PEPTIDES

PROCEEDINGS OF THE FIFTH AMERICAN PEPTIDE SYMPOSIUM

Edited by

Murray Goodman
Department of Chemistry
University of California, San Diego
La Jolla, California 92093

Johannes Meienhofer
Chemical Research Department
Hoffman-La Roche, Inc.
Nutley, New Jersey 07110

A HALSTED PRESS BOOK
John Wiley and Sons
New York   Chichester   Brisbane   Toronto

Distributed by Halsted Press, a division of
John Wiley & Sons, Inc., New York

Library of Congress Cataloging in Publication Data:

American Peptides: Symposium, 5th, University of
California, San Diego, 1977.
Peptides.

Includes indexes.
1. Peptides - Congresses. I. Goodman, Murray,
1928- II. Meinenhofer, Johannes. III. Title

QD431.A1A37 1977 547'.756 77-88855
ISBN 0-470-99384-7

Printed in the United States of America

PREFACE

The Fifth American Peptide Symposium was held on the campus of the University of California, San Diego, from June 20–24, 1977. Representatives from more than twenty nations participated, demonstrating once again the truly international nature of peptide research. There has been an enormous increase in the study of peptides which the Program Committee recognized by bringing speakers to the meeting who created a sense of freshness and timeliness. We were also keenly aware of the need to maintain a balanced program that has become typical of the Peptide Symposia. Therefore we invited leaders in the areas of synthesis, biology, structure and characterization to present their latest findings.

A substantial departure from the format of the preceding meetings was inaugurated with this symposium. We invited only 14 main lectures. All submitted papers were presented in poster sessions. We also included an Open Forum which was designed to bring progress reports of ongoing work to the attention of peptide chemists and biologists.

During this symposium, we also presented the first Alan E. Pierce Award in peptide and chromatographic methodology. The recipient was Professor Miklos Bodanszky who in his award address stressed the uniqueness of each peptide sequence. He demonstrated superbly how peptide chemists must face new and undiminished challenges with each novel structure in the synthesis of complex peptides.

A major biological area of emphasis in the symposium involved brain peptides. The structures of enkephalin, endorphins and other fragments of β-lipotropin were actively discussed. Three of the main lecturers and many poster presentations concentrated on the isolation, characterization, analog syntheses and mode of action of these fascinating molecules. There was by no means complete agreement on the findings reported. Dr. Guillemin, in the concluding remarks of his lecture, took note of this when he indicated that the field is growing rapidly and that a comprehensive picture has not yet evolved.

Our symposium also showed the remarkable advances being made in analytical methodology. Dr. Udenfriend clearly showed that, by use of appropriate labels (radioactive and fluorescent) and modern chromatography, peptide and protein chemistry can be successfully accomplished at the picomole level. These techniques, together with the advances in sequencing and mass spectral analysis, provide us with incredibly sensitive assays of structure and purity.

Conformational studies of peptides, polypeptides and proteins were an important component of the symposium. Four main lectures and numerous posters were devoted to this area of research. From the presentations it is clear that theoreticians and experimentalists are continually learning from each other's results. The former are improving the basis for their potential functions and other ingredients used to predict correct potential energy minima. The latter are unraveling many peptide and protein structures by X-ray crystallography and studies in solution. Utilizing nuclear magnetic resonance, charge-transfer spectra and related measurements, experimentalists are obtaining results which can be used to deduce conformations of peptides in solution. The theoretical studies aid the experimentalists in selecting appropriate measurements needed to propose molecular structures, while the theoreticians are able to refine their calculations on the basis of known structures from X-ray or studies in solutions.

The area of synthesis was well represented. In addition to the Pierce Award address in which Miklos Bodanszky described many exciting challenges remaining for peptide chemists, three main lectures and numerous posters were devoted to descriptions of sophisticated techniques of synthesis including protection, coupling and deprotection. The preparation of important and complex natural peptides was described. In their lectures, Drs. Sakakibara and Merrifield, using classical and solid-phase approaches respectively, showed that constant assessment of side reactions and racemization is necessary to achieve the level of successful syntheses of the natural products and model systems currently of interest. To these important factors, Dr. Feurer added scale-up and other industrial considerations for large-scale peptide synthesis.

As chairman of this year's meeting, I want to take this opportunity to thank the members of the Program Committee who worked closely with me in designing the program and in selecting the submitted papers for poster presentation. A meeting of the size of the Fifth American Peptide Symposium requires substantial funds to be launched successfully. A list of sponsors is contained elsewhere in these proceedings. I want at this point to thank each of the companies, and the Institute of General Medical Sciences of the National Institutes of Health for their generous financial support. There is no way that I can repay my secretaries, Charlotte Beck and Martha Parga, for their unstinting work. My colleague and associate, Dr. Michael Verlander, deserves much praise for his efforts in making sure that everything at the meeting went off according to schedule and without fuss. Much thanks should also go to members of my research group and the Local Committee who helped. Nomi Feldman, as coordinator of the operational aspects of the symposium, performed marvellously and beyond the call of duty. I especially want to acknowledge the efforts and professional insight of my editorial associate, Constance Mullin. To these people and all the attendees I owe a debt of gratitude.

Murray Goodman
La Jolla, California

CONTENTS

Preface iii

ALAN E. PIERCE AWARD LECTURE

MIKLOS BODANSZKY: Peptide Synthesis: An Undiminished Challenge 1

ANALYTICAL TECHNIQUES

SIDNEY UDENFRIEND and STANLEY STEIN: Fluorescent Techniques for Ultramicro Peptide and Protein Chemistry 14

CHARLES W. TODD: Mass Spectral Sequencing of Micro Quantities of Peptides . . . 27

MENACHEM RUBINSTEIN, JANUSZ WIDEMAN and STANLEY STEIN: Isolation and Analysis of Opioid Peptides from Rat Pituitary and Guinea Pig Brain 41

DAVID H. LIVE: Analysis of Fluorescent Labelled Peptides by High Pressure Liquid Chromatography 44

IMRE MOLNÁR and CSABA HORVÁTH: Rapid Analysis of Peptide Mixtures by High Performance Liquid Chromatography with Nonpolar Stationary Phases 48

J. RIVIER, R. WOLBERS and R. BURGUS: Application of High Pressure Liquid Chromatography to Peptides 52

DAVID H. SCHLESINGER, ROBERT JACOBS and DONALD I. HAY: Primary Structure of Protein Inhibitors of Calcium Phosphate Precipitation from Human Salivary Secretions 56

A. W. STRAUSS, A. M. DONOHUE, C. D. BENNETT, J. A. RODKEY and A. W. ALBERTS: Rat Preproalbumin: *In Vitro* Synthesis and Amino Acid Sequence 59

BRAIN PEPTIDES

ROGER GUILLEMIN, FLOYD BLOOM, JEAN ROSSIER, SCOTT MINICK, STEVE HENRIKSEN, ROGER BURGUS and NICHOLAS LING: Recent Physiological Studies with the Endorphins 63

SOLOMON H. SNYDER: Receptors for Putative Neurotransmitter Peptides in the Brain 77

GARLAND R. MARSHALL and FREDRIC A. GORIN: Proposed Opiate Pharmacophore and the Conformation of Enkephalin 84

C. HUMBLET and J. L. DE COEN: Conformational Study of Enkephalin and Analogs in Relation with the Structural Features of Morphine-like Analgesics 88

PETER W. SCHILLER and CHUN F. YAM: Solution Conformation and Opiate Receptor Affinity of [4-Tryptophan]-enkephalin Analogs 92

NICHOLAS LING, SCOTT MINICK, LARRY LAZARUS, JEAN RIVIER and ROGER GUILLEMIN: Structure-Activity Relationships of Enkephalin and Endorphin Analogs 96

R. GLENN HAMMONDS, JR., DAVID PUETT and NICHOLAS LING: Circular Dichroism of Opiate Peptides 100

D. G. SMYTH, B. M. AUSTEN, A. F. BRADBURY, M. J. GEISOW and C. R. SNELL: The Susceptibility of C-Fragment to Endogenous Endopeptidases 103
D. H. COY, P. GILL, A. J. KASTIN, A. DUPONT, L. CUSAN, F. LABRIE, D. BRITTON and R. FERTEL: Synthetic and Biological Studies on Unmodified and Modified Fragments of Human β-Lipotropin with Opioid Activities 107
B. A. MORGAN, J. D. BOWER, K. P. GUEST, B. K. HANDA, G. METCALF and C. F. C. SMITH: Structure-Activity Relationships of Enkephalin Analogs 111
ALAN R. DAY, RICHARD J. FREER and DAVID I. MARLBOROUGH: Synthesis and Pharmacology of Long-acting Enkephalins 114

BIOLOGICAL RECEPTORS AND STRUCTURE-FUNCTION RELATIONSHIPS

DANIEL E. KOSHLAND, JR.: Receptor Interactions in a Sensory System 117
HUGH D. NIALL: Evolution of Structure and Function in Natural Peptides 127
H.-J. FRIESEN, D. BRENDENBURG, C. DIACONESCU, H.-G. GATTNER, V. K. NAITHANI, J. NOWAK, H. ZAHN, S. DOCKERILL, S. P. WOOD and T. L. BLUNDELL: Structure-Function Relationships of Insulins Modified in the Al-Region 136
S. ASWANIKUMAR, B. A. CORCORAN, E. SCHIFFMANN, C. B. PERT, J. L. MORELL and E. GROSS: Peptides with Agonist and Antagonist Chemotactic Activity 141
RICHARD J. FREER and ALAN R. DAY: Synthetic Polypeptides as Chemo-Attractants for Neutrophil Leucocytes: Demonstration of a Specific Receptor for Chemotaxis 146
JAMES T. SPARROW, HENRY J. POWNALL, GERALD F. SIGLER, LOUIS C. SMITH, ANNE K. SOUTAR and ANTONIO M. GOTTO, JR.: The Mechanism of Phospholipid Binding by the Plasma Apolipoproteins . . 149
Z. GRZONKA, E. KOJRO, Z. PALACZ, I. WILLHARDT and P. HERMANN: Tetrazole Analogs of Amino Acids as a Tool in Studies of the Role of Free Carboxyl Group in Biologically Active Systems 153
D. REGOLI, J. BARABE and J. ST-LOUIS: Receptors for Peptide Hormones 157

HORMONES OF THE PITUITARY AND THE HYPOTHALAMUS

CLARK W. SMITH, CHRISTOPHER R. BOTOS and RODERICH WALTER: Conformation-Activity Approach to the Design of Vasopressin Analogs with High and Specific Antidiuretic Activity 161
LESLIE J. F. NICHOLLS, JOSEPH J. FORD, CLAUDE R. JONES, MAURICE MANNING and WILLIAM A. GIBBONS: A Comparison of the Tertiary Structures and Disulfide Bridge Internal Motions of Oxytocin, Arginine-Vasopressin and Their Analogs 165
ROXANNE DESLAURIERS, IAN C. P. SMITH, GEORGE C. LEVY, RONALD ORLOWSKI and RODERICH WALTER: Conformational Studies on [Pro^3, Gly^4]-oxytocin by ^{13}C Nmr 168

B. DONZEL, C. SAKARELLOS and M. GOODMAN: Study of the Conformational and Dynamic Properties of the Luteinizing Hormone-Releasing Factor and Various Analogs using Intramolecular Charge Transfer Complexes 171
PAUL COHEN, ODILE CONVERT, JOHN H. GRIFFIN, PIERRE NICOLAS and CARLO DI BELLO: Carbon-13 Nuclear Magnetic Resonance Studies of the Binding of Selectively ^{13}C-Enriched Oxytocin to its Neurohypophyseal Carrier Protein, Neurophysin I 176
VICTOR J. HRUBY, DIANE M. YAMAMOTO, YOUNG C. S. YANG, and MICHAEL BLUMENSTEIN: Studies of the Interactions of Oxytocin and Arginine Vasopressin with Neurophysins using ^{13}C Nuclear Magnetic Resonance Spectroscopy 179
J. A. EDWARDSON and G. W. BENNETT: Synaptic Interactions Involving Hypophysiotropic Peptides and Neurotransmitters in the Central Nervous System 183
D. SARANTAKIS and J. TEICHMAN: Synthesis of Carbacyclic Analogs of Somatostatin by Combination of Conventional and Solid-Phase Peptide Synthesis Methodology 186
A. S. DUTTA, B. J. A. FURR and M. B. GILES: Synthesis and Biological Activity of α-Aza-Analogs of Luteinizing Hormone-Releasing Factor (LRF) with Potent Antagonist Activity 189
M. FRIDKIN, E. HAZUM, T. BARAM, H. R. LINDNER and Y. KOCH: Hypothalamic and Pituitary LRF-Degrading Enzymes: Characterization, Purification and Physiological Role 193
ADRIANA VIVAS and MARIA ESTER CELIS: Effect of Norepinephrin, Dopamine, and Serotonin on the Inhibition of MSH Release Induced by Pro-Leu-Gly-NH_2 in Pituitaries Incubated *in Vitro* 197
MAURICE MANNING, ANDRAS TURAN, JAYA HALDAR and WILBUR H. SAWYER: Synthesis of Boc-*O*-Benzyl-Homoserine: Its Use in the Synthesis of [4-Homoserine]-oxytocin 201

ANTIBIOTICS, INHIBITORS, TOXINS

PETER PALLAI, TATEAKI WAKAMIYA and ERHARD GROSS: Studies on the Synthesis and Biology of Nisin: Ring A 205
DANIEL H. RICH and ERIC SUN: Kinetics of Inhibition of Pepsin by Pepstatin, Dideoxypepstatin, Acetyl-statin and Acetyl-deoxystatin 209
H. J. CHOU and R. I. GREGERMAN: Preparation and Characterization of a Dextran-Pepstatin Conjugate, A New and Potent Inhibitor of Renin and Pepsin. . . 213
B. F. GISIN, S. KOBAYASHI, D. G. DAVIS and J. E. HALL: Synthesis of Biologically Active Alamethicin 215
GOTFRYD KUPRYSZEWSKI, BERNARD LAMMEK and WITOLD NEUGEBAUER: Tuberculostatic 3-Hydroxyacyl Heptapeptides 218
T. S. ANANTHA SAMY and H. LAZARUS: A Cell Surface Active Antitumor Protein, Neocarzinostatin 221
LYNN H. CAPORALE, BRUCE W. ERICKSON and TONY E. HUGLI: Synthetic Oligopeptides from Human C3a Anaphylatoxin that Mediate the Inflammatory Response 225
HORACIO N. FERNANDEZ and TONY E. HUGLI: Structural and Functional Comparison Between Human C3a and C5a 228
MICHAEL ROSENBLATT, PIERRE D'AMOUR, GINO V. SEGRE and JOHN T. POTTS, JR.: Synthesis of a Fragment of Bovine Parathyroid Hormone, bPTH-(28-48): An Inhibitor of Hormone Cleavage *in Vivo* 232

NORIKAZU NISHINO and JAMES C. POWERS: Peptide Hydroxamates as Inhibitors of Thermolysin and Related Metalloproteases 236
S. I. SALLAY, K. S. L. SRIVASTAVA, S. OROSZLAN and R. V. GILDEN: Synthesis of Antigenic Peptides of Mammalian C-Type RNA Tumor Virus Major Internal Proteins 239
EISUKE MUNEKATA, HEINZ FAULSTICH and THEODOR WIELAND: Exchange of Ala[1] by Diverse Amino Acids in Phalloidin and Toxicological Properties of the New Analogs 243

CONFORMATIONS: THEORY, X-RAY AND RELATED STUDIES

HAROLD A. SCHERAGA: Conformational Energy Calculations on Peptides 246
ETTORE BENEDETTI: Structure and Conformation of Peptides: A Critical Analysis of Crystallographic Data 257
ISABELLA L. KARLE: 3 → 1 and 4 → Intramolecular Hydrogen Bonds in *Cyclo*(Gly-Pro-Gly-D-Ala-Pro) (Crystal Structure Analysis) 274
G. D. SMITH, W. L. DUAX, E. W. CZERWINSKI, N. E. KENDRICK, G. R. MARSHALL and F. S. MATHEWS: The Crystal and Molecular Structure of a Tetrapeptide, the Benzyl Ester of Boc-L-Pro-Aib-L-Ala-Aib. . 277
A. T. HAGLER, B. HONIG and R. SHARON: Theoretical Studies of Protein Folding 280
PETER Y. CHOU and GERALD D. FASMAN: Prediction of Protein Secondary Structure 284
M. MUTTER and P. J. FLORY: Prediction of Cyclization Tendencies of Polypeptides from the Macrocyclization Equilibria 288
F. A. MOMANY, C. Y. BOWERS and J. K. CHANG: On the Use of Conformational Energy Calculations to Show Structural Relationships Between Small Polypeptides which Exhibit Opiate Activity and Those with Pituitary Hormone-Releasing Activity 292
G. MANJULA and C. RAMAKRISHNAN: Studies on the Conformation of Cyclic Tetrapeptides 296
P. K. PONNUSWAMY and P. MANAVALAN: Solvent Accessibilities in Peptides 300
F. FILLAUX, M. H. BARON and C. DE LOZE: The Non-Planar Structure of Self-associated *N*-Methylacetamide 303

CONFORMATIONS: SPECTRAL STUDIES

V. T. IVANOV: Solution Structures of Peptides 307
DAVID COWBURN, ALAN J. FISCHMAN, DAVID H. LIVE, WILLIAM C. AGOSTA and HERMAN R. WYSSBROD: An Approach to the Unequivocal Determination of Peptide Conformations 322
JERRY D. GLICKSON, ROBERT E. LENKINSKI, N. RAMA KRISHNA, DAVID G. AGRESTI and RODERICH WALTER: Development of Nmr and Fluorescence Methods for Determining Peptide Conformations in Solution 325
CLAUDE R. JONES, JOSEPH B. ALPER, MEI-CHANG KUO and WILLIAM A. GIBBONS: Multiple Conformations of the Zwitterionic and Cationic Forms of Enkephalins and Other Peptides 329
KENNETH D. KOPPLE, MEOW-CHAN FENG and ANITA GO: Condensations Products of Met-Val-Gly-Pro-Asn-Gly and Their Conformations 333

A. C. M. PAIVA and L. JULIANO: Conformation of Bradykinin in Aqueous Solution. Electrometric Titration of the Hormone and Related Peptides . . . 337
DANIEL H. RICH and PRADIP BHATNAGAR: Conformational Analysis of Tentoxin and Related Cyclic Tetrapeptides. Isolation and Bioassay of Two Conformers of D-MeAla Tentoxin . . . 340
HORST KESSLER, YVES A. BARA, AXEL FRIEDRICH, WOLFGANG HEHLEIN, PETER KONDOR and MICHAEL MOLTER: Conformational Studies of Cyclic Pentapeptides by Nmr Spectroscopy . . . 343
LILA G. PEASE and CHRISTOPHER WATSON: Conformational and Ion Binding Studies on a Cyclic Pentapeptide: Evidence for β and γ Turns in Solution . . . 346
VINCENT MADISON, JERRY LASKY and BRUCE CURRIE: Asymmetric, Hydrogen-bonded Peptide Complexes . . . 350
HENRY E. AUER, BARBARA E. PAWLOWSKI-KONOPNICKI, YU-CHIH CHEN CHIAO and THOMAS R. KRUGH: Analysis of the Absorption Spectrum of Actinomycin D and the Formation of Its Complexes with Deoxynucleotides . . . 353
C. M. DEBER, P. D. ADAWADKAR and J. TOM-KUN: Carbon-13 Enriched Human Encephalitogenic (EAE) Nonapeptide: Synthesis and Conformational Studies . . . 357
ELISHA HAAS, EPHRAIM KATCHALSKI-KATZIR and IZCHAK Z. STEINBERG: Measurement of the Brownian Motion of Oligopeptide Chain Ends Relative to Each Other . . . 361
M. LECLERC, S. PREMILAT and A. ENGLERT: Conformational Analysis of Adrenocorticotropic Hormone Related to Non-radiative Energy Transfer . . . 364

CONFORMATIONS: POLYPEPTIDES AND PROTEINS

E. PEGGION, A. COSANI, M. PALUMBO and M. TERBOJEVICH: Conformational Aspects of Polypeptide Interactions with Ions . . . 368
CLAUDE LOUCHEUX, MICHEL MORCELLET and CLAUDE FEYEREISEN: Preferential Adsorption and Conformational Transition of Poly[N^5-(3-hydroxypropyl)-L-glutamine] in Water/2-Chloroethanol Mixtures . . . 387
MASAHIRO HATANO, HIROYUKI NOMORI and MASAO YOSHIKAWA: Circular Dichroism of Poly(γ-[2-(9-carbazolyl)-ethyl] -L-glutamate) in Liquid Crystalline and Solid States . . . 390
TULLIO PAGANETTI, VINCENZO RIZZO, ANTONIO BAICI, PETER NEUENSCHWANDER and GIAN PAOLO LORENZI: Conformational Aspects of Protected Co-oligopeptides with Alternating L-Isoleucine and D-Alloisoleucine Residues . . . 395
M. PALUMBO, G. M. BONORA, C. TONIOLO, E. PEGGION and E. STEVENS: Structure of γ-Branched Homo-oligopeptides . . . 399
M. MUTTER, H. MUTTER and E. BAYER: Conformational Studies on Polyoxyethylene-bound Peptides . . . 403
R. S. RAPAKA, KOUJI OKAMOTO, M. M. LONG and D. W. URRY: Syntheses and Properties of Polypeptides and Cross-linked, Insoluble, Peptide Matrices of Elastin . . . 407

R. S. HODGES, S. A. ST-PIERRE and F.-S. TJOENG: Synthetic Sequential Polyheptapeptides as Models for the Two-stranded α-Helical Coiled-coil Structure of Tropomyosin 411
MARIE-H. LOUCHEUX-LEFEBVRE and NICOLE HELBECQUE: Nmr and Circular Dichroism Studies of a Repeating Peptide of a Protein in Human Parotid Saliva H-Gly-Pro_4-OH 415
AVIVA LAPIDOT and CHARLES S. IRVING: Bacterial Cell Wall Peptidoglycan: A Semi-crystalline Matrix or a Random Polymer? An *in Vivo* ^{15}N Nmr Study 419
STEPHEN J. KENNEDY, HENRY R. BESCH, JR., AUGUST M. WATANABE, ROGER W. ROESKE and ALAN R. FREEMAN: Beta-helical Conformations of Peptides and Proteins 423
C. CRANE-ROBINSON, H. HAYASHI, P. D. CARY, T. MOSS, E. M. BRADBURY, L. BÖHN, G. BRIAND, P. SAUTIERE, D. KRIEGER, G. VIDALI, P. N. LEWIS and J. TOM-KUN: The Study of Protein Conformation Using Cleaved Peptides: Application to Histones H3 and H4 427
ANGELO FONTANA and CLAUDIO VITA: Conformational Studies on the Cyanogen Bromide Fragments of Thermolysin 432

SYNTHETIC STUDIES: ADVANCES

SHUMPEI SAKAKIBARA: Solution Synthesis of Complex Peptides by the Maximum Protection Procedure 436
MAX FEURER: Special Features of Large-scale Peptide Synthesis 448
RICHARD A. HOUGHTEN and CHOH HAO LI: Reduction of Methionine Sulfoxide to Methionine in Peptides and Proteins 458
J. KOVACS, Y. HSIEH, K. Y. HUI and S. E. KIM: Further Studies on Racemization in Peptide Synthesis 461
RICHARD G. HISKEY and NORMAN T. BOGGS, III: Studies on Prothrombin. Synthesis of Peptides Containing γ-Carboxyglutamic Acid 465
CHARLES N. C. DREY and GARETH P. PRIESTLEY: Improved Synthesis of Folate Conjugates 468
GERALD SIGLER, WILLIAM FULLER and RAYMOND VAVREK: An Improved Leucine Enkephalin Synthesis 471
F. MARCHIORI, G. BORIN, B. FILIPPI, G. M. BONORA and C. TONIOLO: Preparation and Conformational Analysis of Galline and Clupeines 474
HARI G. GARG and ROGER W. JEANLOZ: Synthesis of Glycotripeptides Containing L-Serine and 2-Acetamido-2-deoxy-β-D-glucopyranosyl Residues 477
FRANK R. N. GURD, WILLIAM H. GARNER, RICHARD D. DIMARCHI and CHI-CHIN WANG: Production of Semisynthetic Myoglobin by Stepwise Fragment Condensation 480
IVAR UGI, GERHARD EBERLE, HEIÑER ECKERT, INGER LAGERLUND, DIETER MARQUARDING, GISELHER SKORNA, REINHARD URBAN, LORENZ WACKERLE and HASSO v. ZYCHLINSKI: The Present Status of Peptide Synthesis by Four-component condensation and Related Chemistry 484

SYNTHETIC STUDIES: SOLID PHASE

R. B. MERRIFIELD, GEORGE BARANY, WESLEY L. COSAND, MARTIN ENGELHARD and SVETLANA MOJSOV: Some Recent Developments in Solid Phase Peptide Synthesis 488

E. ATHERTON and R. C. SHEPPARD: Solid Phase Synthesis on Polyamide Supports: β-Endorphin 503

M. ROTHE, A. SANDER, W. FISCHER, W. MÄSTLE and B. NELSON: Interchain Reactions (Cyclo-Oligomerizations) During the Cyclization of Resin-bound Peptides 506

CHRISTINA BIRR, MARGOT WENGERT-MÜLLER and ANGELIKI BUKU: Total Synthesis on Gel Phase of the Mast-Cell-Degranulating Peptide by Fragment Condensation 510

HARTMUT FRANK, HANSPAUL HAGENMAIER, ERNST BAYER and DOMINIC M. DESIDERIO: Alternating Liquid-Solid Phase Synthesis of Luteinizing Hormone-Releasing Factor 514

GRAHAM MOORE and DENIS MCMASTER: Potassium Cyanide Catalyzed Transesterification: A Mild Procedure for Removing Peptides from the Merrifield Resin with the Protecting Groups Intact 518

J. VAN RIETSCHOTEN, E. PEDROSO MULLER and C. GRANIER: Cysteine Protection in Solid Phase Synthesis of Apamin 522

JAMES BURTON, MICHAEL N. MARGOLIES and EDGAR HABER: Synthesis of an Antibody Fragment: Correcting Loss of Peptide in the Course of Synthesis 525

KUNG-TSUNG WANG and CHI-HUEY WONG: Total Synthesis of Taiwan Cobra Venom Cardiotoxin 528

SYNTHETIC STUDIES: PROTECTION, DEPROTECTION AND COUPLING

ARTHUR M. FELIX, MANUEL H. JIMENEZ and JOHANNES MEIENHOFER: Removal of N^{α}-Benzyloxycarbonyl Groups from Sulfur-containing Peptides by Catalytic Hydrogenolysis in Liquid Ammonia: Stepwise Synthesis of Somatostatin 532

YAKIR S. KLAUSNER, TZVI H. MEIRI and EUGENIA SCHNEIDER: Peptide Synthesis in Aqueous Solution with *o*-Nitro-*p*-Sulfophenyl Esters 536

BORIS WEINSTEIN: 1,4-Dinitro-1,3-butadiene-2,3-dicarboxylic Anhydride, A New Peptide Coupling Agent 539

A. HUBBUCH, W. DANHO and H. ZAHN: Cysteic Acid, a Water Soluble Protecting Group in Peptide Synthesis 540

PETER SIEBER, RUDOLF H. ANDREATTA, KAREL EISLER, BRUNO KAMBER, BERNHARD RINIKER and HANS RINK: The 2-Trimethylsilylethyl Residue, A Selectively Cleavable Carboxylic Acid Protecting Group 543

MASAAKI UEKI, SHIGERU IKEDA and FUMIO TONEGAWA: Phosphinothioyl Groups: New Series of Acid Labile Amino Protecting Groups 546

H. JOSEPH GOREN and MATI FRIDKIN: Base Lability of Reactants and Products of Thiolysis Reaction of N^{im}-2,4-Dinitrophenylhistidine Peptides. An Explanation for the pH Optimum 549

J. RAMACHANDRAN and ELEANOR CANOVA-DAVIS: Synthesis and Use of Photoreactive Sulfenyl Chlorides 553

A. PELLEGRINI and P. L. LUISI: Pepsin Induced Synthesis of Peptide Bonds . . . 556
GILBERTO GOISSIS, BRUCE W. ERICKSON and R. B. MERRIFIELD: Synthesis of Protected Peptide Acids and Esters by Photosolvolysis of 1-Peptidyl-5-bromo-7-nitroindolines . . . 559

OPEN FORUM

P. CRINE, S. BENJANNET, N. G. SEIDAH, M. LIS and M. CHRÉTIEN: *In Vitro* Biosynthesis of β-Endorphin, γ-Lipotropin and β-Lipotropin by the Pars Intermedia of Beef Pituitary Glands . . . 562
E. BRICAS, J. MARTINEZ, D. BLANOT, G. AUGER, M. DARDENNE, J. M. PLEAU and J. F. BACH: The Serum Thymic Factor and its Synthesis . . . 564
CHARLES B. GLASER, ROBERT FALLAT, LUCIJA KARIC, RICHARD STOCKERT and ANATOL G. MORELL: The Role of Carbohydrate in the Structure and Function of Plasma Glycoproteins – Studies on Two Inherited Variants of Alpha-1-Antitrypsin . . . 567
T. CHRISTENSEN, P. VILLEMOES and K. BRUNFELDT: Monitoring with Feedback in Automated Solid Phase Peptide Synthesis . . . 569
M. CHOREV, C. G. WILLSON and M. GOODMAN: A Three-point Model for the Dipeptide Sweetener-Receptor Interaction . . . 572
ARIEH WARSHEL: How Do Enzymes Really Work? . . . 574
MIGUEL A. ONDETTI, EMILY F. SABO, KATHRYN A. LOSEE, HONG SON CHEUNG, DAVID W. CUSHMAN and BERNARD RUBIN: The Use of an Active Site Model in the Design of Specific Inhibitors of Angiotensin-converting Enzyme . . . 576
C. G. WILLSON, M. GOODMAN, J. RIVIER and W. VALE: Topochemically Related Hormone Structures. The Synthesis of Retro-analogs of LRF . . . 579
MICHAEL ROSENBLATT, JANE E. MAHAFFEY and JOHN T. POTTS, JR.: Parathyroid Hormone Inhibitors: Design, Synthesis, and Biological Evaluation . . . 581
ISABELLA L. KARLE: Solvent Channels in the Crystalline Lattices of Antamanide Grown from Polar and Nonpolar Solvents . . . 583
A. T. HAGLER and J. MOULT: Monte Carlo Simulation of Water Structure Around Proteins and Peptides . . . 586
R. ACHER, J. CHAUVET, M. T. CHAUVET and P. CODOGNO: Structure and Conformation of Neurophysins . . . 588
V. RENUGOPALAKRISHNAN and F. JORDAN: Theoretical Studies on Peptide-Solvent Interactions . . . 590
P. L. LUISI, FRANCIS J. BONNER and CH. WALSOE: Spectroscopic and Structural Properties of α-Chymotrypsin in Cyclohexane . . . 591
LEON E. BARSTOW, ROBERT S. YOUNG, EMEL YAKALI, JOHN J. SHARP, JEAN C. O'BRIEN, PHIL BERMAN and HENRY A. HARBURY: The Preparation of Semi-synthetic Horse Heart Cytochrome c . . . 593

COMMITTEES, SPONSORS . . . 596

AUTHOR INDEX . . . 597

SUBJECT INDEX . . . 601

Alan E. Pierce Award Lecture

PEPTIDE SYNTHESIS: AN UNDIMINISHED CHALLENGE

MIKLOS BODANSZKY, *Department of Chemistry, Case Western Reserve University, Cleveland, Ohio 44106*

I would like to express my thanks to the Selection Committee of the American Peptide Symposia for choosing me as the first recipient of the Alan Pierce Award. However, no less gratitude is due to Mr. Roy Oliver of the Pierce Chemical Company. Peptide chemists have received, until now, no real appreciation such as the Pierce Award. This award is a timely recognition of the field of peptide chemistry, a field that requires arduous work, and unrelenting dedication. Last, but not least, I thank our Chairman and host, Professor Murray Goodman, for the invitation that allows a veteran of a quarter century to talk to you on peptide synthesis.

Time restrictions prevent a full presentation of recent advances, and I can mention only a few contributions. These are not necessarily the most significant ones: importance emerges gradually and can be judged only in retrospect. In fact, looking at the impressive array of well-established methods of protection, coupling, etc., one could even question the need for additional procedures. The answer is: yes, we need them, and the reasons for this affirmative answer can be found in the words of Theodor Wieland:[1]

"To the uninitiated, even to the chemist familiar with peptide chemistry only through hearsay, it may appear that nowadays peptide synthesis is mere routine. However, the adept has learned by experience that the individuality of the amino acids, which in the last analysis is one of the elements determining biological specificity, lends to their manipulation in synthesis such diversion, suspense and drudgery as found with no other substances. Thus only can the manifold efforts for development of a methodology of the field be understood. In progressing toward longer and longer, structurally well-defined peptide chains, methods, that are quite trustworthy with small peptides, may prove of little utility. Therefore, it should be possible to choose the right alternative from a large array of procedures. For all these reasons, every synthesis of a complicated polypeptide represents the sum of prolonged intellectual and experimental exertions which, however, often find their reward in the interesting physical, chemical and biological properties of the final product."

Unusual Residues

In addition to the twenty amino acids commonly found in proteins, many unusual residues were identified in microbial peptides, some of these requiring special measures of synthesis. We have to think only of the extreme readiness of phenylglycine to suffer racemization when activated.[2] Not less problematic could be the incorporation of the newly recognized constituent of prothrombin, γ-carboxyglutamic acid.[3] The synthesis, resolution and protection of this amino acid (*1*) has already received growing attention.[4,5] In our own recent work, the incorporation of biocytin[6] (ε-biotinyl-L-lysine), (*2*) presented some problems.

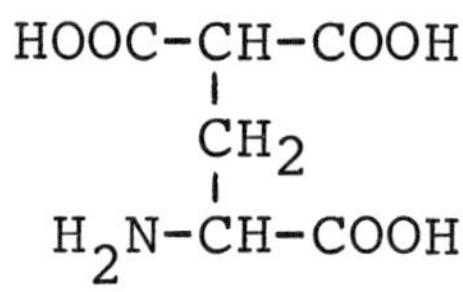

(*1*)

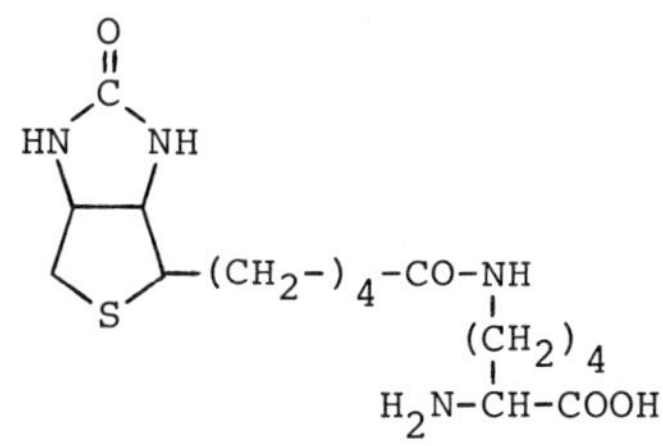

(*2*)

The attempted preparation of biotin *p*-nitrophenyl ester (*5*),[7] with the aid of *p*-nitrophenyl trifluoroacetate,[8] led to a trifluoroacetyl derivative (*4*),[9] which, in turn, could be converted to the desired active ester by alcoholysis. The weak, but definite, nucleophilic character of the urea grouping in biotin (*3*) did not prevent us[10] from obtaining a crystalline pentachlorophenyl ester of Boc-biocytin. [mp. 156° dec., $[\alpha]_D^{25} + 7°$ (*c* 2, DMF)]. Still, the complicating effect of biotin must be taken into consideration during our effort toward the synthesis of a biologically active tryptic fragment of the biotin-containing carboxyl carrier subunit of the transcarboxylase from *Propionibacterium shermanii.*[11]

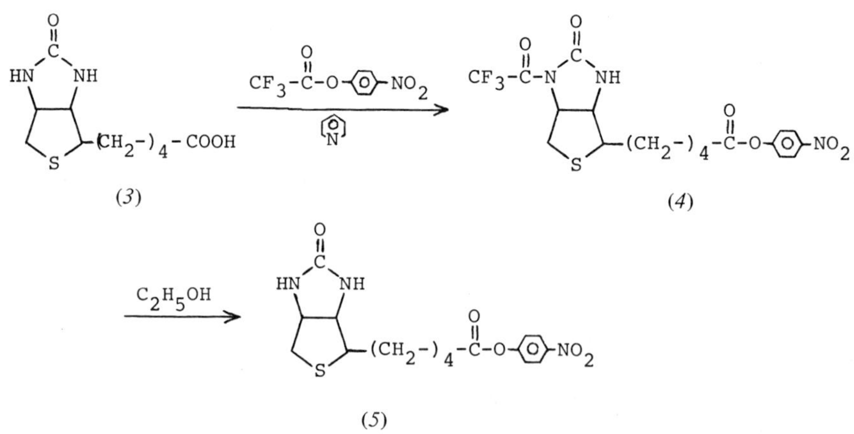

(*3*) (*4*)

(*5*)

Unexpected behavior of an unusual residue occurred in the synthesis of a hormone analog in our laboratory.[12] Replacement of the side-chain amino group of the lysine residue in vasopressin by hydroxyl required the incorporation of ϵ-hydroxynorleucine. When the hydroxy group was left unprotected, the removal of α-amino protecting groups with hydrobromic acid in acetic acid or with trifluoroacetic acid caused considerable cleavage of the peptide bond between ϵ-hydroxynorleucine and the next amino acid, glycine, although the formation of a seven-membered lactone was the driving force of the N→O shift, not a five-membered one, as in the case of homoserine. An old device, acetylation of the hydroxy group, eliminated the problem, but it is obvious that unusual residues in naturally occurring peptides or in their analogs provide new stimuli for the development of novel protecting groups.

Protection and the Removal of Protecting Groups

The introduction of tertiary amine substituents into well-established protecting groups renders them resistant to acids. This idea was generalized in Young's laboratory[13] in Oxford (groups *6–8*), though the underlying principle can be recognized in the earlier proposed[14] isonicotinyloxycarbonyl group (*9*) as well. The isonicotinyloxycarbonyl protection of the ϵ-amino group of lysine residues has already gained practical significance.

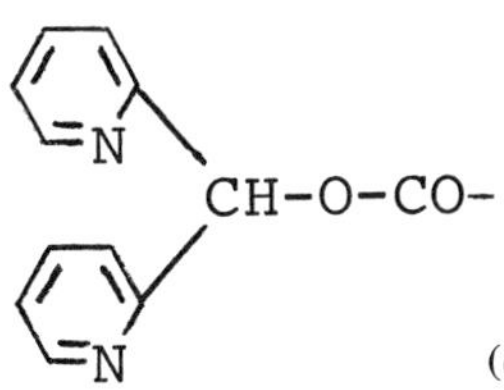

(*6*)

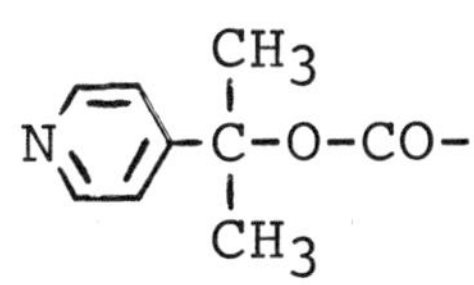

(*7*)

(*8*)

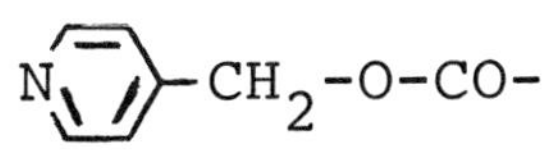

(*9*)

From the attempts toward acid-stable and selectively removable protection of the lysine side chain, the β-methylsulfonylethoxycarbonyl (Msoc) group (*10*) of Tesser[15] is quite promising. The alkali-catalyzed elimination reaction used for the removal of this group is remarkable in its smoothness, but the potential harm to other functions caused even by brief treatment with the calculated amount of

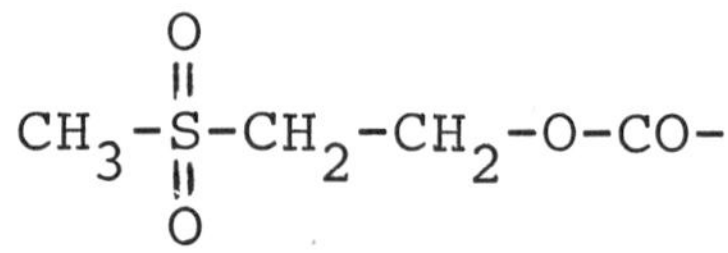

(10)

alkali still requires careful examination. An alternative solution for the stabilization of the benzyloxycarbonyl group against partial removal is substitution of the ring with negative atoms such as Cl[16] or Br[17] or electron-withdrawing groups (NO_2).[18] Such substituted Z-groups remain intact during acidolysis of Boc groups, e.g. with trifluoroacetic acid. On the other hand, unless hydrogenolysis can be used, the negatively substituted Z groups require quite strong acids for their removal. There is a general trend – coming from solid-phase peptide synthesis – for "global deprotection," that is, for the simultaneous removal of all protecting groups with strong acidic reagents. The application of HF,[19] BBr_3,[20] $B(CF_3COO)_3$,[21] CF_3SO_3H,[22] and CH_3SO_3H[23] are all steps in this same direction, as is the potential use of the HF-pyridine complex.[24] Yet, such strong acidic conditions are conducive to previously unobserved side reactions, e.g. the migration of benzyl groups in tyrosine from oxygen into the nucleus. Therefore, some new consideration is due to the "old faithful," the Z-group.[25] Instead of rendering it more acid resistant, one can change the conditions of the cleavage of the Boc group in such a way that the Z-groups on lysine side chains remain essentially intact. Dilution of trifluoroacetic acid with 30% water was proposed by Schnabel et al.,[26] but concern about hydrolysis led us[27] to use a mixture of 70% trifluoroacetic acid and 30% acetic acid, which provides complete removal of Boc groups in about 1/2 hour at room temperature, yet causes only negligible cleavage of side chain Z-groups. A sophisticated approach, reaching selectivity with weaker rather than stronger acids, was proposed by the Ciba-Geigy group.[28] The trityl and biphenylisopropyloxycarbonyl (Bpoc) groups can be selectively removed, each at a distinct, well-defined acidity in 90% trifluoroethanol. The new method found impressive application in a most elegant synthesis of human insulin.[29] An interesting new class of acid labile amino-protecting groups, phosphinamides, was proposed by Kenner et al.[30] Protection of amino acids in the form of diphenylphosphinamides (*11*) protects them against racemization as well.

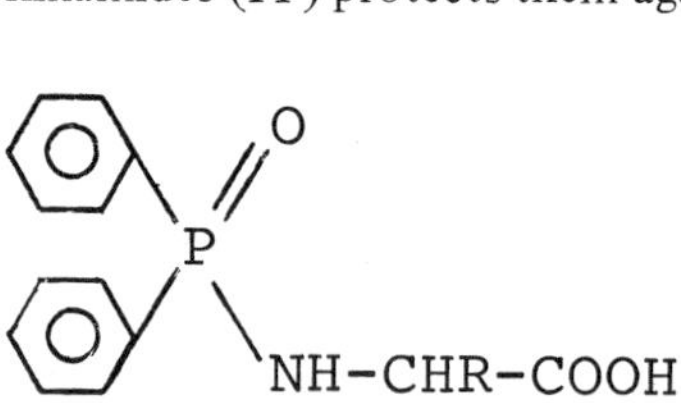

(11)

A similar pattern of development can be recognized in connection with the protection of the phenolic hydroxyl of tyrosine. The "classical" protecting group, *O*-benzyl (*12*),[31] was used without complication, when HBr in AcOH[32] was applied for deprotection (*13*). Later, to prevent the acetylation of serine residues, the acetic acid in the reagent was replaced by trifluoroacetic acid.[33] Subsequently, the already mentioned formation of 3-benzyl-tyrosine[34] was noted (*13* and *14*). (The peptide chemist always sails between Scylla and Charybdis.) Substituted benzyl groups, e.g. the 2,6-dichlorobenzyl group,[17] were introduced, a counter-measure which could diminish the extent of the side

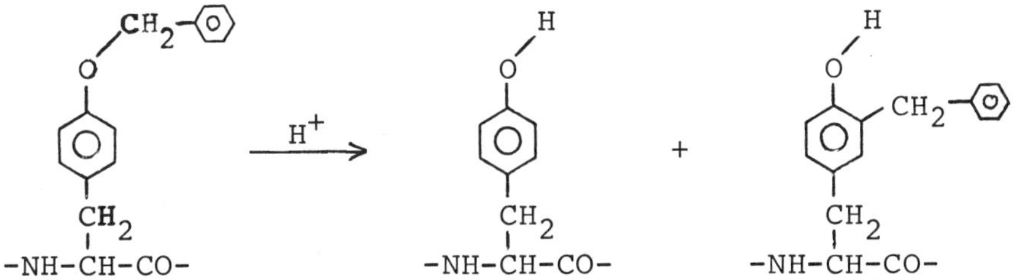

(*12*) (*13*) (*14*)

reaction. An intramolecular mechanism seems to be involved since scavengers were without effect. Yet the solvent dependence of the reaction suggested that a solution of the problem may lie in this direction. We were pleased to observe[35] that only a negligible trace (<0.1%) of 3-benzyltyrosine (*14*) formed when HBr, in a 1:1 mixture of phenol and *p*-cresol, was used for the removal of the benzyl group. The practical application of this reagent is being explored.

The *acid*-catalyzed ring closure in peptides containing β-benzyl aspartyl residues has received new attention: Merrifield[36] replaced this group by the selectively removable phenacyl grouping. In our own work we were more troubled by the *base*-catalyzed formation of succinimide derivatives. For example, in the synthesis of the N-terminal sequence (*15*) of the vasoactive intestinal peptide[37] by stepwise chain lengthening with active esters,[38] a ready ring closure of the

His-Ser-Asp-Ala-Val-Phe-Thr-Asp-Asn-Tyr
1 2 3 4 5 6 7 8 9 10

(*15*)

aspartyl residue in position 8 was noted. The base-catalyzed elimination of the benzyl alcohol might be facilitated by the absence of steric hindrance if the carboxamide of the side chain of the 9-asparagine residue is kept away from the NH group of the same moiety by hydrogen bonding. We are carrying out a search for protecting groups that remain intact during coupling and provide the desired semi-permanent protection, but we also explore the prevention of the ring closure by changing the conditions of the reaction. In a series of experiments, the effects of different bases (tertiary amines) were compared.[39] The hindered base diisopropylethylamine, that was beneficial in other cases,[40] produced nearly as much succinimide byproduct as the unhindered triethylamine. On the other hand, methyldibenzylamine, tribenzylamine and the "proton sponge," 1,8-dimethylaminonaphthalene,[41] greatly reduce the rate of ring closure. Tribenzylamine, perhaps because it is a weak base, diminished the rate of the desired reaction as well. The applicability of methyldibenzylamine and of the "proton sponge" is under investigation.

An alternative solution for the suppression of succinimide ring formation is the acceleration of the rate of the coupling reaction by catalysts that do not enhance the rate of the side reaction. In preliminary experiments, catalysis of active esters with 1-hydroxybenzotriazole[42] gave promising results. No less auspicious were the attempts in which 4-dimethylaminopyridine[43] was adopted for the acceleration of acylation with *p*-nitrophenyl esters.

Activation and Coupling

Efforts toward improved methods of activation of the carboxyl group are continued in many laboratories. The 4-component condensation of Ugi[44] or the reagents that involve acyloxyphosphonium intermediates[45,46] are good examples of such endeavors.

The Bates' reagent[47] (compound *16*)

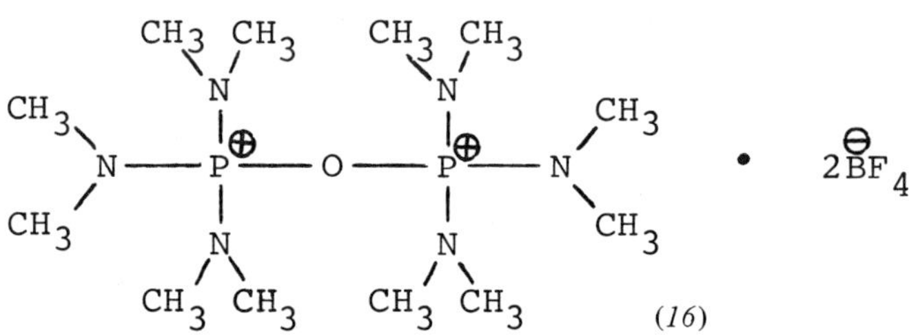

(*16*)

or the reagent (*17*) of Martinez and Winternitz[48]

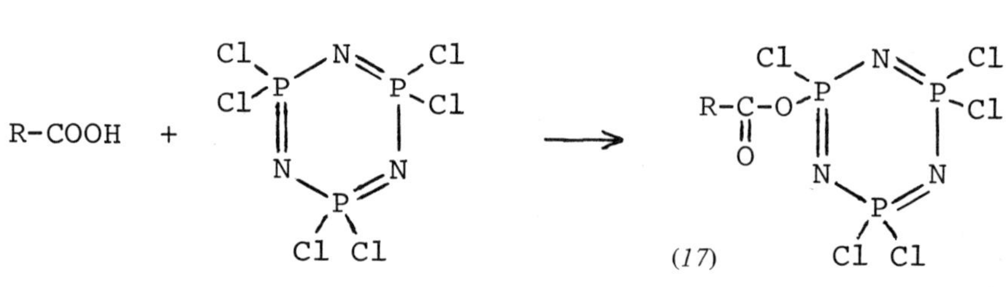

(*17*)

are ingenious and novel, but as yet of unknown significance. It may be interesting to note that well-established reagents such as tri-*p*-nitrothiophenyl phosphite,[49] di-*p*-nitrophenyl sulfite[50] or *p*-nitrophenyl trifluoroacetate,[8] all used in pyridine, also might react via acylpyridinium salts. The recently proposed reactive amides (compound *18*)[51] could indicate a renewed interest in reactive *N*-acyl derivatives.

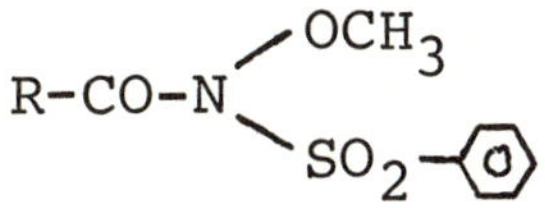

(*18*)

Strategy

Instead of further pursuing novel methods of activation, we should pay attention to the related problem of strategy. Stepwise synthesis of peptides[38] gained new ground by the adaptation of mixed anhydrides, particularly those prepared with isobutyl chlorocarbonate[52] for coupling. Tilak[53] and, more recently, the group of Beyerman[54] carried out syntheses of longer chains in this manner. A brief reexamination of this approach in our laboratory[55] cautions against the uncritical use of the "repetitive excess mixed anhydride (REMA)" method. Some second acylation product could be detected in all experiments (Table I) carried out with mixed anhydrides. The amount of these blocked

Table I. Formation of a Second Acylation Product in the Reaction of *tert*-Butyloxycarbonylamino Acid-isobutylcarbonic Mixed Anhydrides with Amino Acid β-naphthylamides

Boc-amino acid	Amino acid β-naphthylamide	Second acylation product (%)
glycine	glycine	0.5
L-leucine	glycine	0.6
L-phenylalanine	glycine	0.5
O-Bzl-L-serine	glycine	0.7
L-valine	glycine	6.2
L-isoleucine	glycine	8.2
glycine	L-valine	0.5
L-valine	L-valine	3.8
glycine	L-leucine	0.5

peptide byproducts might be acceptable in many cases, but with hindered carboxyl components, such as derivatives of valine and isoleucine, their amount grew to 6 and 8%: clearly reason for concern. Symmetrical anhydrides[56] or active esters[57] should produce – at least in principle – more homogeneous materials.

The expression "*segment condensation*" proposed by Pettit[58] is certainly more descriptive than the term "fragment condensation" used so far. In this strategy, the suppression of racemization by *N*-hydroxysuccinimide (*19*)[59] or by 1-hydroxybenzotriazole (*20*)[60] are major advances. New efficient racemization-suppressing reagents were found, e.g. ethyl 2-hydroxyimino-2-cyano-acetate (*21*).[61] The significance of 1-hydroxybenzotriazole, however, lies in the fact

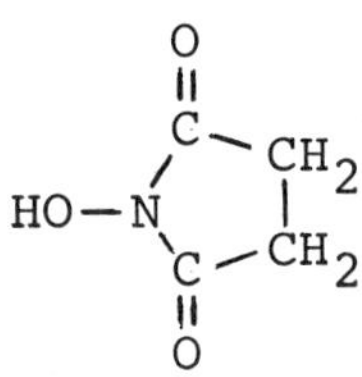

(*19*)

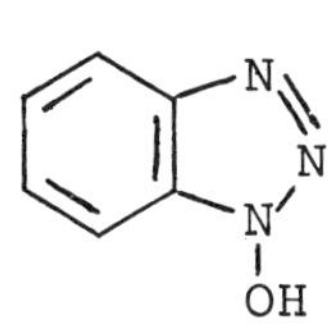

(*20*)

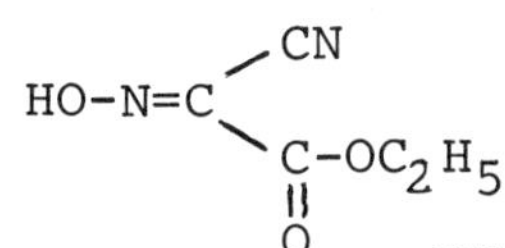

(*21*)

that in addition to its ability to reduce the extent of racemization, it also catalyzes the aminolysis of active esters.[62] It diminishes the danger of acetylation[63] or trifluoroacetylation[64] of amino components used as acetate or trifluoroacetate salts and in some, not yet understood, way it can prevent the extensive hydroxyl *o*-acylation observed in histidine-containing peptides.[65]

While coupling in the presence of these "additives" rendered the potent dicyclohexylcarbodiimide method[66] acceptable for segment condensation, the crucial problem of linking larger segments of a long peptide chain is not yet satisfactorily solved. The classical approach, coupling via azides,[67] is often not as reliable as would be necessary when precious intermediates are involved. *Activation of the carboxyl group* (of the carboxy component) has to be efficient and yet one should avoid overactivation[68] since this will lead to racemization and to other side reactions. The alternative process, *amino-activation*, still awaits discovery. In a search for further solutions, we considered active esters of *peptides* which serve as carboxyl components. Unfortunately, the preparation of active esters generally proceeds through more reactive intermediates, such as *O*-acylisoureas or mixed anhydrides. Yet, it should be possible to prepare active esters via the *activation of the hydroxyl-component.* As an example, the synthesis of *O*-acyl derivatives of benzhydroxamic acid (*22*) by Taschner and his associates[69] can be mentioned. If we consider 1-fluoro-2,4-dinitrobenzene as the activated form of 2,4-dinitrophenol, then the formation of reactive derivatives of carboxyl groups (*23*) during dinitrophenylation[70] is an early, although not practical, example of hydroxyl activation. In our attempts in this direction, the

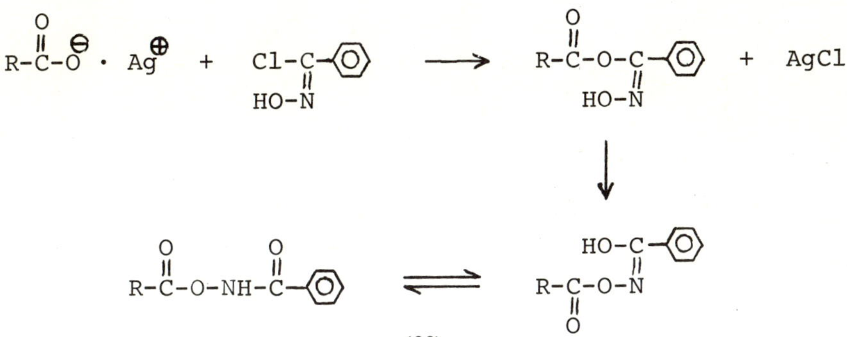

(22)

(23)

reaction of 1-iodo-4-nitrobenzene with the cesium salts of carboxylic acids yielded highly colored solutions, presumably containing Meisenheimer complex type intermediates (*24*):

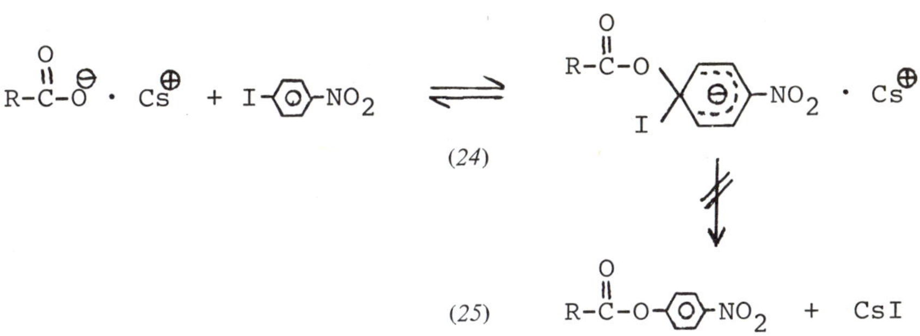

(24)

(25)

We are still looking for a catalyst that would remove the obstacle from the way of these intermediates toward the desired active esters (*25*).

Conclusions

After this brief and, by necessity, superficial survey of the state of peptide synthesis, I am probably expected to look into the future of the field. This would require a better futurologist than I claim to be. Peptide synthesis is about as old as our century. Many of its major advances materialized in the last 25 years and therefore we should assume substantial progress in the last "quarter." Probably larger and more complex peptides will be the targets of syntheses. Beyond this, no generalization is possible. Yet, just because of the individuality of the amino acids and on account of the *individuality of peptide sequences*, we should expect not fewer difficulties and problems, but more. An example of sequence-dependent side reactions, the ring closure of aspartyl peptides to amino succinimide derivatives, is being studied in our laboratory. A recent unexpected observation,[71] the formation of a sulfonic acid derivative during esterification of the phenolic hydroxyl group of tyrosine in the C-terminal 10-peptide of cholecystokinin (CCK),[72] points to the participation of the preceding aspartyl residue in the side reaction. The same operation carried out on the C-terminal 7-peptide of this hormone yielded a biologically highly potent material that contained only the desired *O*-sulfate, but no sulfonic acid derivative. The 7-peptide also has an Aspartyl residue, except that it is more distant from the tyrosine moiety:

```
                                OSO3H
                                |
CCK 24-33      Asp-Arg-Asp-Tyr-Met-Gly-Trp-Met-Asp-Phe-NH2

                                OSO3H
                                |
CCK 27-33                   Tyr-Met-Gly-Trp-Met-Asp-Phe-NH2
```

As you certainly noted, instead of "heptapeptide," I said 7-peptide and replaced "decapeptide" with 10-peptide. I used this opportunity to break with our earlier contention that we are familiar with the Greek numbers. The considerable variations in the nomenclature (e.g. hendecapeptide versus undecapeptide, tetracosapeptide or tetrakosipeptide[73]) of longer peptides suggest that uncertainties in this respect are not limited to our own publications. The use of an arabic numeral preceding the word "peptide" is self-explanatory and simple. The new expressions could be pronounced in the vernacular: ten-peptide in English, dix-peptide in French, zehn-Peptid in German or tiz-peptid in Hungarian, and yet be understood by all of us in the printed publications. I believe we peptide chemists can use any relief, even a small one, such as a simplification in the nomenclature.

Our task will not be easier in the future. As I just demonstrated, the individuality of amino acids and of peptide sequences can produce new surprises. New methods of analysis will reveal side reactions not recognized before. Thus, peptide synthesis still requires uninterrupted vigilance, unrelenting efforts and, particularly, improvements in a methodology which through unequivocal reactions will produce homogeneous final products. I believe that you will share my view about peptide synthesis: its challenge remains undiminished.

Support of our studies by NIH (AM12473) and NSF (CHE7615652) is gratefully acknowledged.

References

1. Wieland, T. (1966) in Foreword to *Peptide Synthesis*, by Bodanszky, M. & Ondetti, M. A., Wiley-Interscience Publ., New York, pp. vii–viii.

2. Bodanszky, M. & Bodanszky, A. (1967) *J. Chem. Soc. Chem. Commun.*, 591–593.

3. Magnusson, S., Sottrup-Jensen, L., Petersen, T. E., Morris, H. R. & Dell, A. (1974) *FEBS Lett.* **44**, 189–193; Stenflo, J., Fernlund, P., Egan, W. & Roepstorff, P. (1974) *Proc. Nat. Acad. Sci. USA* **71**, 2730–2733.

4. Märki, W. & Schwyzer, R. (1975) *Helv. Chim. Acta* **58**, 1471–1477; (1976) *ibid.* **59**, 1591–1592.

5. Boggs, N. T., Gawley, R. E., Koehler, K. A. & Hiskey, R. G. (1975) *J. Org. Chem.* **40**, 2850–2851.

6. Wright, L. D., Cresson, E. L., Skeggs, H. R., Wood, T. R., Peck, R. L., Wolf, D. E. & Folkers, K. (1950) *J. Amer. Chem. Soc.* **72**, 1048.

7. Becker, J. M., Wilchek, M. & Katchalski, E. (1971) *Proc. Nat. Acad. Sci. USA* **68**, 2604–2607; Viswanatha, T., Bayer, E. & Wilchek, M. (1975) *Biochim. Biophys. Acta* **401**, 152–156; McCormick, D. B. (1973) *J. Heterocycl. Chem.* **10**, 235–237.

8. Sakakibara, S. & Inukai, N. (1964) *Bull. Chem. Soc. Japan* **37**, 1231–1232.

9. Bodanszky, M. & Fagan, D. T. (1977) *J. Amer. Chem. Soc.* **99**, 235–239.

10. Bodanszky, M. & Chandramouli, N. (1977) Unpublished.

11. Wood, H. G. (1976) *Fed. Proc.* **35**, 1899–1907.

12. Dreyfuss, P. (1974) *J. Med. Chem.* **17**, 252–255.

13. Coyle, S., Keller, O. & Young, G. T. (1975) *J. Chem. Soc. Chem. Commun.*, 939–940.

14. Veber, D. F., Brady, S. F. & Hirschmann, R. (1972) in *Chemistry and Biology of Peptides*, Proc. 3rd Amer. Pept. Symp., Meienhofer, J., Ed., Ann Arbor Sci. Publ., Ann Arbor, Mich., pp. 315–319.

15. Tesser, G. I. (1975) in *Peptides-1974, Proc. 13th Eur. Pept. Symp.*, Kiryat Anavim, Israel, Wolman, Y., Ed., Wiley, New York-Israel University Press, Jerusalem, pp. 53–56.

16. Ericson, B. W. & Merrifield, R. B. (1972) in *Chemistry and Biology of Peptides*, Proc. 3rd Amer. Pept. Symp., Meienhofer, J., Ed., Ann Arbor Sci. Publ., Ann Arbor, Mich., pp. 191–195.

17. Yamashiro, D., Noble, R. L. & Li, Ch. H. (1972) in *Chemistry and Biology of Peptides*, Proc. 3rd Amer. Pept. Symp., Meienhofer, J., Ed., Ann Arbor Sci. Publ., Ann Arbor, Mich., pp. 197–202.

18. Carpenter, F. H. & Gish, D. T. (1952) *J. Amer. Chem. Soc.* **74**, 3818–3821.

19. Sakakibara, S. & Shimonishi, Y. (1965) *Bull. Chem. Soc. Japan* **38**, 1412–1413.

20. Felix, A. M. (1974) *J. Org. Chem.* **39**, 1427–1429.

21. Pless, J. & Bauer, W. (1973) *Angew. Chem.* **85**, 147–148.

22. Yajima, H., Fujii, N., Ogawa, H. & Kawatani, H. (1974) *J. Chem. Soc. Chem. Commun.*, 107–108.

23. Yajima, H., Kiso, Y., Ogawa, H., Iujii, N. & Irie, H. (1975) *Chem. Pharm. Bull. (Tokyo)* **23**, 1164–1166.

24. Matsuura, S., Niu, C.-H. & Cohen, J. S. (1976) *J. Chem. Soc. Chem. Commun.*, 451–452.

25. Bergmann, M. & Zervas, L. (1932) *Ber.* **65**, 1192–1201.

26. Schnabel, E., Klostermeyer, H. & Berndt, H. (1971) *Justus Liebigs Ann. Chem.* **749**, 90–108.

27. Klausner, Y. S., & Bodanszky, M. (1973) *Bioorg. Chem.* **2**, 354–362.

28. Riniker, B., Kamber, B. & Sieber, P. (1975) *Helv. Chim. Acta* **58**, 1086–1094.

29. Sieber, P., Kamber, B., Hartmann, A., Jöhl, A., Riniker, B. & Rittel, W. (1974) *Helv. Chim. Acta* **57**, 2617–2621.

30. Kenner, G. W., Moore, G. A. & Ramage, R. (1976) *Tetrahedron Lett.*, 3623–3626.

31. Wünsch, E., Fries, G. & Zwick, A. (1958) *Chem. Ber.* **91**, 542–547.

32. Ben-Ishai, D. & Berger, A. (1952) *J. Org. Chem.* **17**, 1564–1570.

33. Preitner, G. & Boissonnas, R. A. (1953) *Helv. Chim. Acta* **36**, 875–886.

34. Iselin, B. (1962) *Helv. Chim. Acta* **45**, 1510–1515.

35. Bodanszky, M. & Bodanszky, A. (1977) Unpublished.

36. Yang, C. C. & Merrifield, R. B. (1976) *J. Org. Chem.* **41**, 1032–1041.

37. Said, S. I. & Mutt, V. (1970) *Science* **169**, 1217–1218; Said, S. I. & Mutt, V. (1972) *Eur. J. Biochem.* **28**, 199–204.

38. Bodanszky, M. (1960) *Ann. N. Y. Acad. Sci.* **88**, 665–668.

39. Bodanszky, M. & Tolle, J. C. (1977) Unpublished.

40. Bodanszky, M. & Bodanszky, A. (1967) *J. Chem. Soc. Chem. Commun.*, 591–593; cf. also Bodanszky, M. & Bath, R. (1968) *ibid.*, 766–767.

41. Alder, R. W., Bowman, P. S., Steele, W. R. P. & Winterman, D. R. (1968) *J. Chem. Soc. Chem. Commun.*, 723–724.

42. König, W. & Geiger, R. (1973) *Chem. Ber.* **106**, 3626–3635.

43. Steglich, W. & Höfle, G. (1969) *Angew. Chem., Int. Ed.* 8, 981.

44. Ugi, I. (1969) Record of Chem. Progress **30**, 289–311; Ugi, I., Aigner, H., Beijer, B., Ben-Efraim, D., Burghard, H., Bukall, P., Eberle, G., Eckert, H., Marquarding, D., Rehn, D., Urban, R., Wackerle, L. & von Zychlinsky, H. (1977) in *Peptides-1976, Proc. 14th Eur. Pept. Symp.*, Loffet, A., Ed., Edition de l'Université de Bruxelles, Brussels, pp. 159–181.

45. Gawne, G., Kenner, G. W. & Sheppard, R. C. (1969) *J. Amer. Chem. Soc.* **91**, 5669–5671.

46. Castro, B. & Dormoy, J. R. (1971) *Bull. Soc. Chim.*, 3034–3036.

47. Bates, A. J., Kenner, G. W., Ramage, R. & Sheppard, R. C. (1973) in *Peptides-1972, Proc. 12th Eur. Pept. Symp.*, Hanson, H. & Jakubke, H. D., Eds., North Holland Publ. Co., Amsterdam, pp. 124–125.

48. Martinez, J. & Winternitz, F. (1975) *Tetrahedron Lett.*, 2631–2632.

49. Farrington, J. A., Kenner, G. W. & Turner, J. M. (1955) *Chem. & Ind.* (London) 1955, 601–602.

50. Iselin, B., Rittel, W., Sieber, P. & Schwyzer, R. (1957) *Helv. Chim. Acta* **40**, 373–387.

51. Nakonieczna, L., Makowski, Z., Berezowska, I. & Taschner, E. (1976) in *Peptides-1976, Proc. 14th Eur. Pept. Symp.*, Loffet, A., Ed., Edition de l'Universite de Bruxelles, Brussels, pp. 117–119.

52. Vaughan, J. R., Jr. & Osato, R. L. (1952) *J. Amer. Chem. Soc.* **74**, 676–678.

53. Tilak, M. A. (1970) *Tetrahedron Lett.*, 849–854.

54. van Zon, A. & Beyerman, H. C. (1973) *Helv. Chim. Acta* **56**, 1729–1740; (1976) *ibid.* **59**, 1112–1126.

55. Bodanszky, M. & Tolle, J. C. (1977) *Internat. J. Pept. Protein Res.*, in press.

56. Weygand, T., Huber, P. & Weiss, K. (1967) *Z. Naturforsch.* **226**, 1084–1085; Birr, Ch., Flor, F., Fleckenstein, P., Wieland, T. (1971) in *Peptides-1971, Proc. 11th Eur. Pept. Symp.*, Nesvadba, H., Ed., North Holland Publ., Amsterdam, pp. 175–184.

57. Bodanszky, M. (1955) *Nature* **175**, 685.

58. Pettit, G. R. (1976) *Synthetic Peptides*, vol. 4, Elsevier, Amsterdam, pp. 22–32.

59. Weygand, F., Hoffmann, D. & Wünsch, E. (1966) *Z. Naturforsch* **21b**, 426–428; Wünsch, E. & Drees, F. (1966) *Chem. Ber.* **99**, 110–120.

60. König, W. & Geiger, R. (1970) *Chem. Ber.* **103**, 788–798.

61. Itoh, M. (1972) in *Chemistry and Biology of Peptides*, Proc. 3rd Amer. Pept. Symp., Meienhofer, J., Ed., Ann Arbor Science Publ., Ann Arbor, Mich., pp. 365–367; (1973) *Bull. Chem. Soc. Japan* **46**, 2219–2221.

62. König, W. & Geiger, R. (1973) *Chem. Ber.* **106**, 3626–3635.

63. Bodanszky, M. (1968) in *Peptides-1968, Proc. 4th Eur. Pept. Symp.*, Bricas, E., Ed., North Holland Publ., Amsterdam, p. 150.

64. Fletcher, G. A., Löw, M. & Young, G. T. (1973) *J. Chem. Soc.*, 1162–1164.

65. Bodanszky, M., Fink, M. L., Klausner, Y. S., Natarajan, S., Tatemoto, K., Yiotakis, A. E. & Bodanszky, A. (1977) *J. Org. Chem.* **42**, 149–152.

66. Sheehan, J. C. & Hess, G. P. (1955) *J. Amer. Chem. Soc.* 77, 1067–1068.

67. Curtius, T. (1902) *Ber.* **35**, 3226–3228.

68. Brenner, M. (1967) in *Peptides, Proc. 8th Eur. Pept. Symp.*, Beyerman, H. C., Van de Linde, A. & Maassen van den Brink, Eds., North Holland Publ., Amsterdam, pp. 1–7.

69. Taschner, E., Rzeszotarska, B. & Lubiewska, L. (1967) Chem. & Ind. (London) 1967, 402–404; cf. also Lubiewska-Nakonieszna, L., Rzeszotarka, B. & Taschner, E. (1970) *Justus Liebigs Ann. Chem.* **741**, 157–166.

70. Heikens, D., Hermans, P. H. & van Velden, F. (1954) *Nature* **174**, 1187–1188.

71. Bodanszky, M. & Natarajan, S. (1977) Unpublished.

72. Mutt, V. & Jorpes, J. E. (1971) *Biochem. J.* **125**, 57–58.

73. Lang, U., Karlaganis, G., Seelig, S., Sayers, G. & Schwyzer, R. (1973) *Helv. Chim. Acta* **56**, 1069–1072.

FLUORESCENT TECHNIQUES FOR ULTRAMICRO PEPTIDE AND PROTEIN CHEMISTRY

SIDNEY UDENFRIEND and STANLEY STEIN,
Roche Institute of Molecular Biology,
Nutley, New Jersey 07110

Isolation and characterization of proteins and peptides is a key step in almost all fields of biology, from enzymology to endocrinology and immunology. The technology introduced over the last twenty-five years, including automated ninhydrin assay, the automated sequenator etc. has made it possible to isolate and characterize those proteins and peptides which are present in relatively high concentration in tissues and are relatively stable, or which can be readily concentrated from large amounts of tissues. However, when one is confronted with a protein or peptide that is present in tissues at concentrations of micrograms per gram or lower, the classical methods can only be applied with difficulty. For example, endocrinologists isolating hormones from the pituitary gland have had to use several hundred thousand glands from slaughterhouse animals. While this approach has been successful it requires an initial outlay for tissue approaching the million-dollar range. The cost of solvents and chemicals is proportionally high.

The use of slaughterhouse materials has other disadvantages, including the problem of controlling proteolytic enzymes. Proteolysis can occur during collection, during transfer, prior to freezing, during thawing, and while working up the tissues and extracts. With slaughterhouse material it is difficult to control the first few steps, resulting in two types of artifacts, 1) the appearance of peptides which are normally not present in the tissues and, 2) the disappearance of larger peptides. Finally, methods which require more than several nanomoles of peptides cannot be applied to the tissues of small laboratory animals usually employed for biological experimentation.

Several years ago fluorescamine (Fluram™) was introduced as a fluorometric reagent for all primary amines[1] and was shown to be useful in assaying amino acids, peptides and proteins in the picomole range.[2] Subsequently, a related reagent, 2-methoxy-2, 4-diphenyl-3(2*H*) furanone (MDPF) was introduced[3] (Fig. 1). Since then, our laboratory and others have found these two reagents to be useful for monitoring all phases of protein and peptide chemistry including isolation, by both classical and chromatographic means, molecular weight determination, amino-acid analysis, fingerprinting and sequencing, as well as peptide synthesis and quantitative assay of tissue extracts. In this review we shall discuss each of these processes and present some applications for which fluorescamine or MDPF may be uniquely suited.

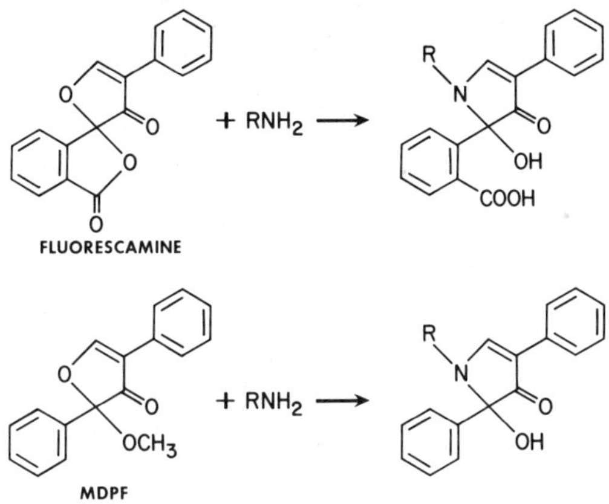

Fig. 1. Fluorogenic reactions of fluorescamine and MDPF with primary amines.

Amino-acid Assay

The properties of fluorescamine make it ideally suited for automated amino-acid analysis.[2] Reaction with amino acids is virtually instantaneous at room temperature and destruction of excess reagent occurs rapidly. Fluorescamine, as well as its hydrolysis products, are nonfluorescent while the amino-acid derivatives are intensely fluorescent. Since picomole quantities are sufficient for assay, small-diameter columns have sufficient capacity and may be used for chromatography. Another important advantage is that fluorescamine is relatively insensitive to ammonia, yielding less than 1% of the fluorescence observed with amino acids. Proline and hydroxyproline can be assayed after oxidation to the corresponding amino aldehydes.[4]

All the above steps have been incorporated into a fully automated amino-acid analyzer which has been described elsewhere.[5,6] Commercial fluorescamine amino-acid analyzers may soon become available.

We have used fluorescamine amino-acid analysis for many problems in which only picomole quantities of peptides or proteins were available (Fig. 2). A unique application of the method has been in the identification of a tissue protein which crossreacts with antibody directed against the enzyme prolyl hydroxylase.[7] The enzyme and the crossreacting protein (CRP) are found in all tissues which form collagen, with the latter always present in excess. Both were purified in microgram quantities from newborn-rat skin. When subjected to gel electrophoresis in sodium dodecyl sulfate (SDS) the enzyme, which exists as a tetramer with a molecular weight of 240,000, was resolved into its two component subunits, a larger one (S_L) about 64,000, and a smaller one (S_s) about 60,000. When subjected to SDS gel electrophoresis CRP migrated to the same position as (S_s)

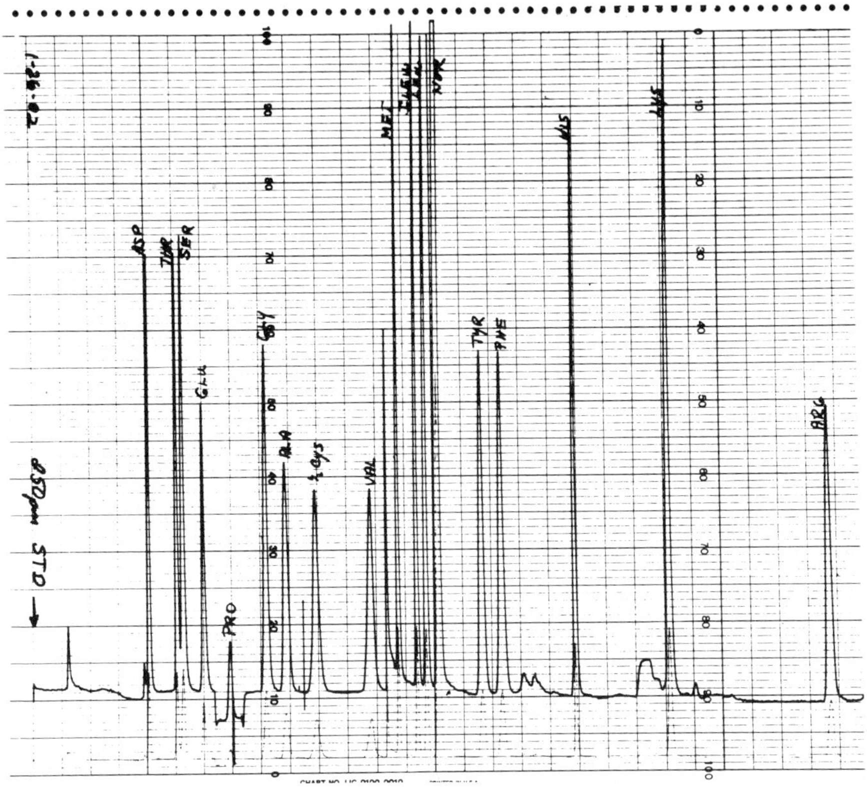

Fig. 2. Amino-acid analysis of a standard mixture (250 pmol of each amino acid). Detection was with fluorescamine. Total run time was 3 h.

indicating both were of the same molecular weight. The stained bands, containing microgram quantities of protein, were cut out, hydrolyzed in acid and subjected to amino-acid analysis as previously described.[8] CRP and S_s were found to have the same amino-acid compositions (within experimental error) providing evidence for their identity. Fluorescamine and MDPF are now being used for fingerprinting microgram quantities of the two peptides with a molecular weight of 60,000 to confirm their identity and to demonstrate for the first time the normal occurrence of an excess of a free subunit along with its enzyme. The requirement of only picomole quantities of the proteins facilitated these studies immensely.

There have been previous attempts to determine the amino-acid composition of protein bands in polyacrylamide gels. However, the large amounts of ammonia released from the gels during hydrolysis made the use of ninhydrin extremely difficult.[9] *o*-Pthalaldehyde, another fluorescent reagent which has been used for amino acid assay at the picomole level[10], would also be unsuitable for this application, since it, unlike fluorescamine yields intense fluorescence with ammonia.

Peptide Purification

The properties of fluorescamine make it ideal for flow systems. Reaction with peptides is rapid and both the reagent and its hydrolysis products are non-fluorescent. When used only for assay, the total column effluent may be directed into a fluorometric detection system, as is done in the amino-acid analyzer. As little as 5 pmol of the dipeptide, carnosine, has been measured by such a procedure.[11] However, for purifying peptides at the nanomole level it is important that only a small portion be used for detection. By employing an automatic stream-sampling valve only a portion of a column effluent is assayed (picomoles) while the remainder (nanomoles) is directed to a fraction collector.[12] Such a system has been used in this laboratory for the chromatographic isolation of nanomole quantities of peptides ranging in size from a few amino-acid residues to over 90. Some applications are described below.

The purity of peptides which are to serve as calibration standards not only for chemical assay, but also for bioassay and radioimmunoassay, must be established. Furthermore, peptides that are to serve as antigenic determinants in the preparation of specific antibodies must be homogeneous to yield meaningful information. Although synthetic peptides can now be made in high purity, impure preparations are still common. For example, synthetic β-endorphin obtained from Dr. C. H. Li was found to be homogeneous by our criteria, while a commercial preparation of the same peptide was not satisfactory (Fig. 3). Impure peptide preparations such as commercial vasopressin, can often be purified by chromatography on a high efficiency column.[13] Radioactive peptides present an even greater problem. They are generally prepared in small amounts and are of high specific activity. Their price is equally high. The laboratories which prepare them must waste relatively large (and therefore costly) amounts to detect and remove impurities. Since few scientists who use them are equipped to monitor their purity, the result is that impure labelled peptides may be used as prime standards in many biological studies. Fluorescamine methodology is ideally suited for the purification of labelled peptides without sacrificing much for assay.

The research in our laboratory has been concerned with the study of biologically active peptides, at present with those possessing opioid activity. To monitor the opioid peptides in column effluents we have set up a competitive binding assay with neuroblastoma X glioma hybrid cells[14] using tritiated leucine-enkephalin as the displaced ligand.[15] Since both the bioassay and the fluorometric procedures are sensitive at the picomole level, we have found it possible to use tissues from relatively few laboratory animals (rats and guinea pigs) instead of slaughterhouse material.[16,17] As noted in the introduction, use of laboratory

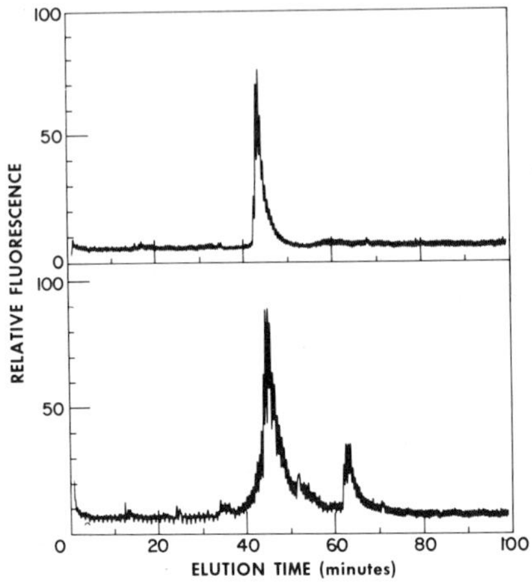

Fig. 3. Comparison of two preparations of synthetic β-endorphin by chromatography on Partisil SCX. Dr. C. H. Li of the Hormone Research Laboratory, University of California at San Francisco, kindly supplied the sample of β-endorphin depicted in the top panel, while the preparation shown in the bottom panel was from a commercial source. The fluorescamine column monitoring system was used for detection.

animals is important for many reasons. Among these are the ability to remove the tissues from the animals rapidly and to homogenize them under conditions where proteolytic degradation is inhibited. Besides generating superfluous peptides that can complicate the isolation procedure, proteolysis can significantly lower the yield of the peptide being sought. Both problems were encountered when we attempted to isolate β-lipotropin from frozen rat pituitary glands from a commercial source. Chromatography of the pituitary homogenate on Sephadex indicated that much of the protein had been degraded to small peptides. The recovery of β-lipotropin was markedly reduced and endorphins were correspondingly increased.[16,17]

In order to investigate the pharmacology and physiology of the opioid peptides it is necessary to use small laboratory animals, especially the rat, which has traditionally been used in studies on analgesia. Thus it was necessary to isolate and characterize the various biologically active peptides from rat tissues, since one might expect to find differences in the primary structures compared to the corresponding peptides which were obtained from large animals. We were able to isolate β-lipotropin from relatively few fresh rat anterior pituitaries.[16,17] An extract of 40 rat glands was fractionated on G-75, then on the strong cation exchanger Partisil SCX[13], and finally on the reverse-phase Lichrosorb RP-18 (Fig. 4). The effluent from each column was monitored with the automated fluorescamine detection system. In addition, aliquots from the collected fractions were treated with trypsin in order to convert the relatively inactive β-lipotropin to highly active cleavage products, which were then determined by

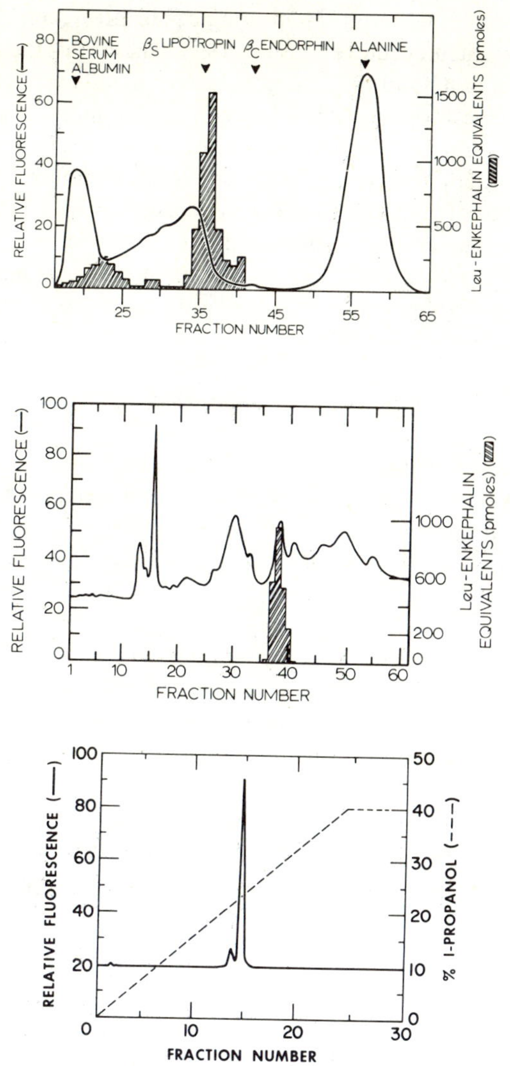

Fig. 4. Chromatographic steps for the purification of rat β-lipotropin (from Ref. 16). The fluorescamine column monitoring system was used for chemical detection. Aliquots of fractions were trypsinized and then tested for opioid activity in the neuroblastoma X glioma binding assay. An extract of 40 rat anterior pituitaries was resolved on Sephadex G-75 (top panel), then on Partisil SCX (middle panel), and finally on Lichrosorb RP-18 (bottom panel). The major peak in the bottom panel represented rat β-lipotropin.

the competitive binding assay. Monitoring of all three columns by the fluorometric and biological procedures consumed less than 20% of the starting material. A yield of 4 nmol of β-lipotropin was obtained from 300 mg of anterior pituitary. Some of this material was used to determine the molecular weight and the amino-acid composition. Similar procedures were used to isolate and characterize rat β-endorphin[17] and are being used in the isolation of opioid peptides from other regions of the brain. We have succeeded in preparing pure methionine-enkephalin from guinea pig striatum and have observed several other peaks of opioid activity in this tissue. The sensitivity of fluorescamine monitoring of columns makes it ideally suited to the isolation and characterization of peptides present in tissues in small quantities.

Elucidation of Primary Structure

These microfluorometric procedures can be useful in elucidating the primary structure of peptides and proteins when only nanomole quantities are available. Picomole amounts have been used for molecular weight determination utilizing comparative gel filtration.[16] Fluorescent procedures have also been used for fingerprinting by a variety of procedures.[18-20] In our laboratory, the cleavage products obtained by enzymatic or chemical degradation of proteins or peptides are resolved by column chromatography using the fluorescence monitoring system. This procedure has been used to demonstrate the identity of rat and camel endorphins.[17] Alternatively, separation of peptides can be achieved by disc gel electrophoresis. For the latter method it has been possible to prelabel proteins or peptides with fluorescamine[21-26] or MDPF[27] and then perform the electrophoresis in the presence of sodium dodecyl sulfate. Application of the latter procedure to the tryptic peptides of aldolase prelabelled with MDPF is shown in Fig. 5. With this methodology the tedious job of staining and destaining is not necessary, since the bands may be visualized with an ultraviolet lamp even during the electrophoresis. Furthermore, because the gels need not be fixed with trichloroacetic acid or acetic acid/methanol, the protein and peptide derivatives

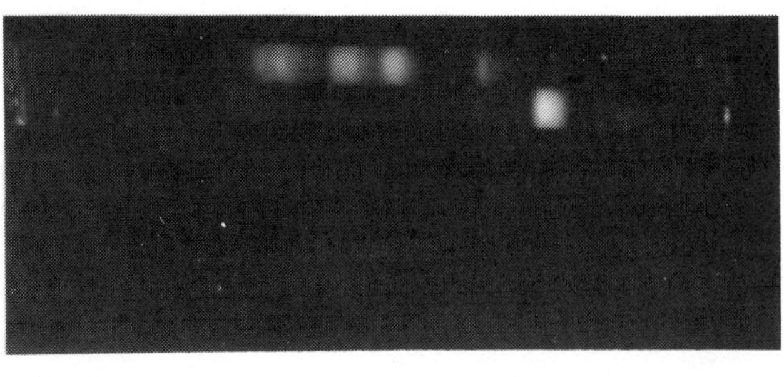

Fig. 5. Polyacrylamide gel electrophoresis in SDS of MDPF-aldolase (bottom) and MDPF-aldolase fragmented with cyanogen bromide (top). Gel top is at the right.

may be recovered from the gel in high yield. This prelabeling technique is especially useful for peptides which cannot be precipitated in the gel (those with a molecular weight of under a few thousand). The MDPF fluorescent label on peptides has been found to be stable even after cleavage at methionyl residues by cyanogen bromide in formic acid.[27] Fluorophor formation, which occurs quantitatively at the ϵ-amino groups of the lysyl residues, should block the activity of trypsin at this site, thereby leading to splitting only at arginine residues. After each chemical or enzymatic cleavage step it would be appropriate to label the newly generated α-amino groups with MDPF.

In our Institute, Dr. C. Y. Lai has been able to hydrolyze the thiazolinone derivatives formed in the Edman degradation and to obtain high yields of the respective amino acids.[28] He is currently utilizing this procedure for sequence analysis. Since the proteins he is using are available in relatively large amounts, he carries out the initial steps with the ninhydrin amino-acid analyzer. With 20 nmol of a peptide he sequences about 10 residues and then employs the fluorescamine amino-acid analyzer because of the decreasing amounts of material. The fluorescamine amino-acid analyzer allows him to sequence an additional 5 residues. It should be possible to utilize the fluorescamine assay exclusively and carry out sequencing with nanomole quantities of peptides.

Obviously fluorescamine is ideally suited for monitoring residues released from peptides by carboxypeptidase or aminopeptidase. A unique approach is to incubate the peptide being sequenced with a dipeptidyl aminopeptidase. MDPF derivatives of the resultant dipeptides are then prepared and from mass spectral analysis it is possible to obtain the composition and sequence of each peptide.[29,30] Repetition of the procedure after a one-step Edman degradation yields the overlapping dipeptides.

Peptide Synthesis

It is important to check for completeness of reaction at each solid-phase step of peptide snythesis and for this application fluorescamine can detect less than 0.5% of uncoupled peptide.[31,32] Alternatively, the nascent peptide may be removed from a small portion of the resin in order to check for homogeneity at intermediate steps. Larger peptides will have accumulated errors, requiring isolation of the final product. The high-efficiency liquid chromatography, which we have introduced, would be suitable for purification, and little of the product will be consumed for column monitoring and amino-acid analysis. Exemplifying the resolving power of high efficiency columns for this purpose is the separation of the hexadecapeptide, α-endorphin, from the heptadecapeptide, γ-endorphin, which has an additional leucine residue at the carboxy terminal (Fig. 6).

Incomplete coupling at each cycle in solid phase synthesis represents an important problem, especially in the preparation of large peptides. Not only is the final product impure, but incorrect peptides, which may contain one or more

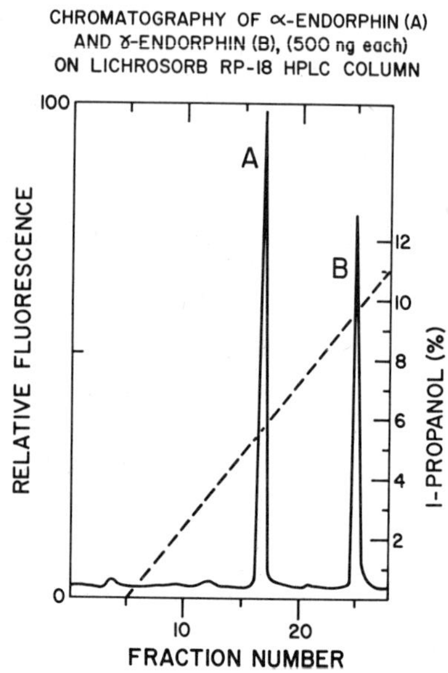

Fig. 6. Reverse-phase column chromatography of α-endorphin and γ-endorphin. The starting and limiting buffers for the linear gradient both contained 1.0 *M* pyridine, 0.5 *M* acetic acid, 10^{-2}% (v/v) thiodiglycol and 10^{-4}% (w/v) pentachlorophenol.

deletions, may be difficult to remove. Reaction with fluorescamine after each coupling step has been used to block further growth of any uncoupled peptide, resulting in a final product that is easier to purify.[33]

Quantitative Analysis of Peptides

Peptides in a column effluent may be quantified by continuously adding fluorescamine with a pump and directing the mixture to the flow-cell of a fluorometer, as is done in the amino-acid analyzer. The monitoring system (see Peptide Purification, above), which takes a portion of the column effluent for detection, also provides quantitation. However, there are advantages to forming the fluorophors prior to resolution of the peptide mixture by column chromatography, in contrast with post-column labeling. Only one pump (for the column) is then required, since the reagent has already been added to the sample. This makes the instrument simpler and more reliable. Since the fluorogenic reagent is not added in a continuous manner, a lower and more stable baseline fluorescence is achieved, thereby extending the lower limit of detection.

Determination of the nonapeptides oxytocin and vasopressin in individual rat posterior pituitaries served as a model study for the prelabeling technique.[34]

For this procedure, a partially purified peptide fraction was prepared as follows: Each pituitary was homogenized in dilute acid and proteins were then precipitated with trichloroacetic acid. Passage through a copper-Sephadex column removed α-amino acids, polyamines and lipoamines. The resultant peptide fraction was treated with fluorescamine and the product resolved on a reverse-phase column with an acetone/water gradient designed for the optimal separation of the derivatives of oxytocin, vasopressin and synthetic vasotocin. The latter had been added to the homogenate as an internal standard (Fig. 7).

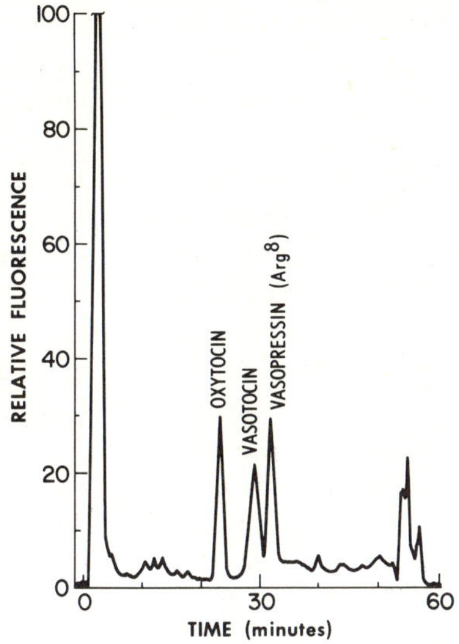

Fig. 7. Reverse-phase column chromatography (Partisil ODS) of a partially purified rat posterior pituitary extract, which had been prelabeled with fluorescamine. The internal standard vasotocin had been added during tissue homogenization.

Specificity and accuracy of the assay were checked in several ways. Fluorophors of synthetic oxytocin and vasopressin co-chromatographed with the tissue-derived nonapeptides. Oxytocin and vasopressin fluorophors in the column effluent were collected and subjected to acid-hydrolysis followed by amino-acid analysis. The expected compositions were obtained, while the negligible quantities of extraneous amino acids indicated the homogeneity of each peak. Quantitation of the peptide fluorophors by amino-acid analysis was in agreement with the fluorescence determinations. Furthermore, aliquots of a pituitary extract were analyzed by a pressor assay specific for vasopressin and the results were in agreement with the fluorescence assay. The chemical assay confirmed previous findings with bioassays and radioimmunoassays that the two nonapeptides are present in equivalent amounts in each pituitary.

Although fluorescamine was used as the prelabeling agent in the initial studies with oxytocin and vasopressin, MDPF is now the reagent of choice for this methodology.[35] Fluorophors derived from MDPF are more stable even at extremes of pH. The more hydrophobic nature of the fluorophors of MDPF, as compared to their fluorescamine analogs [due to a difference of one carboxylic acid group (see Fig. 1)], aids in the reverse-phase chromatography of small peptide-fluorophors, like that produced with the dipeptide carnosine.[35]

The sensitivity of peptide assay by prelabeling is often at or below the picomole level. Fluorometric assays by the prelabeling procedure are currently being developed for the various opioid peptides in tissues. MDPF-fluorophors of methionine-enkephalin and leucine-enkephalin are well resolved on the reverse-phase column and can be measured in the low picomole range.[17] This sensitivity is sufficient for assay of the enkephalins in the brains of individual laboratory animals.

Detection at the picomole level is also convenient for determining the specific activity of synthetic radiolabeled peptides, since little of the precious material need be sacrificed for chemical analysis. Furthermore, both radioactivity and peptide concentration of each peptide-fluorophor peak can be measured and thereby yield specific activity directly. Other regions of the chromatogram reveal both fluorogenic and radioactive impurities in the preparation. Such a procedure has already been applied to [^{3}H]-carnosine prepared enzymatically from either of the labeled component amino acids, histidine and β-alanine.[36] This procedure should lend itself to studies on the *in vivo* turnover of peptides following administration of a radiolabeled precursor amino acid. Measurement of the change in specific activity with time would reveal the anabolic and catabolic rates. In addition, a precursor-product relationship may be ascertained by using pulse-chase type of experiments.

Quantitation by the prelabeling method can achieve a sensitivity comparable to radioimmunoassay and bioassay. An advantage of the chemical assay is its specificity. Radioimmunoassays may be subject to interference by metabolites or other structurally related compounds, while bioassays may involve a complex sequence of events, any step of which could be influenced by unknown factors. A disadvantage of a chemical assay, especially when dealing with a peptide which is a minor component of the tissue, is that preliminary purification is required. However, an added feature is that more than one component can be determined in a single analysis as in the oxytocin and vasopressin assay.[34]

Concluding Remarks

Many of the procedures presented in this article depend on instrumentation which is not yet commercially available. Although the instruments are not unusually complicated, at present, their construction and maintenance demands a substantial part of the investigator's time. Commercial equipment of appropriate design would not only be more reliable, but should also be capable of analysis at lower levels than the laboratory-built instruments. Should commercial instruments become available, fluorescamine and MDPF, as well as other fluorescent reagents, will most likely become widely used, not only because of their inherently greater sensitivity and specificity, but for the resulting economy and other considerations cited above.

Analysis at the picomole level requires special considerations. Losses of peptides due to adsorption becomes readily apparent at these levels. We have minimized this in many ways. We use polypropylene tubes and pipet tips or glassware treated with Siliclad. Peptides are usually stored in dilute hydrochloric acid (0.01 *N*). The antioxidant thiodiglycol and the antimicrobial agent pentachlorophenol are normally added to all buffers used in chromatography. Phenylmethylsulfonyl fluoride, or other protease inhibitors are always added during tissue homogenization, and are generally included in all peptide solutions. In order to minimize background contamination it is often necessary to distill all volatile solvents and reagents, including pyridine, acetic acid, and hydrochloric acid.[34] With these procedures and precautions we have not found it cumbersome to operate at the picomole level. In fact, there are many offsetting features that make the micromethodology far simpler than it appears. We hope that researchers will take advantage of these procedures and that this technology may become routine for the preparation and analysis of peptides.

The unpublished studies on opioid peptides reported here represent work carried out in collaboration with Dr. Menachem Rubinstein; the enkephalin studies were done by Dr. Janusz Wideman. Ms. Louise D. Gerber and Mr. Larry Brink played key roles in all these studies.

References

1. Weigele, M., DeBernardo, S., Tengi, J. P. & Leimgruber, W. (1972) *J. Amer. Chem. Soc.* **94**, 5927–5928.

2. Udenfriend, S., Stein, S., Böhlen, P., Dairman, W., Leimgruber, W. & Weigele, M. (1972) *Science* **178**, 871–872.

3. Weigele, M., DeBernardo, S., Leimgruber, W., Cleeland, R. & Grunberg, E. (1973) *Biochem. Biophys. Res. Commun.* **54**, 899–906.

4. Weigele, M., DeBernardo, S. & Leimgruber, W. (1973) *Biochem. Biophys. Res. Commun.* **50**, 352–356.

5. Stein, S., Böhlen, P., Stone, J., Dairman, W. & Udenfriend, S. (1973) *Arch. Biochem. Biophys.* **155**, 203–212.

6. Felix, A. M. & Terkelsen, G. (1973) *Arch. Biochem. Biophys.* **157**, 177–182.

7. Chen-Kiang, S., Cardinale, G. & Udenfriend, S. manuscript in preparation.

8. Stein, S., Chang, C. H., Böhlen, P., Imai, K. & Udenfriend, S. (1974) *Anal. Biochem.* **60**, 272–277.

9. Houston, L. L. (1971) *Anal. Biochem.* **44**, 81–88.

10. Benson, J. R. and Hare, P. E. (1975) *Proc. Nat. Acad. Sci. USA* **72**, 619–622.

11. Margolis, F., personal communication.

12. Böhlen, P., Stein, S., Stone, J. & Udenfriend, S. (1975) *Anal. Biochem.* **67**, 438–445.

13. Radhakrishnan, A. N., Stein, S., Licht, A., Gruber, K. A. & Udenfriend, S. (1977) *J. Chromatog.* **132**, 552–555.

14. Klee, W. A. & Nirenberg, M. (1974) *Proc. Nat. Acad. Sci. USA* **71**, 3474–3477.

15. Gerber, L. D., Stein, S., Rubinstein, M. & Udenfriend, S., manuscript in preparation.

16. Rubinstein, M., Stein, S., Gerber, L. D. & Udenfriend, S. (1977) *Proc. Nat. Acad. Sci. USA*, *74,* 3052-3055.

17. Rubinstein, M., Wideman, J. & Stein, S. (1977) in *Peptides: Proceedings of the 5th American Peptide Symposium*, Goodman, M. & Meienhofer, J., Eds., John Wiley & Sons, Inc., New York, 41–43.

18. Furlan, M. & Beck, E. A. (1974) *J. Chromatog.* **101**, 244–246.

19. Mendez, E. & Lai, C. Y. (1975) *Anal. Biochem.* **65**, 281–292.

20. Nakai, N., Lai, C. Y. & Horecker, B. L. (1974) *Anal. Biochem.* **58**, 563–570.

21. Eng, P. R. & Parkes, C. O. (1974) *Anal. Biochem.* **59**, 323–325.

22. Pace, J. L., Kemper, D. L. & Ragland, W. L. (1974) *Biochem. Biophys. Res. Commun.* **57**, 482–487.

23. Thompson, E. J., Daroga, B. M., Quick, J. M. S., Shortman, R. C., Hughes, B. P. & Davison, H. N. (1974) *Biochem. Soc. Trans.* **2**, 989–990.

24. Zak, D. L. & Kenney, P. G. (1973) *J. Agr. Food* **22**, 800–802.

25. Stephens, R. E. (1975) *Anal. Biochem.* **65**, 369–379.

26. Rosemblatt, M. S., Margolies, M. H. & Cannon, L. (1975) *Anal. Biochem.* **65**, 321–330.

27. Chen-Kiang, S., Stein, S. & Udenfriend, S. manuscript in preparation.

28. Mendez, E. & Lai, C. Y. (1975) *Anal. Biochem.* **58**, 563–570.

29. Pritchard, D., Schute, W. C. & Todd, C. W. (1975) *Biochem. Biophys. Res. Commun.* **65**, 312–316.

30. Todd, C. W. (1977) in *Peptides: Proceedings of the 5th American Peptide Symposium,* Goodman, M. & Meienhofer, J., Eds., John Wiley & Sons, Inc., New York, 27–40.

31. Felix, A. M. & Jimenez, M. H. (1973) *Anal. Biochem.* **52**, 377–381.

32. Tometsko, A. M. & Vogelstein, E. (1975) *Anal. Biochem.* **64**, 438–443.

33. Felix, A. M., Jimenez, M. H., Vergona, R. & Cohen, M. R. (1975) *Int. J. Pept. Protein Res.* **7**, 11–22.

34. Gruber, K. A., Stein, S., Radhakrishnan, A. N., Brink, L. & Udenfriend, S. (1976) *Proc. Nat. Acad. Sci. USA* **73**, 1314–1318.

35. Wideman, J., Brink, L. & Stein, S. manuscript in preparation.

36. Margolis, F. M., Grillo, M., Wideman, J. & Stein, S., unpublished results.

MASS SPECTRAL SEQUENCING OF MICRO QUANTITIES OF PEPTIDES

CHARLES W. TODD, *Division of Immunology*
City of Hope National Medical Center, Duarte, California, 91010

This presentation will be an overview of the mass spectral approach to peptide sequencing. It is assumed that your presence at this symposium indicates some knowledge and interest in peptide sequencing. Many of you also have familiarity with the application of mass spectrometry to this problem. However, I shall address this talk to those among you who have had limited or no experience with mass spectrometry. For this reason I shall begin with a discussion of the instrumentation and techniques of mass spectrometry. Following that we shall proceed to a consideration of the chemistry involved and the various strategies by which the problem can be approached. Finally we shall conclude with a summary of the present advantages and disadvantages of the mass spectral approach and attempt to visualize probable advances in this field.

The preparation of this overview was greatly simplified by the recent appearance of an excellent review by Arpino and McLafferty.[1] Their bibliography of 251 references would alone use all the space allotted me here. Thus no attempt will be made to do justice to the many individuals who have labored to develop this area of knowledge. Only a few references appearing too late to be included in the review of Arpino and McLafferty[1] are cited here.

A mass spectrometer basically consists of a source in which ions are generated, a means of accelerating the ions, a means of sorting the ions based on their mass (more properly their *m*/*e* ratio), an ion detector (commonly a photo plate or electron multiplier tube), and desirably a data processing system (Figure 1). Details of instrumentation and operation have recently been reviewed by Roboz and Chait.[2]

Mass spectrometers are commonly described in terms of the means by which the ions are sorted. (Arithmetic symbols used are *CF*, centrifugal force; *e*, charge

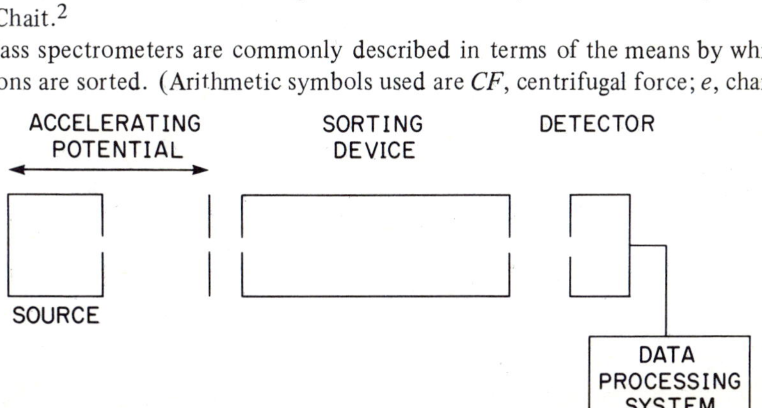

Fig. 1. Basic components of a mass spectrometer.

on electron; E, electric field in electrostatic sector; H, magnetic field; m, mass; r, radius; V, accelerating voltage; v, velocity.) The simplest in concept is the time of flight instrument. Since all singly charged ions possess the same kinetic energy determined by the accelerating voltage ($eV = mv^2/2$), the velocity is inversely proportional to $m^{1/2}$. Thus the time to traverse an evacuated tube from the source to the detector is proportional to $m^{1/2}$. The source is pulsed, and the time of arrival of ions at the detector is measured. Resolution is not high, as it is limited by the initial spread in the kinetic energies of the molecules being ionized and the decay time of the detector system.

Magnetic instruments are based on the principle that a charged particle moving in a magnetic field is deflected by that field in proportion to its charge, velocity, mass, and the strength of the magnetic field (Figure 2). Note that r can be adjusted either by changing the strength of the field, H, or the accelerating voltage, V. Some instruments vary one, while other instruments vary the other.

The ionizing electron beam has a finite width. An ion formed in one region may be subjected to a different accelerating voltage than one formed in another. The difference in kinetic energy will alter the velocity in the magnetic field and thus bring the molecule to focus on the detector as though it had a greater or lesser mass. A device that would allow only molecules with the same kinetic

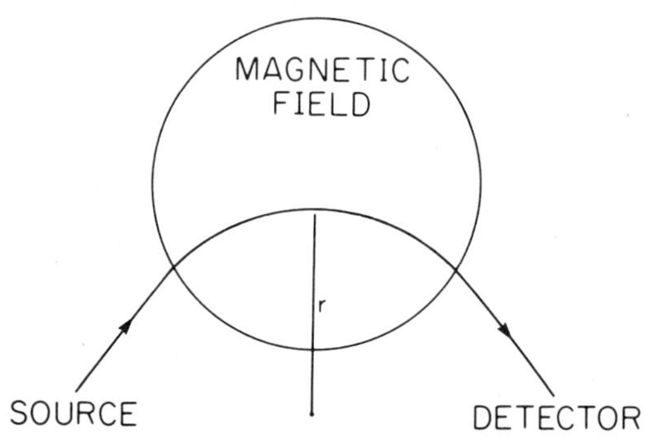

$$CF = Hev = \frac{mv^2}{r}$$

$$\frac{m}{e} = \frac{Hr}{v}$$

$$eV = \frac{mv^2}{r} \qquad v = \sqrt{\frac{2eV}{m}}$$

$$\frac{m}{e} = \frac{Hr}{\sqrt{\frac{2eV}{m}}} = \frac{H^2r^2}{2V}$$

Fig. 2. Principle of magnetic mass spectrometer.

energy to pass would improve the resolution of the instrument. An electrostatic sector is such a device. Ions failing to satisfy the equation for the velocity in Figure 3 will not pass the exit slit.

When the electrostatic sector is placed in series before or after the magnetic sector, one has a double focusing instrument (Figure 4). Such an instrument has increased resolution. This increase in resolution is achieved at the price of decreased sensitivity, since many ions are discarded in the focusing process.

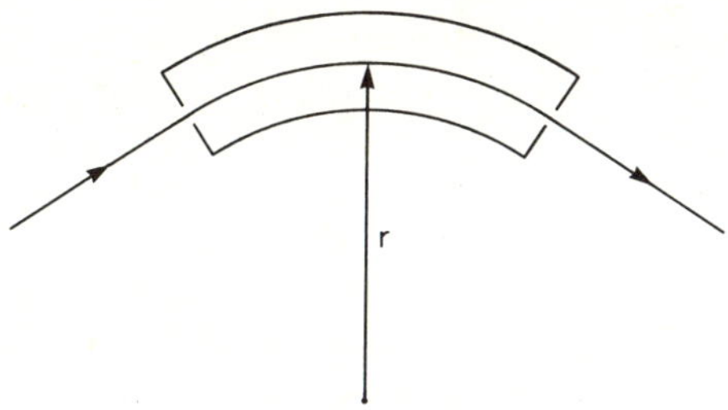

$$CF = \frac{mv^2}{r} = eE$$

$$v = \sqrt{\frac{2eV}{m}}$$

$$\frac{2V}{r} = E \qquad r = \frac{2V}{E}$$

Fig. 3. Principle of the electrostatic sector. The electric field, E, is parallel to the radius.

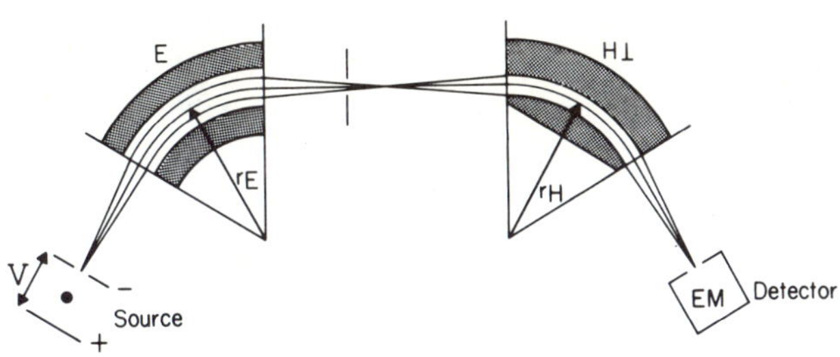

Source

$$eV = \frac{mv^2}{2}$$

$$v = \sqrt{\frac{2eV}{m}}$$

Electric Sector

$$eE = \frac{mv^2}{r_E} = CF$$

$$eE = \frac{m}{r_E}\left(\frac{2eV}{m}\right)$$

$$r_E = \frac{2V}{E}$$

Magnetic Sector

$$Hev = \frac{mv^2}{r_H} = CF$$

$$v = \frac{Hr_H e}{m} = \sqrt{\frac{2Ve}{m}}$$

$$\frac{H^2 r_H^2 e}{m} = 2V$$

$$\frac{m}{e} = \frac{H^2 r_H^2}{2V}$$

Fig. 4. Double focusing mass spectrometer.

Resolution is measured as described in Figure 5. By arbitrary definition resolution is expressed as the increment of masses, Δm, divided by the mass, m, corresponding to the valley between two mass peaks of equal height, h, when the overlap height, Δh, is 10% of h. Resolution is sometimes also expressed as the reciprocal of this figure. Reference to the curve on the left of Figure 5 shows that this valley is lowered by removing ions of higher and lower velocities so that the probability, *P*, curve for velocity distribution is sharpened.

Conventionally mass spectrometers with a resolution of better than 1/5000, or 5000, are referred to as high resolution instruments. Thus an instrument with a resolution of 15,000, such as the du Pont 492, can separate two masses of 150.00 and 150.01 with a 10% overlap in the valley between them. While high resolution and exact mass measurement are not synonymous, such an instrument is capable of determining the mass to 1 part in 200,000. Accordingly the mass of a molecule might be determined as 200.001 with the uncertainty in the 3rd decimal place.

A third way of sorting ions by mass is with a quadrupole filter (Figure 6). The principle of the quadrupole filter is based on alternating electric fields applied to four rods held in square array. To each diagonally paired set of rods a combination of a dc voltage and a radiofrequency (rf) voltage of increasing

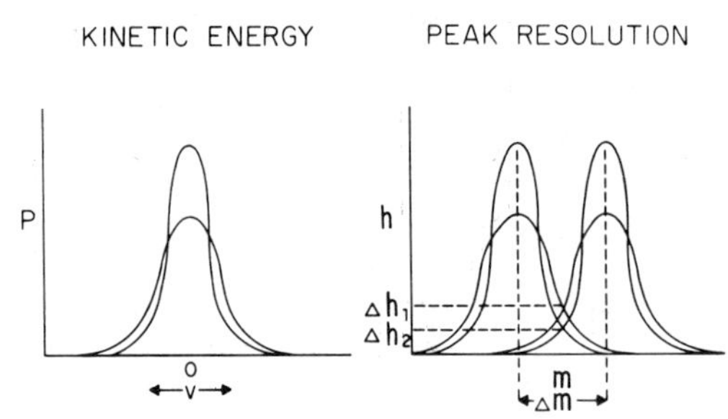

$$\text{Resolution} = \frac{\Delta m}{m} \text{ When } \Delta h = 0.1\,h \text{ (10\% Valley)} = \frac{1}{15{,}000} = \frac{150.01 - 150.00}{150}$$

Fig. 5. Definition of resolution in mass spectrometry.

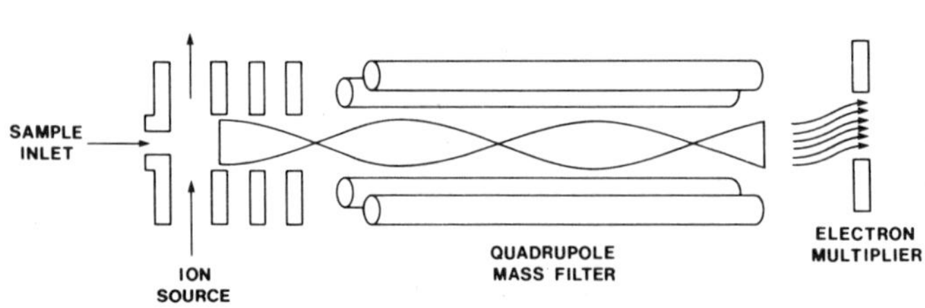

Fig. 6. A quadrupole filter mass spectrometer.

amplitude is applied. One pair receives a positive dc voltage and an rf voltage, and the other pair receives a negative dc voltage and an rf voltage with a 180° phase shift. In any particular voltage ratio and rf voltage, ions whose *m/e* values fall within the trajectory region are transmitted through the quadrupole. Those outside the stable region are neutralized by striking the rods and are pumped away.

Before ions can be sorted, they must be generated. Most mass spectrometry to date has been done on positive ions. The three most common methods of generating positive ions are presented in Figure 7. Less common methods, such as vacuum discharge, thermal ionization, photoionization, and atmospheric pressure ionization are discussed by Roboz and Chait.[2]

In electron ionization (EI) the molecules to be examined are bombarded with electrons, frequently of 70 eV energy. This causes an electron to be ejected when the bombarding electron passes closely enough to the target molecule. A positive ion is thereby generated. Since only about one molecule in 10^5 undergoes ionization, the probability of double ionization is low. Moreover, electron bombardment is usually performed at source pressures of 10^{-5}–10^{-7} torr, so the probability of bimolecular reactions is also low.

The energy of the bombarding electron is about 10 times the energy required to eject an electron from most compounds or to break a covalent bond. Excess energy above that necessary to eject an electron is frequently imparted to the molecule, leaving it in an excited state. This excess energy often causes the molecule to fragment within the time taken to exit from the source (ca. 10^{-5} sec). If it does not and remains intact during the time required for the balance of its journey to the detector (another 10^{-5} sec), it will be recorded as an ion with the mass of the parent molecule, i.e. a molecular ion.

Classical structural studies of organic compounds have been based on controlled fragmentation (degradation). The trick in mass spectrometry, as in the

ELECTRON IONIZATION

$$AB + \epsilon^* \longrightarrow AB^{+\cdot} + \epsilon^* + \epsilon \qquad (M^+)$$
$$AB^{+\cdot} \longrightarrow A\cdot + B^+$$

CHEMICAL IONIZATION

$$CH_4 + \epsilon^* \longrightarrow CH_4^{+\cdot} + \epsilon^* + \epsilon$$
$$CH_4^{+\cdot} + CH_4 \longrightarrow CH_3\cdot + CH_5^+ + C_2H_5^+$$
$$AB + CH_5^+ \longrightarrow ABH^+ + CH_4 \qquad (M+1)^+$$
$$AB + C_2H_5^+ \longrightarrow AB(-H)^+ + C_2H_6 \qquad (M-1)^+$$

FIELD IONIZATION

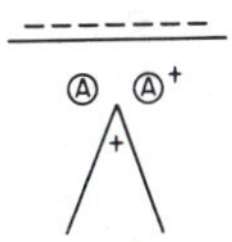

Fig. 7. Methods of generating positive ions.

more classical degradation methods, is to achieve the proper balance between molecular ions and an array of fragment ions of appropriate molecular size. Unfortunately electron bombardment often leads to excessive fragmentation, but fortunately other ionization methods are available.

In chemical ionization (CI) the sample to be examined is bombarded in the presence of an excess of reagent gas, e.g. 1 torr. The reagent gas is preferentially ionized and dissipates its extra energy by bimolecular reactions to generate ions which react chemically with the molecules under study. Normally less excess energy is involved in such reactions, so that fragmentation is less likely to occur. Moreover the amount of fragmentation is under partial control by selecting the reagent gas or the reagent gas mixture. The ions generated in the reagent gas can function either as Bronsted acids or bases and occasionally as Lewis acids. Their relative strengths influence the excess energy imparted to the ionized molecule and thus available for fragmentation. This excess energy transfer is more likely with a Bronsted acid than with a Bronsted base, which, in any event, generates a negative ion. This will be considered later. In functioning as a Bronsted acid, an ion one mass unit greater than the original molecule $(M+1)^+$ is generated. In functioning as a Lewis acid a hydride ion may be extracted to give an $(M-1)^+$ ion or occasionally a complex between the ionized reagent gas and the subject molecule. Chemical ionization does not lead to molecular ions.

Field ionization (FI) can also produce ions by simple loss of electrons to give molecular ions, M^+. The process is analogous to that operating with a lightning rod. In the intense electric field ($\sim 1\ V/\text{Å}$, $10\ kV$ overall) at the tip of a pointed emitter, the electron tunnels from the subject molecule to the positively charged emitter. The tunneling occurs without imparting excess energy to the ionized species.

Thus far all of the ionization methods have required the subject molecule to be present in the vapor phase. This requirement for volatility imposes limitations on the nature of the molecule. In field desorption (FD), the subject molecule is applied directly to the anode before insertion into the mass spectrometer. The applied molecules migrate to the tip, where they are ionized by loss of an electron. The positively charged ion thus generated finds itself in a positive electrical environment and is ejected into the vapor phase with minimal fragmentation.

The requirement for volatility imposes severe limitations on the nature of the peptide samples that can be examined by mass spectrometry. To achieve sufficient volatility the molecular weight must be moderate, and the polarity of the molecule decreased by suitable derivatization. Depending on the nature of the amino acids the upper limit seems to be reached at duodeca- or tetradecapeptides. This limit is also imposed by another practical consideration. Peptides of this size are taxing our present ability to calibrate the mass spectrometer. Moreover, as the size of the peptide increases, it becomes increasingly difficult to preserve sufficient fragments up to the molecular ion needed to provide the requisite structural information. Nonetheless, the prospect of sequencing peptides of such size in a single experiment is enticing and challenging, particularly if one can depend on a computer to digest the information obtained.

A more conservative approach involves prior fragmentation to smaller peptides, which are more readily identified. This approach, as indeed do all approaches to mass spectral sequencing of peptides larger than can be examined in a single scan, requires some method of deducing the structure of the original peptide from the fragments examined.

The extreme example is the use of mass spectrometry as an aid in conventional sequence analysis to identify derivatives, such as the thiohydantoins, of single amino acids. Here the order of generation of these derivatives by Edman degradation provides the ultimate sequence.

The use of enzyme degradation by dipeptidyl aminopeptidase (DAP) proceeds to the next order of complexity. A family of dipeptidyl aminopeptidases exists. The one most commonly used, DAP I, sequentially removes dipeptides starting from the N-terminus. It is blocked by Arg or Lys at the N-terminus and does not cleave on either side of a Pro residue. Another, DAP IV,[3] has the single restriction of not cleaving peptides with Pro in position 3. These apparent limitations can serve as advantages in that the degradation to dipeptides may proceed in more manageable segments. Overlap information to deduce the origin of the dipeptides is obtained by repeating the enzymic hydrolysis after a single Edman degradation or by the addition of a new N-terminal amino acid to the parent peptide.

We have examined fluorescamine derivatives as a means of identifying dipeptides (Figure 8).[4] This derivative usually gives a strong M-18 ion (ketene) with surprisingly little fragmentation. Indeed, as in the example shown, it is often difficult to deduce the order of the amino acids in the dipeptide. In this case the N-terminal amino acid can usually be identified by examination of the hydrolytic product of the fluorescaminated dipeptide. This example illustrates the application of a group which will serve as the primary site of ionization or as the residence site of the excitation energy thus preserving the structure of the derived ion.

Schier et al.[5] identified the ethyl acetoacetate derivative of the partially deuterated methyl ester of the dipeptide by helium CI (Figure 9). This method can be used to examine mixtures of dipeptides without prior separation.

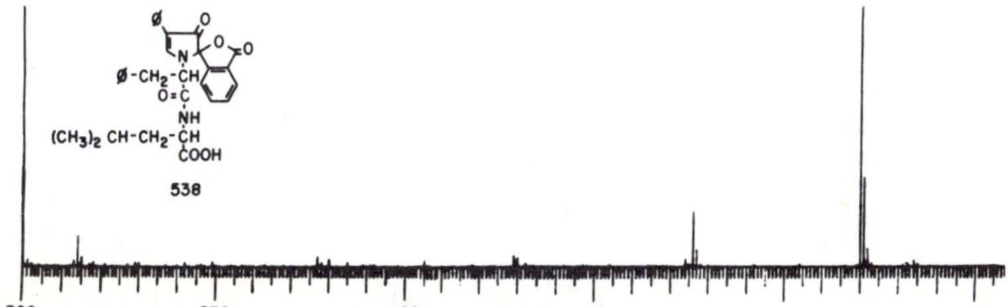

Fig. 8. EI mass spectrum of the fluorescamine derivative of phenylalanyl leucine.

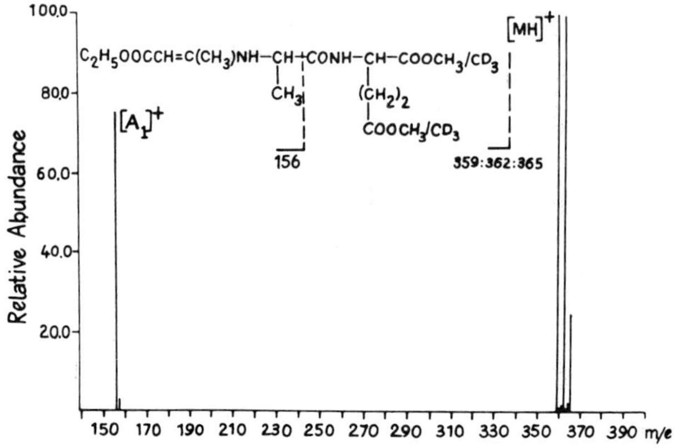

Fig. 9. Helium CI mass spectrum of the ethyl acetoacetate derivative of the partially deuterated methyl ester of alanyl glutamic acid.

For higher peptides, more extensive derivatization is usually necessary to enhance their volatility. It is important to achieve complete replacement of the hydrogens that can reduce volatility through ionic or hydrogen bond formation. Commonly this is done by the methylation method of Hakomori using sodium methylsulfinylmethide. Complications can arise from overmethylation creating sulfonium and ammonium salts or converting glycine to alanine. For detailed discussion of this problem see Arpino and McLafferty.[1]

Such methylated peptides fragment under EI as though they were unzippering from the carboxyl terminus through progressive loss of CO and methyl aldimine residues (Figure 10). The unzippering can also begin at internal peptide bonds. Additionally, a limited number of ions are generated from the C-terminus.

In reconstructing the sequence, one is dealing with a series of ions in which each increment confirms the structure as established by those preceding it. This

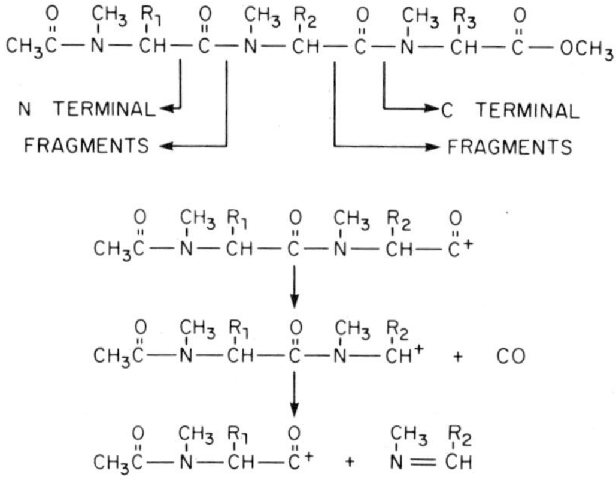

Fig. 10. Fragmentation of methylated peptides by EI.

seems very simple, but there are complications. The intensity of the ions falls off rapidly with size. In addition, many of the constituent amino acids undergo simultaneous chemical alterations. These alterations occasionally aid, but more often complicate, the identification of the amino-acid residues. Familiarity with these alterations is of obvious importance in interpreting the spectra. Due cognizance given to these alterations, the sequence can be reconstructed manually or by computer. Again the review by Arpino and McLafferty[1] should be consulted for details.

Recently increased attention has been given to CI as a means of generating ions for peptide sequencing. As explained earlier, the reagent gas in large excess in gaseous admixture with the peptide derivative can be bombarded by electrons. A variant technique, direct CI, can be used for less volatile peptides. In this method the subject compound coated on the exterior of a solid probe is exposed to previously ionized reagent gas. Ionization favors vaporization. Alternatively, a solution of the subject compound can be sprayed into the ionized reagent gas.

The fragmentation of permethylated acyl peptides by CI with isobutane is similar to that observed by EI, except fragmentation is less.[6] Thus more ions at higher *m/e* values are obtained (Figure 11). The aldimine and N-amide ions usually predominate. A computer program for reconstructing the original sequence from the fragmentation pattern has been written.[7] Although the peptides thus far examined have been short, this method holds considerable promise because of the decreased fragmentation and the increased prevalence of confirmatory ions from C-terminal fragments. As in the case of EI, it is necessary to be aware of degradative reactions characteristic of individual amino-acid residues.

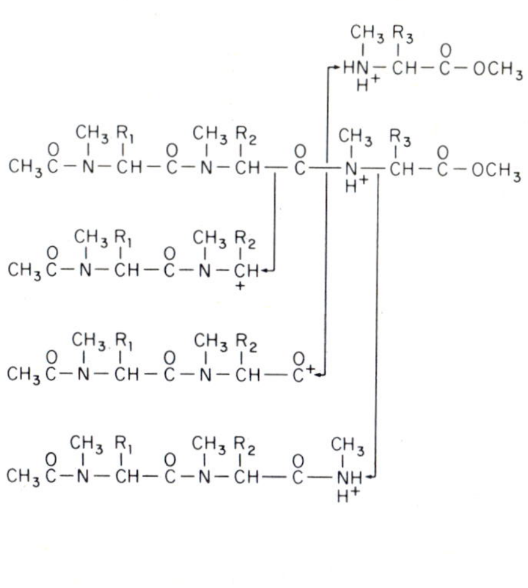

Fig. 11. Fragmentation of methylated peptides by CI.

To extend mass spectrometric examination to peptides larger than can be examined as permethylated derivatives, Biemann has recently revived interest in examination of the polyamino alcohols derived from peptides by reduction with $LiAlD_4$. In this approach the initial peptide, e.g. glucagon with 29 amino acid residues, is subjected to acid hydrolysis to generate a mixture of smaller peptides. The peptides in this mixture are esterified, acylated, and reduced to polyamino alcohols. These are converted to their TMS derivatives. The resultant mixture is examined by gas chromatography coupled with mass spectrometry. The spectra obtained by EI generally provide clear sequence information from abundant fragments derived from both the C- and N-termini of di- to hexapeptide derivatives (Figure 12).

Because of the rapid hydrolysis of amide bonds Glu/Gln and Asp/Asn are not distinguished. Cysteine must be protected, e.g. as aminoethyl Cys, but arginine is converted directly to *N*-methylornithine. Other peculiarities of individual amino acids are discussed by Nau and Biemann.[8] Identification of the individual amino alcohols as well as the reconstruction of the original peptide sequence has been computerized. Important confirmatory evidence in identifying the individual amino alcohols is obtained from the gas chromatographic retention indices, which can be predicted with considerable accuracy by summation of the retention index increments for each amino-acid residue, the acyl and TMS groups.[9]

Thus far field ionization (FI) and field desorption (FD) have not had a major impact on peptide sequencing. Their great promise lies in the inherent ability to circumvent to a considerable degree the volatility problem, which becomes increasingly troublesome as the size of the peptide increases. A major weakness lies in the absence of adequate fragmentation to permit deduction of structure. An example of their potential can be seen in the recent demonstration of a peptide residue of 18 amino acids from underivatized glucagon (29 residues).[10] Since the thermal fragmentation pathways may mimic closely those observed in EI fragmentation, a combination of pyrolysis and FD may provide a new approach, if technical problems of rapid detection can be resolved.

Up to now we have been considering only positive ions. Any negative ions generated during the usual operating mode of a mass spectrometer would be

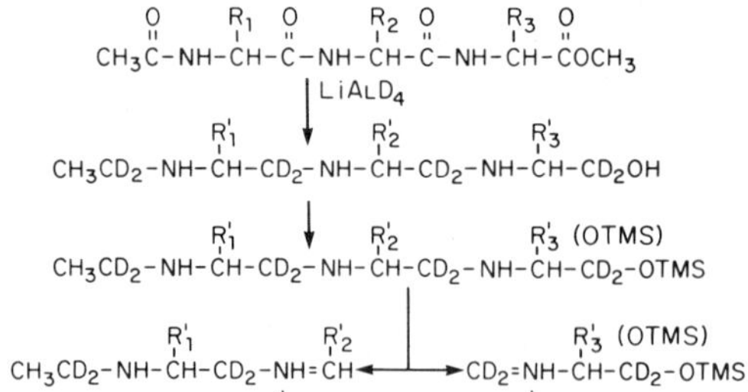

Fig. 12. Fragmentation of polyamino alcohols.

driven in the opposite direction and lost in the source. Obviously the polarity can be reversed. In addition, conditions can be selected to increase the portion of negative ions generated. In pulsed positive-negative ion chemical ionization (PPNICI) mass spectrometry, these conditions have been achieved by Hunt et al.[11] in a manner that permits essentially simultaneous examination of both the positive and negative ions. This method holds considerable promise of providing greater sensitivity in mass spectrometry generally. In addition, increased amounts of high molecular weight negative ions are generated from peptides.

A modified quadrupole mass spectrometer was used, although in principle a magnetic instrument could also be used (Figure 13). The ion source is pulsed with an alternating accelerating voltage at 10 kHz. Ions of identical *m/e*, but different polarity, traverse the quadrupole mass filter with equal facility and exit the filter at the same point. Upon exiting they are deflected to two electron multiplier (EM) detectors, one at a positive, the other at a negative potential. The output from the EM is fed to a light beam oscillograph (LBO), where deflections from each EM are recorded in opposite directions.

Negative ions can be generated in the reagent gas by reaction with free electrons in several ways (Figure 14). In one method CH_4 containing $< 0.1\%$

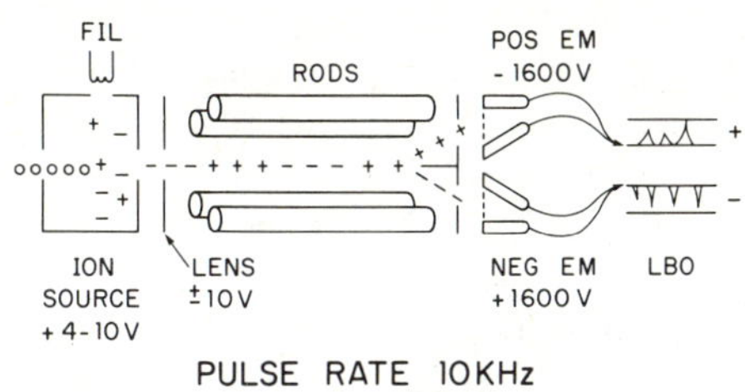

Fig. 13. Pulsed positive-negative ion chemical ionization mass spectrometer.

$AB + e \xrightarrow[\sim 0\,eV]{} AB^{\bullet -}$ RESONANCE CAPTURE

$AB + e \xrightarrow[0-15\,eV]{} A^{\bullet} + B^{-}$ DISSOCIATIVE RESONANCE CAPTURE

$AB + e \xrightarrow[>10\,eV]{} A^{+} + B^{-} + e$ ION PAIR PRODUCTION

$CH_4 + e^* \longrightarrow CH_4^{+\bullet} + e^* + e$

$CH_4^{+\bullet} + CH_4 \longrightarrow CH_5^{+} + CH_3^{\bullet}$

$CH_3ONO + e \longrightarrow CH_3O^{-} + NO$

$M + \underset{\text{BRONSTED BASE}}{CH_3O^{-}} \longrightarrow (M-1)^{-} + CH_3OH$

Fig. 14. Generation of negative ions.

CH_3ONO is used as the reagent gas. Electron bombardment of the CH_4 generates CH_5^+, a strong Bronsted acid, and thermal or near-thermal electrons. These electrons react with the CH_3ONO by dissociative-resonance capture to generate CH_3O^-, a strong Bronsted base. They may also react directly with the subject compound. In the reaction of a Bronsted acid with a Bronsted base, the excess energy resides in the acceptor molecule and can thus lead to fragmentation. Thus in the extraction of a proton from a peptide by a Bronsted base, the excess energy does not reside in the donor (peptide) molecule and fragmentation is less likely.

These principles are exemplified in the PPNICI spectrum of the acetylated permethylated tetrapeptide, methionyl-glycyl-methionyl-methionine (Figure 15). The positive fragments are formed as previously described for CI. The $(M-1)^-$ ion is formed by reaction with CH_3O^-, and negative fragments by dissociative resonance capture. The ion at 385 corresponds to loss of CH_3SH from one of the methionine residues. Additional structural confirmation can be seen by the presence of an ion in the negative ion spectrum 29 mass units (CH_3N) higher than the corresponding ion in the positive ion spectrum, i.e. 188 and 217, 259 and 288, and 403 and 433. Since this technique holds much promise for peptide sequencing, further developments are awaited with interest.

Comparison of mass spectrometry with automated Edman degradation for peptide sequencing reveals advantages and disadvantages for both. While one may attempt to balance off the convenience of one technique against another, normally the average researcher will be governed by what he is equipped to do and his individual capabilities and predilections. The investment in equipment is greater for mass spectrometry, particularly if reliance is placed on computer data processing beyond that normally available with the usual mass spectrometer data systems. This may be offset by the ability of the mass spectrometer to handle a wide variety of problems that can not be approached by a peptide sequencer.

Once techniques are established, the mass spectrometer with appropriate computer facilities can often give answers faster than the Edman sequencer in those

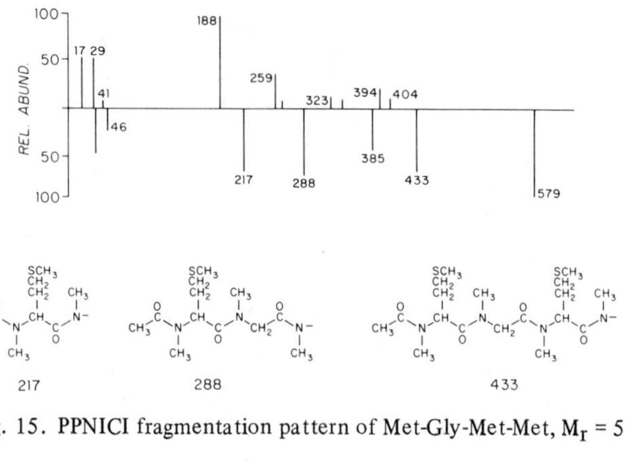

Fig. 15. PPNICI fragmentation pattern of Met-Gly-Met-Met, $M_r = 580$.

applications for which it is suitable. Both are capable of examining approximately the same length of amino-acid sequences in a single experiment, but the Edman approach has been pushed further with highly purified proteins and does not require that the N-terminal peptide sequence to be examined be detached from the balance of the protein. The presence of a blocked N-terminus does not constitute an impediment to the mass spectrometer. While techniques are now available to handle peptides of 10 residues and less by automated Edman sequencing, this size range is well suited for mass spectral sequencing, and with the advent of techniques such as PPNICI mass spectrometry this method for the appropriately equipped laboratory may be the method of choice.

The final choice may well be determined by sensitivity, i.e. the minimum sample required. For either approach the problem has two aspects, the amount of material needed to prepare the sample to be examined and the requisite size of this sample.

Great sensitivity can be attained with automated Edman degradation when the original material can be prepared with radioactive amino acids through appropriate cell culture techniques. This approach avoids the problem of sample loss by adsorption to container walls, since nonradioactive carrier protein may be added.

In mass spectrometry, one is struck by the fact that mass spectrometers are potentially extremely sensitive, but presently very inefficient. In normal scanning only a small fraction of the time is spent at any one mass region and rarely are the ions confined to one region. In multiple specific-ion detection, sensitivity is increased by observing only selected mass regions, but in peptide sequencing one does not know in advance where to look. Photoplate detectors possess the advantage of seeing all regions for as long as one wishes to look, but one is hampered by the absence of facile computer analysis of the results.

As few as 10 ions ($\sim 10^{-22}$ mole) can constitute a peak in a mass spectrum. Some idea of the potential for increased sensitivity is apparent from the fact that of the molecules introduced into the source only one in 10^5 or 10^6 is ionized. Further, of those ionized, only a fraction pass the mass sorting system and arrive at the detector. This is in accord with the statement that $< 10^{-14}$ moles can be detected by single ion monitoring.[1]

Biemann and coworkers[12] have obtained a sequence from as little as 400 nmol of peptide of which only a small portion of derivatized hydrolysate was injected into the gas chromatograph preceding mass spectrometry. He anticipates that the sensitivity could be increased by a factor of 10. Direct coupling of a high pressure liquid chromatograph (HPLC) to the mass spectrometer provides hope of achieving higher sensitivity with derivatized peptide samples not susceptible to gas chromatography. Using this approach, McLafferty et al.[13] have obtained complete sequence information from injection into the HPLC of 10^{-8} g, $< 10^{-10}$ mole, of derivatized oligopeptide even with a 1:100 split on transfer to the mass spectrometer, i.e. $< 10^{-12}$ mole.

Despite the advances in mass spectral sequencing, its use has been limited to relatively few laboratories. A thorough understanding of its complexities is a prerequisite to its successful application.

The research of the author is supported by National Cancer Institute grants CA 16434 and CA 19163 from the National Large Bowel Cancer Program.

References

1. Arpino, P. J. & McLafferty, F. W. (1976) in *Determination of Organic Structures by Physical Methods,* Vol. 6, Nachod, F. C., Zuckerman, J. J. & Randall, E. W., Eds., Academic Press, New York. pp. 1–89.

2. Roboz, J. & Chait, E. (1977) in *Techniques of Chemistry, Vol. I, Physical Methods of Chemistry,* Weissberger, A. & Rossiter, D. W., Eds., John Wiley & Sons, Inc., New York, pp. 129–222.

3. Caprioli, R. M. & Seifert, W. E. (1975) *Biochem. Biophys. Res. Commun.* **64**, 295–303.

4. Pritchard, D. G., Schnute, W. C. & Todd, C. W. (1975) *Biochem. Biophys. Res. Commun.* **65**, 312–316.

5. Schier, G. M., Bolton, P. D. & Halpern, B. (1976) *Biomed. Mass Spectrom.* **3**, 32–40.

6. Mudgett, M., Bowen, D. V., Field, F. H. & Kindt, T. J. (1977) *Biomed. Mass Spectrom.,* **4**, 159–171.

7. Mudgett, M., Sogn, J. A., Bowen, D. V., & Field, F. H. (1977) *Adv. Mass Spectrom.,* in press.

8. Nau, H. & Biemann, K. (1976) *Anal. Biochem.* **73**, 154–174.

9. Nau, H. & Biemann, K. (1976) *Anal. Biochem.* **73**, 139–153.

10. Winkler, H. U., Beuhler, R. J., & Friedman, L. (1976) *Biomed. Mass Spectrom.* **3**, 201–206.

11. Hunt, D. F., Stafford, G. C., Jr., Crow, F. W. & Russell, J. W. (1976) *Anal. Chem.* **48**, 2098–2105.

12. Kelley, H., Förster, H-J., Biemann, K. (1975) *Biomed. Mass Spectrom.* **2**, 313–325.

13. McLafferty, F. W., Knutti, R., Venkataraghavan, R., Arpino, P. J. & Dawkins, B. G. (1975) *Anal. Chem.* **47**, 1503–1505.

ISOLATION AND ANALYSIS OF OPIOID PEPTIDES FROM RAT PITUITARY AND GUINEA PIG BRAIN

MENACHEM RUBINSTEIN, JANUSZ WIDEMAN,
and STANLEY STEIN,
Roche Institute of Molecular Biology, Nutley, New Jersey, 07110

The series of peptides α-, β-, and γ-endorphin and methionine-enkephalin, presumably fragments of β-lipotropin, as well as leucine-enkephalin, have been shown to possess *in vitro* opioid activity[1-5]. Furthermore, β-endorphin has been reported to have potent analgesic activity *in vivo*[6,7]. All of these peptides were characterized after their isolation from many thousands of brains from large animals. It is apparent, however, that elucidation of the biological function, the metabolism, and the pharmacology of this group of peptides requires studies performed on laboratory animals. We wish to present our initial findings on the opioid peptides in the rat, many details of which have already been submitted for publication[8].

A combination of the following three methodologies were employed in our studies. High performance column chromatography was adapted for the resolution of peptides ranging in size up to the 10 kilodalton level[8-10]. Fluorometric detection systems utilizing the reagent fluorescamine were incorporated into an amino-acid analyzer[11] and into an automatic stream-sampling monitor for peptides in column effluents[12]. Finally, a binding assay was developed based on the displacement of [^{3}H] leucine-enkephalin from neuroblastoma X glioma hybrid cells (NG 108-15)[13], which have previously been shown to possess high-affinity opiate receptor sites[14]. Since all three techniques are applicable below the nanomole level, relatively few rat pituitaries were needed for our studies.

Anterior pituitaries from 40 male Wistar rats were removed and frozen within 30 sec after decapitation. Following weighing (300 mg), the tissue was homogenized in 2.5 ml of cold 75% acetone/25% 0.2 *M* hydrochloric acid containing 0.01% thiodiglycol (an antioxidant) and 0.001% phenylmethylsulfonyl fluoride (a protease inhibitor). Acetone was blown off and lipids were extracted into ethyl acetate-ether (3:1 v/v). The aqueous phase was fractionated on a Sephadex G-75 column (1.6 × 40 cm) equilibrated with 1 *M* acetic acid. Opioid activity was detected in regions of the chromatogram corresponding to β-lipotropin, endorphins and enkephalins. In addition, aliquots of fractions from the high molecular weight region (greater than 4,000) were incubated with trypsin prior to carrying out the binding assay. Activity in the presumptive β-lipotropin peak increased about 100-fold. This most likely represented conversion by peptide bond cleavage of the relatively inactive β-lipotropin (IC_{50} = 300 *nM*) to the nonapeptide β-lipotropin 61-69 or to longer endorphins (IC_{50} = 5-9 *nM*)[13]. In addition, trypsin-generated opioid activity was found in a peak estimated to be a peptide of 30 – 50 kilodaltons. This high molecular weight peptide could represent a precursor to β-lipotropin (and presumably the endorphins and methionine-enkephalin), or a precursor to leucine-enkephalin, or a precursor to a still uncharacterized opioid peptide.

The possibility that this last observation was an artifact arising from the binding of a small opioid peptide to a protein was tested by rechromatography on the same column. In the high molecular region all of the (post-trypsinization) activity eluted at the same 30 – 50 kilodalton position. No activity (without trypsinization) was found in the low molecular weight region. Since enough sample was used so that as little as 1% dissociation of activity would have been detected, the existence of a high molecular weight precursor to opioid peptides was confirmed.

The presumptive β-lipotropin peak was further fractionated by cation exchange chromatography on Partisil SCX[9] and finally by partition (reverse-phase) chromatography on Lichrosorb RP-18[10]. As above, column eluates were monitored by the post-trypsinization binding assay, as well as by fluorescamine-detection. Using aliquots of the purified peptide, the molecular weight was confirmed on Biogel P-30 to be close to that of sheep β-lipotropin (10 kilodaltons) and likewise the amino-acid composition resembled known β-lipotropins[8]. A yield of 4 nmol was achieved from 300 mg of tissue, as calculated from the amino-acid analysis. The specific activity in the binding assay (IC_{50} after trypsinization) was found to be equivalent to that of sheep β-lipotropin.

Extraction of anterior pituitaries with 1 *M* acetic acid, instead of acetone/hydrochloric acid, gave a higher recovery of activity in the low molecular weight region of the G-75 chromatogram. Activity was localized at the β-endorphin position and in the salt volume of the G-75 column. Further chromatography of the material in the G-75 salt volume on G-10, followed by reverse-phase chromatography of the G-10 fractions revealed the presence of several active peaks, all in the picomole range. Indeed, the effluent from the initial G-75 fractionation step contained 80 pmoles of β-endorphin and a total of about 800 pmoles (leucine-enkephalin equivalents) in the salt volume, as determined by the binding assay, in comparison to the nanomole quantities of β-lipotropin.

In another experiment, 200 frozen anterior pituitaries from a commercial source were extracted with acetone/hydrochloric acid. Surprisingly, monitoring of the G-75 chromatogram by the binding assay indicated a 30-fold reduction in the β-lipotropin peak and a corresponding increase of the same magnitude in the β-endorphin peak. Furthermore, there was a dramatic shift in the G-75 fluorescence profile towards lower molecular weights. Taking advantage of this fortuitous occurrence, the β-endorphin peak was purified to homogeneity by reverse-phase chromatography. The molecular weight and the amino acid composition were both identical to those of sheep (and camel) β-endorphin[15].

The results of the experiment with the commercial pituitaries once again raises questions about the qualitative and quantitative nature of peptides isolated from slaughterhouse tissues. The use of microwave irradiation, which heat-denatures enzymes within a second[16], might well be the method of choice for killing animals.

A chemical assay has advantages in specificity over bioassay and radioimmunoassay.[17] The steps for this chemical assay are to extract and partially purify the enkephalins, prepare fluorophors with 2-methoxy-2,4-diphenyl-3(2*H*)furanone (MDPF), a structural analog of fluorescamine, and then resolve the fluorophors by reverse-phase column chromatography. As little as 10 pmoles of each enkephalin-fluorophor is readily quantitated. Preliminary assays of enkephalins in guinea pig striatum has shown the feasibility of this method. The values obtained indicate the presence of approximately 150 pmoles of methionine-enkephalin per striatum of a single guinea pig, in agreement with those reported by Smith et al.[18] This amount is more than adequate to permit the use of individual animals for specific chemical assay.

We gratefully acknowledge the excellent technical assistance of Ms. L. D. Gerber and Mr. L. Brink. These studies were carried out with the advice and collaboration of Dr. Sidney Udenfriend.

References

1. Hughes, J., Smith, T. W., Kosterlitz, H. W., Fothergill, L. A., Morgan, B. A. & Morris, H. R. (1975) *Nature (London)* **258**, 577–579.

2. Li, C. H. & Chung, D. (1976) *Proc. Nat. Acad. Sci. USA* **73**, 1145–1148.

3. Teschmacher, H., Opheim, K. E., Cox, B. M. & Goldstein, A. (1975) *Life Sci.* **16**, 1771–1776.

4. Ling, N., Burgus, R. & Guillemin, R. (1976) *Proc. Nat. Acad. Sci. USA* **73**, 3942–3946.

5. Pasternak, G. W., Simatov, R. & Snyder, S. H. (1976) *Molec. Pharm.* **12**, 504–513.

6. Loch, H. H., Tseng, L. F., Wei, E. & Li, C. H. (1976) *Proc. Nat. Acad. Sci. USA* **73**, 2895–2898.

7. Bloom, F., Setal, D., Ling, N. & Guillemin, R. (1976) *Science* **194**, 630–632.

8. Rubinstein, M., Stein, S., Gerber, L. D. & Udenfriend, S. (1977) *Proc. Nat. Acad. Sci., USA* **74**, 3052–3055.

9. Radhakrishnan, A. N., Stein, S., Licht, A., Gruber, K. A. & Udenfriend, S. (1977) *J. Chromatog.* **132**, 552–555.

10. Rubinstein, M. & Stein, S., manuscript in preparation.

11. Stein, S., Böhlen, P., Stone, J., Dairman, W. & Udenfriend, S. (1973) *Arch. Biochem Biophys.* **155**, 202–212.

12. Böhlen, P., Stein, S., Stone, J. & Udenfriend, S. (1975) *Anal. Biochem.* **67**, 438–445.

13. Gerber, L. D., Stein, S., Rubinstein, M. & Udenfriend, S., manuscript in preparation.

14. Klee, W. A. & Nirenberg, M. (1974) *Proc. Nat. Acad. Sci. USA* **71**, 3474–3477.

15. Rubinstein, M., Stein, S. & Udenfriend, S., manuscript in preparation.

16. Clouet, O. H., and Rathner, M. (1976) in *Opiates and Endogenous Opioid Peptides,* Kosterlitz, H. W., Ed., Elsevier/North-Holland Biomed. Press, Amsterdam, The Netherlands, pp. 71–78.

17. Gruber, K. A., Stein, S., Brink, L., Radhakrishnan, A. & Udenfriend, S. (1976) *Proc. Nat. Acad. Sci. USA* **73**, 1314–1318.

18. Smith, T. W., Hughes, J., Kosterlitz, H. W. & Sosa, R. P. (1976) in *Opiates and Endogenous Opioid Peptides,* Kosterlitz, H. W., Ed., Elsevier/North-Holland Biomed. Press, Amsterdam, The Netherlands, pp. 57–62.

ANALYSIS OF FLUORESCENT LABELLED PEPTIDES BY HIGH PRESSURE LIQUID CHROMATOGRAPHY

DAVID H. LIVE, *The Rockefeller University, New York, New York 10021*

Rapid, sensitive and selective means of separating and analyzing peptides have been sought for some time, With the increasing interest in synthesis and purification of biologically active peptides and their analogs, this has become an even more important concern. The development of high pressure (or performance) liquid chromatography (HPLC) with permanently bonded non-polar phase packings has been useful in the separation of a variety of organic compounds. Although applied only sparingly to the analysis of peptides,[1-5] it appears to have great promise. Limits to the detectability of underivitized peptides are on the order of several nanomolar. Gruber et al.,[5] however, have shown that detection of fluorescamine derivitized peptides in HPLC is several orders of magnitude greater, and a considerable degree of selectivity can be obtained from chromatography of pre-derivitized peptides. The derivitization procedure is simple and efficient, so that one can quantitate the chromatogram results. To assess the usefulness of this technique as a chemical assay of the purity of synthetic oxytocin and [Arg^8]-vasopressin (AVP) that are being prepared in our laboratory, we have examined these peptides and a number of their analogs by this method.

Materials and Methods

The chromatographic apparatus used in this work is depicted in Figure 1. Pumps 1 and 2 are Milton Roy Mini-Pumps; the mixing chamber is a 3 × 15 cm

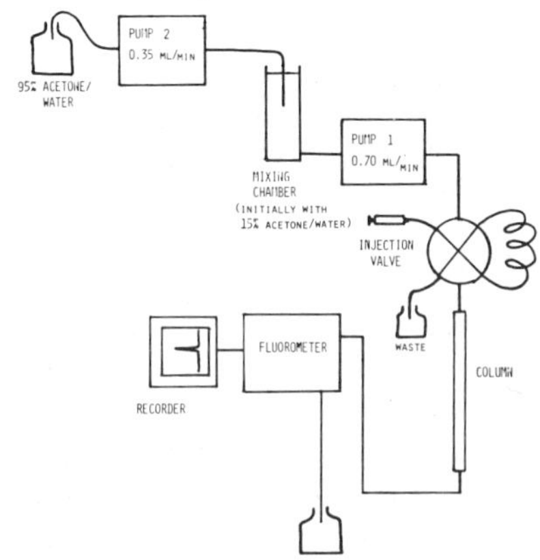

Fig. 1. Schematic diagram of chromatography system.

glass cylinder containing a magnetic stirring bar and supported on a magnetic stirrer. The injection valve used is a Waters model ALC 202 with a 1.5 ml loop. A Whatman Partisil-10 ODS pre-packed column, a Gilson Spectra/Glo fluorometer and an Easterline Angus chart recorder are used. Acetone (reagent grade) was distilled from a glass apparatus before use. Distilled water was additionally purified by passage through a Millipore Milli-Q water purifier. Solvents were made up v/v and contained 0.03% ammonium formate and 0.01% thiodiglycol. Fluorescamine (Fluram, Hoffmann-La Roche) was dissolved in acetone at a concentration of 20 mg/100 ml. Samples for analysis were dissolved in 0.4 ml of 0.046 *M* phosphate buffer (Beckman pH 7 standard buffer was used for the purpose) in a 1.0 × 7.5 cm disposable borosilicate test tube. Fluram solution (0.2 ml) was added and the mixture stirred briefly on a vortex mixer. After 10 min the volume was taken up to 2 ml with water containing 0.03% ammonium formate and 0.01% thiodiglycol, and again subjected to vortex mixing. The sample was then inserted into the loop of the injection valve. Forty-two ml of a 15% acetone/water solution was placed in the mixing chamber, and pump 1 was started. After 10 min pump 2 is started, initiating the linear gradient that runs for 75 min. During the initial 10 min some fluorescence is typically detected, due to contaminants in the derivitization mixture. The pressure on the column varies from about 700 psi initially to about 400 psi at the end of the gradient. At the end of the run the column is washed for 10 min with 95% acetone/water, and reequilibrated for 40 min with 15% acetone/water.

Results and Discsusion

Seventeen peptides including oxytocin, AVP, and 15 analogs were chromatographed in the system described with the results shown in Fig. 2. The resolution of peptides containing rather subtle changes in structure and stereochemistry is quite impressive. The peak widths at half height were on the order of 1 min although some were broader, possibly due to conformational interconversion. The influence of the presence of a free carboxyl group in these peptides is quite clear in the positions of samples 1, 2, and 3, the only ones in this group that possess that function. A similar phenomenon is observed in chromatograms of Asp^1 and Asn^1 angiotensin II. The former elutes quite early, while the latter appears about half-way through the gradient. The presence of carboxyl groups in other peptides examined does not always give rise to early elution. Overall, the system has given highly reproducible results, and has a high degree of sensitivity. A 5 *nM* sample of derivitized oxytocin applied to the column yields a peak with a signal-to-noise ratio of about 30:1. The pulsations of the pump do not contribute in a noticeable way to the noise.

An interesting example of the resolving power of this system is illustrated in Fig. 3, where the chromatogram of one sample of a putative oxytocin is shown. Although the thin layer chromatography (TLC) of this material in three systems

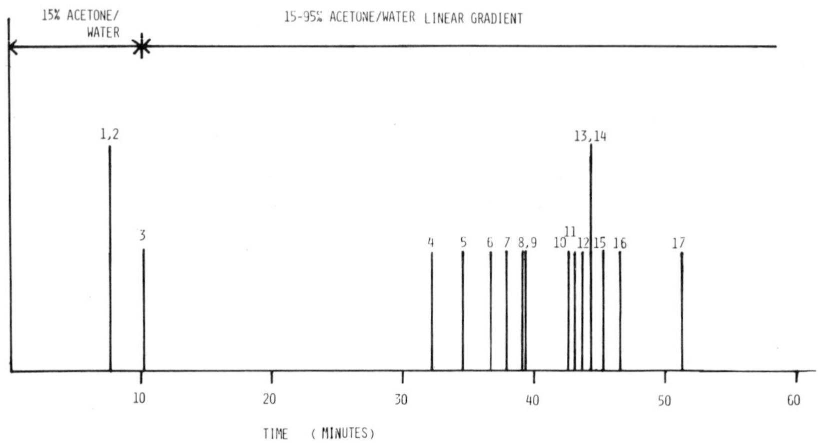

Fig. 2. Elution positions of Fluram derivatives of oxytocin, AVP and 16 analogs in the system described. 1) Des-(Pro-Leu-Gly NH_2)-oxytocin, 2) Des-(Leu-Gly NH_2)-oxytocin, 3) Des-GlyNH_2-oxytocin, 4) Des-Gln-oxytocin, 5) oxytocin, 6) [Phe^3]-oxytocin, 7) [D-Cys^1]-oxytocin, 8) [D-Cys^6]-oxytocin, 9) [1-penicillamine]-oxytocin, 10) [D-Tyr^2]-AVP, 11) [3-cyclohexylalaninc]-oxytocin 12) [di$Ile^{2,4}$]-oxytocin, 13) [Ile^2, Leu^4]-oxytocin, 14) AVP, 15) [di$Leu^{2,4}$]-oxytocin 16) [D-Cys^1]-AVP, 17) [Phe^2, Leu^4]-oxytocin.

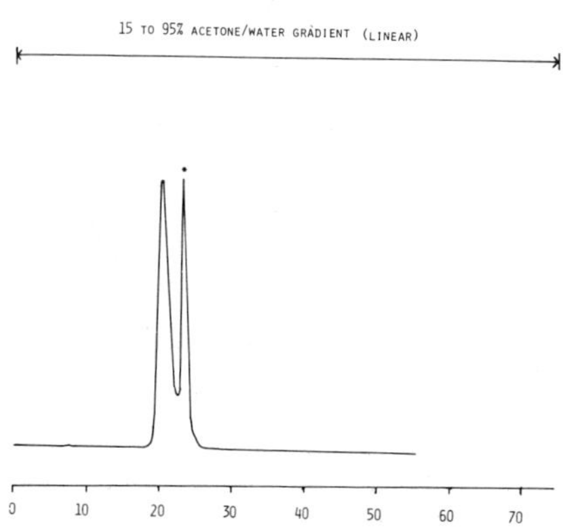

Fig. 3. Chromatogram of putative oxytocin. Peak with star above it is at oxytocin elution position.

with R_f's of 0.34, 0.68, and 0.85, respectively, revealed only a single uniform spot in each that migrated with two other samples of oxytocin, the HPLC clearly shows two species, with the smaller peak co-chromatographing with oxytocin.

We have also examined the products and by-products of peptide synthesis with this method. The species are resolved at least as well as in TLC systems, with the further advantages of HPLC of reproducibility and quantitation.

Investigation of this HPLC methodology indicates that it provides a fairly widely applicable analytical procedure with a high degree of sensitivity and selectivity. It represents a valuable tool in peptide analysis.

I am indebted to Dr. Victor J. Hruby and Dr. R. T. Havran for the analogs of oxytocin and AVP. I wish to thank Dr. Stanley Stein for a number of useful suggestions in the course of this work, and the other members of the Physical Biochemistry Laboratory at The Rockefeller University for their advice and support. This work was supported by NIH Grant AM-2493.

References

1. Tsuji, K., Robertson, J. H. & Bach, J. A. (1974) *J. Chromatog.* **99**, 597–608.
2. Tsuji, K. & Robertson, J. H. (1975) *J. Chromatog.* **112**, 663–672.
3. Hancock, W. S., Bishop, C. A. & Hearn, M. T. W. (1976) *FEBS Lett* **72**, 139–142.
4. Burgus, R. & Rivier, J. (1977) in *Peptides 1976*, Loffet, A., Ed., Editions de l'Université Bruxelles, Bruxelles, pp. 85–94.
5. Gruber, K., Stein, S., Brink, L. Radhakrishnan, A. & Udenfriend, S. (1976) *Proc. Nat. Acad. Sci. USA* 73, 1314–1318.

RAPID ANALYSIS OF PEPTIDE MIXTURES BY HIGH PERFORMANCE LIQUID CHROMATOGRAPHY WITH NONPOLAR STATIONARY PHASES

IMRE MOLNÁR and CSABA HORVÁTH, *Department of Engineering and Applied Science, Yale University, New Haven, CT 06520*

Charcoal, a nonpolar sorbent, had been widely used for the separation of peptides[1] before the advent of ion-exchange chromatography.[2] Recent developments in high performance liquid chromatography revived the interest in the use of nonpolar stationary phases for the separation of biological substances[3,4] by "reversed phase" chromatography, which employs columns packed with 5 or 10-μm porous silica particles having hydrocarbonaceous functions covalently bound to the surface.

This report illustrates the potential of this type of chromatography for the rapid analysis of minute quantities of peptide mixtures. The results suggest that octadecyl-silica columns can be used for fast separation of a wide variety of peptides. By monitoring the column effluent with a UV-detector at 200 nm, the sample components can be analyzed at the subnanomole level without the formation of UV absorbing or fluorescent derivatives.

Experimental

A Perkin-Elmer (Norwalk, Conn.) Model 601 liquid chromatograph with Schoeffel (Westwood, N. J.) FS 770 UV and FS 970 fluorescence detector was used in the gradient elution mode.[5] The sample was introduced with an injection syringe (ES Industries, Marlton, N. J.). The Lichrosorb RP-18, 5-μm columns were supplied by Rainin (Brighton, Mass.). Typical operating conditions are given in the figure legends.

In most experiments 0.1 *M* phosphate buffer, pH 2.1 or 0.5 *M* perchloric acid, pH 0.2 was used as the starting eluent and acetonitrile (Burdick and Jackson, Muskegon, Mich.) as the gradient former. Isocratic elution with the above phosphate buffer was used for the separation of hydrophilic amino acids. The amino acids and peptides were purchased from Sigma (St. Louis, Mo.).

Results and Discussion

In "reversed phase" chromatography solute retention is governed by hydrophobic interactions between the solute and the hydrocarbonaceous ligands at the stationary phase surface. The theoretical basis of this type of chromatography has been analyzed in recent publications.[4,6,7] The increase in the nonpolar moiety of the solute molecule or in the surface tension of the eluent augment solute retention. Conversely, a reduction in the effective surface tension of the eluent or the introduction of polar functions, which can strongly interact with the eluent, into the solute molecules result in a decrease in chromatographic retention.

Hydrophobic amino acids and small peptides can conveniently be separated on presently available nonpolar stationary phases as shown in Fig. 1. Other peptides with hydrophobic residues such as Trp, Leu, Ile, Tyr, or Val can also be readily separated under such conditions.

The technique appears to be generally applicable to the analysis of larger peptides, which usually contain some hydrophobic residues. Fig. 2 illustrates the rapid separation of the tryptic digest of ribonuclease S peptide. The chromatograms in Fig. 3 were obtained with porcine α-melanotropin, which has been purified by conventional methods.[8] The comparison of the two superimposed chromatograms shows that each component contains tryptophan. Only tryptophan has appreciable fluorescence under the conditions employed in this study. Some of the peaks may represent deacetylated or oxidized forms of α-MSH and its conformers.

The conditions described in Figs. 1-3 are not suitable for the chromatographic separation of hydrophilic amino acids and small peptides because of the insufficient retention. However the retention of charged species can be enhanced by adding a counterion having a hydrophobic moiety to the eluent. This technique, using an ionic surfactant, is referred to as "ion-pair" or "soap" chromatography.[9] Figure 4 illustrates that with decyl sulfate in the eluent, the hydrophilic amino acids can be separated.

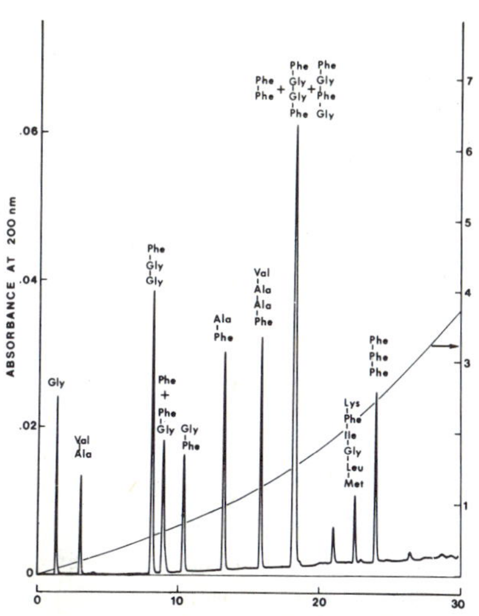

Fig. 1. Chromatogram of small peptides containing hydrophobic amino acids. Column, 5-μm octadecyl-silica, 4.6 mm i.d., 25 cm long; 70°C; flow rate, 2.0 ml/min; inlet pressure, 160 atm; starting eluent, 0.5 *M* perchloric acid, pH 0.2; gradient former, acetonitrile concave gradient in 50 min to 100% at setting 0.3, as illustrated. Sample size approximately 200 ng of each component.

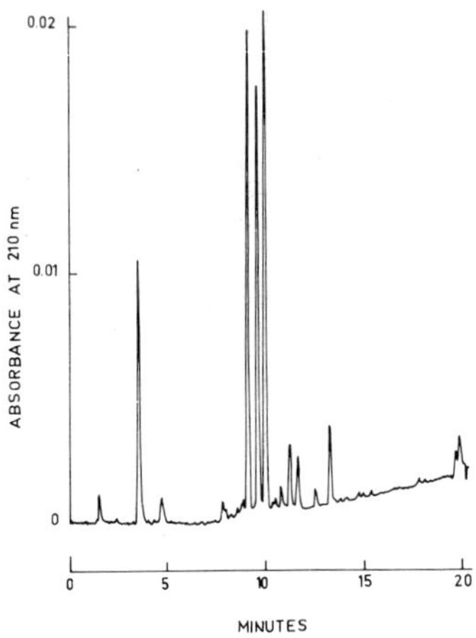

Fig. 2. Chromatogram of the tryptic digest of ribonuclease S-peptide (Lys-Glu-Thr-Ala-Ala-Ala-Lys-Phe-Glu-Arg-Gln-His-Met-Asp-Ser-Ser-Thr-Ser-Ala). Column and operating conditions are the same as those in Figure 1.

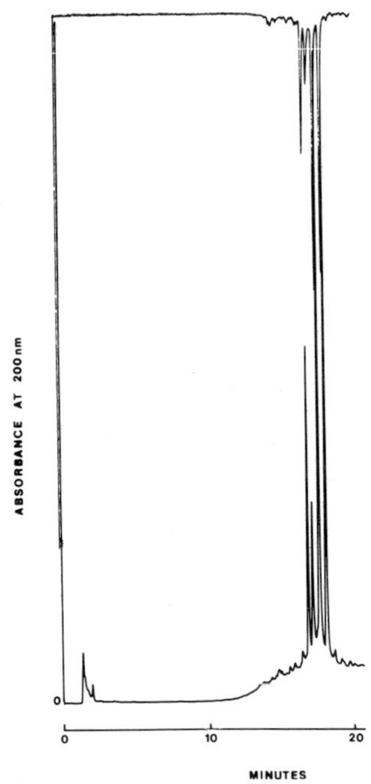

Fig. 3. Chromatogram of porcine α-melanotropin (α-MSH) purified by conventional methods. The amino-acid sequence of the peptide is as follows: CH_3CO-Ser-Tyr-Ser-Met-Glu-His-Phe-Arg-Trp-Gly-Lys-Pro-Val-NH_2. The column and operating conditions are the same as in Fig. 1. The chromatogram at the top has been obtained by tandem monitoring of the effluent with a fluorescence detector. Excitation and emission wavelengths are 280 nm and 340 nm, respectively. Full scale corresponds to a detector output of 0.5 μA at a time constant of 0.5 seconds.

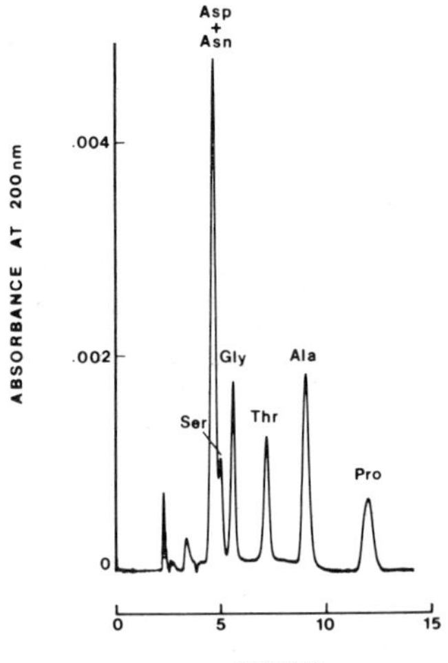

Fig. 4. Chromatogram of polar amino acids on octadecyl-silica with decyl-sulfate in the eluent. Column: 5-μm Lichrosorb ODS, 4.6 mm i.d., 25 cm long; 70°C; inlet pressure, 160 atm; isocratic elution with 0.1 *M* phosphate buffer containing 10^{-3} *M* decyl sulfate; sample size, approx. 100 ng of each component.

This work was supported by grants No. GM 22735 to C. Ambrus and No. GM 20993 from the National Institutes of Health, U. S. Public Health Service. The authors are indebted to S. Lande, N. Asato and J. Rosa for valuable discussions and for the gift of α-MSH and the tryptic digest of the S peptide, respectively, and to P. Molnar for skillful technical assistance.

References

1. Tiselius, A. (1944) in *The Svedberg,* Tiselius, A., Ed., Almquist and Wiksells, Uppsala, pp. 370–371.
2. Moore, S. & Stein, W. H. (1951) *J. Biol. Chem.* **192**, 663–681.
3. Molnar, I. & Horvath, C. (1976) *Clin. Chem.* **22**, 1497–1502.
4. Horvath, C., Melander, W. & Molnar, I. (1976) *J. Chromatogr.* **125**, 129–156.
5. Molnar, I. & Horvath, C. (1977) *J. Chromatogr., Biomed. Appl.* **143**, 391–400.
6. Horvath, C., Melander, W. & Molnar, I. (1977) *Anal. Chem.* **49**, 142–154.
7. Melander, W. & Horvath, C. (1977) *Arch. Biochem. Biophys.*, in press.
8. Lande, S., Lerner, A. B. & Upton, G. V. (1965) *J. Biol. Chem.* **240**, 4259–4263.
9. Knox, J. H. & Laird, R. L. (1976) *J. Chromatogr.* **122**, 17–34.

APPLICATION OF HIGH PRESSURE LIQUID CHROMATOGRAPHY TO PEPTIDES

J. RIVIER, R. WOLBERS and R. BURGUS,
Salk Institute for Biological Studies, La Jolla, California 92037

Introduction

The emphasis of our original report[1] on high pressure liquid chromatography (HPLC) was to demonstrate its power in the resolution of unprotected peptides. These observations were lately confirmed by others.[2,3] A year later it has become obvious that the scope and power of the application of HPLC using non-polar bonded phases for resolution and isolation of synthetic and naturally occurring peptides not only on an analytical but also on a semi-preparative scale have enormously broadened. In this report we show the usefulness of such highly resolutive techniques in separating stereoisomers of peptides of biological interest, for demonstrating their homogeneity and for the study of racemization.

Experimental

Apparatus consisted of Waters Associates Model 204 Liquid Chromatograph, Waters Associates Model UK6 Injector, two Waters Associates Model 6000A Pumps, Waters Associates Model 660 Programmer, Schoeffel Model 770 Multi-wave Length Detector, Infotronics Model 110 Integrator, Linear Instruments Corp. Model 455 Chart Recorder. Three or four Waters Associates 7 mm × 30 cm columns of μBondapak C_{18} were used in series. NH_4OAc or NaH_2PO_4 (0.01 *M*) pH 4.5 buffers were thoroughly degassed and filtered through 0.25μ millipore. Burdick and Jackson glass distilled CH_3CN was degassed. The system was run at room temperature. Flow rate varied from 4 to 6 ml/min. Absorbance was read at 210 nm.

Results

HPLC can be used to demonstrate high purity of a synthetic peptide as well as to show that it is freed from its most likely contaminants. Indeed: a) [D-His^2]-TRF could be isolated on a semi-preparative scale, devoid of TRF which is its most likely contaminant in view of the well documented tendency of histidine to racemize, see Figure 1. b) A baseline separation of somatostatin and [D-Cys^{14}]-somatostatin (which could have been cross contaminated if esterification of Boc-L-Cys(MeOBzl)-OH or Boc-D-Cys(MeOBzl)-OH to the resin had not been free of racemization) has been achieved. The purification steps used, gel filtration and countercurrent distribution, would not have separated quantitatively

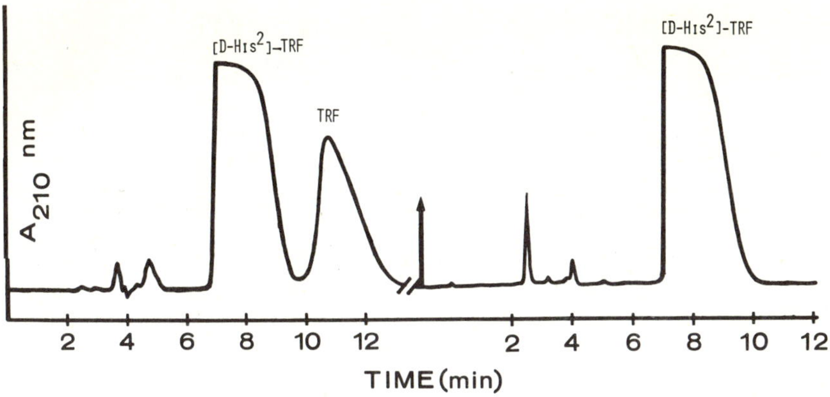

Fig. 1. Semi-preparative purification of [D-His^2]-TRF by HPLC. Conditions: 0.4 × 60 cm column; 1 mg load; 2.5 ml/min; 0.5% CH_3CN/0.01 *M* NH_4OAc, pH 4.5.

the two compounds. (0.7 × 90 cm column; 200 μg each; 6 ml/min, 34% CH_3CN/ 0.01 *M* NH_4OAc pH 4.5; retention times: 23.1 and 26.4 min resp.) c) Fig. 2 shows the potential of HPLC for the study of racemization. It is obvious that diastereoisomeric pairs are well enough resolved to allow in most cases for quantification of less than 0.1% of one conformer contaminating the other and vice-versa. A recent attempt at separating eleven stereoisomeric dipeptide pairs [Phe-(D,L-) Ala, Ser, Thr, Met, Asn, Gln, Leu, His, Phe, Tyr and Trp] led us to study the relative advantages of linear concave or step gradients upon overall resolution. It became evident that a very stringent set of conditions will have to be used for baseline resolution of each and every dipeptide in that group and that better resolution might be reached by the use of stepwise gradients versus continuous gradients.

In a separate experiment L-Phe-D, L-Val, L-Phe-D, L-Ile and L-Phe-D, L-Leu could be separated. Retention times are reported in Table I.

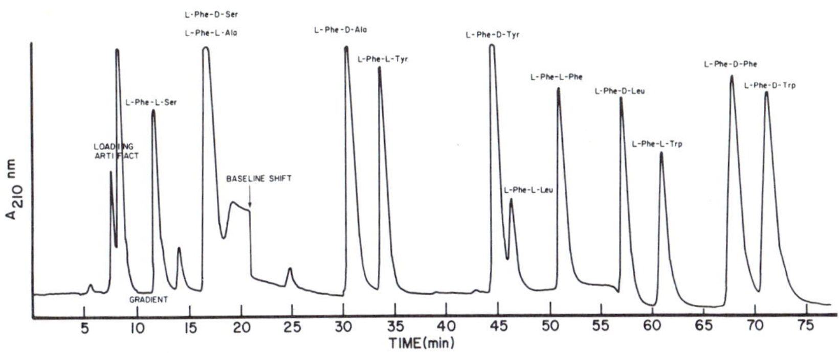

Fig. 2. Analytical separation of diastereoisomeric dipeptides by HPLC. Conditions: 10 min 5% CH_3CN/0.01 *M* NH_4OAc isocratic followed by 40 min linear gradient 5% → 21% CH_3CN; then 21% CH_3CN; 0.7 × 90 cm column; flow rate 6 ml/min.; back pressure: 4500 psi.

Table I. Resolution of L-Phe-L, D-Val, L-Phe-L, D-Ileu and L-Phe-L, D-Leu

Compound	Retention time	Compound	Retention time
L-Phe-L-Val	8.5 min	L-Phe-D-Val	14.8 min
L-Phe-L-Ile	11.8 min	L-Phe-D-Ile	24.7 min
L-Phe-L-Leu	14.1 min	L-Phe-D-Leu	25.7 min

Conditions: 0.7 x 90 cm column. 21% CH_3CN/0.01 M NH_4OAc: isocratic.

Discussion and Prospects

It might be optimistic at the present time to consider HPLC the panacea of the peptide chemist's problems. However, we have been able to demonstrate that it is a most resolutive, quantitative and straightforward means to evaluate the homogeneity of a peptide as exemplified by the baseline separation of [D-His^2]-TRF from TRF, [D-Cys^{14}]-somatostatin from somatostatin, see also Ref. (4). It can also be used for semi-preparative isolation of synthetic and natural peptides. We have also shown that it can resolve diastereoisomeric mixtures of peptides (Fig. 1, Table I). The approach, which is by no means the only one possible, was to couple on solid phase Boc-L-Phe-OH to every common amino acid (in fact a mixture of D,L-amino acids in the ratio of 2 to 1 which by integration allows for immediate identification). After HF cleavage, the free dipeptides (as a racemic mixture) were used as such. Each dipeptide pair was shown to be easily separable allowing for accurate quantitation of less than 0.1% of one conformer contaminating the other and vice-versa. To increase the sensitivity of such a test, one can envision the use of other hydrophobic, UV absorbing, fluorescent or radioactively labelled optically active moieties to be coupled to amino acids. Use of HPLC for isolation of stereoisomeric pairs of protected dipeptides has previously been reported.[5,6] No problem, such as unspecific binding leading to tailing, seems to be attached to the use of gradients, whether stepwise or continuous, as long as the ionic strength of the buffers is sufficient. Disappearance of the buffering capacity of the aqueous phase leads indeed to partial loss of resolution and typically skewed profiles, particularly for products with long retention times. We have not systematically studied the influence of the buffer concentration or composition on resolution or retention times. We have so far used an acetate or phosphate buffer with equal success. We are convinced, however, that other buffers could be used with probably as good if not better results. Addition of a detergent (sodium dodecyl sulfate, courteous suggestion from Dr. Jerry Hawk of Waters Associates Laboratories) in small concentration may help recovery if and when it becomes problematic. More recent experiments show that flow rates are somewhat critical for high resolution, and that maximum flow rates allowed by reasonable back pressures (4500–5000 psi) do not necessarily give the best results.

Conclusions

The time may be near when HPLC could be used as an automated, programmable, very versatile instrument that could not only be used as an amino-acid analyzer (probably after derivatization of the amino acids or of the hydrophobic support), but also as a chromatographic tool to separate peptides and proteins on both an analytical and semi-preparative scale large enough for biological and biochemical research.

We thank R. Ptolemy for his dedicated efforts in the making of the dipeptides used in the racemization test as well as for the running of the HPLC system. Research supported by grants from NIH, AM #18811, HD #09690 and the National Foundation 1-411 to R. Guillemin.

References

1. Burgus, R. & Rivier, J. (1976) in *Peptides 1976*, J. Loffet, Ed., Editions de l'Universite de Bruxelles, pp. 85–94.

2. Hancock, W. S., Bishop, C. A. & Hearn, M. T. W. (1976) *FEBS Lett.* **72**, 139–142.

3. Chang, S-H, Noel, R. & Regnier, F. (1976) *Anal. Chem.* **48**, 1839–1845.

4. Rivier, J. E., Lazarus, L. H., Perrin, M. H. & Brown, M. R. (1977) *J. Med. Chem.*, in press.

5. Gil-Av, E. (1975) in *Peptides 1974*, Y. Wolman, Ed., Israel Universities Press, Jerusalem, pp. 247–256.

6. Furukawa, H., Sakakibara, E., Kamei, A. & Ito, K. (1975) *Chem. Pharm. Bull.* **23**, 1625–1626.

PRIMARY STRUCTURE OF PROTEIN INHIBITORS OF CALCIUM PHOSPHATE PRECIPITATION FROM HUMAN SALIVARY SECRETIONS

DAVID H. SCHLESINGER and ROBERT JACOBS, *Endocrine Unit, Massachusetts General Hospital, Boston, Massachusetts 02114* and DONALD I. HAY, *Forsyth Dental Center and Harvard School of Dentistry, The Fenway, Boston, Massachusetts 02115*

Unlike bone, which exists in a precisely controlled environment, the mineral of the dental enamel is exposed to variable and sometimes hostile conditions. Despite this, the surface enamel is unusually stable, this being attributable to the fact that the fluid surrounding the teeth is supersaturated with respect to the enamel mineral.[1] Recently it has been shown that the supersaturated saliva is stabilized by at least two types of phosphoproteins, which act by inhibiting precipitation of calcium phosphate salts.[2] The complete covalent structure of one of these inhibitors, a tyrosine-rich acidic peptide called statherin, has recently been reported.[3] The other type of inhibitor is a complex group of anionic proline-rich proteins (PRP),[4-6] the four major members of which are structurally related (PRP-I, II, III, IV). These proteins are less potent inhibitors than statherin, but are longer lasting in the oral cavity.[7] A tentative complete sequence of PRP-III has recently been reported.[8]

The purpose of this communication is to report the N-terminal 20 amino acid residues in PRP-I, II, III, and IV, to compare these sequences with that of statherin and to discuss the structure-function relationships of these salivary phosphoproteins in context of their calcium phosphate precipitation-inhibition activities.

Detection, Purification and Structure Determination of Statherin

Statherin was detected in and purified from human parotid saliva by gel filtration and ion exchange chromatography.[3] A "precipitation-inhibition" assay[2] was used, which is based on the ability of statherin to inhibit the precipitation of calcium phosphate from a supersaturated stock solution. Details of the steps leading to the complete sequence elucidation of statherin including enzymatic cleavage, purification of peptide fragments, quantitation during sequence analysis and the sequencing strategy have been recently presented.[3]

Partial Structure Determination of the Anionic Proline-Rich Proteins

Detection and Purification. PRP-I, II, III, and IV was detected in and isolated from human parotid saliva using a dicalcium phosphate dihydrate assay.[7] The principle of the assay is based on the fact that at pH values above 6.2 acidic dicalcium phosphate dihydrate (DCDP) transforms to basic calcium phosphate salts.

The reaction proceeds by dissolution of DCPD, followed by precipitation of basic salts. It is this precipitation step that is blocked by the inhibitors. The PRP's were purified to electrophoretic homogeneity by chromatographic procedures identical to those used for statherin.[3]

Enzymatic Digestion. All four anionic proline-rich proteins were digested with pyroglutamyl-amino peptidase for removal of the N-terminal residue under the following conditions. PRP-I, II, III, and IV (8 mg each) were dissolved in 1.0 ml of 0.1 *M* NH_4HCO_3, pH 8.0 containing 2 μl of β-mercaptoethanol, 20 μl of 0.05 *M* EDTA and digested with the enzyme (0.08 mg) for 24 hr at 37°C.

Automated Liquid Phase Sequence Analysis. Automated sequence analyses of PRP-I, II, III, and IV, after removal of the NH_2-terminal pyroglutamate residue, were performed by the method of Edman and Begg[9] in which a double cleavage Quadrol program and sample preparation were used as previously described for statherin.[3] Identification of the PTH amino acids was carried out by thin layer[10,11] and gas/liquid chromatography.[12]

Results and Discussion

Statherin[3] (Figure 1), an unusually tyrosine-rich acidic phosphoprotein, possesses a highly changed N-terminal sequence through the first one third of the molecule. The COOH terminal 2/3 of statherin is extremely hydrophobic consisting primarily of glutamine, proline and tyrosine in exactly equal amounts.

All four PRP's, in contrast to statherin, are devoid of tyrosine and possess pyroglutamate in the N-terminal position as well as a highly charged N-terminal 11-residue sequence (Figure 1); they lack basic residues. Proline, comprising almost 30% of the residues in the carboxyl-terminal segment of both statherin and the PRP's, is absent from their N-terminal regions.

Similarities in the N-terminal sequences of statherin and the PRP's have led us to speculate that the site of calcium phosphate precipitation-inhibition lies within the N-terminal domains. In fact, the first step by which these salivary proteins inhibit precipitation may be via binding calcium ions. PRP I and III have been shown to bind calcium[13,14] and it seems likely that PRP II and IV will

```
                                                   H2PO3
                                                     |
PRP-I and III:  PCA-Glu-Asp-Leu-Asn-Glu-Asp-Val-Ser-Gln-Glu-Asp-Val-Pro-Leu-Val-Ile-Ser-Asp-Gly-Gly···

PRP-II and IV:  [                     ] Asp [                                                        ]

                    H2PO3 H2PO3
                      |     |
Statherin:      NH2-Asp-Ser-Ser-Glu-Glu-Lys-Phe-Leu-Arg-Arg-Ile-Gly-Arg-Phe-Gly-Tyr-Gly-Tyr-Gly-Pro-

                 21                  25                  30                  35                  40
                -Tyr-Gln-Pro-Val-Pro-Glu-Gln-Pro-Leu-Tyr-Pro-Gln-Pro-Tyr-Gln-Pro-Gln-Tyr-Gln-Gln-

                 41
                -Tyr-Thr-Phe-COOH
```

Fig. 1. Sequences of human salivary and statherin, and the N-terminal parts of PRP I-IV (proline-rich proteins).

possess the property. Equilibrium dialysis studies, presently being conducted, indicate that statherin also binds calcium.

Interestingly, the amino-acid sequence of statherin and the PRP's share similar anionic peptide segments present in other known calcium binding proteins such as the calcium binding protein of bovine intestine,[15] parvalbumen,[16] troponen,[17] the alkali light chain of rabbit skeletal muscle,[18] and the phosphoprotein casein.[19]

An additional property possessed by statherin and the proline-rich proteins is that they selectively adsorb from saliva on to the surfaces of apatitic minerals, such as hydroxyapatite.[20] We have suggested that this property is related to their inhibitory activity, and that the hydroxyapatite binding site of statherin lies in the charged amino-terminal segment.[3]

In conclusion, if it can be shown that precipitation-inhibiting activity is common in mammals and perhaps widespread in the animal kingdom, then it becomes significant as a general biological phenomenon. Initial studies using saliva from monkey, rat and hamster show that inhibitors are also present in these fluids though their nature is undetermined. We offer this concept of "stabilized supersaturation" as new biological property which we suggest is of considerable importance in biological systems in which high concentrations of calcium exist.

This work was supported in part by grants N01-CB-64079 and N01-CB-53868 from the National Cancer Institute and grant DE 03915 from the National Institute for Dental Research.

References

1. Grøn, P. (1973) *Arch. Oral Biol.* **18**, 1385–1392.

2. Grøn, P. & Hay, D. I. (1976) *Arch. Oral Biol.* **21**, 201–205.

3. Schlesinger, D. H. & Hay, D. I. (1977) *J. Biol. Chem.* **252**, 1689–1695.

4. Oppenheim, F. G., Hay, D. I. & Franzblau, C. (1971) *Biochemistry* **10**, 4233–4238.

5. Bennick, A. & Connell, G. E. (1971) *Biochem. J.* **123**, 455–469.

6. Hay, D. I. & Oppenheim, F. (1974) *Arch. Oral Biol.* **19**, 627–632.

7. Hay, D. I. & Grøn, P. (1976) *Proceedings Microbial Aspects of Dental Caries*, Stiles, Loesche and O'Brien, Eds., Sp. Supplement Microbiology Abstracts **1**, 143–150.

8. Bennick, A., Wong, R. & Cannon, M. (1977) *Proceedings Int. Symp. on Calcium Binding Proteins and Calcium Function in Health and Disease*, Corradino, Ed., Elsevier, Amsterdam, The Netherlands, in press.

9. Edman, P. & Begg, G. (1967) *Europ. J. Biochem.* **1**, 80–90.

10. Edman, P. (1970) in *Protein Sequence Determination: A sourcebook of Methods and Techniques*, Needleman, S. B., Ed., Springer-Verlag, New York, pp. 211–266.

11. Schlesinger, D. H., Niall, H. D. & Goldstein, G. (1975) *Biochemistry* **14**, 2214–2218.

12. Pisano, J. J. & Bronzert, T. J. (1969) *J. Biol. Chem.* **244**, 5597–5606.

13. Bennick, A. (1975) *Biochem. J.* **145**, 557–559.

14. Bennick, A. (1976) *Biochem. J.* **155**, 163–166.

15. Huang, W. Y., Cohn, D. V. & Hamilton, J. W. (1975) *J. Biol. Chem.* **250**, 7647–7655.

16. Coffee, C. J. & Bradshaw, R. A. (1973) *J. Biol. Chem.* **248**, 3305–3314.

17. Collins, J. H. (1974) *Biochem. Biophys. Res. Commun.* **58**, 301–304.

18. Tufty, R. M. & Kretsinger, R. H. (1975) *Science* **187**, 167.

19. Mercier, J. C., Grosclaude, F. & Ribadeau-Dumas, B. (1973) *Europ. J. Biochem.* **40**, 323–330.

20. Hay, D. I. (1973) *Arch. Oral Biol.* **18**, 1517–1529.

RAT PREPROALBUMIN: *IN VITRO* SYNTHESIS AND AMINO ACID SEQUENCE

A. W. STRAUSS, A. M. DONOHUE, C. D. BENNETT, J. A. RODKEY and A. W. ALBERTS, *Merck Sharp & Dohme Research Laboratories, West Point, Pa. 19486, and Rahway, N. J. 07065*

Poly(A)-containing RNA from rat liver was prepared from total polysomes by oligo d(T)-cellulose chromatography. Translation of this RNA in wheat germ lysates stimulated incorporation of 19 individual ^{3}H-amino acids and of a mixture of 15 ^{14}C-labeled amino acids into protein. The translation products were immunoprecipitated with antiserum to rat serum albumin (RSA). Electrophoresis of the immunoprecipitate on SDS-polyacrylamide gels revealed a radioactive peak of 70,000 molecular weight (i.e., slightly larger than RSA). Cyanogen bromide fragments of labeled immunoprecipitated cell-free product (ICFP) were identical in size to those of RSA except for the N-terminal fragment which was 2,500 larger. Automated Edman degradations of purified ICFP, labeled with 19 different labeled amino acids, showed that the ICFP had 18 more amino acids at the N-terminus than proalbumin. The ICFP was designated as "preproalbumin." The complete sequence of the pre-piece was found to be:

H-Met-Lys-Trp-Val-Thr-Phe-Leu-Leu-Leu-Leu-Phe-Ile-Ser-Gly-Ser-Ala-Phe-Ser-
1 2 3 4 5 6 7 8 9 10 11 12 13 14 15 16 17 18

Radiolabeling experiments confirmed the sequence of the pro-piece as:

Arg-Gly-Val-Phe-Arg-Arg-
19 20 21 22 23 24

and confirmed the sequence of the first 14 residues of the N-terminus of RSA (cycles 25–38).

Albumin is a major secretory protein synthesized in the liver. In rat serum, this protein contains 583 amino acids in a single peptide chain.[1] *In vivo* studies have shown that albumin is synthesized on membrane-bound polyribosomes.[2] Intracellular albumin exists in two forms, mature serum albumin and proalbumin. Proalbumin purified from rat liver microsomes contains an amino terminal extension of either H-Arg-Gly-Val-Phe-Arg-Arg-serum albumin[3] or H-Gly-Val-Phe-Ser-Arg-serum albumin.[4]

Recently, several groups have reported that mRNAs for secreted proteins translated *in vitro* direct the synthesis of proteins larger than either the secreted, mature proteins or the intracellular proproteins. [Immunoglobulin light chains,[5] MOPC 104E (λ), MOPC-41 (κ), MOPC-321 (κ), lysozyme,[6] ovomucoid,[6] parathormone,[7] prolactin,[8] placental lactogen,[9] proinsulin.[10]] These initial translation products have been called preproproteins and contain an extension (prepiece) of 15 to 29 residues at the amino terminus. It has been postulated[11] that these prepieces function as "signal sequences" which participate in binding nascent chains to membranes and in transporting the proteins across membranes during their synthesis. The prepiece is apparently proteolytically removed during the continued synthesis of the protein.[12]

Materials and Methods

Wheat germ lysate was incubated with poly(A)-containing RNA isolated from rat liver polysomes in the presence of one labeled amino acid plus the other 19 unlabeled amino acids.[13] The translation mixtures were subjected to centrifugation, indirect immunoprecipitation and SDS gel electrophoresis. After elution from the gel, cold rat serum albumin was added and the proteins precipitated with 0.1 *M* HCl in acetone overnight.

Unlabeled rat serum albumin (2–8 mg) and purified preproalbumin (0.01–0.6 picomoles) labeled with various amino acids were subjected to automated, sequential Edman degradation of 35–40 cycles in a Beckman 890C Sequencer using a modification of the program of Bauer et al.[14] The sample containing ^{35}S-cysteine was reduced and alkylated with sodium iodoacetate prior to Edman degradation. Aliquots of the unlabeled phenylthiohydantoin amino acids released were analyzed by gas chromatography and high pressure liquid chromatography[15] and used to calculate repetitive yield (93–96% in all runs).

Results

Preproalbumin has been shown to be the initial product of the translation of albumin mRNA.[13] This study establishes the complete sequence of the 18 residue prepiece, confirms the sequence of the six residue propiece,[3] and of the first 14 residues of the amino terminal of rat serum albumin.[16] For example, when ^{3}H-glycine was incorporated into preproalbumin, ^{3}H-glycine-phenylthiohydantoin was released at step 14 and step 20 during the sequence determination (Figure 1). The repetitive yield determined from the cold RSA included in the sequence determination is also shown, indicating that the amount of glycine recovered at these two steps was as expected for this degradation. Comparison of this prepiece with others whose sequences have been wholly or partially determined (Table I) shows the preponderance of hydrophobic residues in the middle of the chain (outlined) as originally emphasized by Schechter et al.[17] It has been suggested[18] that extensive sequence homology of prepieces of proteins destined for secretion might be found for preproteins of one species or of one tissue. Of the three sequences of rat preproteins now known (Table I), extensive sequence homology is not yet apparent, at least for these preproteins from different tissues of the same species.

The positions of Met, Leu, Arg, Val and Phe in rat preproalbumin have recently been confirmed.[19]

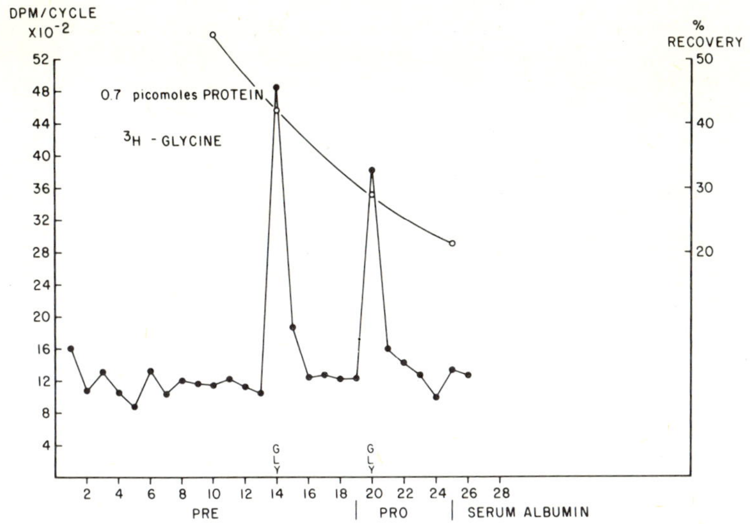

Fig. 1. Illustrative sequence determination, in this case, using 0.7 picomoles of preproalbumin labeled with ^{3}H-glycine.

Table I. Prepieces of Known Sequences

RAT:	Proalbumin	M K W V T -	- - - F L L L L F I	S G S A F S -
MOUSE:	MOPC 1046	M A W I S -	- L I L S L L L F -	P S G A I S -
	MOPC -321	M X T X T -	L L L W V L L L W V	P - X X X T -
	MOPC -41	M D M R A P A Q	I F G F L L L L F -	P - G T R C -
CHICKEN:	Ovomucoid	A M A G	V F V L F S F V L X G F L	P D A A F G -
	Lysozyme	M R S - - -	- L L I L V L C F L	P L A A L G -
COW	Proparathyrin	M M S A K D M V K	V M I V M L A I	X X L A R S D G -
RAT:	Proinsulin	X L K M X	F L F F L K L L X L	X X X X X X X X -
	Prolactin	M -(11 X)	L L L L M M X X L L	X C X X X X X -
HUMAN:	Lactogen	M P-(5 X)	L L L X X X L L	X P X X X X X X X -

Single letter abbreviations: Biochemistry 7, 2703 (1968).

References

1. Judah, J. D. & Quinn, P. S. (1976) *Trends. Biochem. Sci.* **1**, 107–109.
2. Keller, G. H. & Taylor, J. M. (1976) *J. Biol. Chem.* **251**, 3768–3773.
3. Russell, J. H. & Geller, D. M. (1975) *J. Biol. Chem.* **250**, 3409–3413.
4. Urban, J., Inglis, A. S., Edwards, K. & Schreiber, G. (1974) *Biochem. Biophys. Res. Commun.* **61**, 494–501.

5. Burstein, Y. & Schechter, I. (1977) *Proc. Nat. Acad. Sci. USA* **74**, 716–720.

6. Thibodeau, S. N., Gagon, J. & Palmiter, R. (1977) *Fed. Proc.* **36**, 656 (Abstract).

7. Kemper, B., Habener, J. F., Ernst, M. D., Potts, J. T., Jr. & Rich, A. (1976) *Biochemistry* **15**, 15–19.

8. Mauer, R. A., Gorski, J. & McKean, D. J. (1977) *Biochem. J.* **161**, 189–192.

9. Birken, S., Smith, D. L., Canfield, R. E. & Boime, I. (1977) *Biochem. Biophys. Res. Commun.* **74**, 106–112.

10. Chan, S. J., Keim, P. & Steiner, D. F. (1976) *Proc. Nat. Acad. Sci. USA* **73**, 1964–1968.

11. Blobel, G. & Dobberstein, B. (1975) *J. Cell. Biol.* **67**, 835–851.

12. Habener, J. F., Potts, J. T., Jr. & Rich, A. (1976) *J. Biol. Chem.* **251**, 3893–3899.

13. Strauss, A. W., Donohue, A. M., Bennett, C. D., Rodkey, J. A. & Alberts, A. W. (1977) *Proc. Nat. Acad. Sci. USA* **74**, 1358–1362.

14. Brauer, A. W., Margolies, M. N. & Haber, E. (1975) *Biochemistry* **14**, 3029–3035.

15. Rodkey, J. A. & Bennett, C. D. (1976) *Biochem. Biophys. Res. Commun.* **72**, 1407–1413.

16. Bradshaw, R. A. & Peters, T., Jr. (1969) *J. Biol. Chem.* **244**, 5582–5589.

17. Schechter, I., McKean, D. J., Guyer, R. & Terry, W. (1975) *Science* **188**, 160–162.

18. Devillers-Thiery, A., Kindt, T., Scheele, G. & Blobel, G. (1975) *Proc. Nat. Acad. Sci USA* **72**, 5016–5020.

19. Yu, S. & Redman, C. (1977) *Biochem. Biophys. Res. Commun.* **76**, 469–476.

RECENT PHYSIOLOGICAL STUDIES WITH THE ENDORPHINS

ROGER GUILLEMIN, FLOYD BLOOM, JEAN ROSSIER, SCOTT MINICK, STEVE HENRIKSEN, ROGER BURGUS, and NICHOLAS LING, *Laboratories for Neuroendocrinology, The Salk Institute for Biological Studies, La Jolla, California, 92037*

While the core of this lecture will be essentially on recent physiological studies with the endorphins, I will start with a brief introduction now mostly of historical interest, describing the methodology used in our laboratories for the isolation and characterization of the morphinomimetic peptides: α-endorphin and γ-endorphin.

With the availability of the earlier literature demonstrating the existence of specific opiate receptors, as well as that of endogenous ligands for these receptors, most likely of peptide nature, (see *Opiate Narcotics*, A. Goldstein, Ed., Pergamon Press, N. Y., 1975) it became of interest to characterize these endogenous opiate-like peptides, since they might be involved in the physiological mechanisms of control of pituitary functions. In any isolation program of a new biologically active substance, one has two absolute requisites: first, a meaningful bioassay, second, a good supply of starting material.

Assay of Biological Activity. It was decided that morphinomimetic substances would be defined as those substances which would decrease linearly the amplitude of the muscle contractions electrically induced *in vitro* in the myenteric plexus-longitudinal muscle of the guinea pig ileum[1] only if this biological activity could be reversed or prevented by naloxone, a morphine-analog antagonist.

Starting Material. We first confirmed reports by others[2-4] that aqueous extracts of whole brain or of several specific anatomical brain structures (caudate nucleus, hypothalamus) or of the pituitary gland[5] contained naloxone-reversible morphinomimetic activity. Calculations based on simple assumptions relating expected specific activity of the substances to be characterized to their apparent concentration in these extracts indicated that several hundreds of kg of fresh tissues would have to be procured and handled to provide a reasonable chance of characterizing the postulated substances within a reasonable time. Searching through materials available in large quantities from our earlier isolation program of the hypothalamic-hypophysiotropic factors, we found a partially purified extract of (porcine) neurohypophysis-hypothalamus to be considerably enriched in morphinomimetic activity (half-maximal activity in the bioassay at about 10 μg of the dry powder per ml of incubation fluid). This material was used for the isolation of the endorphins; it is an acetic acid/acetone extract corresponding to fraction G of the Kamm procedure[6] performed on tissues of porcine origin consisting of approximately 50% neurohypophysis and 50% pituitary stalk and attached ventral hypothalamus (Pitressin Intermediate, Parke-Davis and Co.). Assays performed in our laboratories showed this material to contain 12 USP units of vasopressin per mg, 4 USP units of oxytocin per mg, 2×10^5 Shizume units of melanotropin (MSH) activity per mg, 1.67 international units of adrenocorticotropin (ACTH) per mg by an *in vitro* corticoidogenic assay. One hundred grams of this material, corresponding to about 1/4 million posterior pituitary-hypothalamus, were used in the purification described below. We now know that the endorphins found in that starting material can not be of posterior pituitary origin; they must come from the ventral hypothalamus and the pars intermedia cells which we know remain attached to the neurohypophysis at time of the gross dissection of the gland at the slaughterhouse.

Purification Scheme. We accepted earlier evidence[2-4] that the morphinomimetic substances of brain origin were polypeptides. The purification sequence finally established from pilot studies consisted of five steps (i to v) as follows: (i) Gel Filtration, on Sephadex G25 equilibrated on 2 *M* acetic acid; (ii) SP-Sephadex chromatography with an elution gradient from 0.001 *M* NH_4OAc, pH 4, to 0.01 *M* NH_4OAc, pH 7, and finally 2 *M* HOAc pH7; (iii) DEAE-Sephadex chromatography with an elution gradient from 0.001 *M* NH_4OAc, pH 7, stepwise to 0.01 *M*, 0.02 *M*, 0.05 *M*, 0.1 *M* NH_4OAc, all at pH 7; (iv) Partition chromatography on Sephadex G50 using the solvent system 1-BuOH/HOAc/H_2O (4:1:5); (v) High pressure liquid chromatography (HPLC); system used for HPLC was 4 mm × 30 cm μBondapak/C18 column using solvent concentration gradient curve 8 computed by a Waters model 660 programmer. Starting buffer A was 20% CH_3CN in 0.01 *M* NH_4OAc, pH 4, and final buffer B was CH_3CN at a program rate of 0-80% B during 20 min. The flow rate was 2.5 ml/min (see Ref. 7).

Fig. 1 shows a typical pattern obtained at step (i). Biological activity as defined above was found essentially in two zones, with the activity of zone 2 totally reversible or statistically so, by naloxone, while that in zone 1 was only reversed 50% in the same conditions, by naloxone. Steps (ii) to (v) were conducted on the lyophilized material from zone 2. Fig. 2 shows the final step (v) of the isolation scheme. It represents the HPLC of one of three active zones obtained at that stage of purification of pool 2. The materials so obtained were called α-endorphin-1, α-endorphin-2, and γ-endorphin. Subsequent studies[7] showed α-endorphin-2 to be most probably an unstable salt form of α-endorphin-1; both had identical amino-acid composition and primary structure. We will thus refer to the single peptide component as α-endorphin.

The primary sequence of α-endorphin and γ-endorphin was obtained by sequential degradation using the Edman method followed by dansylation with [^{14}C] dansyl chloride and identification of phenythiohydantoin derivatives by mass spectrometry. Mass spectrometry of the acetylated peptide fragments obtained after trypsin digestion of the native endorphins was also used to establish

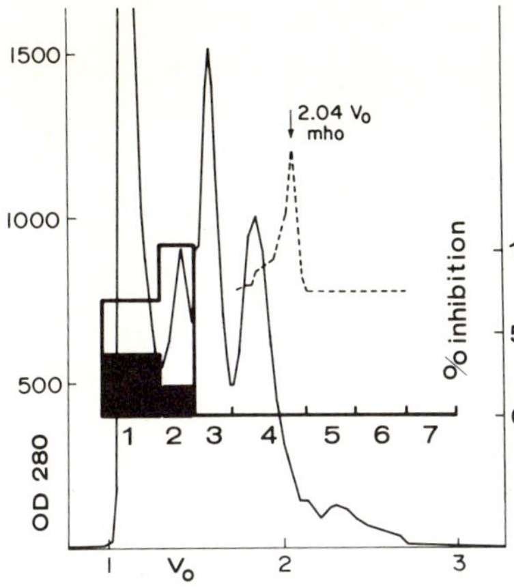

Fig. 1. Sephadex G-25 gel filtration chromatography of pitressin intermediate. Solid line and bar graph show naloxone-reversible biological activity (% inhibition in the myenteric plexus).

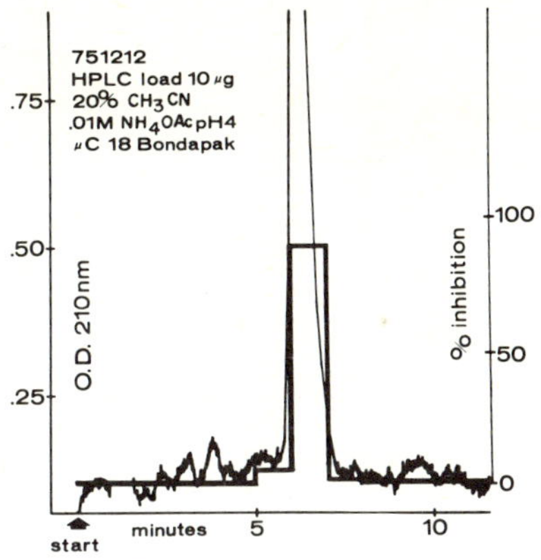

Fig. 2. Analytical HPLC of ovine α-endorphin. Solid line shows naloxone reversible biological activity (% inhibition in the myenteric plexus).

the primary structures. Table I shows the amino-acid analyses of native and synthetic porcine endorphins. Table II shows the dansyl derivatives from the Edman degradation of porcine α-endorphin. Figs. 3–5 show the mass spectra of

Table I. Amino-Acid Analyses of Native and Synthetic Porcine Endorphins

	Native porcine endorphin				Synthetic endorphin		
	α-1[a]	α-2[a]	α-2[b]	γ[a]	α[a]	α[b]	γ[a]
Thr	2.7	2.6	2.4	2.7	2.9	2.0	2.8
Gln	—	—	—	—	—	—	—
Ser	1.6	1.5	1.9[c]	1.6	1.7	1.6[c]	1.8
Glu	2.0	2.0	1.0	1.9	2.0	0.9	2.0
Pro	1.0	1.0	—	1.0	1.0	—	0.9
Gly	1.9	1.9	1.9	2.0	2.1	2.0	2.0
Val	1.0	1.0	1.1[d]	1.0	1.0	1.2[d]	1.0
Met	1.0	1.0	0.9	0.9	1.0	1.0	1.0
Leu	1.1	1.0	1.3[d]	2.0	1.0	1.4[d]	2.0
Tyr	1.0	1.0	0.9	0.9	1.0	1.1	1.0
Phe	1.0	1.0	0.9	1.0	1.0	1.1	1.0
Lys	1.0	1.0	1.0	1.0	1.0	1.0	1.0
NH_3	2.3	1.6	—	2.3	1.2	—	1.3

[a] Hydrolysis in 6 M HCl-0.5% thioglycollic acid.
[b] Enzymatic hydrolysis with papain and leucineaminopeptidase.
[c] Gln does not resolve from Ser under these conditions.
[d] This peak includes a shoulder, probably due to incompletely hydrolyzed peptides.

Table II. Dansyl Derivatives from Edman Degradation of Porcine α-Endorphin (net cpm/Edman cycle)

Dns-amino acid	Edman cycle 0	1	2	3	4	5	6	7	8	9	10	11	12	13	14	15	Control	n*
Tyr/2	*23,921*	0	0	0	301	194	0	66	549	0	13	0	0	240	0	293	1912 ± 571	15
Gly	0	*5,170*	*10,239*	0	0	795	0	0	0	0	0	0	1,991	3,424	206	0	1508 ± 1344	14
Phe	882	590	624	*2,253*	1,303	0	0	0	1,744	0	578	0	0	555	0	0	855 ± 825	14
Met	1	95	61	0	*5,691*	1,771	215	15	0	0	0	0	0	525	0	0	503 ± 100	14
Thr	0	0	0	0	0	*4,987*	1,625	1,118	90	351	62	*4,518*	2,919	2,738	20	*3,121*	1428 ± 1231	10
Ser	0	0	0	0	0	0	*4,695*	3,383	834	*3,590*	1,816	70	889	0	560	12	760 ± 503	12
Glx	0	0	0	0	0	16	0	*12,634*	7,451	1,758	*8,151*	2,603	1,146	1,530	0	0	1046 ± 963	12
Ptc-Lys	—	—	—	—	—	—	—	—	*2,502*	1,038	1,040	0	0	—	160	648	1310 ± 834	6
Pro	0	78	382	364	244	224	0	0	0	0	0	0	*3,814*	716	0	0	604 ± 200	13
Leu	201	0	0	0	0	0	268	695	412	624	748	302	0	*8,298*	1,760	0	1454 ± 546	13
Val	0	0	0	0	0	336	0	7	0	0	0	0	107	1,835	*5,136*	0	1109 ± 543	14
nmol used	1	1	1	1	2	2	2	3	3	3	3	3	4	4	4	4		

Italicized values are those judged to represent newly formed terminal amino acids.
* *n* equals the number of determinations of the control values.

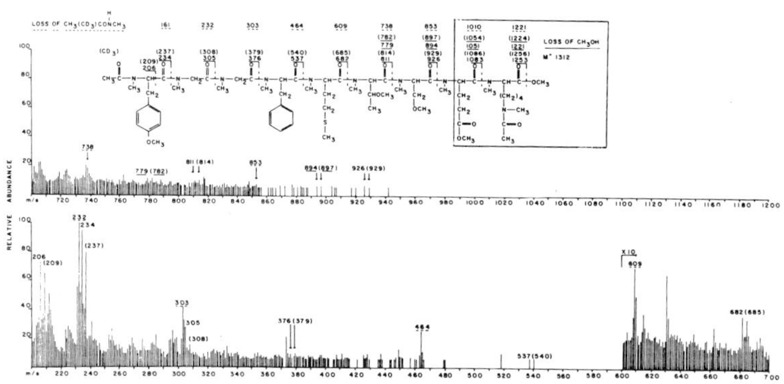

Fig. 3. Mass spectrum of derivatized α-endorphin-1 obtained by fractional vaporization at 280-295°C. The peptide detected has the sequence H-Tyr-Gly-Gly-Gly-Phe-Met-Thr-Ser.

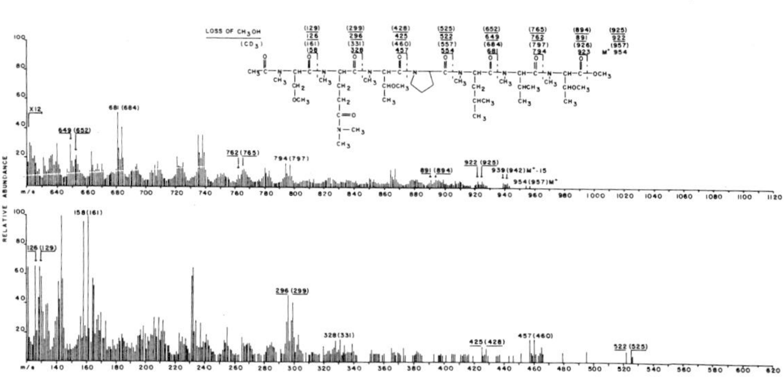

Fig. 4. Mass spectrum of derivatized α-endorphin-1 obtained by fractional vaporization at 231-245°C. The peptide detected has the sequence H-Ser-Gln-Thr-Pro-Leu-Val-Thr-OH.

derivatized fragments of α-endorphin and γ-endorphin. The primary sequence of α-endorphin was established as H-Tyr-Gly-Gly-Phe-Met-Thr-Ser-Glu-Lys-Ser-Gln-Thr-Pro-Leu-Val-Thr-OH. The primary structure of γ-endorphin was established as H-Tyr-Gly-Gly-Phe-Met-Thr-Ser-Glu-Lys-Ser-Gln-Thr-Pro-Leu-Val-Thr-Leu-OH.

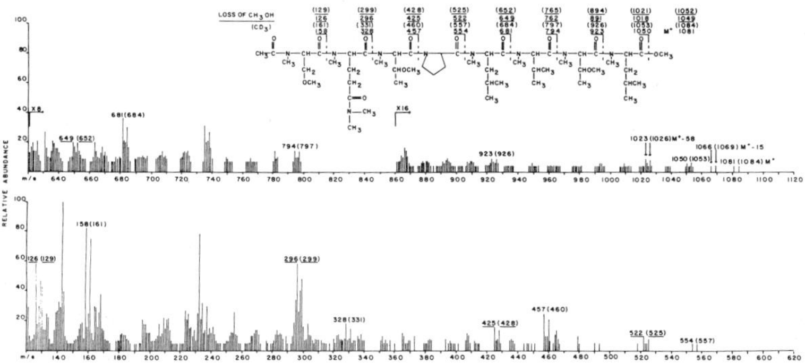

Fig. 5. Mass spectrum of derivatized γ-endorphin obtained by fractional vaporization at 237-246°C. The peptide detected has the sequence H-Ser-Gln-Thr-Pro-Leu-Val-Thr-Leu-OH.

As this work was in progress, Hughes et al.[8] reported that the characterization of porcine *enkephalin*, which had been purified earlier[2], showed it to be the two pentapeptides H-Tyr-Gly-Gly-Phe-Met-OH and H-Tyr-Gly-Gly-Phe-Leu-OH. They also made the remarkable observation that the primary structure of Met^5-enkephalin was identical to the sequence Tyr 61-Met 65 of the hypophysial polypeptide β-lipotropin (β-LPH) isolated by Li et al.[9], while Leu^5-enkephalin shared with β-LPH the sequence Tyr 61-Phe 64. It was obvious that α-endorphin and γ-endorphin have the same amino-acid sequence as β-LPH-(61-76) and β-LPH-(61-77), respectively. The C-fragment[10] of β-LPH [β-LPH-(61-91)] has been also shown to have morphinomimetic activity[10-14]; Li and Chung[14] have named it β-endorphin.

We have found that β-endorphin is by weight the major compound with morphinomimetic activity, reversible with naloxone, in fraction 1 of the first stage of purification of our starting material (see Fig. 1): fraction 1 as in Fig. 1, was further purified by two ion-exchange chromatography and HPLC steps. A material was isolated with morphinomimetic activity which had the same elution characteristics on HPLC as synthetic β-endorphin; its amino-acid composition after hydrolysis was identical to that of porcine β-LPH-(61-91). Though relatively large amounts (ca. 0.5 mg) of that material were available, we decided not to establish the amino-acid sequence; we considered the observations reported here as sufficient proof that this material was indeed β-LPH-(61-91).

Now, almost two years later, we have not found in this starting material any evidence for the presence of either form of the pentapeptide enkephalin. Recently, by the use of a radioimmunoassay we have found evidence of small amounts in fresh brain extracts as well as in fresh pituitary extracts (rat tissues) of one of the two enkephalin pentapeptides; the antiserum available does not distinguish qualitatively between Leu^5-enkephalin and Met^5-enkephalin.

There have been recent reports, from Goldstein's laboratory[15] claiming evidence for the existence in pituitary extracts of substances with morphinomimetic activity that would be different from the already characterized endorphins as fragments of β-lipotropin. To investigate this possibility, we have recently extracted frozen, whole pituitaries from sheep, assaying the fractions obtained for morphinomimetic activity. As shown in Fig. 6, such an extract yielded essentially two zones of biological activity; the first zone following application of the 1 *M*, pH 6.7 NH_4OAc gradient. Upon further purification by HPLC, it yielded a single component with the amino-acid composition of β-endorphin. The second zone after application of that gradient, also yielded a single component with biological activity upon further purification; its amino-acid composition is similar to that of the fragment β-LPH-(61-87) and its biological activity is of the order of the synthetic replicates of that characterized peptide. Thus, we did not obtain any evidence for endorphins of pituitary origin that would not be related to recognized subunits of β-LPH and more specifically of subunits of β-LPH-(61-91).

All the biological studies dealing with the purification scheme described above were conducted with one type of bioassay or another (guinea pig's ileum muscle and myenteric plexus, mouse vas deferens) or several types of binding studies on plasma membrane receptors of a glioma-neuroblastoma cell line cultured *in vitro* or the so-called synaptosome binding assays[12]. Fig. 7 shows one such experiment in which we demonstrated the similarity of binding to rat brain synaptosomes of several of the endorphins[12]. All these biological methods have an inherent limitation. In practice, they do not allow the recognition of a single entity; in other words, all endorphins are active in the synaptosome displacement assay, in the bioassays, etc., though obviously, with different specific activities. To obtain measurements endowed with specificity and also with greater sensitivity than any of the bioassays mentioned, we have recently developed radioimmunoassays for α- and β-endorphin and, as several other laboratories, also for the enkephalin pentapeptides. While these methods have enormous power, they also have their problems and limitations.

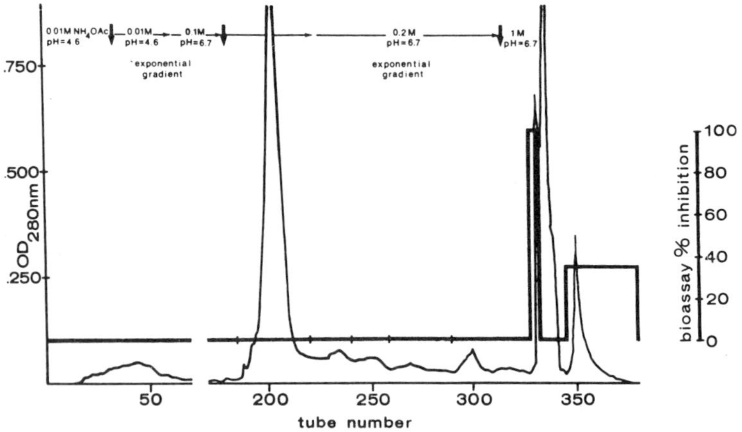

Fig. 6. Carboxymethyl cellulose chromatography of whole ovine pituitary extracts.

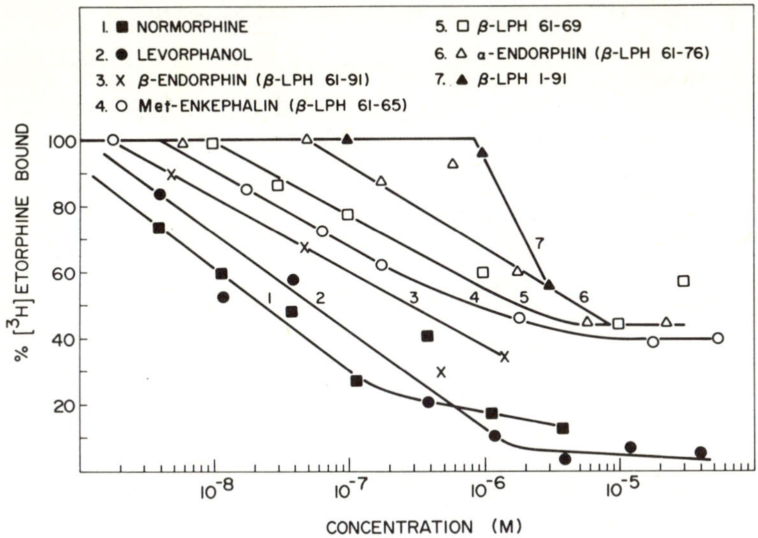

Fig. 7. Competition of the binding of [^{3}H] etorphine by opiates and morphinomimetic peptides. Stock solutions of all compounds were diluted in water containing 1 mg of bovine serum albumin/ml. Each point represents the average value of assays carried out in triplicate.

Tables III and IV show the specificity of two antisera which were raised against α-endorphin and β-endorphin. One of the first observations that were made using the classic methods of immunocytochemistry with these antisera, was that in the pituitary gland of several species studied, α- and β-endorphin were primarily to be found in the cells of the pars intermedia and in a few discrete cells of the adenohypophysis with no evidence of their presence in the pars

Table III. Specificity of α-Endorphin Antiserum RB66-5/10

α-Endorphin	†Y G G F M T S E K S Q T P L V T	100*
α-Endorphin-NH_2	Y G G F M T S E K S Q T P L V T·	5
[Gln^8]-α-Endorphin	Y G G F M T S Q K S Q T P L V T	100
γ-Endorphin	Y G G F M T S E K S Q T P L V T L	0.1
γ-Endorphin-NH_2	Y G G F M T S E K S Q T P L V T L·	0.1
Met^5-Enkephalin	Y G G F M	0
β-LPH-(61-69)	Y G G F M T S E K	0
β-LPH-(70-76)	S Q T P L V T	50
β-LPH-(71-76)	Q T P L V T	15
β-LPH-(72-76)	T P L V T	< 0.05
β-LPH-(73-76)	P L V T	< 0.05
$β_o$-Endorphin	β-LPH-(61-91)	0.05
$β_o$-LPH-(1-91)		0.08

*Numbers show molar ratios x 100 of compound to the left when compared to α-endorphin, for 50% competition of trace binding to the antibody (EB_{50}). Any number different from zero implies that at no less than 3 doses each compound tested gives a binding curve parallel to that of α-endorphin reference standard.
†Peptide sequences shown in the one-letter code of Dayhoff (see Atlas of Protein Sequence and Structure, M.O. Dayhoff, Nat. Biomed. Res. Fdt., Silver Springs, Md., Publ. 1975).
Subscripts o: ovine; p: porcine (Table IV only).

Table IV. Specificity of β-Endorphin Antiserum RB100-10/27

β_o-Endorphin	β-LPH-(61-91)	61 Y G G F M T S E K S	100*
		71 Q T P L V T L F K N	
		81 A I I K N A H K K G Q	
β_p-Endorphin		... A I V K N A H K K G Q	100
[Leu^5]-β_p-Endorphin	[Leu^5]-β-LPH-(61-91)	61 Y G G F L T S E K ...	100
Met^5-Enkephalin	β-LPH-(61-65)		0
α-Endorphin	β-LPH-(61-76)		0
γ-Endorphin	β-LPH-(61-77)		0
δ_o-Endorphin	β-LPH-(61-87)		100
	β-LPH-(62-91)		100
	β-LPH-(66-91)		100
	β-LPH-(78-91)		0
β_o-Lipotropin	β-LPH-(1-91)		100
31K-Precursor			100
1N HOAc Extract (rat)	Whole Pituitary		parallelism
1N HOAc Extract (rat)	Whole Brain		parallelism
Morphine Sulfate			1×10^{-7}

*Legend for Table IV as in Table III

nervosa. These cells of the pars intermedia contain ACTH,[16,17] α-MSH,[16-18] also β-LPH.[18] Recently, using the antisera to α-endorphin and β-endorphin described in Tables III and IV Mains et al.[19] have demonstrated, using the cloned cell line AtT-20 of mouse pituitary origin that β-LPH, β-endorphin, and ACTH (1-39) have a common biosynthetic precursor, a glycoprotein of an approximate molecular weight of 31,000 (referred to later as 31K). This set of observations, along with earlier reports that β-LPH and ACTH were released in similar circumstances[20,21], suggested that in physiological conditions, the biologically active peptides, β-endorphin and ACTH might be secreted simultaneously by the normal pituitary. Indeed, as shown in Fig. 8, we have recently demonstrated that in response to acute stress (instantaneous breaking of the tibia and fibula in rats) there is concomitant elevation of plasma levels of ACTH and plasma levels of

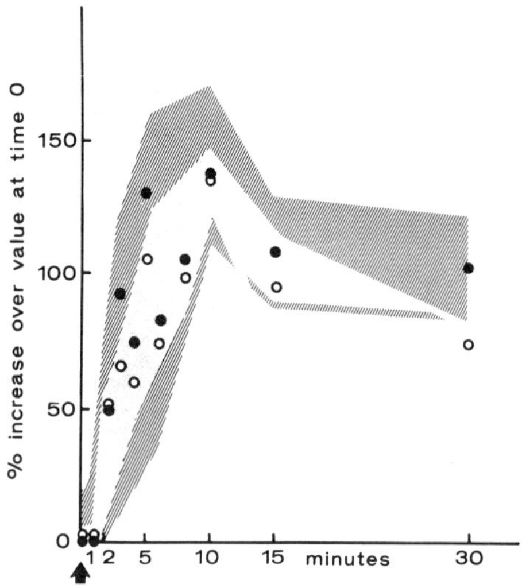

Fig. 8. Plasma levels of adrenocorticotropin (closed circles) and β-endorphin (open circles) measured by radioimmunoassays in trunk blood obtained from rats sacrificed at time intervals shown on the abscissa, following acute stress at time zero. (↑) Solid line shows plasma levels of adrenal corticosterone measured by fluorometry. Shaded areas show confidence limits of the measurements. Correlation coefficient, ρ, between the two populations of ACTH and β-endorphin concentrations is 0.9708 for values of means, (D. F. = 20) and 0.7785 for all individual values (D. F. = 64).

β-endorphin. Moreover, in other studies we have observed a constant parallelism in variations of the pituitary contents in ACTH and β-endorphin as affected by chronic adrenalectomy, and chronic administration of the synthetic glucocorticoid dexamethasone. Also, we have observed that when incubated *in vitro* fragments of the pituitary gland of a patient with an ACTH secreting adenoma secreted similarly elevated amounts of ACTH and β-endorphin; and a pancreatic tumor containing and secreting ACTH was also shown to contain β-LPH and β-endorphin. In preliminary studies, purified hypothalamic extracts (*i.e.* not containing ACTH and/or endorphins) prepared by W. Vale were shown to stimulate simultaneously the secretion of ACTH and β-endorphin by primary monolayer cultures of rat adenohypophyses (unpublished results). Thus we can conclude that in all circumstances in which ACTH is secreted, there is a concomitant secretion of β-endorphin.

The significance of these observations is not clear. We have established (see Fig. 8) that in peripheral blood of the stressed rat it is β-endorphin (as established by gel filtration) that is primarily circulating with smaller amounts of β-LPH. Thus the next question is, what is the significance of these elevated levels of circulating β-endorphin as they appear in response to stress.

Since the early studies of Selye,[22] ACTH has been recognized as the primary pituitary hormone secreted in response to acute stress in all species studied. The teleological proposal has been to relate the acute secretion of ACTH to the corresponding immediate activation of the adrenal cortex for the secretion of glucocorticoids necessary for immediate increase of gluconeogenesis and the ensuing availability of energy rich carbohydrates. In many, though not all species, prolactin and growth hormone may also be released under similar conditions[23]. ACTH, growth hormone and prolactin are pituitary responses to stress, all affecting metabolism. Our results show that β-endorphin is also released in response to acute stress.

There is evidence[24] that peripheral injection of β-endorphin in mice produces analgesia; this was obtained with doses of the peptide leading to plasma concentrations 4 to 5 orders of magnitude greater than those observed here in response to stress. We have failed to observe in rats such central effects of similarly large doses of β-endorphin (up to 20 mk/kg body weight) injected in a peripheral vein. This is in contradistinction to the profound effects (analgesia, catatonia) exerted by small amounts ($\geqslant 0.5\ \mu g$) of β-endorphin when injected directly in the brain and uniformly observed[25]. It may be that our current criteria for assessing possible central effects of the circulating endorphins are naive. Should such effects of the peripherally released β-endorphin eventually be detectable, the results reported here demonstrate that the system necessary for its immediate secretion and availability in response to stress is highly functional and coupled with activation of the ACTH-adrenal cortex axis.

Thus, a holistic response of the organism to stress would involve the immediate secretion of pituitary hormones, some (adrenocorticotropin, growth hormone) involved in somatotropic (metabolic) adaptive reactions, others, (β-endorphin, β-melanocyte stimulating hormone or γ-lipotropin), endowed with neurotropic[25], behavioral or psychotropic adaptive reactions. Suggestive of a vascular pathway from pituitary to brain, recent observations[26] have shown high concentrations of pituitary hormones from all three lobes of the pituitary in the long portal vessels. Functional significance of these observations remains to be proven; there is as yet, no evidence that blood *flows* retrograde from pituitary *into* brain in a functional mechanism able to deliver pituitary peptides. On the contrary, we have evidence that profound variations in the pituitary secretion of β-endorphin are not reflected in concomitant variations of brain levels of β-endorphin. We have reported[27] that several weeks after total hypophysectomy, immunocytochemistry shows neurons and axonal varicosities staining with the fluorescein-coupled antisera to α-endorphin or β-endorphin, similar in numbers and locations to those found in normal intact animals. This is in agreement with an earlier statement by Goldstein[15] reporting identical amounts of opiate-like activity of brain extracts in normal and hypophysectomized rats, as tested by a synaptosomal binding assay. Thus with all this conflicting evidence, we have to search for one or several peripheral targets for the β-endorphin secreted in response to stress in the dynamic fashion and large amounts that we have demonstrated here.

Another series of observations which we have recently made is leading to unexpected questions. We have obtained immune serum in rabbits injected with Leu^5-enkephalin coupled with bovine serum albumin. The antibodies so produced bind Leu^5-enkephalin, and also Met^5-enkephalin though with a sensitivity of 1/30 by weight of the former peptide. This antiserum has no cross reactivity with α-endorphin or β-endorphin. Using this antisera raised against α-endorphin, β-endorphin and Leu^5-enkephalin, we find by immunocytochemistry that there is striking dissociation in the distribution of the two peptides of the brain (see Table V). The same observations are made by radioimmunoassays of extracts of various parts of the brain; those numbers are always related to the immunocytochemical picture. Somewhat to our surprise, we have also found minute but definite amounts of enkephalin in the pituitary as measured by the radioimmunoassay. Some treatments such as chronic administration of morphine will elevate significantly the pituitary content of β-endorphin and, similarly, of enkephalin. Other treatments such as chronic administration of dexamethasone, produce significant decrease of the pituitary content in β-endorphin while at the same time elevating significantly the pituitary content in enkephalin. So far, in the brain no treatment or procedure, such as stress, administration of morphine or dexamethasone, has modified content of various parts of the brain in β-endorphin. The same treatments in the same animals have led in the same extracts of the brain to considerable modifications of the content of enkephalin. At the moment there is no pattern that emerges from these observations; we are still collecting data.

Table V. Immunocytochemical Staining Patterns
Anti-β-Endorphin vs Anti-Enkephalins

	β-Endorphin	Met/Leu – Enkephalin
Thalamus-Periaqueductal G.	++	+++
Amygdala		
Central Nucleus	±	+++
Medial Cortical	++	+
Baso-lateral	++	++
Striatum		
Caudate	0	++
Globus Pallidus	±–+	++++
Spinal Cord	0	++
Pituitary		
Adenohypophysis	+++	+
Intermediate Lobe	++++	±
Neurohypophysis	0	0
Hypothalamus		
Supra-optic Nucl.	±	±
Paraventricular N.	+++	++
Periventricular N.	++++	++++
Arcuate Nucl.	+++	+
Median Eminence	++	++

We and others have recently reported that administration of β-endorphin in one of the lateral ventricles or in the cisterna (studies in rats) will consistently stimulate acutely the secretion of pituitary growth hormone and prolactin. Intravenous administration of β-endorphin leads to secretion of prolactin, but not of growth hormone. We have also reported that β-endorphin does not stimulate the release of the pituitary hormones when added *in vitro*[28]. It is thus most likely that following either intraventricular or intravenous injection that the peptides act somewhere distal to the pituitary to produce these hypophysiotropic effects. Similarly, we have recently observed in collaboration with Weitzman and Fischer (to be published) that administration of β-endorphin in rabbits acutely stimulates the secretion of antidiuretic hormone with an onset identical to that following injection of hypertonic saline but sustained for much greater periods of time (as much as half an hour). Here again, addition of β-endorphin to incubated posterior plus intermediate lobe of rat pituitary does not statistically increase the secretion of vasopressin as measured by radioimmunoassay under conditions in which elevated potassium is highly effective. Thus it is unlikely that the effect of β-endorphin on the secretion of antidiuretic hormone is mediated by the opiate receptors recently described by Simantov and Snyder[29] in the neurohypophysis.

We have recently recognized that administration of β-endorphin by microiontophoresis to neurons throughout the central nervous system will usually produce naloxone-reversible inhibition of the firing rate and will occasionally decrease the amplitude of spontaneous or glutamate-induced firing[30]. A remarkable exception to this statement is the pyramidal neurons of the hippocampus in which on a statistical basis more than 80% of the tested cells are consistently activated by β-endorphin[30]. This activation is also naloxone-preventable or reversible (see Figs. 9 and 10). We have already reported[25] the striking behavioral effects induced in rats by intraventricular administration of doses of the various endorphins. α-Endorphin and β-endorphin produce analgesia of variable duration, but extending from 30 minutes to 4-6 hours, always accompanied by fall in rectal body temperature. α-Endorphin will produce a general tranquilization of the animal with a rapidly appearing wet-dog shake (WDS), an unexpected observation in animals totally naive of administration of opiates or opioid-peptides. β-endorphin will produce the same rapid onset tranquilization (appearing from in 5-15 minutes) also accompanied by the same wet-dog shakes, but rapidly (15-30 minutes) turning into a remarkable state of profound immobilization and catatonia. This is a state in which the animals are totally rigid with

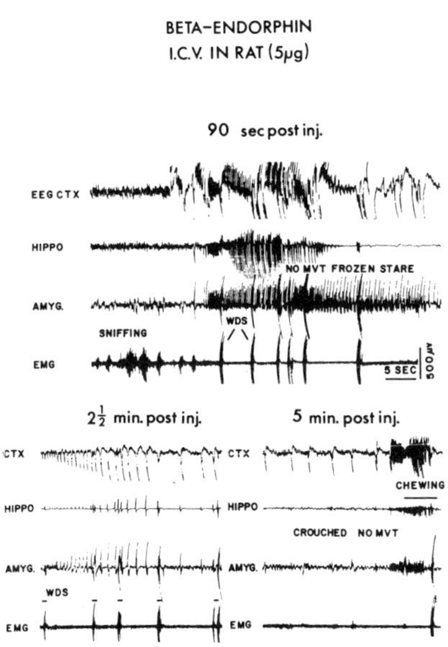

Fig. 9. Injection of β-endorphin (5 μg) in the lateral ventricle of the brain of a rat. Recording with permanently implanted electrodes in the frontal cortex (CTX), the hippocampus (HIPPO), and the amygdala (AMYG.) up to 5 min after injection. EMG is from an electrode in the neck muscles. (WDS = wet-dog shake, MVT = moment).

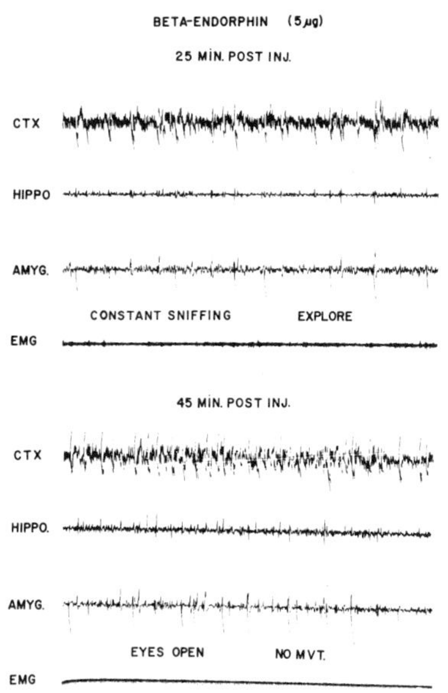

Fig. 10. As in Fig. 9, records at 25 and 40 min post-injection.

shallow abdominal respiration, totally immobile, completely analgetic, total absence of corneal reflex and usually exophthalmic protruding eyeballs. In contradistinction γ-endorphin elevates body temperature and produces a state of agitation of variable degree, occasionally leading to aggressive behavior. All these effects of α-, β-, γ-endorphins are preventable or totally reversible in seconds by intravenous or intraperitoneal injection of naloxone. We have recently studied the effect of β-endorphin with permanently implanted electrodes in the cortex, in the dorsal hippocampus, and in the amygdala. These animals also have a permanently implanted cannula in one of the lateral ventricles and occasionally a catheter in the upper vena cava. Recording in free moving animals is done from the contralateral hemisphere from that in which the cannula is implanted. Following intraventricular administration of β-endorphin, a very consistent, remarkable pattern is observed. Within seconds after administration of the β-endorphin in the contralateral ventricle one observes extremely high amplitude focal seizures in the cortex, the hippocampus and the amygdala, often accompanied by the classical wet-dog shakes. The animals at that time have no analgesia and are freely moving. This high voltage desynchronization will be followed after a few minutes by what appears to be total depolarization of the various areas in which electrodes have been implanted. This picture is broken by spontaneous, multiple ictal episodes as the animal gradually enters the catatonic stage (see Fig. 10) (Soc. Neurosci. Abstr., S. Henriksen et al., 1977). In animals so prepared with the permanently implanted electrodes we have administered up to 20 mgs/kg of β-endorphin by intravenous administration. At no time did we observe any evidence of behavioral effects, analgesia, or any effect on the various electrical recordings. In such animals, with no obvious effect of these enormous doses of β-endorphin administered in a peripheral vein, subsequent injection of a few micrograms of β-endorphin in the lateral ventricle is rapidly followed by the classical syndrome which we have described above and which has been unanimously recognized. These results are totally at variance with those reported by Tseng et al.[24] in which these authors report observing dose-related analgesia in mice injected in a tail vein with β-endorphin.

Obviously, the timing of this lecture in relation to the developing methodology is such that it is impossible to come up with a comprehensive picture. The next few months of exploitation of that powerful methodology should allow us to obtain answers to the many questions that the partial data discussed here are obviously calling for.

Research in our laboratories is presently funded by grants from NIH, Grants, No. AM-18811 and No. HD-09690, W. R. Hearst Foundation (RG) and from NIDA, No. DA-01785 and the A. P. Sloan Foundation (FB).

References

1. Paton, D. M. & Zar, M. A. (1968) *J. Physiol. (London)* **194**, 13–33.

2. Hughes, J., Smith, T., Morgan, B. & Fothergill, L. (1975) *Life Sci.* **16**, 1753–1758.

3. Pasternak, G. W., Goodman, R. & Snyder, S. H. (1975) *Life Sci.* **16**, 1765–1768.

4. Terenius, L. & Wahlstrom, A. (1974) *Acta Pharmacol. Toxicol.* **35**, 55–58.

5. Cox, B. M., Opheim, K. E., Teschemacher, H. & Golstein, A. (1975) *Life Sci.* **16**, 1777-1782.

6. Kamm, O., Aldrich, T. B., Grose, W., Rowe, L. W. & Bugbee, E. P. (1928) *J. Amer. Chem. Soc.* **50**, 563–601.

7. Ling, N., Burgus, R. and Guillemin, R. (1976) *Proc. Nat. Acad. Sci. USA* **73**, 3942–3946.

8. Hughes, J., Smith, T., Kosterlitz, H., Fothergill, L., Morgan, B. & Morris, H. (1975) *Nature* (London) **258**, 577–579.

9. Li, C. H., Barnafi, L., Chretien, M. & Chung, D. (1965) *Nature* **208**, 1093–1094.

10. Bradbury, A., Smyth, D., Snell, C., Birdsall, N. & Hulme, E. (1976) *Nature* (London) **260**, 793–795.

11. Cox, B., Goldstein, A. & Li, C. H. (1976) *Proc. Nat. Acad. Sci. USA* **73**, 1821–1823.

12. Lazarus, L., Ling, N. & Guillemin, R. (1976) *Proc. Nat. Acad. Sci. USA* **73**, 2156–2159.

13. Ling, N. & Guillemin, R. (1976) *Proc. Nat. Acad. Sci. USA* **73**, 3308-3310.

14. Li, C. H. and Chung, D. (1976) *Proc. Nat. Acad. Sci. USA* **73**, 1145-1148.

15. Goldstein, A. (1976) *Science* **193**, 1081–1086.

16. Moriarty, G. C. (1973) *J. Histochem. Cytochem.* **21**, 855–894.

17. Dubois, P., Vargues-Regairaz, H. & Dubois, M. P. (1973) *Z. Zellforsch* **145**, 131–143.

18. Hirata, Y., Matsukura, S., Imura, H., Nakamura, M. & Tanaka A. (1976) *J. Clin. Endocrinol Metab.* **42**, 33–40.

19. Mains, R., Eipper, B. & Ling, N. (1977) *Proc. Nat. Acad. Sci. USA* **74** in press.

20. Abe, K., Nicholson, W. C., Liddle, G. W., Orth, D. N. & Island, D. P. (1969) *J. Clin Invest* **48**, 1580–1585.

21. Gilkes, J. J., Bloomfield, G. A., Scott, A. P., Lowry, P. J., Ratcliffe, J. G., Landon, J. & Rees, L. H. (1975) *J. Clin. Endocrinol. Metab.* **40**, 450–457.

22. Selye, H. (1950) *Stress*, 1 vol., Acta Publ., Montreal, Canada.

23. Glick, S. M., (1968) *Ann. N. Y. Acad. Sci* **148**, 471–477.

24. Tseng, L. F., Loh, H. H. & Li, C. H. (1976) *Nature (London)* **263**, 239–243.

25. Bloom, F., Segal, D., Ling, N. & Guillemin R. (1976) *Science* **194**, 630–634.

26. Krivoy, W. & Guillemin, R. (1961) *Endocrinology* **69**, 170–172.

27. Guillemin, R., Ling, N., Lazarus, L., Burgus, R., Minick, S., Bloom, F., Nicoll, R., Siggins, G., & Segal, D. (1977) Ann. N. Y. Acad. Sci. (in press).

28. Rivier, C., Vale, W., Ling, N., Brown, M. & Guillemin R. (1977) *Endocrinology* **100**, 238–241.

29. Simantov, R. & Snyder, S. H. (1977) *Brain Res.* **124**, 178–184.

30. Nicoll, R., Siggins, G., Ling, N., Bloom, F. & Guillemin, R. (1977) *Proc. Nat. Acad. Sci. USA* **74** (in press).

RECEPTORS FOR PUTATIVE NEUROTRANSMITTER PEPTIDES IN THE BRAIN

SOLOMON H. SNYDER, *Departments of Pharmacology and Experimental Therapeutics and Psychiatry and Behavioral Sciences, Johns Hopkins University School of Medicine, Baltimore, Maryland 21205*

Peptide hormones and most neurotransmitters exert their physiological effects by first attaching to receptor sites on the plasma membranes of target cells. These receptors can be labeled and assayed by means of the binding of radioactive transmitters, peptides or antagonist drugs. Though these procedures are, in principle, simple, they can pose difficult technical challenges in execution. The number of such receptors in target tissues is finite, with densities in the nanomolar range. Yet, most radiolabeled ligands employed to assay the receptors are chemically reactive agents which can bind via ionic and non-ionic mechanisms to numerous chemical groups on biological membranes. The number of such "non-specific" binding sites usually greatly exceeds those of the physiological receptors. Success in labeling receptors takes advantage of properties common to most receptors, such as their high affinity for the ligand. Employing ligands of high specific activity, low concentrations can be utilized which should favor binding to specific receptor sites. Extensive but rapid washing procedures will tend to remove nonspecifically bound ligands selectively.

Utilizing the principles described above, several workers have identified binding of ligands associated with peptide hormones and neurotransmitter receptors. Direct binding of insulin, glucagon, thyrotropin, ACTH and other hormones has been observed. For neurotransmitters, the transmitter molecule itself often has a relatively low affinity for its receptors and so more potent antagonist drugs have often been employed. The most extensively investigated example is the nicotinic acetylcholine receptor of the electric organ of certain invertebrate fish such as the electric eel or Torpedo. Snake toxins such as *Naja naja* cobra toxin or α-bungarotoxin are potent, almost irreversible nicotinic cholinergic antagonists and bind fairly selectively to receptor sites. In our laboratory neurotransmitter receptors in the brain have been investigated through the reversible binding of drugs and transmitter molecules.

The Opiate Receptor

The biological function of the opiate receptor appears to mediate the neurotransmitter actions of opioid peptides in the brain, the enkephalins and endorphins. The opiate receptor has been labeled using tritiated forms of opiate agonists or antagonists. Affinities of numerous opiates for binding sites parallel, in general, their relative analgesic properties. A direct comparison of biological and binding potency, requisite evidence to show that binding involves the biologically meaningful receptor, must be obtained in a single tissue. In the case of opiates, pharmacological potency can be assessed in smooth muscle systems, since the ability of opiates to block electrically induced contractions in these tissues parallels their analgesic potencies. These tissues, such as the guinea pig ileum and mouse vas deferens thus appear to possess pharmacological opiate receptors with properties similar to those of the brain. The potencies of opiate agonists and antagonists in competing with ^{3}H-opiates for binding sites in the guinea pig ileum correlate extremely closely with pharmacological activity in the same tissue specimens.[1]

Studies of the influence of ions on opiate receptor binding have provided insight into possible mechanism for synaptic actions of opiates and opioid peptides. Low concentrations of sodium, as little as 1 *mM*, enhance receptor binding of antagonists while decreasing the binding of agonists.[2] The influence of sodium is selective, since while lithium, whose properties resemble those of sodium, can reproduce these effects somewhat, other monovalent cations, such as rubidium, cesium and potassium, fail to discriminate agonists and antagonists. These findings suggest that the opiate receptor may exist in two interconvertible states with respective preferences for binding of agonists and antagonists. Sodium apparently binds preferentially to the "antagonist" state. One might speculate that synaptic effects of opiate agonists occur by altering the relative proportions of agonist and antagonist receptors, which changes the tightness of attachment of sodium to the receptor in turn altering the membrane's permeability to sodium. This model, based on biochemical data, finds support in recent neurophysiological studies. Zieglgänsberger et al.[3] have shown that opiates and opioid peptides inhibit neuronal firing without hyperpolarizing cells. Instead they block excitatory actions of glutamate and acetylcholine through selective effects at synaptic sodium channels.

The influence of sodium on opiate-receptor binding also has practical consequences. One can predict whether a drug is an agonist or antagonist by the effect sodium has on its receptor interactions. One measures the binding of a drug such as the antagonist ^{3}H-naloxone in the presence or absence of sodium. Nonradioactive agonists become substantially less potent in competing for receptor binding when sodium is added to incubation media while pure antagonists show no loss in potency. Drugs with both agonist and antagonist properties are affected in an intermediate fashion. Such mixed agonist-antagonists have important therapeutic properties. Certain of these agents elicit analgesia, a function of their agonist effects. But, perhaps because of their antagonist propensities, they are markedly less addicting than conventional opiate agonists. These mixed agonist-antagonists are the most promising of all drugs as relatively addiction-free pain relievers. While it is difficult to identify such agents in screening procedures in intact animals, opiate receptor interactions afford a simple, sensitive, inexpensive yet highly reliable screen.

Autoradiographic studies have permitted a localization of opiate receptors at specific microscopic sites throughout the central nervous system[4,5] (Fig. 1).

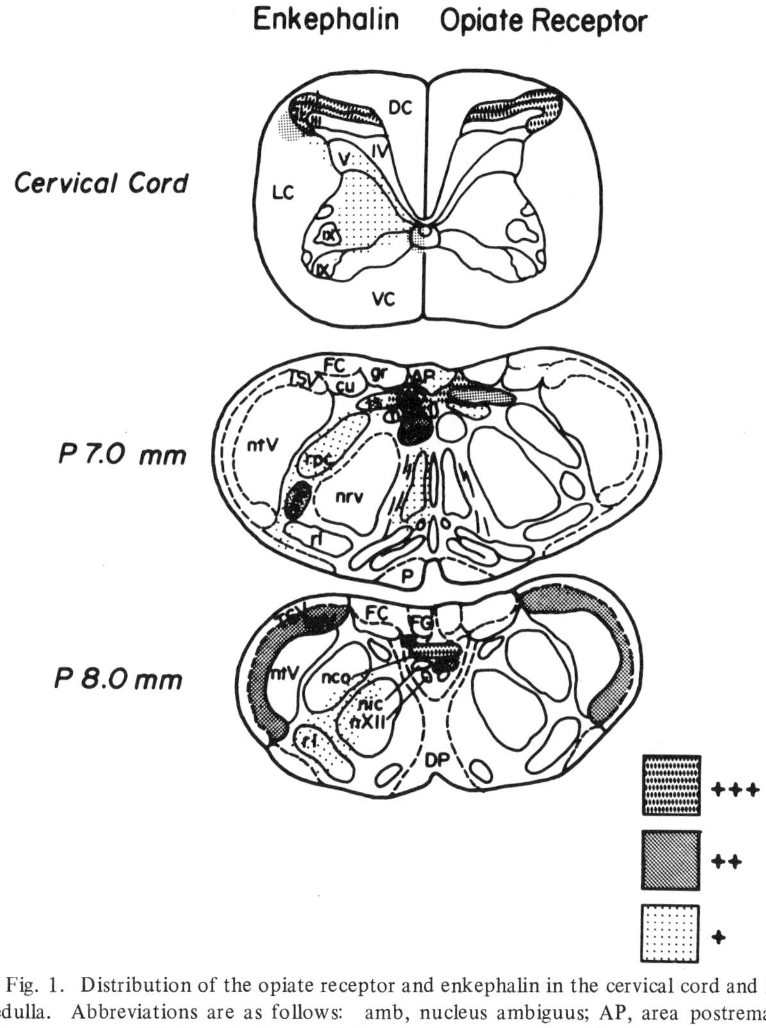

Fig. 1. Distribution of the opiate receptor and enkephalin in the cervical cord and lower medulla. Abbreviations are as follows: amb, nucleus ambiguus; AP, area postrema; cu, nucleus cuneatus; DC, dorsal column; DP, decussatio pyramidis; FC, fasciculus cuneatus; FG, fasciculus gracilis; gr, nucleus gracilis; io, nucleus olivaris inferior; LC, lateral column; nco, nucleus commissuralis; nic, nucleus intercalatus; nrv, nucleus reticularis medullae oblongatae pars ventralis; nts, nucleus tractus solitarius, ntV, nucleus tractus spinalis nervi trigemini; nX, nucleus originis dorsalis vagi; nXII, nucleus originis nervi hypoglossi; P, tractus corticospinalis; rl, nucleus reticularis lateralis; rpc, nucleus reticularis parvocellularis; sgV, substantia gelatinosa trigemini; ts, tractus solitarius; TSV, tractus spinalis nervi trigemini; VC, ventral column. Adapted from Simantov et al.[12]

Receptors are highly concentrated in areas associated with the integration of pain perception, emotional regulation and visceral reflexes associated with opiate action. For instance, receptors are highly concentrated in the substantia gelatinosa of the spinal cord and the trigeminal nucleus, the first way stations in sensory perception and areas where the analgesic effects of opiates are in part mediated. The periaqueductal grey of the brainstem, which plays a major role in drug and electrically induced analgesia, is enriched in opiate receptors. The locus coeruleus, which contains the highest concentration of norepinephrine cells in the brain, is thought to influence affective behavior and has a high density of opiate receptors as does the amygdala, which is part of the emotion-regulating limbic system. Vagal nuclei in the brainstem, such as nucleus tractus solitarius, and nucleus ambiguus, participate in sensory visceral reflexes such as cough and respiratory reflexes, and have high concentrations of opiate receptors. Perhaps the well known cough suppressive effects of opiates as well as the sometimes lethal depression of respiration elicited by these drugs are mediated through these nuclei.

The dramatic features of the opiate receptor suggested that there might exist an endogenous morphine-like substance in the brain. By monitoring the effects of brain extracts on smooth muscle contractions, Hughes et al.[6] isolated and characterized from pig brain the two pentapeptide enkephalins, methionine-enkephalin, (Tyr-Gly-Gly-Phe-Met)-enkephalin, and leucine-enkephalin, (Tyr-Gly-Gly-Phe-Leu)-enkephalin. Independently, by measuring the ability of brain extracts to compete for opiate receptor binding, Simantov and Snyder[7] subsequently identified the same two pentapeptides in calf brain. The amino acid sequence of Met-enkephalin is contained within the 91 amino acid pituitary peptide β-lipotropin, a fragment of which, amino acids 61-91 (β-endorphin) has potent opiate activity.[8-10]

Development of antisera to the enkephalins has permitted their measurement by radioimmunoassay as well as their localization in the central nervous system by immunohistochemistry[11,12] (Fig. 1). Sites most enriched in enkephalin correspond fairly closely to those with high concentrations of opiate receptors. Enkephalin is contained within neurons. Enkephalin cell bodies are enriched in the same areas which have the highest density of enkephalin-containing nerve terminals. This suggests that most enkephalin is contained in short neurons restricted to defined areas of the central nervous system.

Enkephalins display the same pharmacological properties as opiates, causing analgesia, influencing smooth muscle contractions, and binding to the opiate receptor. Many enkephalin analogs have been synthesized with a view to protecting enkephalin from very rapid proteolytic degradation. Certain of these are thousands of times more potent than parent enkephalins (R. Miller, personal communication). Though enkephalins and endorphins do appear to have addictive potential similar to that of morphine in rats,[13-15] it is possible that certain enkephalin analogs will demonstrate useful therapeutic properties in man.

Angiotensin and Neurotensin

There appear to be a large number of other peptides with possible central nervous functions. In the interest of brevity, we have focused on the enkephalins, though Substance P is particularly well characterized as a sensory transmitter candidate. There is also substantial evidence for a role of angiotensin II and neurotensin in the brain. Direct administration of small doses of angiotensin II into the brain can influence cardiovascular reflexes and drinking behavior. Renin and angiotensin-converting enzyme exist in the brain. Recently we detected angiotensin II receptor binding in brain tissue similar to levels of receptor binding in the adrenal cortex, the peripheral tissue most enriched in angiotensin receptors.[16] Indeed, in a screen of various tissues, only brain and adrenal cortex demonstrated detectable levels of angiotensin II receptor binding. Substrate specificity of receptors in the brain and adrenal cortex is quite similar (Fig. 2). Taken together these data suggest a natural role for angiotensin in the brain.

Angiotensin receptor binding sites in the brain have about one order of magnitude greater affinity for angiotensin and its analogs than those in the adrenal cortex.[16] The dissociation constant for angiotensin II at calf and rat brain receptors is about 0.2 *nM*. This high affinity, as well as the striking selectivity of the angiotensin receptor of brain tissue, has facilitated development of a radioreceptor assay for endogenous angiotensin-like material (J. P. Bennett, Jr. and S. H. Snyder, manuscript in preparation).

Neurotensin is a tridecapeptide whose amino-acid sequence[17] is <Glu-Leu-Tyr-Glu-Asn-Lys-Pro-Arg-Arg-Pro-Tyr-Ile-Leu-OH). Neurotensin was isolated from hypothalamic extracts as an apparent byproduct of the isolation of substance P. It was detected as a substance which in small doses could elicit hypotension, increased vascular permeability, pain sensation, increased hematocrit, cyanosis, morphine-inhibitable stimulation of ACTH secretion, increased LH secretion, increased FSH secretion, hyperglycemia and a variety of smooth muscle effects including contraction of the rat uterus and guinea pig ileum.

Like enkephalin, somatostatin and Substance P, neurotensin is localized only to the central nervous system and gut.[18] Radioimmunoassay reveals marked regional differences in neurotensin levels throughout the brain.[18-21] In calf brain the highest concentrations occur in the hypothalamus and basal ganglia with much lower values in cerebellum and white matter. In the cerebral cortex, there are pronounced variations with highest levels in the parahippocampal gyrus. Subcellular fractionation studies indicate a localization of neurotensin to synaptosomal fractions, which are enriched in pinched-off nerve terminals.[19]

Using ^{125}I-neurotensin, it has been possible to detect specific receptor binding.[22] The dissociation constant of neurotensin is about 3 *nM*. There is a single population of binding sites with a Hill coefficient of 1.2, indicating the absence of cooperativity. The density of binding sites in rat cerebral cortex is about 3 pmol/g wet weight, similar to that of several neurotransmitter receptors.

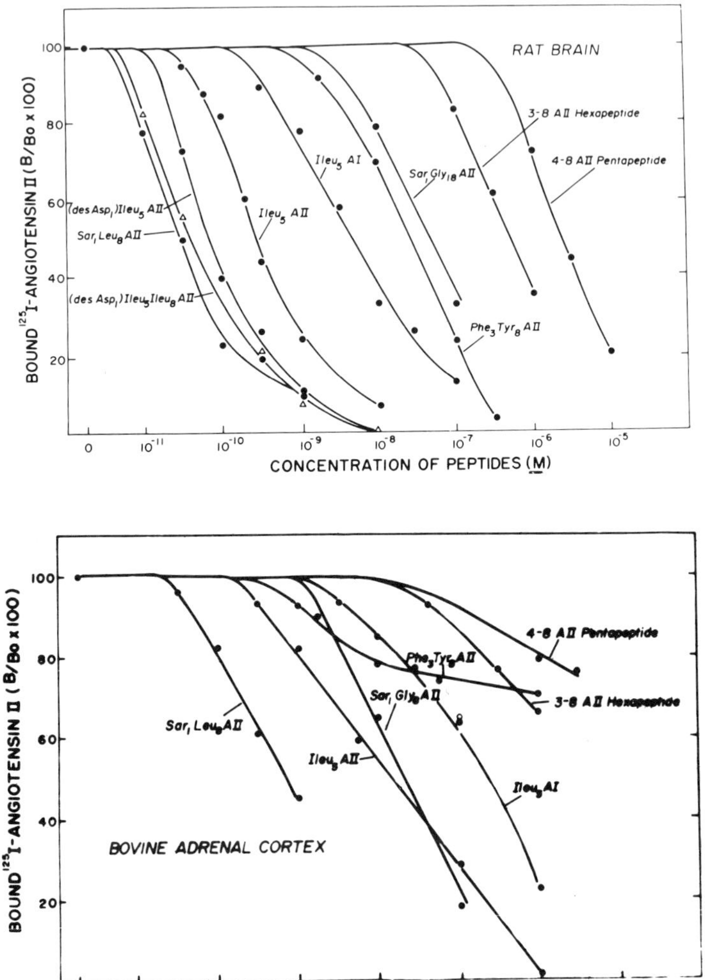

Fig. 2. Displacement of ^{125}I-angiotensin II bound to rat brain (upper) and bovine adrenal cortex (lower) membranes by angiotensin peptides. ^{125}I-Angiotensin II (0.05 *nM*) was incubated with membranes and bound radioactivity in the presence of increasing concentrations of various angiotensin peptides was assayed by filtration. B_O = Binding in the absence of unlabeled peptides, B = binding in presence of unlabeled peptides. Adapted from Bennett and Snyder.[16]

Marked regional variations exist in neurotensin binding which parallel, in part, variations in endogenous neurotensin. Highest binding is found in the dorsomedial thalamus, parahippocampal cerebral cortex (which possesses the highest endogenous levels of the cerebral gyri) and hypothalamus. Lowest levels occur in the cerebellum and brainstem. The strongest evidence that the binding sites involve physiological neurotension receptors emerges from an examination of the relative abilities of five partial sequence fragments of neurotensin to compete for binding. Their relative potencies correspond fairly well to their relative activities in a number of peripheral systems.

In summary, we have reviewed data regarding receptor binding and other evidence compatible with central nervous system roles for enkephalin, angiotensin II and neurotensin as possible neurotransmitters. Similar evidence exists for substance P, somatostatin, TRF, vasoactive intestinal peptide and perhaps even gastrin. Since these peptides were discovered fortuitously in most cases, it is conceivable that numerous other peptide transmitter candidates exist in the brain. This multiplicity of transmitters greatly enlarges the scope of problems and promises to elucidate neurotransmission in the central nervous system.

References

1. Creese, I. & Snyder, S. H. (1975) *J. Pharm. Exptl. Ther.* **194**, 205–219.
2. Pert, C. B. & Snyder, S. H. (1974) *Mol. Pharmacol.* **10**, 86–879.
3. Zieglgänsberger, W., Fry, J. P. & Herz, A. (1976) in *Opiates and Endogenous Opioid Peptides*, Kosterlitz, H. W., Ed., North Holland, Amsterdam, pp. 231–238.
4. Pert, C. B., Kuhar, M. J. & Snyder, S. H. (1976) *Proc. Nat. Acad. Sci. USA* 73, 3729–3733.
5. Atweh, S. & Kuhar, M. J. (1977) *Brain Res.* **124**, 53–67.
6. Hughes, J., Smith, T. W., Kosterlitz, H. W., Fothergill, L., Morgan, B. A. & Morris, H. R. (1975) *Nature* **258**, 577–579.
7. Simantov, R. & Snyder, S. H. (1976) *Proc. Nat. Acad. Sci. USA* 73, 2515–2519.
8. Bradbury, A. M., Smyth, D. G., Snell, C. R., Birdsall, N. J. M. & Hulme, E. C. (1976) *Nature* **260**, 793–795.
9. Cox, B. M., Goldstein, A. & Li, C. H. (1976) *Proc. Nat. Acad. Sci. USA* 73, 1821–1823.
10. Ling, N. & Guillemin, R. (1976) *Proc. Nat. Acad. Sci. USA* 73, 3308–3310.
11. Elde, R., Hökfelt, T., Johannson, O. & Terenius, L. (1976) *Neurosciences* **1**, 349–355.
12. Simantov, R., Kuhar, M. J., Uhl, G. & Snyder, S. H. (1977) *Proc. Nat. Acad. Sci. USA* 74, 2167–2171.
13. VanRee, J. M., DeWied, D., Bradbury, A. F., Hulme, E. C., Smyth, D. G. & Snell, C. R. (1976) *Nature* **264**, 792–794.
14. Wei, E. & Loh, H. (1976) *Science* **193**, 1262–1263.
15. Tseng, L.-F., Loh, H. H. & Li, C. H. (1976) *Proc. Nat. Acad. Sci. USA* 73, 4187–4189.
16. Bennett, J. P., Jr. & Snyder, S. H. (1976) *J. Biol. Chem.* **251**, 7423–7430.
17. Carraway, R. & Leeman, S. E. (1975) *J. Biol. Chem.* **250**, 1907–1918.
18. Carraway, R. & Leeman, S. E. (1976) *J. Biol. Chem.* **251**, 7045–7052.
19. Uhl, G. R. & Snyder, S. H. (1977) *Eur. J. Pharmacol.* **41**, 89–91.
20. Uhl, G. R. & Snyder, S. H. (1976) *Life Sci.* **19**, 1827–1832.
21. Kobayashi, R., Brown, M. & Vale, W. (1977) *Brain Res.* **126**, 584–588.
22. Uhl, G. R., Bennett, J. P., Jr. & Snyder, S. H. (1977) *Brain Res., 130,* 299-313.

PROPOSED OPIATE PHARMACOPHORE AND THE CONFORMATION OF ENKEPHALIN

GARLAND R. MARSHALL and FREDRIC A. GORIN,
Department of Physiology and Biophysics, Washington University School of Medicine, St. Louis, Missouri 63110

Two pentapeptides, the enkephalins, are endogenous to the mammalian central nervous system and possess opioid activity. There have been several proposals comparing the structural similarity of the enkephalins with the opiate compounds of the morphine and oripavine families. Unfortunately, there is much controversy regarding the identification and topography of the chemical substituents responsible for the biological activity of the opiates. It has been suggested that different sites exist on the narcotic receptors for binding different chemical groups corresponding to the different classes of narcotics.[1] However, more than one class of narcotic receptor mediating the opiates' analgesic effect may exist.[2] This latter possibility appears to be discounted in such *in vitro* preparations as rat neostriatum[3] and the neuroblastoma-glioma hybrid cell line NG108-15.[4] For this reason, the enkephalins were structurally compared with those opiates shown to bind in a competitive fashion and inhibit adenylate cyclase in the NG108-15 cell line.

Determination of the Opiate Pharmacophore

Using a molecular graphics system,[5] the crystallographic structures of morphine hydrochloride, and 7-α-(1-(R)-hydroxyl-1-methylbutyl-)-6,14-*endo*ethenotetrahydrothebaine (THT) were compared with that of 1-β-2′-hydroxy-2,9-dimethyl-5-phenyl-6,7-benzomorphan (GPA1657). This molecule was simulated using the skeleton of morphine with the C-ring replaced by a 5-substituted phenyl ring positioned to correspond with published nmr data.[6] Comparison of the three opiates demonstrated spatial correspondence of their phenolic A-rings, nitrogen atoms of their D-rings, and C5 and C6 atoms of their hydrocarbon C-rings.

Earlier proposals suggested that opiate activity required the correct spatial configuration of a phenolic ring and a tertiary amine as observed in morphine. These substituents appear to be necessary but insufficient requirements since tyramine is inactive as an opiate. In these semi-rigid opiates, we propose that atoms of the C-ring provide the additional third site necessary to define an essential pharmacophore.

Fitting Enkephalin to this Pharmacophore

One assumes that the phenolic side chain and nitrogen terminus of Tyr^1 of enkephalin correspond to the 3-hydroxy-phenethylamine moiety of the opiate pharmacophore. Chemical modification of the phenolic group of Tyr^1 reduces the potency of enkephalin, and [Phe^1]-Met^5-enkephalin is only weakly active.[7] Additionally, desamino-Tyr^1, Met^5-enkephalin is biologically inactive.[8] The tetrapeptide is the minimal structural unit capable of binding to brain tissue and eliciting a biological response.[7,9] Correspondence between the essential C-ring atoms and the phenyl ring of Phe^4 was assumed.

A minimum of ten rotatable bonds is required to fit the tetrapeptide fragment of enkephalin to the proposed opiate pharmacophore. To reduce the conformational freedom at residue two, the potent analog [D-Ala^2]-Met^5-enkephalinamide was chosen for study.[10] The side chain torsion angles of Tyr^1 were fixed so that the phenolic ring and the amino terminus of the first residue spatially corresponded to the aromatic A-ring and nitrogen of the D-ring of morphine, respectively. Two target points were positioned relative to these groups to correspond spatially to C5 and C6 of morphine's C-ring. The remaining eight torsional bonds of the backbone of the peptide and side chain of Phe^4 were systematically explored with 31° increments.[11] A single sterically allowed conformation was found with the *meta* carbon of the aromatic ring of Phe^4 corresponding to C6. A summary of the torsional angles for this conformation is shown in Table I.

Consistency with Structure-Activity Relationship Data

The combinatorial nature of the conformational search required a coarse 31° scan. It is of interest, however, to see if the torsion angles of the single conformation of enkephalin found to fit the proposed opioid pharmacophore are consistent with structure-activity relationship data of the enkephalins. Based upon the Ramachandran plots in Figure 1, predictions for the *in vitro* activity of *N*-methyl, alpha-methyl, and D,L-amino-acid analogs of enkephalin can be made (Table II). Side-chain orientation of the first and fourth residues is important

Table I. Torsional Angle Data of a Conformation of Enkephalin Found to Fit the Proposed Opiate Pharmacophore

χ_1 (Tyr^1) = -106	ψ_1 (Tyr^1) = 129	ϕ_3 (Gly^3) = -118
χ_2 (Tyr^1) = -163	ϕ_2 (D-Ala^2) = 160	ψ_3 (Gly^3) = 98
χ_1 (Phe^4) = -87	ψ_2 (D-Ala^2) = -87	ϕ_4 (Phe^4) = -87
χ_2 (Phe^4) = -56		

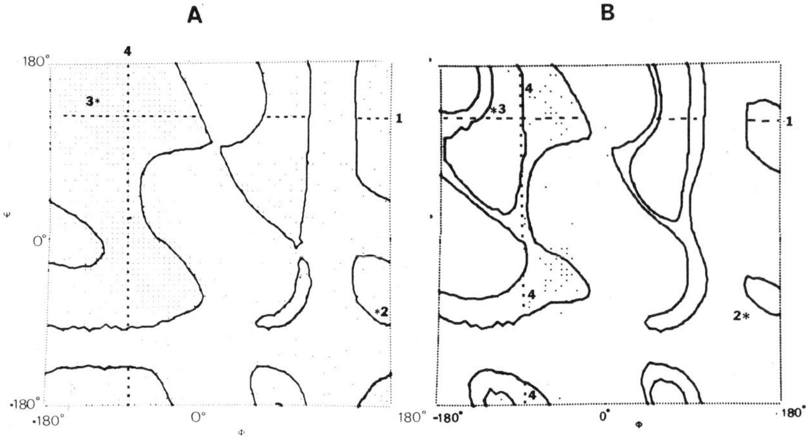

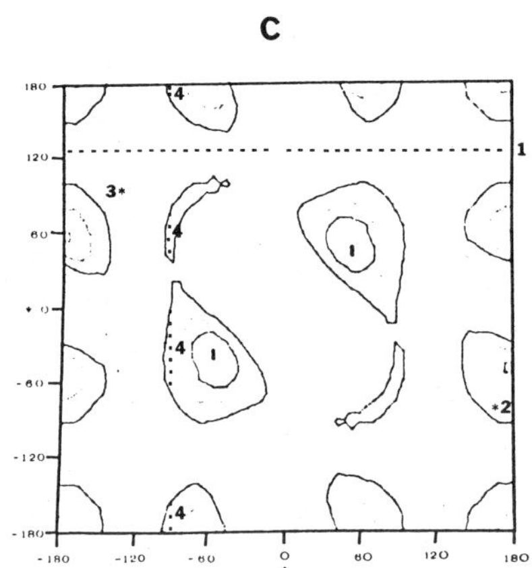

Fig. 1. Kitaigorodsky potential energy plots for acetyl-L-alanine methylamide,A; acetyl-*N*-methyl-L-alanine methylamide,B; and acetyl-aminoisobutyric acid methylamide,C. 1 = ψ_1 (129°), Tyr[1]; 2 = ϕ_2,ψ_2 (160°,–87°), [D-Ala[2]]; 3 = ϕ_3,ψ_3 (-118,98), Gly[3]; 4 = ϕ_4 (–87) Phe[4].

Table II. Predicted Biological Activity of Enkephalin Analogs Based Upon Their Backbone Conformations

Analog	Prediction	Analog	Prediction
[D-Tyr^1]	inactive	[Aib^3]	inactive
[D-Ala^2]	assumed[10]	[Pro^3]	inactive
[L-Ala^2]	active[9]	[N-Me-L-Ala^3]	active
[Aib^2]	active	[N-Me-D-Ala^3]	inactive
[Pro^2]	inactive	[Sar^3]	active
[N-Me-L-Ala^2]	inactive	[D-Phe^4]	active
[N-Me-D-Ala^2]	active	[Phe (αMe)^4]	active[12]
[Sar^2]	active	[β-phenyl-Pro^4]	active
[L-Ala^3]	active[9]	[N-Me-L-Phe^4]	active
[D-Ala^3]	inactive	[N-Me-D-Phe^4]	active

and cannot be predicted in this fashion. Analogs of enkephalin with D-amino acid, *N*-methyl, and alpha-methyl substitutions aid in defining the backbone conformation of enkephalin at the receptor.

Conclusion

X-ray crystallographic structures of conformationally constrained opiate compounds were examined to identify and spatially localize the minimal number of chemical substituents essential for biological activity in an *in vitro* cell hybrid line where binding and biological efficacy could be measured in the same system. SAR data of the enkephalins were used to identify those chemical substituents of the pentapeptides which correspond to the atoms defining an essential opiate pharmacophore.

References

1. Portoghese, P. S. (1965). *J. Med. Chem.* **8**, 609–616.
2. Lord, A. H., Waterfield, A. A., Hughes, J. & Kosterlitz, H. W. (1977). *Nature* **267**, 495-499.
3. Minneman, K. P. & Iverson, L. L. (1976). *Nature* **262**, 313–314.
4. Klee, W. A. & Nirenberg, M. (1976). *Nature* **263**, 609–613.
5. PDP 11/40-based molecular display system similar to that of D. D. Hodges, D. H. Nordby, G. R. Marshall, *Abst. 169th Natl. ACS Meeting Comp-7* (1975).
6. Yohoyama, N., Block, F. B. & Clarke, F. H. (1970). *J. Med. Chem.* **13**, 488–492.
7. Morgan, B. A., Smith, C. F. C., Waterfield, A. A., Hughes, J. & Kosterlitz, H. W. (1976). *J. Pharm. Pharmacol.* **28**, 660–661.
8. Büscher, H. H., Hill, R. C., Romer, D., Cardinaux, F., Closse, A., Hauser, D. & Planos, J. (1976). *Nature* **261**, 523–425.
9. Terenius, L., Wahlstrom, A., Lindeberg, G., Karlsson, S. & Ragnarsson, U. (1976). *Biochem. Biophys. Res. Commun.* **71**, 175–177.
10. Pert, C. B., Pert, A., Chang, J. & Fong, B. T. W. (1976). *Science* **194**, 330–332.
11. Bosshard, H. E., Barry, C. D., Fritsch, J. M., Ellis, R. A. & Marshall, G. R. (1972). *Proc. 1972 Summer Simulation Conf.* **1**, 58.
12. Gorin, F., Marshall, G. R. unpublished data.

CONFORMATIONAL STUDY OF ENKEPHALIN AND ANALOGS IN RELATION WITH THE STRUCTURAL FEATURES OF MORPHINE-LIKE ANALGESICS

C. HUMBLET, *Laboratoire de Chimie Moléculaire Structurale, Facultés Universitaires de Namur, 61, rue de Bruxelles, B-5000 – NAMUR, Belgium* and J. L. DE COEN, *Laboratoire de Chimie Biologique, Université de Bruxelles, 67, rue des Chevaux, B-1640 – RHODE St GENESE, Belgium*

Introduction

The isolation from mammalian brain[1,2] of two pentapeptides, named enkephalin (Tyr-Gly-Gly-Phe-X, X=Met or Leu) with morphine-like activity has stimulated a number of studies. Numerous analogs have been synthesized and their biological activity and affinity assessed by the opiate-receptor-binding assay.[3] These studies suggest that the enkephalins bind to the same receptor site as morphine. In order to establish on a quantitative basis, the structural similarities between the peptides and the opiates, we undertook a theoretical conformational analysis of Met^5- and Leu^5-enkephalin and three analogs: [D-Ala^2]-, [D-Ala^3]-, and [D-Phe^4]-Met^5-enkephalin.

The first analog has been reported[3] to be ten times more active than the native peptide, however its binding affinity for the receptor is only 1.2 times that of Met^5-enkephalin. The two other D-analogs are totally inactive although [D-Ala^3]-Met^5-enkephalin retains some affinity for the receptor (~0.1). According to our calculations[4] obtained for Met^5-enkephalin the conformational state of this peptide in solution can be described as an equilibrium between extended and folded structures. The folded conformers contain β-turns between Gly^2-Gly^3 and/or Gly^3-Phe^4.

Using another theoretical approach, based on extensive use of a minimization technique, Isogai et al.[5] have found, in agreement with our results, that "many compact conformations, including those containing various standard bends, were of comparable energy". However, these authors have found, in contradiction to us, that "one conformation has a potential energy about 5 kcal/mole below that of the large group of compact conformations". It is probably the different weight attributed to the strength of the hydrogen bond in the two works, which explains the discrepancy between these theoretical results. The energy parameters in our study were chosen so as to simulate an aqueous environment with a low contribution for hydrogen bonding (~-2.0 kcal/mole).

Results and Discussion

Following the same strategy as that previously applied to Met^5-enkephalin, we calculated for each analog the relative probabilities of several hundreds of distinct main-chain conformers. These calculations take into account the statistical contribution arising from the great number of arrangements that the side chains can assume. This is the most important feature of our strategy: the backbone is fixed into various, quite distinct, probable conformations; while the side chains are kept mobile and can express their different potentialities of interaction.

Each analog appears to have its own distribution of conformers at the equilibrium. That these distributions differ from that of Met^5-enkephalin is clearly shown in Fig. 1 where the relative probabilities for each analog are reported (when greater than 1%). Four, among the five most probable conformations of Met^5-enkephalin, appear also for Leu^5-enkephalin, though populated to a different extent. As far as the three D-analogs are concerned, it appears that the number of populated conformers which are similar to those of the native peptide is two for D-Ala^2, one for D-Ala^3 and zero for D-Phe^4. These results suggest a decreasing similarity in the overall topology of the three analogs which could be correlated with the decreasing affinity of these molecules for the receptor. On the other hand, the fact that conformer number 4 for Met^5-enkephalin appears to be twice as populated in the case of D-Ala^2 suggests that this structure is probably more efficient for binding to the receptor. This conformer appears neither for D-Ala^3 nor for D-Phe^4 which are known to bind poorly to the receptor.

The calculation of the mean values of several intramolecular distances, from the distribution of conformers shown in Fig. 1 indicates that D-Ala^2 is slightly more compact than the native peptides ($\bar{d}_{(C_1^\alpha - C_5^\alpha)}$ = 8.0 Å instead of 8.6 Å for both Met^5- and Leu^5-enkephalin, while D-Ala^3 and D-Phe^4 are a little more extended ($\bar{d}_{(C_1^\alpha - C_5^\alpha)}$ = 9.0 Å and 9.5 Å). The same observation is apparent for the mean distances between the centers of the aromatic rings of the Tyr and Phe residues which are: 8.7 Å for Met^5, 8.6 Å for Leu^5, 8.4 Å for D-Ala^2, 9.8 Å for D-Ala^3 and 9.5 Å for D-Phe^4. On the other hand, the mean distance $\bar{d}_{(N-O)}$ between the amino terminal and the oxygen of the hydroxyl group of Tyr is quite similar for the five derivatives ($\bar{d}_{(N-O)} \cong 7.6$ Å). This results from the preference in every case of a *trans* conformation for the $\chi_{(C\alpha - C\beta)}$ angle of the side chain. In morphine, the corresponding distance is 7.1 Å.

In order to compare the results of our analysis with spectroscopic data measured in solution, we calculated the mean values of various pmr coupling constants. Each analog is characterized by a given set of coupling constants. The set corresponding to Met^5-enkephalin is reported in Table II in comparison with experimental values[6] measured in DMSO. It can be seen that a reasonable agreement is obtained excepted for $\bar{J}_{(NH-C^\alpha H)}$ of Met and one of the two $\bar{J}_{(C^\alpha H-C^\beta H)}$ of the Tyr side chain. These discrepancies can not be fully evaluated at the present stage of our calculations.

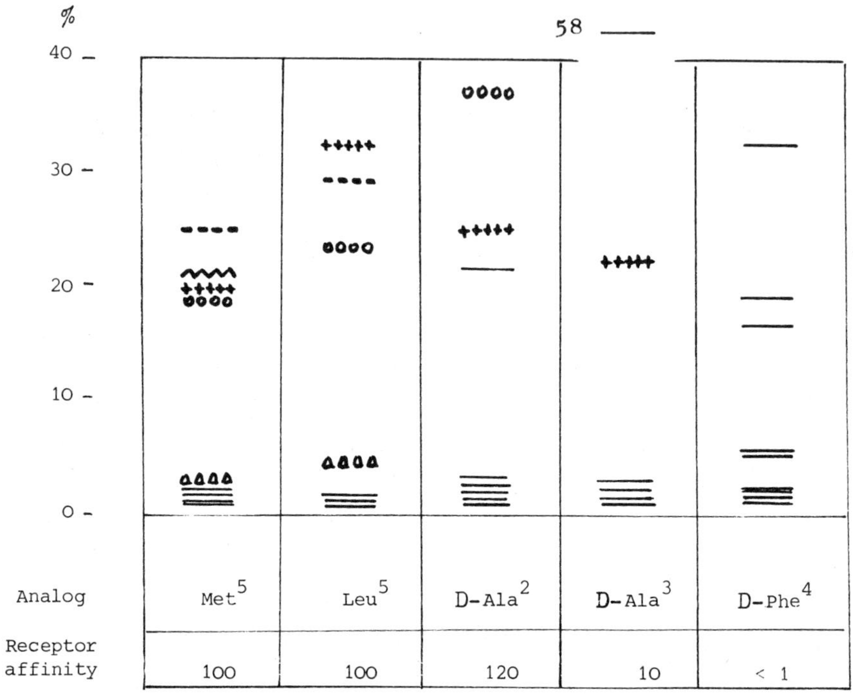

Fig. 1. Low energy conformers of enkephalin derivatives in equilibrium at 25°C. Each line represents a different conformation of the backbone and the probabilities reported along the vertical axis were calculated taking into account the various populated side-chain arrangements. The five most probable conformations of Met^5-enkephalin are identified by decorated lines. The ϕ, ψ angles corresponding to these structures are listed in Table I.

Table I. Torsional ϕ, ψ Angles of the Five Most Probable Conformers of Met^5-enkephalin (see Figure 1)

Conformer	ϕ_1	ψ_1	ϕ_2	ψ_2	ϕ_3	ψ_3	ϕ_4	ψ_4	ϕ_5	ψ_5
- - - -	60	140	-80	80	-80	80	-80	80	-140	140
ΛΛΛΛ	60	140	-80	80	-180	180	-80	80	-140	140
+ + + +	60	140	80	-80	-60	-60	-160	160	-140	140
o o o o	60	140	80	-80	-80	80	-80	80	-140	140
△△△△	60	140	-180	180	-80	80	-80	80	-140	140

Conclusions

Theoretical analysis indicates several interesting features of the conformational properties of Met^5-enkephalin and four of its analogs.

Native Met^5 and Leu^5-enkephalin exhibit a similar structure. The most probable conformers of the very active D-Ala^2 derivative correspond to models with a turn at the level of residues 2 and 3, which are also populated for the native peptides. Only one of these two conformers is also present for D-Ala^3, which is known to be inactive. The populated conformers of the inactive D-Phe^4 analog are quite different from those of the natural peptides.

Table II. Calculated Mean Values of Vicinal Coupling Constants (Hz) for Met[5]-enkephalin Compared with Experimental Data[6] Measured in DMSO (in parenthesis)

Residue	$\bar{J}_{(NH-C^{\alpha}H)}$		$\bar{J}_{(C^{\alpha}H-C^{\beta}H)}$	
1 Tyr			3.2 (2.9)	12.6 (5.8)
2 Gly	6.3	6.0		
3 Gly	5.8 (5.2)$^+$	5.0		
4 Phe	7.3 (8.6)		3.4 (3.8)	9.6 (9.9)
5 Met	10.2 (7.4)		5.4 (7.4)$^+$	10.4

\+ Mean values

Our calculations strongly suggest a structural analogy between the Tyr amino terminal and a corresponding part of morphine.[7] However, our results indicate that this similarity alone can not explain the biological activity of the enkephalins because it is also apparent in the case of the inactive derivatives. The spatial arrangement of other parts of the peptide could be equally important; for example, the positions of the side chains of Phe[4], Met[5] or Leu[5] relative to the Tyr[1] residue could play a crucial role.

Conformational analysis of other enkephalin derivatives, as well as a detailed study of the structural similarities between them, morphine and other opiates, for which the crystal structure is known, has been undertaken in our laboratory. Such studies should help to assess the structure-function relationship which has only tentatively been presented here.

We thank Drs. M. Koch, E. Ralston and S. Wodak for stimulating discussions. The calculations were performed at the Centre de Calcul de l'Université Libre de Bruxelles. This work was supported by a grant from the Ministère de la Politique Scientifique. C. Humblet is Boursière de l'Institut pour l'Encouragement de la Recherche Scientifique dans l'Industrie et l'Agriculture, J. L. De Coen is Chercheur Qualifié du Fonds National de la Recherche Scientifique.

References

1. Hughes, J., Smith, T. W., Kosterlitz, J. W., Fothergill, L. A., Morgan, B. A. & Morris, H. R. (1975) *Nature* **258**, 577–579.

2. Simantov, R. & Snyder, S. H. (1976) *Proc. Nat. Acad. Sci. USA* **73**, 2515–2519.

3. Coy, D. H., Kastin, A. J., Schally, A. V., Morin, O., Caron, N. G., Labrie, F., Walker, J. M., Fertel, R., Bernton, G. G. & Sandman, C. A. (1976) *Biochem. Biophys. Res. Commun.* **73**, 632–638.

4. De Coen, J. L., Humblet, C. & Koch, M. H. J. (1976) *FEBS Lett.* **73**, 38–42.

5. Isogai, Y., Nemethy, G. and Sheraga, H. A. (1977) *Proc. Nat. Acad. Sci. USA* **74**, 414–418.

6. Garbay-Jaureguiberry, C., Roques, B. P., Oberlin, R., Anteunis, M. and Lala, A. K. (1976) *Biochem. Biophys. Res. Commun.* **71**, 558–565.

7. Bradbury, A. F., Smyth, D. G. & Snell, C. R. (1976) *Nature* **260**, 165–166.

SOLUTION CONFORMATION AND OPIATE RECEPTOR AFFINITY OF [4-TRYPTOPHAN]-ENKEPHALIN ANALOGS

PETER W. SCHILLER and CHUN F. YAM, *Laboratory of Chemical Biology and Polypeptide Research, Clinical Research Institute of Montreal, 110 Pine Avenue West, Montreal, Quebec H2W 1R7, Canada.*

Numerous studies with rigid, morphine-derived analgesics have permitted a description of the opiate receptor topography in terms of binding sites.[1,2] The enkephalin-opiate receptor system is thus uniquely suited for studying the conformational requirements of a polypeptide-receptor interaction. Studies with the morphine-derived 7α-(1(R)-hydroxy-1-methylphenylalkyl)-6,14-*endo*-ethenotetrahydrooripavines[1] indicate that optimal interaction with the receptor is brought about by three-point attachment involving the tertiary nitrogen, the phenol ring and the phenyl ring, whereby the correct intramolecular distance (~ 10Å) between the two aromatic rings is essential for activity. In the case of Met^5-enkephalin (H-Tyr-Gly-Gly-Phe-Met-OH) evidence has been presented that the α-amino group, the tyrosyl residue and possibly the aromatic side chain in position 4 are the corresponding moieties interacting with the opiate receptor and that an appropriate conformation exists which accommodates these groups in a similar spatial disposition as it is present in the potent 7α-(1(R)-hydroxy-1-methyl-3-phenylpropyl)-6,14-*endo*-ethenotetrahydrooripavine (PEO)[3]. The effect of substitutions in positions 2 and 3 with other non-functional amino acids on opiate receptor affinity[4,5] suggests that the two glycine residues play the role of a spacer maintaining the correct spatial disposition of the two aromatic rings. Replacement of phenylalanine by tryptophan yielded an active analog[3] which permits the measurement of the intramolecular distance between the fluorescent side-chain residues in positions 1 (phenol) and 4 (indole) by measurement of excitation energy transfer. The present paper describes the determination of this interesting conformational parameter in two analogs of [Trp^4]-Met^5-enkephalin.

Opiate Receptor Affinities

Opiate receptor affinities of the analogs synthesized by the solid-phase method were determined by displacement of ^{3}H-naloxone essentially according to the method described by Pert et al.[4] Incubation for 1 hr at 0°C eliminated the possibility of proteolytic peptide degradation and permitted the determination of true, molar affinities. [Trp^4]-Met^5-enkephalin displays approximately the same affinity as its parent compound Met^5-enkephalin (Table I). Replacement of glycine in position 2 of [Trp^4]-Met^5-enkephalin with D-alanine results in a 3-fold reduction of affinity and substitution with L-alanine in the same position lowers the affinity to a mere 2%. Analogous substitutions in Met^5-enkephalin

Table I. Relative Opiate Receptor Affinities of Enkephalin Analogs

Analog[a]	ϕ_D^o	E	R_o [Å]	r [Å]
Tyr—Gly—Gly—Trp—Met	0.027	0.70	10.7	9.3
Tyr—D-Ala—Gly—Trp—Met	0.022	0.62	10.3	9.5
Tyr—L-Ala—Gly—Trp—Met	0.026	0.81	10.6	8.3

[a]Peptide concentration = 3 x 10^{-5}M; solvent = H_2O.

had given rise to parallel reductions in affinity[4] (Table I) and it thus appears that tryptophan is an equivalent substitute for phenylalanine in position 4. The set of [Trp^4]-Met^5-enkephalin analogs is therefore well suited for a study of the relationship between opiate receptor affinity and solution conformation.

Conformational study by energy transfer experiments

Radiationless transfer of excitation energy from a donor fluorophore to an acceptor chromophore is distance-dependent[6] and its quantitative evaluation permits the calculation of intramolecular distances in biopolymers. The determination of the intramolecular distance between the donor Tyr^1 and the acceptor Trp^4 was performed according to a recently published procedure[3,7]. The intramolecular donor-acceptor separation, r, is related to the transfer efficiency, E, by Eq. (1).

$$r = (E^{-1}-1)^{1/6} R_o \qquad (1)$$

where the so-called Förster critical distance, R_o, is related to various measurable spectroscopic parameters by Eq. (2), which permits its computation:

$$R_o = \left[8.79 \times 10^{-28}(\chi^2/n^4)\, \phi_D^o J_{AD}\right]^{1/6} \text{[cm]} \qquad (2)$$

where χ = orientation factor, n = refractive index, ϕ_D^o = donor fluorescence quantum yield in absence of transfer and J_{AD} = overlap integral. The two difficult parameters to determine are χ^2 and ϕ_D^o. The results of recent ^{1}H nmr studies of enkephalin indicate that both the tyrosine residue in position 1 and the phenylalanine side chain in position 4 enjoy considerable orientational freedom[8,9] and it is thus reasonable to assume random donor-acceptor orientation ($\chi^2 = 2/3$). The ϕ_D^o values needed for the computation of the R_o values of the [X^2,Trp^4]-Met^5-enkephalin analogs were ideally determined with the corresponding [X^2,Phe^4]-Met^5-enkephalin analogs (Table II) using tyrosine as a reference ($\phi_{Tyr} = 0.14$). The same values[3] of n and J_{AD} were used for all three analogs. Normalization at 370 nm of the fluorescence emission spectra of the

Table II. Intramolecular Tyr-Trp Distances in [Trp⁴]-Met⁵-Enkephalin Analogs

Analog[a]	ϕ_D^o	E	R_o [Å]	r [Å]
Tyr—Gly—Gly—Trp—Met	0.027	0.71	10.5	9.1
Tyr—D-Ala—Gly—Trp—Met	0.022	0.62	10.2	9.4
Tyr—L-Ala—Gly—Trp—Met	0.026	0.81	10.4	8.2

[a]Peptide concentration = 3×10^{-5} M; solvent = H_2O.

[X^2,Trp^4] Met^5-analogs obtained with excitation at 270 and 293 nm and subsequent subtraction of the spectra permitted the determination of the donor quantum yields in the presence of transfer, ϕ_D. The transfer efficiencies were then calculated according to Eq. (3).

$$E = 1 - \left(\phi_D/\phi_D^o\right) \quad (3)$$

Parameters and resulting Tyr-Trp distances, r, are listed in Table II.

Discussion

The tyrosine fluorescence quantum yielded of Met^5-enkephalin (ϕ = 0.027) in H_2O is within the range of values normally observed with tyrosyl peptides[10]. Since hydrogen bonding of the phenolic hydroxyl with carbonyl groups quenches fluorescence completely[10], this finding is incompatible with models of enkephalin which propose a hydrogen bond between the tyrosyl hydroxyl and a carbonyl group of the peptide backbone[11,12]. The measured intramolecular distance between the aromatic rings in positions 1 and 4 of Met^5-enkephalin suggests some type of a folded conformation and is compatible with a β_1-bend model stabilized by a hydrogen bond between the amino group of methionine and the carbonyl group of glycine in position 2. A Tyr-Trp separation of 9–11 Å is observed in the latter model. Evidence for the existence of a preferred solution conformation and in favor of the 5 → 2 hydrogen-bonded model has recently been obtained from nmr experiments[8,9,12]. Since studies with the rigid PEO-compounds indicate a distance of ~ 10 Å between the receptor binding sites of the phenol ring and the phenyl group, it appears that the distance between the aromatic rings in the solution conformation is similar to that in the receptor-bound conformation. This argument is based on the likely, but not yet proven assumption, that the side chain in position 4 interacts with the second lipophilic binding site. The Tyr-Trp separations obtained with the [D-Ala^2]- and [L-Ala^2]-analogs of [Trp^4]-Met^5-enkephalin (Table II) indicate that no drastic change in solution conformation is associated with the drop in activity observed with these compounds. As in the case of the oripavines[1], it appears that subtle changes in the distance between the two rings produce a large effect on affinity.

This work was supported by an operating grant (MA-5655) of the Medical Research Council of Canada.

References

1. Lewis, J. W., Bentley, K. W. & Cowan, A. (1971) *Ann. Rev. Pharmacol.* **11**, 241–270.

2. Feinberg, A. P., Creese, I. & Snyder, S. H. (1976) *Proc. Nat. Acad. Sci. USA* **73**, 4215-4219.

3. Schiller, P. W., Yam, C. F. & Lis, M. (1977) *Biochemistry* **16**, 1831–1838.

4. Pert, C. B., Bowie, D. L., Fong, B. T. W. & Chang, J. K. (1976) In *Opiates and Endogenous Opioid Peptides,* Kosterlitz, H., Archer, S., Simon, E. J. & Goldstein, A., Eds., North-Holland, Amsterdam, pp. 79–86.

5. Terenius, L., Wahlström, A., Lindeberg, G., Karlsson, S. & Ragnarsson, U. (1976) *Biochem. Biophys. Res. Commun.* **14**, 587–603.

6. Förster, T. (1948) *Ann. Phys.* **2**, 55–75.

7. Schiller, P. W. (1975) in *Biochemical Fluorescence: Concepts*, vol. 1, Chen, R. F. & Edelhoch, H., Eds., Dekker, New York, pp. 285–303.

8. Garbay-Jaureguiberry, C., Roques, B. P., Oberlin, R., Anteunis, M. & Lala, A..K. (1976) *Biochem. Biophys. Res. Commun.* **71**, 558–565.

9. Jones, C. R., Gibbons, W. A. & Garsky, V. (1976) *Nature (London)* **262**, 779–782.

10. Cowgill, R. W. (1976) in *Biochemical Fluorescence: Concepts*, vol. 2, Chen, R. F. & Edelhoch, H., Eds., Dekker, New York, pp. 441–486.

11. Isogai, Y., Némethy, G. & Scheraga, H. A. (1977) *Proc. Nat. Acad. Sci. USA* **74**, 414–418.

12. Khaled, M. A., Long, M. M., Thompson, W. D., Bradley, R. J., Brown, G. B. & Urry, D. W. (1977) *Biochem. Biophys. Res. Commun.* **76**, 224–231.

STRUCTURE-ACTIVITY RELATIONSHIPS OF ENKEPHALIN AND ENDORPHIN ANALOGS

NICHOLAS LING, SCOTT MINICK, LARRY LAZARUS, JEAN RIVIER and ROGER GUILLEMIN, *The Salk Institute for Biological Studies, Laboratories for Neuroendocrinology La Jolla, California 92037*

Endogenous peptides with morphinomimetic activity have been characterized as the pentapeptides[1,2] Met-enkephalin (Tyr-Gly-Gly-Phe-Met-OH) and Leu-enkephalin (Tyr-Gly-Gly-Phe-Leu-OH) in brain and as α-, β-, γ-, δ-endorphins in the pituitary of various mammals.[3-7] What is remarkable is that, with the exception of Leu-enkephalin, all the other five opioid peptides have primary structures corresponding to those of β-lipotropin[8] (β-LPH) subunits (61-65), (61-76), (61-91), (61-77) and (61-87), respectively. With an overall objective of finding more potent agonists, antagonists and maybe eventually non-addictive analgesics, we have carried out a systematic study involving the substitution of various amino acid residues in the enkephalins and the endorphins to assess the relative importance contributed by each residue.

All peptide analogs were synthesized by solid phase methodology[9] and purified as previously described.[10] Relative potencies of the synthetic analogs (on an equimolar basis with Met-enkephalin = 100) were measured by their ability to inhibit the electrically induced contractions of the guinea pig ileum-*myenteric plexus.*[11]

To find out the relative importance of the various side-chain groupings from the enkephalin sequence, each residue of Met-enkephalin was successively replaced by L-alanine (Table I). Only [Ala^2]- and [Ala^5]-enkephalin showed appreciable activity, which implies that the phenolic group at Tyr^1 and the phenyl group at Phe^4 are required either for binding to the receptor or for maintaining a favorable conformation of the peptide. A series of D-alanine substituted analogs was also prepared (Table I) and the D-Ala^2 substitution yielded compounds with much higher activity than the respective parent peptides. This could be due to the stabilization of a favorable conformation by D-alanine or to the inhibition of enzymatic degradation.

To introduce flexibility to the peptide backbone, we have successively replaced each of the optically active residues of Met-enkephalin with glycine (Table I). Only [Gly^5]-enkephalin showed some activity which indicates that the stereochemistry of the 5th residue is not critical for biological activity.

Along the strategy of altering the backbone conformation and inhibiting enzyme degradation but keeping the side-chain groupings intact, a series of analogs of the enkephalins was prepared in which each of the optically active

Table I. Relative Potencies of L-Alanine, D-Alanine, Glycine and D-Isomer Substituted Analogs of Enkephalin

Analogs	Rel. Potency	(95% Confidence Limits)
Ala-Gly-Gly-Phe-Met-OH	0*	
Tyr-Ala-Gly-Phe-Met-OH	6.4	(3.7 - 10.0)
Tyr-Gly-Ala-Phe-Met-OH	< 1	
Tyr-Gly-Gly-Ala-Met-OH	0	
Tyr-Gly-Gly-Phe-Ala-OH	2.9	(2.4 - 3.5)
D-Ala-Gly-Gly-Phe-Met-OH	0	
Tyr-D-Ala-Gly-Phe-Met-OH**	316	(191 - 673)
Tyr-D-Ala-Gly-Phe-Leu-OH**	230	(197 - 272)
Tyr-Gly-D-Ala-Phe-Met-OH	4	(2 - 8)
Tyr-Gly-Gly-D-Ala-Met-OH	< 1	
Tyr-Gly-Gly-Phe-D-Ala-OH	6.5	(5.5 - 8.1)
Gly-Gly-Gly-Phe-Met-OH	0	
Tyr-Gly-Gly-Gly-Met-OH	0	
Tyr-Gly-Gly-Phe-Gly-OH	< 1	
D-Tyr-Gly-Gly-Thr-Met-OH	< 1	
Tyr-Gly-Gly-D-Phe-Met-OH	< 1	
Tyr-Gly-Gly-Phe-D-Met-OH	10.5	(9.2 - 11.9)
Tyr-Gly-Gly-Phe-D-Leu-OH	15.0	(13.4 - 16.9)

* 0 activity is assumed when the highest dose (100 μg) given did not depress the contractions of the guinea pig ileum-myenteric plexus.

** These analogs have been published by Hambrook et al.[12], Pert et al.[13] and Miller et al.[14].

residues was replaced by its D-isomer (Table I). Only the D-Met^5 and D-Leu^5 analogs showed activity with the D-Leu being more potent than the D-Met.

From these studies it can be concluded that only the 2nd and 5th positions of enkephalin can be safely manipulated to yield compounds with potent activity. For this reason a series of D-Ala^2 and 5th position substituted analogs of enkephalin was synthesized (Table II). The most potent compounds have Met-NH_2, Leu-NH_2, D-Leu-OH, D-Phe-OH, (N^αMe)Leu-NH_2 at the 5th position.

Table III shows the relative potencies of some endorphin analogs. It is interesting to note that extension of the sequence beyond Met-enkephalin, which is β-LPH-(61-65), towards the COOH-terminal of β-LPH produces a gradual drop of biological activity until it reaches the 87th residue. However, incorporation of the last four residues *in toto* produces an enormous increase in activity. As in

Table II. Relative Potencies of Enkephalin Analogs Substituted at Positions 2 and 5

Analog	Rel. Potency	(95% Confidence Limits)
Tyr-D-Ala-Gly-Phe-Met-NH_2*	528	(492 - 569)
———————Leu-NH_2	528	(491 - 570)
———————D-Met-NH_2	246	(207 - 297)
———————D-Leu-NH_2	406	(341 - 484)
———————D-Met-OH	419	(382 - 461)
———————D-Leu-OH**	684	(619 - 753)
———————D-Ala-OH	205	(178 - 237)
———————D-Lys-OH	328	(288 - 374)
———————D-Phe-OH	522	(452 - 607)
———————D-Pro-OH	32	(28 - 38)
———————($\underline{N}^{\alpha}$Me)Leu-OH	387	(291 - 397)
———————($\underline{N}^{\alpha}$Me)Leu-NH_2	711	(671 - 753)

* This analog has been published by Pert et al.[13]
** This analog has been synthesized by Wellcome Research Lab.

Table III. Relative Potencies of Endorphin Analogs

Analog		Rel. Potency	(95% Confidence Limits)
β-LPH-(61-68)		67	(61 - 75)
β-LPH-(61-69)		60	(38 - 85)
β-LPH-(61-76)	α-endorphin	36	(19 - 53)
β-LPH-(61-77)	γ-endorphin	23	(11 - 37)
β-LPH-(61-79)		37	(29 - 45)
β-LPH-(61-87)	δ-endorphin	48	(38 - 60)
β_p-LPH-(61-91)*	β_p-endorphin	450	(281 - 966)
β_o-LPH-(61-91)*	β_o-endorphin	450	(281 - 966)
	α-endorphin amide	72	(47 - 102)
	$[Leu^5]$-α-endorphin	10	(4 - 26)
	$[D\text{-}Ala^2]$-α-endorphin	37	(22 - 49)
	$[D\text{-}Ala^2, D\text{-}Leu^5]$-α-endorphin	1.7	(0.3 - 2.7)
	γ-endorphin amide	41	(27 - 53)
	$[Leu^5]$-γ-endorphin	8.2	(7.7 - 8.6)
	$[D\text{-}Ala^2]$-γ-endorphin	37	(21 - 49)
	$[D\text{-}Ala^2, D\text{-}Leu^5]$-γ-endorphin	2.8	(2.7 - 2.9)
	$[Leu^5]$-β-endorphin	75	(50 - 107)
	β_p-LPH-(66-91)	0	
	β_p-LPH-(62-91)	0	

*Subscripts o = ovine, p = porcine.

the case of the enkephalins, substitution of an amide group at the COOH-terminus or D-alanine at the 2nd position of α- and γ-endorphins increases their potencies while replacement of the methionine residue with leucine lowers their activities. This leucine modification also applies to β-endorphin. In contrast to the enkephalins, replacement of the 2nd and 5th positions of α- and γ-endorphins with D-alanine and D-leucine, respectively, does not increase their potencies. This finding might imply that, at least for the α- and γ-endorphins, cleavage to the enkephalins is necessary for bioactivity. The des-Tyr^1 and des-Tyr^1 . . . Met^5 analogs of β-endorphin are completely inactive which implies that the tyrosine and the Tyr-Gly-Gly-Phe-Met sequence respectively are required for activity of β-endorphin.

The authors wish to thank M. Mercado, L. Koski, F. Luansing, E. Calonge, R. Kaiser, R. Schroeder, L. Yang and R. Wolbers for their excellent technical assistance. This research is supported by NIH Grant No. HD-09690 and AM-18811, and the W. R. Hearst Foundation.

References

1. Hughes, J., Smith, T., Kosterlitz, H., Fothergill, L., Morgan, B. & Morris, H. (1975) *Nature* **258**, 577–579.

2. Simantov, R. & Snyder, S. (1976) *Life Sci.* **18**, 781–788.

3. Guillemin, R., Ling, N. & Burgus, R. (1976) *C. R. Acad. Sci. Paris*, Series D **282**, 783–785.

4. Ling, N., Burgus, R. & Guillemin, R. (1976) *Proc. Nat. Acad. Sci. USA* **73**, 3942–3946.

5. Bradbury, A., Smyth, D. & Snell, C. (1975) in *Peptides: Chemistry, Structure, Biology*, Walter, R. & Meienhofer, J., Eds., Ann Arbor Science, Ann Arbor, pp. 609–615.

6. Li, C. H. & Chung, D. (1976) *Proc. Nat. Acad. Sci. USA* **73**, 1145–1148.

7. Chretien, M., Benjannet, S., Dragon, N., Seidah, N. G. & Lis, M. (1976) *Biochem. Biophys. Res. Commun.* **72**, 472–478.

8. Li, C. H., Barnafi, L., Chretien, M. & Chung, D. (1965) *Nature* **208**, 1093–1094.

9. Merrifield, R. B. (1963) *J. Amer. Chem. Soc.* **85**, 2149–2154.

10. Ling, N. (1977) *Biochem. Biophys. Res. Commun.* **74**, 248–255.

11. Kosterlitz, H. & Watt, A. (1968) *Br. J. Pharmac.* **33**, 266–276.

12. Hambrook, J. M., Morgan, B. A., Rance, M. J. & Smith, C. F. C. (1976) *Nature* **262**, 782–783.

13. Pert, C. B., Bowie, D. L., Fong, T. W. & Chang, J. K. (1976) in *Opiates & Endogenous Opioid Peptides*, Kosterlitz, H., Ed., North-Holland Publishing Co., Amsterdam, pp. 79–86.

14. Miller, R. J., Chang, K., Cuatrecasas, P. & Wilkinson, S. (1977) *Biochem. Biophys. Res. Commun.* **74**, 1311–1317.

CIRCULAR DICHROISM OF OPIOID PEPTIDES

R. GLENN HAMMONDS, JR., DAVID PUETT, *Department of Biochemistry, Vanderbilt University, Nashville, TN 37232,* and NICHOLAS LING, *Neuroendocrinology Laboratory, The Salk Institute, La Jolla, CA 92037*

Several peptides which exhibit morphine-like biological actions have recently been identified[1-3]. Each of these is a subsequence of the 91 residue pituitary hormone, β-lipotropin (β-LPH). In this communication the CD spectra of four such peptides; methionine-enkephalin (β-LPH 61–65), α-endorphin (β-LPH 61–66), γ-endorphin (β-LPH 61–67), and β-endorphin (β-LPH 61–91), are reported and their conformational implications discussed.

The peptides were synthesized by a solid phase procedure[4] and assayed by the depression of electrically induced contractions of the myenteric plexus-longitudinal muscle of guinea pig ileum[5]. Sedimentation equilibrium studies of the four opiate peptides in 0.1 *M* KCl, 1 *mM* bicine, pH 8 demonstrated monodispersity in the concentration range employed for spectral measurements. Diffusion constants were also determined for the three endorphins by the approach to equilibrium method[6] and all were found to be between 1.9 and 2.0 $\times 10^{-6}$ cm^2/sec.

Figure 1 shows the CD spectra of all four peptides in 0.1 *M* KCl, 1 *mM* bicine pH 8.0, 25°C. The spectrum of Met^5-enkephalin is characterized by a positive band at 218 nm and a shoulder at 225 nm, characteristic of aromatic chromophores. Neither 6 *M* GdmCl nor 75% TFE appreciably alters the magnitude of this spectrum (Table I). The spectra of α-, β-, and γ-endorphin are all negative in the region 210–250 nm, with γ slightly more negative than α, and β considerably more negative than γ. There is a well-defined minimum at 225 nm in the spectrum of β-endorphin. The strong protein denaturant 6 *M* GdmCl considerably

Table I. Mean-Residue Ellipticity (deg·cm^2/dmole) at 220 nm of Opioid Peptides in Various Solvents

Peptide	Buffer	6 *M* GdmCl[a]	75%TFE[a]
Met^5-enkephalin	4140	3816	3461
α-endorphin	-1,215	-100	-2,970
γ-endorphin	-1,275	-210	-2,940
β-endorphin	-2,170	-110	-17,250

a) GdmCl, guanidinium chloride; TFE, trifluoroethanol

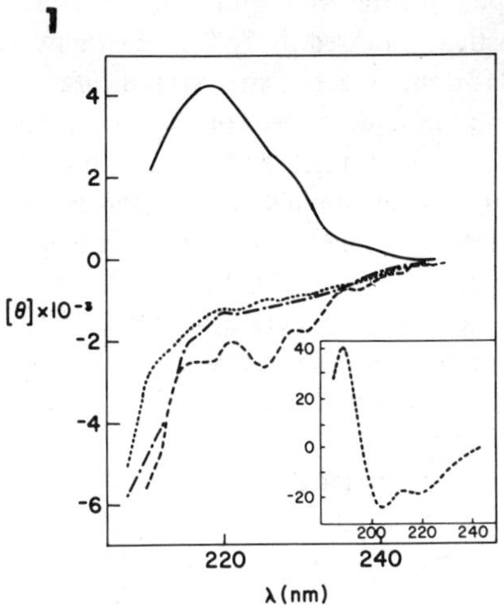

Fig. 1. Far UV CD spectra of opioid peptides. Spectra were recorded at 25°C in 0.1 *M* KCl, 1 *mM* Bicine, pH 8.0. The insert shows the CD of β-endorphin at 25°C in 75% TFE. $[\theta]$ = mean residue ellipticity. See Fig. 2 for curve labels.

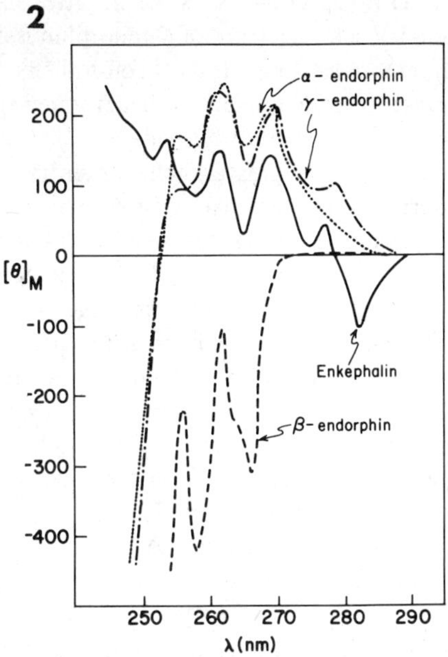

Fig. 2. Near UV CD spectra of opioid peptides. Spectra were recorded as in Fig. 1. $[\theta]_m$ = molar ellipticity.

diminishes the ellipticity at 220 nm of the endorphins, but has a small effect on the spectrum of Met^5-enkephalin (Table I). If β-turns or β-sheets were present in Met^5-enkephalin in aqueous solution, as has been recently suggested[7,8], one might expect CD changes in the spectral region of the peptide chromophore upon disruption of this structure. Although such changes were not observed, the relatively small magnitude of the β-turn spectrum mitigates against detection in the presence of other chromophores.

The decrease in ellipticity seen in the endorphins with 6 *M* GdmCl apparently reflects a change in the average conformation of peptide chromophore, since Met^5-enkephalin is unaffected. Unconstrained least-squares analysis[9] of the spectra of the endorphins gives quite good results (by the sums test) for β-endorphin, indicating 10% of α-helix and 10% of β-sheet. Analysis, leaving out the β-sheet reference spectra, gave clearly inferior sums test and residuals. Since the error in this method is (generously) placed at 5–10%, there is at most a just detectable amount of secondary structure in β-endorphin. The spectra of α-endorphin and γ-endorphin cannot be adequately decomposed into the reference spectra employed. However, their smaller ellipticity, and the lack of some features of the spectra of β-endorphin make it appear likely that they have even less α-helix or β-sheet than β-endorphin.

TFE (75%) increases the negative ellipticity of the endorphins (Table I). The far UV CD spectrum of β-endorphin is drastically altered by 75% TFE (Figure 1, inset), suggesting a typical coil-to-helix transition. Spectral analysis indicates the presence of about 50% α-helicity in β-endorphin under these conditions. Hollasi et al.,[10] have recently made a similar observation. Thus, β-endorphin appears to have a much greater helix potential than Met^5-enkephalin or the shorter endorphins; at least part of this may arise simply from the longer chain length.

Figure 2 shows the near UV CD spectra of Met^5-enkephalin and the endorphins. The spectra of Met^5-enkephalin and α- and γ-endorphin are similar, although not identical; this suggests that the common aromatic chromophores, Tyr^1 and Phe^4, exist in slightly different chemical environments. The spectrum of β-endorphin is considerably different from that of the other peptides. It has fewer observed transitions and is negative in sign rather than positive. This may be due to development in β-endorphin of a conformation in which the aromatic residues are shielded from the solvent, or it may be due to cancellation by the additional chromophore, Phe^{18}.

In summary, the far UV CD results suggest that the three endorphins possess limited secondary structure in aqueous solution. However, the α-helix potential is quite high for β-endorphin, and it is of great importance to ascertain if this coil-helix transition is involved in receptor recognition and interaction. The near UV CD spectra of Met^5-enkephalin, α- and γ-endorphin, and β-endorphin show significant differences. A plausible explanation for these differences is that some chain folding, perhaps involving a limited domain, occurs in β-endorphin. This would be consistent with the reported resistance of β-endorphin to peptidase action[11].

This research was supported by NIH (Research Grants HD-09690 and AM-15838; RCDA AM-00055 to D.P.) and in part by the VU Research Council.

References

1. Hughes, J. (1975) *Nature* **258**, 577–579.
2. Guillemin, R., Ling, N. & Burgess, R. C. (1976) *C. R. Acad. Sci. Paris, Ser. D*, 783–785.
3. Cox, B. M., Goldstein, A. & Li, C. H. (1976) *Proc. Nat. Acad. Sci. USA* **73**, 1821–1823.
4. Ling, N. (1977) *Biochem. Biophys. Res. Commun.*, **74**, 248–255.
5. Paton, D. M. & Zars, M. A. (1968) *J. Physiol.* **194**, 13–33.
6. Van Holde, K. E. & Baldwin, R. L. (1958) *J. Phys. Chem.* **62**, 734–743.
7. Jones, C. R., Gibbons, W. A. & Garsky, V. (1977) *Nature* **262**, 778–782.
8. Khaled, M. A., Long, M. M., Thompson, W. D., Bradley, R. J., Brown, B. & Urry, D. W. (1977) *Biochem. Biophys. Res. Commun.* **76**, 224–231.
9. Hammonds, R. G. (1977) *Eur. J. Biochem.* **74**, 421–424.
10. Hollosi, M., Kajtar, M. & Gráf, L. (1977) *FEBS Lett* **74**, 185–189.
11. Geisow, M. J. & Smyth, D. G. (1977) *Biochem. Biophys. Res. Commun.* **75**, 625–629.

THE SUSCEPTIBILITY OF C-FRAGMENT TO ENDOGENOUS ENDOPEPTIDASES

D. G. SMYTH, B. M. AUSTEN, A. F. BRADBURY, M. J. GEISOW, C. R. SNELL, *National Institute for Medical Research, Mill Hill, London NW7 1AA, England*

The first isolation of lipotropin C-Fragment (residues 61-91, recently called β-endorphin) and of C′-Fragment (61-87) was described two years ago at the IVth American Peptide Symposium in New York.[1] Since that time the presence of C-Fragment in pituitary has been confirmed,[2-4] the peptide has been found in brain[5] and it has been shown to exhibit a range of activities.[6,7] Characteristic of C-Fragment is the long duration of its effects, which contrast with the weak and short-lasting activities exhibited by N-terminal stretches of the C-Fragment sequence (γ- and α-endorphins, 61-77 and 61-76, and methionine enkephalin, 61-65). If the duration of action is related to stability, C-Fragment should be highly resistant to attack by brain proteases and the shorter peptides easily degraded. This has been demonstrated in a study of the proteolysis of C-Fragment by soluble enzymes and by membrane bound proteases from brain.

The N-terminal tyrosine was released from C-Fragment by aminopeptidase M with great difficulty; at an enzyme-substrate ratio of 1:60 (w/w) pH 7.4, 37°C, less than 10 per cent of the terminal residue was liberated in 8 hours. From C′-Fragment the terminal tyrosine was again released slowly but rapid cleavage occurred with the pentapeptide 61-65 and tridecapeptide 61-73. Similar results were obtained on incubation of this series of peptides with a membrane bound protease present in a particulate fraction of rat brain (Fig. 1).

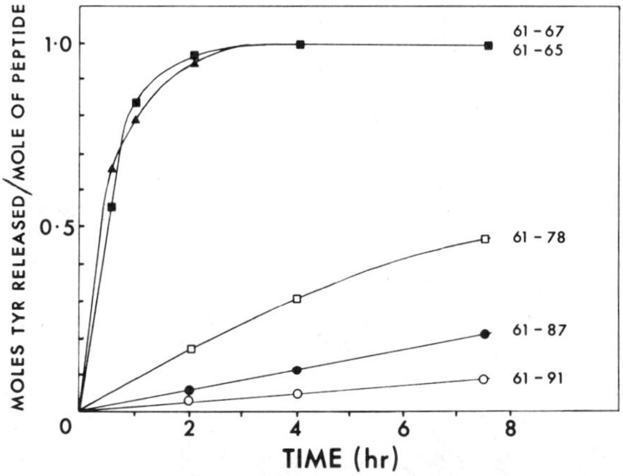

Fig. 1. Release of tyrosine from a series of peptides related to C-Fragment by a membrane-bound protease from brain.

The unusual stability exhibited by C-Fragment may be attributed to a folding of the peptide chain which renders the N-terminal residue less accessible to the enzyme. This is consistent with a study of the chiroptic properties of C-Fragment, which have indicated an atypical environment for the tyrosine residue.[8] The conformational properties are not retained by des-tyrosine-C-Fragment since the rate of release of the penultimate glycine from this peptide by aminopeptidase M was similar to the release of glycine from des-tyrosine-methionine-enkephalin. Thus the N-terminal tyrosine residue appears to play an important role in maintaining the conformation of C-Fragment.

The C-terminal residues of C-Fragment are highly resistant to the action of carboxypeptidase A (CpAse A) and of a corresponding enzyme in brain synaptosomes.[9] Under physiological conditions of pH and ionic strength, less than 1% of the C-terminal glutamine was released when C-Fragment was incubated for 4 hours with CpAse A (1:200, mole enzyme/mole of peptide). This was a surprising result, as C-terminal glutamine is normally released with ease. Since a high resistance was exhibited also by the hexapeptide Ala-Tyr-Lys-Lys-Gly-Gln (Fig. 2) the C-terminal stability of C-Fragment, like that of the synthetic peptide, can be seen to be a property of the primary structure and not the consequence of a specific conformation which would render the terminal residue less accessible. Under the same conditions of digestion glutamine was released from the pentapeptide Ala-Tyr-Lys-Gly-Gln fifty times more rapidly. Thus C-Fragment is resistant to the action of both NH_2-peptidase and carboxypeptidase. It seems likely, therefore, that the termination of its biological action is initiated by an endopeptidase.

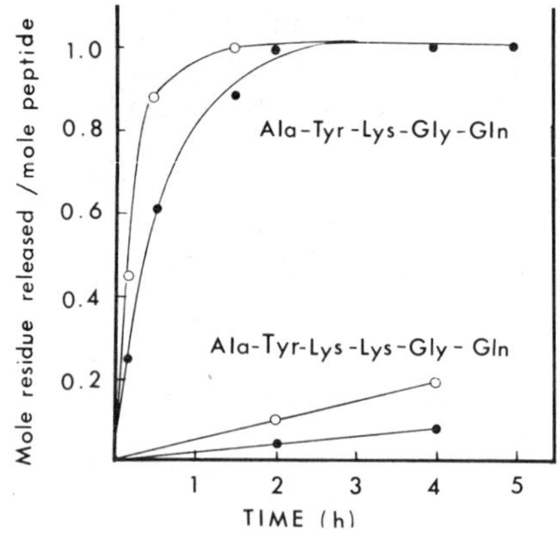

Fig. 2. Release of glutamine and glycine from Ala-Tyr-Lys-Lys-Gly-Gln and Ala-Tyr-Lys-Gly-Gln by carboxypeptidase A. —○—○— Gln, —●—●—Gly

Mild digestion of C-Fragment by different endopeptidases led to specific cleavage in the central section of the peptide chain. Rennin, chymotrypsin, *Armillaria mellea* protease and trypsin-catalyzed hydrolysis at positions 77-78,

78-79, 78-79 and 79-80, respectively, and the corresponding N-terminal peptides were isolated in high yield.

```
61              65              70
Tyr-Gly-Gly-Phe-Met-Thr-Ser-Glu-Lys-Ser-Gln-Thr-Pro-Leu-

75                  80                  85                  91
Val-Thr-Leu-Phe-Lys-Asn-Ala-Ile-Val-Lys-Asn-Ala-His-Lys-Lys-Gly-Gln
```

↑ rennin (Leu-Phe) ↑ chymotrypsin, *Armillaria* protease (Phe-Lys) ↑ trypsin (Lys-Asn)

Under mild conditions the proteases failed to cleave the N-terminal section at sites that would be expected to be susceptible. Chymotrypsin did not attack the Phe-Met bond (positions 64-65 of lipotropin), *Armillaria* and staphylococcal proteases gave negligible cleavage at Glu-Lys (68-69), trypsin exhibited a strong preference for Lys-Asn (79-80) rather than Lys-Ser (69-70), and rennin did not attack Phe-Met (64-65) or Leu-Val (74-75). The resistant section of C-Fragment thus extends from position 61 through position 75. When the digestions were performed under more vigorous conditions or when hexacitraconyl C-Fragment or the heptapeptide 61-67 was used as substrate, cleavage took place at the less reactive sites in the N-terminal region. It appears that C-Fragment adopts a preferred conformation which places residues 76-79 in a relatively exposed environment and renders other residues less accessible.

Incubation of C-Fragment at pH 7.4 with washed membranes from rat brain led to specific cleavage at Leu-Phe (77-78), forming γ-endorphin, and there was in addition a small amount of methionine-enkephalin (Fig. 3). The products were generated in high yield, however, only in the presence of bacitracin ($10^{-4}M$) or at lower pH. The heptapeptide 61-67, which is unable to maintain a preferred conformation, was rapidly cleaved by the membranes with the formation of methionine-enkephalin and threonylserine even at neutral pH. On incubation of C-Fragment with slices of rat striatum, the same series of specific products was formed (Fig. 4).[10]

It is clear that a preferred conformation of C-Fragment directs the route by which the peptide is degraded: the conformational properties impart stability to the N-terminal region and this is retained to a degree in the corresponding region of the endorphins. The formation of γ- and α-endorphin therefore does not require the action of an enzyme with a unique specificity nor does their survival in brain depend on packaging within the protected environment of a vesicle. Methionine-enkephalin appears to be formed by an endopeptidase with a higher pH optimum than that which forms γ-endorphin from C-Fragment but the pentapeptide is vulnerable to exopeptidases and is rapidly destroyed. It remains to be seen whether the endorphins and enkephalin are elaborated to perform a specific

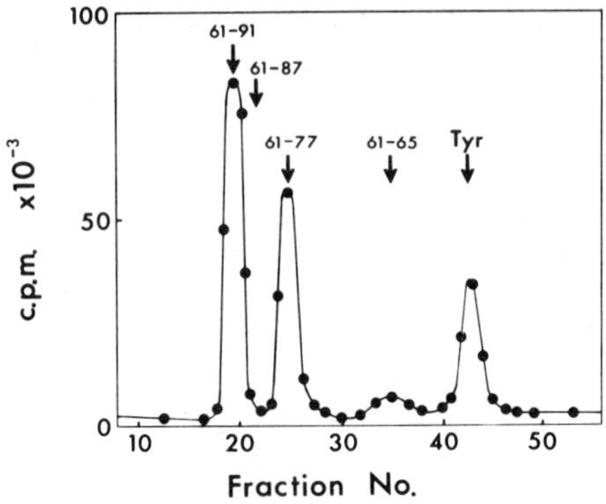

Fig. 3. Gel filtration of the products of digestion of (^{125}I) C-Fragment by membrane bound proteases from brain on Sephadex G-50.

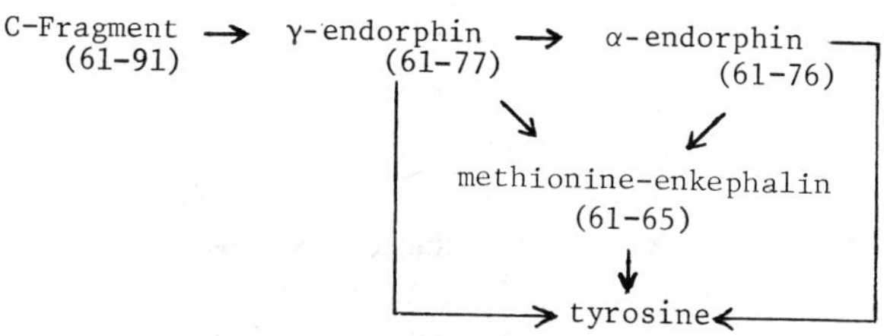

Fig. 4. Degradation of C-Fragment by extracellular enzymes in striatal slices.

physiological function or whether they are simply intermediate peptides which form transiently during the extracellular degradation of C-Fragment.

References

1. Bradbury, A. F., Smyth, D. G. & Snell, C. R. (1975) Proc. 4th Amer. Peptide Symposium, Walter, R. & Meienhofer, J., Eds., Ann Arbor Sci. Publ., Ann Arbor, pp. 609–615.
2. Li, C. H. & Chung, D. (1976) *Proc. Nat. Acad. Sci. USA* 73, 1145–1148.
3. Graf, L., Barat, E. & Patthy, A. (1976) *Acta Biochem. Biophys. Acad. Sci. Hung.* **73**, 3942–3946.
4. Guillemin, R., Ling, N. & Burgus, R. (1976) *C. R. Acad. Sci. Paris, Ser. D.* **282**, 783–785.
5. Bradbury, A. F., Feldberg, W. S., Smyth, D. G. & Snell, C. R. (1976) in *Opiates and Endogenous Opioid Peptides, Proc. Int. Narcotics Research Club*, Kosterlitz, H. W., Ed., Elsevier/North Holland Biomedical Press, pp. 9–17.
6. Feldberg, W. & Smyth, D. G. (1976) *J. Physiol.* **260**, 30-31.
7. Feldberg, W. & Smyth, D. G. (1977) *Brit. J. Pharmac.* **60**, 445-454.
8. Bayley, P. B., Snell, C. R. & Smyth, D. G., (1977) *Biochem. Soc. Trans.* **5**, 683-686.
9. Geisow, M. J. & Smyth, D. G. (1977) *Biochem. Biophys. Res. Commun.* **75**, 625-629.
10. Symth, D. G. & Snell, C. R. (1977) *Febs Lett* **78**, 225-228.

SYNTHETIC AND BIOLOGICAL STUDIES ON UNMODIFIED AND MODIFIED FRAGMENTS OF HUMAN β-LIPOTROPIN WITH OPIOID ACTIVITIES

D. H. COY, P. GILL, A. J. KASTIN, *Department of Medicine, Tulane University School of Medicine and VA Hospital, New Orleans, Louisiana 70112;* A. DUPONT, L. CUSAN, F. LABRIE, *Laboratory of Molecular Endocrinology, University of Laval, Quebec, Canada;* D. BRITTON, R. FERTEL, *Department of Pharmacology, Ohio State University School of Medicine, Ohio*

Several large peptides with morphinomimetic activity have recently been found[1,2] in hypothalamic and pituitary tissue and appear to be derived from the C-terminal region of β-LPH. All have the Met^5-enkephalin sequence[3] at their N-terminus and range in size from the 16 and 17 residue peptides, β-LPH(61-76) (α-endorphin) and β-LPH(61-77) (γ-endorphin), to the largest of the series, β-LPH(61-91) (β-endorphin) which has been isolated from human, camel, sheep, and pig pituitaries. We undertook the synthesis of all of these peptides in order to evaluate their biological properties in a wide range of assay systems. We were also interested in preparing their [D-Ala^2]-analogs, since this substitution in the enkephalins results in a massive increase in opioid activities.[4] The physiological precursor of Leu^5-enkephalin[3] still remains to be discovered, however, it seems probable that it originates from a β-LPH or β-endorphin peptide with Leu in the relevant position. [Leu^5]-β_h-endorphin was, therefore, also prepared in order to examine its properties.

Synthesis

The endorphins were assembled on the appropriate amino acid-substituted, 1% cross-linked Merrifield resins which were prepared by Gisin's $CsHCO_3$ method. The substitutions were about 0.5 mmole per g. In the automated solid-phase synthesis, low (25%, 15 min) and high (50%, 15 min) concentrations of TFA/ CH_2Cl_2 were used in consecutive deprotection steps in an effort to promote complete removal of α-Boc-groups by subjecting the resin to varying degrees of swelling. The following side-chain protecting groups were used: Tyr, 2-bromo-carbobenzoxy; Lys, 2-chlorocarbobenzoxy; Ser, Thr, and Glu, benzyl. Boc-amino acids were coupled in the presence of diisopropylcarbodiimide (DIC) which, unlike the commonly used dicyclohexylcarbodiimide, forms a far more soluble urea.[5] Boc-Asn and Gln were coupled with the addition of 1 equivalent of 1-hydroxybenzotriazole and DIC. All reactions appeared to proceed as far as they were able within 30 min and were monitored by the Kaiser ninhydrin test.

Where necessary, re-coupling was carried out in DMF using 1.5 equivalents of the pre-activated, symmetrical anhydrides,[6] formed from a 2:1 molar mixture of the appropriate Boc-amino acid and DIC. In this situation, the complete solubility of diisopropylurea in DMF or CH_2Cl_2 containing a few drops of DMF avoids filtration of moisture-sensitive anhydrides before use. In the infrequent event of incomplete re-coupling, free amino groups were terminated by reaction with 5% *N*-acetylimidazole in CH_2Cl_2. Completed peptide resins, from which the N-terminal Boc-groups had been removed, were cleaved and deprotected by a standard treatment with 10% anisole/HF.

Purification

Both α- and γ-endorphins and their [D-Ala^2]-analogs were purified initially by gel filtration on Sephadex G-50 in 0.2 *M* AcOH. The γ-endorphins were subjected to a final purification by partition chromatography on Sephadex G-50 using *n*-butanol/acetic acid/water (4:1:5) and the α-endorphins to partition chromatography in the system *n*-butanol/*n*-propanol/acetic acid/water (7:1:2:10). Yields were in the region of 20–30% and they gave satisfactory amino-acid analyses and were homogeneous by silica gel TLC in 5 solvent systems and reversed phase HPLC. The following $[\alpha]_D$ values were obtained in 0.2 *M* AcOH: α-endorphin, -65° (*c*, 0.56; 27°); [D-Ala^2]-α-endorphin, -59° (*c*, 0.56; 27°); γ-endorphin, -88° (*c*, 0.59; 28°); [D-Ala^2]-γ-endorphin, -100° (*c*, 0.28; 28°).

Initial elution of the β-endorphins on Sephadex G-25 in 2 *M* AcOH gave exclusion volume fractions containing roughly 75% pure peptide. This material was then subjected to partition chromatography on Sephadex G-50 in the system 0.1 *M* acetic acid/*n*-butanol/pyridine (11:5:3) followed by ion-exchange elution on microcrystalline CM-cellulose using a linear gradient of NH_4OAc buffers (0.1 *M* at pH 6.0 to 0.4 *M* at pH 7.0). Final peptides were obtained in 10–15% yields and gave correct amino-acid analyses after acid and enzymatic hydrolysis. Digestion with trypsin gave the expected 5 major peptide fragments by TLC which were readily separated by chromatography on silica gel and analyzed to give the expected amino-acid ratios. TLC on silica gel in 5 solvent systems revealed that the β-endorphins were homogeneous at loads up to 30 μg. HPLC was carried out on a column (0.4 × 30 cm) of C_{18} μ-Bondapak under isocratic [0.01 *M* NH_4OAc, pH 4.0: CH_3CN containing 1% *N*-ethylmorpholine (45:55)] or linear gradient conditions (25% to 75% of buffer 2). Sharp elution profiles were obtained as measured by absorption at 220 nm and representative traces for β_h-endorphin and [Leu^5]-β_h-endorphin are shown in Fig. 1. The following $[\alpha]_D$ values were obtained in 0.2 *M* AcOH: β_h-endorphin, -83° (*c*, 0.65; 26°); [Leu^5]-β_h-endorphin, -78° (*c*, 0.55; 26°); [D-Ala^2]-β_h-endorphin, -82° (*c*, 0.49; 26°).

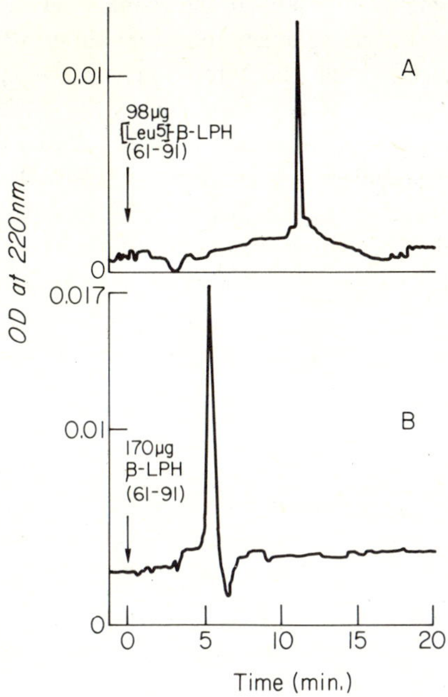

Fig. 1. HPLC of [Leu^5]-β_h-endorphin and β_h-endorphin on a column of μ-Bondapak C-18 (0.4 × 30 cm) using 0.01 *M* NH_4OAc/CH_3CN containing 1% *N*-ethylmorpholine for elution under the gradient (A) or isocratic (B) conditions described in the text.

Biological Activities

Peptides were assayed *in vitro* for their abilities to bind to partially purified brain opiate receptors and to inhibit electrically-evoked contractions of mouse vas deferens. The experimental conditions have been described previously.[4] In the vas deferens bioassay, α-, γ-, and β_h-endorphin were less effective than Met^5-enkephalin with molar ID_{50} values of 5×10^{-8}, 1.5×10^{-8}, and 3.8×10^{-8}, respectively, compared to a value of 1.1×10^{-8} for the pentapeptide. Their [D-Ala^2]-derivatives were roughly 10 times more active, having the following values: α-endorphin, 1.8×10^{-9}; γ-endorphin, 1.7×10^{-9}; β_h-endorphin, 4.2×10^{-9}. Thus, they were somewhat less active than [D-Ala^2]-enkephalin (1.2×10^{-9}).[4] [Leu^5]-β_h-endorphin (1.8×10^{-8}) was about twice as active as β_h-endorphin itself. In terms of binding affinity to opiate receptors, β_h-endorphin and its [Leu^5]- and [D-Ala^2]-analogs were approximately 30% as active as Met^5-enkephalin and α- and γ-endorphin and their [D-Ala^2]-analogs exhibited only 3% binding affinity. *In vivo* β_h-endorphin was 2,000–4,000 times more effective than Met^5-enkephalin in releasing radioimmunoassayable growth hormone and prolactin after intracerebral injection and its [D-Ala^2]- and [Leu^5]-analogs were similarly active. For comparison, [D-Ala^2]-enkephalinamide was as potent on a weight basis. Although α- and γ-endorphin were many times more potent than Met^5-enkephalin *in vivo*, they were still much less active than β_h-endorphin.

Apparently, extensions to the C-terminus of Met^5- or Leu^5-enkephalin in varying degrees adversely affect binding affinities to opiate receptors but greatly increase *in vivo* activity presumably by protecting the N-terminus from enzymatic attack. In [D-Ala^2]-enkephalin, this function is performed by the D-Ala-residue and its incorporation into the larger peptides results in little increase in *in vivo* activity since the active part of the molecule is already protected. The [D-Ala^2]-endorphins possibly have increased activity in the vas deferens system because of the presence of a different set of degrading enzymes or different conformational or structural requirements for receptor binding in the vas deferens.

We would like to thank Dr. David Huang for performing the HPLC investigation. Supported in part by NIH grant DA 01806.

References

1. Li, C. H. & Chung, D. (1976) *Proc. Nat. Acad. Sci. USA* 73, 2515–2519.
2. Ling, N., Burgus, R. & Guillemin, R. (1976) *Proc. Nat. Acad. Sci. USA* 73, 3942–3946.
3. Hughes, J., Smith, T. W., Kosterlitz, H. W., Fothergill, L. A., Morgan, B. A. & Morris, H. R. (1975) *Nature* **258**, 577–579.
4. Coy, D. H., Kastin, A. J., Schally, A. V., Morin, O., Caron, N. G., Labrie, F., Walker, J. M., Fertel, R., Berntson, G. G. & Sandman, C. A. (1976) *Biochem. Biophys. Res. Commun.* 73, 632–638.
5. Sarantakis, D., Teichman, J., Lien, E. L. & Fenickel, R. L. (1976) *Biochem. Biophys. Res. Commun.* 73, 336–342.
6. Hemmasi, B. & Bayer, E. (1974) *Hoppe Seyler's Z. Physiol. Chem.* **355**, 481–489.

STRUCTURE-ACTIVITY RELATIONSHIPS OF ENKEPHALIN ANALOGS

B. A. MORGAN, J. D. BOWER, K. P. GUEST, B. K. HANDA, G. METCALF and C. F. C. SMITH, *Reckitt & Colman Ltd., Pharmaceutical Division, Hull, HU8 7DS, England*

Introduction

The discovery that enkephalins can act as ligands for opiate receptors[1] and subsequent work confirming their opiate-like activity has opened up a new field of medicinal chemistry with exciting prospects. In order to explore the relationship between structure and activity, we have synthesized a large number of peptides related to the enkephalins and assessed their biological effects. In this study we compare the pharmacological effects exerted by a selection of the analogs.

Chemistry

The analogs were synthesized by solution methods utilizing a strategy similar to that previously described.[2] The purified peptides were lyophilized from 0.1 *M* HCl and then water before biological testing. Care was taken to minimize racemization of the *N*-alkyl tyrosine derivatives. Boc-MeTyr(Bu^t)-OH was synthesized by methylation of Boc-Tyr(Bu^t)-OH according to the method of Benoiton.[3] The *N*-cyclopropylmethyl (*c*PrMe) derivative was synthesized by the route shown in Scheme 1.

$$\text{H-Tyr(Bu}^t\text{)-OMe} \xrightarrow{\text{Na}_2\text{CO}_3\text{/KI/}c\text{PrMeBr}} c\text{PrMeTyr(Bu}^t\text{)-OMe} \xrightarrow{\text{(i) NaOH, (ii) HCl}}$$

$$\text{Cl}^-\cdot\text{H}^+\text{-}c\text{PrMeTyr(Bu}^t\text{)-OH} \xrightarrow{\text{BocN}_3\text{/TMG/DMF}} \text{Boc-}c\text{PrMeTyr(Bu}^t\text{)-OH}$$

Scheme 1. Synthesis of N^α-cyclopropylmethyl-*O*-*t*-butyltyrosine.

Biological Assay Procedures

In vitro methods: Mouse vas deferens (MVD); the method used was essentially that of Hughes et al.[4] Guinea pig ileum (GPI); a method derived from that described by Kosterlitz and Watt was used.[5] In both assays relative agonist potencies were calculated by comparison with Met[5]-enkephalin on the same tissue.

In vivo methods: Rat tail flick assay; the method used was essentially that of Janssen et al.[6] Water at 55° ± 1°C was used as the nociceptive stimulus. Animals were tested prior to and at various times after intraventricular administration (10 μl) to assess the duration of the effects produced. Animals which failed to withdraw their tails within 30 seconds were considered completely analgesed. Results are expressed as a percentage of the 30 second cut off time so that a typical control response of 3 seconds corresponds to 10% analgesia. Rat catalepsy; the method of Blane et al.[7] was used.

Results and Discussion

Following our discovery[8] that the first metabolic step for the enkephalins was cleavage of the Tyr-Gly bond we found, like others, that substitution of a D-amino acid at position 2 or *N*-alkylation of tyrosine was effective in increasing stability. The effects of these modifications in combination with variations at the C-terminus resulted in an interesting series of changes in activity (Table I).

In vitro tests indicated that *N*-methylation of tyrosine (*4,5*) resulted in an increase in GPI potency and a decrease in MVD potency. Alkylation of the tyrosine nitrogen with a cyclopropylmethyl group (*3*) resulted in the reduction of agonist potency. The selection of a cyclopropylmethyl group was not arbitrary; substitution of groups such as cyclopropylmethyl and allyl on the nitrogen of opiate "agonists" such as morphine or oxymorphone dramatically alter the profile of the drug, inducing varying degrees of antagonist character. In the MVD preparation peptide (*3*) showed no ability to antagonize the effects of the agonist etorphine. This result is in contrast to the suggestion that [*N*-allyl-Tyr^1, D-Ala^2]-Met^5-enkephalin has partial agonist character as measured by the "sodium index" receptor-binding technique.[9] Substitution by D-Ala at position 2 (*6,7,8,9*) increased potency in both assay procedures. Interesting effects were observed by combining D-Ala^2 substitution with increasing degrees of truncation at the C-terminus. Removal of the terminal carboxyl group as in (*10*) increased potency in the GPI and halved potency in the MVD. Removal of the C-terminal alkyl moiety as in (*11*) increased GPI potency but greatly decreased MVD potency. Removal of the carboxamide group from this tetrapeptide gave (*12*) with increased potency and selectivity in the GPI; a picture which was unaffected by methylation of the C-terminal amide as in (*13*).

Effects following intraventricular administration in rats were also investigated. The MeTyr^1 (*4,5*) and D-Ala^2 pentapeptides (*6,7,8,9*) caused strong naloxone reversible antinociception and catalepsy usually lasting 2–3 hours. The tetrapeptide (*10*) did not cause antinociception at 100 μg but did induce catalepsy. The tripeptide (*12*) caused strong naloxone reversible antinociception and catalepsy lasting up to 5 hours.

Various conformations have been proposed for the enkephalins, including a Tyr^1 $\rightarrow$ Phe^4 β-turn, a Gly^2 $\rightarrow$ Met^5 β-turn, and an inverse γ-turn. Obviously the truncated analogs (*12,13*) cannot form a Gly^2 $\rightarrow$ (Met/Leu)^5 β-turn and the tertiary amide (*13*) cannot form the Tyr^1 $\rightarrow$ Phe^4 β-turn hydrogen bond; however

Table I. Enkephalin Analogs: Results of Pharmacological Tests

No.	Structure	In vitro		In vivo (Rat i.c.v.)			
		MVD†	GPI††	Dose µg (n)*	Analgesia		Catalepsy**
					Extent**	Duration+	
1	Tyr-Gly-Gly-Phe-Met-OH	100	100	1000 (5)	40	0.25	40
2	MeTyr-Gly-Gly-Phe-Met-OH	20	100	NT			
3	*c*PrMeTyr-Gly-Gly-Phe-Met-OH	0.8	1.5	NT			
4	MeTyr-Gly-Gly-Phe-Met-NH_2	10	525	200 (5)	100	3	100
5	MeTyr-Gly-Gly-Phe-Met-NH-Pr	25	470	100 (5)	100	3	100
6	Tyr-D-Ala-Gly-Phe-Met-OH	500	625	250 (5)	72	2	80
7	Tyr-D-Ala-Gly-Phe-Met-NH_2	190	800	100 (5)	88	2	100
8	Tyr-D-Ala-Gly-Phe-Met-NH-Pr	520	480	100 (3)	76	3	100
9	Tyr-D-Ala-Gly-Phe-Leu-NH_2	160	500	100 (3)	100	2	100
10	Tyr-D-Ala-Gly-Phe-$NHCH_2CH_2CHMe_2$	44	940	100 (5)	20	-	80
11	Tyr-D-Ala-Gly-Phe-NH_2	6	515	NT			
12	Tyr-D-Ala-Gly-$NHCH_2CH_2Ph$	2.3	440	100 (5)	83	4	100
13	Tyr-D-Ala-Gly-$NMeCH_2CH_2Ph$	7.5	530	100 (5)	35	3	100
14	Morphine	3	100	100 (10)	88	4	50

† Results expressed as percentage of methionine-enkephalin (Met^5-enkephalin); ID_{50} Met^5-enkephalin 32.5 ± 3 *nM*

†† Results expressed as percentage of Met^5-enkephalin; ID_{50} Met^5-enkephalin 174 ± 21 *nM*

* Minimum dose tested which caused appreciable analgesia

** Peak effects (see test procedures)

\+ hours

all the compounds have the structural features necessary to form an inverse γ-turn. It would be premature to allot unique importance to a single conformation; however, it is tempting to draw attention to the correlation between GPI potency and the ability to form an inverse γ-turn. In any event, the data presented in Table I, provide further evidence for the occurrence of differing types of opiate receptor; however, the relevance of these differences to the varying types of activity attributed to opiate drugs in man remains unresolved.

It is a pleasure to acknowledge the contributions of M. Botham, I. Guest, D. Hagues, S. Howe, T. Robinson and A. Wilson.

References

1. Hughes, J., Smith, T. W., Kosterlitz, H. W., Fothergill, L. A., Morgan, B. A. & Morris, H. R. (1975) *Nature* **258**, 577–579.

2. Bower, J. D., Guest, K. P. & Morgan, B. A. (1976) *J. Chem. Soc. Perkin Trans. I.* 2488–2492.

3. McDermott, J. R. & Benoiton, N. L. (1973) *Can. J. Chem.* **51**, 1915–1919.

4. Hughes, J., Kosterlitz, H. W. & Leslie, F. M. (1975) *Brit. J. Pharmac.* **53**, 371–381.

5. Kosterlitz, H. W. & Watt, A. J. (1968) *Brit. J. Pharmac. Chemother.* **33**, 266–276.

6. Janssen, P. A. J., Niemegeers, C. J. E. & Dony, J. G. H. (1963) *Arzneim Forsch.* **13**, 266–276.

7. Blane, G. F., Boura, A. L. A., Fitzgerald, A. E. & Lister, R. E. (1967), *Brit. J. Pharmac. Chemother.* **30**, 11–22.

8. Hambrook, J. M., Morgan, B. A., Rance, M. J. & Smith, C. F. C. (1976) *Nature* **262**, 782–783.

9. Pert, C. B. (1977) in *Centrally acting peptides*, Hughes, J., Ed., Macmillan, London (in press).

SYNTHESIS AND PHARMACOLOGY OF LONG-ACTING ENKEPHALINS

ALAN R. DAY, RICHARD J. FREER, *Department of Pharmacology, Medical College of Virginia, Richmond, Virginia 23298* and DAVID I. MARLBOROUGH, *The John Curtin School of Medical Research, Canberra City, Australia*

The enkephalins and endorphins belong to a class of opiate peptides derived from β-lipotropin.[1] All of the peptides contain the fragment 61-65 of β-lipotropin. Unlike β-endorphin, the enkephalins are metabolized rapidly by tissue enzymes.[2] Along with other researchers we have sought to introduce features into the molecule of enkephalin which would retard enzymatic degradation but still retain biological potency. The first compound of this type [D-Ala^2]-enkephalinamide was described by Pert et al.[3] Following their lead we have synthesized [D-Ala^2, decarboxy-Nle^5]-enkephalin (I) and [N^α-MeTyr^1, decarboxy-Nle^5]-enkephalin (II).

Methods

Both peptides I and II were prepared by the classical solution methods outlined in Figure 1.

The method selected was the mixed anhydride procedure outlined by Beyerman[4]. Excess mixed anhydride acylating component was removed by reaction with 3-dimethylamino-1-propylamine. Each intermediate was isolated and checked routinely by thin layer chromatography and electrophoresis where appropriate. All had correct amino-acid and elemental analyses. Free peptides were isolated after treatment of the fully protected sequence with HF/anisole.

Bioassay of the peptides versus morphine sulphate was determined in the stimulated guinea pig ileum assay as described by Harris et al.[5] Mouse tail-flick assay was carried out as outlined by Pedigo et al.[6] The rats used in this study were implanted with cannulae into the peri-aqueductal gray area. Peptides were delivered in pH 7.4 saline solution (1 μl) and tail-flick latency determined 20 min later. In the cross-tolerance study mice were made tolerant to morphine sulphate by twice-daily subcutaneous injection of the narcotic over a period of eleven days. The test dose of narcotic or peptide was given intraventricularly four hours after final subcutaneous injection of morphine sulphate. The resistance to enzymatic degradation of the peptides was assessed by incubation of the peptide with rat brain homogenate at 37°, 25° and 0°C. Aliquots of the mixture were removed periodically, boiled for 2 min and assayed on the electrically stimulated guinea pig ileum.

Results

Some pharmacological properties of peptides I and II have been tabulated in Table I.

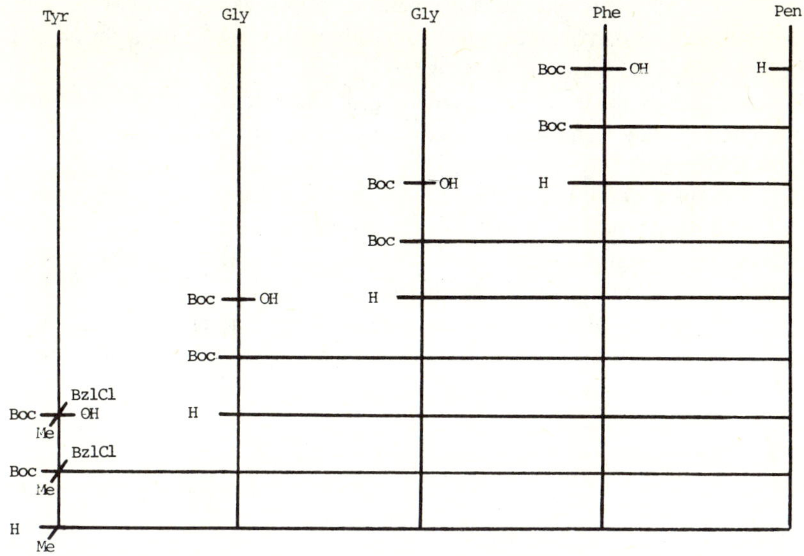

Figure 1. Synthesis of Enkephalin Analogs (Pen is an abbreviation for *n*-pentylamine).

As can be seen Met⁵-enkephalin was equipotent with morphine sulphate in the stimulated ileum assay. Peptides I and II showed significantly better activity, particularly peptide I which is 6-7 fold more active than the narcotic. The *in vivo* assay (mouse tail-flick) demonstrates that morphine sulphate is approximately 3X more potent than peptide I and 6X more potent than peptide II on a molar basis. The duration of effect of the peptides was 30 min at a dosage of 4-6 μg. Administration of naloxone (1 mg/kg, sc) 15 min prior to intraventricular

Table I. Pharmacological properties of [D-Ala^2, decarboxy-Nle^5]-enkephalin (I) and [N^{α}-MeTyr^1, decarboxy-Nle^5]-enkephalin (II) in guinea pig ileum, mouse tail-flick (upper) and morphine tolerant rats (lower).

Compound	Guinea Pig Ileum (ED_{50})	Mouse Tail-Flick Assay (% Antinociception)				Mouse Tail-Flick (ED_{50} @ 15 min)
		Dose	15min	30min	60min	
I	6.1×10^{-8} M	4 μg	71	31	8	2.2 μg
II	2.0×10^{-7} M	6 μg	79	68	20	4.2 μg
Met^5-enkephalin	2.4×10^{-7} M	16 μg	9	21	12	-
Morphine Sulphate	3.1×10^{-7} M	NOT DETERMINED				0.36 μg

	Percent Antinociception	
	Drug Naive	Chronic Morphine
I (16 μg)	100	39
Morphine Sulphate (16 μg)	100	30

injection of the peptides caused a decrease in antinociception to control saline levels. Mice which had been treated chronically with narcotic for a period of 11 days showed cross tolerance to peptide I.

Figure 2 shows the effect of incubating peptides I, II and Met^5-enkephalin with crude rat brain homogenates and assaying aliquots of the incubate on the stimulated ileum.

At 37°C, peptides I and II are completely stable for at least a period of 90 min. In contrast, Met^5-enkephalin is rapidly degraded at 37°C ($t_{1/2} \sim 4$ min) while the rate of degradation is somewhat less at 25°C. At 0°C Met^5-enkephalin is stable over a period of at least 60 min (data not shown).

Figure 3 demonstrates the behavior of peptides I, II and morphine when administered directly into the peri-aqueductal gray area of rats. In this system

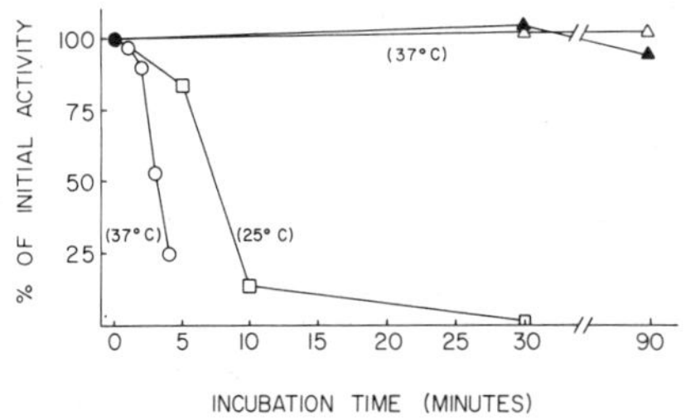

Fig. 2. Stability of [D-Ala^2, decarboxy-Nle^5]-enkephalin (▲), [N$^\alpha$-MeTyr^1, decarboxy-Nle^5]-enkephalin (△) and Met^5-enkephalin [(○–○) and (□–□)] in crude rat brain homogenates at 37° and 25°C.

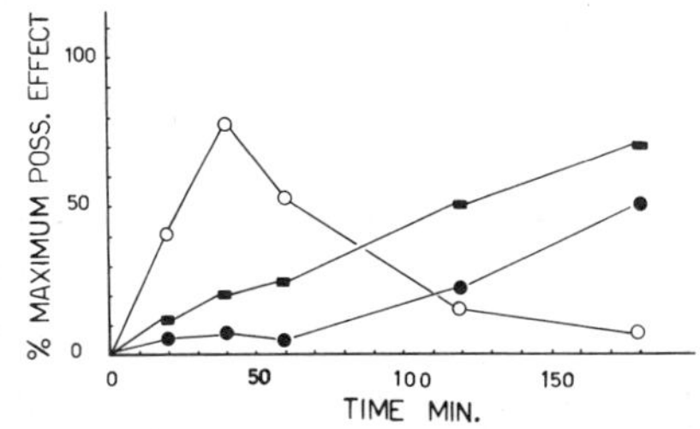

Fig. 3. Antinociceptive activity of morphine (○), [D-Ala^2, decarboxy-Nle^5]-enkephalin (■–■) and [N$^\alpha$-MeTyr^1, decarboxy-Nle^5]-enkephalin (●–●) after direct application into the peri-aqueductal gray area in rats.

the onset of action of morphine is clearly more rapid and less prolonged than that of the peptides. The reason for this is not clear but perhaps indicates a transport phenomenon.

The authors are grateful to Drs. W. L. Dewey, J. Carney and J. A. Rosecrans for biological testing. Supported by USPHS Grants DA-01647 and DA-00296.

References

1. Hughes, J., Kosterlitz, H. W., Fothergill, L. A., Morgan, B. A. & Morris, H. R. (1975) *Nature* **258**, 577–579.

2. Marks, N., Grynbaum, A. & Neidle, A. (1976) *Biochem. Biophys. Res. Commun.* **74**, 1552–1559.

3. Pert, C. B., Pert, A., Chang, J. K. & Fong, B. T. W. (1976) *Science* **194**, 330–332.

4. Beyerman, H. C. (1972) in *Chemistry and Biology of Peptides,* Meienhofer, J., Ed., Ann Arbor Science, Michigan, pp. 351–357.

5. Harris, L. S., Dewey, W. L., Howes, J. F., Kennedy, J. S. & Pars, H. (1969) *J. Pharm. Exp. Ther.* **169**, 17–22.

6. Pedigo, N. W., Dewey, W. L. & Harris, L. S. (1975) *J. Pharm. Exp. Ther.* **193**, 845–852.

RECEPTOR INTERACTIONS IN A SENSORY SYSTEM

DANIEL E. KOSHLAND, JR., *Department of Biochemistry*
University of California, Berkeley, California 94720

Unity and diversity are characteristics which we identify with biological systems. Once any system has been thoroughly delineated, it is quite easy to perceive the essential unities and the planned diversities. In the early stages of research, however, the diversity of species can obscure an essential unity, and repetitious similarities can obscure a functionally significant diversity. In the area of receptor protein interaction, the interplay between unity and diversity is particularly challenging. The discovery of peptide communicators, such as the enkephalins, peptide-P, sleep factor, etc. suggest a class of substances somewhere between neurotransmitters and hormones. Are we looking at totally different classes of messenger molecules selectively evolved to have separate properties, or a single unity in which there will be a continuous gradation between classes previously considered to be quite distinct? In this situation the studies which we have been doing on a simple sensory system may be helpful.

Bacterial Sensing

The chemotactic system of bacteria involves a sensing system which is important to the survival of these species[1,2]. It tells the bacterium to migrate in a direction favorable to its survival and away from situations threatening its survival. It does this by a sensory system which involves receptors, a central processing system, and a motor response[3-7]. The outside environment communicates with this system by means of chemoeffectors which become bound to receptor proteins. Thus the system has analogy to neurotransmitters as well as hormone receptors.

In Figure 1 is shown a schematic representation of the current knowledge of the chemotactic signalling system of the bacteria. A variety of receptors is present in the membrane of the bacterium, each of which is specific for one or a few chemicals. One receptor, for example, binds ribose and allose but no other compounds so far detectable. Another binds aspartic acid and glutamic acid preferentially. Thus, the individual receptor molecules have specificities similar to those of enzymes and receptors in other species.[3] The chemoeffector binding to the receptor induces a conformational change leading to the transmission of a signal to a central processing system[8]. Though the system is central, indicated by the additivity of signals in relation to the final response, the complexity of the processing system is indicated by the fact that nine genes have been identified[9] and their gene products apparently are required for correct interpretation of the message[9-12]. The signal which emerges from the processing system controls flagella function in such a way that the tumbling frequency of the flagella is

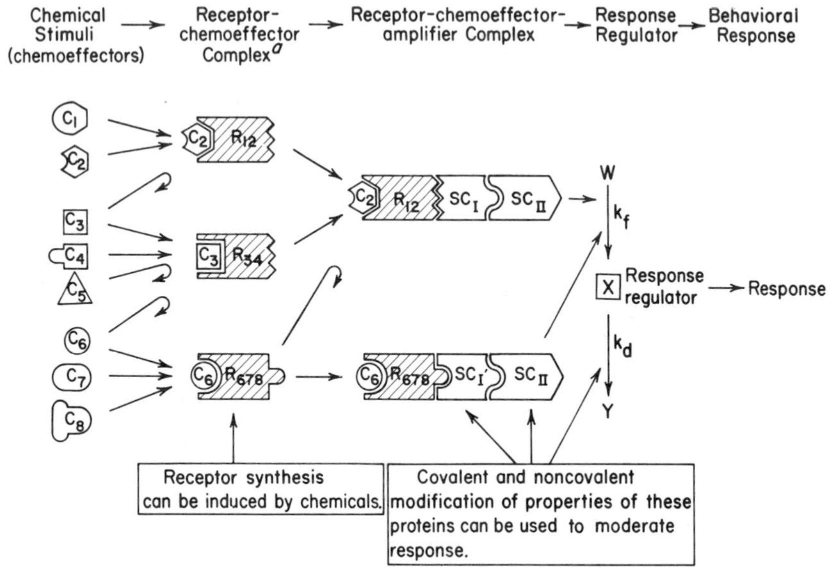

Fig. 1. Schematic representation of general signalling systems. A series of stimuli interact in various ways with individual receptors. Compounds C_1 and C_2 bind to receptor 12, but not to receptors 34 or 678. Compound C_5 binds to no receptors, and hence cannot be detected by the organism. On binding to the receptor, an induced conformation change occurs such that R_{12} and R_{34} receptors are attracted to signal components I. This means that the receptor chemoeffector complexes will compete with each other and can limit responses if the number of SC_I molecules is significantly smaller than the number of R_{12} and R_{34} receptors. Chemoeffectors C_6, C_7, and C_8 are focused through a separate processing machinery via $SC_{I'}$ by the specificity of R_{678} for $SC_{I'}$ and not SC_I. Signal components I and I′ can then interact with other signalling components of the general system, which may or may not be similar to each other and are designated by SC_{II}. Ultimately the signal from this system interacts with one of the two steps of the response regulator system, here designated as being formed from W in a k_f step and being decomposed to Y in a step designated k_d. The effects of the two chemoeffectors may be positive or negative, depending on whether they increase or decrease the rates of the k_f or k_d steps. Favorable effects (increase of attractant, decrease of repellent) reinforce each other and are inhibited by unfavorable effects (decrease of attractant or increase of repellent). The level of tumbling regulator then determines the behavioral response in the same way that a thermostat regulates a furnace. The receptor proteins can be induced by chemicals or be constitutive. The properties of the various enzymes and receptors in the system can be altered by covalent or noncovalent modification leading to enhanced or subdued sensory responses. The level of X relative to a threshold controls the sensory response.

modified[13,14] to control migration towards favorable stimuli, usually nutrients, and away from unfavorable ones, usually toxic substances.

Receptor – Chemoeffector Interactions of the Chemotaxis System

Since the features of such a system have been described in detail elsewhere in a number of reviews[3-7], I shall concentrate here on certain features of the receptor-chemoeffector interaction which may be of particular relevance to other systems:

(*i*) The bacterium controls those aspects of the environment to which it will respond by the number and specificity of the receptors on its surface. Individual receptors have, in most cases, rather sharply defined specificities but are usually not absolutely specific. For example, in Table I are shown the specificities of the

Table I. Binding of Sugars to Galactose and Ribose Receptors of *Salmonella typhimurium*

Saccharide	Dissociation Constant ($\underline{M}$)
Galactose Receptor	
Galactose	2×10^{-7}
Glucose	10^{-7}
Arabinose	4×10^{-5}
Lactose	6×10^{-4}
Fucose	6×10^{-3}
Methyl galactoside	Not detectable
Ribose	No binding
Allose	No binding
Ribose Receptor	
Ribose	3×10^{-7}
Allose	3×10^{-4}
1,5-anhydroribitol	Not detectable
Ribose-1-P	Not detectable
D-arabinose	Not detectable
2-deoxy-D-ribose	Not detectable
D-xylose	Not detectable
3-deoxy-D-ribose	Not detectable
3-o-methyl-D-ribose	Not detectable

galactose receptor of *Salmonella typhimurium* and the ribose receptor of the same species[15,16]. The ribose and galactose receptors in *E. coli* have been identified and have similar specificities[17,18]. Ribose receptors seem to bind only two chemicals of those tested so far, and in this case, allose binds 1,000 times less effectively than ribose. Galactose and glucose both bind very tightly to the galactose receptor and other sugars bind weakly or not at all. Thus the specificities of the receptors determine which compound will be detected at low concentrations, which require high concentrations, and which compounds will be ignored completely.

In the bacteria there are approximately 20 receptors for attractants and possibly almost as many for repellents and each receptor responds to several compounds[4]. Hence a wide spectrum of environmental stimuli can be processed by this single cell. Man shows a similarly wide spectrum of responses and also is incapable of detecting certain chemicals, sound frequencies, wave lengths of light, etc.

(*ii*) The intensity of the bacterial response varies enormously from receptor to receptor. The dissociation constant of the sugar or amino acid from the receptor is not in itself any indication of the behavioral response of the organism. It merely indicates the range of concentrations of the chemoeffector molecule over which the sensory system will be sensitive. For example, the response to serine of *Salmonella* is thousands of times greater than the response to ribose when each of their receptor molecules are half saturated.[4,6] The precise reason for this particular difference in quantitative response is partially understood. In the case of the galactose receptor it is known that there are ten times as many ribose receptors as galactose receptors per cell and the response to ribose is proportionately greater[19]. This suggests that the numbers of receptor molecules are a major factor in the intensity of the response. Such a conclusion is further supported by the finding that a number of the receptors are induced[3]. Obviously there will be no response in the mature bacterium unless the growth conditions lead to the induction of the particular receptor molecules.

In addition there appears to be a second factor relating to the efficiency in which the receptor transmits its signal further. The response to serine is so much higher than to ribose that it would require 10^7 receptor molecules per cell, if numbers alone count. Hence the response of a particular receptor must involve both the number of receptors and the efficiency of its interaction with the signalling system.

(*iii*) The receptors react by an induced conformational change[8] and in the case of the periplasmic protein do so by an induced association generated by the conformational change[19]. This was demonstrated by competition between two receptor molecules, the galactose and glucose receptor, both of which compete for a common signal component (cf. Figure 2). Such a competitive interaction of receptor molecules has been postulated in higher species[20]. It could be demonstrated in the bacterial case by virtue of the fact that the receptors could

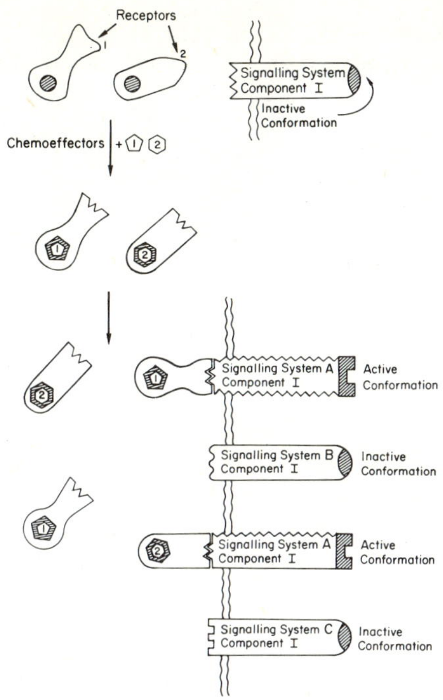

Fig. 2. Floating receptor model. Receptors are initially in conformations that are not attracted to component I, but are induced into new conformations by the chemoeffectors. As a result, individual chemoeffector-receptor complexes are induced to encounter and associate with the first component of the signalling system. If one binds there, it induces a conformation change, which activates the signalling system and begins a signal that can be amplified in a cascade process. If two receptor-chemoeffector complexes compete for the same site, one stimulus can diminish or completely block another.

be isolated in pure form and mutants were available to test the interactions between the receptors.

(*iv*) All the receptors tested so far have a dual function[21]. In some cases their interactions suggest that the protein which serves as a receptor can serve in two different pathways. For example, it has been shown that the galactose-binding protein can serve as the first component of the galactose-transport system and also as the first component of the galactose-chemotaxis system[22]. Since mutants are available which can selectively eliminate either transport or chemotaxis, it is clear that this common protein diverges in function at this initial stage.

In addition to this kind of dual use of a common protein, it appears that perturbation of the chemotactic sensing system can be obtained by other modifications which would appear not to be natural physiological functions. For example, a blue-light effect has been observed to generate tumbling[23]. This effect has the action spectrum of a flavin. Moreover, it was established that the blue-light effect perturbs the electron-transport system and is not observed in the absence of such electron transport. The fact that the bacteria have no physiological reason to respond to blue light and in fact require light intensities far above those encountered in the environment[24] strongly suggests that this is not a natural receptor. What seems obvious is that a perturbation of the electron transport system causes changes in the cellular milieu which is detected by the chemotaxis system. Similarly it has recently been shown that changes in the membrane potential are perceived in the same way that attractants and repellents are perceived by the biological system[25-27]. Furthermore, it has been shown that the MgCa-ATPase is a receptor for chemotaxis at concentrations of magnesium and calcium which are much higher than the organism is ever likely to encounter.[21]

These results all lead to a suggestion that the bacterial sensing system is performing a role perhaps not unlike that of pain and pleasure in higher organisms[7]. Just as pain and pleasure provide incentives to remove ourselves from toxic conditions, or to seek more advantageous surroundings, so the tumbling frequency of bacteria urges its migration to more favorable environments and away from dangerous ones. Both man and bacteria have evolved systems to provide surveillance of the general physiological health of the organism and to respond so as to optimize the proper functioning of all its component parts.

(v) The receptor competition mentioned above allows the sensing system to make a judgment when two different chemicals are providing the same function, in this case providing a carbon source. If the environment has large concentrations of one carbon source which is perfectly adequate for the needs of the bacterium, a gradient of a second carbon source might cause it to migrate away from the first one. Thus it would leave a completely adequate and efficient source of carbon to respond to a lower concentration of a compound which duplicates its already adequate nutrient needs. By having the receptors compete with each other for a common signalling system the bacterial system prevents this kind of inefficiency.

In a larger sense, however, this competition as shown in Figure 2 provides a second general function. Competition between two sugars at a single active site requires common structural features. If, however, two compounds which provide the same values to the organism have quite different structural features, it would be difficult to arrange such a judgmental kind of competition. By having the receptors rather than the chemoeffectors compete with each other for a site, any two compounds can be designed to compete. Each of the chemoeffectors can bind to highly specific sites on their individualistic proteins. The protein has a common region which interacts with the signal components and hence the protein serves an adapter function to create competition which would be difficult to devise by both sugars binding to a single active site. That this system may not be unique to bacteria is indicated by the findings that noradrenalin and acetylcholine compete in some tissues and not in others. Clearly the device of having common signalling components in certain cells and not in others could serve this function in any cell.

This finding also suggests a caution. In many cases overall behavioral properties are used to delineate the properties of a receptor. For example, odor is frequently delineated by behavioral responses of a wide variety of rather similar chemical compounds which are deduced to bind to a common receptor. In many cases these chemical compounds are no more different or no more similar than ribose is to galactose. The deduction that the chemoeffectors bind to a common protein is clearly erroneous in the case of the ribose and galactose responses in chemotaxis and may be equally misleading in the responses of hormones and of odoriferous compounds in higher species.

(*vi*) The receptor response is integrated by a temporal sensing system, one type of which is shown very schematically in Figure 3. The response regulator is stimulated by the chemoeffector receptor complex and undergoes a transient increase followed by an adaptation response which returns it to its former level. The behavioral response is controlled not only by the signal generated, but by the time-decay characteristic of the response regulator. These various stimuli cause increases and decreases in the rate constants of the enzymes leading to the formation and degradation of the response regulator (in the chemotaxis case, a

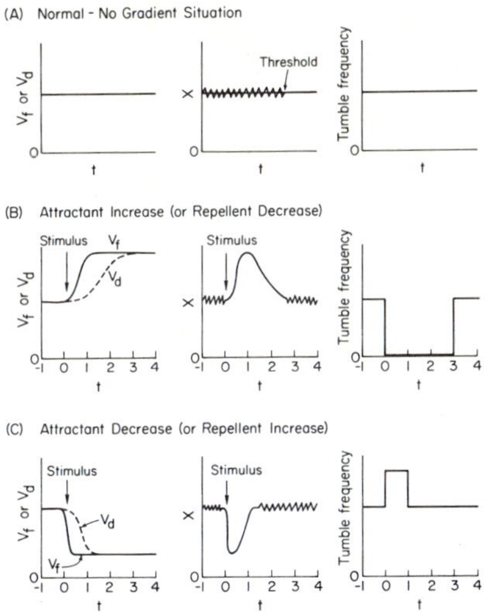

Fig. 3. Response of wild-type bacterium to attractants and repellents, as explained by a response (tumble) regulator model. The variation over time for the enzyme activities, the level of tumble regulator and the tumbling frequency is shown for three situations. (A) In absence of a gradient, $V_f = V_d$ are constant over time, and X (the tumble regulator) concentration varies around the threshold in a Poissonian manner. The tumble frequency remains essentially constant. (B) Sudden increase in attractant increases the rate of V_f faster than V_d, leading to a transient increase in concentration of X and a transient decrease in tumbling frequency. Repellent decrease gives the same effect. (C) Sudden decrease in repellent decreases the rate of V_f more rapidly than V_d, leading to a decrease in concentration of X and a transient increase in tumbling frequency.

regulator of tumbling frequency). Hence the pool level of the chemical or membrane potential which serves as the response regulator provides a convenient method for integrating a variety of signals.

The response regulator levels can be altered by genetic mutation of any of the enzymes of the central processing system, by nutrients which alter the level of compounds affecting this central processing system, and by growth conditions which induce the presence or absence of certain key proteins. It seems quite simple to extend this principle to higher species and to allow one cell to have one type of response regulator which then generates a signal, stimulating a second cell, which may have a similar or different response regulator. Each cell may have a different repertoire of receptors or of enzymes processing these receptors so that its time-decay characteristic is sufficiently different to provide different kinds of judgments of multiple stimuli. In this case there is a unity in the basic mechanism of a response and at the same time a mechanism for diversity by simply varying numbers and specificities of individual proteins.

Unity and Diversity in Signalling Systems

Unity and diversity is apparent in the comparison of receptor interaction in these biological species and what is known of higher systems. Firstly, the response of the cell to stimuli including an initial response and an adaptation phenomenon appears common to most signalling systems. Secondly, the responsiveness of an individual cell is determined by the repertoire of receptors on its surface. The specificity of the receptor, their numbers and their affinities for the remaining components of the signalling system would appear to be a very common feature of each cell. Thirdly, an induced conformational change which triggers the response, now demonstrated in a bacterial system, appears to be common to all receptors. Fourthly, the concept of an induced association of receptors has been indicated in the bacterial and mammalian systems. However, some receptor-signal component systems are associated before binding such as the MgCa-ATPase system. Thus both options are available. Fifthly, the output of the cell, which could be control of flagella tumbling, an action potential, or the release of a chemical, can in each case be the accumulative effect of a central processing system involving enzymes which respond to the interaction of a number of receptors. If such a response regulator does control individual cells to provide an integrated response to a variety of stimuli, this would be a most interesting unity in all signalling systems.

The potential of the system for diversity is also impressive. In bacteria it has been shown that a variety of receptors exists in an individual cell but different species have different repertoires of receptors. Similarly, cells in higher systems could utilize differences in the variety of receptors, their numbers and their affinities for the second components of the signalling system as a means of obtaining cellular diversity. Secondly, the adapter function of having two receptors compete for the same component of a signalling system can easily be modified in different cellular types. Not only can different chemicals be made to compete or enhance the activity of each other in different cells, but it is quite simple to eliminate entirely the presence of certain receptors in certain cells or to allow their presence only if induced by appropriate growth conditions, hormones, or peptides in the circulating medium. Thirdly, the time characteristics of different cells can be quite different with only minor modification either in the number of enzyme molecules or in their types. Thus cells responding to neurotransmitters could be designed with adaptive responses which adapt in a very short time in order to provide an extremely rapid and localized signal from one cell to the next. Enkephalin-type molecules could well be of intermediate time spans designed in such a way that they spread over a moderately larger network of cells because they exist for longer periods in the medium. Finally, hormones could operate on a longer time schedule from either of the former two classes so that they traverse far greater distances in the organism. Pheromones might be the ultimate in this classification since they extend information between organisms and hence are changed by dilution. Further, the response times within the cell can also be adapted to the particular needs of that cell.

The diversity is created in these cases by the time constants for the survival of the chemoeffector molecules in the milieu and the time constants of the processing system within the cell. Fundamentally similar mechanisms with different rate constants provide an enormous diversity. In the case of higher species there have been indications that receptors are themselves modified or translocated to less accessible regions on reaction. No such engulfing or degradation of receptors has yet been established in the bacterial system. The possibility exists that this represents an additional mechanism in the higher species, a mechanism which has not yet been uncovered in the bacteria. Further some types of adaptation which have been found in bacteria can explain a number of mammalian phenomena without invoking covalent destruction of receptors.

In conclusion it appears that both in the bacteria and in higher species, receptors trigger responses through common principles which show a unity in all species and a variety of differing rate constants, affinities, and receptor quantities which allow diversities in the behavioral responses.

This work was supported by a United States Public Health Service Grant #AM 09765 and a National Science Foundation Grant #BMS 71-0133A03.

References

1. Englelmann, T. W. (1881) *Pfluegers Arch. Gesamte Physiol. Menschen Tiere* **25**, 285–292.
2. Pfeffer, W. (1883) *Ber. Dtsch. Bot. Ges.* **1**, 524–533.
3. Adler, J. (1975) *Annu. Rev. Biochem.* **44**, 341–356.
4. Adler, J. (1969) *Science* **166**, 1588–1597.
5. Berg, H. C. (1975) *Annu. Rev. Biophys. Bioeng.* **4**, 119–136.

6. Koshland, D. E., Jr. (1977) in *Advances in Neurochemistry*, vol. 2, Agranoff, B. W. & Aprison, M. H., Ed., Plenum Press, pp. 277–341.

7. Koshland, D. E., Jr. (1977) *Science* **196**, 1055–1063.

8. Zukin, R. S., Hartig, P. R. & Koshland, D. E., Jr. (1977) *Proc. Nat. Acad. Sci. USA* **74**, 1932–1936.

9. Warrick, H., Taylor, B. & Koshland, D. E., Jr. (1977) *J. Bacteriol.* **130**, 223–231.

10. Armstrong, J. B. & Adler, J. (1969) *J. Bacteriol.* **97**, 156–161.

11. Parkinson, J. F. (1974) *Nature (London)* **252**, 317–319.

12. Silverman, M., Matsumura, P., Hilmen, M. & Simon, M. (1977) *J. Bacteriol.* **130**, 877–887.

13. Macnab, R. M. & Koshland, D. E., Jr. (1972) *Proc. Nat. Acad. Sci. USA* **69**, 2509–2512.

14. Berg, H. C. & Brown, D. A. (1972) *Nature (London)* **239**, 500–504.

15. Zukin, R., Strange, P., Heavey, L. & Koshland, D. E., Jr. (1977) *Biochemistry* **16**, 381–386.

16. Aksamit, R. & Koshland, D. E., Jr. (1974) *Biochemistry* **13**, 4473–4478.

17. Anraku, Y. (1968) *J. Biol. Chem.* **243**, 3116–3128.

18. Hazelbauer, G. L. & Adler, J. (1971) *Nature (London) New Biol.* **230**, 101–104.

19. Strange, P. G. & Koshland, D. E., Jr. (1976) *Proc. Nat. Acad. Sci. USA* **73**, 762–766.

20. Cuatrecasas, P. (1974) *Annu. Rev. Biochem.* **43**, 169–214.

21. Zukin, R. & Koshland, D. E., Jr. (1976) *Science* **193**, 405–408.

22. Ordal, G. W. & Adler, J. (1974) *J. Bacteriol.* **117**, 517–526.

23. Macnab, R. & Koshland, D. E., Jr. (1974) *J. Mol. Biol.* **84**, 399–406.

24. Taylor, B. L. & Koshland, D. E., Jr. (1975) *J. Bacteriol.* **123**, 557–569.

25. Ordal, G. W. & Goldman, D. J. (1975) *Science* **189**, 802–804.

26. Jong, M. H., Van der Drift, C. & Vogels, G. D. (1976) *Arch. Microbiol.* **111**, 7–11.

27. Miller, J. B. & Koshland, D. E., Jr. (1977) *Proc. Nat. Acad. Sci. USA* in press.

EVOLUTION OF STRUCTURE AND FUNCTION IN NATURAL PEPTIDES

HUGH D. NIALL, *Howard Florey Institute, University of Melbourne, Parkville, Australia 3052*

Naturally occurring peptides, small and large, subserve an impressive array of biological functions. Perhaps their major role in multicellular organisms is to act as chemical messengers that are released from one cell or cell type and diffuse or are transported to other cells to modulate their activity. How, and whence, this communications system has evolved is of considerable interest. A better understanding of the evolutionary process could also be of practical value in allowing us to recognize patterns in the organization of hormonal messengers and possibly to discover new regulatory peptides or new functions for those we already know.

Requirements for Information Transfer

Let us first consider what are the requirements for a cell-cell communications system. The dimension of *time* is involved, since cells must be able to regulate the behavior of other cells both over the short term (e.g., in minute-by-minute adjustments of parameters such as blood glucose or blood electrolyte levels) and over the long term (e.g., in the coordination of growth and reproduction). Likewise the dimension of *space* is involved, since cells must be able to influence both their nearest neighbors and cells that are a meter or more distant. In fact, an outline is starting to emerge of three different functional modes of communication.

1) Cells may direct a particular chemical message only to one other cell or to a very limited number. The clearest example of this is neurotransmission.

2) Cells may release substances which have local regulatory effects in the same or neighboring tissues but which do not, by and large, influence distant tissues. Prostaglandins in the reproductive system and the incompletely characterized substances involved in wound-healing, seem to act in this way.

3) Cells may release substances into the blood stream to act at distant sites – i.e., the so-called classical hormones.

The role of peptides as classical hormones has long been known. More recently it has been appreciated that there is a diverse array of peptides acting in the central nervous system (e.g., thyrotropin-releasing factor (TRF), luteinizing hormone-releasing factor (LRF), somatostatin, substance P, and opiate related peptides). It is thought that they act as local tissue regulators and in some instances probably also as neurotransmitters. Control of sleep, memory, perceptions, emotions and appetitive drives are possible functions for these peptides, in what could, by analogy, be called the "Central Endocrine System."

Apart from the temporal and spacial requirements of a cell-cell communications system, there are certain constraints and requirements of the actual molecules used for transmission of the chemical signals. I have reviewed these requirements elsewhere.[1] Briefly, they are high information content, thermodynamic stability, ease of replication in large numbers with a low copying error, ability to cross at least one cell membrane, and destructibility. At the other end of the system, the molecules receiving the signals (receptors) have their own set of requirements. One of these is that their information content (reflected in the complexity of receptor structure) has to be greater than that of the hormone. The latter has only to carry a message, while the receptor must not only receive this message (requiring a complementary fit with the hormone) but also pass it on through a second interaction with other target cell components. It is not surprising, therefore, to find that both steroid and peptide hormone receptors are complex, high-molecular weight entities with allosteric properties.

Without analyzing in further detail the required characteristics of hormone and receptor molecules, it can be seen that polypeptides are well-suited to both roles. It is equally clear that receptors as well as hormones are subject to evolutionary change through mutations in the appropriate gene or genes.

Mechanisms of Hormone Evolution

It is apparent from the above considerations that one cannot simply consider the evolution of peptide hormones solely in terms of amino-acid sequences of the known hormones. What has evolved is a complex system involving biosynthesis of both hormone and receptor, their intracellular processing and their transport to the site of action (Fig. 1). The whole process must be under the control of multiple genes, all of which are subject to mutation with a consequent alteration in the system. Since multiple recognition events are involved (see asterisks in Fig. 1) the polypeptides that are hormone or receptor precursors must contain all the structural features necessary for recognition by the cellular modules responsible for transport, activation, storage, etc. Those portions of a peptide hormone molecule that are apparently not needed for activation of target cells (e.g., the C-terminal regions of parathyroid hormone and ACTH) may be required for interactions during biosynthesis or secretion. In the case of the parathyroid hormone, the "biologically inactive" region comprising residues 35-84 is relatively highly conserved when the sequences of the bovine, porcine and human hormones are compared. This in itself is presumptive evidence for an important role for this region, since there is reason to believe that non-functional or less critical regions of peptide sequence fairly rapidly accumulate point mutations in the divergence of species. Thus the C-peptide of insulin and the fibrinopeptides show substantial sequence differences between human and bovine structures, presumably because their actual presence and perhaps chain length are important, but the nature of the amino acids comprising them is not at all

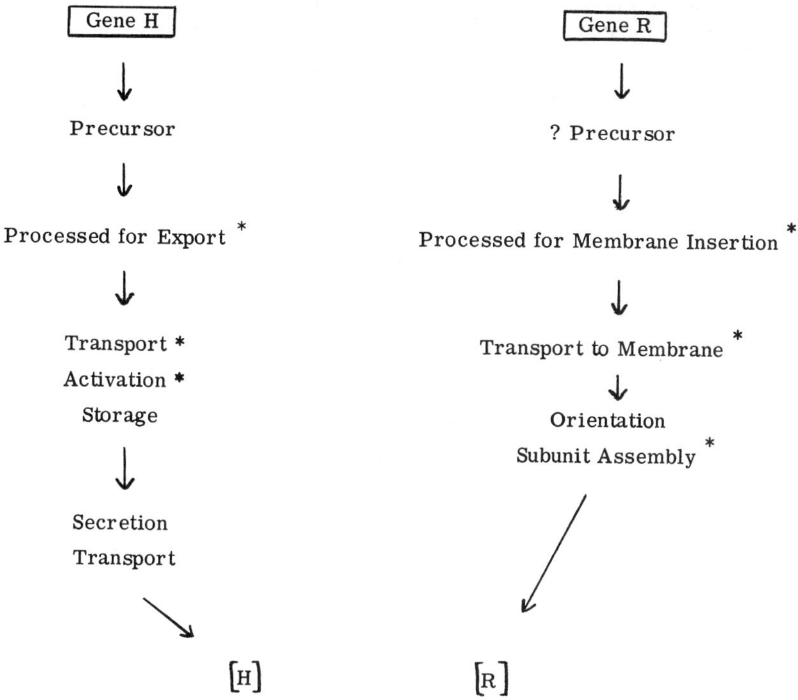

Fig. 1. Events in hormone action. * Denotes a step where there must be recognition of a specific structural feature of the hormone or receptor molecule or their precursors.

critical. From this kind of indirect argument, it seems likely that the C-terminal region of the parathyroid hormone either is important during biosynthesis or secretion, or conceivably has its own separate hormonal function, not as yet discovered.

Origin of Peptide Hormones

From what we do know of peptide sequences of hormones, it seems that most of the contemporary peptide hormones evolved from a limited set of ancestral polypeptides. This argument is based on the extensive sequence homology not only between hormones with related functions (e.g., prolactin and placental lactogen) but also between hormones with no known shared function (e.g., placental lactogen and secretin). The main mechanism by which they arose is also fairly clear – i.e., by gene duplication. This is an advantageous method, since the redundant copy of the gene is free to undergo mutation while the original gene continues to regulate biosynthesis of the original hormone. When a new and useful hormone appears, natural selection will ensure survival of the mutated gene. Presumably new hormone receptors also evolve mainly by gene duplication, though we have as yet no structural information to confirm this.

It is not clear at present which ancestral polypeptides have evolved into current regulatory peptides and hormones. Adelson[2] has proposed, on the basis of what must be regarded as doubtful or marginally significant sequence homologies, that peptide hormones have evolved from primitive digestive enzymes. This would make them the products of endodermal cells. Pearse and Pollak,[3] on the other hand, have very persuasively argued for an origin of many peptide hormones from neuroectodermal cells that have migrated from the neural crest into tissues that then develop an endocrine function. These cells can be recognized by specific cytochemical staining techniques. They are termed "APUD" cells, a shorthand name based on their possession of amine precursors and decarboxylase activity. The thyroid C cells that secrete calcitonin are the best example of this cell type. Their neuroectodermal origin has been most convincingly demonstrated, providing strong support for the theory of Pearse and Pollak.[3]

Probably more important than the cell of origin of peptide hormones is the common mode of biosynthesis that seems to be shared by all secreted proteins. The initial gene product has an N-terminal sequence of 15-30 residues that are largely hydrophobic in character. (See Strauss et al.[4] for further details and references.) This region (known as the "prepro" sequence) attaches to the membrane of the rough endoplasmic reticulum as translation is taking place on adjacent polysomes. In some manner not yet understood, the growing peptide chain is then "pulled" through the membrane. The hydrophobic extension is then cleaved off by one or more proteases and the functional protein or proprotein is transported via the Golgi region to be packaged, activated if necessary, and secreted as required. The whole process is directed by the "prepro" sequence which must act as a code indicating "export from cell." Membrane proteins (such as hormone receptors) and proteins used for the cell's internal economy probably have their own handling codes, but these are not yet known (Fig. 2).

This coding system is very relevant to the evolution of peptide hormones. Any new hormone that evolves has to be secreted in order to have any influence on the course of events. This makes it quite unlikely that hormones would evolve from proteins that are not normally secreted from the cell, since they would almost certainly lack the appropriate structural features required for secretion. This perspective makes it much more understandable that gene duplication has been such a fertile provider of new hormones, since any new gene product resulting from a mutation in the redundant gene has already solved the problems of packaging and secretion. These hormones have evolved from preexisting hormones, and, more distantly, from other secreted proteins. Whether these ancestral molecules were digestive enzymes,[2] or neurosecretory products[3] is perhaps of less relevance. Very possibly both events have occurred.

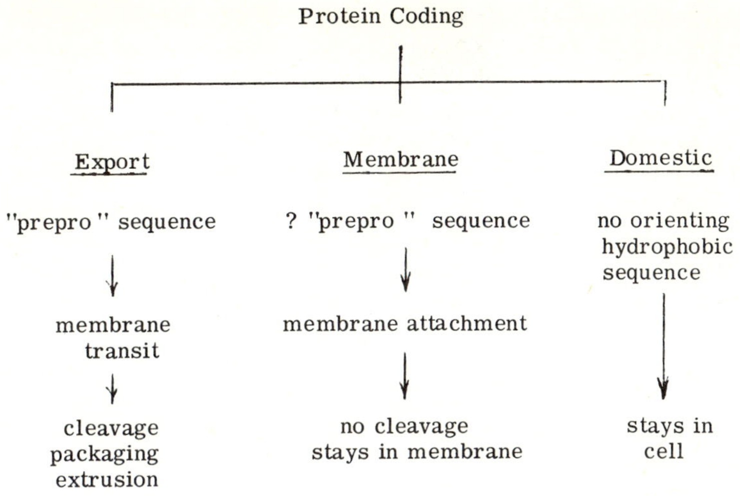

Fig. 2. Theoretical model for structural coding of proteins during biosynthesis.

Chemistry of Relaxin: Homology with Insulin

Relaxin is a peptide hormone of ovarian origin. Originally described by Hisaw,[5] it has the property of widening the interpubic ligament during pregnancy in mammals through its actions on connective tissue. Relaxin preparations also have been shown to inhibit spontaneous contractions in the uterus. Thus, it has an important function in preparing the birth canal for parturition and possibly also in influencing the actual timing of that event, since serum relaxin levels have been shown to fall sharply just prior to delivery.[6]

Recently, in collaboration with Dr. Bryant-Greenwood and her group at the University of Hawaii, we have determined the amino-acid sequence of porcine relaxin, using material purified according to the procedure of Sherwood and O'Byrne.[7] Our proposed sequence (shown in Fig. 3) has been recently reported;[8] another group has independently proposed a relaxin sequence differing in some respects from our own.[9]

A-chain

H-Arg-Met-Thr-Leu-Ser-Glu-Lys-Cys-Cys-Gln-Val-Gly-Cys-Ile-Arg-Lys-Asp-Ile-Ala-Arg-Leu-Cys-OH

B-chain

PCA-Ser-Thr-Asn-Asp-Phe-Ile-Lys-Ala-Cys-Gly-Arg-Glu-Leu-Val-Arg-Leu-Trp-Val-Glu-Ile-Cys-Gly-Ser-Val-Ser-Thr-Trp-Gly-Arg-OH

Fig. 3. The amino-acid sequence of porcine relaxin.

A most unexpected and interesting finding was a clear structural homology between relaxin and insulin. Relaxin, like insulin, consists of A- and B-chains linked in an identical fashion by 2 disulfide bridges. The relaxin A-chain has an additional intrachain bridge linking residues 8 and 13. All six half-cystines can be exactly aligned with those in insulin. There is conservation of virtually all the residues of the hydrophobic core of insulin; these are known from crystallographic studies. In insulin these are the half-cystine residues found at A-6 and A-11, together with A-2 isoleucine, A-16 leucine, with B-11 and B-15 leucines. At several other positions there is conservation of the hydrophilic or hydrophobic nature of the residue. These findings[8] suggest that the folding of the relaxin chains may occur in a similar way to that in insulin, and that the overall tertiary structure of the two hormones may be quite similar. In fact our collaborators on the X-ray crystallography of relaxin, Drs. Neil Isaacs and Guy Dodson of Leeds University, England, have built a model of the relaxin structure by fitting the relaxin side chains to the backbone of the 2-Zn insulin structure. This was achieved with almost no distortion. Of course, this is purely a hypothetical model, and some revision will almost certainly be needed in the light of accurate crystallographic data.

A detailed comparison between the relaxin and insulin structures is outside the scope of the present discussion. However, what can be said is that relaxin and insulin are closely related in an evolutionary sense. There is conservation of primary structure (i.e., amino-acid sequence homology), secondary structure (e.g., the B-chain α-helix), disulfide pattern and tertiary folding. Since we have preliminary evidence for a precursor to relaxin of the approximate size of proinsulin, it seems that the biosynthetic mechanisms that lead to correct folding are also conserved. The disposition and nature of surface residues however is quite different, and as far as we know at present there is no immunological or biological cross-reactivity between the two hormones.

One possible mode of evolution of relaxin and related peptides is shown in Fig. 4. One can postulate that a series of gene duplications took place in a gene coding for an ancestral peptide, perhaps one that had growth-regulatory properties. Gene products appeared corresponding to proinsulin, prorelaxin, and probably also to peptides with nonsuppressible insulin-like activity; i.e., NSILA.[10] Nerve growth factor (not shown in Fig. 4) is probably derived from a further gene duplication.[11]

With further evolution, proteolytic cleavage mechanisms appeared to generate the relaxin and insulin structures. Relaxin, like guinea pig, coypu, and hagfish insulins, lacks the Zn coordinating histidine residue at position B-10. Possibly a later mutation gave rise to this property in most insulins.

An alternate possibility which must at least be considered is that the similarities between relaxin and insulin are not based upon a common genetic ancestor but have evolved by a convergent evolutionary process. There may be only a limited number of options for carrying out successful chain folding to give a

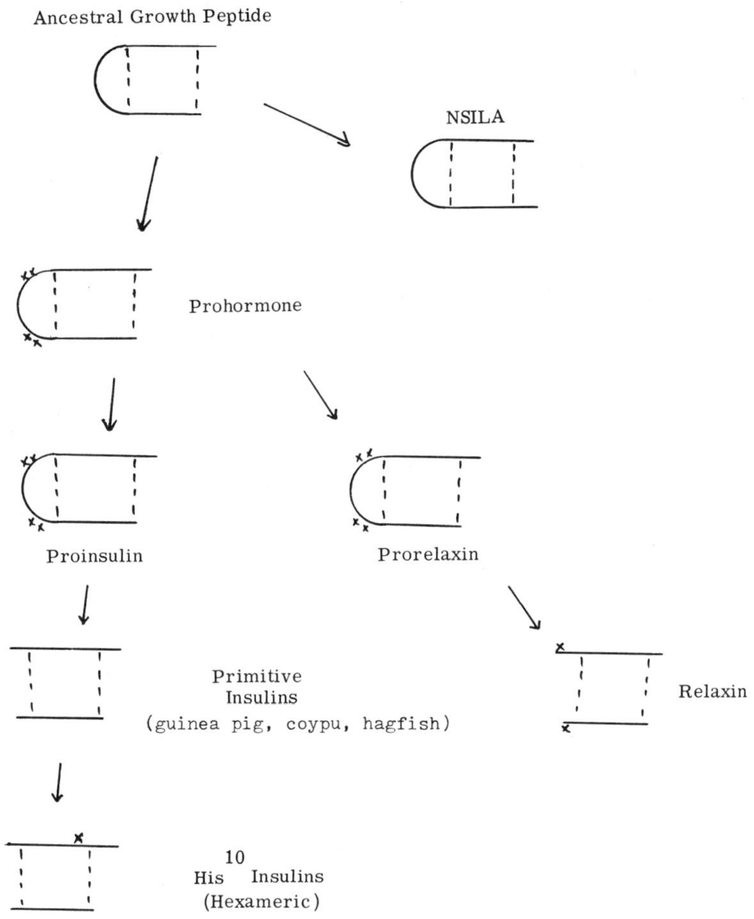

Fig. 4. Hypothesis for evolution of relaxin and related peptides.

structure of this sort. The use of an orienting C peptide sequence and a particular positioning of disulfide bridges may be a solution to a particular problem that has been arrived at through two independent evolutionary routes. However, on balance, the first hypothesis looks, to me at least, more attractive.

It will be of great interest to determine the sequence of relaxins from other species. Also of interest are experiments currently underway in our laboratory to form relaxin-insulin hybrids between the natural chains of each parent molecule. An intriguing question is whether the relaxin A-chain, for example, can be made to couple to the insulin B-chain. If this can be achieved, even in low yield, it will be possible to study the immunological and biological properties of the hybrid. In passing I would like to propose the names "insulaxin" (for insulin A- plus relaxin B-chains) and "relaxulin" (relaxin A- plus insulin B-chains) for these rather enticing hybrids.

Future Prospects

As more polypeptide structures become known, there will certainly be discovered further surprising relationships such as the relaxin-insulin story I have outlined or the fascinating coevolution of peptides with ACTH, MSH and endorphin activities, as most elegantly narrated by Guillemin et al.[12] A problem in the study of evolution through analysis of contemporary molecules is that we are, perforce, dealing only with evolutionary success stories. The failures have vanished from the scene. If we are to understand fully the way in which evolutionary forces work, it will be necessary to look at the molecular lesions in hormones or receptors produced by unfavorable mutations. Some of these are accessible through the study of clinical syndromes. However, there is clearly a need for other model systems. The synthesis of multiple analogs is one approach – the main one used so far. However, it is quite empirical, and very costly in time and effort. It seems likely that in the future much greater use will be made of computer-graphic analysis. This will allow the effect of hypothetical mutational events (amino-acid deletions, additions or substitutions for example) to be superimposed on the known three-dimensional structure of the polypeptide. With improvements in our ability to predict secondary and tertiary structure of peptides from their sequence, it should be possible to see in a very graphic way the effect on function of localized changes of structure, without the need for actual synthesis of the molecule. Of course, work along these lines has been progressing, though quite slowly, for some years already.

Another future prospect is the insertion of genes coding for polypeptide hormones into bacteria with subsequence biosynthesis of hormones or hormone precursors. The actual techniques of insertion have already been worked out. To induce transcription and translation of the inserted gene is another matter, however, and even if a preprohormone molecule is synthesized, the bacteria will lack the cellular apparatus for its proteolytic activation and secretion. Nevertheless, these problems will in time also be solved. It will then be possible to subject the bacteria to mutagenic agents, generating a whole series of altered hormones as hormone precursors. This should allow us to appreciate more clearly the kind of constraints that have operated during evolutionary history.

References

1. Niall, H. D. (1976) in *Protein Structure and Evolution*, Fox, J. L., Dehl, Z. and Blazij, A., Eds., Marcel Dekker, New York, pp. 429–438.

2. Adelson, J. W. (1971) *Nature* **229**, 321–324.

3. Pearse, A. G. E. & Pollak, J. M. (1972) in *Endocrinology 1971, Proceedings of the 3rd International Symposium*, Taylor, S., Ed., Heinemann, London, pp. 145–152.

4. Strauss, A. W., Donohue, A. M., Bennett, C. D., Rodkey, J. A. & Alberts, A. W. (1977) in *Peptides: Proceedings of the 5th American Peptide Symposium,* Goodman, M. & Meienhofer, J., Eds., John Wiley & Sons, Inc., New York, 59–62.

5. Hisaw, F. L. (1926) *Proc. Soc. Exp. Biol. Med.* **23**, 661–663.

6. Sherwood, O. D., Chang, G., Bevier, G. W. & Dziuk, P. J. (1975) *Endocrinology* **97**, 834–837.

7. Sherwood, O. D. & O'Byrne, E. M. (1974) *Arch. Biochem. Biophys.* **160**, 185–196.

8. James, R., Niall, H., Kwok, S. & Bryant-Greenwood, G. (1977) *Nature* **267**, 544–546.

9. Schwabe, C., McDonald, J. K. & Steinetz, B. G. (1977) *Biochem. Biophys. Res. Comm.* **75**, 503–510.

10. Rinderknecht, E. & Humbel, R. E. (1976) *Proc. Nat. Acad. Sci. USA* **73**, 4379–4381.

11. Frazier, W. A., Hogue-Angeletti, R. & Bradshaw, R. A. (1972) *Science* **176**, 482–488.

12. Guillemin, R., Bloom, F., Rossier, J., Minick, S., Henriksen, S., Burgus, R. & Ling, N. (1977) in *Peptides: Proceedings of the 5th American Peptide Symposium,* Goodman, M. & Meienhofer, J., Eds., John Wiley & Sons, Inc., New York, 63–76.

STRUCTURE-FUNCTION RELATIONSHIPS OF INSULINS MODIFIED IN THE A1-REGION

H. -J. FRIESEN, D. BRANDENBURG, C. DIACONESCU, H. -G. GATTNER, V. K. NAITHANI, J. NOWAK and H. ZAHN;
Deutsches Wollforschungsinsitut a.d. TH Aachen, Federal Republic of Germany, S. DOCKERILL, S. P. WOOD and T. L. BLUNDELL;
Birkbeck College, London University, London WC 1E 7HX, United Kingdom

Synthesis, CD, X-ray, *in vivo* and *in vitro* biological properties of a series of Gly^{A1} substituted insulins have been reported.[1,2] With increasing size of uncharged substituents (acetyl, Boc, dimethylformylthiazolidinecarbonyl) *in vitro* biological activities and receptor binding decreased, while the native conformation and ability to dimerize were disturbed.[2] However, the relative importances of size, flexibility, polarity and especially charge of the substituent for conformation and biological properties are still unclear. We report herein the synthesis of several bovine insulin derivatives covering these variables and their crystallization and preliminary X-ray studies and *in vitro* bioactivities.

Preparations

Derivatives were characterized by cellulose acetate electrophoresis (pH 2.2, pH 8.6), paper electrophoresis (pH 2.2) after sulfitolysis or performic acid oxidation[3], UV spectra, amino-acid analysis[4] and end-group determination.[5] They are considered to be pure within relevant limits. Experimental conditions for pH 3[3,6], pH 7.3[6,7], pH 8.1[6], pH 8.4[3], pH 8.6[6] and pH 8.8[6] ion exchange chromatography have been described elsewhere. A1-(Boc-Gly)-, A1-(Boc-Trp)-, A1-Gly- and A1-Trp-insulin were prepared by reacting native insulin with activated esters (-ONp, -ONSu) in DMSO. Subsequent purification by pH 3 and pH 8.4 ion exchange chromatography yielded the Boc-Gly- and Boc-Trp-derivatives. After TFA treatment the Gly- and Trp-derivative were obtained.

Preparation of additional derivatives (acetyl-, palmitoyl-, Tfa-, Boc-, citraconyl-(Ct), succinyl-insulin) from native insulin is outlined in Fig. 1. Preparation of derivatives via B1-Msc-insulin is outlined in Fig. 2. Preparation of A1-Mtc-insulin via B1,B29-Boc_2-insulin was not attempted as the Mtc-group (Mtc = methylthioethyloxycarbonyl) is sensitive to TFA (A. Hubbuch & H. -J. Friesen, unpubl.). Several derivatives were synthesized via B1,B29-Boc_2-insulin (Fig. 3) which was prepared from A1-Tfa-insulin in gram amounts and in pure form.[6] In coupling neutral amino acids or Glu, unreacted A1 amino groups were trifluoroacetylated

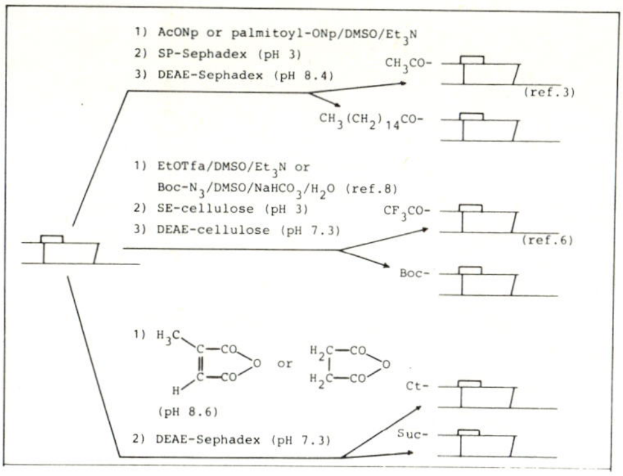

Fig. 1. Preparation of Gly^{A1}-substituted insulins from native insulin.

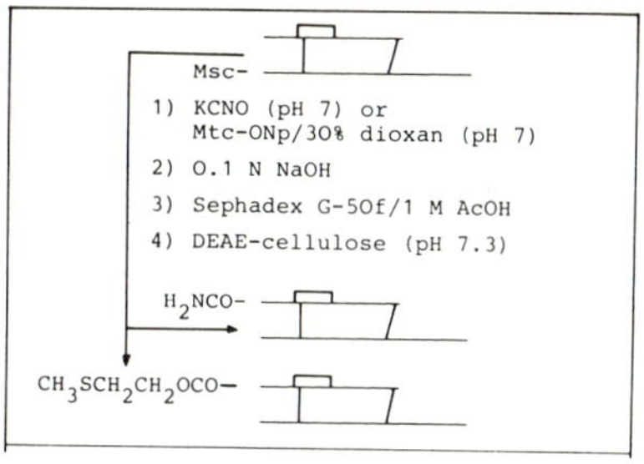

Fig. 2. Preparation of Gly^{A1}-substituted insulins via B1-Msc-insulin conditions for carbamylation similar to Ref. 9; prep. of Mtc-ONp, Ref. 11; prep. of B1-Msc-insulin and removal of Msc-groups, Ref. 10.

to enable separation of insulin not aminoacylated at A1 and derivative after removal of Boc-groups (no charge difference otherwise except Glu). The A1 amino group was converted to a guanido function by reaction with guanyl-dimethylpyrazole[14] similar to Bodanszky's procedure for the preparation of Arg^{8}-vasopressin.[15] After TFA-treatment, products were purified by preparative isoelectric focusing[12] (PIEF) in 6 *M* urea or ion exchange chromatography. Arg was coupled at pH 10–10.5 by reaction with NCA-Arg-hydrobromide.[16]

Crystallization

Crystallization was attempted under conditions for rhombohedral 2 Zn crystals[17] except for A1-Ct-insulin, which would not withstand the acidic pH during crystallization (Ct = citraconyl). Only Lys-Arg-, Arg-Arg- and Arg-Arg-Arg-insulin did not form crystals in temperature-dependent experiments.[18] Arg-, Lys- and guanido-insulin showed very little temperature dependence of solubility near the turbidity point and formed small crystals. Acetyl-, Tfa- and

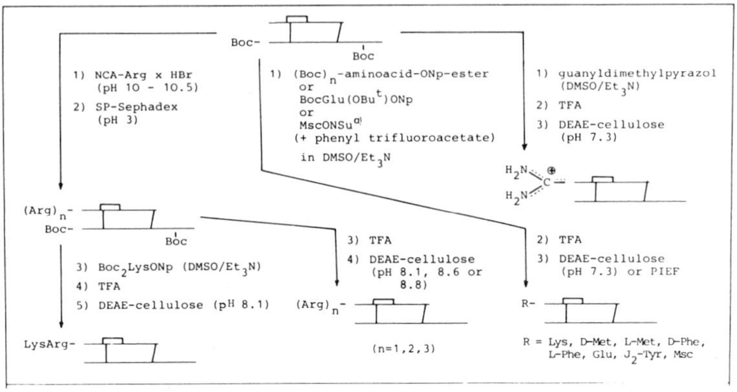

[a]prep. of Msc-ONSu, Ref. 13.

Fig. 3. Preparation of Gly^{A1}-substituted insulins via B1,B29-Boc$_2$-insulin reaction conditions, Ref. 6 (detailed publ. in prep.).

carbamoyl-insulin formed small crystals, too, as a consequence of low solubility as with bovine insulin.[2] D-Phe, Phe, D-Met-, Met-, Mtc-, Boc- and especially Glu-, Pro-, Suc-, Msc- and biotinyl-insulin[19] showed strong temperature dependence of solubility. We suspect, that polar or negatively charged substituents are favorable for obtaining large crystals.

X-ray Studies

Diffraction patterns of D-Phe-, L-Met-, Glu-, Lys-, Boc-, guanido- and acetyl-insulin were similar. Crystals were isomorphous with rhombohedral 2 Zn insulin (spacegroup R3). Patterns of D-Phe-, Met-, Msc-, Glu- and Boc-insulin were difficult to distinguish by eye. Intensity changes in the guanido- are even less pronounced and resemble that of acetyl-insulin. These preliminary data, as well as a 3Å difference Fourier map of the D-Phe-insulin, indicate conformations similar to the Boc-insulin[2] for the other derivatives investigated.

Influence of Size, Flexibility, Polarity and Charge on *In Vitro* Activity

In vitro activities were determined in the free fat cell assay.[20,21] Values are shown in Fig. 4. In the series of neutral, nonflexible substituents activity decreases with increasing size of the substituent (acetyl – Boc, 37.5 – 19%; compare Ref. 2). Aminoacyl-derivatives with flexible side chain show the activity to be dependent on charge, rather than size of the adduct (Arg, Lys: 40–45%; Glu: 22%). Aminoacyl-derivatives with a neutral side chain show little correlation of the size of substituent or conformation at the α-C atom with activity (Trp and diiodo-Tyr have side chains with restricted flexibility compared to the others).

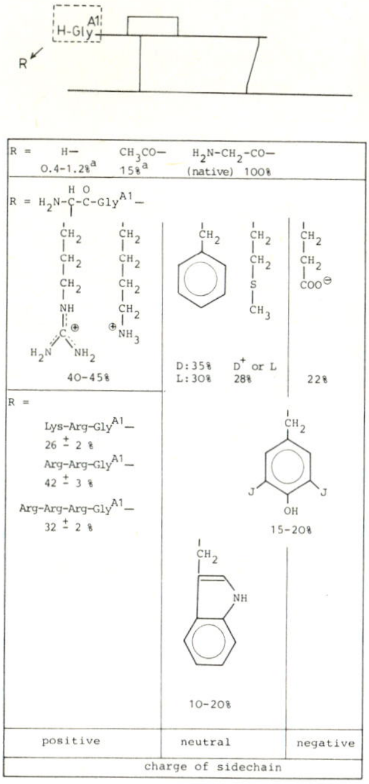

Fig. 4a. *In vitro* activities of des-Gly^{A1}-, des-amino^{A1}-insulin and Gly^{A1}-aminoacylated insulins carrying free functional groups at the substituent.
$^{+}$activity calculated from a preparation containing 10% native insulin assuming an additive effect for derivative and native.
[a]values taken from Ref. 21.

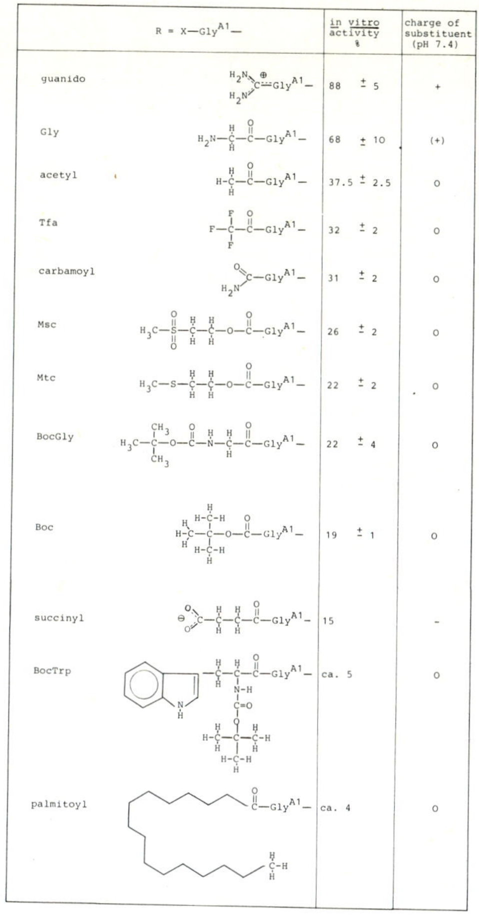

Fig. 4b. *In vitro* activities of Gly^{A1}-substituted insulins carrying Gly, blocked or no amino groups at the N-terminus of the A-chain.
[a]values taken from Ref. 21.
Results of one or more preparations and one or more tests (confidence limits, n⩾9) shown

Further addition of positive charges (Lys-Arg-, Arg-Arg-Arg-insulin) seems to have a relatively favorable effect on activity. The isosteric A1-guanido- and A1-carbamoyl-insulin clearly show the importance of a positive charge for activity (88 and 31%), as do the Suc- and Ct-insulin in comparison with these (both ca. 15%).

It is unclear whether large hydrophobic substituents with flexible or fixed structure (palmitoyl, Boc-Trp) influence activity by direct steric hindrance and/or secondary effects (disturbance of hormone conformation essential for activity).

The authors thank the "Deutsche Forschungsgemeinschaft" (AZ Za 5/28, SFB 113), "Bundesministerium für Forschung und Technologie," "Land NRW" and the Science Research Council for financial support and EMBO for a short term fellowship to H.-J. F., Mrs. Croon for secretarial help and Mr. Arns for drawings.

References

1. Geiger, R. (1976) *Chemiker Zeitung* **100,** 111–129.

2. Pullen, R. A., Lindsay, D. G., Wood, S. P., Tickle, I. J., Blundell, T. L., Wollmer, A., Krail, G., Brandenburg, D., Zahn, H., Gliemann, J. & Gammeltoft, S. (1976) *Nature* **259,** 369–373.

3. Brandenburg, D., Gattner, H. -G. & Wollmer, A. (1972) *Z. Physiol. Chem.* **353,** 599-617.

4. Spackman, D. H., Stein, W. H. & Moore, S. (1958) *Anal. Chem.* **30,** 1190-1206.

5. Hartley, B. S. (1970) *Biochem. J.* **119,** 805–822.

6. Friesen, H. -J. (1976) Dr.rer.nat.thesis, TH Aachen, Germany.

7. Lindsay, D. G. & Shall, S. (1969) *Biochem. J.* **115,** 587–595.

8. Geiger, R., Schöne, H. -H. & Pfaff, W. (1971) *Z. Physiol. Chem.* **352,** 1487–1490.

9. Lindsay, D. G., Loge, O., Losert, W. & Shall, S. (1972) *Biochem. Biophys. Acta* **263,** 658–665.

10. Geiger, R., Geisen, K., Summ, H. -D. & Langner, D. (1975) *Z. Physiol. Chem.* **356,** 1635–1649.

11. Kunz, H. (1976) *Chem. Ber.* **109,** 3693–3706.

12. Radola, B. J. (1974) *Biochim. Biophys. Acta* **386,** 181–195.

13. Tesser, G. I. & Balvert-Geers, I. C. (1975) *Int. J. Pep. Protein Res.* **7,** 295–305.

14. Habeeb, A. F. S. A. (1972) in *Methods in Enzymology,* Vol. XXV, McCormick, D. B. & Wright, L. D., Eds., Acad. Press, New York, pp. 558–566.

15. Bodanszky, M., Ondetti, M. A., Birkhimer, C. S. & Thomas, P. L. (1964) *J. Amer. Chem. Soc.* **86,** 4452–4459.

16. Weinert, M., Kircher, K., Brandenburg, D. & Zahn, H. (1971) *Z. Physiol. Chem.* **352,** 719–724.

17. Schlichtkrull, J. (1956) *Acta Chem. Scand.* **10,** 1455–1458.

18. Blundell, T. L. & Johnson, L. N. (1976) *Protein Crystallography* Acad. Press, New York.

19. Friesen, H.-J., Hofmann, K. & Finn, F. M., manuscript in preparation.

20. Moody, A. J., Stan, M. A., Stan, M. & Gliemann, J. (1974) *Horm. Met. Res.* **6,** 12–16.

21. Gliemann, J. & Gammeltoft, S. (1974) *Diabetologia* **10,** 105–113.

PEPTIDES WITH AGONIST AND ANTAGONIST CHEMOTACTIC ACTIVITY

S. ASWANIKUMAR, B. A. CORCORAN, E. SCHIFFMANN, *Connective Tissue Section, LDBA-NIDR,* C. B. PERT, *Section on Biochemistry, APB-NIMH,* J. L. MORELL and E. GROSS, *Section on Molecular Structure, RRB-NICHD National Institutes of Health Bethesda, Maryland 20014*

Formyl-L-Phe-D-Leu-L-Phe-D-Leu-L-Phe-OH is a potent chemoattractant for neutrophils. *tert*-butyloxycarbonyl-L-Phe-D-Leu-L-Phe-D-Leu-L-Phe-OH is an antagonist to this type of chemotaxis.

N-Formylation of peptides with the sequence L-Met-X- (X = amino-acid residue with an aromatic or other hydrophobic side chain) gave products with chemotactic activity.[1] A detailed study[2] of structure-activity relationships resulted in the synthesis of chemoattractants of high potency ($10^{-11}M$). In subsequent investigations it was found that there is not an absolute requirement for methionine.[3] Methionine may be replaced by norleucine or norvaline with insignificant loss in chemotactic activity.[4]

Since the most potent chemotactically active peptides known to date have in common the sequence Leu-Phe (or Phe, followed by other amino-acid residues),[2-4] the peptide R-L-Phe-D-Leu-L-Phe-D-Leu-L-Phe (R = H) and its *N*-formylated, *N*-acetylated, and *N*-*tert*-butyloxycarbonylated derivatives [R = HCO–, H_3CCO–, $(CH_3)_3COCO$–] were examined for their chemotactic potential. Peptides of bacterial origin with established antibiotic properties containing the sequence Leu-Phe, (gramicidin S and tyrocidin) or Ile-Phe (bacitracin) were also screened for the potential to act as attractants for neutrophils.

Boc-L-phenylalanyl-D-leucyl-L-phenylalanyl-D-leucyl-L-phenylalanine (I) was synthesized by the *N*-hydroxysuccinimide ester method beginning with Boc-L-phenylalanine. The theoretical weight gain occurred at each step and amino-acid analysis indicated quantitative coupling. The final product showed one major spot and minor contaminants in thin layer chromatography (TLC; benzene/acetic acid = 7:1). Crystallization from ethyl acetate/petroleum ether gave a material pure by amino-acid analysis (0.4 mg: phenylalanine, 1.40 μmol; leucine, 0.87 μmol) and TLC in the solvent system given above.

Formyl-L-phenylalanyl-D-leucyl-L-phenylalanyl-D-leucyl-L-phenylalanine. Formic acid-*N*-hydroxysuccinimide ester was prepared by treatment of formic acid with *N*-hydroxysuccinimide and DCC in ethyl acetate (1 h, 0°C; 30 min, r.t.). After removal of the urea by filtration and the addition of petroleum ether crystallization occurred. The acetic acid-*N*-hydroxysuccinimide ester was prepared by the same procedure

Crystalline Boc-L-Phe-D-Leu-L-Phe-D-Leu-L-Phe-OH (80 mg) was allowed to stand in 10 ml of trifluoroacetic acid (TFA) at room temperature for 60 min. Following the removal of TFA under vacuum, the salt was dissolved in 30% ethanol/water. Upon addition of ammonium hydroxide to neutrality the free peptide (II) precipitated. One half of the product was allowed to react (30 min; r.t.) with a ten-fold excess of *N*-hydroxysuccinimidyl formate in 70% tetrahydrofuran/water containing 10 equivs. of $NaHCO_3$. After evaporation of THF and dilution with water the solution was acidified to pH 2.0 by adding hydrochloric acid and extracted with ethyl acetate. The peptide obtained after solvent evaporation showed the correct amino-acid analysis and weight recovery of amino acids (0.30 mg: Phe, 1.06 μmol; Leu, 0.72 μmol) and the absence of the free amino group upon reaction with phenyl isothiocyanate. The remaining portion of II was acetylated following the procedure given for *N*-formylation using only a two-fold excess of *N*-hydroxysuccinimidyl acetate. The isolated product also showed the absence of free amino groups and the correct amino-acid analysis and weight recovery (0.35 mg: Phe, 1.30 μmol; leucine, 0.93 μmol).

Formyl-L-norleucyl-L-leucyl-L-*p*-^{3}H-phenylalanine was prepared from formyl-L-norleucyl-L-leucyl-L-*p*-phenylalanine employing a catalytic replacement technique.[4] Activated rabbit neutrophils were obtained 12 to 14 hours after the intraperitoneal injection of 0.1% glycogen in phosphate buffered saline.[5]

Chemotactic activities were measured[6] by counting in a hemocytometer the cells adhering to the underside of a micropore filter separating two chambers, the upper one containing the cells, the lower the attractant. In the standard binding assay[3] neutrophils (4.4×10^6 cells) briefly treated with 0.1 *mM* tosyl-L-phenylalanyl chloromethane (TPCK; Calbiochem, Gaithersubrg, Maryland) were incubated at 0°C for 1 h in 2 ml of Gey's balanced salt solution[7] containing 50,000 cpm of HCO-HN-L-norleucyl-L-leucyl-L-*p*-tritio-phenylalanine (1.5 *nM*). Cells were rapidly filtered on glass fiber filters and the filters counted in a liquid scintillation counter with an efficiency of 40%.

Gramicidin S was purchased from Calbiochem, Gaithersburg, Maryland, bacitracin from Sigma, St. Louis, Missouri, tyrocidin from Serva Chemicals, New York, New York.

Formyl-L-Phe-D-Leu-L-Phe-D-Leu-L-Phe-OH is a potent chemoattractant for neutrophils, the potency of the acetyl derivative is lower by approximately 3 powers of 10, the $^{\alpha}N$-unprotected peptide shows the lowest activity (Figure 1). The Boc-protected peptide is an inhibitor of chemotaxis; the concentration to affect 50% inhibition (ID_{50}) of cellular migration is $8 \times 10^{-7} M$ (Table I). The effect of II and its *N*-protected derivatives on the specific binding of the labeled attractant formyl-L-norleucyl-L-leucyl-L-*p*-^{3}H-phenylalanine is shown in Figure 1. The curves are essentially parallel indicating that the peptides interact at the same receptor site as the labeled ligand. The calculated ID_{50} values of Table I are in agreement with the biological activities.

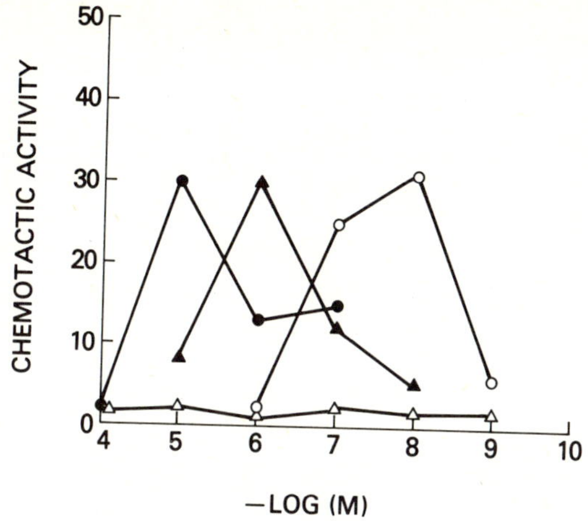

Figure 1. Chemotactic activity of neutrophils in response to H-L-Phe-D-Leu-L-Phe-D-Leu-L-Phe-OH (•), HCO-HN-L-Phe-D-Leu-L-Phe-D-Leu-L-Phe-OH (○), H_3CCO-HN-L-Phe-D-Leu-L-Phe-D-Leu-L-Phe-OH (▲), and $(H_3C)_3$ COCO-HN-L-Phe-D-Leu-L-Phe-D-Leu-L-Phe-OH (△). The results are expressed as the average number of neutrophils in 10 fields at a magnification of X970 for triplicate samples. The SEM did not exceed 10% for values above 10.

Table I. Chemotactic Activities of Synthetic Peptides and Peptide Antibiotics

Peptide	Chemotaxis $ED_{50}(M)^a$	Specific Binding $ID_{50}(M)^b$	Inhibition of Chemotaxis $ID_{50}(M)^c$
HCO-L-Met-L-Leu-L-Phe-OH	7×10^{-11}	3.3×10^{-10}	—
H-L-Phe-D-Leu-L-Phe-D-Leu-L-Phe-OH	4×10^{-6}	1×10^{-5}	—
Ac-L-Phe-D-Leu-L-Phe-D-Leu-L-Phe-OH	5×10^{-7}	4×10^{-7}	—
HCO-L-Phe-D-Leu-L-Phe-D-Leu-L-Phe-OH	4×10^{-9}	2×10^{-8}	—
t-Boc-L-Phe-D-Leu-L-Phe-D-Leu-L-Phe-OH	—	7×10^{-7}	8×10^{-7}
Gramicidin S	3×10^{-6}	5×10^{-5}	—
Tyrocidin	4×10^{-6}	2×10^{-5}	—
Bacitracin	2×10^{-5}	9×10^{-5}	—

Standard conditions were employed as described in Methods for chemotaxis and binding assays. The values are means of triplicate samples varying less than 10%.

[a] ED_{50} is defined as that concentration of attractant giving a half-maximal chemotactic response.

[b] ID_{50} is defined as that concentration of peptide causing 50% inhibition of specific binding of HCO-L-Nle-L-Leu-L(^{3}H) Phe-OH on neutrophils.

[c] ID_{50} is defined as that concentration of peptide, not itself an attractant, causing 50% inhibition of leucotactic response to the standard potent attractant HCO-L-Met-L-Leu-L-Phe-OH.

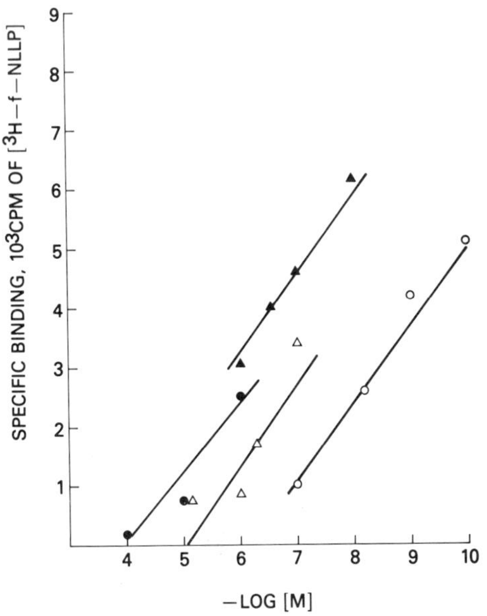

Figure 2. Inhibition of specific binding of HCO-L-Nle-L-Leu-L-(^{3}H)Phe-OH to neutrophils by H_2N-L-Phe-D-Leu-L-Phe-D-Leu-L-Phe-OH (●), HCO-HN-L-Phe-D-Leu-L-Phe-D-Leu-L-Phe-OH (○), H_3CCO-HN-L-Phe-D-Leu-L-Phe-D-Leu-L-Phe-OH (▲), and $(H_3C)_3$COCO-HN-L-Phe-D-Leu-L-Phe-D-Leu-L-Phe-OH (△). ID_{50} for binding was estimated from these plots of specific binding versus concentration of peptide. The values are means of triplicate samples varying less than 10%.

Gramicidin S, tyrocidin, and bacitracin are chemotactically active at concentrations of ~ 10 μM. The ability of the peptide antibiotics to displace the specific binding of a chemoattractant is closely related with their chemotactic potencies (Table I).

Specific proteases have been invoked to participate in the chemotactic response.[8] When the effect on neutrophils of II, its *N*-acyl analogs, and the antibiotics was tested in the presence of protease inhibitors, similar observations were made (Table II). Tosyl-L-phenylalanyl-chloromethane (TPCK), an inhibitor of chymotrypsin, was more effective in inhibiting chemotaxis than $^{\alpha}N$-tosyl-L-lysyl chloromethane (TLCK), an inhibitor of trypsin. If proteolytic cleavage takes place indeed, that of the substrates with alternating L and D configurations will deserve further attention.

The response of leucocytes to the *N*-formylated pentapeptide lends additional support to the role attributed in chemotaxis to the first member in the family of acyl groups. It remains to be seen why replacement of the formyl hydrogen by the tertiary butoxy group changes the picture drastically, this derivative displaying activity that is *antagonistic* to the response of neutrophils to chemoattractants.

Table II. Effect of Specific Protease Inhibitors on the Chemotactic Response to Peptides

Peptide	Addition to Cells	Inhibition of Chemotaxis[a] %
Bacitracin, 0.1mM	TPCK[b], 0.1mM	100
Bacitracin, 0.1mM	TLCK[c], 0.1mM	49
Gramicidin S, 0.01mM	TPCK, 0.1mM	100
Gramicidin S, 0.01mM	TLCK, 0.1mM	62
Tyrocidin, 0.01mM	TPCK, 0.1mM	100
Tyrocidin, 0.01mM	TLCK, 0.1mM	0
HCO-L-Phe-D-Leu-L-Phe-D-Leu-L-Phe-OH, 10nM	TPCK, 0.1mM	84
HCO-L-Phe-D-Leu-L-Phe-D-Leu-L-Phe-OH, 10nM	TLCK, 0.1mM	0
HCO-L-Phe-D-Leu-L-Phe-D-Leu-L-Phe-OH, 10nM	BTEE[d], 0.1mM	100
HCO-L-Phe-D-Leu-L-Phe-D-Leu-L-Phe-OH, 10nM	TAME[e], 0.1mM	0

[a]Assays for chemotaxis were carried out as described in Methods

[b]Tosyl-L-phenylalanyl chloromethane

[c]αN-Tosyl-L-lysyl chloromethane

[d]Benzoyl-L-tyrosine ethyl ester

[e]Tosyl-L-arginine methyl ester

References

1. Schiffmann, E., Corcoran, B. A. & Wahl, S. M. (1975) *Proc. Nat. Acad. Sci.* USA **72**, 1059–1062.

2. Showell, H. J., Freer, R. J., Zigmond, S. H., Schiffmann, E., Aswanikumar, S., Corcoran, B. A. & Becker, E. L. (1976) *J. Exp. Med.* **143**, 1154–1169.

3. Aswanikumar, S., Corcoran, B., Schiffman, E., Day, A. R., Feer, R. J., Showell, H. J., Becker, E. L. & Pert, C. (1977) *Biochem. Biophys. Res. Commun.* **74**, 810–817.

4. Day, A. R., Showell, H. J., Becker, E. L., Schiffmann, E., Corcoran, B., Aswanikumar, S., Pert, C., Radding, J. A., & Freer, R. J. (1977) *FEBS Lett,* in press.

5. Tempel, T. R., Snyder, R., Jordan, H. V. & Morgenhagen, S. E. (1970) *J. Peridontol.* **41**, 3/71–12/80.

6. Schiffmann, E., Showell, H. J., Corcoran, B. A., Ward, P. A., Smith E. & Becker, E. L. (1975) *J. Immunol.* **114**, 1831–1837.

7. Gey, G. O. & Gey, M. K. (1936) *Am. J. Cancer,* **27**, 45–76.

8. Aswanikumar, S., Schiffmann, E., Corcoran, B. A. & Wahl, S. M. (1976) *Proc. Nat. Acad. Sci.* USA **73**, 2439–2442.

SYNTHETIC POLYPEPTIDES AS CHEMO-ATTRACTANTS FOR NEUTROPHIL LEUCOCYTES: DEMONSTRATION OF A SPECIFIC RECEPTOR FOR CHEMOTAXIS

RICHARD J. FREER and ALAN R. DAY, *Department of Pharmacology, Medical College of Virginia, Richmond, VA 23298*

A variety of substances is reportedly chemotactic for neutrophil leucocytes.[1] Unfortunately, most are either incompletely characterized or too complex to allow a detailed analysis of the nature of the interaction with the neutrophils. However the recent observation of Schiffmann et al.[2] that the acylated amino acid *N*-formyl-methionine and some *N*-formyl dipeptides were chemotactic allowed a systematic and detailed analysis of the structural requirements for chemotaxis in a series of small molecular weight polypeptides.[3] The most active compound identified was *N*-formyl-Met-Leu-Phe-OH ($7 \times 10^{-11} M$). The results of a detailed analysis of position 1 (*N*-formyl-Met), as well as the synthesis of an intrinsically radiolabeled analog for binding studies, is reported herein.

Peptides were prepared by either standard solid-phase[4] or classical[5] methods. Formylation was carried out at first by the method of Sheehan and Yang[6] and later by DCC coupling directly on the resin or in solution. The synthesis of the radiolabeled analog, *N*-formyl-norleucyl-leucyl-(*p*-^{3}H)-phenylalanine has been described.[7] All peptides were homogeneous by thin-layer chromatography, high voltage electrophoresis and showed the expected amino-acid analyses. Biological activities were determined by Drs. Becker, Schiffmann and Pert.

Results and Discussion

Our previous study[3] had clearly demonstrated that there existed on the surface of the neutrophil leucocyte a receptor site which appeared to initiate both chemotaxis and lysosomal enzyme (lysozyme and β-glucuronidase) release. One of the requirements for a highly active compound appeared to be the presence of an *N*-formyl-methionine residue in the 1 position. The results of a detailed analysis of this position indicate that while the formyl group is essential, variations in the side chain are somewhat better tolerated (Table I). As can be seen acetylation (III) or removal of the α-amino group (II) yields relatively inactive analogs (~0.05%) as does the replacement by the ethyl group (IV). Therefore simply blocking or removing the amino group is not sufficient to confer enhanced chemotactic potency, indicating a unique and essential role for the formyl group.

Variations in the methionine side chain also produced changes in biological activity (Table I). Specifically the following were noted.

Table I. Position 1 Analogs of N^{α}-Formyl-Met-Leu-Phe-OH

No.	Structure	Chemotaxis (ED_{50})	Lysosomal Enzyme Release (ED_{50})	
			Lysozyme	β-Glucuronidase
I	CHO-Met-Leu-Phe-OH	7×10^{-11} M	2.4×10^{-10}	2.6×10^{-10}
II	deamino-Met-Leu-Phe-OH	1.9×10^{-7}	1.1×10^{-6}	2.8×10^{-6}
III	Acetyl-Met-Leu-Phe-OH	2.0×10^{-7}	1.4×10^{-6}	1.7×10^{-6}
IV	2-Ethyl-hexanoyl-Leu-Phe-OH	2.9×10^{-7}	1.6×10^{-6}	1.9×10^{-6}
V	CHO-Gly-Leu-Phe-OH	4.3×10^{-6}	7.6×10^{-6}	1.0×10^{-5}
VI	CHO-Ala-Leu-Phe-OH	ND	3.6×10^{-5}	4.1×10^{-5}
VII	CHO-Abu-Leu-Phe-OH	6.5×10^{-7}	1.8×10^{-6}	2.3×10^{-6}
VIII	CHO-Nva-Leu-Phe-OH	ND	1.3×10^{-8}	1.6×10^{-8}
IX	CHO-Nle-Leu-Phe-OH	6.6×10^{-10}	1.5×10^{-9}	1.9×10^{-9}
X	CHO-Eth-Leu-Phe-OH	4.3×10^{-10}	1.2×10^{-9}	1.5×10^{-9}
XI	CHO-Cys(Me)-Leu-Phe-OH	ND	8.2×10^{-8}	7.5×10^{-8}
XII	CHO-Phe-Leu-Phe-OH	1.1×10^{-7}	5.7×10^{-7}	8.5×10^{-7}
XIII	CHO-Leu-Leu-Phe-OH	ND	4.8×10^{-8}	5.2×10^{-8}
XIV	CHO-Ile-Leu-Phe-OH	ND	2.0×10^{-9}	2.1×10^{-9}
XV	CHO-Val-Leu-Phe-OH	ND	7.6×10^{-8}	8.6×10^{-8}

1. The presence of a S atom in the side chain was beneficial but not essential (compare I and IX) for good activity.

2. In the aliphatic series, biological potency increased with increasing side-chain length (compounds V thru IX).

3. Aliphatic side chains with β or γ branching showed much reduced activity (compare VIII to XV; compare IX to XIII and XIV).

The marked structure-activity requirements exhibited for the chemotactic response in neutrophils suggested that these compounds were interacting with a specific receptor site on the neutrophil. To investigate this directly a radiolabeled analog was prepared.[7] The compound prepared was N^{α}-formyl-Nle-Leu-*p*-Cl-Phe-OH which was then subjected to dehalotritiation to give the intrinsically labeled (14 Ci/mmole) ligand. Norleucine was chosen since the analog has good activity (~10%) and would not pose any problem during the catalytic tritiation. The resulting radiolabeled peptide was homogeneous in several thin-layer chromatography systems and was stable for at least 9 mo at -20°C.

Using this reagent, a specific, saturable binding site was demonstrated on the surface of the neutrophil leucocyte.[8] Scatchard analysis indicated a single binding site with a K_D of $1.5 \times 10^{-9} M$ and an R_{max} of 10^5 binding sites per cell. The data presented in Figure 1 show that the specific binding is almost absolutely correlated (r=0.99) with both chemotaxis and lysosomal enzyme release. Interestingly, a naturally occurring chemotactic factor from *E. coli* will also interact with this binding site while the chemotactic fragment of activated C_5 will not.

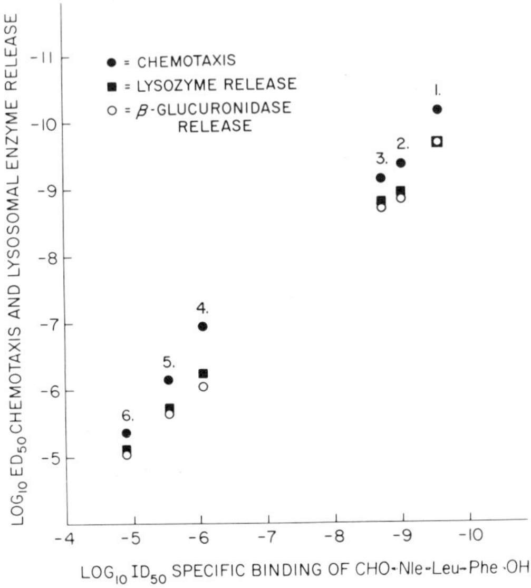

Fig. 1. Correlation between inhibition of specific binding of CHO-Nle-Leu-*p*-^{3}H-Phe-OH versus chemotactic and lysosomal enzyme releasing activity of several peptides. 1=CHO-Met-Leu-Phe-OH, 2=CHO-Eth-Leu-Phe-OH, 3=CHO-Nle-Leu-Phe-OH, 4=CHO-Phe-Leu-Phe-OH, 5=CHO-Met-Leu-Arg-OH and 6=CHO-Gly-Leu-Phe-OH.

Previous data,[7,8] as well as that presented here, indicate that the chemotactic response of the neutrophil leucocyte to the synthetic peptides is mediated by a highly specific receptor site. Binding data using a ^{3}H labeled analog substantiate the structure-activity data.

The authors are grateful to Drs. E. L. Becker, E. Schiffmann, and C. B. Pert for biological data (supported by NIDR Contract DE-62494).

References

1. Wilkinson, P. C. (1974) *Chemotaxis and Inflammation*, Churchill Livingston, Edinburgh, London, p. 54.
2. Schiffmann, E., Corcoran, B. & Wahl, S. (1975) *Proc. Nat. Acad. Sci. USA* **72**, 1059–1062.
3. Showell, H. J., Freer, R. J., Zigmond, S. H., Schiffmann, E., Aswanikumar, S., Corcoran, B. & Becker, E. L. (1976) *J. Exptl. Med.* **143**, 1154–1169.
4. Stewart, J. M. & Young, J. D. (1969) *Solid-Phase Peptide Synthesis*, W. H. Freeman, San Francisco.
5. Tilak, M. A. (1970) *Tetrahedron Lett.* **11**, 849–854.
6. Sheehan, J. C. & Yang, D. D. H. (1958) *J. Amer. Chem. Soc.* **80**, 1154–1158.
7. Day, A. R., Showell, H. J., Becker, E. L., Schiffmann, E., Corcoran, B., Aswanikumar, S., Pert, C. B., Radding, J. A. & Freer, R. J. (1977) *FEBS Lett.* **77**, 291–294.
8. Aswanikumar, S., Becker, E. L., Corcoran, B., Day, A. R., Freer, R. J., Pert, C. B. & Schiffmann, E. (1977) *Biochem. Biophys. Res. Commun.* **74**, 810–817.

THE MECHANISM OF PHOSPHOLIPID BINDING BY THE PLASMA APOLIPOPROTEINS

JAMES T. SPARROW, HENRY J. POWNALL, GERALD F. SIGLER, LOUIS C. SMITH, ANNE K. SOUTAR, and ANTONIO M. GOTTO, JR., *Department of Medicine, Baylor College of Medicine, The Methodist Hospital, Houston, Texas 77030*

Previously, we have reported[1] the synthesis and purification of a synthetic protein having the sequence of human plasma apolipoprotein apoLP-C-I (Fig. 1) and have demonstrated that the synthetic material binds to phospholipid and activates lecithin:cholesterol acyltransferase (LCAT) to the same extent as native apoLP-C-I. Upon binding to phospholipid, both synthetic and native apolipoprotein C-I show increases in α-helicity as measured by circular dichroism and show shifts in the intrinsic tryptophan fluorescence maximum, indicative of the tryptophan being placed in a more hydrophobic environment. Using the Chou-Fasman[2] rules to predict protein structure, CPK – space filling models of the sequence of apolipoprotein C-I were built. We observed that two faces occur on the helical segments: one a non-polar face and the other a polar face.[3] These segments occur between residues 7-14, 18-28 and 33-53. Using the hydrophobicity assignments of Bull and Breese[4], the mean residue hydrophobicity index of each segment is, respectively, -1,081, -952 and -880 cal/residue. We have shown previously that synthetic amphipathic peptides having low hydrophobicity indices do not bind to phosphatidylcholines, whereas those with higher indices do.[5]

We now report the synthesis and purification of four fragments of apoLP-C-I and their ability to bind phosphatidylcholine and activate LCAT. The fragments corresponding to sequence positions 39-57, 32-57, 24-57, 17-57 were synthesized by solid phase methodology on an improved resin previously reported[6]. The peptides were cleaved from the resin with anhydrous HF/anisole and purified by gel filtration on Bio-Gel P-10 and by ion exchange chromatography on Sulfopropyl Sephadex-25. Polyacrylamide gel electrophoresis at both acidic and basic pH showed the peptides to be homogeneous, and each peptide had the expected amino-acid analysis (Table I). With vesicles of dimyristoylphosphatidylcholine-cholesterol, 20 μg of the three fragments stimulated LCAT activity by 50, 60 and 100%, respectively of the values found for apolipoprotein C-I

H-Thr-Pro-Asp-Val-Ser-Ser-Ala-Leu-Asp-Lys-Leu-Lys-Glu-Phe-Gly-Asn-Thr-Leu-Glu-

Asp-Lys-Ala-Arg-Glu-Leu-Ile-Ser-Arg-Ile-Lys-Gln-Ser-Glu-Leu-Ser-Ala-Lys-Met-Arg-

Glu-Trp-Phe-Ser-Glu-Thr-Phe-Gln-Lys-Val-Lys-Glu-Lys-Leu-Lys-Ile-Asp-Ser-OH

Fig. 1. The amino acid sequence of apolipoprotein C-I.

Table I. Amino Acid Analyses of the Purified Synthetic Peptides.

Amino Acid	AA 39-57	AA 32-57	AA 24-57	AA 17-57	Peptides I	II	III	IV
Asp	1.0 (1)	1.1 (1)	1.0 (1)	2.0 (2)	- (0)	- (0)	- (0)	- (0)
Thr	0.9 (1)	0.9 (1)	0.9 (1)	1.7 (2)	- (0)	- (0)	- (0)	- (0)
Ser	1.8 (2)	3.7 (4)	4.6 (5)	4.8 (5)	5.7 (6)	5.7 (6)	5.8 (6)	7.7 (8)
Glu	4.0 (4)	4.7 (5)	6.7 (7)	8.0 (8)	2.0 (2)	2.1 (2)	2.1 (2)	2.1 (2)
Ala	- (0)	0.9 (1)	1.0 (1)	2.0 (2)	- (0)	1.9 (2)	1.0 (1)	- (0)
Val	0.9 (1)	1.0 (1)	0.8 (1)	1.0 (1)	0.9 (1)	0.9 (1)	0.9 (1)	0.9 (1)
Met	- (0)	1.1 (1)	0.9 (1)	1.1 (1)	- (0)	- (0)	- (0)	- (0)
Ile	1.0 (1)	1.0 (1)	2.5 (3)	2.6 (3)	- (0)	- (0)	- (0)	- (0)
Leu	1.1 (1)	2.0 (2)	2.9 (3)	3.9 (4)	2.0 (2)	2.0 (2)	2.0 (2)	4.0 (4)
Tyr	- (0)	- (0)	- (0)	- (0)	1.0 (1)	- (0)	- (0)	1.0 (1)
Phe	1.8 (2)	2.0 (2)	2.0 (2)	2.2 (2)	0.9 (1)	1.0 (1)	1.1 (1)	1.0 (1)
Lys	3.9 (4)	4.9 (5)	5.9 (6)	7.0 (7)	2.0 (2)	2.1 (2)	1.9 (2)	2.0 (2)
Arg	0.9 (1)	0.9 (1)	2.0 (2)	3.1 (3)	- (0)	- (0)	- (0)	- (0)
Trp	ND (1)	ND (1)	ND (1)	ND (1)	ND (1)	- (0)	ND (1)	ND (1)

(Fig. 2). The lipid binding properties of the four fragments were studied by fluorescence and circular dichroic spectroscopy. Changes in the ellipticity at 222 nm are observed for the three larger fragments, while the smaller fragment shows minimal changes (Table II). Changes in the intrinsic tryptophan fluorescence maximum were also observed for these larger fragments (Table II). The lipid-peptide complexes were subjected to density gradient ultracentrifugation and the majority of the larger peptides was found associated with the phospholipid band (Fig. 3).

From these studies we conclude that the 39-57 fragment does not contain the necessary requirements for binding to phosphatidylcholine, whereas the addition of seven residues causes the synthetic peptide to bind phospholipid. We

Table II. Fluorescence and Circular Dichroic Spectral Changes of Synthetic Peptides with Phospholipid.

PEPTIDE	$\Delta\lambda_{MAX}$	$[\Theta]_{222}$ PEPTIDE ALONE	$[\Theta]_{222}$ PEPTIDE PLUS PHOSPHOLIPID
AA_{39-57}	0	-1,914	-1,914
AA_{32-57}	12 NM	-3,829	-11,104
AA_{24-57}	11 NM	-4,427	-10,034
AA_{17-57}	12 NM	-7,452	-14,664
I	9.5 NM	-3,210	-7,782
II	---	-2,355	-2,107
III	5.5 NM	-3,625	-5,870
IV	19 NM	-5,931	-24,218

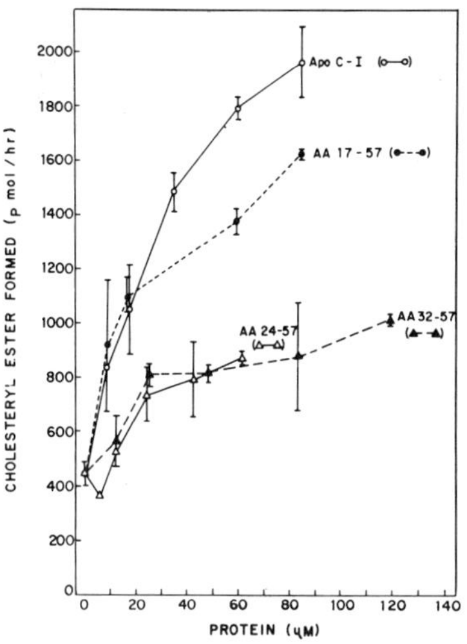

Fig. 2. Lecithin cholesterol acyltransferase activation by synthetic fragments of apolipoprotein C-I.

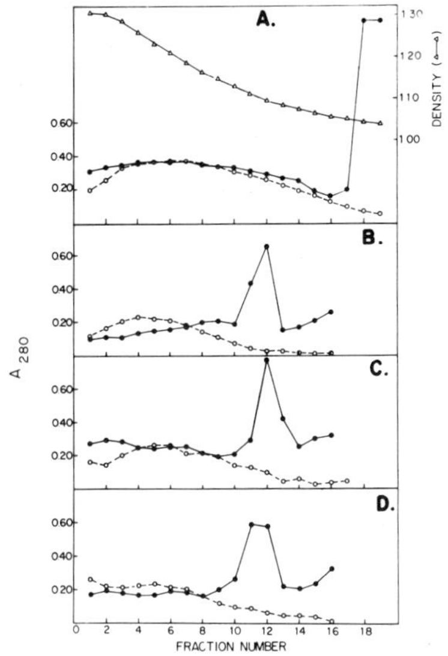

Fig. 3. Density gradient profiles A: 39–57; B: 32–57; C: 24–57; D: 17–57 o–o Peptide; •–• Peptide plus phospholipid.

feel that the minimum sequence necessary for binding to be approximately 16 to 20 residues and that an α-helix is necessary which is amphipathic and contains a hydrophobic face with a high mean residue hydrophobicity. Using these criteria for binding, we have designed several peptides and studied their ability to bind to phosphatidylcholine. The sequences of these peptides are shown in Figure 4. The peptides were synthesized on the improved resin and were cleaved with anhydrous HF/anisole and purified by gel filtration on Bio-Gel P-10 and by ion exchange chromatography on DEAE-cellulose. Each of the peptides had the expected amino-acid analysis (Table I) and showed the expected number of tryptic fragments by high voltage electrophoresis and paper chromatography. Peptide IV showed a single band on polyacrylamide gel electrophoresis; peptides I, II and III diffused too rapidly from the gels. Peptides I and IV bind phospholipid

Val-Ser-Ser-Leu-Lys-Glu-Tyr-Trp-Ser-Ser-Leu-Lys-Glu-Ser-Phe-Ser
I
Val-Ser-Ser-Leu-Lys-Glu-Ala-Ala-Ser-Ser-Leu-Lys-Glu-Ser-Phe-Ser
II
Val-Ser-Ser-Leu-Lys-Glu-Ala-Trp-Ser-Ser-Leu-Lys-Glu-Ser-Phe-Ser
III
Val-Ser-Ser-Leu-Leu-Ser-Ser-Leu-Lys-Glu-Tyr-Trp-Ser-Ser-Leu-Lys-Glu-Ser-Phe-Ser
IV

Fig. 4. The amino-acid sequence of model peptides for phospholipid binding. Mean residue hydrophobicities are I, –1092; II, –852, III, –965; and IV, –1120.

in a manner similar to the human apolipoproteins; i.e., they show a shift in the intrinsic tryptophan fluorescence maximum and an increase in ellipticity at 222 nm in the circular dichroic spectrum (Table II). Density gradient ultracentrifucation of the peptide-phospholipid mixtures demonstrated that some of peptide I and the majority of peptide IV were associated with the phosphatidylcholine (Fig. 5). From these results, we feel that these peptides demonstrate that the important criteria for binding of the apolipoproteins to phospholipids is the ability to form a stable α-helix, the amphipathic nature of the helix formed, and the hydrophobicity of the non-polar face.

This material was developed by the Atherosclerosis, Lipids and Lipoprotein Section of the National Heart and Blood Vessel Research and Demonstration Center, Baylor College of Medicine, a grant supported research project of the National Heart, Lung and Blood Institute, National Institutes of Health, Grant #HL-17269. L. C. Smith and H. J. Pownall are Established Investigators of the American Heart Association. This research was partially funded by the Texas Affiliate, AHA.

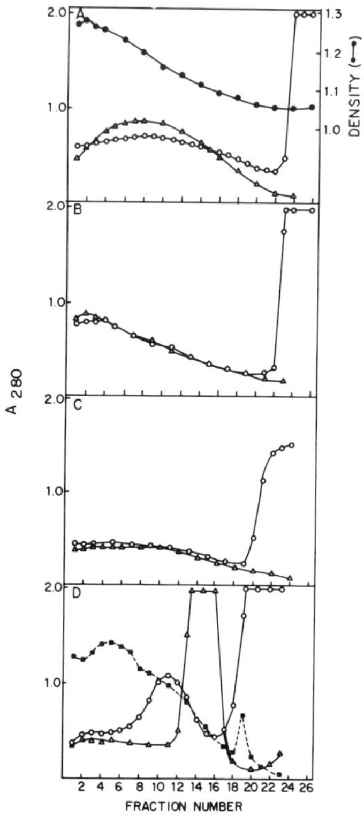

Fig. 5. Density Gradient Profiles. A: Peptide I; B: Peptide II; C: Peptide III; D: Peptide IV. A, B, C: Δ–Δ Peptide alone; o–o – peptide plus phospholipid; D: ■–■ peptide alone; peptide plus (Δ–Δ) dimyristoyl- and (o–o) palmitoyl-oleoyl-phosphatidylcholine.

References

1. Sigler, G. F., Soutar, A. K., Smith, L. C., Gotto, A. M. & Sparrow, J. T. (1976) *Proc. Nat. Acad. Sci. USA* **73**, 1422–1426.

2. Chou, P. Y. & Fasman, G. D. (1974) *Biochemistry* **13**, 211–245.

3. Segrest, J. P., Jackson, R. L., Morrisett, J. D. & Gotto, A. M. (1974) *FEBS Lett.* **38**, 247–253.

4. Bull, H. B. & Breese, K. (1974) *Arch Biochem Biophys* **191**, 665–670.

5. Sparrow, J. T., Morrisett, J. D., Pownall, H. J., Jackson, R. L. & Gotto, A. M., (1975) in *Peptides, Chemistry, Structure and Biology*, Walter, R. and Meienhofer, J., Eds., Ann Arbor Science, Ann Arbor, pp. 597–602.

6. Sparrow, J. T. (1976) *J. Org. Chem.* **41**, 1350–1353.

TETRAZOLE ANALOGS OF AMINO ACIDS AS A TOOL IN STUDIES OF THE ROLE OF FREE CARBOXYL GROUP IN BIOLOGICALLY ACTIVE SYSTEMS

Z. GRZONKA, E. KOJRO, Z. PALACZ, *Institute of Chemistry, University of Gdańsk, 80-952 Gdańsk, Poland,*
I. WILLHARDT and P. HERMANN,
Physiologisch-chemisches Institut, Martin-Luther-Universität, 402-Halle/Saale, G.D.R.

The 5-substituted tetrazolyl group (T) is isofunctional and isosteric to the carboxyl group.[1] As a result of these properties it can be useful in studies of biologically active systems in which the presence of free carboxyl group is indispensable for activity.

The present studies served to analyze: (a) the susceptibility of the tetrazole analogs of amino acid and peptide derivatives to the proteolytic enzymes; and (b) the chiroptical properties of this class of compounds. The successful synthesis of the γ-tetrazole analog of L-glutamic acid is also reported.

Susceptibility of Tetrazole Derivatives to Proteolytic Enzymes

Leucineaminopeptidase splits the peptide bond between amino acid and tetrazole (T) analog of amino acid with one order lower velocity compared to leucine amide (Table I). On the other hand, the tetrazole derivatives are not substrates of those enzymes which require the presence of free carboxyl group, i.e. of acylase I, and carboxypeptidase A. Experiments were undertaken to study whether Z-Gly-PheT acts as an effector of the enzymatic hydrolysis of Z-Gly-Phe-OH, which is the typical substrate of carboxypeptidase A. To our surprise Z-Gly-PheT acts as an activator of the competitive type. The results agree well with those of others[2-4] on the activation of the enzymatic hydrolysis of Z-Gly-Phe-OH by certain amino-acid derivatives which are not substrates of carboxypeptidase A.

Chiroptical Properties of Tetrazole Analogs of Amino Acids

The vast majority of compounds carrying the tetrazolyl group linked directly to the chirality center was found to possess a negative Cotton effect in the CD curves over the 196–219 nm range (Table II). Only the analogs of alanine (AlaT), and phenylalanine (PheT) display a positive Cotton effect within that

Table I. Susceptibility of Tetrazole Derivatives to Proteolytic Enzymes

Substrate	Specific activity M hydrolysis/min·mg enzyme
Leucineaminopeptidase	
H-Ala-GlyT	0.73
H-Phe-GlyT	4.45
H-Phe-Gly-GlyT	hydrolyzed
$H\text{-}Leu\text{-}NH_2$	18.73
Acylase I	
ClAc-PheT	not hydrolyzed
Ac-LeuT	not hydrolyzed
ClAc-Phe-OH	1.11
Carboxypeptidase A	
Z-Gly-PheT	not hydrolyzed
ClAc-PheT	not hydrolyzed
ClAc-PheT*	not hydrolyzed
ClAc-Phe-OH	0.32

* incubation in the presence of Zn^{+2}

range indicated. According to UV spectra this CD peak can be assigned to the $n \to \pi^*$ transition in the tetrazolyl moiety. It shows a characteristic shift depending on the solvent polarity (Fig. 1). Temperature studies show no correlation between CD bands and conformational equilibria. The CD peak of PheT might arise from superposition of the $n \to \pi^*$ transition of the tetrazolyl group and that of the ^1L_a one of the phenyl ring. More likely it results from the interaction of the two chromophores. The β-analog of Asp and the γ-analog of Glu and their derivatives (see Table II) show similarity of chiroptical properties to those of other α-amino acids.[5] Free amino acids are characterized by a positive Cotton effect in the CD curves at about 200 nm which is due to the $n \to \pi^*$ transition in the carboxyl group.

Synthesis of γ-Tetrazole Analog of L-Glutamic Acid

Syntheses of optical active tetrazole analogs of aspartic acid, and the α-analog of glutamic acid have been described[6,7], using 1,3-dipolar cycloaddition of NH_4N_3 or $Al(N_3)_3$ to the appropriate nitriles. However, as we confirmed the Z-Abu(CN), *2*, does not form the γ-tetrazole analog of glutamic acid by this

Table II. Circular Dichroism of Tetrazole Analogs of Amino Acids and Their *N*-Benzyloxycarbonyl Derivatives

Compound	CD* $[\theta]_{max}$ deg·mol^{-1}cm^2	λ_{max} nm	Compound	CD* $[\theta]_{max}$ deg·mol^{-1}cm^2	λ_{max} nm
AlaT	+28	212	PheT	+4600	210
Z-AlaT	-44	219		+38	249
AspT	-490	208	Z-PheT	-13050	203
Z-AspT	-520	215		-105	254
AsnT	-450	201	CysT(Bzl)	-3320	211
Z-AsnT	-1010	214		+940	230
Z-ValT	-2800	201	Z-CysT(Bzl)	-10050	208
ProT	-1270	201		+765	230
Z-ProT	-13600	196			
LeuT	-1390	197			
Z-LeuT	-1740	200			
GluT	-1550	199	$AspT^{\beta}$	+4750	199
Z-GluT	-3700	198	$Z\text{-}AspT^{\beta}$	-840	227
GlnT	-1850	199			
Z-GlnT	-1740	202	$AspT^{\beta}\text{-}NH_2$	+2280	202
Z-MetT	-3440	205	$Z\text{-}AspT^{\beta}\text{-}NH_2$	-3600	217
LysT	-2200	198			
Z-LysT(Z)	-1790	200	$GluT^{\gamma}$	+4720	200
			$Z\text{-}GluT^{\gamma}$	+2710	215

* in 0.1 N HCl (free tetrazole analogs of amino acids), or in methanol (N-benzyloxycarbonyl derivatives).

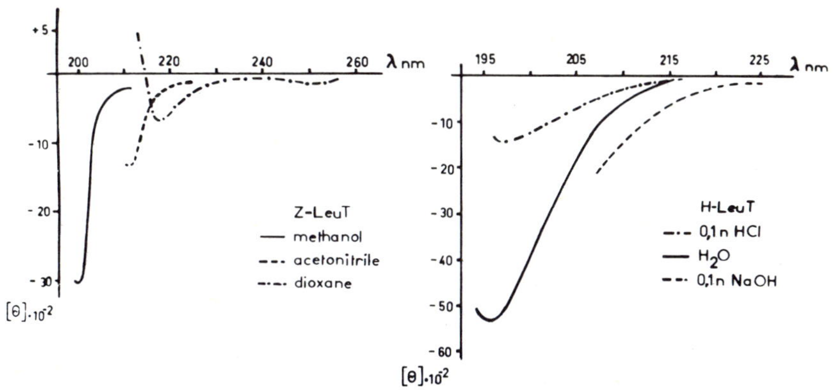

Fig. 1. Solvent dependence of CD spectra.

method. Recently, Sisido et al.[8] described a new method of synthesis of 5-substituted tetrazoles by cycloaddition of trialkyltin azides to nitriles followed by splitting of the N-Sn bond. Using this method we synthesized γ-(5-tetrazolyl)-α-amino-L-butyric acid (Scheme I). Starting with Z-Gln-OMe, *1*, the carboxamide group was dehydrated with $C_6H_5SO_2Cl$ in pyridine to the nitrile, *2*, which was refluxed with an equimolar amount of $Bu_3Sn\text{-}N_3$ in THF for 70 h. The tin derivative, *3*, obtained was transformed to tetrazole, *4*, by splitting the N-Sn bond by HCl/ether. The carboxyl protecting group was removed by alkaline hydrolysis and the benzyloxycarbonyl group by the action of HBr/AcOH. The structure of this new amino acid was confirmed by IR, nmr and CD spectra. Moreover, the γ-(5-tetrazolyl)-α-amino butyryl residue was incorporated into model dipeptides as a N- or C-terminal residue.

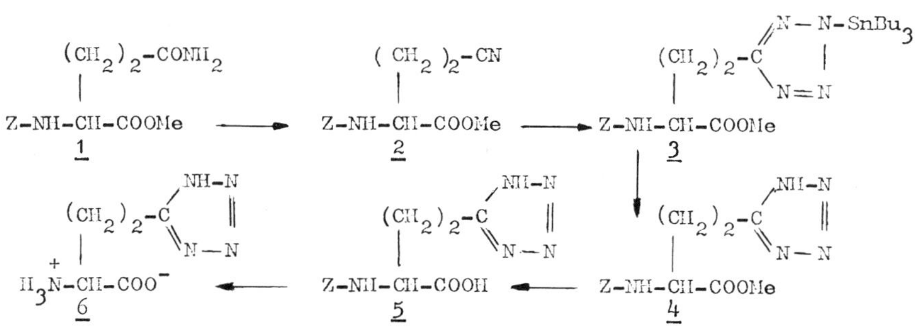

Scheme I. The synthesis of γ-(5-tetrazolyl)-α-amino-L-butyric acid.

References

1. Morley, J. S. (1968) *Fed. Proc.* **27**, 1314–1317.
2. Whitaker, J. R., Menger, F. & Bender, M. L. (1966) *Biochemistry* **5**, 386–392.
3. Davies, R. C., Auld, D. S. & Vallee, B. L. (1968) *Biochem. Biophys. Res. Commun.* **31**, 628–633.
4. Schechter, I. & Zazepizki, E. (1971) *Eur. J. Biochem.* **18**, 469–473.
5. Poloński, T. (1975) *Tetrahedron* **31**, 347–352, and references cited herein.
6. Grzonka, Z. (1972) *Roczniki Chem.* **46**, 1265–1273.
7. Grzonka, Z. (1973) *Roczniki Chem.* **47**, 1401–1406.
8. Sisido, K., Nabika, K., Isida, T. & Kozima, S. (1971) *J. Organomet. Chem.* **33**, 337–346.

RECEPTORS FOR PEPTIDE HORMONES

D. REGOLI, J. BARABE, and J. ST-LOUIS, *Department of Physiology and Pharmacology, Medical School, University of Sherbrooke, Sherbrooke, Que, Canada*

The comparison of the affinities of agonists and antagonists, combined with the evaluation of other parameters (e.g. the onset and the duration of action) can help to establish whether antagonists bind to the same receptor site(s) as the agonists. Such a knowledge is essential in choosing the general approach for the analysis and the interpretation of the pharmacological findings, namely the "one receptor site concept" elaborated by Ariens[1] or the "two receptor sites concept" recently discussed by Ariens and Beld[2].

Vasoactive peptides such as angiotensin II (ATII) and bradykinin (BK) have been extensively tested in isolated vascular smooth muscles, which respond to the two agents with rapid and stable changes of tone, that can be measured easily and precisely[3,4]. ATII and BK share with neurotransmitters and with biologically active amines the ability of stimulating several smooth muscles via the activation of specific receptors[3,4]; therefore a direct comparison between the two groups of agents is feasible. Peptides are larger and bigger than amines and contain a variety of chemically reactive groups, both hydrophilic and hydrophobic. Because of their size, their charges, and the presence of various chemical groups, it is generally admitted that: a) Peptides do not pass through the cell membrane[5,6]. Some workers have shown that receptors for ATII[7], insulin[8] and ACTH[9] are located in the cell membrane; b) Peptides presumably interact with several sites of the receptor protein. Such multiple interactions, possibly of different chemical nature, are expected to prolong the duration of the peptide-receptor complex. If this complex remains active throughout its full existence, affinity should be high. The affinity constants of peptide hormones, including those of ATII[10] are generally higher than those of amines[11] and approach those of antagonists for neurotransmitters and amines.

Results of studies with ATII and BK in smooth muscle of the rabbit aorta support some of these concepts. Rabbit aorta strips, suspended in oxygenated Krebs medium at 37°C, were contracted with submaximal concentrations of noradrenaline (NA), of various angiotensins and kinins. At the plateau of contraction, the agents were washed out and the time required for the tissue to relax by 50% (RT_{50}) was measured. In other experiments, the apparent affinities of the various agents were calculated from concentration-response curves and are expressed as pD_2 (-log of the concentration of agent producing 50% of the maximum response) according to van Rossum[12]. The results presented in Table I indicate that, in parallel with the decrease of the pD_2 value, the time to relax by 50% (RT_{50}) is comparatively shorter. The findings with ATII, its analogs and NA have been discussed in a recent paper.[13] A good direct correlation has been

Table I. Intrinsic Activities (α^E), Affinities (pD_2) and Time to 50% Relaxation (RT_{50}) of NA, Angiotensins and Kinins in Rabbit Aorta Strips.

Agent	α^E	pD_2	RT_{50}	Agent	α^E	pD_2	RT_{50}
Asp^1-ATII	1.0	8.86	7.1 ± 0.3*	BK	1.0	6.40	4.0 ± 0.2*
β-Asp^1-ATII	1.0	8.34	5.4 ± 0.3	Lys-BK	1.0	7.73	31.1 ± 4.4
Asn^1-ATII	1.0	8.11	5.2 ± 0.2	Met-Lys-BK	1.0	7.84	36.5 ± 3.3
des-Asp^1-ATII	1.0	7.62	3.3 ± 0.3				
NA	1.0	7.28	1.3 ± 0.1				

* means ± S.E. of 8-12 determinations

shown to exist between the pD_2 values of angiotensins and their respective RT_{50}. It has also been demonstrated that the rate of diffusion of the various agents out of the biophase is not determinant for the differences in the rate of relaxation[13]. Thus the rate of relaxation can be taken as an accurate measure of the rate of dissociation of the drug-receptor complex, and the complex NA-receptor appears to have a much shorter life than the complex ATII-receptor. Similar results have been obtained with kinins. The RT_{50} of this group of peptides increase in parallel with the pD_2 values. Moreover, the findings with kinins indicate that the addition of a positive charge (e.g. Lys) at the *N*-terminal increases affinity: in fact, Lys-BK and Met-Lys-BK are 21 and 28 times more active than BK. Metabolic degradation should not be responsible for the difference, because BK is rather resistant to aminopeptidases and the inhibition of converting enzyme with high concentrations of the Squibb nonapeptide (SQ 20881) does not increase significantly the affinity of BK for the aortic receptors. The reason why Met-Lys-BK is more active than Lys-BK is still unknown.

While the N-terminal portion seems to play a major role for the affinity of both ATII[14] and BK, the C-terminal portion is definitely involved in the process of stimulation (intrinsic activity: α^E). Gradual change from agonist to antagonist via intermediate dualistic compounds have been observed with a stepwise change of the structure of C-terminal Phe, in both ATII and des-Arg^9-BK (D-BK). (This octapeptide fragment of BK is approximately 6 times more potent than the nonapeptide and it has therefore been used, instead of BK, to explore the chemical features required for the activation of the aortic receptors.) The results presented in Table II indicate that optimum stimulation of receptors for both peptides is provided by the presence of Phe in position 8. Replacement of this residue with a) Tyr gives full agonists with lower affinity; b) cyclohexylalanine (Cha) and Nle gives partial agonists in the ATII and an antagonist ([Nle^8]-D-BK) in the D-BK series; c) Leu gives antagonists; d) other residues containing aliphatic side chains shorter than that of Leu give weaker antagonists; and, e) esterification of the terminal COOH does not influence the intrinsic activity of ATII[15] but reduces slightly that of D-BK.

Table II. Intrinsic Activities (α^E) and Affinities (pD_2 and pA_2) of ATII, des-Arg^9-BK and their Analogs Containing Various Residues at the C-Terminal Position. The Pharmacological Parameters Have Been Measured in Rabbit Aorta Strips.

Agent	α^E	pD_2	pA_2*	Agent	α^E	pD_2	pA_2*
ATII	1.0	8.86	-	des-Arg^9-BK (D-BK)	1.0	7.20	-
[Tyr^8]-ATII	1.0	7.94	-				
[Nle^8]-ATII	0.55	8.60	8.44	[Tyr^8]-D-BK	1.0	4.39	-
[Cha^8]-ATII	0.52	8.55	8.30	[Phe-OMe^8]-D-BK	0.82	6.81	-
[Leu^8]-ATII	0	-	8.60	[Leu^8]-D-BK	0	-	6.75
[Val^8]-ATII	0	-	7.96	[Leu-OMe^8]-D-BK	0	-	6.66
[Ala^8]-ATII	0	-	7.22	[Nle^8]-D-BK	0	-	6.31
				[Ala^8]-D-BK	0	-	3.95

* pA_2: -log of the concentration of antagonist that reduces the effect of a double dose of agonist to that of a single dose.

Any substitution of the other residues of ATII and, to some extent, of des-Arg^9-BK, is accompanied by variable change of affinity, but not of intrinsic activity (see Regoli et al.[16] for a review on ATII). From the data of Table II, it is evident that the C-terminal residue of both peptides is primarily involved in the simulation of receptors and contributes, to some extent, also to the binding. The most important finding presented in Table II is the similarity of the affinities of ATII/[Leu^8]-ATII and of D-BK/[Leu^8]-D-BK. The peptide antagonists have chemical structures, molecular sizes, and physico-chemical properties practically identical to those of the agonists. In this respect peptides differ from neurotransmitters and amines, which, according to Ariens and Beld[2], have little or no chemical relationship with their respective antagonists, and show much lower affinities than the antagonists. It is therefore conceivable that peptide agonists and antagonists have similar onset and duration of action. Experiments were performed to elucidate this point.

The phases of the myotropic effect (development of contraction, plateau of contraction, relaxation) are easily measured with agonists. The average time required by aortic strips to reach the plateau of contraction in the presence of ATII and D-BK and the time to relax by 100% after washout of the two peptides were used in the experimental protocol designed to evaluate the onset and the duration of action of antagonists. It was found that the onset of action of [Leu^8]-ATII ($4.8 \times 10^{-8} M$) and of [Leu^8]-D-BK ($1.1 \times 10^{-6} M$) is fairly rapid (5 min or less) and corresponds to the time required by ATII ($4.4 \times 10^{-9} M$) and D-BK ($9.8 \times 10^{-8} M$) to elicit full contractions of the rabbit aorta. The antagonism by [Leu^8]-ATII is reversible in 30 min[17] and that by [Leu^8]-D-BK in 15-20 min, as illustrated in Fig. 1.

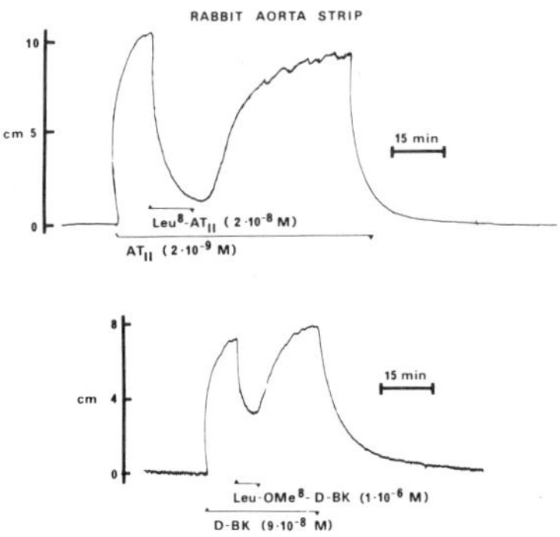

Fig. 1. Onset and duration of action of antagonists for ATII and D-BK in rabbit aorta strips.

These results suggest that agonists and antagonists of the ATII and BK series bind to the same receptor site(s). Thereby, the second criterion proposed by Schild[18] for the classification of different receptor types (namely the affinity constants of competitive antagonists) is applicable to ATII and BK. Moreover, competitive antagonists of ATII and BK could possibly become useful tools for binding assays and for the isolation of receptors. Finally, the interpretation of experimental findings with ATII, BK and hopefully other peptide hormones can be based on the "one receptor concept" of Ariens[1] and on classical receptor theories.

References

1. Ariens, E. J. (1964) Molecular Pharmacology, Academic Press, London.
2. Ariens, E. J. & Beld, A. J. (1977) *Biochem. Pharmacol.* **26**, 913–918.
3. Rioux, F., Park, W. K. & Regoli, D. (1973) *Can. J. Physiol. Pharmacol.* **51**, 665–672.
4. Regoli, D., Barabé, J. & Park, W. K. *Can. J. Physiol. Pharmacol.* (in press).
5. Cuatrecasas, P. (1974) *Ann. Rev. Biochem.* **43**, 169–214.
6. Schwyzer, R. (1974) *Pure Appl. Chem.* **37**, 299–314.
7. Devynck, M. A., Pernollet, M. G., Meyer, P., Fermandjian, S. & Fromageot, P. (1973) *Nature New Biol.* **245**, 55–58.
8. Cuatrecasas, P. (1974) *Biochem. Pharmacol.* **23**, 2353–2361.
9. Lang, U., Karlaganis, G., Vogel, R. & Schwyzer, R. (1974) *Biochemistry,* **13**, 2626–2633.
10. Peach, M. J. (1977) *Physiol. Rev.* **57**, 313–370.
11. Kahn, C. R. (1976) *J. Cell. Biol.* **70**, 261–286.
12. Van Rossum, J. M. (1968) *Rec. Adv. Pharmacol.* **2**, 99–134.
13. St. Louis, J., Regoli, D., Barabé, J. & Park, W. K., *Can. J. Physiol. Pharmacol.* (in press).
14. Regoli, D., Rioux, F., Park, W. K., Choi, C. (1974) *Can. J. Physiol. Pharmacol.* **52**, 39–49.
15. Rioux, F., Park, W. K. & Regoli, D. (1975) *Can. J. Physiol. Pharmacol.,* **53**, 383–391.
16. Regoli, D., Park, W. K. & Rioux, F. (1974) *Pharmacol. Rev.* **26**, 69–123.
17. Rioux, F. (1974) Ph.D. Thesis, University of Sherbrooke, Sherbrooke, Que.
18. Schild, H. O. (1973) in *Drug Receptors,* Rang. H. P. Ed., University Park Press, Baltimore, pp. 29–36.

CONFORMATION-ACTIVITY APPROACH TO THE DESIGN OF VASOPRESSIN ANALOGS WITH HIGH AND SPECIFIC ANTIDIURETIC ACTIVITY

CLARK W. SMITH, CHRISTOPHER R. BOTOS and RODERICH WALTER, *Department of Physiology and Biophysics, University of Illinois Medical Center, Chicago, Illinois 60612*

Recently a model of the biologically active conformation of vasopressin bound to its antidiuretic receptor was developed.[1] This model is principally based on the preferred backbone conformation of vasopressin determined in dimethylsulfoxide,[2] but includes the stacking interaction of the aromatic side chains of the Tyr-Phe sequence observed for vasopressin in aqueous medium[3,4] (Fig. 1). In vasopressin – and very likely in many other peptide hormones – it is the side chains of residues located in the corner positions of folds or bends, such as the corner residues in the two β-turns of neurohypophyseal hormones (positions 3, 4, 7 and 8), which are most exposed. It is these residues which are readily available for intermolecular interactions that contain the "binding elements"[5] and contribute to hormone-receptor binding.

It should be possible to enhance selectively a particular biological activity by introducing modifications in the side chains of these residues. The important contributions of the side chains in the corner positions 3 and 8 are well documented and the contribution of position 4 is emerging as a result of work from several laboratories. In an initial effort to test the vasopressin model, the remaining corner position, Pro^7, which has been least investigated, has been subjected to structural changes designed to enhance binding to the antidiuretic receptor.

Recently we began investigations at these corner positions of the β-turns by making substitutions with residues which possess double bonds in their side chains in the hope that such deformable electron clouds with their ability to undergo π-π or other nonbonded interactions may enhance receptor binding of the resultant analogs (provided the steric fit at the receptor is correct).

With these considerations in mind we have synthesized [7-(3,4-dehydroproline)]-Arg-vasopressin and its deamino analog. The protected nonapeptide precursors were synthesized by the Merrifield method of solid phase synthesis[6] using a scheme of deprotection, neutralization and coupling described previously.[7] *t*-Butyloxycarbonyl-3,4-dehydroproline was the generous gift of Dr. J. Meienhofer, Hoffmann-La Roche, Nutley, N. J. The protecting groups were removed by reduction with Na in liquid NH_3[8,9] and cyclization was accomplished by oxidation with ICH_2CH_2I.[10] The analogs were purified by gel filtration on Sephadex, ion exchange chromatography on carboxymethylcellulose,[11] and partition chromatography.[12] The purity and identity of the final product was verified by TLC in several solvent systems, amino-acid analysis, and elemental

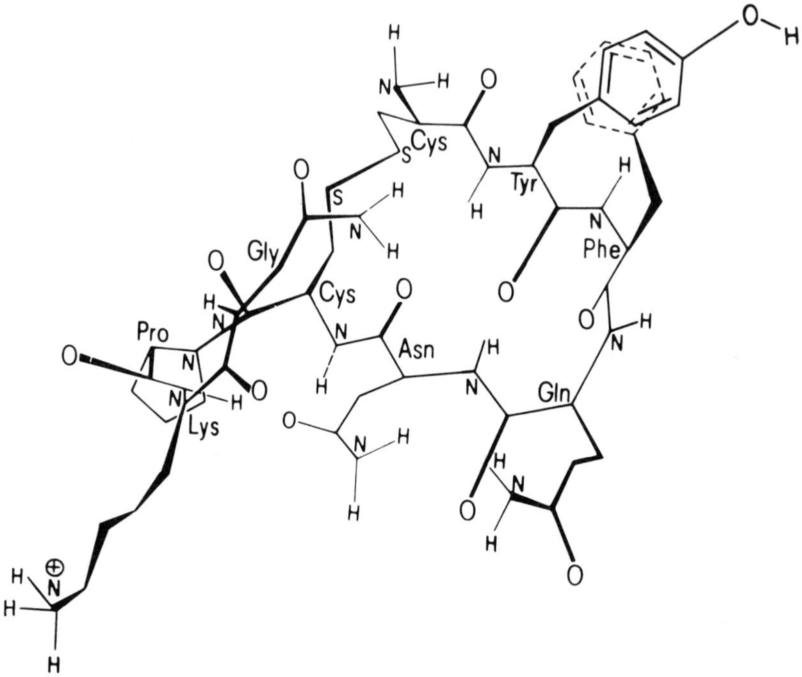

Fig. 1. Preferred average conformation of lysine vasopressin in aqueous media. One surface of the biologically active model is featureless and hydrophobic. Amino-acid side chains of residues along the rim of this surface of vasopressin extending from residue 3 via the hydrocarbon portion of Glu[4] all the way around through Cys[6] and Pro[7] to the hydrocarbon portion of the side chain of residue 8 are proposed for initial recognition by the antidiuretic receptor. The other surface of the molecule is overlayed by the glutamine and asparagine side chains as well as the C-terminal tripeptide; the latter bends in the direction of the primary amino group of Cys[1]. The lysine (or arginine) side chain is shown here in an extended conformation. However, in the model it is approximately parallel to the peptide backbone in the region occupied by the asparagine residue, but above the β-pleated sheet of the 20-membered ring of vasopressin. The carboxamide group of Asn[5] and the basic moiety of residue 8 (the "active elements")[5] in this hydrophilic cluster are proposed to act in concert on the receptor to initiate the sequence of events which lead to the antidiuretic response.

analysis. The antidiuretic potencies were determined using anesthetized male rats according to the method of Jeffers et al.[13] as modified by Sawyer.[14] This method yields antidiuretic activities in terms of the intensity of antidiuresis based on the maximal depression of urine flow after injections, and has been shown to yield figures least subject to errors introduced by differing durations of antidiuretic actions.[15] The pressor potencies were determined using anesthetized male rats as described in the U. S. Pharmacopeia.[16] Either the four-point assay design of Schild[17] or matches were used as compared to U. S. P. posterior pituitary reference standard.

The results are summarized in Figure 2. Immediately apparent is the nearly two-fold increase of antidiuretic potency over the highest reported value for Arg-vasopressin (AVP)[18] as a result of the introduction of the double bond into the pyrrolidine ring of proline. This is accompanied by a nearly 60% reduction in pressor potency. The deletion of the N-terminal amino group further dissociates these activities and enhances the antidiuretic potency to a value of 3310 ± 240 (mean ± SEM) making it the most potent antidiuretic compound to date.

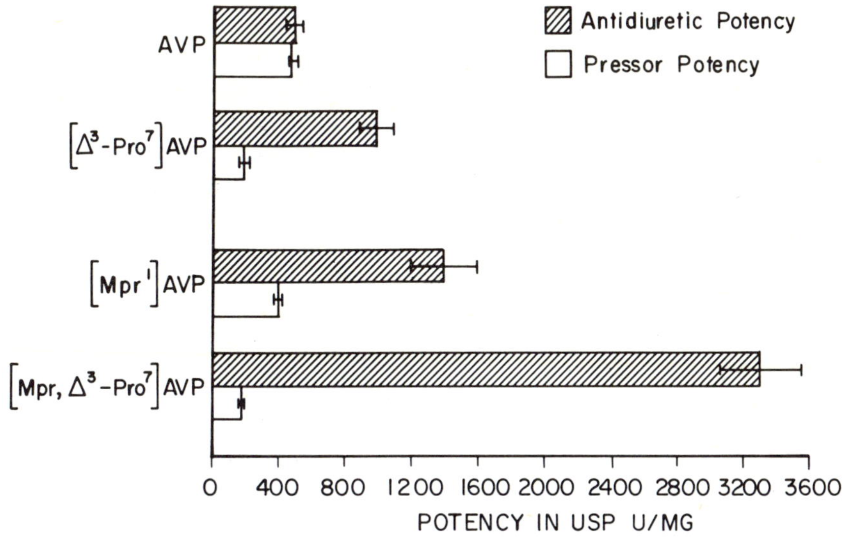

Fig. 2. Antidiuretic and pressor potencies of vasopressin analogs. Values are given as mean ± SEM. AVP potencies are those of Meienhofer et al.[18] and [Mpr^1]AVP potencies are those of Huguenin et al.[19] AVP = Arg-vasopressin; Mpr = β-mercaptopropionic acid.

These data would seem to support the contention that introducing a deformable electron cloud into a binding element should increase the affinity of the peptide for the receptor; although, in vivo potency data alone cannot distinguish between affinity effects and other factors such as resistance to enzymatic degradation.

The authors wish to thank Mr. G. Skala and Ms. S. Chan for their help in performing the bioassays. This study was supported in part by USPHS grant AM-18399 and a Pharmaceutical Manufacturer's Foundation Research Starter Grant.

References

1. Walter, R., Smith, C. W., Mehta, P. K., Boonjarern, S., Arruda, J. A. L. & Kurtzman, N. A. (1977) in *Disturbances in Body Fluid Osmolality*, Andreoli, T. E., Grantham, J. and Rector, Jr., F. C. Eds., Am. Physiol. Soc., Bethesda, pp. 1–36.
2. Walter, R., Ballardin, A., Schwartz, I. L., Gibbons, W. A. & Wyssbrod, H. R. (1974) *Proc. Nat. Acad. Sci. USA* **71**, 4528–4532.
3. Deslauriers, R. & Smith, I. C. P. (1970) *Biochem. Biophys. Res. Commun.* **40**, 179–185.
4. Von Dreele, P. H., Brewster, A. I., Dadok, J., Scheraga, H. A., Bovey, F. A., Ferger, M. F. & du Vigneaud, V. (1972) *Proc. Nat. Acad. Sci. USA* **69**, 2169–2173.
5. Walter, R. (1977) *Fed. Proc.* **36**, 1872–1878.
6. Merrifield, R. B. (1963) *J. Amer. Chem. Soc.* **85**, 2149–2154.
7. Smith, C. W. & Ferger, M. F. (1976) *J. Med. Chem.* **19**, 250–254.
8. Sifferd, R. H. & du Vigneaud, V. (1935) *J. Biol. Chem.* **108**, 753–761.
9. du Vigneaud, V., Ressler, C., Swan, J. M., Roberts, C. W. & Katsoyannis, P. G. (1954) *J. Amer. Chem. Soc.* **76**, 3115–3121.
10. Weygand, F. & Zumach, G. (1962) *Z. Naturforsch. B* **17**, 807–810.
11. Larsson, L.-E., Melin, P. & Ragnarsson, U. (1976) *Int. J. Pept. Protein Res.* **8**, 39–44.
12. Yamashiro, D. (1964) *Nature* **201**, 76–78.
13. Jeffers, W. A., Livezey, M. M. & Austin, J. H. (1942) *Proc. Soc. Exp. Biol. Med.* **50**, 184–188.
14. Sawyer, W. H. (1958) *Endocrinology* **63**, 694–698.
15. Sawyer, W. H., Acosta, M., Balaspiri, L., Judd, J. & Manning M. (1974) *Endocrinology* **94**, 1106–1115.
16. "The Pharmacopeia of the United States of America" (1970) 18th revision, Mack, Eston, Pa., p. 771.
17. Schild, H. O. (1942) *J. Physiol. (London)* **101**, 115–130.
18. Meienhofer, J., Trzeciak, A., Havran, R. T. & Walter, R. (1970) *J. Amer. Chem. Soc.* **92**, 7199–7202.
19. Huguenin, R. L., Sturmer, E., Boissonnas, R. A. & Berde, B. (1965) *Experientia* **21**, 68–69.

A COMPARISON OF THE TERTIARY STRUCTURES AND DISULFIDE BRIDGE INTERNAL MOTIONS OF OXYTOCIN, ARGININE-VASOPRESSIN AND THEIR ANALOGS.

LESLIE J. F. NICHOLLS, JOSEPH J. FORD, CLAUDE R. JONES, *Department of Biochemistry, College of Agricultural and Life Sciences, University of Wisconsin-Madison, Madison, WI 53706,* MAURICE MANNING, *Department of Biochemistry, Medical College of Ohio at Toledo, P. O. Box 6190, Toledo, OH 43614,* and WILLIAM A. GIBBONS, *Department of Biochemistry, College of Agricultural and Life Sciences, University of Wisconsin-Madison, Madison, WI 53706.*

Although there are 162 possible conformations for the cystine bridge of a given neurohypophyseal hormone[1] (NHH), there have been few experimental studies of their existence or biological importance; often the bridge is considered to be fixed.[2] Experimental evidence is equally unavailable for secondary conformational averaging at the other amino-acid residues of the ring, although the tail is considered highly mobile.[3]

The effect of amino-acid substitution (principally at positions 3, 4, 7, and 8) on conformation and function have been largely restricted to the β-turn model;[4] the effect on the disulfide bridge and other residues remote from the substitution site has been neglected. We provide experimental evidence for the existence of multiple secondary conformations and for quantitation of multiple tertiary conformations in a number of NHH and postulate that individual NHH conformers are separately responsible for each different biological activity viz. binding to different receptors and to the different proteases which form and cleave them. The statistical weights of each contributing disulfide bridge conformation vary with pH, solvent, and analog; each has a different ring size which modulates and, in turn, is affected by the conformation of the other residues i.e. correlations may exist.[1]

Results and Discussion

Multiple disulfide conformations and ring sizes. The cysteine $\langle {}^3J_{\alpha\beta} \rangle$ values for a number of NHH are shown in Table I; they depend strongly on solvent, pH, and temperature, providing evidence for considerable mobility about the C^{α}–C^{β} bonds.[1,5] This means that the ring size and conformation vary with rotation around these bonds. Therefore different secondary conformations of the NHH

Table I. $\langle {}^3J_{\alpha\beta} \rangle$ Values for a Number of NHH Half Cysteine Residues

Molecule	Solvent	Cys[1]	Cys[6]
Oxytocin	D_2O(pH 7.4)	4.7; 8.0	3.4; 9.1
Oxytocin	DMSO	6.6; 7.5	6.1; 7.3
Oxytocin	D_2O(pH 3.8)	5.4; 5.8	4.0; 9.4
Arginine-vasopressin	D_2O(pH 3.8)	4.2; 5.3	2.8; 10.0
Arginine-vasotocin	D_2O(pH 3.8)	4.4; 5.6	3.5; 9.7
D-Arginine-vasopressin	D_2O(pH 3.8)	3.7; 5.5	3.5; 9.5

coexist. The corollary is that the physical parameters are averages over all the coexisting secondary conformations.

Assuming frozen rotation the $\langle {}^3J_{\alpha\beta} \rangle$ values yield the possible χ_1 angles. The value, thus calculated, is usually eclipsed, making the concept of frozen rotamers unlikely. Both crystallography[6] and theory[7] also predict near-classical values.

The Raman spectra of a wide range of disulfide compounds show the same three disulfide frequencies.[8] Analysis of this data by Sugeta et al.[8] predicts a three-fold potential for rotation about the C^β-S bonds, consistent with extensive averaging of disulfide bridge conformations. It has been reported[9] that the disulfide bridge of oxytocin is considerably more mobile than that of penicillamine-oxytocin, but that even this is not rigid.

The values in Table I indicate that substituting an amino acid at position 3 and/or 8 affects the conformational preferences at non-adjacent positions e.g. 1 and 6. Further, this finding has important biological consequences; the changes in disulfide conformational averaging between oxytocin and its congeners are largely one of degree and not kind, individual activities perhaps being reflected in one specific cystine (secondary) conformation, rather than all being related to a single secondary conformation.[4] Calculations predict several oxytocin backbone conformations of similar stability.[7] We thus have a strong case (a) for multiple disulfide bridge conformations in the NHH, (b) that co-existing disulfide conformers are the same for each molecule and (c) that variation in the statistical weights of these conformers leads to the variation of $\langle {}^3J_{\alpha\beta} \rangle$ with solvent, sequence, pH, and temperature—these variations can lead to considerable changes in relative ring sizes.

The existence of multiple tertiary conformations is confirmed by the dependence of side chain $\langle {}^3J_{\alpha\beta} \rangle$ values on solvent, pH, and temperature.[1,3,5] They may be quantified by rotamer analysis; the product of the individual rotamer populations gives the statistical weights of the co-existing conformers.[10] We have demonstrated this;[10] our data for the NHH indicate that no individual tertiary conformation has a statistical weight greater than 0.02 compared to the fact that

12 individual gramicidin S tertiary conformations have a combined statistical weight of 0.60. Therefore extensive conformational averaging of the NHH occurs.

Conclusions

Conformational averaging of the NHH depends on pH, temperature and sequence. There is extensive evidence for the existence of multiple secondary (particularly disulfide bridge) and tertiary solution conformations. It is likely that the co-existing conformers do not vary with these parameters but that their relative statistical weights do. We postulate a three-fold potential for rotation about the cysteine C^{α}-C^{β} bonds but it is not clear whether the rotational states are classical. This rotation has considerable effect on the ring size. The statistical weights of co-existing tertiary conformations can be obtained from rotamer analysis. Relative differences in principal biological functions can be related to differing statistical weights of NHH conformers.

This work was supported in part by research grants from the National Science Foundation (BMS 72 23819) and the National Institutes of Health (AM18604 and HD06351). C. R. J. was supported by an Enzymology Postdoctoral Training Grant (#IT32AM070490).

References

1. Nicholls, L. J. F., Jones, C. R. & Gibbons, W. A. (1977) *Biochemistry* **16,** 2248–2254.

2. Boicelli, C. A., Bradbury, A. F. & Feeney, J. (1977) *J. Chem. Soc. Perkin II,* 477–482.

3. Feeney, J., Roberts, G. C. K., Rockey, J. H. & Burgen, A. S. V. (1971) *Nature New Biol.* **232,** 108–110.

4. Stahl, G. L. & Walter, R. (1977) *J. Med. Chem.* **20,** 492–495 and references therein.

5. Feeney, J. (1975) *Proc. R. Soc. Lond.* **A345,** 61–72.

6. Rosenfield, R. (1974) Ph.D. Thesis, State University of New York at Buffalo.

7. Kotelchuck, D., Scheraga, H. A. & Walter, R. (1972) *Proc. Nat. Acad. Sci. USA* **69,** 3629–3633.

8. Sugeta, H., Go, A. & Miyazawa, T. (1973) *Bull. Chem. Soc. Jap.* **46,** 3407–3411.

9. Meraldi, J. P., Hruby, V. J. & Brewster, A. I. R. (1977) *Proc. Nat. Acad. Sci. USA* **74,** 1373–1377.

10. Kuo, M. C., Jones, C. R. & Gibbons, W. A. (1977), Biochemistry, submitted for publication.

CONFORMATIONAL STUDIES ON [Pro^3,Gly^4]-OXYTOCIN BY ^{13}C NMR.

ROXANNE DESLAURIERS and IAN C. P. SMITH, *Division of Biological Sciences, National Research Council, Ottawa, Canada K1A OR6,* GEORGE C. LEVY, *Department of Chemistry, Florida State University, Tallahassee, Florida 32306,* RONALD ORLOWSKI and RODERICH WALTER, *Department of Physiology and Biophysics, University of Illinois Medical Center, Chicago, Ill. 60680.*

A major problem in analysis of the conformational flexibility of peptides using ^{13}C spin-lattice relaxation times (T_1) is determining the relative contributions of overall and segmental motion to the observed relaxation times. Even in the case of cyclic amino acids and peptides, internal motions or pseudorotations can occur which are reflected in the observed T_1 values. Proline and proline-containing peptides provide good examples of a cyclic residue which undergoes internal flexion at a rate which is sufficiently rapid to affect the T_1 values. In the case of oxytocin, the cyclic portion of the peptide backbone could possibly be undergoing a conformational averaging process. We have investigated the ^{13}C nuclear magnetic resonance (nmr) characteristics of an oxytocin analog in which substitutions were made to favor steric restriction of the peptide backbone. [Pro^3,Gly^4]-oxytocin is an analog for which a type II β-turn[1,2] has been found by ^{1}H nmr spectroscopy[3], in agreement with theoretical predictions.

Herein we report data showing that, although the -Pro-Gly- sequence introduced in positions 3 and 4 of oxytocin is expected to favor a β-turn and thereby possibly diminish the frequency of oscillations within the ring, we observe *increased* motion for the α-carbons of the glycyl residue relative to the other α-carbons in the cyclic moiety. Secondly the introduction of a prolyl residue into the ring portion of oxytocin produces conformational heterogeneity as manifest by the presence of both *cis* and *trans* isomers about the -Tyr-Pro- peptide bond.

Our investigations were carried out at 68 MHz and 25 MHz. ^{13}C nmr spectra were obtained using previously described techniques[4,5]. Chemical shifts are reported in parts per million (ppm) downfield from external tetramethylsilane (TMS), using 1.5 ml samples containing 100 mg of peptide in D_2O. pH values are uncorrected meter readings in D_2O, and were adjusted using CD_3COOH.

Results and Discussion

The assignment of the ^{13}C spectrum of [Pro^3,Gly^4]-oxytocin, Table I, was based on comparison with the previously assigned spectrum of oxytocin at the same pH[6]. The values in parentheses are "extra" resonances which have been

Table I. ^{13}C Chemical Shifts and Spin Lattice Relaxation Times of [Pro^3,Gly^4]-Oxytocin in D_2O.

Residue		pH=3.5		pH=9.0		NT_1
		M	m	M	m	
Cys^1	αCH	52.9		53.1		143
	βCH$_2$	40.9		45.5	(46.0)	176
	C=O	167.8		175.4		-
Tyr^2	αCH	54.2	(55.0)	54.1	(54.9)	145
	βCH$_2$	36.4	(37.1)	36.6		220
	γC	128.5		128.5	(127.9)	-
	δCH	131.9		132.0		210
	εCH	116.5	(116.7)	116.6	(117.0)	222
	ζC	155.7		156.0		-
	C=O	172.9		171.6		-
Pro^3	αCH	63.3		63.2		142
	βCH$_2$	29.8	(32.2)	29.7	(31.9)	233
	γCH$_2$	26.1	(22.6)	26.2	(22.5)	263
	δCH$_2$	48.9	(47.7)	48.9	(48.1)	200
	C=O	175.1		175.1		-
Gly^4	αCH$_2$	43.5	(42.5)	43.7	(42.6)	200
	C=O	175.1		175.1		-

Residue		pH=3.5	pH=9.0		NT_1
		M	M	m	
Asn^5	αCH	50.8	50.9	(51.2)	155
	βCH$_2$	37.7	37.2	(37.8)	180
	γC=O	175.5	175.7		-
	C=O	173.1	172.5		-
Cys^6	αCH	52.0	52.2	(51.9)	150
	βCH$_2$	39.3	40.3		150
	C=O	170.6	170.4		-
Pro^7	αCH	61.6	61.9		175
	βCH$_2$	30.2	30.4		327
	γCH$_2$	25.2	25.8		407
	δCH$_2$	48.9	48.9		200
	C=O	175.3	175.2		-
Leu^8	αCH	53.6	53.6		185
	βCH$_2$	40.6	40.6		293
	γCH	25.3	25.3		310
	δCH$_3$	23.1	23.3		1590
	δCH$_3$	21.7	21.7		1435
	C=O	176.0	176.1		-
Gly^9	αCH$_2$	43.1	43.2		327
	C=O	174.9	174.9		-

Chemical shifts are reported in parts per million downfield from external tetramethylsilane (TMS).

M = major conformer
m = minor conformer

Assignments of carbonyl carbons are tentative.

NT_1 values are measured in milliseconds, N is the number of directly bonded hydrogens, pH 3.5.

assigned to the presence of a *cis* -Tyr-Pro- peptide link[7,8]. The relative ratio of the *cis* and *trans* isomers is *ca.* 1:3. The effect of the *cis-trans* isomerism about the -Tyr-Pro- peptide bond was manifest on the tyrosyl residue which demonstrated doubling of the α, β and ε resonances with chemical shifts of 55.0, 37.1 and 116.7 ppm. There is also, at 68 MHz, doubling of the α carbon resonance of the asparaginyl residue. The intensity ratio of the minor and major peaks in both cases is also 1:3. In the carbonyl carbon region at 68 MHz, there is definite doubling of the resonances assigned to the cystine residue. Other resonances may be doubled elsewhere in the spectrum which are not detectable due to their weak intensities or to overlap with other resonances in the spectrum of the major component. The spectra from the two prolyl and two glycyl residues in [Pro^3,Gly^4]-oxytocin are quite distinct. The α-carbons of the prolyl-3 and prolyl-7 residues differ by 1.5 ppm, those of the glycyl-4 and glycyl-9 residues by 0.4 ppm. The assignment of the glycyl-4 α-carbon resonance was verified by measuring the spectrum of [Leu^2, Pro^3, α,α-$^2H_2$-Gly^4]-oxytocin. Deuteration at the α carbon of the glycyl residue in position 4 effectively causes the disappearance of this α-carbon resonance due to the coupling between carbon and the deuterons and to the long relaxation time of this non-protonated carbon. We also obtained a spectrum of [Pro^3]-oxytocin in order to determine whether the *cis-trans* isomerism was manifest and if so, to what extent the neighboring glycyl-4 residue influenced this ratio in [Pro^3,Gly^4]-oxytocin. When comparing the relative ratios of *cis:trans* isomers in [Pro^3]-oxytocin and [Pro^3,Gly^4]-oxytocin, it can be seen that the sterically more crowded [Pro^3]-oxytocin shows a greater percentage of *cis* isomer (45%).

The results of the spin-lattice relaxation time (T_1) measurements performed at 68 MHz on the major (*trans*) isomer of [Pro^3,Gly^4]-oxytocin are given in Table I. The general trend of the relaxation times follows what has been observed in oxytocin[9] itself: a restricted backbone in the cyclic portion of the peptide and increasing segmental motion of the acyclic terminal tripeptide. In the cyclic portion we found a larger NT_1 value for the α-carbon of the glycyl residue than those for the α-carbons of the other amino acids in the 20-membered ring. Comparing the NT_1 values observed for similar residues in the peptide, the glycyl-4 and prolyl-3 residues are more restricted than their congeners in positions 9 and 7. This is consistent with the relative ratios of T_1 values within the two prolyl residues; that in position 3 reflects less intracyclic mobility, possibly as a result of through-bond, if not through space, constraint from the neighboring Tyr residue and the cyclic peptide backbone. We see from the data in Table I, that although the -Pro-Gly- sequence may favor formation of a β-turn, the formation of such a turn does not restrict the intrinsic flexibility at the α-carbon of the glycyl residue.

References

1. Urry, D. W. & Ohnishi, M., (1970) in *Spectroscopic Approaches to Biomolecular Conformation,* Urry, D. W., Ed., American Medical Association, Chicago, Ill., pp. 263–300.

2. Chandrasekaran, R., Lakshminarayanan, A. V., Pandya, U. V. & Ramachandran, G. N., (1973) *Biochim. Biophys. Acta* **303**, 14–27.

3. Wyssbrod, H. R., Ballardin, A., Gibbons, W. A., Roy, J., Schwartz, I. L. & Walter, R. (1975) in *Peptides: Chemistry, Structure, Biology,* R. Walter and J. Meienhofer, Eds., Ann Arbor, Sci. Publ., Ann Arbor, Mich., pp. 815–822.

4. Lyerla, J. R. & Levy, G. C. (1974) in *Topics in Carbon-13 NMR Spectroscopy*, Vol. 1, G. C. Levy, Ed., Wiley-Interscience, New York, pp. 79–148.

5. Deslauriers, R., Levy, G. C., McGregor, W. H., Sarantakis, D. & Smith, I. C. P. (1975) *Biochemistry* **14**, 4335–4343.

6. Walter, R., Prasad, K. U. M., Deslauriers, R. & Smith, I. C. P. (1973) *Proc. Nat. Acad. Sci., USA* **70**, 2086–2090.

7. Dorman, D. G. & Bovey, F. A., (1973) *J. Org. Chem.* **38**, 2379.

8. Bystrov, V. F., (personal communication).

9. Deslauriers, R., Smith, I. C. P. and Walter, R. (1974) *J. Amer. Chem. Soc.* **96**, 2289–2291.

STUDY OF THE CONFORMATIONAL AND DYNAMIC PROPERTIES OF THE LUTEINIZING HORMONE-RELEASING FACTOR AND VARIOUS ANALOGS USING INTRAMOLECULAR CHARGE TRANSFER COMPLEXES

B. DONZEL, C. SAKARELLOS and M. GOODMAN,
Department of Chemistry, University of California, San Diego, La Jolla, California 92093

Introduction

Our earlier work[1,2] involved charge transfer (CT) complexes to show that the folding of the luteinizing hormone-releasing factor (LRF) is influenced by the nature of the amino acid at position six. The observed intramolecular CT-intensity between [Nva^8(Nic$^+$)] and Trp^3 moieties is highest in the [D-Ala^6] analog as compared to the Gly^6 and [L-Ala^6] analogs. (Nic$^+$ is the abbreviation used for nicotinamide.) This trend is in qualitative agreement with the calculated[3] relative stabilities of β^{II}-turns containing an L- or a D-amino acid residue at the third position of the turn. We now present further CT results on analogs in which we have altered the positions of the CT donor-acceptor moieties or changed other parts of the LRF sequence (see Table I).

Table 1. LRF and Charge-Transfer Labelled Analogs

	1	2	3	4	5	6	7	8	9	10
	pGlu^1	His^2	Trp^3	Ser^4	Tyr^5	Gly^6	Leu^7	Arg^8	Pro^9	Gly^{10}-NH$_2$
(I)			Trp^3			D-Ala^6		Nva^8 (Nic$^+$)		
(II)		Trp^2	His^3			D-Ala^6	Nva^7 (Nic$^+$)	Ala^8		
(III)			Trp^3			L-Ala^6		Nva^8 (Nic$^+$)		
(IV)			Trp^3			D-Ala^6	Nva^7 (Nic$^+$)	Ala^8		
(V)			Trp^3	Glu^4 ┐		D-Ala^6	┘ Orn^7	Nva^8 (Nic$^+$)		
(VI)		des-His^2	Trp^3					Nva^{8+} (Nic$^+$)		
(VII)			Trp^3					Nva^8 (Nic$^+$)	Pro^9-NHEt	des-Gly^{10}
(VIII)		des-His^2	Trp^3					Nva^8 (Nic$^+$)	Pro^{9}NHEt	des-Gly^{10}
(IX)			Trp^3					Nva^8 (Nic$^+$)		

Evidence for a Preferential Folding Sequence in the LRF Molecule

The two analogs I and II have the same number of residues between the side chains carrying the acceptor and donor moieties. An approximately 35% higher CT intensity in I than in II (Figs. 1, 2) indicates that the backbone folding occurs preferentially around the central tetrapeptide sequence Ser-Tyr-Gly-Leu. The intramolecular CT spectra of III and II are almost identical (Fig. 2). Using this value as a reference characteristic for the flexibility of the LRF molecule, we can deduce that the conformational effect in I accounts for a 53% increase of the total population of the folded conformers.

The acceptor and donor moieties in IV are separated by three residues only, but would be located in opposite directions in a hypothetical planar extended structure. However, the observed intramolecular CT-effect in IV is approximately 34% higher than in II (which indicates that the peptide chain is highly flexible). It must be emphasized that the intramolecular CT effect does not reach the level of I in spite of the fact that IV contains one less residue between the acceptor and the donor moieties. We interpret this to indicate that compound I is preferentially folded around residue six.

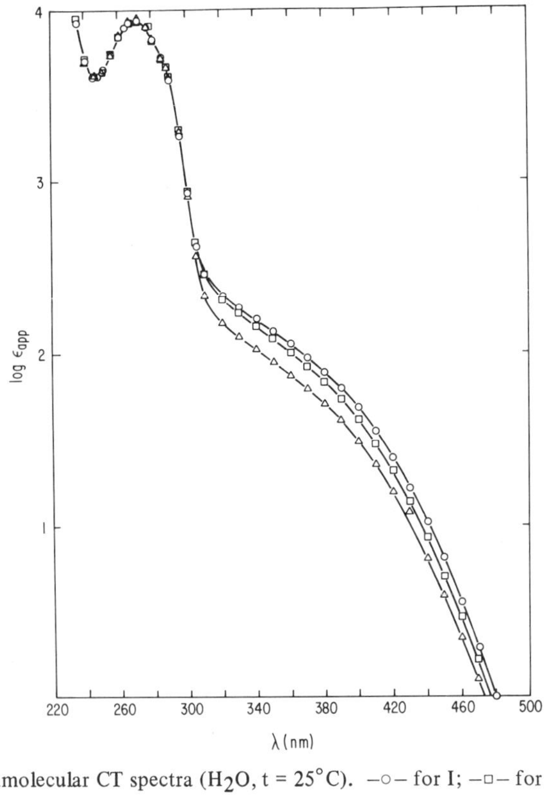

Fig. 1. Intramolecular CT spectra (H_2O, t = 25°C). –○– for I; –□– for IV; –△– for II.

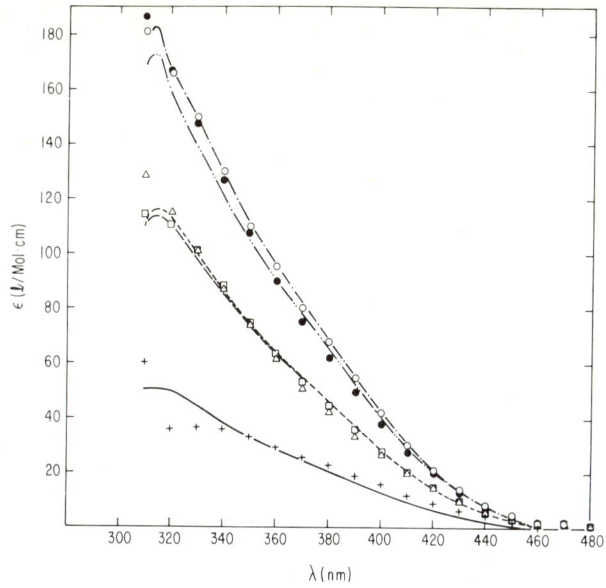

Fig. 2. Comparison between the experimental and simulated[4] intramolecular CT spectra. —·— Simulation and ○ experimental for I; —··— Simulation and ● experimental for IV; - - - - Simulation and △ experimental for III; ——— Simulation and □ experimental for II; —— Simulation and + experimental for V.

Synthesis and Conformation of [Glu^4,D-Ala^6,Orn^7,Nva^8(Nic$^+$)]-LRF·Cl$^-$ (V)

Side chain-side chain cyclization can be envisaged as a method to stabilize secondary structures of peptides.[5] In an attempt to determine the conformational effect of such a ring formation, we prepared the analog V (Schemes 1, 2) in which the side chains of the fourth and seventh residues of LRF (i.e., the central tetrapeptide sequence) are covalently linked through an amide bond. Under optimal conditions of cyclization, a 65% yield of pure, cyclic, monomeric material was obtained. The very low CT intensity of V (20%) compared to I (Fig. 2) at 380 nm indicates that the formation of the 17-membered ring has strongly perturbed the folded structure of the peptide backbone. It seems likely that the geometric constraints of the ring force the entire cyclic moiety into a nearly planar structure. These steric constraints greatly outweigh the stabilization energy of a regular, hydrogen-bonded β-pleated sheet structure.

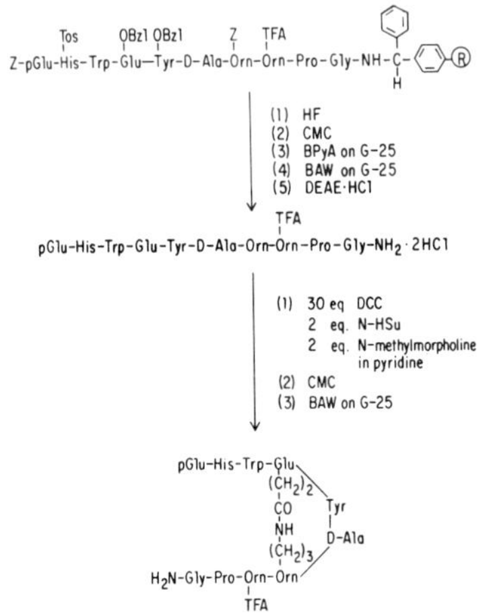

Scheme 1. Synthesis of [Glu4,D-Ala6,Orn7,Orn8(TFA)] LRF.

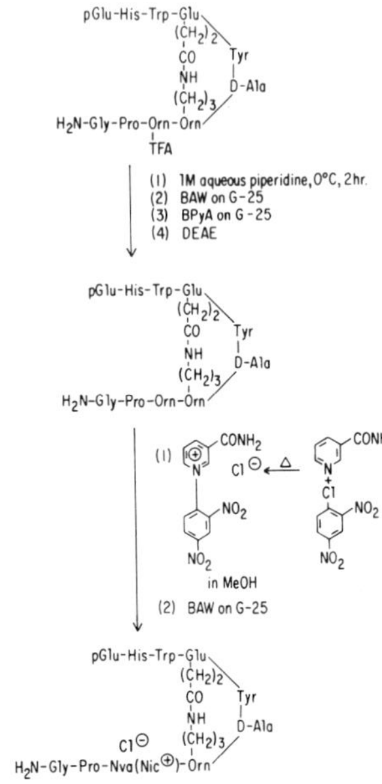

Scheme 2. Synthesis of [Glu4,D-Ala6,Orn7,Nva8(Nic^+Cl^-)]-LRF.

Study of the Specific Influence of Residues Two and Ten Upon the Folding Probability of the LRF-molecule

In addition to position 6, positions 2 and 10 in LRF were found to be most critical for biological activity.[6,7] In order to determine whether these modifications influence the folding probability of the peptide backbone, we have prepared the three CT labeled analogs, VI, VII and VIII.

At wavelengths longer than 350 nm, the analogs VII and VIII show similar difference spectra (Fig. 3). At 380 nm, the CT intensities are slightly higher (about 8%) than in the reference analog IX. The analog VI, on the other hand, shows a slightly lower charge transfer effect (about 8%) as compared to reference analog IX. We can conclude that the introduction of an ethylamide group at the tenth position produces a measurable increase of the backbone folding, whereas the deletion of the histidyl residue at position two decreases folding to a similar extent. The two modifications do not seem to be conformationally interrelated, since the CT intensities in VII and VIII are the same.

From these studies we have gained further insight into the conformational characteristics of the LRF molecule, providing strong evidence for a preferential folding sequence. We examined the structure of a cyclic analog as well as the influence of positions 2 and 10 upon the folding probability. Our goal is to place

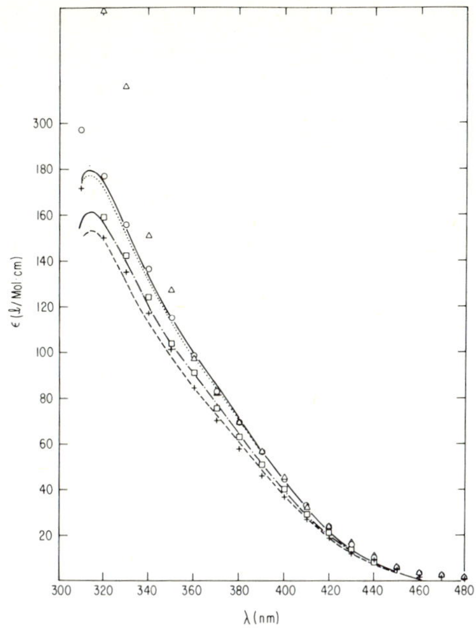

Fig. 3. Comparison between the experimental and simulated[4] intramolecular CT spectra. —— simulation and ○ experimental for VIII; . . . . . simulation and △ experimental for VII; –•– simulation and □ experimental for IX; - - - - simulation and + experimental for VI.

these studies on a quantitative base and to gain information concerning the active conformation of the hormone.

The authors gratefully acknowledge financial support by a grant from the National Institutes of Health (USPHS AM 15410).

References

1. Donzel, B., Gilon, C., Blagdon, D., Erisman, M., Burnier, J., Goodman, M., Rivier, J. & Monahan, M. (1975) in *Peptides: Chemistry, Structure and Biology*, Walter, R. & Meienhofer, J., Eds., Ann Arbor Science Publ., Inc., Ann Arbor, Mich., pp. 863-869.

2. Donzel, B., Rivier, J. & Goodman, M. (1977) *Biochemistry* **16**, 2611-2618.

3. Nemethy, G. & Printz, M. P. (1972) *Macromolecules* **5**, 755-758.

4. Deranleau, D. A. & Schwyzer, R. (1970) *Biochemistry* **9**, 126-134.

5. Donzel, B., Rivier, J. & Goodman, M. (1977) *Biopolymers*, in press.

6. Vale, W., Rivier, C., Brown, M., Leppaluoto, J., Ling, N., Monahan, M. & Rivier, J. (1976) *Clin. Endocrin.* **5**, Suppl., 261s-272s.

7. Fujino, M., Kobayashi, S., Obayashi, M., Shinagawa, S., Fukuda, T., Kitada, C., Nakayama, R., Yamazaki, I., White, W. F. & Rippel, R. H. (1972) *Biochem. Biophys. Res. Commun.* **49**, 863-869.

CARBON-13 NUCLEAR MAGNETIC RESONANCE STUDIES OF THE BINDING OF SELECTIVELY ^{13}C-ENRICHED OCYTOCIN TO ITS NEUROHYPOPHYSEAL CARRIER PROTEIN, NEUROPHYSIN I.

PAUL COHEN, ODILE CONVERT, JOHN H. GRIFFIN*, PIERRE NICOLAS, *Université Pierre et Marie Curie, Paris, France; *Scripps Clinic and Research Foundation, La Jolla, California 92037;* and CARLO DI BELLO, *Universita di Padova, Padova, Italy*

Carbon-13 enriched hormones offer a new approach for studying conformational and microdynamical properties of neurohypophyseal peptides bound to their physiological carrier proteins, the neurophysins[1-3]. Our initial studies involving the ^{13}C-enrichment of residues 3 and 9 of ocytocin demonstrated that residue 3 was involved in interactions between ocytocin and neurophysin while residue 9, the C-terminal glycinamide, was not affected[1,3]. In order to characterize further the potential involvement of each residue of ocytocin in the binding reaction, ocytocin molecules containing 85% ^{13}C-enrichment in the amino-acid residues at positions 7 and 8 have been synthesized and studied using ^{13}C nmr spectroscopy[4]. The chemical shift and spin lattice relaxation time (Table I) of each individual carbon atom of proline and leucine were measured on ^{13}C Pro^7- and ^{13}C Leu^8-ocytocin when each hormone was free in solution or bound to

Table I. NT_1 Values in Milliseconds (±15%) for ^{13}C Atoms of Pro^7- and Leu^8-ocytocins* (from Convert et al.[4]).

Carbon atom	Ocytocin (7 mM) alone	Ocytocin (7 mM)+0.28 eq. of neurophysin I
Pro^7 C^{α} (CH)	143	127
Leu^8 C^{α} (CH)	167	159
Pro^7 C^{β} (CH_2)	286	264
Leu^8 C^{β} (CH_2)	252	254
Pro^7 C^{γ} (CH_2)	304	290
Leu^8 C^{γ} (CH)	300	309
Pro^7 C^{δ} (CH_2)	178	144
Leu^8 C^{δ_1} (CH_3)	1340	1164
Leu^8 C^{δ_2} (CH_3)	1668	1612

* The given value of NT_1 is the mean value obtained from analysis of each signal of the multiplets arising from ^{13}C - ^{13}C coupling.

bovine neurophysin (Figs. 1 and 2). There was no indication that these nuclear magnetic resonance parameters of the two residues are affected by hormone binding[4] (Table I). The observed chemical shift values showed that the Pro^7 peptide bond remains in the *trans* conformation upon complex formation[4]. Based on the observed T_1 values, it is concluded that the segmental mobility of the entire tripeptide tail of the nonepeptide hormone is unaffected upon binding

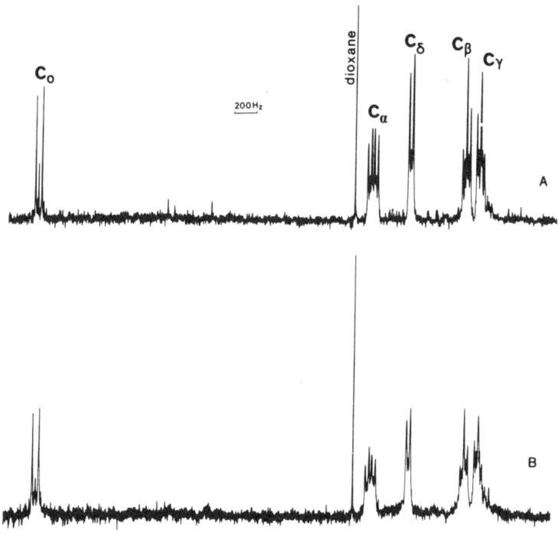

Fig. 1. ^{13}C nmr spectra of 85% ^{13}C enriched Pro^7-ocytocin A. Peptide alone at 7 *mM* (91,000 transients) B. same sample + 0.28 equivalents of neurophysin 1 (91,000 transients) (from Convert et al.[4]).

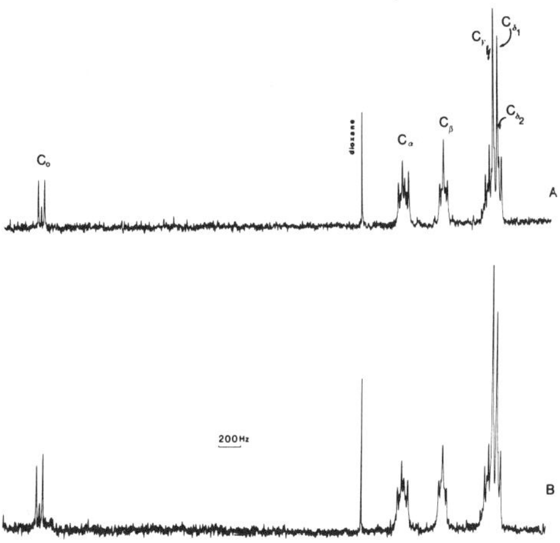

Fig. 2. ^{13}C nmr spectra of 85% ^{13}C enriched Leu^8-ocytocin A. Peptide is 7 *mM* (50,000 transients) B. same sample + 0.28 equivalents of neurophysin 1 (50,000 transients) (from Convert et al.[4]).

to neurophysin and that no large conformational rearrangement occurs in that region of the ligand in this process (Table I). These results[4], together with previous spectroscopic and thermodynamic data on the binding process[2,5] and with the proposed models for the preferred conformation of ocytocin in solution[6,7] allow the description of a more detailed picture for the association of this hormone to neurophysin. (See Fig. 3.[4]) This scheme, albeit oversimplified, should prove useful in the design and interpretation of future experiments.

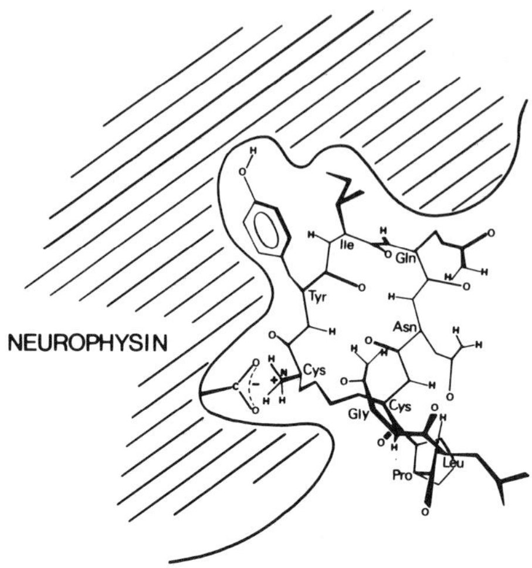

Fig. 3. A possible schematic representation of ocytocin binding to the high affinity binding site of a neurophysin protomer on the dimeric molecule[8]. The parallel diagonal lines in the neurophysin are meant to indicate a hydrophobic region of the protein molecule. The peptide backbone of ocytocin has been represented[6,7]. Its N-terminal tripeptide portion is presumed to be involved in interactions with the binding site while the C-terminal tripeptide tail is represented free from any direct bonds with the protein (from Convert et al[4]).

We gratefully acknowledge the helpful encouragement of Dr. Pierre Fromageot, Service de Biochimie, C. E. N., Saclay, who kindly provided the ^{13}C enriched amino acids for these studies. The help of Professor Basselier, Laboratoire de Chimie Organique Structurale, Université P. et M. Curie, Paris, in making available the nmr spectrometer is acknowledged. This is publication number 1351 from the Department of Immunopathology, Scripps Clinic and Research Foundation.

References

1. Griffin, J. H., DiBello, C., Alazard, R., Sala, E. & Cohen, P. (1975) in *Peptides: Chemistry, Structure, and Biology,* Walter, R. and Meienhofer, J., Eds., Ann Arbor Science, Ann Arbor, Mich. pp. 823–828.

2. Cohen, P., Camier, M., Wolff, J., Alazard, R., Cohen, J. & Griffin, J. H. (1975) *Ann. N. Y. Acad. Sci.* **248**, 463–479.

3. Griffin, J. H., DiBello, C., Alazard, R., Nicolas, P. & Cohen, P. (1977) *Biochemistry,* **16**, in press.

4. Convert, O., Griffin, J. H., DiBello, C., Nicolas, P. & Cohen, P. (1977) *Biochemistry* **16**, in press.

5. Breslow, E. (1975) in *Ann. N. Y. Acad. Sci.* **248**, 423–441.

6. Walter, R., Schwartz, I. L., Darnell, J. H. & Urry, D. W. (1971) *Proc. Nat. Acad. Sci. USA* **68**, 1355–1358.

7. Glickson, J. D., Rowan, R., Pitner, P. T., Dadok, J., Bothner-By, A. A. & Walter, R. (1976) *Biochemistry* **15**, 1111–1119.

8. Nicolas, P., Camier, M., Dessen, P. & Cohen, P. (1976) *J. Biol. Chem.* **251**, 3965–3971.

STUDIES OF THE INTERACTIONS OF OXYTOCIN AND ARGININE VASOPRESSIN WITH NEUROPHYSINS USING ^{13}C NUCLEAR MAGNETIC RESONANCE SPECTROSCOPY

VICTOR J. HRUBY, DIANE M. YAMAMOTO, YOUNG C. S. YANG, *Department of Chemistry, University of Arizona, Tucson, Az. 85721,* and MICHAEL BLUMENSTEIN, *Department of Biochemistry and Pharmacology, Tufts University School of Medicine, Boston, Ma. 02111*

The interactions of oxytocin and arginine vasopressin (AVP) with the neurophysins have been studied by many investigators using a variety of techniques[1-3]. We have been studying these interactions using ^{13}C nuclear magnetic resonance (nmr) in conjunction with solid-phase peptide synthesis. Our basic approach has been to synthesize the hormones enriched to 90% ^{13}C in specific positions, and then to examine the ^{13}C nmr spectrum of an equimolar (or other) mixture of neurophysin and enriched hormone and to observe the enriched resonance, which stands out above the background due to the natural abundance ^{13}C (1.1%) of the protein. Several syntheses of amino acids and hormones labeled in various positions have been completed and the hormone-protein interactions studied.

Results

In Fig. 1 are shown spectra of neurophysin I (NPI) and [1-hemi-L-[2-^{13}C]-cystine]-oxytocin. Upon binding to the protein, the resonance due to the enriched position is broadened and is shifted upfield by 2.7 ppm. The position of the resonance is independent of temperature and concentration, but the linewidth is highly dependent on these parameters, varying from about 20 hz at 11 mg/ml protein and 37°C to about 150 hz at 65 mg/ml protein and 22°C (at high concentrations the enriched peak is not clearly resolved from the natural abundance background). Upon addition of excess hormone to the solution, separate peaks due to both free and bound hormone are observed, indicating a slow rate of exchange ($< 15\ sec^{-1}$) for the hormone between neurophysin and the bulk solution.

Fig. 2 shows spectra of both [1-hemi-D- and [1-hemi-L-[1-^{13}C] cystine, 8-arginine]-vasopressin and neurophysin II(NPII). The resonance due to the enriched carbon of the natural diastereomer moves upfield by several parts per million as the hormone binds to the protein. In contrast, with the D-diastereomer the position of the enriched peak is unaffected by the presence of neurophysin. Titration curves of both diastereoisomers in the presence and absence of NPII are shown in Fig. 3. Data were not obtained below pH 6.5 due to precipitation of protein. Data with labeled oxytocin was identical to that obtained with labeled AVP.

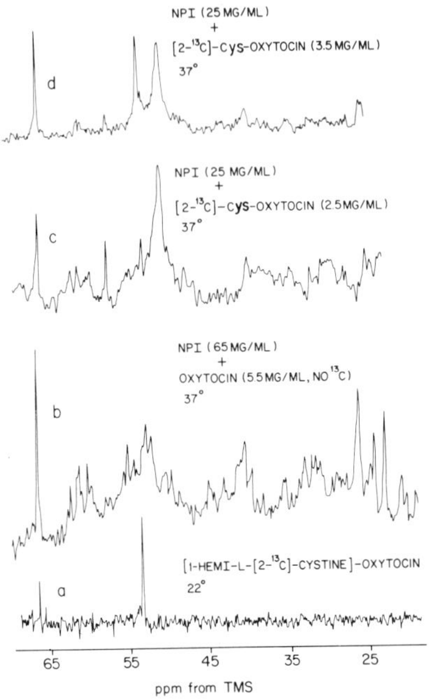

Fig. 1. ^{13}C nmr spectra of [1-hemi-[2-^{13}C] cystine]-oxytocin at 67.9 MHz, pH 6.6 and various conditions.

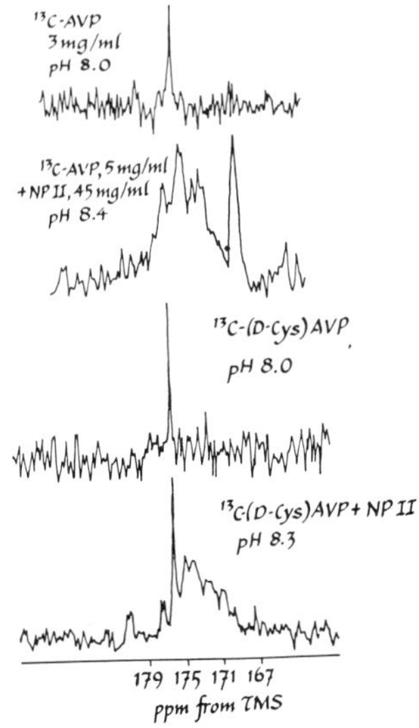

Fig. 2. ^{13}C nmr spectra at 25° of [1-hemi-L(D)-[1-^{13}C] cystine, 8-arginine]-vasopressin at 15.1 MHz and under various conditions.

Experiments with [2-[2-^{13}C] tyrosine]-oxytocin are shown in Fig. 4. Here, binding to neurophysin causes a downfield shift of 2.2 ppm. As was the case with [1-hemi-[2-^{13}C] cystine]-oxytocin, addition of excess hormone leads to two resonances, indicative of the slow exchange condition for hormone-protein interaction.

Discussion

For all three of the labeled hormone derivatives, binding to neurophysin was accompanied by substantial chemical shift changes. For [1-hemi-[1-^{13}C] cystine]-oxytocin, the total shift change can be explained by postulating that the α-amino group of half-cystine-1 is protonated throughout the observed pH range (6.5–8.9). This indicates that the pKa of this group, which is 6.3 in the free hormone, is raised to a value of above 9 upon binding to neurophysin. The raised pKa is consistent with the interactions of this amino group with a negatively charged group on the protein. Furthermore, the fact that the pKa increase is so large supports the hypothesis[1] that the amino group is in a region which is inaccessible to solvent in the hormone-protein complex. The fact that no chemical shift change is seen from pH 6.5–8.9 also means that the apparent pKa of about 7

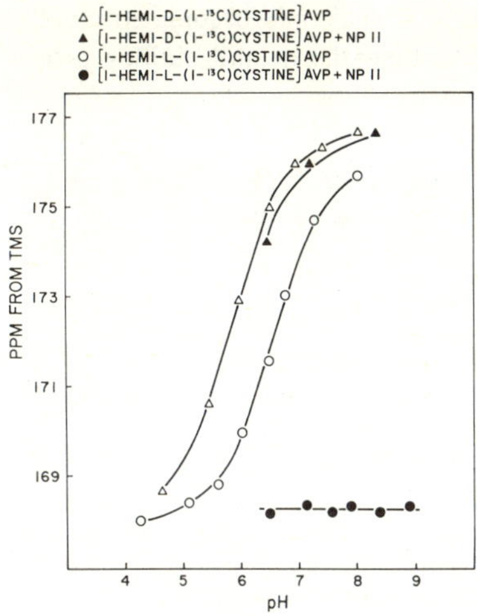

Fig. 3. Plot of the chemical shift vs. the pH of [1-hemi-L(D)-[1-^{13}C]-cystine, 8-arginine]-vasopressin.

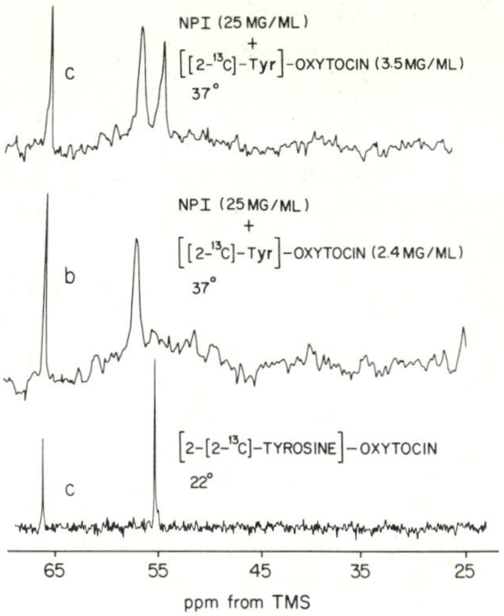

Fig. 4. ^{13}C nmr spectra at pH 6.6 of [2-[2-^{13}C] tyrosine] oxytocin at 67.9 MHz and under various conditions.

which influences the protein-hormone association constant[4] is most likely due solely to the pKa at the α-amino of free hormone.

The raising of the pKa of the α-amino group can account for at most a 1 ppm upfield shift of the α-carbon of half-cystine-1. The additional upfield shift (almost 2 ppm) could be due to a number of other factors such as the electric field effect of a nearby charge, a change in conformation of the peptide backbone, or the movement of this carbon into a medium which has a dielectric constant which differs from that of the bulk H_2O solution. Similar reasons could account for the observed shift of the α-carbon of tyrosine-2 (in free oxytocin the position of this resonance is independent of pH). Currently it is difficult to choose among the above possibilities, but hopefully studies with hormones enriched in other positions, as well as model studies will allow us to do so.

The three enriched positions all yielded data indicating a slow rate of exchange of the hormone between NPI or NPII and the bulk solution. From the data with the carbonyl enriched peptides (in which our linewidth measurements were most accurate) we can set an upper limit of 5 sec^{-1} for the exchange rate at pH 7.5, 25°C. This is somewhat slower than the rate of 18 sec^{-1} found by Pearlmutter and McMains[5]. The off rate may be dependent on neurophysin concentration as well as buffer conditions and our experiments were performed at concentrations an order of magnitude greater than those of Pearlmutter and in a different buffer. Other experiments[3,6] which we have conducted with [9-[2-^{13}C] glycinamide]-oxytocin yielded data indicating an exchange rate of greater than 1000 sec^{-1}. We believe that the exchange process sensed in the experiments with the glycinamide enriched hormone was a weak microscopic interaction between neurophysin and the hormone, rather than the overall binding process sensed by the labeled carbon in the other three positions.

This work was supported by the U. S. Public Health Service (VJH, MB, National Magnet Lab) and the National Science Foundation (VJH, National Magnetic Lab).

References

1. For a review see Breslow, E. (1974) *Advan. Enzymol.* **40**, 272–333.

2. For a review see Cohen, P., Camier, M., Wolff, J., Alazard, R., Cohen, J. S. & Griffin, J. H. (1975) *Ann. N. Y. Acad. Sci.* **248**, 463–479.

3. For more recent references see Blumenstein, M. & Hruby, V. J. (1976) *Biochem. Biophys. Res. Commun.* **68**, 1052–1058.

4. Camier, M., Alazard, R., Cohen, P., Pradelles, P., Morgat, J. L. & Fromageot, P. (1973) *Eur. J. Biochem.* **32**, 207–214.

5. Pearlmutter, A. F. & McMains, C. (1977) *Biochemistry* **16**, 628–633.

6. Blumenstein, M. & Hruby, V. J., manuscript submitted.

SYNAPTIC INTERACTIONS INVOLVING HYPOPHYSIOTROPIC PEPTIDES AND NEUROTRANSMITTERS IN THE CENTRAL NERVOUS SYSTEM

J. A. EDWARDSON and G. W. BENNETT, *Department of Physiology, St. George's Hospital Medical School, London SW17 ORE, England*

Hypophysiotropic peptides such as thyrotrophin-releasing factor (TRF) and somatostatin have a widespread distribution in nerve-endings throughout the central nervous system. In the hypothalamus their functions regulate pituitary secretion but elsewhere in the brain they appear to be involved in synaptic processes, although their precise role and mechanisms of action remain to be established. The present report describes studies with nerve-endings (synaptosomes) isolated from various brain regions. The peptidergic terminals which contain such factors may have functional connections with nerve-endings which release acetylcholine (ACh), monoamines or the putative amino-acid neurotransmitters.

Synaptosomes are formed during the homogenization of brain tissue under controlled conditions and can be isolated using differential and density gradient centrifugation. Rapid re-sealing of the pre-synaptic membrane occurs, retaining cytosol, mitochondria, and secretory granules within a miniature cell-like structure which lacks a nucleus but retains many metabolic and secretory properties of the nerve-ending *in situ*. Thus, when incubated in suitable media, synaptosomes respire at a high rate; accumulate K^+ and extrude Na^+; and synthesize proteins, phospholipids and neurotransmitters (see Ref. 1 for references). Synaptosomes contain a high proportion of the neurosecretory peptides present in brain tissue and these can be released in a calcium-dependent manner from the nerve-endings by depolarizing stimuli such as elevated K^+ levels or electrical field stimulation. This has been shown[1-3] for corticotrophin-releasing factor (CRF), vasopressin, prolactin release-inhibiting factor, TRF and the gonadotrophin-releasing factor (LRF).

A direct action of neurotransmitters on the neurosecretory nerve-endings of the median-eminence of sheep and rat has been shown in studies with synaptosomes isolated from this region.[3,4] Acetylcholine (ACh) at concentrations as low as $10^{-11}M$ stimulated release of CRF in amounts similar to those elicited by depolarizing agents such as elevated K^+ levels or electrical field stimulation. The action of ACh was blocked by the muscarinic blocker, atropine, and also dopamine at $10^{-8}M$. The monoamine also blocked release of CRF induced by electrical field stimulation. In contrast, dopamine caused the release of both TRF and LRF from median-eminence synaptosomes, while ACh was without effect.[4] Also, serotonin blocked the release of TRF in such preparations but had no effect on release of LRF or CRF. These data suggest that neurotransmitters in the median-eminence may specifically modulate the release of hypothalamic hormones from the peptidergic nerve terminals and thus regulate pituitary function.

The above findings suggested that similar interactions might be demonstrated with synaptosomes from other brain regions and in relation to other neurotransmitters. Suspensions of synaptosomes were prepared from the brain stem and cerebral cortex of rats and sheep. Such preparations release the physiologically active amino acids Glu, Asp, Gly and γAbu (γ-aminobutyric acid) in a preferential manner in response to depolarizing stimuli.[5] The effects of a variety of cerebral peptides on such mechanisms were studied using a sensitive chromatographic procedure[5] to measure amino-acid release. Figure 1 shows that TRF at $10^{-8}M$ added to brain stem synaptosomes caused a 6-fold increase in the K^+ induced release of glycine, and to a lesser extent the release of glutamic acid, aspartic acid and γ-aminobutyric acid but *not* other amino acids. In the absence of elevated K^+ levels, TRF had no effect on the basal release of amino acids. That the effect was on release rather than re-uptake mechanisms was confirmed in experiments using ^{3}H-labelled Glu and Gly; TRF did not influence uptake by synaptosomes but, after suitable pre-incubation, facilitated release in response to

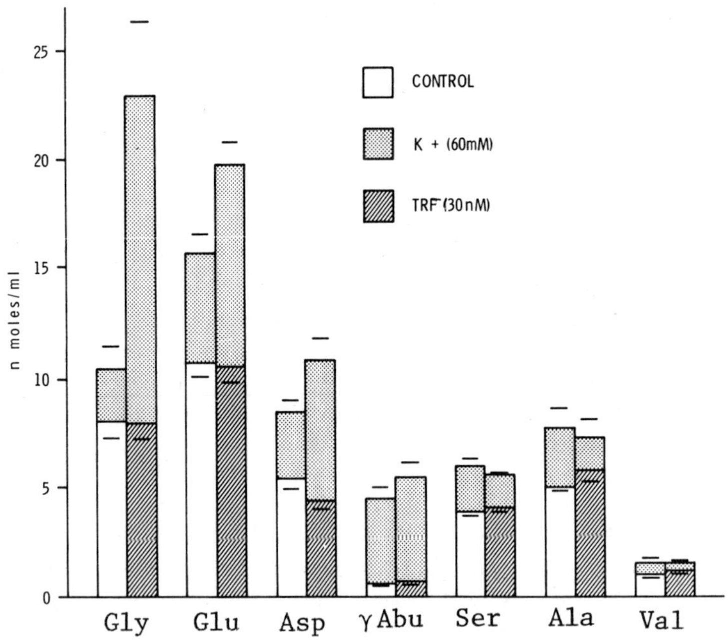

Fig. 1. Effect of TRF on the potassium induced release of amino acids from sheep brain stem synaptosomes.

K^+. Other peptides including LRF, somatostatin (SRIF) and β-endorphine were without effect. Enhanced release of neurotransmitter amino acids in response to K^+ was also observed when TRF was added to suspensions of nerve endings isolated from rat cerebral cortex. The effect was most marked with the excitatory neurotransmitters, glutamic and aspartic acids.

Synaptosome preparations have been used to investigate the effects of peptides and amino-acid neurotransmitters on TRF release. It has been shown that a significant increase in the release of TRF was observed when somatostatin ($10^{-8}M$) or glycine ($10^{-4}M$) was added to suspensions of brain stem synaptosomes, while addition of LRF, αMSH, or glutamic acid showed no such increase. In view of the widespread inhibitory actions of these two substances on neuronal function, it is likely that their apparent stimulatory effect on TRF release is due to the inhibition of release of an undetermined inhibitory factor which acts on the TRF-containing nerve terminal.

These findings are consistent with a role of extra-hypothalamic TRF and somatostatin in neurotransmission and indicate the existence of presynaptic monoamine-peptide, amino acid-peptide and even peptide-peptide interactions at neurosecretory nerve-terminals in the central nervous system. At present there is no ultrastructural evidence for direct axo-axonic or other 'point-to-point' synapses between peptidergic and other types of nerve-ending, even in the hypothalamic median eminence where these have been carefully looked for. Thus, more diffuse functional connections may exist, where the specificity is determined by the distribution of receptors.

Synaptosomes are generally accepted to consist of presynaptic rather than postsynaptic elements. Interactions between peptides and amino acids at the postsynaptic level are also likely to occur, but such problems will require the application of other, more sophisticated experimental procedures. However, our findings indicate that TRF may act presynaptically to increase the evoked release of amino-acid neurotransmitters, and may explain the potent behavioral effects of hypothalamic hormones.

This work was supported by a Programme Grant from the Medical Research Council. Some of this work was carried out in collaboration with Dr. S. L. Jeffcoate, Miss D. Holland and Miss N. White and with the excellent technical assistance of Mrs. J. Pennington.

References

1. Bennett, G. W. & Edwardson, J. A. (1975) *J. Endocr.* **65**, 33–44.
2. Edwardson, J. A., Bennett, G. W. & Bradford, H. F. (1972) *Nature* **240**, 554–556.
3. Bennett, G. W., Edwardson, J. A., Holland, D., Jeffcoate, S. L. & White, N. (1975) *Nature* **257**, 323–325.
4. Edwardson, J. A. & Bennett, G. W. (1974) *Nature* **251**, 425–427.
5. Bradford, H. F., Bennett, G. W. & Thomas, A. J. (1973) *J. Neurochem.* **21**, 495–505.

SYNTHESIS OF CARBACYCLIC ANALOGS OF SOMATOSTATIN BY COMBINATION OF CONVENTIONAL AND SOLID-PHASE PEPTIDE SYNTHESIS METHODOLOGY

D. SARANTAKIS and J. TEICHMAN, *Research Division, Wyeth Laboratories Inc., Radnor, Pennsylvania, 19087*

The synthesis of cyclic peptides is usually accompanied by mediocre yields during the cyclization step[1]. A convenient and rapid method for the construction of the linear suitably protected precursors of the cyclic peptides would be of great advantage for their routine synthesis.

It has been known that fragments of protected polypeptides can be removed from the Merrifield-type polystyrene resin (hydroxymethyl polystyrene type) by hydrazinolysis in excellent yield and without racemization[2]. These fragments further could be used for either direct cyclization by the azide method[3] or after coupling to a desired amino-acid derivative or peptide. The advantages of the combination of the solid phase techniques and conventional procedures have been discussed by Merrifield[4].

We have employed the above methodology for the synthesis of carbacyclic (I) and bicyclic (II) analogs of somatostatin.

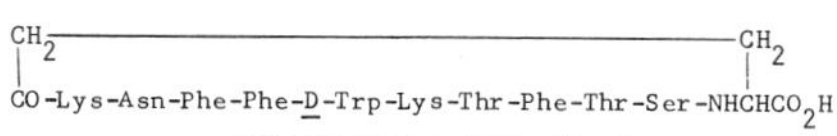

(I)

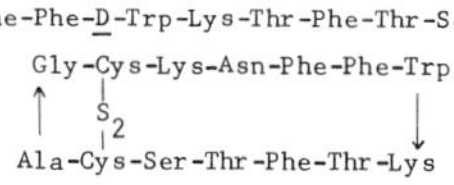

(II)

Carbacyclic analogs (III) of somatostatin have been prepared by conventional methods[5]. Our approach as shown in Scheme 1, utilizes the intermediate hydrazide (Ib) which was obtained by treatment of the peptide-resin (Ia) with 50

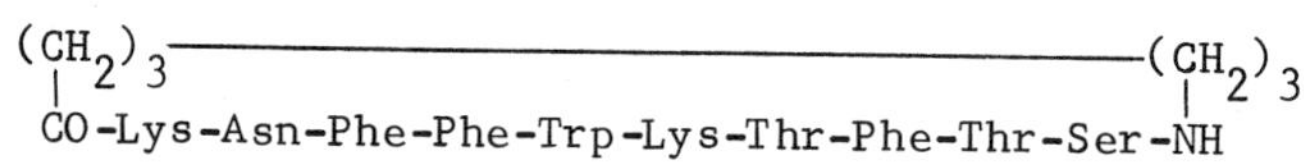

(III)

equivalents of anhydrous hydrazine in DMF/*n*-butanol for 3 hours. The hydrazide (Ib) was purified by chromatography through Sephadex LH 20 with DMF as solvent, then coupled with H-Glu(O-Bu^t)-OBzl by the azide method of Honzl and Rudinger[6] to afford (Ic) which was again purified by chromatography on Sephadex LH 20 with DMF as elution solvent. The undecapeptide was treated with TFA in the presence of anisole to afford the partially deprotected intermediate (Id) which was cyclized to (Ie) by the method of Wieland et al.[7]. The product was not purified but it was globally deprotected by hydrogenolysis in the presence of Pd-C to give a mixture of the monomeric and higher polymeric

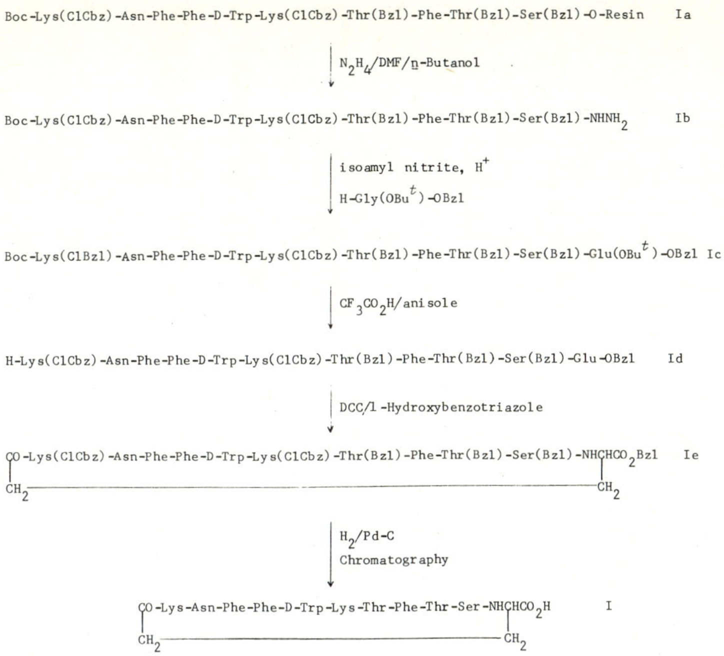

Scheme 1. Synthesis of the carbacyclic somatostatin analog (I).

species. Purification was carried out by repeated gel filtrations through Sephadex G-25 to obtain (I) in an average yield of 5%, based on the deprotected crude product. The synthesis of (II) (Scheme 2) was accomplished through the intermediate (IIb) which was obtained from (IIa) by hydrogenolysis as for (Ib). The hydrazide (IIb) was too insoluble to be chromatographed, therefore it was purified by digestion with warm methanol and repeated precipitations from DMF/H_2O. The N-terminal amino group was deprotected by TFA in the presence of anisole and cyclization was carried out in high elution by the azide method[8] to afford (IId). The final deprotection was carried out by treatment with liquid anhydrous HF to afford (IIe) which was oxidized by air in rather high dilution to the bicyclic compound (II). Purification of (II) was carried out by repeated gel filtration through Sephadex G-25.

Scheme 2. Synthesis of the Bicyclo-somatostatin analog (II).

References

1. Schröder, E. & Lübke, K. (1965) *The Peptides* Vol. I, Academic Press, New York, pp. 271–286.

2. Ohno, M. & Anfinsen, G. B. (1967) *J. Amer. Chem. Soc.* **89**, 5994–5995.

3. Kopple, K. D. (1972) *J. Pharm. Sci* **61**, 1345–1356.

4. Merrifield, R. B. (1973) in *Chemistry of Polypeptides,* Katsoyannis, P. G., Ed., Plenum Press, New York, pp. 335–361.

5. Veber, D. F., Strachan, R. G., Bergstrand, S. J., Holly, F. W., Homnick, C. F., Hirschmann, R., Torchiana, M. & Saperstein, R. (1976) *J. Amer. Chem. Soc.* **98**, 2367–2369.

6. Honzl, J. & Rudinger, J. (1961) *Coll. Czech. Chem. Commun.* **26**, 2333–2334.

7. Wieland, T., Birr, C. & Flor, F. (1969) *Ann. Chem.* **727**, 130–137.

8. Mazur, R. H. & Schlatter, J. M. (1964) *J. Org. Chem.* **29**, 3213–3216.

SYNTHESIS AND BIOLOGICAL ACTIVITY OF α-AZA-ANALOGS OF LUTEINIZING HORMONE-RELEASING FACTOR (LRF) WITH POTENT ANTAGONIST ACTIVITY

A. S. DUTTA, B. J. A. FURR and M. B. GILES,
Imperial Chemical Industries Limited, Pharmaceuticals Division, Alderley Park, Macclesfield, Cheshire, U.K.

Antagonists of LRF have been prepared by omitting the histidine residue from the 2 position[1] and by certain other modifications in the 2, 3 and 6 positions.[2-4] We have previously argued that antagonists of biologically active peptides may arise by α-aza-replacement.[5,6] Aza-amino-acid replacement in a peptide may change the overall conformation of the molecule, and in cases where this change affects only the intrinsic activity but not the affinity of the peptide, the resulting analogs may have inhibitory properties. Applied to LRF we first synthesized analogs containing α-aza-amino acids in positions 6 and 10. These analogs were somewhat less active than LRF in inducing ovulation in androgen-sterilized constant-oestrus rats and were considerably less active than LRF in releasing LH and FSH in immature male rats.[5] We now report that when α-aza-change is combined with changes in positions 2 and 6 some of the resulting compounds have high antagonist activity. The following α-aza-analogs (*1–9*) were synthesized (Scheme 1). In addition two non-aza-analogs (*10, 11*) were synthesized for comparison purposes.

The synthetic routes to the aza-analogs are described in Figures 1 and 2. The tri- or tetrapeptide fragments, except the ones containing an aza-amino acid, were prepared by stepwise coupling procedures. Coupling of these fragments by the azide method gave the final products which were purified by Sephadex LH-20

Glu-His-Trp-Ser-Tyr-Gly-Leu-Arg-Pro-Gly-NH_2	(LRF)
Glu-Trp-Ser-Tyr-Azgly-Leu-Arg-Pro-Gly-NH_2	(*1*)
Glu-Trp-Ser-Tyr-Azala-Leu-Arg-Pro-Gly-NH_2	(*2*)
Glu-Trp-Ser-Tyr-Azgly-Leu-Arg-Pro-NHC_2H_5	(*3*)
Glu-Trp-Ser-Tyr-Azala-Leu-Arg-Pro-NHC_2H_5	(*4*)
Glu-Trp-Ser-Tyr-Gly-Leu-Arg-Pro-Azgly-NH_2	(*5*)
Glu-Trp-Ser-Tyr-D-Ala-Leu-Arg-Pro-Azgly-NH_2	(*6*)
Glu-Trp-Ser-Tyr-D-Phe-Leu-Arg-Pro-Azgly-NH_2	(*7*)
Glu-D-Phe-Trp-Ser-Tyr-D-Phe-Leu-Arg-Pro-Azgly-NH_2	(*8*)
Glu-D-Trp-Trp-Ser-Tyr-D-Phe-Leu-Arg-Pro-Azgly-NH_2	(*9*)
Glu-D-Phe-Trp-Ser-Tyr-D-Phe-Leu-Arg-Pro-Gly-NH_2	(*10*)
Glu-D-Phe-Trp-Ser-Tyr-D-Ala-Leu-Arg-Pro-Gly-NH_2	(*11*)

Scheme 1

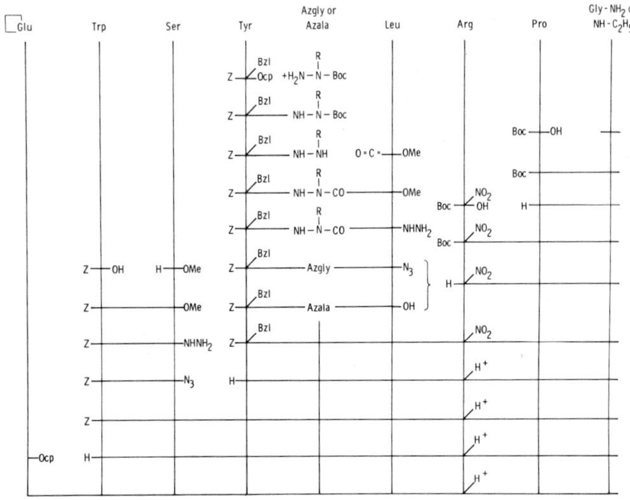

Fig. 1. Synthesis of des-His^2-[Azgly^6]-LRF, (*1*); des-His^2-[Azala^6]-LRF, (*2*); des-His^2-[Azgly^6, Pro-ethylamide^9]-des-Gly-NH_2^{10}-LRF, (*3*); and des-His^2-[Azala^6, Pro-ethylamide^9]-des-Gly-NH_2^{10}-LRF, (*4*). Z = benzyloxycarbonyl; Boc = *t*-butoxycarbonyl; Bzl = benzyl; Ocp = 2,4,5-trichlorophenyl ester; Azgly = -NH-NH-CO-; Azala = -NH-N(Me)-CO-.

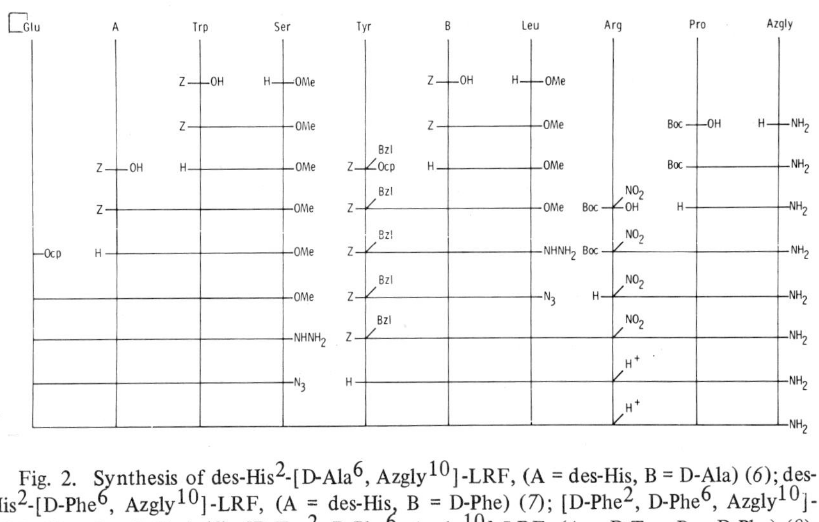

Fig. 2. Synthesis of des-His^2-[D-Ala^6, Azgly^{10}]-LRF, (A = des-His, B = D-Ala) (*6*); des-His^2-[D-Phe^6, Azgly^{10}]-LRF, (A = des-His, B = D-Phe) (*7*); [D-Phe^2, D-Phe^6, Azgly^{10}]-LRF, (A = B = D-Phe) (*8*); [D-Trp^2, D-Phe^6, Azgly^{10}]-LRF, (A = D-Trp, B = D-Phe) (*9*).

column chromatography using dimethylformamide as solvent and partition chromatography on Sephadex G-25 using *n*-butanol/acetic acid/water (4:1:5) and *n*-butanol/acetic acid/water/pyridine (5:1:5:1).

All analogs were tested in androgen-sterilized constant-oestrus rats for their ability (a) to induce ovulation (agonist activity), and (b) to inhibit the ovulation induced by LRF (0.5 μg/rat) (antagonist activity). In these experiments the rats were killed three days after injection and the ovaries were examined for the presence of corpora lutea. With one exception, none of the analogs showed agonist activity at doses up to 2 mg/rat. The exception was des-His^{2}-[Azala^{6}, Pro-ethylamide^{9}]-des-Gly-NH_2^{10}-LRF (*4*), which induced ovulation in 2 out of 3 rats at a dose of 2 mg/rat (but not at lower doses). The antagonist activity of the analogs is summarized in Table I. Des-His analogs containing the azaglycyl or azalanyl residue in position 6 were inactive but when azaglycine was substituted in position 10, the resulting compounds showed antagonist activity. Des-His^{2}-[Azgly^{10}]-LRF blocked ovulation completely at a dose of 250 μg/rat.

The corresponding non-aza-compound, des-His^{2}-LRF has been reported to have a very weak antagonist activity in an *in vitro* test system. Our most active compound, [D-Phe^{2}, D-Phe^{6}, Azgly^{10}]-LRF, was fully active at a dose of 15 μg/rat (i.v. or s.c.). It also blocked ovulation completely when injected (100 μg/rat, i.v.) 60 minutes before LRF (0.5 μg/rat) and partially when injected 120 minutes before LRF. The activity of [D-Phe^{2}, D-Phe^{6}, Azgly^{10}]-LRF was compared with [D-Phe^{2}, D-Phe^{6}]-LRF and [D-Phe^{2}, D-Ala^{6}]-LRF in our test system. These two compounds were fully active at a dose of 62.5 and 125 μg/rat respectively.

Further work on the synthesis of analogs containing azalanine in position 10 and D-Trp in positions 2 and 6 is in progress.

Table I. Effects of α-Aza-analogs of LRF on the Response to LRF in Androgen-sterilized Constant-oestrus Rats

No.	Compound	Dose (μg/rat)	Response (No. ovulating / No. treated)
1	Des-His^{2}-[Azgly^{6}]-LRF	1000	3/3
2	Des-His^{2}-[Azala^{6}]-LRF	1000	3/3
3	Des-His^{2}-[Azgly^{6},Pro-ethylamide^{9}]-des-Gly-NH_2^{10}-LRF	500	3/3
4	Des-His^{2}-[Azala^{6},Pro-ethylamide^{9}]-des-Gly-NH_2^{10}-LRF	500	3/3
5	Des-His^{2}-[Azgly^{10}]-LRF	250 125	0/3 2/6
6	Des-His^{2}-[D-Ala^{6},Azgly^{10}]-LRF	2000 500	2/3 3/3
7	Des-His^{2}-[D-Phe^{6},Azgly^{10}]-LRF	250 125	0/3 2/3
8	[D-Phe^{2},D-Phe^{6},Azgly^{10}]-LRF	15.6 7.8	0/3 1/3
9	[D-Trp^{2}-D-Phe^{6},Azgly^{10}]-LRF	125 62.5	0/3 2/3
10	[D-Phe^{2}, D-Phe^{6}]-LRF	125 62.5	0/3 1/3
11	[D-Phe^{2}, D-Ala^{6}]-LRF	250 125	0/3 2/3

References

1. Vale, W., Grant, G., Rivier, J., Monahan, M., Amoss, R., Blackwell, R., Burgus, R. & Guillemin, R. (1972) *Science* **176**, 933–934.

2. Yardley, J. P., Foell, T. J., Beattie, C. W. & Grant, N. H. (1975) *J. Med. Chem.* **18**, 1244–1247.

3. Nishi, N., Coy, D. H., Coy, E. J., Arimura, A. & Schally, A. V. (1976) *J. Reprod. Fert.* **48**, 119–124.

4. Humphries, J., Wan, Y. P., Folkers, K. & Bowers, C. Y. (1976) *Biochem. Biophys. Res. Commun.* **72**, 939–944.

5. Dutta, A. S., Giles, M. B., Furr, B. J. A. & Morley, J. S. (1976) *Clin. Endocrin.* **5**, **Suppl**, 291S–298S.

6. Dutta, A. S. & Morley, J. S. (1976) *Peptides 1976, Proceedings of the 14th European Peptide Symposium*, Loffet, A., Ed., Editions de l'Universite de Bruxelles, Belgium, 517–522.

HYPOTHALAMIC AND PITUITARY LRF-DEGRADING ENZYMES: CHARACTERIZATION, PURIFICATION AND PHYSIOLOGICAL ROLE

M. FRIDKIN, E. HAZUM, T. BARAM, H. R. LINDNER and Y. KOCH, *Department of Organic Chemistry and of Hormone Research, The Weizmann Institute of Science, Rehovot, Israel*

Peptidases capable of rapidly degrading and inactivating luteinizing hormone-releasing factor (LRF), also called LHRH, have been found in the hypothalamus[1,2] and the anterior pituitary[1,3] of the rat. As we have previously established, the enzyme of hypothalamic (HT) origin preferentially cleaves LRF at the Gly^6-Leu^7 bond.[1] We are now reporting that an enzyme found in the anterior pituitary (AP) attacks the peptide at the same site and in addition, the pituitary contains an enzyme activity that appears to open the N-terminal pyroglutamyl ring.

Enzyme preparations were obtained by homogenizing anterior pituitaries with phosphate buffered saline, pH 6.9 followed by centrifugation at 100.000 X g and collection of supernatant fraction. LRF (1 mg; a gift from N.I.H.) together with 1.5 μCi of <Glu-^{3}H-LRF (New England Nuclear, Boston) was incubated with this supernatant (3 mg protein) for 10 min at 37°C and the reaction was terminated by boiling for 3 min. Analysis of the enzymatic breakdown products (Fig. 1) was performed by high voltage paper electrophoresis at pH 1.9 (2.5% formic acid/7.5% acetic acid) for 60 min at 60 V/cm.

The various regions (marked I-VII) were eluted from the paper with 10% acetic acid and the eluate subjected to amino-acid analysis. The slowest radioactive component (I) was fluorescent (Trp), Pauli-positive, ninhydrin-negative and contained Glu (1.2), His (1.0), Ser (1.0), Tyr (0.97), Gly (0.80), indicating the structure of <Glu-His-Trp-Ser-Tyr-Gly-OH. The second radioactive band (II) was fluorescent (Trp), Pauli and ninhydrin positive. Amino-acid composition

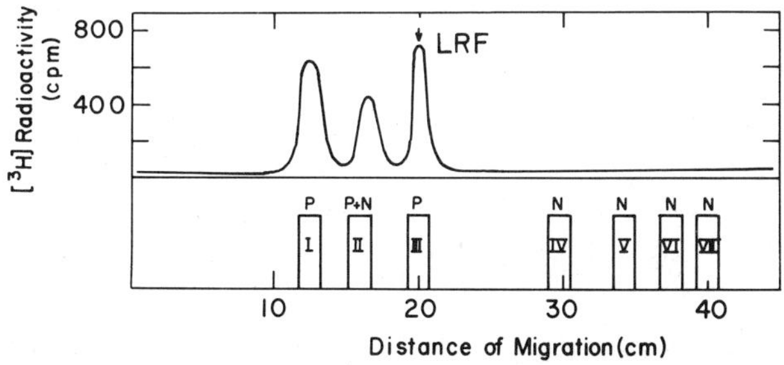

Fig. 1. Electropherogram of the degradation products of LRF. P., Pauli positive; N., Ninhydrin positive.

was Glu (1.25), His (1.0), Ser (0.74), Tyr (0.88), Gly (1.1), indicating the structure of H-Glu-His-Trp-Ser-Tyr-Gly-OH. The third radioactive component (III) was fluorescent, Pauli positive and identical in its amino-acid composition and mobility to intact LRF. Of the four non-radioactive but ninhydrin-positive components, only region V had a composition compatible with being a fragment of LRF. It contained Leu (0.92), Arg (1.06), Pro (1.01), Gly (1.00), suggesting that it represents the C-terminal sequence H-Leu-Arg-Pro-Gly-NH_2.

The enzymic inactivation of LRF by both the hypothalamic and the anterior pituitary enzymes was temperature dependent with an optimum of 37°C (Fig. 2), and pH-dependent with an optimum at pH = 7.6 (Fig. 3). The neutral pH-optimum suggests that the enzyme(s) concerned are not lysozomal cathepsins. K_m values were 8.7×10^{-7} *M* and 6.4×10^{-7} *M* for the enzymes of the AP and HT, respectively.

LRF degradation by both AP and HT enzyme preparations can be inhibited by diisopropylfluorophosphate (10^{-6} *M*), *p*-chloromercuribenzoate (10^{-5} *M*) and by *N*-ethylmaleimide (10^{-5} *M*). These results may indicate the presence of serine and/or thiol in the vicinity or at the active site of the degrading enzymes. The enzymes can also be inhibited by bacitracin (10^{-6} *M*) as well as by crude preparations of Kallikrein inactivator (Trasylol; a gift from Bayer A.G., Frankfurt; Refs. 1, 4). Both enzymes were not inhibited by the metal-chelating agents 1,10-phenanthroline (10^{-3} *M*) and ethylenediamine tetraacetic acid (10^{-3} *M*), suggesting that there are no metal requirements for enzymic activity.

The inhibitory component of Trasylol was utilized for affinity-purification of both HT and AP enzymes. Trasylol (4350 KIE/mg) was fractionated on a BioRex-70 ion-exchange column and the enzyme-inhibiting fraction was coupled to Sepharose by the CNBr procedure.[5] The conjugate was used for affinity-chromatography of the enzymes as illustrated in Fig. 4. A nine- to twelve-fold increase in the specific activities of the AP and HT enzymes, respectively, was thus achieved. Bacitracin was also coupled to Sepharose either directly or through two different extension arms: 1) – NH $(CH_2)_2$-COC_6H_4-*p*-N=N-bacitracin, 2) – NH $(CH_2)_5$-CO-NH-CH (COOH)-$CH_2C_6H_4$-*p*-N=N-bacitracin. The three conjugates which contained ~25 μmol bacitracin/ml, failed to separate the active

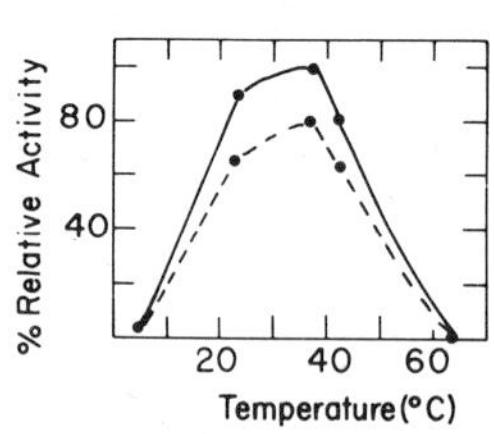

Fig. 2. Temperature dependence of enzyme activity. •—• pituitary enzyme; •- -• hypothalamic enzyme.

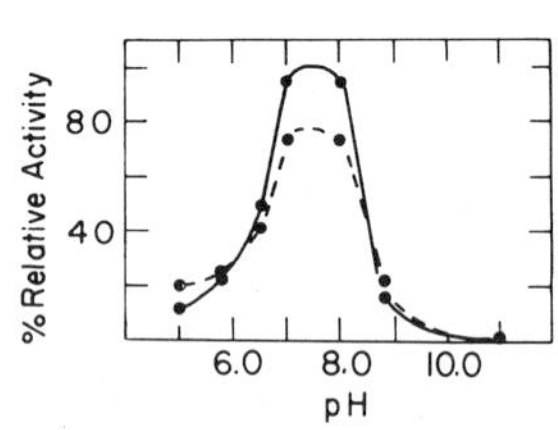

Fig. 3. pH dependence of enzymic activity. •—• pituitary enzyme; •- -• hypothalamic enzyme.

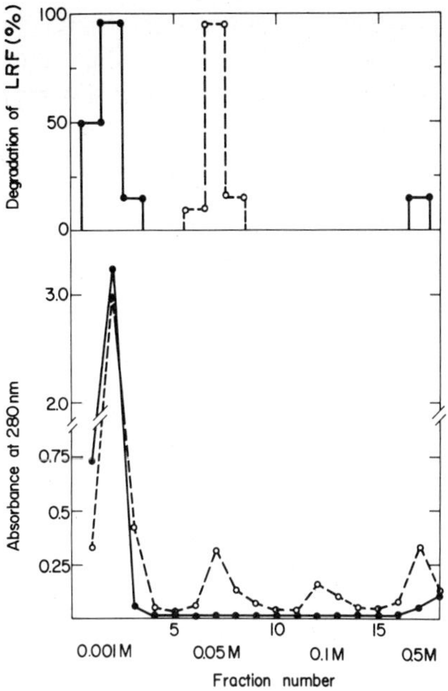

Fig. 4. Elution patterns of LRF-degrading enzyme preparations on affinity and control columns. Preparations were applied to 75 × 5 mm columns and eluted with a discontinuous gradient of phosphate buffer (0.001–0.5 *M*; pH 6.8); flow rate 0.25 ml/min; fractions of 1 ml were collected. Control resin was prepared by a similar coupling of myosin. ○- -○, absorbance and degrading-activity of Trasylol affinity column eluate; ●—●, absorbance and degrading activity of control column eluate.

enzyme fraction from the bulk of the accompanying proteins, using affinity chromatographic techniques, though the two conjugates with extension-arms inhibited the degradation of LRF by both HT and AP preparations.

Since the degrading enzymes are present at the sites of release (hypothalamus) and of action (pituitary) of LRF, we examined the possibility that they may have a physiological role in the fine control of the amount of the hormone at the receptor site. Several synthetic analogs of LRF, modified at positions Gly^6 and Leu^7, have been reported to possess enhanced luteinizing hormone-releasing activities.[6,7] This suggested a relationship between the resistance of these analogs to LRF-degrading enzymes and their enhanced biological potency. Analogs modified at the Gly^6 or at both the Gly^6 and Leu^7 residues have a higher resistance to the LRF-degrading enzymes;[3] the relative resistance of the analogs is: [D-Ala^6, (N^α-Me)Leu^7]-LRF > [D-Trp^6]-LRF > [D-Leu^6]-LRF > [D-Ala^6] LRF > LRF. [D-Leu^6, (N^α-Me)Leu^7]-LRF is as resistant as [D-Ala^6, (N^α-Me)-Leu^7]-LRF to enzymic attack.

This ranking order corresponds, generally, to the known order of biological potencies of these LRF analogs.[6,7] [D-Ala^6, (N^α-Me)Leu^7]-LRF, however, is exceptional in that, though immune to enzymic attack, it is only as active as [D-Ala^6]-LRF. This discrepancy may stem from impaired interaction of the analog with its receptor due to steric hindrance by the N^α-methyl group.

Further indication of the physiological role for the pituitary enzyme in determining the effective concentration of LRF at the receptor site is provided by the finding that enzyme activity in the pituitary is influenced by estrogen (Fig. 5). Ovariectomy causes a 50% decrease, while estradiol replacement restores enzymic activity to pre-castration levels. Thus degradation may act as a quenching mechanism that regulates the concentration of this neurohormone at the pituitary receptor-sites.

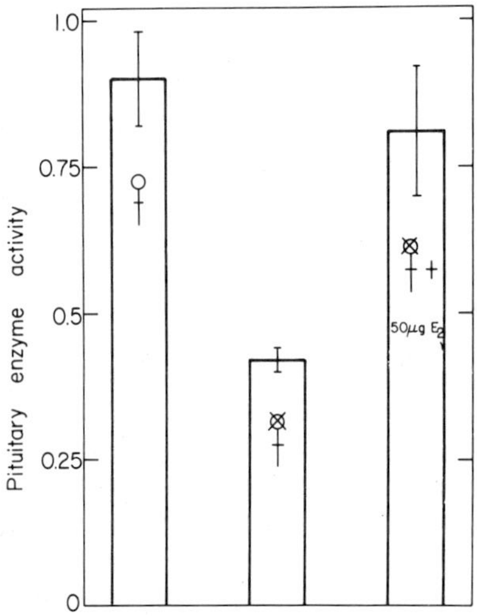

Fig. 5. Changes in LRF-degrading enzyme activity in the pituitary. Enzyme activity is defined as the reciprocal of the amount of tissue (mg) degrading 60% of 500 ng LRF in 15 min at 37°C. ♀, pituitaries taken from intact female; ⚢, pituitaries taken 2 weeks after ovariectomy; ⚢ + E_2, pituitaries taken from ovariectomized rats 3 days after a s.c. injection of 50 μg of estradiol benzoate.

Supported by the Ford Foundation and the Population Council Inc., N. Y.

References

1. Koch, Y., Baram, T., Chobsieng, P. & Fridkin, M. (1974) *Biochem. Biophys. Res. Commun.* **61**, 95–103.

2. Marks, N. & Stern, F. (1974) *Biochem. Biophys. Res. Commun.* **61**, 1458–1463.

3. Koch, Y., Baram, T., Hazum, E. & Fridkin, M. (1977) *Biochem. Biophys. Res. Commun.* **74**, 488–491.

4. McKelvy, J. F., LeBlanc, P., Laudes, C., Perrie, S., Grimm-Jorgensen, Y. & Kordon, C. (1976) *Biochem. Biophys. Res. Commun.* **73**, 507–515.

5. Porath, J., Axen, R. & Ernback, S. (1967) *Nature* **215**, 1491–1492.

6. Ling, N. & Vale, W. (1975) *Biochem. Biophys. Res. Commun.* **63**, 801–806.

7. Coy, D. H., Vilchez-Martinez, J. A., Coy, E. J. & Schally, A. V. (1976) *J. Med. Chem.* **19**, 423–425.

EFFECT OF NOREPINEPHRIN, DOPAMINE, AND SEROTONIN ON THE INHIBITION OF MSH RELEASE INDUCED BY Pro-Leu-Gly-NH_2 IN PITUITARIES INCUBATED *IN VITRO*

ADRIANA VIVAS and MARIA ESTER CELIS,
Instituto de Investigación Médica Mercedes y Martin Ferreyra Casilla de Correo 389 – 5000 Córdoba, Argentina

Introduction

Melanotropins released from the vertebrate pars intermedia are under inhibitory control of the hypothalamus.[1] Both direct neuronal innervation, as well as hypothalamic inhibiting and releasing factors, are implicated in the regulation of the pars intermedia function. The tripeptide Pro-Leu-Gly-NH_2 was found to inhibit MSH (melanocyte-stimulating hormone) release *in vivo* and *in vitro.*[2,3] On the other hand, the release of MSH which occurs spontaneously from incubated glands was inhibited by catecholamines (norepinephrin, NE, and dopamine, DA).[4] Since both the tripeptide and catecholamines act at the pituitary level, we shall discuss available data and provide information for our *in vitro* studies on the regulation of pars intermedia function.

Materials and Methods

Male rats were decapitated and the pituitary divided in halves. Both halves of the same pituitary were incubated at 37°C in minimum essential medium, using a Dubnoff metabolic shaker. After 30 min of incubation, each half was placed in 1 ml of fresh buffer and compounds to be tested were added to one hemipituitary, whereas the other half served as control. The MSH concentration in the medium was determined by bioassay and expressed as control percentage.

Results

As can be seen in Table I various doses of NE inhibited MSH release, and the most effective dose was 1.25 ng/ml. In contrast DA showed an inhibitory action only at a dose of 1.25 ng/ml; all the other doses tested were ineffective in affecting MSH release. However when DA (50–500 ng/ml) and Pro-Leu-Gly-NH_2 (20 ng/ml) were both present, the inhibition of MSH release was greater than the effect produced by Pro-Leu-Gly-NH_2 alone. An inhibitory effect was also observed when 20 ng/ml were incubated with 50 or 500 ng/ml of NE. These results

Table I. Effect of Dopamine (DA) or Norepinephrin (NE) on the Inhibition of MSH Release Induced by Pro-Leu-Gly-NH_2 (PLG) in Incubated Rat Hemipituitaries

	Dose, ng/ml					
	0	1.25	5	50	500	5,000
DA	-2.2±3.7(7)	-41.5±12.81(4)	-23.0±16.0(9)	-10.5±11.0 (9)	- 1.2±9.3 (4)	-13.7±17.7 (7)
NE	-2.2±3.7(7)	-64.5± 5.51(4)	-12.0±13.7(9)	-25.6± 8.5(5)	-25.7±6.21(4)	-37.0± 4.51(4)
DA plus 20PLG	-53.0±3.7(7)	-47.5-14.0 (3)	-30.0± 4.0(9)	-87.0±13.02(5)	-80.0±5.22(9)	-40.0± 7.2 (6)
NE plus 20PLG	-49.1±1.2(7)	-27.0±23.0 (4)	-45.0± 3.5(10)	-66.0± 1.62(5)	-68.0±12.02(5)	-43.2± 2.3 (6)

The MSH activity of the incubation medium of the treated hemipituitary is expressed as percentage change of that of the control incubated half. In parentheses number of experiments. Mean ± SEM.

1 0.05 compared with hemipituitary incubated in catecholamine-free medium.

2 0.025 compared with PLG alone.

indicate that DA at certain concentrations may potenciate the inhibiting effect of Pro-Leu-Gly-NH_2 on MSH secretion whereas the action of NE and Pro-Leu-Gly-NH_2 are additive.

Pimozide was used as a dopaminergic-receptor blocker and phenoxybenzamine as a α-blocker to find out whether the synergistic effect of DA on the inhibition exerted by Pro-Leu-Gly-NH_2 is implemented through an α-adrenergic or a dopaminergic receptor. The results showed that pimozide does not block the synergistic effect between DA and Pro-Leu-Gly-NH_2; whereas phenoxybenzamine blocks this action and the hemipituitary incubated with phenoxybenzamine plus Pro-Leu-Gly-NH_2 gives similar results as an hemipituitary containing only the Pro-Leu-Gly-NH_2 (Fig. 1). The same α-blocker was used to study the effect of NE, which was found to be incapable of inhibiting MSH release when phenoxybenzamine was present in the hemipituitary (Fig. 1). Neither pimozide nor phenoxybenzamine can act alone in the release of MSH into the medium. Adrenaline showed no action on the release of MSH into the incubation medium in any of the concentrations tested (1.25; 5; 50; 500 and 5000 ng/ml).

When propranolol, a β-adrenergic blocker, was analyzed alone, it showed an inhibiting effect on MSH release in all the concentrations used. When propranolol and Pro-Leu-Gly-NH_2 were added together at low doses of the β-blocker ($10^{-8}M$), the inhibition of the release of MSH was similar to that of Pro-Leu-Gly-NH_2 alone; but in high concentrations ($10^{-6}M$ and $10^{-5}M$) it increased the inhibition of MSH release (Fig. 2). Serotonin tested at the same concentrations as the other amines was ineffective in inducing changes in MSH release; but in the high concentration of $10^{-5}M$ it blocked the inhibitory effect of Pro-Leu-Gly-NH_2 on the MSH release.

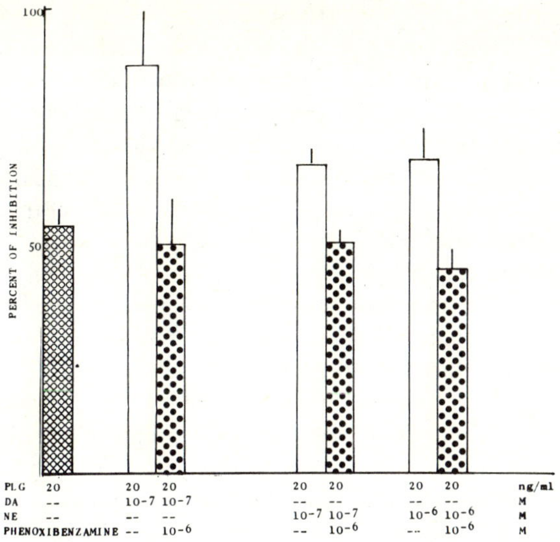

Fig. 1. Effect of phenoxibenzamine on the synergistic effect of DA and on the additive effect of NE in the inhibition of MSH exerted by Pro-Leu-Gly-NH_2 (PLG). Vertical lines indicate SEM. In parenthesis number of experiments.

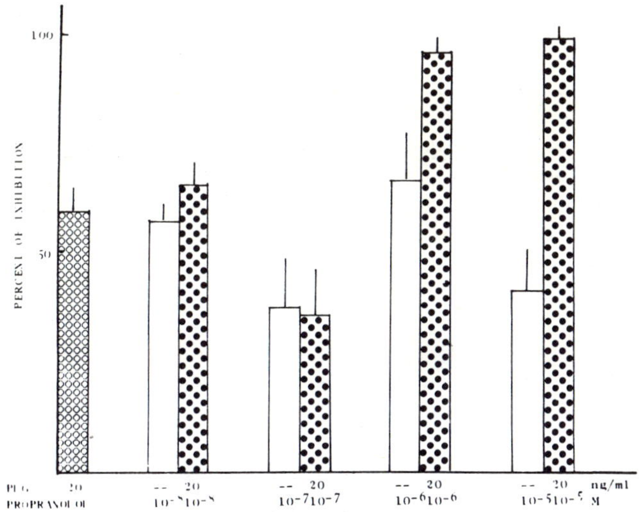

Fig. 2. Effect of propranolol on the MSH release by pituitaries incubated *in vitro* in presence and in absence of Pro-Leu-Gly-NH_2 (PLG).

Discussion

The results obtained with NE are in agreement with the work of Bower et al.[4] Besides this direct effect on the pars intermedia; catecholamines have been shown to inhibit MSH release in the rat by an action at hypothalamic level via the release of melanotropin release-inhibiting factor.[5] A similar mechanism has been proposed for prolactin control.[6] The synergistic effect that in some doses DA has on the inhibition exerted by Pro-Leu-Gly-NH_2 is blocked by phenoxybenzamine which in turn also blocks the additive effect of NE on Pro-Leu-Gly-NH_2 inhibition. These results suggest that an α-adrenergic receptor is implicated in this action. Epinephrine in our experiments does not show any action on MSH release; other authors found inhibition or release of MSH according to the dose used.[4] Propranolol showed inhibition of MSH release when it was tested alone. This effect could be a consequence of propranolol inhibition of Ca^{++} effluxes.[7] Besides, propranolol has an additive effect on Pro-Leu-Gly-NH_2 in high doses. Serotonin was ineffective in inducing any change in MSH release, contrary to what occurs *in vivo* where it does induce MSH release.[8] Serotonin in high doses ($10^{-5}M$) blocked the inhibitory effect of Pro-Leu-Gly-NH_2.

Supported by the Consejo Nacional de Investigaciones Cientificas y Técnicas of Argentina.

References

1. Etkin, W (1967) in *Neuroendocrinology,* Vol. 2, Martini, L. & Ganong, W. F., Eds., Academic Press, New York, pp. 261–282.
2. Celis, M. E., Taleisnik, S. & Walter, R. (1971) *Proc. Nat. Acad. Sci. USA* **68**, 1428–1433.
3. Vivas, A. & Celis, M. E. (1977) *Acta Physiol. Latinoamer.* In press.
4. Bower, A., Hadley, M. E. & Hruby, V. J. (1974) *Science* **184**, 70–72.
5. Taleisnik, S., Tomatis, M. E. & Celis, M. E. (1972) *Neuroendocrinology* **10**, 235–245.
6. Greibrokk, T., Hansen, J., Knudser, R., Yiu-Kuen, L., Folkers, F. & Bower, C. Y. (1975) *Biochem. Biophys. Res. Commun.* **67**, 338–344.
7. Seeman, P. (1972) *Pharmacol. Rev.* **24**, 583–655.
8. Taleisnik, S., Celis, M. E. & Tomatis, M. E. (1973/74) *Neuroendocrinology* **13**, 327–338.

SYNTHESIS OF Boc-*O*-BENZYL-HOMOSERINE: ITS USE IN THE SYNTHESIS OF [4-HOMOSERINE]-OXYTOCIN

MAURICE MANNING, ANDRAS TURAN, JAYA HALDAR and WILBUR H. SAWYER, *Medical College of Ohio, Department of Biochemistry, Toledo, Ohio 43699; College of Physicians and Surgeons of Columbia University, Department of Pharmacology, New York, N. Y. 10032.*

The synthesis of [4-homoserine]-oxytocin was originally proposed in 1970 to test an hypothesis presented at the 2nd American Peptide Symposium in Cleveland.[1] This hypothesis attempted to explain the remarkable two-fold enhancement of oxytocic activity of [4-threonine]-oxytocin relative to oxytocin.[2]

The incorporation of homoserine into oxytocin proved extremely difficult. The difficulty stemmed from the well-known tendency of γ-hydroxy amino acids to lactonise under acidic conditions.[3-5] Thus homoserine (Hse) and its N^{α}-substituted derivatives readily form the corresponding lactones. To date homoserine has been incorporated into only one synthetic peptide.[6] The method described did not involve derivatization of the OH group. Our attempts to use this approach to synthesize [Hse^4]-oxytocin were unsuccessful. We now report the synthesis of N^{α}-*tert*-butyloxycarbonyl-*O*-benzyl-L-homoserine (Boc-Hse (OBzl)) and its successful incorporation into position 4 in oxytocin to give [Hse^4]-oxytocin. Boc-Hse (OBzl) was obtained by an adaptation of a recently reported method used for the preparation of Boc-*O*-benzyl-serine[7] (Fig. 1). It was incorporated into the required protected nonapeptide by the Merrifield solid phase method[8] following previously published procedures.[2,9] Deblocking with sodium in liquid ammonia[10], cyclization[11], and purification by gel filtration on Sephadex G-15 by the two-step procedure involving 50% AcOH and 0.2 *M* AcOH as eluants respectively as previously described[12,13] afforded the desired [Hse^4]-oxytocin. The pharmacological properties of this peptide were evaluated by previously described procedures.[14,15] A more complete description of its synthesis and pharmacological properties will be reported elsewhere.[16]

Results and Discussion

Synthetic Aspects. Boc-L-Homoserine was obtained as an oil, free from lactone, by a modification of the DMSO method.[17] It was characterized as the dicyclohexylammonium (DCHA) salt. Storage as the DCHA salt was also necessary to prevent spontaneous lactone formation. Lactone formation was further avoided by the direct conversion of the Boc-homoserine DCHA salt into the sodium salt with aqueous sodium chloride in DMF prior to the benzylation step. The desired

L-Homoserine

↓ Boc-Azide:Et_3N

Boc-Homoserine (oil)

↓ Dicyclohexylamine

Boc-Homoserine DCHA (80%)

↓ M.P. 148-149° $[\alpha]^{23}_D$ -2.6 (C, 2, DMF)
NaCl

Boc-Homoserine Na

↓ Sodium Hydride
Bzl—Br

Boc-O-Benzyl-Homoserine (oil) (48%)

↓ Cyclohexylamine

Boc-O-Benzyl-Homoserine CHA salt (99%)

M.P. 145-147°; $[\alpha]^{23}_D$ +3° (C, 3 methanol)

Fig. 1. Synthesis of Boc-Hse (OBzl)

Boc-*O*-benzyl-homoserine, obtained as an oil, was characterized as its cyclohexylammonium (CHA) salt. To our knowledge this is the first successful synthesis of homoserine derivative suitable for general use in peptide synthesis.

Pharmacological Aspects. The oxytocic and antidiuretic properties of [Hse^4-serine]-oxytocin are presented in Table I. [Hse^4]-oxytocin is only one fourth as active as oxytocin in the oxytocic assay system.

Background to Hypothesis. In comparing the oxytocic activities of all known 4-substituted analogs of oxytocin,[1,2] it became clear that the nature of the side chain at position 4 plays a critical role in determining the degree of oxytocic activity of a given analog. The critical structural features are: 1) the length of the side chain; 2) the degree of branching; 3) the possession of both lipophilic and hydrophilic character; 4) the nature of the hydrophilic substituent.

Hypothesis. In satisfying factor 3, both glutamine and threonine have equal amounts of lipophilic character as measured by the number of methylene and methyl groups. Thus, it was postulated that the greater effectiveness of threonine relative to glutamine is related solely to factor 4 i.e. the β-OH group is more effective than the γ-$CONH_2$ group in helping to elicit the oxytocic response.

Support of Hypothesis. [4-Serine]-oxytocin has twice the oxytocic potency of [4-asparagine]-oxytocin (Table I). This is apparently due to the greater effectiveness of the β-OH as compared to the β-$CONH_2$ group.

Test of Hypothesis. This was based on the assumption that the β- and γ-positions at position 4 are equivalent as far as the effectiveness of the OH group is concerned. Since glutamine and homoserine differ only with respect to a $CONH_2$/OH interchange at their respective γ-carbons, [4-homoserine]-oxytocin would, if both the hypothesis and the assumption upon which it was based were correct, be twice as potent as oxytocin.

Table I. Pharmacological Activities (Units/mg ± Standard Errors) of [4-Homoserine]-Oxytocin, Oxytocin and Related Peptides

	Peptide	Rat Oxytocic (no Mg++)	Rat Antidiuretic
1)	[4-Homoserine]-oxytocin[a]	125±13	0.24 ±0.03
2)	Oxytocin[b]	520±12	4.0 ±0.8
3)	[4-Threonine]-oxytocin[b]	923±95	1.8 ±0.3
4)	[4-Serine]-oxytocin[c]	195±30	0.06 ±0.01
5)	[4-Asparagine]-oxytocin[d]	108±29	0.044±0.005

[a]Present communication and reference[16]; [b]Reference[13]; [c]Reference[18]; [d]Reference[19]

Conclusion. It is clear from the data presented in Table I that our attempt to explain the high oxytocic potency of [Thr^4]-oxytocin was too simplistic in ignoring from the beginning the distinction between β and γ substituents. The fact that [Hse^4]-oxytocin is much less active than oxytocin does not invalidate the original hypothesis. The hypothesis was based on an incorrect assumption. This peptide was not, in retrospect, the best peptide to test the hypothesis with. To test the hypothesis properly would require the incorporation at position 4 of β-methyl asparagine. It now appears that the greater effectiveness of threonine relative to glutamine at position 4 is due to a combination of factors 2 and 4 i.e. to the degree of branching and to the nature of the hydrophilic substituent. It is also clear that the relative effectiveness of the OH group vis-a-vis the $CONH_2$ group on position 4 side chains, in helping to elicit the oxytocic response, depends very much on whether it is attached to a β- or a γ-carbon.

This work was supported in part by research grants from the National Institute of Child Health and Human Development (No. HD 06351), the National Institute of Arthritis, Metabolism and Digestive Disease (No. AM 01940), and National Heart and Lung Institute (No. HL 12738). The authors wish to thank Ms. Cindy Licata for invaluable assistance in the preparation of the manuscript.

References

1. Manning, M. & Sawyer, W. H. (1972) in *Progress in Peptide Research* (Proceedings of the Second American Peptide Symposium) Lande, S., Ed., Gordon and Breach Publishers, New York, pp. 79–87.
2. Manning, M., Coy, E. & Sawyer, W. H. (1970) *Biochemistry* 9, 3925–3930.
3. Armstrong, M. D. (1949) *J. Amer. Chem. Soc.* 71, 3399–3402.
4. Greenstein, J. P. & Winitz, M. (1961) *Chemistry of the Amino Acids,* John Wiley and Sons, Inc., New York, Vol. 3, pp. 2612–2616.
5. Flavin, M. & Slaughter, C. (1965) *Biochemistry* 4, 1370–1375.
6. Morley, J. S. & Smith, J. M. (1968) *J. Chem. Soc.,* 726–733.
7. Sugano, H. & Miyoshi, M. (1976) *J. Org. Chem.* 41, 2352–2353.
8. Merrifield, R. B. (1963) *J. Amer. Chem. Soc.* 85, 2149–2154.

9. Manning, M. (1968) *J. Amer. Chem. Soc.* **90,** 1348–1349.

10. du Vigneaud, V., Ressler, C., Swan, J. M., Katsoyannis, P. & Roberts, G. W. (1954) *J. Amer. Chem. Soc.* **76,** 3115–3121.

11. Hope, D. B., Murti, V. V. S. & du Vigneaud, V. (1962) *J. Biol. Chem.* **237,** 1563–1566.

12. Manning, M., Wuu, T. C. & Baxter, J. W. M. (1968) *J. Chromatogr.* **38,** 396–398.

13. Manning, M., Coy, E. J. & Sawyer, W. H. (1971) *Experientia* **27,** 1372–1374.

14. Munsick, R. A. (1960) *Endocrinology* **66,** 451–457.

15. Sawyer, W. H. (1958) *Endocrinology* **63,** 694–698.

16. Turan, A., Manning, M., Haldar, J. & Sawyer, W. H. (1977) *J. Med. Chem.* (in press).

17. Stewart, J. M. & Young, J. D. (1969) in *Solid Phase Peptide Synthesis,* W. H. Freeman, San Francisco, CA, pp. 29–30.

18. Guttman, St. & Berde, B. (1963) *Helv. Chim. Acta* **46,** 1625–1636.

19. Jaquenoud, P. A. & Boissonnas, R. A. (1962) *Helv. Chim. Acta* **45,** 1601–1607.

STUDIES ON THE SYNTHESIS AND BIOLOGY OF NISIN: RING A

PETER PALLAI, TATEAKI WAKAMIYA, and ERHARD GROSS,
Section on Molecular Structure Reproduction Research Branch
National Institute of Child Health and Human Development
National Institutes of Health Bethesda, Maryland 20014

The heterodetic pentacyclic peptide nisin[1] (Figure 1) and its N-terminal fragments (residues 1–21) affect neonatal and neoplastic tissue by inducing fetal resorption and inhibiting the growth of tumor tissue. Trypsin-derived fragment T_1 (residues 1–12) displays similar activities. The peptide 1–7 (Figure 1) with one residue each of lanthionine (Ala-*S*-Ala), dehydroalanine (Dha), and dehydrobutyrine (Dhb) is predicted to also display the biological activities indicated. In order to test this hypothesis and to provide the first cyclic peptide for the stepwise condensation of Rings A (residues 3–7), B (residues 8–11), and C (residues 13–19), precursors have been provided for synthetic routes. These precursors offer a maximum of necessary flexibility (for cyclization via peptide bond or sulfide-bridge formation) to assemble Ring A with dehydroalanine in position 5 (Figure 1). With these precursors there is also the possibility of incorporating radiolabel for the study of the mode of action of nisin, nisin fragments, and their analogs in the physiological environment.

Precursors for Cyclization of Ring A via Peptide Bond Formation (Route A)

The suitably protected tripeptide consisting of residues 4–6 of nisin (Figure 1) was condensed with differently protected lanthionine of appropriate chirality to establish feasible routes for the cyclization to Ring A.

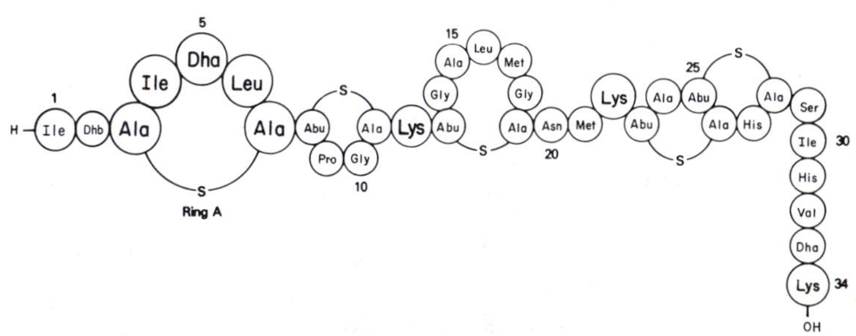

Fig. 1. The structure of nisin. Ring A: residues 3–7; Abu = aminobutyric acid; Dha = dehydroalanine; Dhb = dehydrobutyrine (β-methyldehydroalanine); Ala-*S*-Ala = lanthionine; Abu-*S*-Ala = β-methyllanthionine.

Boc-L-Isoleucyl-dehydroalanyl-L-leucine (I; Boc-L-Ile-Dha-L-Leu-OH) was prepared by stepwise condensation using *N*-hydroxysuccinimide esters for activation (Figure 2). Boc-L-Isoleucine *N*-hydroxysuccinimide ester was allowed to react with *S*-methyl-L-cysteine in tetrahydrofuran (THF)/water in the presence of excess $NaCHO_3$ under nitrogen to give Boc-L-isoleucine-*S*-methyl-L-cysteine (II) in 86% yield. II was converted to the *N*-hydroxysuccinimide ester and coupled to L-leucine methyl ester. Boc-L-isoleucyl-*S*-methyl-L-cysteinyl-L-leucine methyl ester (III) was isolated in 89% yield. Fluorosulfonic acid methyl ester[2] was employed to form the sulfonium salt of III for the conversion of the cysteine residue to dehydroalanine via triethylamine-catalyzed β-elimination. Boc-L-isoleucyl-dehydroalanyl-L-leucine-methyl ester (IV) was secured in 92% yield [thin layer chromatography (TLC): R_f = 0.48 in benzene/acetic acid = 7:1; Anal. calcd for $C_{21}H_{37}N_3O_6$: C, 58.99; H, 8.72; N 9.83; Found:C, 58.91, H, 8.67; N 9.78.– Hydrolysis of IV with 1.5 equivs. of NaOH in aqueous THF (100 min., r.t.) gave I in 96% yeild, mp 85–86°C; TLC single spots in several solvent systems; amino-acid analysis (residue ratios): Ile, 0.99; Leu, 1.0; NH_3, 1.09. Nmr [$CDCl_3$ internal standard tetramethyl silane (TMS)] dehydroalanine, β-protons: 5.36 and 6.41 ppm; amide proton, 8.59 ppm.

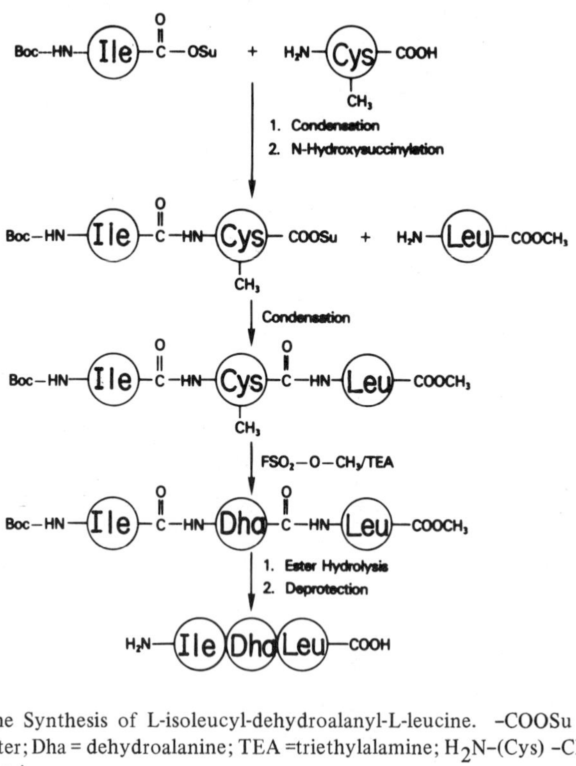

Fig. 2. The Synthesis of L-isoleucyl-dehydroalanyl-L-leucine. –COOSu = *N*-hydroxysuccinimide ester; Dha = dehydroalanine; TEA =triethylalamine; H_2N–(Cys) –CH_3; –COOH = *S*-methyl-L-cysteine.

Differently Protected Lanthionines were required to investigate synthetic routes (cf. Figure 3 for one approach chosen) for the *meso*-form of lanthionine (chiralities seen for this amino acid in nisin). In substitution reactions with precursors (*N*-protected β-haloalanine and *N,S*-disubstituted cysteine derivatives) of the L-configuration, chirally homogeneous L-lanthionine was obtained directly; it was not necessary to resort to the resolution of optical isomers. The same is not true for *meso*-lanthionine. Here the chiralities to be faced lead to a situation that is more prone to side reactions causing alterations in chirality.

Precursor for Cyclization via Sulfide-Bridge Formation (Route B)

A pentapeptide containing residues 4 through 6 of nisin and protected dehydroalanine and cysteine methyl ester in the terminal positions was used in studies aiming at the cyclization of Ring A via addition of the sulfhydryl group across the double bond in the unsaturated amino acid. Resolution will be required for the chirally inhomogenous product to be expected.

Boc-Dehydroalanyl-L-isoleucyl-dehydroalanyl-L-leucyl-*S*-benzoyl-L-cysteine methyl ester (V) was derived from the reaction of Boc-*S*-methyl-L-cysteinyl-L-isoleucyl-*S*-methyl-L-cysteine with L-leucyl-*S*-benzoyl-L-cysteine methyl ester (*N*-hydrooxysuccinimide ester activation) followed by the conversion of the cysteine residues to dehydroalanine. The latter was accomplished by forming the disulfonium salt with methyl fluorosulfonate[2] and catalyzing the β-elimination reaction with triethylamine compound. V was isolated as a powder in 72% yield. Amino-acid analysis (residue ratios): cysteine, 1.01; isoleucine, 1.10, leucine, 1.00, ammonia, 1.98. Nmr (δ DMSO-d_6/TMS): 5.20; 5.36; 5.98; 6.30 (each 1 H, s, β-protons dehydroalanine residues); 7.32 and 8.67 (2 × 1 H, s, amide proton

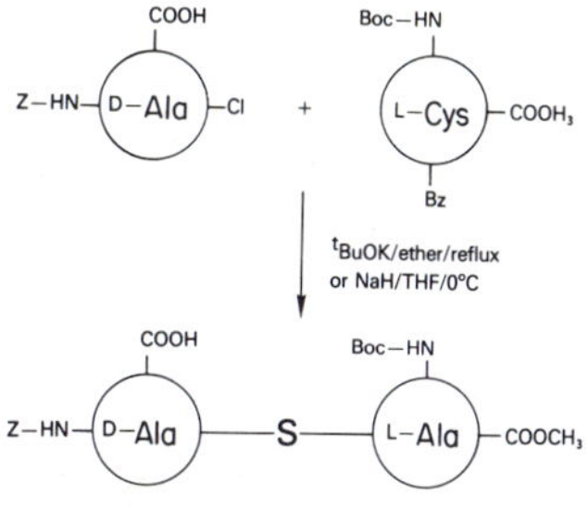

Fig. 3. The synthesis of *meso*-lanthionine. ^{t}BuOK = potassium-*tert*-butylate; THF = tetrahydrofuran; Z-HN-(D-Ala)-Cl; –COOH = *N*-benzyloxycarbonyl-β-chloro-D-alanine; Boc-HN-(L-Cys)-Bz; –$COOCH_3$ = *N-tert*-butyloxycarbonyl-*S*-benzoyl-L-cysteine methyl ester; (D-Ala)-*S*-(L-Ala) = *meso*-lanthionine.

dehydroalanine residues); 7.39 (2 H, t, *meta*-positions benzoyl); 7.53 (1 H, t, *para*-position benzoyl) 7.86 (2 H, d, *ortho*-positions benzoyl).

Analytical scale reactions to condense I with lanthionine (mixture of *meso*- and L- and/or D-forms; protected as shown in Figure 4) were carried out successfully thus setting the stage to explore cyclization to Ring A via peptide bond formation.

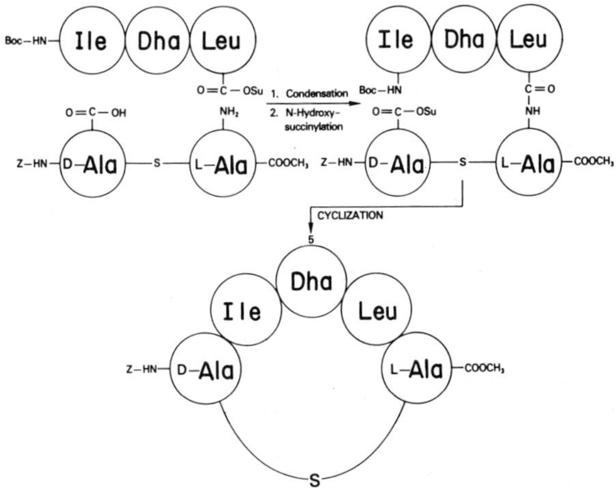

Fig. 4. Condensation of (Boc-L-isoleucyl-dehydroalanyl-L-leucine with *meso*-lanthionine (here shown to form the peptide bond between residues corresponding to position 6 and 7 of nisin) and cyclization to Ring A via Route A. Dha = dehydroalanine; -COOSu = *N*-hydroxysuccinimide ester; (D-Ala)-S-(L-Ala) = *meso*-lanthionine.

References

1. Gross, E. & Morell, J. L. (1971) *J. Amer. Chem. Soc.* **93**, 4634–4635.
2. Rich, D. H. & Tam, J. P. (1975) *Tetrahedron Lett.* 211–212.

KINETICS OF INHIBITION OF PEPSIN BY PEPSTATIN, DIDEOXYPEPSTATIN, ACETYL-STATIN AND ACETYL-DEOXYSTATIN.

DANIEL H. RICH and ERIC SUN, *School of Pharmacy, University of Wisconsin, Madison, WI 53706*

Pepstatin (*1*) is a tight-binding ($K_i \cong 10^{-10}$ *M*), inhibitor of several acid proteases including pepsin, renin and cathepsin D. We have shown that the hydroxyl group in the third residue of pepstatin is essential for the low K_i, because dideoxypepstatin (*2*), which lacks this hydroxyl group is a much weaker inhibitor of porcine pepsin ($K_i = 2 \times 10^{-7}$ *M*).[1]

In order to obtain a better understanding of the mechanism of inhibition of pepsin by pepstatin, which may be a transition-state-analog inhibitor,[2] and to clarify the function of the essential hydroxyl group in the third residue, we have carried out detailed kinetics of the inhibition of pepsin by pepstatin (*1*), dideoxypepstatin (*2*), acetyl-statin (*3*) and acetyl-deoxystatin (*4*) using the heptapeptide *5* as the pepsin substrate. See Scheme 1.

```
Iva-Val-Val-Sta-Ala-Sta   (1)   where Sta = 3-hydroxy-4-amino-6-
                                            methyl heptanoic acid
Iva-Val-Val-DeoxySta-Ala-DeoxySta   (2)

Ac-Sta   (3)

Ac-DeoxySta   (4)   where DeoxySta = 4-amino-6-methyl heptanoic acid

Phe-Gly-His-Phe(NO2)-Phe-Ala-Phe-OMe   (5)
```

Scheme 1.

The following results were obtained. Inhibitors *2, 3* and *4* were found to be simple competitive inhibitors of pepsin with K_i respectively equal to 2×10^{-7} *M*, 1.2×10^{-4} *M* and 3×10^{-3} *M*. In contrast, inhibition by pepstatin (*1*) was found to be non-competitive as judged by a Lineweaver-Burke plot of this system at steady-state.[3] A non-competitive inhibition mechanism would be inconsistent with the transition-state-analog-inhibitor hypothesis.

Close inspection of the inhibition kinetics revealed an important difference between the rate at which maximum inhibition is produced by the inhibitors. Figure 1A shows V_i for hydrolysis of *5* in the absence of inhibitor [I] and Figure 1B shows V_i in the presence of *2*. In both cases V_i is linear for 10–75 sec. Figure 1C shows V_i when enzyme and *1* are added to *5 without pre-equilibration*. V_i is clearly non-linear in the first 0–60 sec and inhibition of the enzyme is steadily increasing. If this concentration of *1* is preequilibrated with pepsin for 3–5 min no hydrolysis occurs and the enzyme is completely inhibited (Figure 1D).

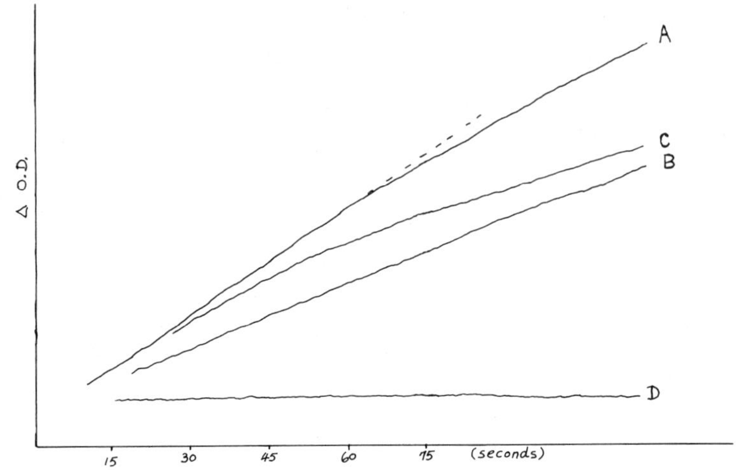

Fig. 1. Initial velocity (V_i) for hydrolysis *5*. A. [I] = 0; B. Dideoxypepstatin = 3 × 10^{-7} *M*; C. Pepstatin, 2 × 10^{-8} *M*, not pre-equilibrated; D. Pepstatin, 2 × 10^{-8} *M*, pre-equilibrated 5 min.

The data in Figure 1 establish that the binding of *1* to pepsin does not reach steady-state for at least 2–3 min. We propose that the inhibition follows Scheme 2 where the first-order approach to equilibrium k_2/k_{-2} is slow requiring 3–5 min for completion. Since

$$E + I \underset{k_{-1}}{\overset{k_1}{\rightleftharpoons}} EI \underset{k_{-2}}{\overset{k_2}{\rightleftharpoons}} \text{'EI'} \qquad \text{where } K_i = 10^{-10} = \frac{k_{-2}}{k_2} \cdot \frac{k_{-1}}{k_1}$$

Scheme 2.

k_2 is slow and K_i = 10^{-10} *M*, then k_{-2} must be much smaller and because 'EI' is not present initially Scheme 2 can be simplified to Scheme 3.

$$E + I \underset{k_{-1}}{\overset{k_1}{\rightleftharpoons}} EI \xrightarrow{k_2} \text{'EI'}$$

Scheme 3.

Scheme 3 is frequently used to determine the kinetic parameters of irreversible inhibitors[4] and can be used to analyze the pepstatin results by plotting the reciprocal pseudo-first order rate constant (k_{app}) vs. the reciprocal of inhibitor concentration [I] according to Eq (1):

$$\frac{1}{k_{app}} = \frac{k_{-1}/k_1}{k_2[I]} + \frac{1}{k_2} \qquad (1)$$

The V_i for hydrolysis of *5* at increasing concentrations of *1* is shown in Figure 2. This data was analyzed graphically to determine k_{app} (Figure 3). Following Eq (1), a plot of $1/k_{app}$ vs $1/[I]$ (Figure 4) gives the following constants: $k_2 = 0.022$ sec^{-1}, $k_{-1}/k_1 = 1.22 \times 10^{-8}$ *M* and $T_{1/2} = 31$ sec.

Because of the close structural similarity between *1* and *2*, which is a pure competitive inhibitor,[1] it is probable that pepstatin *1* is also a competitive inhibitor of pepsin. This could be tested by measuring V_i (using a high k_{cat} substrate) before the relaxation process proceeds; this work is in progress. However, assuming competitive inhibition by *1*, it is interesting to compare the effect on K_i produced by adding a hydroxyl group to acetyl-deoxystatin *4* and to predict the K_i of pepstatin in the absence of other effects. Addition of the hydroxyl

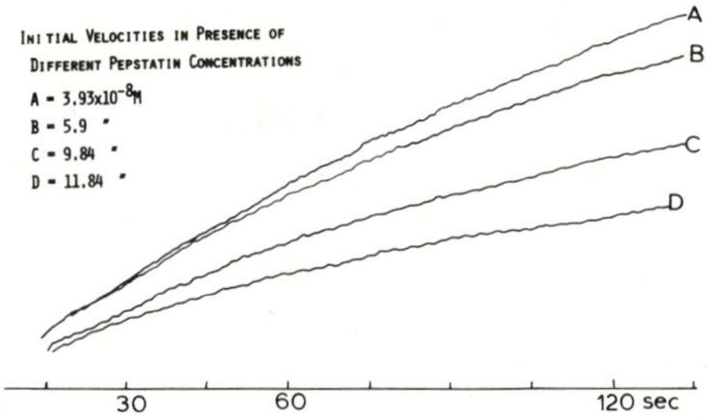

Fig. 2. Initial velocity for hydrolysis of *5* in presence of different pepstatin concentrations.

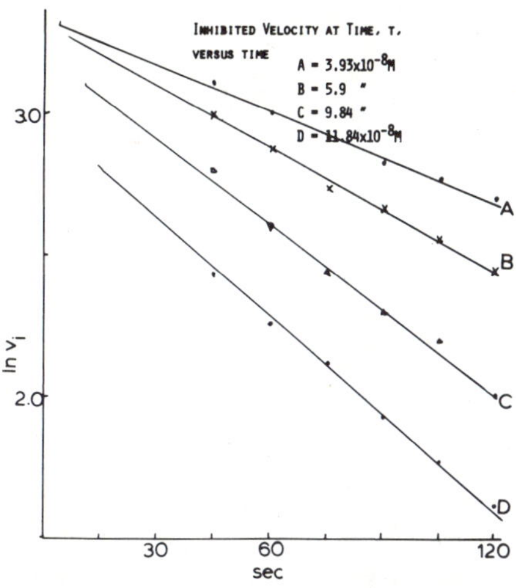

Fig. 3. Inhibited velocity at time, *T*, vs. time in the presence of different pepstatin concentrations, A, B, C, and D.

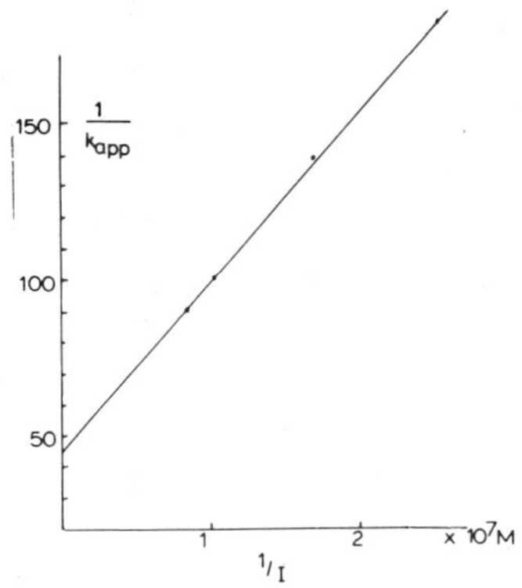

Fig. 4 Plot $1/k_{app}$ vs. $1/[I]$.

group to *4* decreases K_i by 18-fold. Both inhibitors are competitive and therefore the factor of 18 can be used to estimate the contribution of this hydroxyl group to inhibition when placed in the enzyme's active site. Addition of the hydroxyl group to dideoxypepstatin *2* would decrease the K_i by 18 to 1.17×10^{-8} *M*. This value agrees closely with k_{-1}/k_1 (1.22×10^{-8} *M*) derived for Scheme 2.

The inhibition of pepsin by pepstatin *1* is described by Scheme 2. The value for k_{-1}/k_1 is in accord with that expected from addition of a hydroxyl group to *2*. The difference between k_{-1}/k_1 (10^{-8} *M*) and K_i (10^{-10} *M*) is due to the slow first order process EI ⇋ 'EI'. The magnitude of k_2 is not unusually small for an enzymatic process and is comparable to the rate constant for a conformational change.[5] If the conformation of 'EI' resembles the transition-state for hydrolysis of substrate then *1*, but not *3*, may be a transition-state analog inhibitor of pepsin.

We thank the Graduate School of the University of Wisconsin-Madison for partial support of this work.

References

1. Rich, D. H., Sun, E. & Singh, J. (1977) *Biochem. Biophys. Res. Commun.* **74**, 762–767.

2. Marciniszyn, J. P., Hartsuck, J. A. & Tang, J. J. N. (1976) *J. Biol. Chem.* **251**, 7088–7094.

3. McKown, M. M., Workman, R. J. & Gregerman, R. I. (1974) *J. Biol. Chem.* **249**, 7770–7774.

4. Tipton, K. F. (1973) *Biochem. Pharmacol.* **22**, 2933–2941.

5. Hammes, G. G. & Schimmel, P. R. (1970) in *The Enzymes*, Vol. II, Boyer, P. D., Ed., Academic Press, New York, p. 67–114.

PREPARATION AND CHARACTERIZATION OF A DEXTRAN-PEPSTATIN CONJUGATE, A NEW AND POTENT INHIBITOR OF RENIN AND PEPSIN.

H. J. CHOU & R. I. GREGERMAN, *Gerontology Research Center, Clinical Physiology Branch, National Institute on Aging, National Institutes of Health, Baltimore City Hospitals, Baltimore, MD. 21224.*

Pepstatin, (isovaleryl-L-valyl-L-valyl-4-amino-3-hydroxy-6-methylheptanoyl-L-alanyl-4-amino-3-hydroxy-6-methylheptanoic acid, M_r 685) is a pentapeptide obtained from *Streptomyces.* The compound is a potent specific inhibitor of several acid proteases (pepsin, cathepsin D, renin). The effect of this peptide on the production of angiotensin I by inhibition of renin both *in vivo*[1] and *in vitro*[2] has been described. These results suggest that pepstatin might be useful in evaluating the role of the renin-angiotensin system under various physiologic and pathologic conditions. Pepstatin could also be of potential clinical usefulness in the differential diagnosis of hypertension and the detection of renovascular hypertension.

Experimental use of pepstatin has been precluded by the very low water solubility of the peptide and its short duration of action, apparently due to its rapid clearance from the circulation. We had previously manipulated the solubility characteristics of the tetradecapeptide renin substrate by coupling the peptide to polyglutamic acid.[3] Subsequently, synthetic polymers have been used as carriers for several drugs in order to prolong the duration of their activity.[4,5] In order to circumvent both the problem of limitation of dosage due to solubility and duration of action, we have coupled pepstatin through its C-terminal carboxyl to several high-molecular-weight, water-soluble polymers, since the C-terminal carboxyl seems to be not essential for pepstatin's effectiveness. Such conjugates would be expected to maintain their potent inhibitory effects. The conjugates would presumably also exert an *in vivo* effect of longer duration of action than pepstatin itself and would be expected to approach that of the survival time of the polymer in the circulation.

The most promising pepstatin-conjugate polymer we have prepared (Scheme 1) is one in which the peptide is coupled to high-molecular-weight dextran, a relatively non-allergenic, soluble polysaccharide.

An essential feature in the success of the synthesis is the use of low concentrations of cyanogen bromide to avoid irreversible precipitation of the polysaccharide during the activation reaction. Coupling of pepstatin to ethylene diamine dextran conjugate is performed in a 1:1 mixture of pyridine and 0.1 *M* sodium phosphate, pH 5.5, with 1-ethyl-3-(3-dimethylaminopropyl)-carbodiimide as the coupling catalyst. In this solvent system all reactants are soluble. Removal of unreacted components is accomplished by prolonged dialysis. After coupling of pepstatin to the dextran, an excess of glycine is added to saturate unreacted imidocarbonate groups, thus preventing the initially soluble conjugate from becoming insoluble during freeze-drying of the preparation.

(i) dextran (T-40) + cyanogen bromide $\xrightarrow{25°}$ activated dextran

(ii) activated dextran + $NH_2(CH_2)_2NH_2$ $\xrightarrow{4°}$ $NH_2(CH_2)_2$-NH-dextran

(iii) $NH_2(CH_2)_2$ NH-dextran + pepstatin

$$\xrightarrow[\text{25°, 1-ethyl-3-(3-dimethylaminopropyl)-carbodiimide}]{\text{pyridine in 0.1 M sodium phosphate pH 5.5}}$$

pepstatin-NH-$(CH_2)_2$-NH-dextran

Scheme 1. Preparation of the pepstatin-conjugate polymer.

Amino-acid analysis indicates that 1 mg of polymer conjugate contains 20 μg of pepstatin. Control experiments without added carbodiimide shows absence of pepstatin, indicating that the peptide does not become non-specifically adsorbed by dextran during preparation. The solubility of dextran-pepstatin conjugate is the same as that of uncoupled dextran.

The conjugate was tested for its ability to inhibit the acid proteases (human renin and porcine pepsin) using the labeled polymeric substrate assay.[3] The polymer conjugate maintains the same potent inhibitory effects as pepstatin itself toward human renin and porcine pepsin (Fig. 1, A, B). The pepstatin conjugate also shows inhibitory effects on human renin using partially purified hog renin substrate or human renin substrate with determination of the angiotensin I product by radioimmunoassay[6]. Evaluation of the effectiveness of the pepstatin conjugate *in vivo* is in progress.

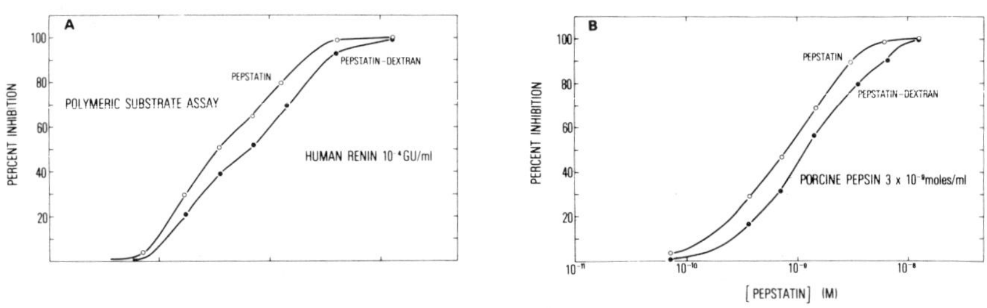

Fig. 1. Inhibition of (A) human renin (10^{-4}GU/ml) and (B) porcine pepsin (3×10^{-9} mol/ml) by pepstatin and pepstatin-dextran conjugate.

References

1. McKown, M. M., Workman, R. J. & Gregerman, R. I. (1974) *J. Biol. Chem.* **249**, 7770–7774.

2. Miller, R. P., Puper, C. J., Wilson, C. W. & DeVito, E. (1972) *Biochem. Pharmacol.* **21**, 2941–2944.

3. Bath, N. W. & Gregerman, R. I. (1972) *Biochemistry* **11**, 2845–2852.

4. Havron, A., Weiner, B. Z. & Zilkha, A. (1974) *J. Med. Chem.* **17**, 770–772.

5. Ringsdorf, H., Ritter, H. & Rolly, H. (1976) *Makromol. Chem.* **177**, 741–746.

6. Chou, H. J. & Gregerman, R. I. Submitted.

SYNTHESIS OF BIOLOGICALLY ACTIVE ALAMETHICIN

B. F. GISIN, S. KOBAYASHI, *Department of Biochemistry, The Rockefeller University, New York, N. Y. 10021*, D. G. DAVIS, *Department of Chemistry, Adelphi University, Garden City, N. Y. 11530*, and J. E. HALL, *Department of Physiology and Pharmacology, Duke University Medical Center, Durham, N. C. 27710*

Alamethicin (ALA) is a peptide antibiotic[1] that modifies natural[2] and artificial[3-7] membranes. In artificial lipid bilayer membranes it induces voltage-dependent conductances[4] that are due to ion-conducting channels[5-9] similar to those observed in nerve membranes. The cyclic structure that was first proposed for ALA by Payne et al.[10] was found to be incorrect and instead an open chain sequence was proposed by Martin and Williams[11] and Jung et al.[12] The latter structure (ALA-α, Fig. 1) was synthesized but the product was different from the main component of natural ALA.[13] One difference was the acidity of the carboxyl groups. Synthetic ALA-α had a pK_a' of 4.4 compared to 5.2 for natural ALA (Table I).[13]

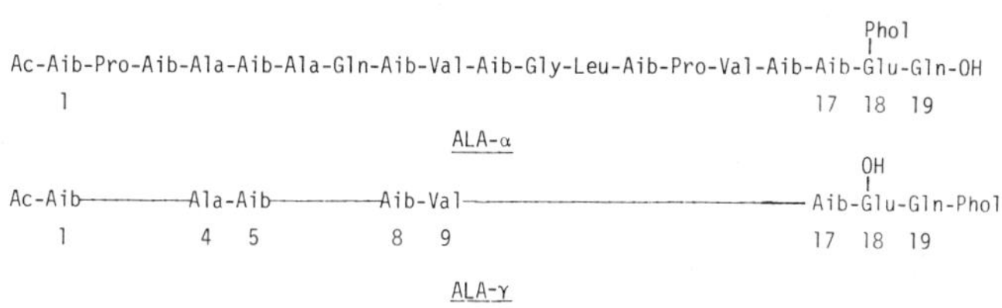

Fig. 1. Aib = α-aminoisobutyric acid, Phol = L-Phenylalaninol, Ac = acetyl. ALA-α, structure proposed in Ref. 11 and 12. ALA-γ, structure of synthetic, biologically active alamethicin (this paper).

Table I. Dissociation Constants of Glu Derivatives

Compound	Position of Carboxyl Group	pK_a' a)
Boc-Glu(OBzl)-OH	α	4.4
Boc-Glu(Phol)-OH	α	4.4 b)
synt. ALA-α	α	4.4
Boc-Glu(OH)-OBzl	γ	5.1
Boc-Glu(OH)-Phol	γ	4.9 b)
synt. ALA-γ	γ	5.3
nat. ALA	γ (?)	5.2

a) Apparent dissociation constants determined with glass electrode in $EtOH/H_2O$ (3:7, v/v) at 25°. Estimated experimental error, ± 0.1 pK units.

b) $EtOH/H_2O$ (2:8, v/v).

This suggested that the position of the phenylalaninol (Phol) residue may have been misassigned in previous structural work. For that reason the isomer of ALA-α in which the Phol residue was linked to the C-terminal α-carboxyl group of Gln[19] rather than to the γ-carboxyl group of Glu[18] was synthesized. This peptide, ALA-γ (γ for γ-carboxyl group), is the subject of the present contribution.

The synthesis of ALA-γ was performed by the solid-phase method[14] using the stepwise fragment condensation approach applied to the synthesis of ALA-α.[13] In the final condensation step, an *N*-acetyl 17-residue fragment was coupled to the resin-bound dipeptide derivative H-Glu(ORes)-Gln-Phol and the unreacted portion of the large fragment was washed out. This scheme was used in order to avoid the potentially difficult task of having to separate unreacted 17-residue peptide from the 19-residue target peptide. Coupling was effected by DCC/HOBt in CH_2Cl_2/DMF over 40 hr. Purification, after cleavage with HF, was on a Sephadex G-25 column in EtOH/H_2O (1:1). The amino-acid composition of a hydrolyzate of the product was in agreement with the expected values (in parentheses): Glu 2.91 (3), Gly 1.01 (1), Ala 1.92 (2), Val 1.95 (2), Leu 1.01 (1), Aib 8.48 (8), Phol 0.87 (1).

Synthetic ALA-γ co-chromatographed with the main component of natural ALA on silicagel G thin layer plates in six different solvent systems including three that clearly separated ALA-γ and ALA-α. The methyl esters of natural ALA and of ALA-γ co-chromatographed in five systems in two of which the methyl ester of ALA-α migrated differently. The pK_a' (Table I) of ALA-γ was within experimental error the same as for the natural product but expectedly different from that of synthetic ALA-α.

The 220-MHz proton nmr spectrum of natural ALA contained all of the signals present in the spectrum of synthetic ALA-γ at the same intensities and chemical shifts. There were additional signals in the natural ALA spectrum that were absent in the ALA-γ spectrum. This is consistent with the presence of heterogeneities in the natural sample that were not removed by the purification techniques used. (Crude natural ALA was purified on an anion exchange resin followed by Sephadex G-25 chromatography.) A likely possibility is the presence[10] of Aib^6-alamethicin to the extent of 15–30 mol-% and of other homologues in low concentration.

The antibiotic activity was assayed using *Bacillus subtilis* in liquid cultures containing varying concentrations of peptide. Synthetic ALA-γ was 60–80% as potent as natural ALA in this assay (Fig. 2) while ALA-α had little activity ($<$10%). In lipid bilayer membranes the specific activity of ALA-γ was 20–30% that of natural ALA (Fig. 3).

In conclusion, synthetic ALA-γ is indistinguishable from the main component of natural ALA in chromatographic behavior and acidity. It is very potent in antibiotic assays and in inducing voltage-dependent conductances in lipid bilayer membranes. This may mean that ALA-γ is similar to natural ALA in structure.

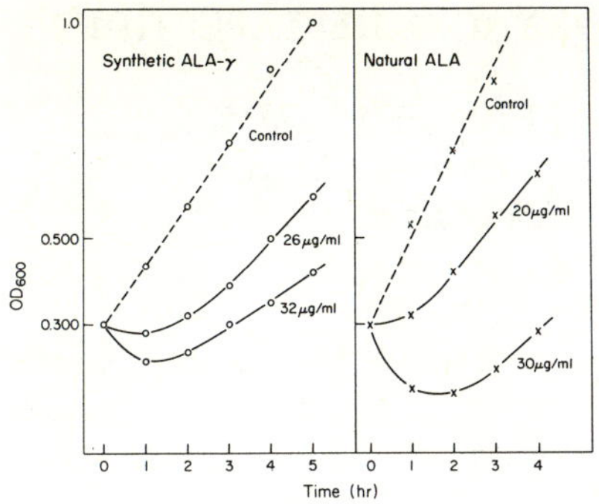

Fig. 2. Growth inhibition of *Bacillus subtilis 168* in liquid culture by natural alamethicin and by synthetic ALA-γ.

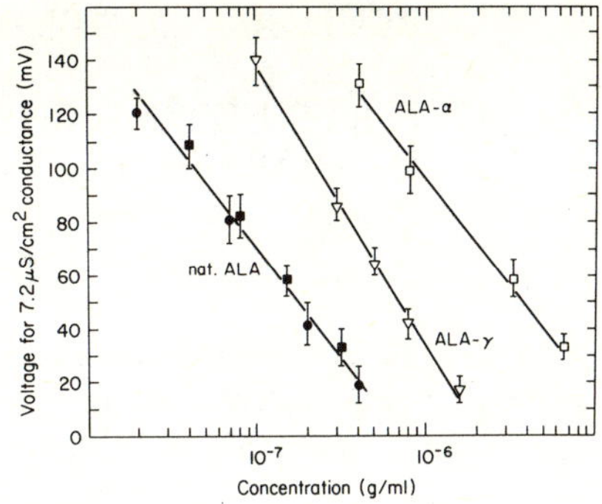

Fig. 3. Effect of concentration of natural and synthetic alamethicins on the voltage required to produce a conductance of 7.2 $\mu S/cm^2$ in lipid bilayer membranes made of phosphatidyl ethanolamine/monoolein (2:1).

Alternatively, ALA-γ may represent just one component of natural ALA with other components being more active.

We wish to thank Dr. Z. K. Borowska for assistance with the antibiotic assays and Dr. R. B. Merrifield for reading the manuscript. Samples of natural alamethicin were kindly provided by Dr. G. B. Whitfield of the Upjohn Company. Nmr spectra were taken at the facilities at The Rockefeller University (NSF Grant BMS 74-12247) and at Carnegie Mellon University (NIH Grant RR-00292). This work was supported by NIH Grants HL-17961 and HL-19951.

References

1. Meyer, C. E. & Reusser, F. (1967) *Experientia* **23**, 85–86.
2. Jung, G., Dubischar, N., Leibfritz, D., Ottnad, M., Probst, H. & Stumpf, Ch. (1975) in *Peptides 1974*, Y. Wolman, Ed., John Wiley & Sons, N. Y., p. 345–354.
3. Lau, A. L. Y. & Chan, S. (1974) *Biochemistry* **13**, 4942–4948.
4. Mueller, P. & Rudin, D. O. (1968) *Nature* **217**, 713–719.
5. Eisenberg, M., Hall, J. E. & Mead, C. A. (1973) *J. Membrane Biol.* **14**, 143–176.
6. Boheim, G. (1974) *J. Membrane Biol.* **19**, 277–303.
7. Gordon, L. G. M. & Haydon, D. (1975) *Phil. Trans. R. Soc. Lond. B.* **270**, 433–447.
8. Hall, J. E. (1975) *Biophys. J.* **15**, 934.
9. Baumann, G. & Mueller, P. (1974) *J. Supramolec. Struc.* **2**, 538–557.
10. Payne, J. W., Jakes, R. & Hartley, B. S. (1970) *Biochem. J.* **117**, 757–766.
11. Martin, D. R. & Williams, R. J. P. (1976) *Biochem. J.* **153**, 181–190.
12. Jung, G., Dubischar, N. & Leibfritz, D. (1975) *Eur. J. Biochem.* **54**, 395–409.
13. Gisin, B. F., Kobayashi, S. & Hall, J. E. (1977) *Proc. Nat. Acad. Sci. USA* **74**, 115–119.
14. Merrifield, R. B. (1963) *J. Amer. Chem. Soc.* **85**, 2149–2154.

TUBERCULOSTATIC 3-HYDROXYACYL HEPTAPEPTIDES

GOTFRYD KUPRYSZEWSKI, BERNARD LAMMEK and WITOLD NEUGEBAUER, *Institute of Chemistry, University of Gdańsk, Poland*

In 1951 Kochi[1,2] isolated a tuberculostatic antibiotic and named it esperin. Further investigations of the antibiotic[3-7] revealed that it presumably consisted of a mixture of a few compounds of the following structure:

$$\text{R-}\underset{\lfloor\_\_\_\_\_\_\_\_\_\_\_\_\_\_\_\_\_\_\_\_\_\_\_\_\_\_}{\text{CH}}\text{-CH}_2\text{-CO-Glu-Leu-Leu-Val-}\underset{\text{O}\_\_\_\_\rfloor}{\text{Asp}}\text{-Leu-Leu} ,$$

where R is $CH_3(CH_2)_9$, $CH_3(CH_2)_{10}$ and $CH_3(CH_2)_{11}$.

Accordingly, the molecules contain one residue each of 3-hydroxy acid, L-glutamic acid, L-aspartic acid and L-valine, as well as two residues each of L-leucine and D-leucine. The C-terminal position has been shown to be occupied by D-leucine.

As the sequence of the peptide chain of esperin was determined by mass spectrometry,[7] the position of the second D-leucine residue has not yet been elucidated. The configuration of the 3-hydroxy acid residues also remains unknown.

A mild hydrolysis of the lactone bond of esperin affords the appropriate 3-hydroxyacyl heptapeptide referred to as esperic acid. Analogs of esperic acid, similar to esperin itself, have been found to exhibit tuberculostatic activity. Were both the position of one of the D-leucine and the configuration of the 3-hydroxy acid residue to remain unknown, eighteen analogs of esperic acid would be expected. The following structures can be assigned to these analogs:

$$CH_3\text{-}(CH_2)_{\underline{n}}\text{-}\underset{\text{OH}}{\text{CH}}\text{-}CH_2\text{-CO-Glu-Leu-D-Leu-Val-Asp-Leu-D-Leu} \quad \text{I}$$

$$CH_3\text{-}(CH_2)_{\underline{n}}\text{-}\underset{\text{OH}}{\text{CH}}\text{-}CH_2\text{-CO-Glu-D-Leu-Leu-Val-Asp-Leu-D-Leu} \quad \text{II}$$

$$CH_3\text{-}(CH_2)_{\underline{n}}\text{-}\underset{\text{OH}}{\text{CH}}\text{-}CH_2\text{-CO-Glu-Leu-Leu-Val-Asp-D-Leu-D-Leu} \quad \text{III}$$

Ia: $n = 9$, (R); Ib: $n = 9$, (S); Ic: $n = 10$, (R); Id: $n = 10$, (S); Ie: $n = 11$, (R); If: $n = 11$, (S); IIa: $n = 9$, (R); IIb: $n = 9$, (S); IIc: $n = 10$, (R); IId: $n = 10$, (S); IIe: $n = 11$, (R); IIf: $n = 11$, (S); IIIa: $n = 9$, (R); IIIb: $n = 9$, (S); IIIc: $n = 10$, (R); IIId: $n = 10$, (S); IIIe: $n = 11$, (R); IIIf: $n = 11$, (S). The configurations of the 3-hydroxy acid residues are given in parentheses.

The syntheses of the eighteen analogs of esperic acid have been carried out in this work. First, racemic 3-hydroxytridecanoic, 3-hydroxytetradecanoic and 3-hydroxypentadecanoic acids were obtained by employing the Reformatsky method.[8,9] The racemates were then resolved into enantiomers.[10,11]

Nine analogs of esperic acid (Ic, Id, If, IIc, IId, IIf, IIIc, IIId and IIIf) were obtained by aminolysis of the *N*-hydroxysuccinimide esters of appropriate 3-hydroxy acids with heptapeptides synthesized by the Merrifield method. The 3-hydroxyacyl heptapeptides obtained were purified by column chromatography.

The nine remaining analogs (Ia, Ib, Ie, IIa, IIb, IIe, IIIa, IIIb and IIIe) were obtained by classical methods employed in peptide chemistry. *N*-Ethoxycarbonyl-2-ethoxy-1,2-dihydroquinoline (EEDQ) and the azide of the diphenyl ester of phosphoric acid (DPPA) were used to synthesize the peptide bonds. The amino groups were protected by *t*-butyloxycarbonyl residues and the carboxyl group of the C-terminal D-leucine was protected by converting it to the *p*-nitrobenzyl ester (ONb). The γ-carboxyl group of L-glutamic acid and the β-carboxyl group of L-aspartic acid were protected by converting them to the benzyl esters. The N-terminal amino acid was acylated with 3-hydroxy acids by using their active *N*-hydroxysuccinimide esters (see Scheme 1). Final products were purified by column chromatography.

The eighteen chromatographically homogeneous 3-hydroxyacyl heptapeptides were identified by quantitative assessment of their amino-acid composition, elemental analyses, IR spectra and determination of their molecular weights by mass spectrometry.

Seven (Ia, Ic, Id, Ie, If, IIIa and IIIe) of the eighteen 3-hydroxyacyl heptapeptides were found to exhibit tuberculostatic activity. On this basis one can assume that esperin constitutes a group of compounds with structures corresponding to the seven active 3-hydroxyacyl heptapeptides.

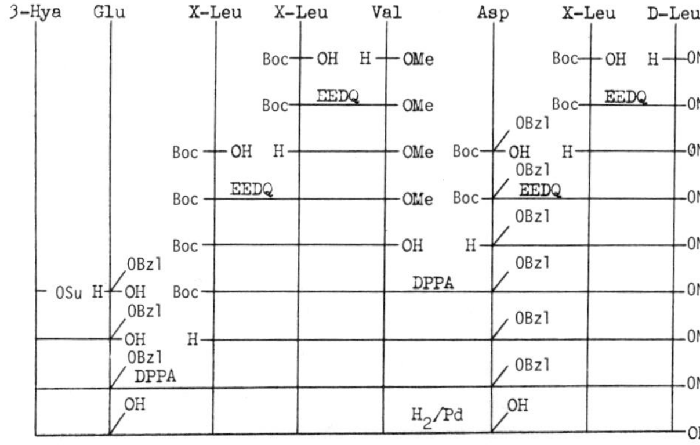

Scheme 1. A schematic diagram of the syntheses of analogs of esperic acid. X-Leu stands for D- or L-leucine; 3-Hya stands for 3-hydroxy acid.

References

1. Kochi, M. (1951) Jap. pat. 1145; (1952) *Chem. Abstr.* **46**, 1159.

2. Kochi, M. (1951) Jap. pat. 2497; (1953) *Chem. Abstr.* **47**, 690.

3. Ogawa, H. & Ito, T. (1951) *J. Agr. Chem. Soc. Jap.* **24**, 191–196.

4. Ogawa, H. & Ito, T. (1952) *J. Agr. Chem. Soc. Jap.* **26**, 432–434.

5. Ito, T. & Ogawa, H. (1959) *Bull. Agr. Chem. Soc. Jap.* **23**, 536–547.

6. Ovchinnikov, Yu. A., Ivanov, V. T., Kostetsky, B. V. & Shemyakin, M. M. (1966) *Tetrahedron Lett.*, 5285–5290.

7. Thomas, D. W. & Ito, T. (1969) *Tetrahedron*, 1985–1990.

8. Reformatsky, S. (1887) *Chem. Ber.* **20**, 1210–1211.

9. Rieke, R. D. & Uhm, S. J. (1975) *Synthesis*, 452–453.

10. Hiramoto, H., Okada, K., Nagai, S. & Kawamoto, H. (1971) *Chem. Pharm. Bull. Jap.* **19**, 1308–1314.

11. Lammek, B., Neugebauer, W. & Kupryszewski, G. (1976) *Roczniki Chem.* **50**, 997–1000.

A CELL SURFACE ACTIVE ANTITUMOR PROTEIN, NEOCARZINOSTATIN

T. S. ANANTHA SAMY and H. LAZARUS,
Sidney Farber Cancer Institute, Harvard Medical School,
44 Binney Street, Boston, MA 02115

Neocarzinostatin (NCS), an antitumor protein isolated from *Streptomyces carzinostaticus*[1], is a single chain polypeptide with 109 amino-acid residues. The amino-acid sequence of NCS has been established[2]. NCS is highly cytotoxic for mammalian cells *in vitro*. Although it inhibits DNA synthesis and induces DNA strand scissions *in vitro* and *in vivo*[3-5], its precise mechanism of action is not known.

The two free amino groups in NCS (α-NH_2 of Ala[1] and ϵ-NH_2 of Lys[20]) were reacted with 3-(4-hydroxyphenyl)propionic acid *N*-hydroxysuccinimide ester in 0.1 *M* borate buffer, pH 8.5 at 5° C. The reaction went to completion with two moles of active ester reacting with one mole of protein. The bis-amino substituted protein, bis[Ala[1], Lys[20]]-3-(4-hydroxyphenyl)propionamide-NCS (BisHPP-NCS), was purified by chromatography on a CM-cellulose column with 0.05 *M* acetic acid and acetate buffer gradient, pH 3–5. Electrometric titration indicated the absence of free amino groups in BisHPP-NCS; this was evident in the increased mobility on electrophoresis at pH 8.6 (Figure 1) and the low isoelectric point, pI 3.1 *vs.* 3.3 native NCS. The BisHPP-NCS was biologically active in growth inhibition of *Sarcina lutea* and of human lymphoid leukemic (CCRF-CEM) cells, in vitro (Table I). It was as potent as native NCS in the L1210 and P388 mouse leukemias. These results suggested that the two amino groups are not essential for biological activity of NCS.

After ascertaining the nonfunctional role of the amino groups, NCS was immobilized by coupling to the *N*-hydroxysuccinimide ester derivative of agarose (Affigel 10, BioRad) in 0.1 *M* phosphate buffer, pH 7.5. The coupling was

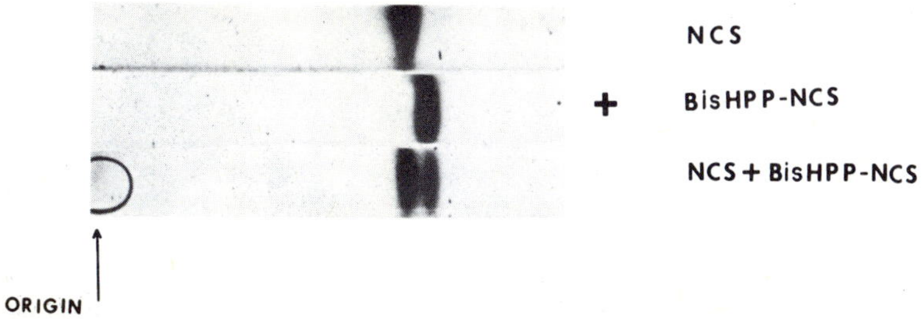

Fig. 1. Acrylamide gel electrophoresis of NCS and BisHPP-NCS at pH 8.6.

Table I. Growth Inhibitory Activity of NCS and BisHPP-NCS Against *S. lutea* and CCRF-CEM Cells *in vitro.*

	INHIBITORY ACTIVITY	
	Sarcina Lutea	CCRF-CEM Cells
	MIC μg/ml	ID_{50} μg/ml
Native NCS	0.05	0.022
BisHPP-NCS	0.05	0.039

carried out at 5° C for 15 h. The beads were then extensively washed with deionized water and buffers of pH 4.0, 7.5 and 8.6. The agarose immobilized NCS (AG-NCS) was found to contain 10–20 μg NCS per mg agarose (determined by amino-acid analysis). Acrylamide gel electrophoresis, crossed-immunoelectrophoresis, immunodiffusion and agar gel diffusion tests against *S. lutea*, all showed that there was no free NCS present in the AG-NCS preparation (Figures 2A, B & C). These tests are sensitive enough to detect 1% free NCS, if present, in the AG-NCS preparation. AG-NCS was 15–40% biologically active; several different preparations inhibited the growth of CCRF-CEM cells *in vitro* at an ID_{50} 6–15 $\times$ 10^{-9} *M* (Figure 3). The immobilized NCS also, inhibited the incorporation of ^{3}H-TdR into DNA in these cells. Since agarose immobilized NCS (bead size, 75–150 μ) cannot penetrate the leukemic cell (size 10–15 μ), it is suggested that NCS exerts its toxic effect in these cells by interaction with cell-surface receptor(s).

We recognize that there may be release of free ligand into solvent after coupling of the protein to a solid matrix. To test whether there was any such release on storage over a period of time, we prepared an AG-NCS derivative containing covalently bound ^{14}C-alanine. The AG-NCS-^{14}C-alanine complex was stored frozen for 4 weeks and at specified time intervals aliquots were

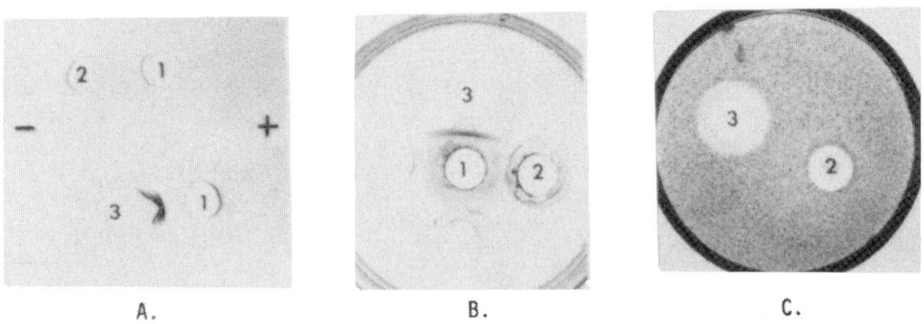

Fig. 2. Tests for the presence of any unbound NCS in AG-NCS preparation: A. Counter-immunoelectrophoresis at pH 8.6, B. immunodiffusion and C. growth inhibition with *S. lutea;* (1) rabbit anti-NCS serum, (2) AG-NCS and (3) NCS.

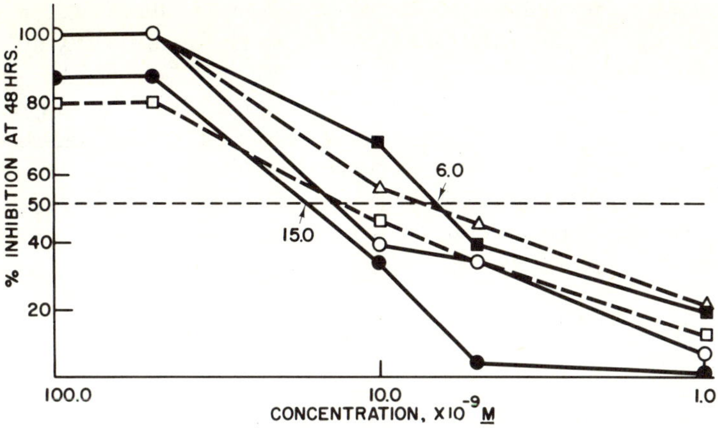

Fig. 3. Growth inhibitory activity of 5 different preparations of AG-NCS against CCRF-CEM cells *in vitro*.

withdrawn, centrifuged and the supernatant solution was analyzed for ^{14}C-radioactivity and for NCS content by a radioimmunoassay procedure developed in this laboratory.[6] No detectable quantities of ^{14}C-label or NCS were found. We also tried to detect the release of NCS during incubation of AG-NCS with leukemic cells. CCRF-CEM cells (2×10^6 cells/ml) were incubated with AG-NCS (2 μg/ml on NCS basis) at 37° C for 3 h. At the end of the incubation period three components of the system (cells, AG-NCS and media) were separated by sedimentation over a Ficoll-Hypaque gradient. The inhibitory activity of each fraction was determined by the addition of fresh media and/or cells. Only AG-NCS was active. The spent media did not kill the fresh cells, indicating that NCS was not released during incubation with leukemic cells. This data clearly indicates that AG-NCS is cytotoxic and is able to inhibit DNA synthesis by interaction with the cell membrane.

The mechanism by which NCS induces cytotoxicity is not clear. Experiments are in progress in our laboratory to elucidate the cytotoxic mechanism in leukemic cells. We envisage the following possibilities: (a) initial binding of NCS to cell-surface receptor(s) and later transmission of cytotoxic effect to the target site through some conformational change on the membrane as in the case of colicins[7], (b) changes in intracellular cyclic AMP levels leading to the inhibition of DNA synthesis, and (c) changes in membrane properties of the cell. In lymphoid cell lines up to 2% of cellular DNA is associated with cytoplasmic membranes[8]. The function of this species of DNA is not clear and whether NCS has any specific relation with this DNA species, we do not know.

The surface-active property of NCS may have implications both in the mechanism of DNA synthesis, in general, and cancer pharmacology in particular. Many anticancer agents are small molecular-weight (less than 2,000) compounds which react with nuclear DNA directly or inhibit nucleic acid or protein synthesis by interaction with intracellular constituents. NCS, on the other hand, can inhibit DNA synthesis without getting into the cell. Induction of toxicity by such a mechanism is unique for antitumor agents. If NCS could be coupled to a highly characterized tumor-specific antibody, then one could selectively transport NCS to tumors. This might result in increased selectivity for tumor cells and decreased toxicity for normal cells. A mechanism of inhibition of DNA synthesis in mammalian cells by a relatively large molecular-weight compound, like NCS, with signals originating from the cell surface, would represent a novel mechanistic aspect in studies relating to metabolism of cellular DNA.

Supported by NCI Grants CA17305 and CA06516.

References

1. Ishida, N., Miyazaki, K., Kumagai, K. & Rikimaru, M. (1965) *J. Antibiot.* **18**, 68–72.
2. Meienhofer, J., Maeda, H., Glaser, C. B., Czombos, J. & Kuromizu, K. (1972) *Science* **178**, 875–876.
3. Beerman, T. A. & Goldberg, I. H. (1974) *Biochem. Biophys. Res. Commun.* **59**, 1254–1261.
4. Tatsumi, K., Nakamura, T. & Wakisaka, G. (1974) *Gann* **65**, 59–461.
5. Ohtsuki, K. & Ishida, N. (1975) *J. Antibiot.* **28**, 143–148.
6. Samy, T. S. A. & Raso, V. (1976) *Cancer Res.* **36**, 4378–4381.
7. Nomura, M. (1964) *Proc. Nat. Acad. Sci. USA* **52**, 1514–1521.
8. Hall, M. R., Meinke, W., Goldstein, D. A. & Lerner, R. A. (1971) *Nature, New Biol.* **234**, 227–229.

SYNTHETIC OLIGOPEPTIDES FROM HUMAN C3a ANAPHYLATOXIN THAT MEDIATE THE INFLAMMATORY RESPONSE

LYNN H. CAPORALE and BRUCE W. ERICKSON,
The Rockefeller University, New York, New York 10021,
and TONY E. HUGLI, *Department of Molecular Immunology, Scripps Clinic and Research Foundation, La Jolla, California 92037*

Activation of the complement (C) system of human serum by antigen-antibody complexes or certain polysaccharides causes cleavage of protein C3 into two biologically active fragments, the 9000-dalton anaphylatoxin, C3a, and the 171,000-dalton protein, C3b.[1] Human C3a is a potent mediator of acute inflammation, causing degranulation of mast cells, contraction of smooth muscle, and increased vascular permeability.[2] These biological activities are lost, however, when the C-terminal arginine is enzymatically removed from the polypeptide.

The synthetic[3] octapeptide corresponding to the C-terminus of human C3a exhibits the biological activities and specificity of natural[2] C3a. The octapeptide is 2% as active as C3a on a molar basis (about 20% on a weight basis). A few nanomoles of the octapeptide cause contraction of guinea pig ileum and increased vascular permeability in human skin. This paper deals with the contributions of peptide length, the side chains of Ser^{71} and His^{72}, and positively charged groups near the N-terminus to the activities of human C3a-(70-77)-octapeptide.

The synthetic C3a oligopeptides listed in Fig. 1 were assembled by the stepwise solid-phase method[4] and were purified by gel filtration, ion-exchange chromatography, and, when necessary, by preparative thin-layer chromatography. They were chromatographically homogeneous and gave the correct amino-acid ratios. Their inflammatory activities were measured by their ability to contract smooth muscle from guinea pig ileum and to increase vascular permeability in human skin.

```
70  71  72  73  74  75  76  77             70  71  72  73  74  75  76  77

Ala-Ser-His-Leu-Gly-Leu-Ala-Arg (1)    Ala-Ala-Ala-Leu-Gly-Leu-Ala-Arg (7)
    Ser-His-Leu-Gly-Leu-Ala-Arg (2)        Ala-Ala-Leu-Gly-Leu-Ala-Arg (8)
        His-Leu-Gly-Leu-Ala-Arg (3)            Ala-Leu-Gly-Leu-Ala-Arg (9)
            Leu-Gly-Leu-Ala-Arg (4)         Ac-His-Leu-Gly-Leu-Ala-Arg (10)
                Gly-Leu-Ala-Arg (5)             Ac-Leu-Gly-Leu-Ala-Arg (11)
                    Leu-Ala-Arg (6)    Asn-Lys-Pro-Leu-Gly-Leu-Ala-Arg (12)
```

Fig. 1. Synthetic peptides based on the C-terminus of C3a anaphylatoxin. Numbering corresponds to the C3a sequence.[2]

The relative molar activities of these peptides in the smooth muscle assay are compared in Fig. 2. The C-terminal 13-residue peptide was previously observed[3] to be no more active than octapeptide *1*. Shortening the octapeptide, however, did result in a progressive loss of activity. For example, tetrapeptide *5* is about 100 times less active than pentapeptide *4*. Peptides containing five or more residues desensitized ileal smooth muscle to contraction by human C3a and vice versa, as illustrated in Fig. 3. This effect was specific for C3a because smooth muscle desensitized to C3a was still contracted by human C5a anaphylatoxin and histamine. Prior treatment with the antihistaminic agent chlorpheniramine blocked contraction by C3a and the synthetic peptides but did not block contraction caused by bradykinin.

The presence of residues 70 to 72 produced a 4-fold increase in smooth muscle activity. The hydroxyl group of Ser^{71} and the imidazole ring of His^{72} evidently do not contribute to the increase in activity because analogs *7* to *9* bearing alanine at positions 71 and 72 were as active as the corresponding human peptides. The acetylated hexapeptide *10* was as active as hexapeptide *3*. Furthermore, the possibility that a positively charged group is required at either the amino terminus or at the histidine side chain was eliminated because the acetylated pentapeptide *11* was as active as the free pentapeptide *4*. Finally, certain residues in positions 70 to 72 will not produce the 4-fold increase in activity. For example, the octapeptide *12* corresponding to the C-terminus of pig[5] C3a was only slightly more active than the pentapeptide *4*. The 4-fold difference in smooth muscle activity between the human octapeptide *1* and pentapeptide *4* may be due to a conformational state involving the peptide backbone of residues 70 to 72 that is not as available to the pig octapeptide *12*.

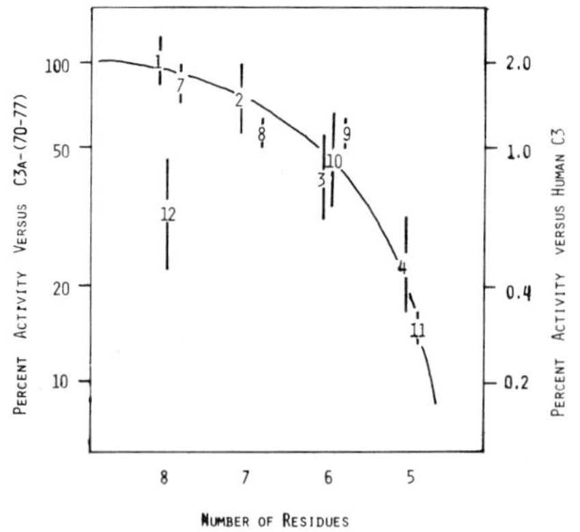

Fig. 2. Relative molar smooth muscle activity for ten C3a peptides. Peptide numbers correspond to those in Fig. 1. Bars represent the minimum effective dose range.

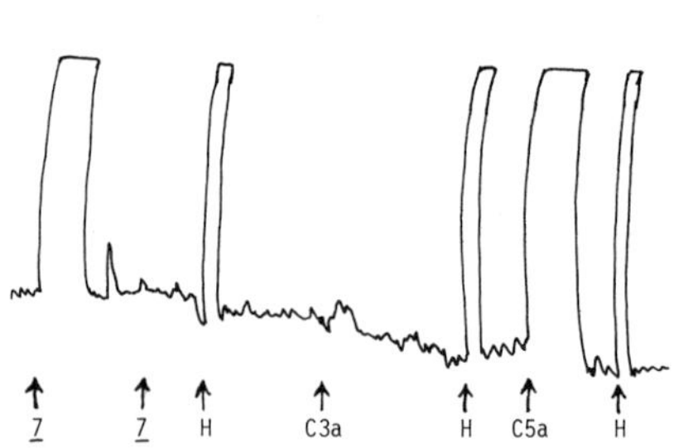

Fig. 3. Smooth muscle specificity of octapeptide *7*. After contraction by *7*, ileum responded to complement peptide C5a and histamine (H) but not to C3a or more *7*.

The second C3a bioassay involved the increase in vascular permeability of human forearm skin as measured by the dose-dependent diameter of the wheal-and-flare reaction. As shown in Fig. 4, the peptide dose necessary to produce half-maximal wheal increased progressively as the length of the C3a peptides decreased. These results are qualitatively similar but quantitatively different from the smooth muscle results. Addition of Leu^{73} to tetrapeptide *5* increased the ileum activity 100-fold but the skin activity only 10-fold. In contrast, the addition of residues 70 to 72 increased the ileum activity only 4-fold but increased the skin activity 10-fold. Interestingly, even the C-terminal arginine residue appears active in sufficient quantity. Thus the tetrapeptide *5* is 10^2 times more active, the octapeptide *1* is 10^4 times more active, and C3a is 10^6 times more active than free arginine in producing wheal-and-flare reaction in human skin. As before, the hydroxyl group of Ser^{71} and imidazole group of His^{72} are not essential for this activity because the alanine-containing analogs *7* and *8* were equally as active as the human C3a peptides *1* and *2*, respectively.

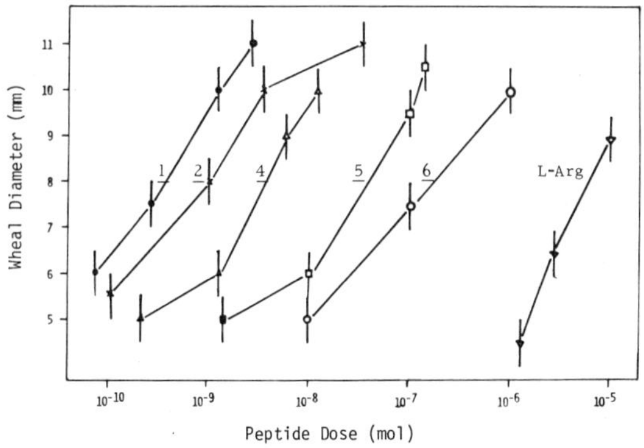

Fig. 4. Increase in vascular permeability. Code numbers refer to Fig. 1. Dose for half-maximal response increased as the peptide length decreased.

We thank Ms. Freda Kerman, Mr. Vincent Claps, and Mr. Peter Tippett for excellent technical assistance. This research was supported in part by grants HL 16411, HL 19795, and HL 20220 from the U. S. Public Health Service, by Grants in Aid 74-844 and 74-864 from the American Heart Association, and by funds contributed by the New York and California Heart Associations. T. E. H. is an Established Investigator (72-175) of the American Heart Association.

References

1. Hugli, T. E. (1975) in *Proteases and Biological Control*, Reich, E., Rifkin, D. B., and Shaw, E., Eds., Cold Spring Harbor Laboratory, Cold Spring Harbor, New York, pp. 273–290.

2. Hugli, T. E. (1975) *J. Biol. Chem.* **250**, 8293–8301.

3. Hugli, T. E. & Erickson, B. W. (1977) *Proc. Nat. Acad. Sci. USA* **74**, 1826–1830.

4. Erickson, B. W. & Merrifield, R. B. (1976) in *The Proteins*, Neurath, H. & Hill, R. L., Eds., Academic Press, New York, 3rd Edn., Vol. II, pp. 255–527.

5. Corbin, N. C. & Hugli, T. E. (1976) *J. Immunol.* **117**, 990–995.

STRUCTURAL AND FUNCTIONAL COMPARISON BETWEEN HUMAN C3a AND C5a

HORACIO N. FERNANDEZ and TONY E. HUGLI,
Scripps Clinic and Research Foundation, Department of Molecular Immunology, La Jolla, California 92037

In the course of complement activation, components C3 and C5 are sequentially cleaved, each giving rise to a large fragment which integrates into the cytolytic cascade, and a small, biologically active fragment which is released into the fluid phase (Fig. 1). These small split products (C3a and C5a) may participate in the inflammatory reaction following complement activation through their ability to induce smooth muscle contraction, increase vascular permeability and, in the case of C5a, to induce a directional migration of polymorphonuclear leukocytes and macrophages.[1] Removal of a C-terminal arginyl residue from either C3a or C5a abrogates all of the activities except for chemotaxis. Therefore, indigenous serum carboxypeptidase serves as an effective control enzyme for these nascent polypeptides when they become generated in circulation.

Human and porcine C3a isolated from complement-activated sera have a molecular weight[2] of 9,000, while human C5a exhibits a higher apparent molecular weight of 15,000 to 16,000 when estimated by gel filtration or on SDS

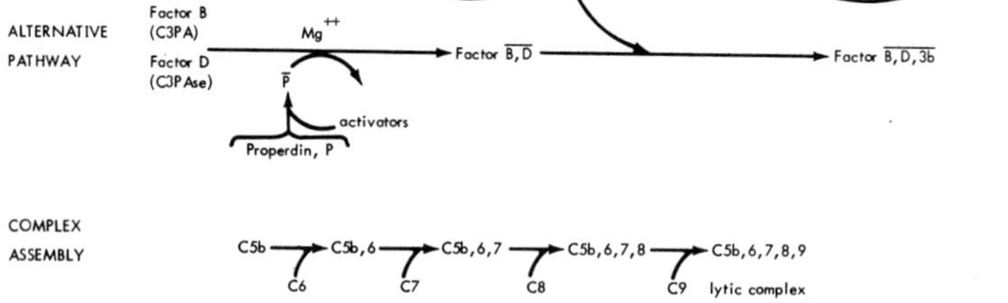

Fig. 1. A simplified scheme for the activation of complement (C). Activators such as agar, complex polysaccharides (inulin and zymosan) and immune complexes initiate the cascade via certain recognition components. Activation then proceeds through several orderly and selective proteolytic steps involving the intermediate complement components. Following these enzymatic conversions a self-associative complex is formed composed of the late components C5b to C9. The complex which is formed may be cytolytic when constructed on a cellular membrane surface.

polyacrylamide gels.[3] Unlike the C3a fragments, human C5a contains a sizable carbohydrate moiety consisting of mannose, galactose, glucosamine and sialic acid. The molecular weight of the C5a polypeptide moiety was estimated by quantitative C-terminal analysis to be about 8,500 which corresponds favorably with the size of C3a molecules. The polypeptide portion of human C5a is also similar in size to that reported for the porcine C5a molecule,[4] which is claimed to be free of carbohydrates.

Determination of the primary structure of human C5a afforded a comparison between this structure and the primary sequences of human and porcine C3a anaphylatoxins.[5,6] Originally, a tentative manual alignment was performed for the N-terminal 25 residues in human C5a and C3a based on the locations of a tyrosine and two half-cysteine residues which occur in both sequences at approximately identical positions. As shown in Figure 2, from a manual alignment of the two structures it was obvious that a marked similarity exists. Statistical evaluation of the degree of similarity was performed using the Align 5 computer program as previously described.[7] This comparison of the N-terminal 25 residues of C3a and C5a provided a value removed 4.08 standard deviations from a mean random score (Table I). Ancestral relationships between C3a and C5a were strongly implied from these preliminary data. Extending our comparison to include the entire C5a anaphylatoxin structure indicated that a highly significant structural similarity exists throughout the entire sequence of the two polypeptide chains. A total of 9 gaps were introduced in order to optimize the identities between C3a and C5a including a 2-residue indentation at the N-terminus of C5a.

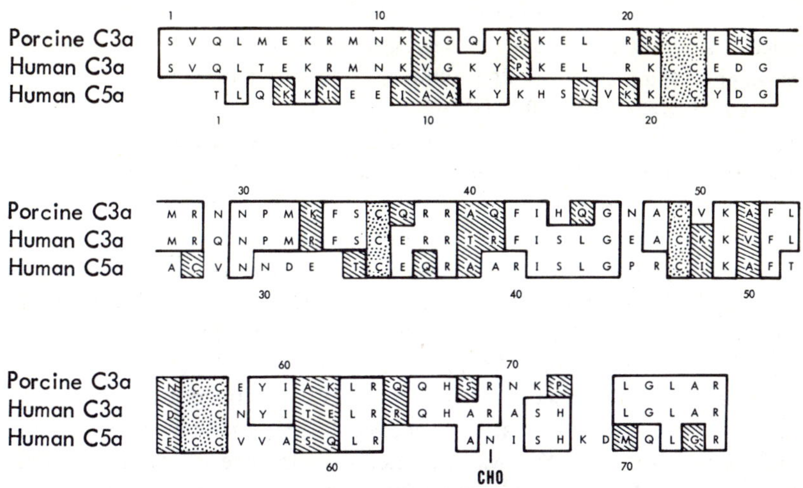

Fig. 2. The complete primary structures of porcine C3a and human C3a and C5a are compared. Alignments were made to optimize the identity between these three sequences. Identities have been signified by outlining the residues ☐. All single base mutational changes were indicated by line stipple ▨ and the six constant cysteine positions were distinguished by dot stipple ▩.

Table I. Evaluation of the Primary Structural Comparison Between C3a and C5a

A. Extent of similarity obtained from a manual alignment of C3a and C5a:

Comparison	Identities*	Identities plus single base mutations	(P)†
Human C3a and human C5a	36	59	0.01% <p < 0.1%
Human C3a and porcine C3a	66	86	p < 0.01%

B. Statistical evaluation # of the similarity obtained from a computer-generated alignment of C3a and C5a:

	Real Score [R]	Mean Random Score [M]	Random S.D. [SD]	Alignment Score [(R-M)/SD]
Human C3a and human C5a	191	151	9.81	4.08
Human C3a and porcine C3a	270	156	11.0	10.5

* Gaps introduced in either sequence were considered non-identities.

† The alignment comparison proposed by Moore and Goodman[8] was used to estimate the probability (P) that the aligned sequences belong to a random population. A penalty of 3 MMD (nucleotide differences per codon) was imposed for each insertion or deletion.

\# The N-terminal regions (residues 1-25) in human C5a, human C3a and porcine C3a were compared employing the Align 5 computer program of Barker and Dayhoff. The best-fit manual sequence alignment for this region was identical to the computer-generated alignment.

These gaps and indentations must also compensate for the net difference in chain length between C3a (77 residues) and C5a (74 residues). The percentage of identical residues was 36% and, if the residue identities are computed including replacements which could result from a single step mutational event, this value was increased to 59% (see Table I).

The alignment statistics for the human C5a and C3a sequences as determined by the criteria proposed by Moore and Goodman[8] indicates a very slight probability that these sequences belong to a random population, in fact less than a 0.1% chance. Thus the similarity previously evidenced in the N-terminal portion of these peptides actually exists throughout the entire length of the C3a and C5a molecules. Further support for a common genetic origin of the anaphylatoxins is supplied by the positioning of six half-cysteine residues which are located at nearly identical sites in both C3a and C5a. In addition, there are two relatively uncommon Cys-Cys sequences in each anaphylatoxin. Integrity of the disulfide bonds was essential for stabilization of the secondary structure and expression of full activity[9] in C3a. Hence it was predicted from the linear distribution of half-cysteine residues in C5a that arrangement of the three disulfide bonds will be found identical to those in C3a. Both human C3a and C5a have very similar and characteristic circular dichroism spectra indicating a major contribution from alpha helical structure, possibly involving as many as 40% of the residues. These data taken together would then suggest that these two molecules must share highly similar conformations.

Evidence of a common genetic ancestry for the anaphylatoxins also implies a single genetic origin for the parent molecules C3 and C5. This possibility affords some interesting suggestions concerning the phylogenetic development of the complement system. Complement factors C2, C4 and factor B have already been shown in man to be associated via genetic linkage with the HLA complex of the major histocompatibility system.[10] Previously, no evidence of genetic linkage or support for any other form of genetic relationship between C3 and C5 have been detected. Chemical evidence for C3 and C5 being related structurally poses the question of: Are these components related functionally, and in what order did these proteins evolve? These chemical and biological comparisons then set the stage for answering some of the basic questions concerning the genetic origins of the complex complement system.

This is publication number 1349 from Scripps Clinic and Research Foundation, supported by a National Institutes of Health Program Project Grant from the National Heart and Lung Institute (HL 16411), HL 20220, and AHA 74-864 from the American Heart Association. Dr. Hugli is the recipient of an Established Investigatorship from the American Heart Association (72-175).

References

1. Snyderman, R. & Pike, M. (1977) in *Comprehensive Immunology*, vol. 2, Day, N. K. & Good, R. A., Eds., Plenum Publishing Corp., New York, pp. 159–181.

2. Hugli, T. E., Vallota, E. H. & Müller-Eberhard, H. J. (1975) *J. Biol. Chem.* **250**, 1472–1478.

3. Fernandez, H. N. & Hugli, T. E. (1976) *J. Immunol.* **117**, 1688–1692.

4. Lieflander, M., Dielenberg, D., Schmidt, G. & Vogt, W. (1972) *Hoppe-Seyler's Z. Physiol. Chem.* **353**, 385–392.

5. Hugli, T. E. (1975) *J. Biol. Chem.* **250**, 8293–8301.

6. Corbin, N. C. & Hugli, T. E. (1976) *J. Immunol.* **117**, 990–995.

7. Fernandez, H. N. & Hugli, T. E. (1977) *J. Biol. Chem.* **252**, 1826–1828.

8. Moore, G. W. & Goodman, M. (1977) *J. Mol. Evol.* **9**, 121–130.

9. Hugli, T. E., Morgan, W. T. & Müller-Eberhard, H. J. (1975) *J. Biol. Chem.* **250**, 1479–1483.

10. Rosen, F. & Alper, C. A. in *Advances in Human Genetics*, Harris, H. & Hirschhorn, K., Eds., Plenum Press, New York, in press.

SYNTHESIS OF A FRAGMENT OF BOVINE PARATHYROID HORMONE, bPTH-(28–48): AN INHIBITOR OF HORMONE CLEAVAGE *IN VIVO*

MICHAEL ROSENBLATT, PIERRE D'AMOUR, GINO V. SEGRE, and JOHN T. POTTS, JR., *Endocrine Unit, Department of Medicine, Massachusetts General Hospital and Harvard Medical School, Boston, Massachusetts 02114*

Bovine parathyroid hormone (bPTH), an 84-amino-acid single-chain peptide, undergoes metabolic conversion *in vivo*. The principal cleavage positions lie between amino-acid residues 33 and 34, and 36 and 37.[1] The anatomical locale, kinetics and physiological role of hormonal cleavage is unknown. However, since structure–activity studies[2] reveal the minimum region necessary for biological activity to be the sequence 2–27, biologically active fragments of PTH may be generated by the cleavage process, or may be involved in PTH action on target organs. Cleavage, causing activation of the hormone, might be an important control in one or more PTH actions, perhaps an abnormally functioning control in disorders of parathyroid regulation. Alternatively, peripheral cleavage of the hormone may be principally catabolic, i.e., the pathway of hormone removal.

One approach to investigate the significance of cleavage of the hormone is to inhibit the enzymes responsible for the process. We synthesized a region of the hormone molecule likely to be a substrate for the cleavage enzymes (but lacking inherent hormonal activity) so that the peptide could be administered in large quantity. Accordingly, a 21-amino-acid sequence of the hormone, bPTH-(28–48), containing the cleavage sites of native bPTH[3] (Fig. 1) was selected for synthesis by the Merrifield solid-phase technique.[4]

The major synthetic product by gel filtration and CM-cellulose chromatography yielded the theoretical amino-acid composition (Table I) and was found to be homogeneous by thin layer chromatography and by polyacrylamide-gel isoelectric focusing. Sequence analysis was performed to quantitate contamination by deletion-containing error peptides.[5] Although extractive losses prevented analysis of more than the N-terminal 1/3 of the peptide, the region examined was 97% pure.

Of particular interest was the finding that approximately 1/3 of the synthetic product was peptides of lower molecular weight than the desired bPTH-(28–48). These side-products were isolated in the same purification systems and identified by amino-acid analysis (Table I) and sequence analysis as the COOH-terminal fragments: bPTH-(38–48) and bPTH-(39–48). It is unlikely that such fragments arise at the time of hydrogen fluoride cleavage because complementary N-terminal fragments were not found. More likely, these side-products result

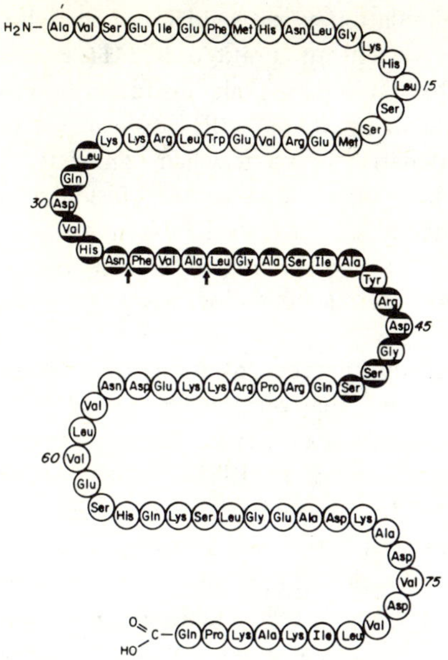

Fig. 1. (left) The amino-acid sequence of bovine parathyroid hormone. Shaded residues indicate the sequence of bPTH-(28–48). Arrows indicate the principal sites of cleavage.

Table I. Amino-Acid Content after Acid Hydrolysis of Synthetic bPTH-(28–48) and Side Products of the Synthesis, bPTH-(38–48) and bPTH-(39–48)

Amino Acid	bPTH-(28—48) expected	obtained	bPTH-(38—48) expected	obtained	bPTH-(39—48) expected	obtained
Histidine	1	1.0	0	< 0.1	0	< 0.1
Arginine	1	1.0	1	1.2	1	1.0
Aspartic Acid	3	3.2	1	1.4	1	1.1
Serine	3	2.8	3	2.9	3	2.6
Glutamic Acid	1	1.1	0	< 0.1	0	< 0.1
Glycine	2	2.1	2	1.7	1	1.2
Alanine	3	3.1	2	2.0	2	2.0
Valine	2	1.9	0	< 0.1	0	< 0.1
Isoleucine	1	0.9	1	0.9	1	1.0
Leucine	2	2.0	0	< 0.1	0	< 0.1
Tyrosine	1	1.0	1	1.0	1	1.1
Phenylalanine	1	1.0	0	< 0.1	0	< 0.1

from premature termination of peptide chains for sequence-dependent physical or chemical reasons.[6,7] The discovery and identification of these peptide side-products emphasizes the need to purify and evaluate thoroughly by multiple techniques a synthetic peptide before it is used in a biological application.

The synthetic fragment bPTH-(28–48) was assessed for PTH-like activity *in vitro* in the rat renal cortical adenylyl-cyclase assay.[8] bPTH-(28–48) demonstrated no agonist activity to a maximum concentration of $1.35 \times 10^{-4} M$. The hormone fragment also failed to inhibit stimulation by native bPTH of adenylyl-cyclase activity when tested to a maximum concentration ratio of bPTH-(28–48) to native bPTH of 500 to 1. However, the synthetic peptide was found to possess the biological property that suggested its synthesis: bPTH-(28–48) is an effective inhibitor of *in vivo* cleavage of native hormone when administered simultaneously with ^{125}I-labeled native hormone. Intravenous infusion of a large molar excess of bPTH-(28–48) inhibited cleavage to the extent that the formation of hormone fragments in treated rats was decreased by 75% of that seen in control rats, and there was a prolongation (two- to four-fold) of the disappearance of intact hormone (Fig. 2).

Further studies are necessary to determine optimal concentrations and delivery methods for cleavage inhibition. If it is found that very high molar ratios of bPTH-(28–48) to native hormone are required to block cleavage, design of more effective analogs will be undertaken, such as analogs of bPTH-(28–48) bearing modifications that prolong survival in blood, or containing D-amino acids at positions 33 and 34 which might prolong survival at the site of enzymic cleavage. Ultimately, the presently reported fragment, bPTH-(28–48), or its analogs should prove valuable in furthering understanding of the metabolism, mode and control of action of parathyroid hormone, as well as the problem of heterogeneity of circulating parathyroid hormone.

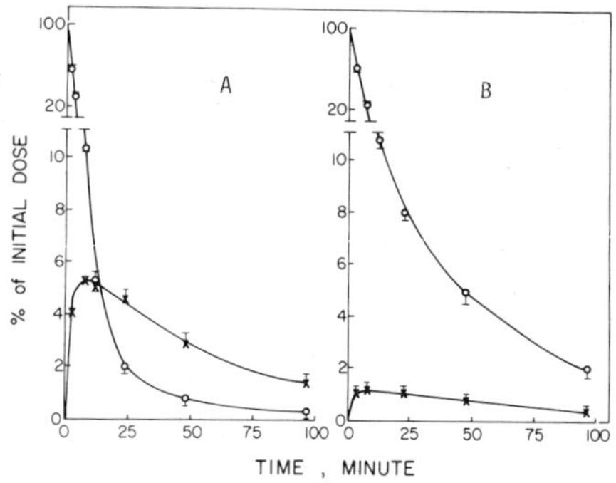

Fig. 2. Cleavage Chromatogram: Disappearance of intact ^{125}I-labeled bPTH (○) and appearance and disappearance of ^{125}I-labeled COOH-terminal fragments (X). (A) and (B) represent the results in control rats and in rats treated with bPTH-(28–48). Each value is the mean ± SEM of results obtained in three independent studies.

This work was supported in part by grants AM04501 and AM11794 from the National Institute of Arthritis, Metabolism, and Digestive Diseases (NIAMDD), a grant from the John A. Hartford Foundation, Inc., and contract NAS 9-11011 with the National Aeronautics and Space Administration. M.R. is a recipient of a Research Fellowship from the Charles A. King Trust and a National Research Service Award. G.V.S. is a recipient of a Research Career Development Award from the NIAMDD.

References

1. Segre, G. V., Niall, H. D., Habener, J. F. & Potts, J. T., Jr. (1974) *Amer. J. Med.* **56**, 774–784.

2. Tregear, G. W., Van Rietschofen, J., Greene, E., Keutmann, H. T., Niall, H. D., Reit, B., Parsons, J. A. & Potts, J. T. (1973) *Endocrinology* **93**, 1349–1353.

3. Rosenblatt, M., Segre, G. V. & Potts, J. T., Jr. (1977) *Biochemistry* **16**, 2811–2816.

4. Merrifield, R. B. (1969) *Adv. Enzymol.* **32**, 221–296.

5. Tregear, G. W. (1975) in *Peptides 1974: Proceedings of the 13th European Peptide Symposium*, Y. Wolman, Ed., New York, N. Y., Wiley, pp. 177–189.

6. Erickson, B. W. & Merrifield, R. B. (1976) in *The Proteins*, Neurath, H. & Hill, R. H., Eds., New York, N. Y., Academic Press, 3rd ed., vol. 2, pp. 255–257.

7. Hancock, W. S., Prescott, D. J., Vagelos, P. R. & Marshall, G. R. (1973) *J. Org. Chem.* **38**, 774–781.

8. Marcus, R. & Aurbach, G. D. (1969) *Endocrinology* **85**, 801–810.

PEPTIDE HYDROXAMATES AS INHIBITORS OF THERMOLYSIN AND RELATED METALLOPROTEASES

NORIKAZU NISHINO and JAMES C. POWERS, *School of Chemistry, Georgia Institute of Technology, Atlanta, Georgia 30332*

Metalloproteases are a family of proteolytic enzymes characterized by the presence of an essential active site zinc atom. Members of the family include widely studied enzymes such as thermolysin[1] and carboxypeptidase A, and less well understood enzymes such as collagenase, acrolysin and the angiotensin-converting enzyme[2]. Metalloproteases are involved in a number of important physiological processes including fertilization, kinin release and arthritis. Specific and potent inhibitors for these enzymes should be valuable for the study of their biological role and in addition may find practical application in the control of certain diseases.

Relatively few inhibitors of metalloproteases have been reported. All of the enzymes are inhibited by zinc chelating agents such as 1,10-phenanthroline, and by thiol compounds such as cysteine. Thermolysin is specifically inhibited by phosphoramidon (Rhamnose-Pi-Leu-Trp-OH)[3], a fermentation product; and a series of mercaptoalkanoyl derivatives of proline has been recently reported to be specific inhibitors of the angiotensin-converting enzyme[2]. In this paper we report the synthesis of a number of peptide hydroxamates which are specific inhibitors of several metalloproteases. In designing these compounds we sought to utilize specific substrate binding sites which are different in various members of the metalloprotease family and to include a functional group which would coordinate strongly to the active site zinc atom which is present in all members of the family.

Thermolysin was used as the model system in our initial experiments. This enzyme hydrolyzes the peptide bond on the imino side of hydrophobic amino acids, in particular isoleucine, leucine and phenylalanine. Z-Gly-Leu-NH_2 is a typical small synthetic substrate. X-ray crystallographic studies[4] have shown that binding of such a substrate to thermolysin involves coordination of the carbonyl oxygen of the scissile peptide bond of the substrate to the zinc atom of the enzyme and interaction of the side chain of the hydrophobic amino acid residue with the S_1' binding subsite[5] of the enzyme (*1*). (See Scheme 1.)

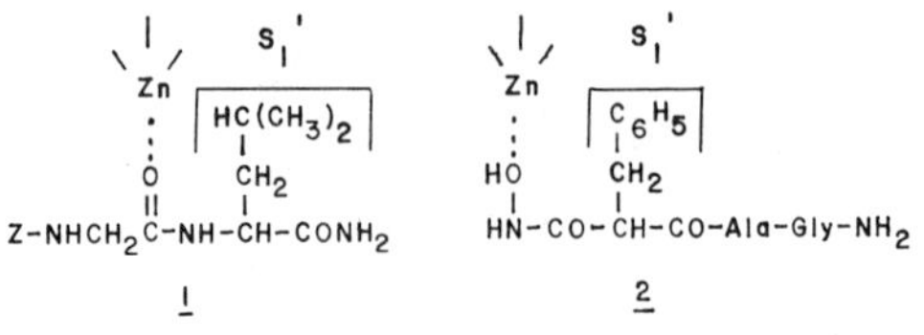

Scheme 1.

Therefore we have initially investigated compounds containing a leucine residue to interact with thermolysin's substrate binding site.

The peptide hydroxamates synthesized are given in Table I. Various *N*-protected peptides were converted to the hydroxamates by reaction of hydroxylamine with the *N*-hydroxysuccinimide esters. The benzylmalonyl derivatives were prepared by acylation of peptide amines with $BzlONHCOCH(CH_2C_6H_5)CO_2H$ followed by removal of the *O*-benzyl group by catalytic hydrogenation. All the hydroxamates were noncompetitive inhibitors of thermolysin using furylacryloyl-Gly-Leu-NH_2 as a substrate. K_I values were determined by Dixon plots and are listed in Table I. The hydroxamates were not cleaved by thermolysin.

A number of interesting conclusions can be reached upon examination of the data. All of the simple hydroxamates are bound much more tightly to thermolysin than are the corresponding substrates. The K_M for the thermolysin hydrolysis of Z-Gly-Leu-NH_2 is 21,000 μM at pH 7.0[6] compared to a K_I of 13 μM for Z-Gly-Leu-NHOH at pH 7.2, a difference of 1,600 fold. Both the *N*-Me and *O*-Me derivatives of Z-Gly-Leu-NHOH are bound much less tightly, confirming the involvement of the hydroxamate moiety and suggesting coordination of this group to the zinc atom of thermolysin. The side-chain specificity of the inhibitors is shown by the fact that Z-Gly-Leu-NHOH is bound 72 times more tightly than Z-Gly-Gly-NHOH, which suggests that the leucyl side chain is interacting with the primary substrate binding site (S_1') of thermolysin.

Several aspects of the mode of binding of hydroxamates to thermolysin are not yet clear. If the leucyl side chain of Z-Gly-Leu-NHOH is interacting with S_1' subsite and the hydroxamate is coordinating to the zinc atom, then the D-isomer

Table I. Inhibition of Thermolysin by Peptide Hydroxamates.[a]

Hydroxamates	K_I (μM)
Z-Gly-L-Leu-NHOH	13
Z-Gly-D-Leu-NHOH	59
Z-Gly-L-Leu-N(CH_3)OH	2200
Z-Gly-L-Leu-$NHOCH_3$	No inhibition
Z-Gly-Gly-NHOH	940
Z-L-Leu-NHOH	10
Z-Gly-Gly-L-Leu-NHOH	39
Z-Gly-Gly-D-Leu-NHOH	250
HONH-DL-CO-CH($CH_2C_6H_5$)-$COOC_2H_5$	20
HONH-DL-CO-CH($CH_2C_6H_5$)-CO-L-Ala-Gly-NH_2	0.66
HONH-DL-CO-CH($CH_2C_6H_5$)-CO-L-Ala-Gly-$NHC_6H_4NO_2$	0.43[b]

[a]0.1 *M* Tris/HCl, pH 7.2, 2% DMF, 25°C. [b]I_{50} value.

would be expected to bind more tightly than the L-isomer. However, in both cases where both isomers were examined, the L-isomer was a better inhibitor. An alternate binding mode is to place the leucyl side chain in the S_1 subsite of enzyme with the hydroxamate coordinating to the zinc atom. This would explain the optical specificity and the S_1 subsite does appear to be somewhat hydrophobic. Further discussion of the binding mode of inhibitors such as Z-Gly-Leu-NHOH must await the results of an X-ray crystallographic study which is currently underway in the laboratory of Brian Matthews at the University of Oregon.

Benzylmalonyl derivatives were specifically designed to bind in a manner analogous to substrate binding (see *2* in Scheme 1) and were in fact the best inhibitors in the series. Extension of the peptide chain yields lower K_I values, evidence that the inhibitors are interacting with thermolysin's extended substrate binding region.

Another purpose for designing specific enzyme inhibitors is to utilize them in the purification of specific enzymes by affinity chromatography. In order to demonstrate the usefulness of peptide hydroxamate inhibitors in this regard, we attached DL-HONHCOCH($CH_2C_6H_5$)CO-Ala-Gly-OH to aminopropyl agarose. This resin was used for the purification of thermolysin and the separation of neutral proteases A and B from *Bacillus subtilis* (Fig. 1). Neutral proteases A and B have been successfully separated previously using a different adsorbent.[7] The hydroxamate column was quite specific since chymotrypsin and carboxypeptidase A were not retained on the column.

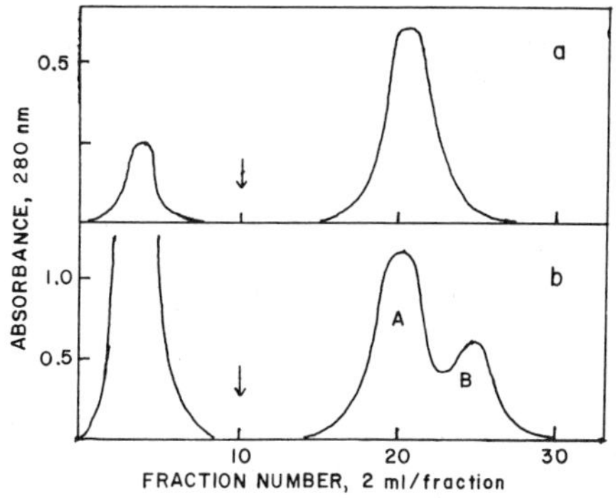

Fig. 1. Purification of thermolysin and neutral proteases A and B by affinity chromatography. (a) Thermolysin was placed on the column and washed with 0.1 *M* Tris/HCl pH 7.2, 0.01 *M* $CaCl_2$ and eluted with 0.1 *M* Tris/HCl pH 9.0, 0.1 *M* $CaCl_2$. (b) Neutral proteases A and B were separated by a linear gradient using 0.1 *M* Tris/HCl pH 8.0, 0.05 *M* $CaCl_2$ and 0.1 *M* Tris/HCl pH 9.0, 0.1 *M* $CaCl_2$.

This research was supported by grants HL 18679 and RR 07024-11 (Biomedical Research Support Grant Program) from the NIH. We would like to express our appreciation to Dr. J. Feder for the sample of crude *B. subtilis* neutral protease.

References

1. Matsubara, H. & Feder, J. (1971) in *The Enzymes*, Vol. 3, Boyer, P. D., Ed., Academic Press, New York, pp. 765–786.
2. Ondetti, M. A., Rubin, B. & Cushman, D. W. (1977) *Science* **198**, 441–444.
3. Suda, H., Aoyagi, T., Takeuchi, T. & Umezawa, H. (1973) *J. Antibiotics* **26**, 621–623.
4. Kester, W. R. & Matthews, B. W. (1977) *Biochemistry*, in press.
5. Schechter, I. & Berger, A. (1967) *Biochem. Biophys. Res. Commun.* **27**, 157–162.
6. Morihara, K. & Tsuzuki, H. (1970) *Eur. J. Biochem.* **15**, 374–380.
7. Pangburn, M. K., Burstein, Y., Morgan, P. H., Walsh, K. A. & Neurath, H. (1973) *Biochem. Biophys. Res. Commun.* **54**, 371–379.

SYNTHESIS OF ANTIGENIC PEPTIDES OF MAMMALIAN C-TYPE RNA TUMOR VIRUS MAJOR INTERNAL PROTEINS

S. I. SALLAY, K. S. L. SRIVASTAVA, *Department of Medicinal Chemistry, School of Pharmacy, Purdue University, West Lafayette, IN 47907 and Department of Chemistry, Indiana-Purdue University, Fort Wayne, IN 46805,* S. OROSZLAN and R. V. GILDEN, *Frederick Cancer Research Center, Frederick, MD 21701.*

Detection of virus-specific products by immunologic means has proven invaluable for studies of C-type RNA tumor viruses. Their most useful antigenic reactivity has been associated with the ~ 30,000 dalton major internal protein, carrying a mosaic of about six distinct species-specific and interspecies determinants, referred to synonymously as the group-specific antigen, gs-antigen or p30.[1] This protein is an important marker for a variety of studies relating to inter-viral relationships, natural history and the correlation of C-type RNA viruses to cancer. The gs-protein can be readily detected in adult, embryonic and tumor tissues even though infectious virus cannot be demonstrated.[2] It accounts for ~ 25-30% of the virion protein mass and is the major component of the nucleoid protein coat of the various C-type RNA viruses. They are homologous, single-chain proteins, assembled by 251 to 268 amino-acid residues, containing only two cysteine units.[3]

Recently N-terminal-sequence analysis of the first twenty to thirty amino acids of hamster, mouse, rat, cat, RD-114, gibbon ape, baboon and woolly monkey C-type RNA tumor virus p30s were determined by Oroszlan et al.[3,4] Each gs-antigen starts with a Pro-Leu-Arg conserved sequence. A hypervariable 4–10 sequence of amino acids follows. Finally, a longer conserved region extends from residues 11 to 24 of the p30 antigens. As exceptions, the gibbon ape and woolly monkey p30s possess an insertion of a total of seven amino-acid residues between the 6–7 and 8–9 sequences of hypervariable region.[4] Limited tryptic digest of gs-antigens have demonstrated the existence of a homologous polypeptide core of these protein molecules.[5]

Since the 11–21 fragment of the conserved region (*1*, Scheme 1) is common for all mammalian gs-antigens so far investigated, the strategy of its synthesis was designed to use it as a "prefabricated building block" which allows further extensions at its termini while the side-chain protections remain intact. This undecapeptide also promised great utility for use in immunoassays of C-type viral activity, having two of the six or seven tyrosine residues of the total gs-antigen.[3] The specific activity of ^{125}I-labeling achievable for this undecapeptide is approximately one magnitude larger than that of the intact p30 molecule.

The undecapeptide (*1*) was synthesized[6] through intermediates *2* → *12*. The fully protected undecapeptide (*11*) was purified on a silicagel column; m.p.

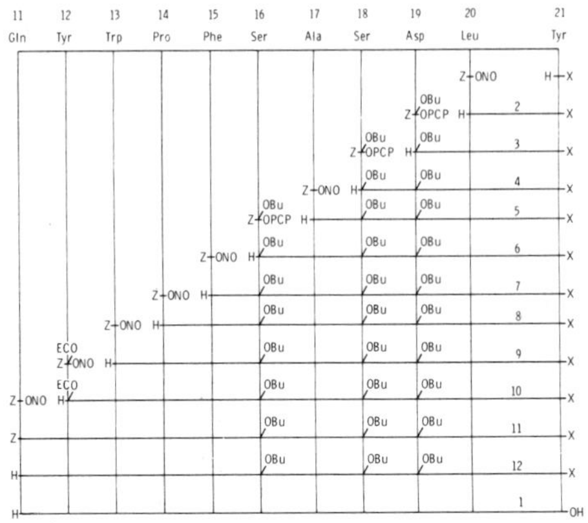

Scheme 1. X-OCH_3; ECO=C_2H_5-O-CO-

198-200°C. The TFA-salt of *1* was isolated through high-voltage electrophoresis and used for immunological studies after it was covalently attached to bovine serum albumin (BSA). The BSA-*1* conjugate forced antibody production in goat and the immune serum in micro-complement fixation tests reacted not only with the synthetic antigen (BSA-*1*), but also with the native p30s from all mammalian C-type RNA viruses. Specific binding of both internally (^{14}C) and externally (^{125}I) labeled p30s to the antibody against BSA-*1* was also shown in radioimmunoassays. This has been the first demonstration of a C-type virus synthetic antigen.[7]

The preliminary immunological findings encouraged us to extend the 11-21 undecapeptide through its amino terminus to the 1-21 segments of baboon and cat RNA tumor virus p30 antigens. First, a nonapeptide-hydrazide (*21*), representing the 1-9 sequence of the baboon virus gs-protein was synthesized[8] from a common precursor: tripeptide-Z-hydrazide (*13*, Scheme 2). The two fragments, *15* and *19*, were condensed to the nonapeptide derivative (*20*). The latter was converted to the Boc-nonapeptide hydrazide (*21*, m.p. 255-260°C) ready for fragment condensation with the undecapeptide derivative (*12*).

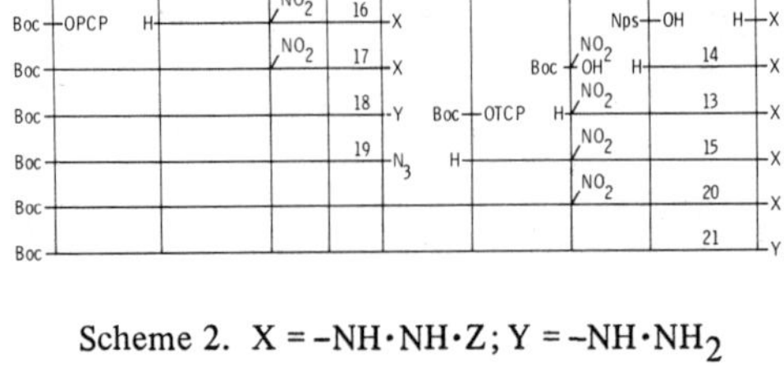

Scheme 2. X = –NH·NH·Z; Y = –NH·NH_2

A third sequence, the 1-10 segment of cat RNA tumor virus gs-antigen also was synthesized (Scheme 3). The final product, Boc-decapeptide hydrazide (*31*, m.p. 268-272°C) was purified by liquid chromatography. The chromatograms of the crude and purified fractions of *31* are shown on Figures 1 and 2.

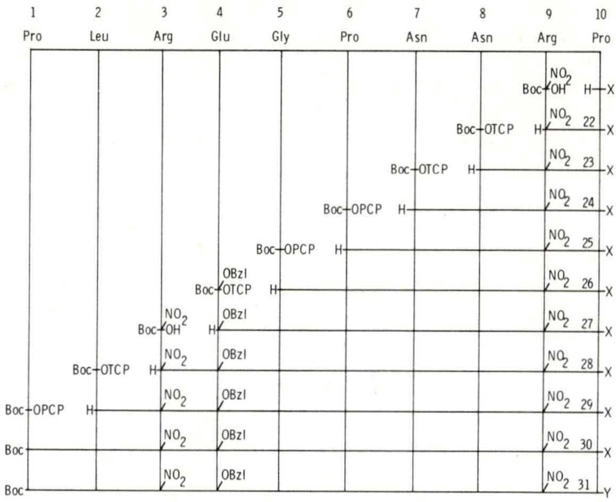

Scheme 3. X = –NH·NH·Z; Y = –NH·NH$_2$

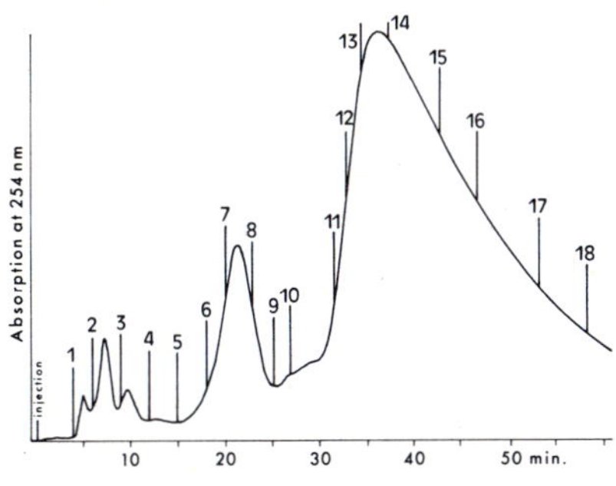

Fig. 1. LC of crude *31* (40 mg) on Bondapak C_{18} column/4 feet. CH_3CN/H_2O = 1:4, + 1% AcOH.

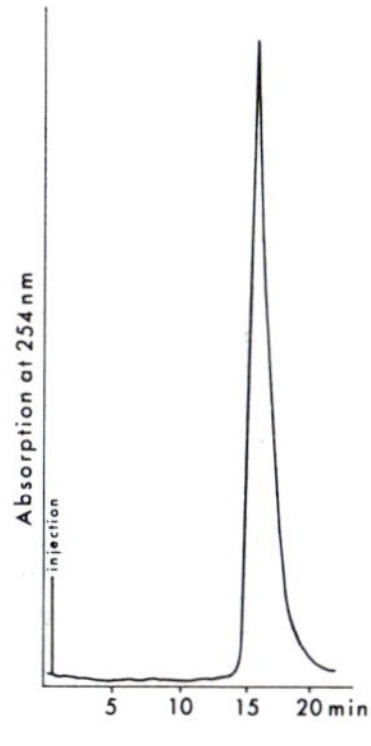

Fig. 2. LC of pure *31* (fractions 13–15 of Fig. 1) on μ Bondapak C_{18} column/1 foot. CH_3CN/H_2O = 1:4, + 1% Ac OH.

Boc-OSu[9] (m.p. 98–100°C) was found to be an excellent reagent for *t*-butyloxycarbonylation throughout the synthesis. The reagent was obtained in ~ 70% yield and is marketed by Regis Chemicals, Morton Grove, Illinois. Active esters were used for the coupling reactions except in cases of Arg and Thr residues which were incorporated by the mixed anhydride technique. The amino-acid analysis of the products described herein were carried out on a Beckman Model-121 amino acid analyzer. Waters Associates' liquid chromatograph Model ALC/GPC 244 was used for purification.

The excellent technical assistance of Dr. William F. Erbelding and the amino-acid analysis of Dr. Louis Henderson is greatly appreciated. This work was supported by Contract N01-CP-53530 from the Special Virus Cancer Program, National Institutes of Health – National Cancer Institute, Bethesda, MD. One of us (S. I. S.) was also generously supported by Thermtron Products, Inc. and Allen County Cancer Society, Fort Wayne, IN. This paper is dedicated to Professor R. B. Woodward of Harvard University on the occasion of his 60th birthday.

References

1. Oroszlan, S., Bova, D. & Gilden, R. V. (1975) *Immunochemistry* **12**, 61–66.
2. Gilden, R. V. & Oroszlan, S. (1972) *Proc. Nat. Acad. Sci. USA* **69**, 1021–1025.
3. Oroszlan, S., Copeland, T., Summers, M. R., Smythers, G. & Gilden, R. V. (1975) *J. Biol. Chem.* **250**, 6232–6239.
4. Oroszlan, S., Copeland, T., Smythers, F., Summers, M. R. & Gilden, R. V., (1977) *Virology* 77, 413–417.
5. Oroszlan, S., Summers, M. R., Foreman, C. & Gilden, R. V. (1974) *J. Virol.* **14**, 1559–1574.
6. Sallay, S. I., Oroszlan, S. & Gilden, R. V. (1975) First Chemical Congress of the North American Continent, Mexico City, Abstract BMPC #18.
7. Oroszlan, S., Gilden, R. V. & Sallay, S. I. (1975) First Chemical Congress of the North American Continent, Mexico City, Abstract BMPC #14.
8. Sallay, S. I., Srivastava, K. S. L., Oroszlan, S. & Gilden, R. V., 172nd ACS National Meeting, San Francisco, 1976; Abstract MEDI #77.
9. Frankel, M., Ladkany, D., Gilon, D. & Wolman, Y. (1966), *Tetrahedron Lett.*, **39**, 4765–4768.

EXCHANGE OF Ala[1] BY DIVERSE AMINO ACIDS IN PHALLOIDIN AND TOXICOLOGICAL PROPERTIES OF THE NEW ANALOGS

EISUKE MUNEKATA, HEINZ FAULSTICH and THEODOR WIELAND, *Max-Planck-Institut für medizinische Forschung, Abteilung Naturstoff-Chemie, Heidelberg, West Germany.*

For the purpose of extensive studies on the correlations between chemical structure and biological activity of toxic cyclopeptides from the poisonous mushroom *Amanita phalloides*[1], new analogs were derived from naturally occurring phalloidin (*1*) by means of following reactions:

1) Acid hydrolysis of the peptide bond between γ, δ-dihydroxyleucine (position 7) and alanine (position 1),

2) cleavage of Ala[1] by Edman degradation,

3) coupling with Boc-protected amino acid,

4) removal of Boc-group with CF_3COOH and

5) opening of γ-lactone ring and cyclization to bicyclicpeptide[2] (Scheme 1).

In addition, the bicyclic hexapeptide, des-Ala[1]-phalloidin (*6*), and the bicyclic octapeptide, *endo*-Ala[1a]-phalloidin (*8*) were prepared as shown in Scheme 2.

The yields of the cyclization reactions, Rf-values of the analogs in tlc and toxicities in white mice are compiled in Table I. Val[1] -(*2d*) and Leu[1] -analogs

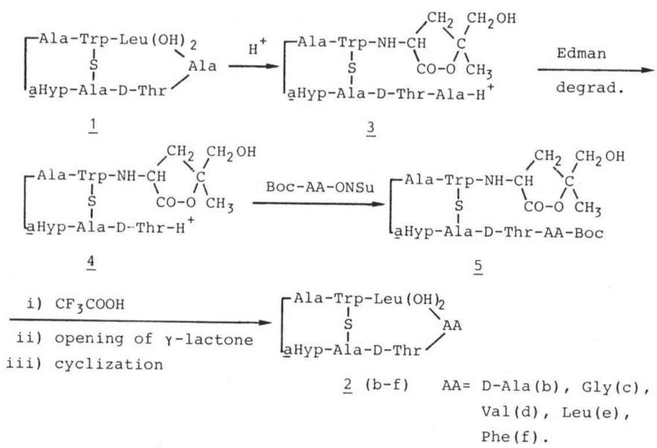

Scheme 1.

(*2e*) exhibit essentially identical toxicological and spectroscopical properties. Gly[1]-analog (*2c*) possesses slightly reduced toxicity although the CD curve and UV-difference spectrum[3] of the bound complex with rabbit muscle actin are significantly different. The Phe[1]-analogue (*2f*) shows a normal CD curve and

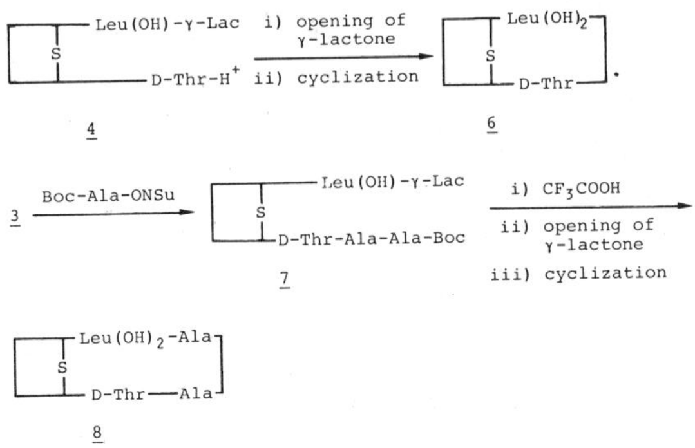

Scheme 2.

Table I. Yields of Cyclization, Rf-Values (on Silica Gel tlc Plates, Kieselgel 60 F_{254} Merck, in $CHCl_3$: CH_3OH: H_2O; 65:25:4; by vol.) and Toxicities (LD_{50} mg/kg in white mice) of Phalloidin Analogs.

Amino acid in position 1		Yield of cyclization (%)	Rf-value	Toxicity
Ala[a]	2a	8.7	0.31	2.0
D-Ala	2b	6.6	0.32	-[b]
Gly	2c	22.0	0.30	7.5
Val	2d	2.4	0.45	2.5
Leu	2e	2.7	0.48	2.5
Phe	2f	2.2	0.48	20.0
des-Ala	6	7.7	0.32	-[b]
Ala_2	8	28.0	0.40	-[b]

[a] Substance obtained by recyclization of secophalloidin[2].

[b] Tested in doses up to 30mg/kg.

UV-difference spectrum with actin, being nevertheless about tenfold less poisonous than the natural product. D-Ala[1] -analogue (*2b*), des-Ala[1]-phalloidin (*6*) and *endo*-Ala[1a]-phalloidin (*8*) have no toxicity and the spectra of these derivatives are evidently abnormal.

The results of present work can be summarized as follows:

1) To be toxic, the bicyclic peptide must consist of seven amino acids, since the hexapeptide *6* and the octapeptide *8* are non-toxic.

2) The methyl group of Ala[1] may be replaced by an isopropyl- or isobutyl-group without loss of toxicity. However, toxicity is reduced by substitution of

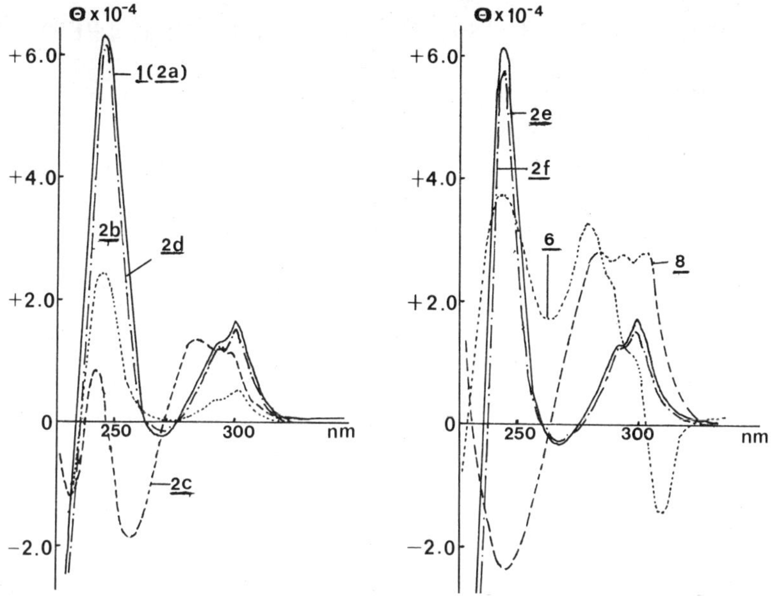

Fig. 1. CD-spectra of phalloidin *1*(*2a*) and its analogues in H_2O.

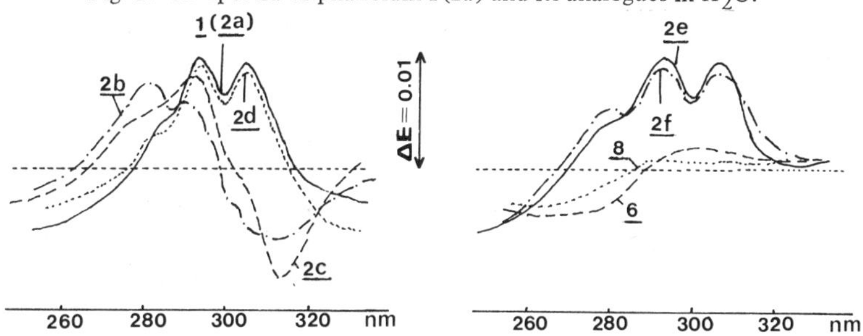

Fig. 2. UV-difference spectra of bound complex of actin with phalloidin *1*(*2a*) and its analogues in Trisbuffer (pH 7.4).

the methyl group by either hydrogen or the benzyl group.

3) Change of configuration at Ala[1] completely removes the toxic properties of the cyclic peptide.

References

1. Wieland, Th. & Wieland, O. (1972) in *Microbial Toxin,* vol. 8, Kadis, S., Ciegler, A. & Ajl, S. J., Eds., Academic Press, New York, p. 249.

2. Munekata, E., Faulstich, H. & Wieland, Th. (1977) *Angew. Chem.* **89**, 274–275. (*Angew. Chem. Int. Ed. Engl.* **16**, 267–268).

3. Wieland, Th., de Vries, J. X., Schäfer, A. & Faulstich, H. (1975) *FEBS Lett.* **54,** 73–75.

CONFORMATIONAL ENERGY CALCULATIONS ON PEPTIDES

HAROLD A. SCHERAGA, *Department of Chemistry, Cornell University, Ithaca, New York 14853*

Introduction

Empirical energy calculations are carried out to determine the low-energy conformations of polypeptides and proteins and to elucidate the interactions which lead to the low-energy structures. The computational strategies for small peptides differ from those for proteins, and primarily the former will be considered in this paper. The reader is referred to a recent review for a similar discussion of the computational problem for globular proteins.[1]

Since a solution of a small peptide in equilibrium with its surroundings at constant temperature and pressure is considered to have a minimum Gibbs free energy with respect to conformational changes, it is necessary to have procedures to compute the conformational energy of the peptide and to minimize it by varying the conformation. These procedures have been reviewed elsewhere.[1-5] The essential ingredients of such procedures are: (i) values of the bond lengths and bond angles of the amino-acid residues, (ii) a method to generate an arbitrary conformation of the polypeptide chain, (iii) empirical potential energy functions for interactions between various parts of the polypeptide chain, in any arbitrary conformation, (iv) a method to include the effect of solvation (hydration), (v) an algorithm to minimize the conformational energy (i.e., the potential energy plus free energy of solvation), and finally (vi) a method to compute the entropy of libration of the structure around its minimum-energy conformation. These procedures are discussed briefly below.

While there will no doubt be improvements in potential functions, parameters, and procedures for energy minimization, we have considerable information in these areas – even though refinements are needed. The major difficulty remaining, in order to compute the most stable conformation of a polypeptide or protein, is that which arises from the existence of many minima in the multidimensional conformational energy space (the multiple-minimum problem). In this paper, we will show how this problem has been surmounted for small open-chain oligopeptides, cyclic peptides, and regular-repeating-sequence polypeptide analogs of fibrous proteins. The current efforts to solve this problem for *globular* proteins are reviewed elsewhere.[1]

ECEPP (Empirical Conformational Energy Program for Peptides)

This computer program has been deposited with the Quantum Chemistry Program Exchange (QCPE), together with extensive documentation to aid in its use. From a critical survey of the structural literature on amino acids and peptides, a standard geometry (bond lengths and bond angles) for all of the naturally occurring amino-acid residues was deduced.[6] The program employs matrix multiplication methods to generate the atomic coordinates for any desired conformation of a polypeptide, as defined by the dihedral angles for rotation about the single bonds of the backbone and side chains. These dihedral angles (the independent variables in this problem) are designated as ϕ and ψ for rotation about the N-C$^{\alpha}$ and C$^{\alpha}$-C$'$ bonds, respectively, of the backbone, and χ^1, χ^2, etc., for rotation about the single bonds of the side chains. (The peptide group is usually fixed in the planar *trans* conformation, except for peptide bonds preceding proline, where both *trans* and *cis* conformations are considered.) The justification for keeping the bond lengths and bond angles fixed, and the peptide groups in the planar conformation, has been presented elsewhere.[7]

ECEPP then computes the conformational energy of the specified conformation, consisting of torsional energies for most side-chain single bonds (in the form of periodic functions), and interaction energies for all atom pairs whose interatomic distance is capable of variation. The torsional energy parameters were obtained from the experimental torsional barriers of small molecules, and the parameters for interatomic interactions from calculations on crystal structures of small molecules.[6]

The relations used to compute the interaction energy for atom pair i,j are:

(1) Electrostatic:

$$U_{el} = 332.0\, q_i q_j / D r_{ij} \tag{1}$$

where q_i and q_j are partial atomic charges in electronic charge units, D is the dielectric constant, r_{ij} is the internuclear distance in Angstrom units, and 332.0 is the factor to convert to energy units of kcal/mole,

(2) Nonbonded (modified Lennard-Jones 12-6 potential):

$$U_{nb} = F A^{k\ell} / r_{ij}^{12} - C^{k\ell} / r_{ij}^{6} \tag{2}$$

where the coefficients $A^{k\ell}$ and $C^{k\ell}$ are assigned specific values for each combination of atom types k and ℓ, and F has a value defined by the interaction type,

(3) Hydrogen bonded [substituted for (2) in appropriate cases]:

$$U_{hb} = A'_{HX} / r_{HX}^{12} - B_{HX} / r_{HX}^{10} \tag{3}$$

where A'_{HX} and B_{HX} are specific coefficients for the different combinations of donors and acceptors.

It should be emphasized that the partitioning of the total conformational energy into components, as discussed above, is an artificial procedure. It is only the *total* conformational energy that is relevant. Therefore, the components of the total energy must be self-consistent, i.e., they must pertain to a *total* energy function that is based on experimental data. For this reason, it makes no physical sense to compare the individual components of one energy function with those of another.

Though not yet in the ECEPP documentation supplied to QCPE, hydration is included by means of a solvent-shell model.[8-10] It is also possible to include the effect of ionic strength on polypeptide conformation, as has been done in the computation of the effect of variation in pH and ionic strength on the helix-coil transition in poly(L-lysine) in aqueous salt solution.[11]

A variety of minimization procedures can be used to minimize the conformational energy.[12-19] Finally, the matrix of second derivatives of the conformational energy provides a measure of the librational entropy.[20-22]

Preference for the *Trans* Conformation in Peptide Groups

It is of interest, at this point, to consider the well-known preference for the *trans* conformation in peptide groups, and for the possible existence of the *cis* conformation in peptide bonds preceding proline. This question was examined[23] for blocked dipeptides (those with *N*-acetyl and *N'*-methyl amide end groups), and it was found that the instability of the *cis* conformation in Gly-Gly comes primarily from interactions of the C_1^α and H_1^α atoms with the C_2^α and H_2^α atoms, and also from favorable interactions present in the *trans* form which are disallowed in the *cis* form, and from conformational entropy. The instability of the *cis* form in Gly-Pro is much less than in Gly-Gly because unfavorable interactions of the type $H_1^\alpha \cdots H_2^\alpha$ and $C_1^\alpha \cdots C_2^\alpha$ present in the *cis* conformation of Gly-Gly are present in *both* the *cis* and *trans* forms of Gly-Pro. The instability of the *cis* conformation in Gly-Pro arises mainly from the change in electrostatic energy caused by the restricted rotation about the N-C^α bond of Pro. This conclusion is supported by experimental observations.[24] Ramachandran and Mitra[25] also computed the *cis*/*trans* ratio for several dipeptides, and obtained *cis* contents that were much larger than those computed by Zimmerman and Scheraga[23] or observed by nmr spectroscopy in *nonpolar* solvents by Wüthrich et al.[26] and by Stimson et al.[24] Also, an electron diffraction study of *N*-methyl acetamide in the gas phase[27] gave no evidence for any *cis* conformation. Thus, local interactions account for the preference for the *trans* over the *cis* conformation in short peptides. In polar solvents, the *cis* content in X-Pro peptide groups is higher than in nonpolar solvents.[24] The observation of *cis* X-Pro peptide bonds in some proteins[28] is probably due to an increment in free energy of stabilization from long-range interactions. However, while such long-range interactions may stabilize the *cis* conformations of X-Pro peptide groups, they are apparently not strong enough to lead to *cis* conformations in peptide groups not containing proline.

Approximations to ECEPP

In order to speed up computations in the *initial* stages of the calculations on a polypeptide or protein, various approximations are used.

The first of these makes use of "united atoms" to represent aliphatic and aromatic -CH, $-CH_2$, and $-CH_3$ groups, thereby reducing the number of units between which the interaction energies have to be computed. The united-atom version of ECEPP is known as UNICEPP (United Atom Conformational Energy Program for Peptides). It was parameterized[29] to provide the same relative conformational energies as ECEPP.

Second, a united-residue approximation[30] was introduced to reduce the required number of computed interactions still further. When residues are far apart, they contribute to the overall (attractive) interaction energy of the protein. However, the nonbonded and electrostatic interaction energies between two such residues can be well approximated by an interaction between two points, rather than by the time-consuming procedure of calculating the interaction energies between every pair of atoms in the two residues.

Finally, for some applications, the backbone and side chains of all residues can each be represented by spheres of appropriate radii,[31] thereby further reducing the required computation time.

The utility of these approximate methods lies mainly in its applicability to proteins. However, even for small peptides, considerable savings of computer time can be achieved by the UNICEPP and spherical-representation approximations. The united-residue approximation is not applicable to oligopeptides because of their small size.

In the *final* stages of the computation, the approximate procedures are abandoned, and ECEPP is used.

Previous Systems Considered

These techniques have been applied to a variety of systems with considerable success.[1,4] These include: the conformational preferences of single amino acids, open-chain peptides, and cyclic peptides; the conformational properties of randomly-coiled polypeptides; calculation of the preferred type (e.g., α or ω) and right- or left-handedness of helical polypeptides; calculation of the thermodynamic parameters for helix-coil transitions in poly(amino acids), and for helix-helix transitions (e.g., the interconversion between polyproline I and II); the computational energetic refinement of the X-ray structures of proteins; computations of structures of homologous proteins (e.g., the calculation of the structure of α-lactalbumin from that of lysozyme, and the structures of several neurotoxic trypsin and chymotrypsin inhibitors from that of bovine pancreatic trypsin inhibitor); and computations of the structural features of enzyme-substrate complexes.

In order to discuss some specific examples, it is convenient to classify polypeptides as small open-chain structures, cyclic structures, and synthetic regular-repeating-sequence analogs of fibrous proteins. A fourth category, globular proteins, is considered elsewhere.[1]

Multiple-Minimum Problem

While mathematical procedures exist to pass from one local minimum to another in conformational space,[32-36] they consume too much computer time. Therefore, alternative methods are required.

The approach to circumvent the multiple-minimum problem differs for the various categories of polypeptides mentioned above. It has been possible to surmount this problem in the first three categories because the number of variables is small, and appropriate strategies can be used so that the conformational space can be covered adequately (even though still requiring much computer time) to locate the global minimum. Energy minimization is carried out from many appropriately-selected starting conformations, and the global minimum is then identified from a listing of the various local minima.

For a (blocked) single amino-acid residue, i.e., one in the *N*-acetyl-*N′*-methyl amide form, the conformational space can be searched completely, and all local minima within any desired range of the global one determined. For the blocked 20 naturally occurring amino acids, there are ~7 backbone minima, each of which generally includes a large number of side-chain minima.[37] For purposes of illustration, let us assume that there are ~20 low-energy backbone-side-chain combination conformations for each blocked amino acid. If, as a strategy to solve the multiple-minimum problem for an oligopeptide, we were to start with all combinations of low-energy single-residue conformations, and minimize the energy of each one, we would have to consider 400, 8000, 1.6×10^5, 3.2×10^6, and 6.4×10^7 energy minimizations for di-, tri-, tetra-, penta- and hexapeptides, respectively, and 10^{130} for a protein of 100 residues. Thus, such a complete exploration of conformational space generally becomes impractical for structures larger than a di- or tripeptide. In special cases where proline (with its restricted pyrrolidine ring) is involved, larger structures can be handled. Therefore, to treat oligopeptides, it is necessary to devise strategies (which may differ from one peptide to another, depending upon its composition) to sample the conformational space only partially, with a likelihood that one has not overlooked the region in which the global minimum occurs.

The simplest type of oligopeptide, the blocked dipeptide, has been treated by starting with conformations that include all combinations of low-energy single-residue minima and several standard bend structures, and then minimizing the conformational energy.[38,39] The molecules considered were a series of X-Pro, Pro-X,[38] X-Ala, Ala-X,[39] X-Gly, Gly-X,[39] and X-Ser, Ser-X[39] dipeptides, where X was a variety of amino-acid residues. The conformations of most of these dipeptides were found to be determined primarily by *intra*-residue interactions. Also, the librational entropy was computed and found to influence the relative stabilities of some minima. Because of the existence of many low-energy minima for each dipeptide, such systems are best regarded as statistical ensembles of low-energy structures. The probabilities of occurrence of bends in most of the blocked dipeptide sequences were found to correlate well with experimental X-ray data on bends in globular proteins.[40]

For small open-chain peptides, the conformational space that should be explored can be limited by breaking the oligopeptide into blocked single-residue and di- and tripeptide fragments, whose computed low-energy conformations can be used to build up larger structures.[38,39,41,42] The computed low-energy conformations of single residues can be supplemented by those conformations that are found to occur frequently in the known X-ray structures of proteins. For cyclic peptides such as gramicidin S that possess symmetry, the symmetry itself and the requirement to close the ring restrict the conformational space.[43] In both open and cyclic peptides, various types of β-bends are also considered as starting conformations. Finally, when available, the generally-incomplete conformational information from nmr studies can be used to obtain additional starting conformations. Variations of these strategies have been used, for example, for the tripeptide *p*-Glu-His-Pro-NH_2,[44] the blocked tetrapeptide Thr-Asp-Gly-Lys,[41] and the pentapeptide Tyr-Gly-Gly-Phe-Met,[42] the cyclic hexapeptide cyclohexaglycine[45,46] and the cyclic decapeptide gramicidin S,[43] and synthetic models of collagen.[47,48] Some of these calculations are discussed below.

Different approaches have been taken to try to solve the multiple-minimum problem for globular proteins. These are discussed elsewhere.[1]

Small Open-Chain Structures

We will describe some recent computations on enkephalin,[42] as an example of a small open-chain structure. The amino-acid sequence of this pentapeptide is either Tyr-Gly-Gly-Phe-Met or Tyr-Gly-Gly-Phe-Leu, both of which are biologically active. In order to treat these pentapeptides, a variety of selection strategies was used. The starting conformations included several regular repeating conformations of the pentapeptide, various forms of chain reversals (bends), combinations of conformations that are of low energy for single amino acid residues or dipeptides, and compact conformations obtained by model building.[42] Energy minimizations were then carried out for these structures. Side-chain orientations were varied systematically to test the effect of possible side chain-backbone interactions.

Over 50 conformations (indicated in Fig. 1 of Ref. 42) were found within an energy range of 11 kcal/mole above the global minimum. The global minimum energy conformation differed by up to 5 kcal/mole from the other conformations. It is characterized by a type II′ β-turn centered on Gly-3 and Phe-4 and the folding of the Tyr-1 side chain against one side of the fold, with the Tyr-OH group hydrogen bonded to the backbone. This backbone conformation permits several side-chain conformations of Phe and Met (or Leu) with differing energies. Some of these conformations are of very low energy and form part of a low-energy group. In this group, two slightly different orientations of the Tyr side chain are possible, in which the OH group forms a hydrogen bond either (A) with the C=O of Gly-3, or (B) with the C=O of Phe-4 which lie close to each other. Changing from one to the other involves a small movement of the Tyr side chain, a change in the orientation of the planar peptide group between Tyr-1 and Gly-2 (the latter causing changes in the values of ψ_1 and ϕ_2), as well as smaller changes in the dihedral angles of Gly-2. However, the overall folding of the molecule, including backbone and side chains, is not altered significantly.

In Met-enkephalin, conformation (A) is lower in energy by 1.5 kcal/mole than (B) and, therefore, was considered as the major form of the molecule.[42] In Leu-enkephalin, the two conformations are much closer in energy. In fact, (B) is lower, but only by 0.2 kcal/mole. Therefore, both forms are about equally probable. In both peptides, the side chains of Phe and Met can adopt several different orientations for the same backbone structure without large increases in energy. These low-energy conformations are consistent with published nmr parameters for Met-enkephalin in dimethylsulfoxide.[49-51]

The analog of Met-enkephalin in which Gly-2 is replaced by D-alanine is reported to be at least as active as Met-enkephalin itself.[52] This suggests that the molecule bound to the receptor must take up a conformation which is of low energy for a D-alanyl residue in position 2, irrespective of what is the most stable conformation in solution. The lowest-energy conformation computed for Met-enkephalin in free solution[42] cannot accommodate this substitution. However, the computations indicate that about 40% of a large group of compact conformations with relative energies in the range of 5 to 10 kcal/mole can accommodate a D-alanine in position 2. Most of these conformations involve various types of bends centered at Gly-2 and Gly-3. The conformation suggested by Momany[53] for the D-alanyl analog also has an energy falling within this range. The facts that D-Ala can be substituted for Gly-2, with retention of activity, but that L-Ala cannot be, suggest (but by no means prove) that, in binding to its receptor, enkephalin can undergo a conformational change, using the binding energy to compensate for the increase in conformational energy. If the nature of the receptor were known, it would be possible to compute the conformation of the enkephalin-receptor complex, as has been done for enzyme-substrate complexes of lysozyme and oligosaccharides.[54,55]

Cyclic Peptides

As an example of a cyclic peptide, we cite the calculations on gramicidin S,[43] a cyclic decapeptide. This peptide has C_2 symmetry, and appropriate procedures[56] were used to generate closed rings with C_2 symmetry. To overcome the multiple-minimum problem, a variety of strategies were developed to insure an adequate coverage of conformational space; i.e., an appropriate number (viz., 10,541) of starting conformations was selected (for subsequent energy minimization) to make sure that no region of conformational space (in which the global minimum might be) would be overlooked. The computed low-energy structure agrees with the available nmr data, which are cited in Ref. 43. Recent nuclear Overhauser measurements[57] provide additional confirmation of this structure.

Synthetic Polypeptide Models of Fibrous Proteins

Collagen is an example of a fibrous protein, whose triple-stranded helical structure is the same as that of some synthetic regular-repeating poly-tripeptides of the general type poly(Gly-X-Y), where X and/or Y can be Pro or Hyp. Calculations on poly(Gly-Pro-Pro)[47] led to a coiled-coil triple-stranded structure – and showed how and why inter-chain interactions alter the conformation of the single-stranded structure and favor the coiled-coil triple-stranded structure over the single-stranded one. The calculated structure was subsequently shown to agree (within an RMS deviation of 0.3 Å for all non-hydrogen atoms) with that of Okuyama et al.,[58] based on a single-crystal X-ray structure of $(\text{Gly-Pro-Pro})_{10}$. These calculations have been extended[48] to include poly(Gly-Pro-Hyp), poly(Gly-Pro-Ala) and poly(Gly-Ala-Pro); preliminary results[48] indicate that the first two form coiled-coil triple-stranded structures, like poly(Gly-Pro-Pro), whereas the third forms a parallel-chain structure, all of these results being in agreement with experiment.[59-61]

In the case of these collagen-like repeating poly-tripeptides, the (energetic) multiple-minimum problem could be overcome because the structure is regular; i.e., the conformation of each chain could be generated by symmetry operations from, e.g., the structure of the Gly-Pro-Pro unit. This unit, in turn, is sufficiently small so that its conformational space can be covered adequately (as in the case of small oligopeptides) to select numerous starting conformations (for subsequent energy minimization), and to ensure that the region containing the global minimum would not be missed.

Globular Proteins

Though not considered in this paper, for the sake of completeness it should be mentioned that, as indicated above, the number of combinations of low-energy single-residue minima is much too large (for a protein of 100 residues) to be able to adopt the strategies used for systems with a small number of independent variables. Therefore, a different approach is required. For this purpose, approximate procedures are used to reduce the required computer time and to lead to an approximate structure of the protein, i.e., one whose conformational energy would lie in the potential energy well of the native protein. Then, "exact" procedures, such as those involving ECEPP (or UNICEPP) are used to locate the minimum of that particular potential well. The progress that has been made in attempting to locate such a minimum-energy conformation of a globular protein has been reviewed elsewhere.[1]

Concluding Remarks

In contrast to globular proteins, the multiple-minimum problem is less severe for small oligopeptides and/or synthetic analogs of fibrous proteins. However, even for these simpler systems, it is necessary to observe some essential precautions in order to obtain valid results. First of all, the selection strategies for the energy-minimization procedure should cover the conformational space properly. Secondly, when several minima lie close to the global one, the librational entropy should be included in the computation of the statistical weight to identify the most stable structure properly. Thirdly, proper statistical mechanical averaging should be carried out over all low-energy structures before comparing the computed conformational properties of the system with experimental ones. Finally, though it sometimes has been omitted, the effect of the solvent on the calculated structures should be included in the computations.

This work was supported by research grants from the National Science Foundation (PCM75-08691), and from the National Institute of General Medical Sciences of the National Institutes of Health (GM-14312).

References

1. Némethy, G. & Scheraga, H. A. (1977) *Quart. Revs. Biophys.*, in press.
2. Ramachandran, G. N. & Sasisekharan, V. (1968) *Adv. Protein Chem.* **23**, 283–437.
3. Scheraga, H. A. (1968) *Adv. Phys. Org. Chem.* **6**, 103–184.
4. Scheraga, H. A. (1974) in *Peptides, Polypeptides, and Proteins*, Blout, E. R., Bovey, F. A., Goodman, M. & Lotan, N., Eds., Wiley Interscience, New York, pp. 49–70.
5. Anfinsen, C. B. & Scheraga, H. A. (1975) *Adv. Protein Chem.* **29**, 205–300.
6. Momany, F. A., McGuire, R. F., Burgess, A. W. & Scheraga, H. A. (1975) *J. Phys. Chem.* **79**, 2361–2381.
7. Scheraga, H. A. (1977) in *Frontiers in Physico-Chemical Biology*, Pullman, B., Ed., Academic, New York, in press.
8. Gibson, K. D. & Scheraga, H. A. (1967) *Proc. Nat. Acad. Sci. USA* **58**, 420–427.
9. Hopfinger, A. J. (1971) *Macromolecules* **4**, 731–737.
10. Hopfinger, A. J. (1973) *Macromolecules* **6**, 423–437.
11. Hesselink, F. T., Ooi, T. & Scheraga, H. A. (1973) *Macromolecules* **6**, 541–552.
12. Davidon, W. C. (1959) U.S. AEC Research and Development Report, ANL-5990.
13. Fletcher, R. & Powell, M. J. D. (1963) *Computer J.* **6**, 163–168.
14. Fletcher, R. & Reeves, C. M. (1964) *Computer J.* **7**, 149–154.
15. Powell, M. J. D. (1964) *Computer J.* **7**, 155–162.
16. Pennington, R. H. (1965) *Introductory Computer Methods and Numerical Analysis*, Macmillan, New York, p. 236.

17. Zangwill, W. I. (1967) *Computer J.* **10**, 293–296.
18. Pearson, J. D. (1969) *Computer J.* **12**, 171–178.
19. Fletcher, R. (1970) *Computer J.* **13**, 317–322.
20. Gō, N. & Scheraga, H. A. (1969) *J. Chem. Phys.* **51**, 4751–4767.
21. Gō, N., Gō, M. & Scheraga, H. A. (1974) *Macromolecules* 7, 137–139.
22. Gō, N. & Scheraga, H. A. (1976) *Macromolecules* 9, 535–542.
23. Zimmerman, S. S. & Scheraga, H. A. (1976) *Macromolecules* 9, 408–416.
24. Stimson, E. R., Zimmerman, S. S. & Scheraga, H. A., *Macromolecules,* in press.
25. Ramachandran, G. N. & Mitra, A. K. (1976) *J. Mol. Biol.* **107**, 85–92.
26. Wüthrich, K., Grathwohl, C. & Schwyzer, R. (1974) in *Peptides, Polypeptides and Proteins*, Blout, E. R., Bovey, F. A., Goodman, M. & Lotan, N., Eds., Wiley Interscience, New York, pp. 300–307.
27. Kitano, M., Fukuyama, T. & Kuchitsu, K. (1973) *Bull. Chem. Soc., Japan* **46**, 384–387.
28. Huber, R. & Steigemann, W. (1974) *FEBS Lett.* **48**, 235–237.
29. Dunfield, L. G., Burgess, A. W. & Scheraga, H. A., *J. Phys. Chem.*, to be submitted.
30. Pincus, M. R. & Scheraga, H. A. (1977) *J. Phys. Chem.*, **81**, 1579–1583.
31. Tanaka, S. & Scheraga, H. A. (1975) *Proc. Nat. Acad. Sci. USA* **72**, 3802–3806.
32. Gibson, K. D. & Scheraga, H. A. (1969) *Proc. Nat. Acad. Sci. USA* **63**, 9–15.
33. Gibson, K. D. & Scheraga, H. A. (1970) *Comput. Biomed. Res.* **3**, 375–384.
34. Crippen, G. M. & Scheraga, H. A. (1969) *Proc. Nat. Acad. Sci. USA* **64**, 42–49.
35. Crippen, G. M. & Scheraga, H. A. (1971) *Arch. Biochem. Biophys.* **144**, 453–461, 462–466.
36. Crippen, G. M. & Scheraga, H. A. (1973) *J. Computational Phys.* **12**, 491–497.
37. Zimmerman, S. S., Pottle, M. S., Némethy, G. & Scheraga, H. A. (1977) *Macromolecules* **10**, 1–9.
38. Zimmerman, S. S. & Scheraga, H. A. (1977) *Biopolymers* **16**, 811–843.
39. Zimmerman, S. S. & Scheraga, H. A., *Biopolymers*, submitted.
40. Zimmerman, S. S. & Scheraga, H. A. (1977) *Proc. Nat. Acad. Sci. USA*, in press.
41. Simon, I., Némethy, G. & Scheraga, H. A., *Macromolecules*, to be submitted.
42. Isogai, Y., Némethy, G. & Scheraga, H. A. (1977) *Proc. Nat. Acad. Sci. USA* **74**, 414–418.
43. Dygert, M., Gō, N. & Scheraga, H. A. (1975) *Macromolecules* 8, 750–761.
44. Burgess, A. W., Momany, F. A. & Scheraga, H. A. (1975) *Biopolymers* **14**, 2645–2647.
45. Gō, N. & Scheraga, H. A. (1973) *Macromolecules* **6**, 525–535.
46. Gō, N. & Scheraga, H. A. (1974) *Macromolecules* 7, 148.
47. Miller, M. H. & Scheraga, H. A. (1976) *J. Polymer Sci. (Polymer Symposia)* **54**, 171–200.
48. Miller, M. H. & Scheraga, H. A., *Macromolecules*, to be submitted.
49. Garbay-Jaureguiberry, C., Roques, B. P., Oberlin, R., Anteunis, M. & Lala, A. K. (1976) *Biochem. Biophys. Res. Commun.* **71**, 558–565.
50. Jones, C. R., Gibbons, W. A. & Garsky, V. (1976) *Nature* **262**, 779–782.
51. Bleich, H. E., Cutnell, J. D., Day, A. R., Freer, R. J., Glasel, J. A. & McKelvy, J. F. (1976) *Proc. Nat. Acad. Sci. USA* **73**, 2589–2593.
52. Pert, C. B., Pert, A., Chang, J. K. & Fong, B. T. W. (1976) *Science* **194**, 330–332.
53. Momany, F. A. (1977) *Biochem. Biophys. Res. Commun.* **75**, 1098–1103.
54. Pincus, M. R., Zimmerman, S. S. & Scheraga, H. A. (1976) *Proc. Nat. Acad. Sci. USA* **73**, 4261–4265.
55. Pincus, M. R., Zimmerman, S. S. & Scheraga, H. A. (1977) *Proc. Nat. Acad. Sci. USA*, **74**, 2629–2633.

56. Gō, N. & Scheraga, H. A. (1973) *Macromolecules* **6**, 273–281.

57. Rae, I. D., Stimson, E. R. & Scheraga, H. A. (1977) *Biochem. Biophys. Res. Commun.*, **77**, 225-229.

58. Okuyama, K., Tanaka, N., Ashida, T. & Kakudo, M. (1976) *Bull. Chem. Soc., Japan* **49**, 1805–1810.

59. Andreeva, N. S., Esipova, N. G., Millionova, M. I., Rogulenkova, V. N., Tumanyan, V. G. & Shibnev, V. A. (1970) *Biofizika* **15**, 198–205.

60. Traub, W. & Yonath, A. (1967) *J. Mol. Biol.* **25**, 351–355.

61. Segal, D. M. & Traub, W. (1969) *J. Mol. Biol.* **43**, 487–496.

STRUCTURE AND CONFORMATION OF PEPTIDES: A CRITICAL ANALYSIS OF CRYSTALLOGRAPHIC DATA

ETTORE BENEDETTI, *Istituto Chimico, Universita di Napoli, Via Mezzocannone 4, 80134 Napoli, Italy*

The derivation of information from a comparison of peptide crystal structures, even if complicated by intricate hydrogen bonding networks, may provide useful tests for model energy functions and parameters as well as give indications of trends in conformations for backbone and side-chain structures. In this paper we present results on the geometry of the peptide bond and the conformation of the peptide backbone and side-chains as derived from the analysis of the published crystal structures.

Geometry of the Peptide Unit

The dimensions of the peptide unit[1] and its revised values[2,3] have been widely used. From the structures reported in the literature we have isolated a sample of recent and more accurate studies of amino-acid derivatives having both ends blocked with hydrophobic groups and di-, tri-, or higher peptides presenting at least once the residue shown below in parentheses:

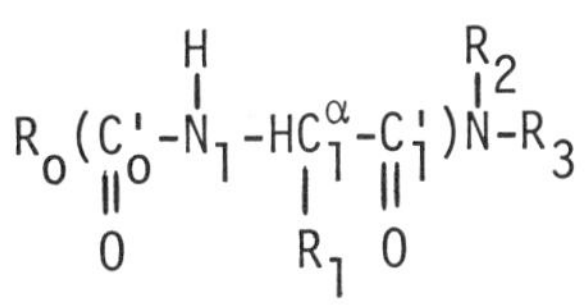

The sample does not include any of the peptides used for the average by the previous authors;[1-3] none of the chosen peptides present both terminals in the ionic form (NH_3^+ or COO^-), so that the derived geometrical parameters will leave aside any possible influence of such terminals. For the description of the geometry of the peptide unit and conformation of the peptide chain the conventions proposed by the IUPAC-IUB Commission on Biochemical Nomenclature[4] is followed. Table Ia, b lists the geometry of the 34 crystal structures considered in the analysis. Figure 1 reports the calculated weighted averages of bond lengths and angles. Bond lengths present smaller dispersion about the mean values than that observed for bond angles. The loss in coherence for the angular values of the same geometrical portion of the residue must be ascribed to the influence of different molecular environments in the crystal structures (small bond angle deformations are easily achieved without involving appreciable amounts of energy).

The geometry for the *trans* unit differs slightly with respect to the corresponding values of Ramachandran et al.,[3] (lower values) indicating that, within

Table Ia. Bond Distances in Peptides from Crystal Data (E.s.d.'s in unit of the last significant figure)

Peptide	R_0–C_0'	C'=O	C'–N	N–H	N–C^{α}	C_1^{α}–C_1'	N_2–R_2, R_3	Ref.
N-Ac-Gly-NH-iPr	1.508(11)	1.263(10)	1.305(11)	0.89(11)	1.449(9)	1.570(13)	1.478(12)	5
		1.182(9)	1.354(9)	1.22(13)				
N-Piv-Gly-NH-iPr	1.540(5)	1.230(4)	1.335(5)	0.91(5)	1.466(4)	1.511(5)	1.473(5)	5
		1.236(4)	1.324(4)	0.93(5)				
N-(Diethylacetyl)-Gly-	1.53(2)	1.24(1)	1.34(1)	—	1.45(2)	1.53(1)	1.53(2)$^{+}$	5
NH-iPr		1.21(1)	1.33(2)					
N-Ac-DL-Val-NMe$_2$	1.493(10)	1.232(8)	1.351(8)	1.10(7)	1.444(8)	1.536(7)	1.474(8)	5
	1.425(10)	1.234(8)	1.340(8)	1.01(8)	1.453(8)	1.533(7)	1.468(8)	
		1.219(7)	1.353(7)	1.468(8)*				
		1.227(7)	1.355(7)	1.457(7)*				
N-Ac-DL-Leu-NHEt	1.543(11)	1.228(9)	1.346(10)	1.02(8)	1.440(9)	1.509(8)	1.515(12)$^{+}$	5
		1.247(9)	1.345(9)	1.20(11)				
N-Ac-DL-Leu-NHMe	1.558(12)	1.246(8)	1.298(10)	—	1.435(9)	1.540(8)	1.440(9)	6
		1.243(8)	1.322(8)					
N-Ac-DL-pseudo-Leu-N-Me$_2$	1.502(5)	1.220(4)	1.313(5)	0.95(5)	1.453(4)	1.502(5)	1.447(6)	5
		1.238(4)	1.343(5)	1.453(5)*				
N-Ac-DL-Phe-NHMe	1.488(15)	1.225(9)	1.348(10)	—	1.436(11)	1.509(12)	1.456(15)	7
		1.263(10)	1.310(11)					
N-Ac-L-Phe-NHMe	1.488(9)	1.245(9)	1.311(8)	—	1.447(8)	1.521(9)	1.472(9)	8
		1.245(9)	1.319(9)					
N-Ac-L-Pro-NH$_2$	1.505(3)	1.239(2)°	1.351(4)°	1.468(3)°	1.479(4)°	1.530(3)	0.75(6)	9
		1.234(2)	1.326(3)				0.78(4)	
N-Ac-L-Pro-NHMe	1.490(4)	1.245(4)°	1.337(4)°	1.476(4)°	1.472(4)°	1.530(4)	1.453(4)	10
		1.231(4)	1.316(4)	0.87(5)				
N-Ac-L-Tio-Pro-NH$_2$	1.499(5)	1.248(3)°	1.357(4)°	1.422(6)°	1.455(6)°	1.536(5)	0.76(5)	9
		1.213(5)	1.308(5)				0.96(4)	
N-Ac-L-Met-N-Me$_2$	1.496(8)	1.239(6)	1.332(6)	0.96(7)	1.449(7)	1.519(8)	1.492(8)	5
		1.229(8)	1.350(8)	1.464(9)*				
N-Piv-(N'-Me)-L-Gln-NHMe	1.530(8)	1.236(7)	1.336(7)	1.01(7)	1.466(6)	1.533(7)	1.450(9)	5
		1.232(6)	1.328(7)	0.81(8)		1.512(7)	1.469(7)	
		1.227(6)	1.332(7)	0.97(6)				
N-Piv-L-Ser-NHMe	1.527(5)	1.228(4)	1.322(5)	0.83(4)	1.448(5)	1.512(5)	1.427(7)	5,11
		1.231(5)	1.333(5)	0.85(4)				
N-Ac-L-Tyr-NHMe	1.469(3)	1.231(3)	1.345(3)	—	1.452(3)	1.533(3)	1.455(3)	12
		1.233(3)	1.311(3)					
N-Ac-DL-Trp-NHMe	1.504(5)	1.242(4)	1.318(4)	—	1.448(3)	1.522(4)	1.461(4)	13
		1.230(3)	1.325(4)					
N-Ac-L-His-NHMe	1.501(7)	1.238(5)	1.323(6)	—	1.447(5)	1.519(6)	1.455(8)	14
		1.250(5)	1.325(5)					
N-Piv-L-Pro-Gly-NMe$_2$	1.534(10)	1.226(8)°	1.345(8)°	1.466(11)°	1.476(9)°	1.524(9)	1.471(12)	5
		1.225(9)	1.346(9)	1.16(9)	1.426(9)	1.535(10)		
		1.216(9)	1.355(9)	1.463(11)*				
t-Boc-Gly-L-Pro-OH	—	1.203(7)	1.344(7)	0.89	1.445(7)	1.510(7)	—	15,16
		1.226(7)°	1.332(7)°	1.472(7)°	1.449(7)°	1.512(7)		
t-Boc-Gly-L-Pro-OBzl	—	1.214(7)	1.322(7)	0.88	1.429(7)	1.512(7)	—	16
		1.234(7)°	1.326(7)°	1.462(7)°	1.452(7)°	1.512(7)		
t-Boc-L-Pro-Gly-OH	—	1.223(7)°	1.346(8)°	1.471(7)°	1.469(5)°	1.520(5)	—	17
		1.228(6)	1.340(5)	1.0(1)	1.447(7)	1.520(6)		
Cbz-Gly-L-ProOH	—	1.209(10)	1.320(10)	—	1.444(10)	1.520(10)	—	18
		1.217(10)	1.333(10)°	1.475(10)°	1.475(10)°	1.524(10)		
N-iBu-L-Pro-L-Ala-NHiPr	1.60(2)	1.23(1)°	1.33(1)°	1.46(1)°	1.48(1)°	1.50(1)	1.49(1)	5
		1.27(1)	1.32(1)		1.44(1)	1.51(1)		
		1.25(1)	1.33(1)					
N-iBu-L-Pro-D-Ala-NH-iPr	1.51(2)$^{+}$	1.27(1)°	1.34(1)°	1.47(1)°	1.45(1)°	1.49(1)	1.52(1)$^{+}$	5
		1.23(1)	1.36(1)		1.46(1)	1.46(1)		
		1.26(1)	1.33(1)					
N-Ac-L-Pro-D-Lac-NHMe	1.515(14)	1.250(9)°	1.362(12)°	1.490(11)°	1.449(11)°	1.544(12)	1.459(16)	5,19
		1.241(9)	1.320(15)	1.04(10)				
N-Ac-L-Pro-L-Lac-NHMe	1.508(13)	1.261(10)°	1.331(11)°	1.485(11)°	1.474(11)°	1.519(11)	1.480(12)	5,19
		1.240(9)	1.338(10)	1.09(11)				
N-Ac-L-Phe-L-Tyr-OH	1.507(3)	1.231(3)	1.335(3)	0.87(2)	1.442(3)	1.528(2)	—	20
		1.233(2)	1.336(3)	0.89(2)	1.454(2)	1.531(2)		
t-Boc-Sar-Gly-OBzl	—	1.217(4)	1.343(4)	1.456(4)	1.452(4)	1.520(4)	—	21
		1.214(4)	1.340(4)		1.440(4)	1.507(4)		
t-Boc-Gly-L-Ala-OH	—	1.219(5)	1.347(5)	0.969(63)	1.454(4)	1.538(5)	—	22
		1.241(5)	1.326(5)	0.909(62)	1.454(4)	1.521(5)		
Cbz-Gly-Pro-Leu-OH	—	1.210(5)	1.334(5)	0.88(5)	1.439(5)	1.516(5)	—	23
		1.235(5)°	1.340(5)°	1.473(5)°	1.473(5)°	1.528(5)		
		1.227(5)	1.333(5)	0.86(5)	1.454(5)	1.522(5)		
t-Aoc-L-Pro-L-Pro-L-Pro-	—	1.219(8)°	1.336(8)°	1.465(8)°	1.445(8)°	1.536(8)	—	24
OH		1.221(8)°	1.329(8)°	1.471(8)°	1.455(8)°	1.511(8)		
		1.245(8)°	1.330(8)°	1.482(8)°	1.456(8)°	1.531(8)		
t-Boc-(L-Pro)$_3$-L-Pro-	—	1.217(12)°	1.336(12)°	1.463(12)°	1.466(12)°	1.535(12)	—	25
OBzl		1.211(12)°	1.358(12)°	1.471(12)°	1.432(12)°	1.517(12)		
		1.230(12)°	1.310(12)°	1.502(12)°	1.443(12)°	1.561(12)		
		1.230(12)°	1.315(12)°	1.506(12)°	1.460(12)°	1.516(12)		
t-Boc-(S-Bzl)-L-Cys-Gly-	—	1.201(9)	1.339(9)	—	1.452(9)	1.531(10)	—	26
OMe		1.223(8)	1.348(9)		1.461(10)	1.488(12)		

(°) The underlined values are significantly larger or smaller than the "best value". Distances preceding a proline residue.

(*) Values not included in the mean because of a tertiary amide.

(+) Disorder is present: the corresponding values are not included in the calculation.

Abbreviations: iPr = isopropyl, Piv = pivalyl, iBu = isobutyl, Tio = thiozolidine.

Table Ib. Bond Angles in Peptides from Crystal Data

Peptide	O–C'–N	C'–N–C^{α}	C'–N–H	H–N–C^{α}	N–C^{α}–C'	C^{α}–C'–O	C^{α}–C'–N	Ref.
N-Ac-Gly-NH-iPr	122.0(7) 126.4(7)	121.1(7)	147 (6) 99 (4)	91 (6) 134 (4)	110.1(6)	121.8(7)	111.2(7)	5
N-Piv-Gly-NH-iPr	120.7(3) 123.9(3)	122.1(3)	123 (3) 119 (3)	115 (3) 116 (3)	111.9(3)	121.1(3)	114.9(3)	5
N(Diethylacetyl)-Gly-NH-iPr	122 (1) 125.0(9)	116.9(8)			109.9(8)	123 (1)	111.5(8)	5
N-Ac-DL-Val-NMe$_2$	121.1(6) 122.3(6) 121.5(5)* 121.3(5)*	122.0(5) 122.1(5)	124 (4) 118 (4) 125.3(5)* 125.4(5)*	113 (4) 104 (4) 118.2(5)* 116.9(5)*	109.3(4) 107.4(4)	119.9(5) 120.7(5)	118.6(5) 118.0(5)	5
N-Ac-DL-Leu-NHEt	121.9(7) 121.3(7)	120.6(7)	129 (5) 117 (5)	110 (5) 120 (5)	111.5(5)	123.6(6)	115.1(6)	5
N-Ac-DL-Leu-NHMe	122.7(7)	120.8(6)	—	—	112.0(6)	120.5(6)	116.7(6)	6
N-Ac-DL-pseudo-Leu-NMe$_2$	122.6(4) 120.8(3)	122.7(3)	116 (3) 126.4(4)*	121 (3) 113.9(4)*	107.8(3)	119.5(3)	119.7(3)	5
N-Ac-DL-Phe-NHMe	122.2(7) 122.2(8)	123.1(7)			109.6(7)	118.1(8)	119.7(8)	7
N-Ac-L-Phe-NHMe	122.3(6)	125.0(6)	—	—	109.2(5)	119.8(6)	117.8(6)	8
N-Ac-L-Pro-NH$_2$	121.5(7)° 123.1(7)	120.7(7)°	125.8(7)°	113.1(7)°	114.2(7)	117.9(7)	119.0(7)	9
N-Ac-L-Pro-NHMe	120.2(3)° 124.4(3)	121.4(3)°	125.6(3)°	112.2(3)°	114.3(3)	117.6(3)	117.9(3)	10
N-Ac-L-Tio-Pro-NH$_2$	119.6(7)° 124.0(7)	121.0(7)°	121.3(7)°	115.9(7)°	115.3(7)	118.6(7)	117.4(7)	9
N-Ac-L-Met-NMe$_2$	122.7(5) 122.3(5)	123.6(4)	110 (4) 125.2(5)*	125 (4) 116.5(5)*	109.3(4)	120.0(5)*	117.8(5)*	5
N-Piv(N'-Me)-Gln-NHMe	121.2(5) 122.3(5) 123.9(5)	120.3(4)	128 (4) 99 (5) 115 (4)	111 (4) 136 (4) 121 (4)	110.2(4)	122.7(4) 122.7(5)	115.1(5) 113.4(4)	5
N-Piv-L-Ser-NHMe	120.5(3) 121.7(3)	121.8(3)	127 (3) 109 (3)	110 (3) 127 (3)	111.8(3)	123.6(3)	114.6(3)	5,11
N-Ac-L-Tyr-NHMe	122.4(3) 123.6(3)	124.5(3)	—	—	113.7(3)	118.1(3)	118.1(3)	12
N-Ac-L-Trp-NHMe	123.7(3) 122.8(3)	124.9(2)	—	—	109.9(2)	121.5(2)	115.7(2)	13
N-Ac-L-His-NHMe	122.2(4) 122.3(4)	121.4(3)	—	—	111.7(3)	121.6(4)	116.1(4)	14
N-Piv-L-Pro-Gly-NMe$_2$	119.4(6)° 124.5(6) 124.5(7)*	117.2(5)° 120.1(6)	130.0(6)° 118 (4) 125.6(7)*	112.2(5)° 119 (4) 117.3(7)*	111.3(5) 110.7(6)	122.9(6) 120.7(6)*	112.6(5) 114.8(6)*	5
t-Boc-Gly-L-Pro-OH	125.0(5) 122.1(5)°	124.5(5) 120.8(5)°	126.6(5)°	112.5(5)°	108.0(5) 110.6(5)	120.6(5)	117.4(5)	15,1
t-Boc-Gly-L-Pro-OBzl	124.5(5) 121.8(5)°	120.4(5)	126.8(5)°	112.6(5)°	111.9(5) 110.8(5)°	121.8(5) 121.3(2)	116.4(5)	16
t-Boc-L-Pro-Gly-OH	125.1(2)° 123.4(2)	123.6(2)° 120.6(2)	120.8(2)°	114.5(2)°	112.5(2) 110.7(2)		115.1(2)	17
Cbz-Gly-L-ProOH	125.5(7) 121.9(7)°	120.7(7) 120.5(7)°	— 126.5(7)°	— 112.8(7)°	111.2(7) 109.5(7)	122.9(7)	115.2(7)	18
N-iBu-L-Pro-L-Ala-NHiPr	122.3(10)° 124.3(9) 122.3(9)	118.7(8)° 124.9(8)	129.5(9)°	111.7(8)°	112.6(8) 114.4(9)	119.1(8) 120.7(9)	116.6(8) 116.7(9)	5
N-iBu-L-Pro-D-Ala-NHiPr	119.7(9)° 120.5(8) 120.7(9)	120.3(8)° 123.6(7)	128.0(9)°	111.7(8)°	110.8(7) 112.9(8)	122.9(8) 119.1(9)	116.6(7) 120.1(8)	5
N-Ac-L-Pro-D-Lac-NHMe	120.7(8)° 124.4(9)	119.1(8)°	126.2(8)°	114.4(7)°	110.6(7) 110.2(7)	116.3(0)	119.4(8)	5,19
N-Ac-L-Pro-L-Lac-NHMe	120.9(7)° 123.3(7)	120.7(6)°	125.9(6)°	112.2(6)°	115.2(7) 111.2(6)	118.4(7)	118.2(7)	5,19
N-Ac-L-Phe-L-Tyr-OH	122.1(2) 121.8(2)	123.3(2) 121.0(2)	—	—	111.8(2) 111.1(2)	122.6(2)	115.3(2)	20
t-Boc-Sar-Gly-OBzl	124.4(3)° 123.9(3)	120.7(3) 122.9(3)	—	—	113.1(3) 112.1(3)	123.4(3) 125.9(3)	112.6(3)	21
t-Boc-Gly-L-Ala OH	123.3(1) 122.2(1)	121.5(1) 124.2(1)	124.6(4) 125.3(5)	113.6(4) 110.3(5)	112.2(1) 106.4(1)	121.9(1) 122.5(1)	115.7(1)	22
Cbz-Gly-Pro-Leu-OH	126.0(8) 120.6(8)° 122.8(8)	120.7(8) 125.1(8)° 121.8(8)	122.0(8)*	112.9(8)*	112.9(8) 110.5(8) 112.9(8)	123.2(8) 122.0(8)	116.2(8) 115.3(8)	23
t-Aoc-L-Pro-L-Pro-L-Pro-OH	125.0(4)° 120.4(4)° 121.6(4)°	123.5(4)° 118.4(4)° 121.0(4)°	120.8(4)° 128.3(4)° 127.2(4)°	113.8(4)° 113.1(4)° 111.8(4)°	111.3(4) 110.2(4) 109.9(4)	121.4(4) 120.4(4)	118.2(4) 118.0(4)	24
t-Boc-(L-Pro)$_3$-L-Pro-OBzl	124.3(7)° 121.4(7)° 120.3(7)° 122.3(7)°	123.6(7)° 118.4(7)° 122.1(7)° 118.7(7)°	120.8(7)° 126.4(7)° 125.8(7)° 128.0(7)°	115.3(7)° 114.8(7)° 111.9(7)° 113.2(7)°	110.8(7) 110.5(7) 109.1(7) 112.9(7)	122.0(7) 120.9(7) 121.3(7) 121.3(7)	116.4(7) 118.9(7) 116.4(7)	25
t-Boc-(S-Bzl)L-Cys-Gly-OMe	126.6(7) 124.0(6)	121.1(6) 119.9(6)	—	—	108.7(6) 110.1(7)	120.3(6)	115.7(6)	26

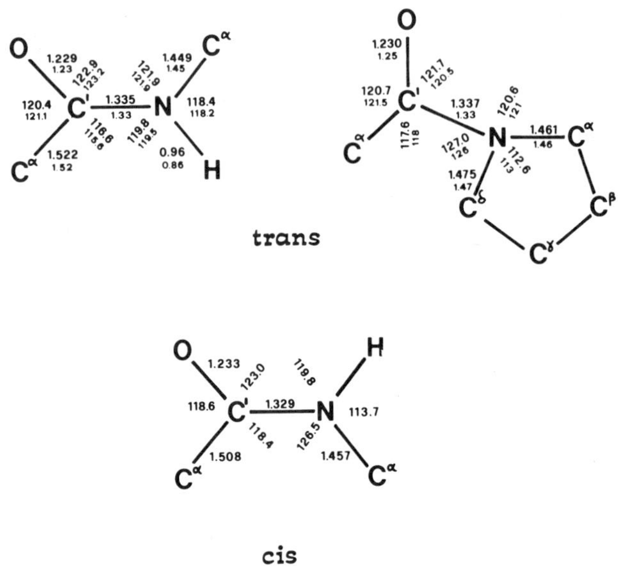

Fig. 1. Revised geometry of the *trans* peptide unit, the *trans* proline unit and the *cis* unit. E.s.d.'s of lengths and angles are 0.004–0.008 Å (0.014 Å for N-H) and 0.5°–1.5° respectively. Values involving H atoms are from neutron data (Ref. 5). Lower values from Refs. 3 and 27.

the accuracy of the measurements, the values are not greatly influenced by the presence of the ionic terminals on one or both sides of the peptide linkage. The larger differences between the two sets of values, found for bond angles, must be ascribed to packing forces such as the system of hydrogen bonds observed in the crystals. We also find no appreciable difference in the bond angle C′-Cα-N between glycyl residues and the other amino-acid residues, at variance with previous indications.[3] The mean value of this angle obtained from all the data with the exception of glycyl residues is 111.1°, while the glycyl residues present an average value of 110.4°.

The values for the geometry of a *trans* bond preceding a proline residue present appreciable differences; the mean values (Figure 1) obtained from the proline peptides reported in Table Ia, b are compared with those recently proposed by Ashida and Kakudo[27] (lower values). The results show differences with respect to the *trans* unit primarily for the nitrogen atom, because of the restrictions imposed by the presence of the pyrrolidine cyclic system. It should be mentioned that the values reported in Figure 1 pertain to the planar peptide unit even if their derivation has been carried out from crystal structures where non-planar peptide units occur. The procedure given by Kolaskar et al.[28] is suggested for the cases when non-planar distortion at the N atom is needed.

The dimensions of the *cis* peptide unit derived from the geometry observed in cyclic dipeptides (diketopiperazines) is given in Figure 1. Only 8 structures of linear peptides present *cis* peptide linkages, i.e.: Cbz-Gly-Pro-Leu-OH[23] (racemic mixture), *t*-Boc-L-Pro-Gly-OH,[17] *t*-Aoc-(L-Pro)$_3$OH,[24] *t*-Boc-(L-Pro)$_4$-OBzl,[25] *t*-Boc-Sar-Gly-OBzl,[21] *t*-Boc-L-Pro-OH,[30] *t*-Boc-Tio-Pro-OH[30] and *t*-Boc-D-Val-OH.[31] In all published structures presenting the residue X-L-Pro-Y, where X=*t*-Boc, *t*-Aoc, the urethane bond is always in the *cis* configuration; also in peptides such as X-Sar, where X=*t*-Boc or an amino-acid residue, the same situation is verified as in *t*-Boc-Sar-Gly-OBzl,[21] actinomycin D,[32] *cyclo*(Sar)$_4$,[33] *cyclo*(Gly-Sar$_3$),[34] *cyclo*(Gly-Sar)$_2$,[34] etc. The only case of *cis* peptide bond for which no tertiary nitrogen atom is involved (that is Pro or Sar residues), is *t*-Boc-D-Val-OH. In this compound, the *cis* configuration at the urethane-CONH-linkage is fixed by crystal packing forces which are governed by H-bonds and the unique type of van der Waals interactions.

Cis peptide bonds involving proline residues show significant differences in the angles involving the nitrogen atom: the trend observed is that C$'$-N-C$^\alpha$ shows larger value, while C$'$-N-C$^\delta$ shows smaller values than the corresponding values for the *trans* form of the same peptide linkage. The steric repulsion between the C$^\alpha$ atom (of the preceding residue) and the C$^\alpha$ atom of the prolyl residue in the *cis* form is bigger than that between the oxygen atom of the carbonyl group and the C$^\alpha$ atom of the prolyl residue in the *trans* form.

Molecular Conformation

The solid state molecular conformation of cyclic dipeptides, listed in Table II will be considered first, since their conformation presents the highest constrictions for torsional angles, followed by cyclic peptides with larger number of residues in the ring and the linear peptides, listed in Table III. Finally, a statistical account of the side chains conformational characteristics will be given.

Cyclic Peptides. Solution as well as solid state studies of cyclic dipeptides have shown that the 2,5-piperazinedione ring (DKP) must be considered rather flexible. The conformations observed in crystal structures of cyclic dipeptides can now be rationalized in four groups: planar conformation, boat conformations with C$^\beta$ atoms in the axial or in the equatorial positions and chair conformation. These structures are characterized by the six values of the ϕ, ψ and $\omega(\phi_1,\psi_1,\omega_1,\phi_2,\psi_2,\omega_2)$ torsional angles.[4] In addition, the dihedral angle between the two amide planes might be taken as a direct measure of the folding of the diketopiperazine ring,[52] (β values in the Hooker's notation). Planar structures have all six values of torsional angles close to zero and consequently $\beta \sim 0°$. Boat conformations with axial or equatorial substituents on C$^\alpha$ atoms are distinguished by opposite values for the ϕ and ψ angles: positive ϕ values (and negative ψ values) characterize boat conformation with axial C$^\beta$ substituents, with $\beta<0$, while negative ϕ values (and positive ψ value) characterize boat conformations with equatorial C$^\beta$ substituents with $\beta>0$. The chair conformation, instead, is characterized by an alternating sequence of positive and negative signs for the

Table II. Conformational Parameters for Cyclic Dipeptides

Cyclic Dipeptide	ψ_1	φ_1	ω_1	ψ_2	φ_2	ω_2	(52) β	DKP'ring Conforma tion	C^β Substit. Position	Ref.
$\underline{c}$(Gly)$_2$	1	- 1	1	- 1	1	- 1	0	Planar	-----	35
$\underline{c}$(-L-Ala-D-Ala)	3	- 3	3	- 3	3	- 3	0	Planar	-----	36
$\underline{c}$(L-Ala)$_2$	21	-32	8	27	-26	1	+29	Boat	Equatorial	36
$\underline{c}$(L-MeAla-L-Ala)	-19	26	- 9	-23	30	- 5	-22	Boat	Axial	37
$\underline{c}$(L-MeAla)$_2$	-18	24	0	-27	33	-10	-25	Boat	Axial	38
$\underline{c}$(DL-MeAla$_2$)	-17	20	1	-26	29	- 8	-12	Boat	Axial	38
$\underline{c}$(L-Cys)$_2$- Acetic Acid	-20	37	-17	-17	34	-14	25	Boat	Axial	39
$\underline{c}$(L-Pro-Gly)	38	-44	0	33	-37	7	+38	Boat	Equatorial	40
$\underline{c}$(L-Pro-L-Leu)	34	-41	6	34	-41	6	+37	Boat	Equatorial	41
$\underline{c}$(L-Trp-Gly)	- 1	- 4	2	1	4	- 3	0	Planar	-----	42
$\underline{c}$(L-Tyr-Gly)	-13	19	- 4	-10	16	- 7	-15	Boat	Axial	43
$\underline{c}$(L-Thr-L-His)·2H_2O	-11	7	5	-12	7	5	-10	Boat	Axial	44
$\underline{c}$(L-Pro-D-Phe)	15	-17	3	17	-19	1	+16	Boat	Equat.Axial	45
$\underline{c}$(L-Pro)$_2$	37	-38	1	36	-37	- 1	+38	Boat	Equatorial	46
$\underline{c}$(L-HyPro)$_2$	37	-39	1	37	-39	1	+41	Boat	Equatorial	47
$\underline{c}$(L-MePhe)$_2$	-11	21	- 7	-15	25	-12	-19	Boat	Axial	38
$\underline{c}$(L-MePhe-D-MePhe)	-13	14	-14	13	-14	14	---	Chair	-----	38
$\underline{c}$(L-Tyr-L-Ser)	6	-12	5	5	-10	6	0	Planar	-----	43
$\underline{c}$(Sar)$_2$	6	- 7	7	- 6	7	- 7	---	Chair	-----	48
$\underline{c}$(Sar-L-Val)	-14	21	-12	-24	32	- 0	-23	Boat	Axial	49
$\underline{c}$(L-MeVal)$_2$	-33	36	- 1	-36	40	- 4	-41	Boat	Axial	38
$\underline{c}$(L-MeVal-D-MeVal)	-17	18	-19	17	-18	19	---	Chair	-----	38
$\underline{c}$(L-Val-D-Val)	11	-13	- 1	8	- 9	3	+11	Boat	Equatorial	50
$\underline{c}$(Gly-Sar)·H_2O	3	- 2	- 5	- 2	4	1	0	Planar	-----	51
$\underline{c}$(Gly-N-iPr·Gly)	18	-26	7	29	-26	7	+19	Boat	Equatorial	51

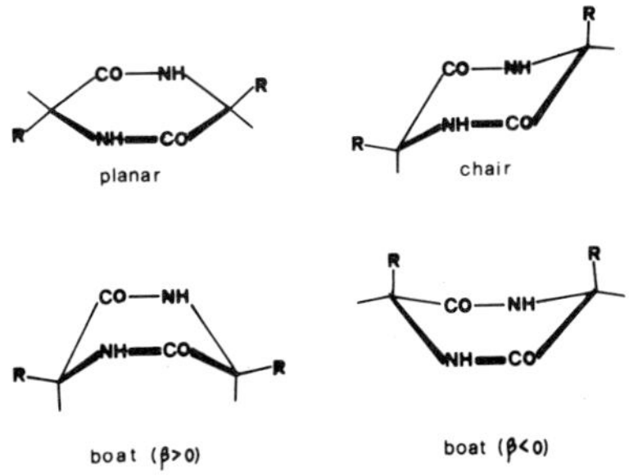

values of the six torsional angles. (This conformation cannot be described utilizing the Hooker's β parameter.) As shown in Table II the ϕ and ψ values may vary in the range -40°, +40°; the highest values observed occur when prolyl or *N*-substituted amino acid residues are present. Distortion from planarity of the amide bond may be as great as 20°: deviations $|\Delta\omega|$ up to 15° are quite probable, since the energy increase for them is only of the order of 0.5 kcal/mol.[28]

Correlations can be made between amino-acid composition and the preferred conformation of the cyclic peptide. Dipeptides with no aromatic side chains are less affected by attractive interactions between substituents and DKP ring: one might expect the DKP ring to minimize ring strain by assuming one (or both)

Table III. Conformation Angles φ and ψ for Linear Peptides*

Peptide	φ_1	ψ_1	Ref
N-Ac-Gly-NHiPr	−78	160	5
N-Piv-Gly-NHiPr	−112	142	5
N-(Diethylacetyl)-Gly-NHiPr	−169	175	5
N-Ac-Gly-NHMe	−71	159	53
N-Ac-L-Ala-NHMe	−84 −88	159 155	54
N-Ac-DL-Ala-NHMe	−76 −80	150 152	54
N-Ac-DL-Leu-NHMe	−94	140	6
N-Ac-DL-Leu-NHEt	−91	144	5
N-Ac-L-Leu-NHiPr	−90	158	5
N-Ac-L-Met-NMe$_2$	−126	162	5
N-Ac-DL-Met-NHMe	−117	109	8
N-Ac-L-Pro-NH$_2$	−80	−14	9
N-Ac-L-TioPro-NH$_2$	−85	−9	9
N-Ac-L-Pro-NHMe	−76	−16	10
N-Ac-DL-Phe-NHMe	−105	108	7
N-Ac-L-Phe-NHMe	−107	104	8
N-Ac-DL-Val-NMe$_2$	−90 −92	122 123	5
N-Ac-L-Val-NHMe	−1[illegible]8	113	8

Peptide	φ_1	ψ_1	φ_2	ψ_2	Ref
N-Ac-L-Trp-NHMe	−104	141			13a
N-Ac-L-Trp-OMe	−66	157*			13b
N-Ac-L-His-NHMe	−72	156			14
N-Ac-L-Tyr-NHMe	−80	−3			12
N-Piv-L-Ser-NHMe	−92	151			5,11
N-Piv-(N'-Me)-L-Gln-NHMe	−72	165			5
Cbz-L-Leu-OpNO$_2$Bzl	−93	49*			55
N-Tosyl-L-Pro-L-HypOH-H$_2$O	−122	157	−50	138*	56
Chloro-acetyl-Gly-Gly-OH	−74	154	159	−171*	57
t-Boc-Sar-Gly-OBzl	72	−154	81	174*	21
t-Boc-Gly-L-Pro-OH	172	178	−70	−26*	15,16
t-Boc-Gly-L-Pro-OBzl	−109	164	−76	−23*	16
t-Boc-L-Pro-Gly-OH	−61	147	−67	162*	17
Cbz-Gly-L-Pro-OH	−108	176	−69	−32*	18
N(9-β-D-Ribofuranosylpurin-6-yl)-Gly-L-Ala-OH·1.5 H$_2$O	−107	−142	−62	161*	58
t-Boc-(S-Bzl)-L-Cys-Gly OMe	−124	103	153	170*	26
L-Ala-L-Ala-OH			−113	103*	60
L-Ala-L-Ala-OH·HCl			−154	161*	61

Peptide	φ_1	ψ_1	φ_2	ψ_2	φ_3	ψ_3	φ_4	ψ_4	Ref
L-Pro-L-Phe-OMe·HBr			−97	−42*					62
L-Pro-L-Leu-Gly-NH$_2$			−61	128	73				63
Gly-L-Phe-Gly-OH·H$_2$O			−126	134	84	179*			59
N-iBu-L-Pro-D-Ala-NHiPr	−62	137	96	3					5
N-Piv-L-Pro-Gly-NMe$_2$	−71	157	−76	175					5
N(halo-Ac)-L-Phe-L-Phe-OEt	−102 −100	106 110	−122 −118	53* 40*					64
N-Ac-L-Phe-L-Tyr-OH	−110	153	−50	43*					20
t-Boc-Gly-L-Ala-OH	−85	55	16	7*					22
L-Leu-L-Pro-Gly-OH			−68	162	−177	180*			65
DL-Leu-Gly-Gly-OH			−167	−172	−165	−177*			66
L-Ala-L-Ala-L-Ala-OH·$^1/_2$ H$_2$O			−146 −156	145 150	−147 −160	172* 144*			67
Cbz-Gly-L-Pro-L-Leu-OH	−79	174	−71	145	−89	38*			23
N-iBu-L-Pro-L-Ala-NHiPr	−59	136	66	14					5
t-Aoc-L-Pro-L-Pro-L-Pro OH	−59	155	−99	139	−73	166*			24
p-BrCbz-Gly-L-Pro-L-Leu-Gly-OH	91	170	−58	−33	−104	8			68
(S-Bzl)-L-Cys-L-Pro-L-Leu-Gly-NH$_2$			−66	−29	−115	13			70
o-BrCbz-Gly-L-Pro-L-Leu Gly-L-Pro-OH	−97	−165	−65	−27	−105	8	−117	178	69
t-Boc-L-Pro-L-Pro-L-Pro L-Pro-OBzl	−68	152	−70	165	−61	152	−58	−38	25

* the values marked with an asterisk refer to a torsional angle at the C-terminal involving the oxygen atom of the hydroxyl group.

boat conformations, and interactions between side chains might favor one boat form over the other. However, the lessened importance of intramolecular interactions with the DKP ring apparently makes the various DKP ring conformations very close in energy and other effects such as close packing, hydrogen bonding and intermolecular stacking interactions became more important in fixing the conformation. Cyclic dipeptides with one aromatic substituent show a definite preference for this substituent to fold back over the DKP ring as a result of a direct interaction between the two rings, particularly if the second amino-acid residue is not very bulky or if it has the opposite configuration. This is verified for *cyclo*(L-Tyr-Gly),[43] *cyclo*(L-Pro-D-Phe),[45] *cyclo*(L-Tyr-L-Ser),[43] *cyclo*(L-MePhe-D-MePhe)[38] and *cyclo*(L-MePhe)$_2$.[38] The only exception is the extended conformation of the tryptophan side chain in the crystal of *cyclo*(L-Trp-Gly). The presence of an L-Pro residue in the DKP, because of the constraints due to the pyrrolidine ring, forces the conformation of the cyclic dipeptide toward the boat form with the equatorial position of the C^β atoms ($\beta > 0$) if the other residue has the same configuration.

For *N*-methylated diketopiperazines[38] the amide bonds are more easily deformed; optically active isomers assume a boat conformation with C^β atoms in the axial position while *meso* isomers assume the chair structure as a preferred conformation.

In cyclic dipeptides because of the steric requirements of the six-membered ring, the two amide groups are forced to be *cis* ($\omega \cong 0°$). In the case of tripeptides, cyclization can take place only if all three units are in the *cis* configuration. In higher cyclic peptides, i.e. with four or more units, one or more *cis* bonds are found. As the number of peptide units increases the more stable *trans* configuration is generally more common. The occurrence and positions of *cis* amides for cyclic peptides is given in Table IV.

Table IV. Occurrence of *cis* Amide Bonds in Cyclic Peptides

Peptide	No. of residues in ring	No. of atoms in ring	Conformation	CIS Peptide bonds	Ref
c(Pro)$_3$	3	9	ccc	Pro, Pro, Pro	71
c(L-Pro-L-Pro-L-Hyp)	3	9	ccc	Pro, Pro, Hyp	71
c(D-Hyiv-L-MeIle-D-Hyiv-L-MeLeu)	4	12	tctc	MeIle, MeLeu	72
c(Sar)$_4$	4	12	tctc	Sar, Sar	33
c(Gly-Sar$_3$)	4	12	tctc	Sar, Sar	34
c(L-Ala-Sar$_3$)	4	12	tctc	Sar, Sar	34
c(D-Ala-Sar$_3$)	4	12	tctc	Sar, Sar	34
c(Gly-Sar-Gly-Sar)	4	12	tctc	Sar, Sar	34
c(L-MeAla-L-Leu-D-MePhe-Gly) (dihydrotentoxin)	4	12	tctc	Leu, Gly	73
c(Sar)$_5$	5	15	ccctt	Sar, Sar, Sar	74
c(L-Ala-Sar$_4$)	5	12	ccctt	Sar, Sar, Sar	75
c(Sar)$_6$	6	18	cctcct	Sar, Sar, Sar, Sar	76
c(Sar)$_7$	7	21	tttcccc	Sar, Sar, Sar, Sar	77
Ilamycin B$_1$	7	21	tctttct	MeLeu, MeLeu	78
c(Sar)$_8$	8	24	ccttcctt	Sar, Sar, Sar, Sar	79
Li$^+$-Antamanide	10	30	tcttttcttt	Pro, Pro	80
Na$^+$ [Phe^4, Val^6]-antamanide	10	30	tcttttcttt	Pro, Pro	81
Cyclosporin A	11	33	tttttttttct	MeLeu	82

The *cis* state in tri- or higher cyclic peptides occurs when a substitution on the nitrogen atom is present, i.e., Pro, Hyp, Sar, MeLeu, MeIle, with a single exception in dihydrotentoxin,[73] where, most unexpectedly, the *cis* configuration occurs for the secondary amide bonds involving the Gly and L-Leu residue and not for the L-MeAla and D-MePhe residues. In all cases listed the C=0 and N-R bonds point outwards with respect to the cyclic structure.

In the two known cases of cyclic tripeptides, i.e., *cyclo*(L-Pro)$_3$ and *cyclo*(L-Pro-L-Pro-L-Hyp)[71] the overall conformation of the ring system is the same. The backbone conformation in the crystal does not possess a true three-fold symmetry; the relative rigidity of these tripeptides must be sought in the intrinsic conformation of the peptide chain itself enhanced by the presence of the pyrrolidine rings.

Six structures of cyclic tetrapeptides have been elucidated, one of which presents all *trans* peptide bonds, i.e., *cyclo*(Me$_2$Gly-L-Phe-D-Pro-LX),[83] and five of them present in position 2 and 4 *cis* peptide bonds. The conformational features of six independent molecules in the structures of *cyclo*(Sar)$_4$,[33] *cyclo*(Gly-Sar$_3$),[34] *cyclo*(D,L-Ala-Sar$_3$),[34] *cyclo*(Gly-Sar)$_2$[34] and dihydrotentoxin,[73] are strikingly similar in spite of the differences in the chemical structure, confirming the earlier conclusion drawn from nmr data, on the existence of only one conformer at least for the Sar-containing cyclic peptides. In these latter structures no evidence is found for transannular interactions. The reason for the remarkable identity of all six ring conformations must therefore lie in the intrinsic conformation of the peptide chain itself. The same conclusion was obtained in the case of the two pentapeptide *cyclo*(Sar)$_5$[74] and *cyclo*(Ala-Sar$_4$)[75] and for *cyclo*(Sar)$_n$ with n=6,7 and 8,[76,77,79] both in solution and in the solid state from the observation of conformational homogeneity. Increasing the number of peptide units in the ring, the flexibility of the ring and consequently the number of conformers, increase even though intra-ring interaction such as hydrogen bond or ion binding may stabilize preferentially one conformation. As an example, we may consider the crystal structures of uncomplexed valinomycin, a cyclic dodecadepsipeptide *cyclo*(D-Val-L-Lac-L-Val-D-Hyv)$_3$ in two crystalline modification, P1[89] and P2$_1$,[90] and its K$^+$ complex,[91] Figure 2. While the uncomplexed valinomycin molecule in both crystalline modification shows approximately the same conformation stabilized by all possible NH . . . 0 intramolecular hydrogen bonds (four of the 4 → 1 type (conformation II) and two of the 5 → 1 type), the K$^+$ complex shows a different conformation, with the same six H-bonds. Octahedral coordination at the K$^+$ ion by six carbonyl oxygen atoms takes place, requiring the opening of the interior of the structure with large changes in some conformational angles. While the uncomplexed molecule does not maintain the possible threefold symmetry but an approximate center of symmetry which relates residues i with i+6, the K$^+$ complex shows a pseudo S$_6$ symmetry.

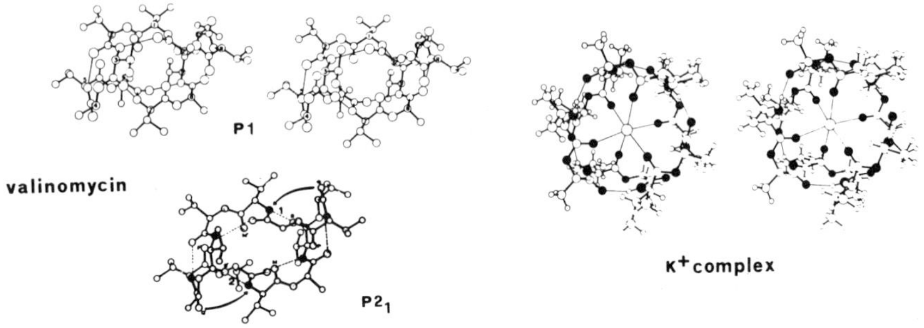

Fig. 2. The uncomplexes valinomycin in space groups P1 and $P2_1$ and its K^+ complex.

Linear Peptides. Table III lists the conformational angles ϕ and ψ for each residue in 55 linear peptides. The majority of them falls within the allowed regions. In particular, Pro residues concentrate in two regions centered around $\phi = -70$, $\psi = -25$ and $\phi = -65$, $\psi = -150°$, while Gly residues seem to spread out over a larger range of values for these angles as a consequence of the greater freedom caused by the absence of substitution on C^α. The other amino-acid residues all fall within the allowed regions with few exceptions represented by those residues presenting ionic terminals or involved in particular intramolecular hydrogen bonds cases for which the residues may be forced in an "unusual" conformation.

Various types of intramolecular hydrogen bonds have been proposed[92] and found in peptides. An H-bond of an amino-acid residue of sequence number m and C=0 of a residue of sequence number n is designated $m \rightarrow n$ so that for example in a system of three linked peptide units the $2 \rightarrow 2$, the $2 \rightarrow 3$, the $3 \rightarrow 1$ (or $4 \rightarrow 2$) and the $4 \rightarrow 1$ are intramolecularly H-bonded conformations. The ϕ and ψ angles involved in each of such conformations experimentally observed in linear or cyclic peptides are given in Table V. Figure 3 gives one experimental model for the residues involved in each of the seven types of intramolecular NH . . . . 0 bond as observed in crystalline peptides. Mainly nmr techniques have been used to assess the existence of NH . . . . 0 intramolecular hydrogen bonds in solution. For most of the structures determined in the solid state by X-ray, the spectroscopic results have been confirmed. However, there are cases in which complete disagreement or a partial agreement occur between results in solution and in the solid state. An example is given by the crystal structure of *cyclo*(L-Ala-L-Pro-D-Phe)$_2$[93] which presents all *trans* amide linkages. Two intramolecular hydrogen bonds of the type $4 \rightarrow 1$ (conformation II) were predicted from nmr data.[94] In the solid state, the experimental structure confirms partially the solution results, showing a C2 symmetry, but the expected hydrogen bonds are not found even if the conformational angles involving the peptide chain are consistent with the values accepted for β-turns. The peptide is involved in

Table V. Intramolecular NH . . . 0 Bonds

Substance	Residues 2	3	4	2	2	3	3	4	4	Ref.
A. 3 → 1 TRANS, Axial										
Dihydrochlamydocin	$(CH_3)_2$Gly			+72	−65					83
	D-Pro			+82	−73					
B. 3 → 1 TRANS, Equatorial										
Cyclosporin A	L-Ala			−92	+64					82
C. 4 → 1 TRANS, Type I										
c(Gly)$_6$	Gly	Gly		−69	−30	− 94	+11			84
	Gly	Gly		+69	+33	+ 92	− 7			
c(Gly$_4$−2D−Ala)	D-Ala	D-Ala		+66	+15	+131	−31			85
Li$^+$ antamanide	L-Ala	L-Phe		−69	−13	− 84	− 6			80
	L-Phe	L-Phe		−79	−13	− 90	+ 8			
S-benzyl-L-Cys-L-Pro-L-Leu-Gly-NH_2	L-Pro	L-Leu		−66	−29	−115	+13			70
p-BrCbz—Gly-L-Pro-L-Leu-Gly-OH	L-Pro	L-Leu		−58	−27	−104	+ 8			68
o-BrCbz—Gly-L-Pro-L-Leu-Gly-L-Pro-OH	L-Pro	L-Leu		−65	−27	−105	+ 8			69
N-Ac-L-Pro-L-Lac-$NHCH_3$	L-Pro	L-Lac*		−55	−22	− 81	−11			86
D. 4 → 1 TRANS, Type II										
Ferrichrome A	L-Ser	Gly		−57	+132	+ 82	− 1			87
L-Pro-L-Leu-Gly-NH_2	L-Leu	Gly		−61	+128	+ 72				63
N-Ac-L-Pro-D-Lac-$NHCH_3$	L-Pro	D-Lac*		−62	+140	+ 91	− 8			88
valinomycin (P1)	L-Val	D-Hyv*		−63	+129	+ 96	− 3			89
	L-Val	D-Hyv*		−67	+130	+ 82	+ 3			
	D-Val	L-Lac*		+63	−134	− 74	− 6			
	D-Val	L-Lac*		+60	−135	− 98	+14			
N-i Bu — L-Pro-L-Ala-NH-iPr	L-Pro	L-Ala		−59	+136	+ 66	+14			5
N-i Bu — L-Pro-D-Ala-NH-iPr	L-Pro	D-Ala		−62	+137	+ 96	+ 3			5
E. 4 → 1 CIS										
Ilamycin B_1	L-Try	L-MeLeu		−86	+117	−128	+99			78
	L-Ala	L-MeLeu		−61	+126	−121	+38			
F. 5 → 1 TRANS										
Valinomycin (P1)	L-Lac*	L-Val	D-Hyv*	−74	− 6	−108	+78	+146	−11	89
	D-Hyv*	D-Val	L-Lac*	+82	+ 3	+108	−69	−164	+23	

Notes: *O replaces NH

Section C Both L or both D or both Gly.
Section D L,D or D,L or L, Gly.
Section E Residue 3 N-substituted.

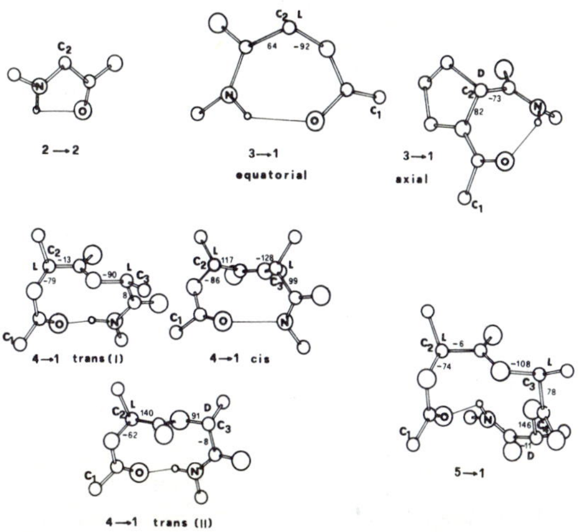

Fig. 3. Types of intramolecular H-bonds as found in crystalline peptides.

extensive H bonds with solvents of crystallization. Because of crystal packing forces, it is always a possibility that the structure of a peptide in the crystal is not the same as that in solution or in the biological systems; however, data have accumulated for several systems, which in the majority of the cases show closely similar molecular conformations in both states. Crystallization of higher peptides is a serious problem mainly because of the large number of almost isoenergetic conformers present. The ease of crystallization of certain peptides, for example cyclic peptides, with respect to linear peptides must be ascribed to the reduced number of conformers in solution so that then the molecule in the two states may exhibit the same conformation intrinsic to the peptide itself.

Side-Chain Conformations

Among the amino acids presenting a side-chain extending beyond the C^β atom, only those residues sufficiently numerous in crystal-structure analyses will be considered. The results concerning the conformations in the solid state for the side-chains of Ser, Val, Leu, Phe, Tyr and Pro are summed up in the hystograms reported in Figure 4.

In the 14 crystal structures in which the side-chain of serine is reported, the χ_1 angle assumes values centered at about 60° in 10 cases, while the -60° and 180° conformations are each represented in two cases, indicating a slight preference for the oxygen atom in the γ position to achieve a conformation close to the nitrogen atom to the preceding amide bond. However, the conformation of this side chain may be influenced to a certain extent by the hydrogen bond formation in which the oxygen atom could be involved.

In valyl residues the two torsional angles $\chi_{1,1}$ and $\chi_{1,2}$ in more than 80% of the cases examined (~40 independent residues) assume values centered and about -60°, 180°, respectively, for the L configuration of the α-carbon, and opposite values for a D configuration. Only in very few cases the conformation, described with $\chi_{1,1} \cong -60$ and $\chi_{1,2} \cong 60°$, is observed. A correlation between the side-chain angular values and the backbone angular values, is that a $\chi_{1,1}$ value close to -60° corresponds, usually, to two groups of conformations for the backbone chains with $\phi \cong -65°$, $\psi \cong 130°$ and $\phi \cong -110°$, $\psi \cong 70°$, respectively.

In the leucyl residues a χ_1 value of about 60° is not observed: such a value would result in severe steric interactions between the substituents on C^β and C^γ and the backbone chain. For the three angular values defining the conformation the leucyl side chains χ_1, $\chi_{2,1}$ and $\chi_{2,2}$ only two combinations of angles are observed, centered at about -60°, -60°, 180° and 180°, 180°, 60°. In addition, only small deviations from these values are found (at most ± 20°). The first combination is represented in the majority of the crystal structures examined. The values -60, 180° for the two angles $\chi_{2,1}$ and $\chi_{2,2}$ are the same as those observed in valyl residues for $\chi_{1,1}$ and $\chi_{1,2}$ pertaining both to an isopropyl group. The χ_1 ($\cong$ -60°) conformation for the side chain of leucine residues seems

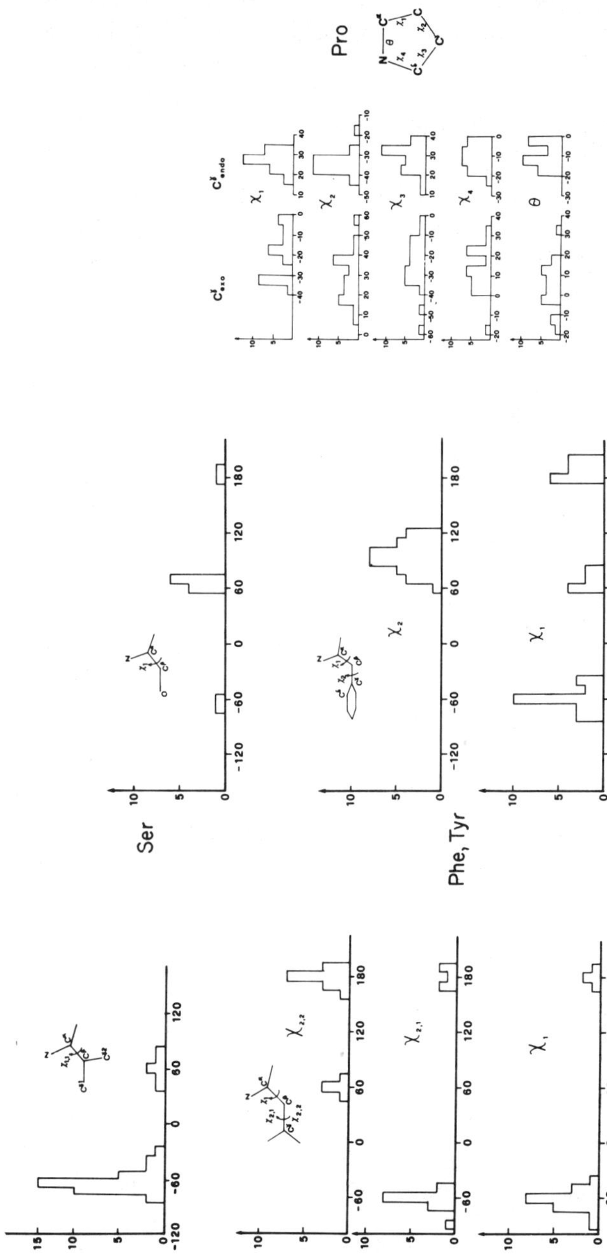

Fig. 4. Hystograms for the side-chain conformational angles of Val, Ser, Leu, Phe, Tyr and Pro residues.

to be correlated mainly with two groups of backbone conformations defined by $\phi \cong -80°$, $\psi \cong 140°$ and $\phi \cong -120°$, $\psi \cong 10°$. The other conformation ($\chi_1 \cong 180$) cannot be associated significantly with a particular backbone conformation.

For the aromatic side chains of phenylalanine and tyrosine, as indicated by a recent study,[96] three major classes of conformers can be singled out from the crystal structures in which these residues are present. These conformations are defined by the three possible values of the χ_1 angular parameter: 60°, 180°, -60°, while the χ_2 angle seems to show only one value centered at about 90°, even if it shows a broad dispersion around the mean. In more than 60 aromatic residues examined, containing either phenylalanine or tyrosine, the statistical distribution of the three sets of values for χ_1 and χ_2 is about 20% : 35% : 45% for the conformations with $\chi_1 \cong 60°$, $\chi_1 \cong 180°$, $\chi_1 \cong -60°$, respectively. Furthermore, for structures in which $\chi_1 \cong 180°$ a discontinuity has been observed in the dependence of ψ from χ_2: two classes of conformers with $\chi_1 \cong 180°$ can be distinguished; the first for χ_2 values smaller than 85° one finds $\psi > -50°$ and the second for χ_2 values larger than 85° one finds $\psi < -50°$.

In prolyl residues, among the possible conformations observed for the rather flexible pyrrolidine ring (with puckering at the various atoms of the five-membered ring), those presenting γ-puckering are by far the more common as also confirmed by minimum energy calculations.[97] These conformations denoted as C^γ-*exo* (or conformation A) and C^γ-*endo* (or conformation B) have been characterized by the following ranges for the sequence of $\chi_1, \chi_2, \chi_3, \chi_4$ and θ dihedral angles: $-30° < \chi_1 < 0°$, $15° < \chi_2 < 50°$, $-30° < \chi_3 < -15°$, $5° < \chi_4 < 25°$, $-15° < \theta < 15°$ and $20° < \chi_1 < 35°$, $-40° < \chi_2 < -30°$, $20° < \chi_3 < 35°$, $-20° < \chi_4 < -5°$, $-15° < \theta < 15°$ for the C^γ-*exo* and C^γ-*endo* conformations, respectively. In the hystograms reported in Figure 4, these angular values show a considerable spread about the mean value, which is larger for the C^γ-*endo* than for the C^γ-*exo* conformation, indicating a greater rigidity for the latter.

References

1. Corey, R. B. & Pauling, L. (1953) *Proc. Roy. Soc. (London)* **B141**, 10–20.

2. Marsh, R. E. & Donohue, J. (1967) *Adv. Protein Chem.* **22**, 235–255.

3. Ramachandran, G. N., Kolaskar, A. S., Ramakrishnan, C. & Sasisekharan, V. (1974) *Biochem. Biophys. Acta* **359**, 298–302.

4. IUPAC-IUB Commission on Biochemical Nomenclature (1970) *Biochemistry* **9**, 3471-3479.

5. Aubry, A. (1976) Ph.D. Thesis, Universite de Nancy I; C.N.R.S., n, A.O.12.610.

6. Ichikawa, T. & Iitaka, Y. (1969) *Acta Cryst.* **B25**, 1824–1833.

7. Harada, Y. & Iitaka, Y. (1974) *Acta Cryst.* **B30**, 726–730.

8. Harada, Y. & Iitaka, Y. (1977) *Acta Cryst.* **B33**, 247–249.

9. Benedetti, E., Christensen, A., Gilon, C., Fuller, W. & Goodman, M. (1976) *Biopolymers* **15**, 2523–2534.

10. Matsuzami, T. & Iitaka, Y. (1971) *Acta Cryst.* **B27**, 507–516.

11. Aubry, A., Protas, J., Marraud, M. & Neel, J. (1976) *Acta Cryst.* **B32**, 2749–2754.

12. Cotrait, M. & Bideau, J. P. (1973) *Cryst. Struct. Comm.* **2**, 111–114; (1974) *Acta Cryst.* **B30**, 1024–1028.

13. (a) Harada, Y. & Iitaka, Y. (1977) *Acta Cryst.* **B33**, 244–247; (b) Cotrait, M. & Barrans, Y. (1974) *Acta Cryst.* **B30**, 510–513.

14. Harada, Y. & Iitaka, Y. (1977) *Acta Cryst.* **B33**, 250–252.

15. Benedetti, E., Palumbo, M., Bonora, G. M. & Toniolo, C. (1976) *Macromolecules* **9**, 417–420.

16. Marsh, R. E., Venkatesan, K. & Murthy, K. (1977) *J. Amer. Chem. Soc.* **99**, 1251–1256.

17. Benedetti, E., Pavone, V., Toniolo, C., Bonora, G. M. & Palumbo, M. (1977) *Macromoles*, submitted for publication.

18. Tanaka, I., Kozima, T., Ashida, T., Tanaka, N. & Kakudo, M. (1977) *Acta Cryst.* **B33**, 116–119.

19. Lecomte, C., Aubry, A. & Protas, J. (1974) *Acta Cryst.* **B30**, 1992–1996 and 2343–2348.

20. Stenkamp, R. E. & Jensen, L. H. (1973) *Acta Cryst.* **B29**, 2872–2878.

21. Itoh, H., Yamane, T., Ashida, T., Sugihara, T., Imanishi, Y. & Higashimura, T. (1976) *Acta Cryst.* **B32**, 3355–3357.

22. Gadret, M., Leger, J. M. & Carpy, A. (1977) *Acta Cryst.* **B33**, 1067–1071.

23. Yamane, T., Ashida, T., Shimonishi, K., Kakudo, M. & Sasada, Y. (1976) *Acta Cryst.* **B32**, 2071–2076.

24. Kartha, G., Ashida, T. & Kakudo, M. (1974) *Acta Cryst.* **B30**, 1861–1866.

25. Matsuzaki, T. (1974) *Acta Cryst.* **B30**, 1029–1036.

26. Kashino, S., Ashida, T. & Kakudo, M. (1974) *Acta Cryst.* **B30**, 2074–2076.

27. Ashida, T. & Kakudo, M. (1974) *Bull. Chem. Soc. Japan* **47**, 1129–1133.

28. Kolaskar, A. S., Lakshminarayanan, A. V., Sarathy, K. P. & Sasikharan, V. (1975) *Biopolymers* **14**, 1081–1094.

29. Ramachandran, G. N. & Sasisekaran, V. (1968) *Adv. Protein Chem.* **23**, 283–437.

30. (a) Benedetti, E., Ciajolo, M. R. & Maisto, A. (1974) *Acta Cryst.* **B30**, 1783–1788; (b) Robert, F. (1976) *Acta Cryst.* **B32**, 2367–2369.

31. Toniolo, C., Palumbo, M. & Benedetti, E. (1976) *Macromolecules* **9**, 420–424.

32. Jain, S. C. & Sobel, H. M. (1972) *J. Mol. Biol.* **68**, 1–20.

33. Groth, P. (1970) *Acta Chem. Scand.* **24**, 780–790.

34. Declercq, J. P., Germain, G., van Meerssche, M., Debaerdemacker, T., Dale, J. & Titlestad, K. (1975) *Bull. Soc. Chim. Belg.* **84**, 275–287.

35. Corey, R. B. (1938) *J. Amer. Chem. Soc.* **60**, 1598–1604; Delgeilh, R. & Marsh, R. E. (1959) *Acta Cryst.* **12**, 1007–1014.

36. Benedetti, E., Corrandini, P., Goodman, M. & Pedone, C. (1969) *Proc. Nat. Acad. Sci. USA* **62**, 650–652; Benedetti, E., Corradini, P. & Pedone, C. (1969) *Biopolymers* **7**, 751–764; Benedetti, E., Corradini, P. & Pedone, C. (1969) *J. Phys. Chem.* **73**, 2891–2895; Sletten, E. (1970) *J. Amer. Chem. Soc.* **92**, 172–177.

37. Filhol, A. & Timmins, P. A. (1976) *Acta Cryst.* **B32**, 3116–3118.

38. Benedetti, E., Marsh, R. E. & Goodman, M. (1976) *J. Amer. Chem. Soc.* **98**, 6676–6684.

39. Mez, H. C. (1974) *Cryst. Struct. Comm.* **3**, 657–660.

40. Von Dreele, R. B. (1975) *Acta Cryst.* **B31**, 966–970.

41. Karle, I. L. (1972) *J. Amer. Chem. Soc.* **94**, 81–84.

42. Morris, A. J., Geddes, A. J. & Sheldrick, B. (1974) *Cryst. Struct. Commun.* **3**, 345–348.

43. Lin, C. F. & Webb, L. E. (1973) *J. Amer. Chem. Soc.* **95**, 6803–6811.

44. Cotrait, M., Ptak, M., Busetta, B. & Heitz, A. (1976) *J. Amer. Chem. Soc.* **98**, 1973–1976.

45. Ramani, R., Venkatesan, K., Marsh, R. E. & Hu Kung, W. J. (1976) *Acta Cryst.* **B32**, 1051–1056.

46. Benedetti, E., Goodman, M., Marsh, R. E., Rapoport, H. & Musich, J. A. (1975) *Cryst. Struct. Commun.* **4**, 641–645.

47. Karle, I. L., Ottenheym, H. C. J. & Witkop, B. (1974) *J. Amer. Chem. Soc.* **96**, 539–543.

48. Groth, P. (1969) *Acta Chem. Scand.* **23**, 3155–3162.

49. Timmins, P. A. (1975) *Acta Cryst.* **B31**, 2561–2565.

50. Benedetti, E., Goodman, M., Easter, D. C. & Marsh, R. E. unpublished results.

51. Brucker, S. (1974) *Gazz. Chim. Ital.* **104**, 1077–1086.

52. Hooker, T. M., Bayley, P. M., Radding, W. & Schellman, J. A. (1974) *Biopolymers* **13**, 549–566.

53. Iwasaki, F. (1974) *Acta Cryst.* **B30**, 2503–2505.

54. Harada, Y. & Iitaka, Y. (1974) *Acta Cryst.* **B30**, 1452–1459.

55. Coiro, V. M., Mazza, F. & Mignucci, G. (1974) *Acta Cryst.* **B30**, 2607–2613.

56. Sabesan, M. N. & Venkatesan, K. (1971) *Acta Cryst.* **B27**, 1879–1883.

57. Rao, S. T. & Mallikarjunan, M. (1973) *Cryst. Struct. Commun.* **2**, 257–260.

58. Narayanan, P., Berman, H. M. & Rousseau, R. (1976) *J. Amer. Chem. Soc.* 98, 8472–8475.

59. Marsh, R. E. & Glusker, J. P. (1961) *Acta Cryst.* **14**, 110–116.

60. Fletterick, R. J., Tsai, C. & Hughes, R. E. (1971) *J. Phys. Chem.* 7, 918–922.

61. Tokuma, Y., Ashida, T. & Kakudo, M. (1969) *Acta Cryst.* **B25**, 1367–1373.

62. Sabesan, M. N. & Venkatesan, K. (1961) *Z. fur Krist.* **134**, 230–242.

63. Reed, L. L. & Johnson, P. L. (1973) *J. Amer. Chem. Soc.* **95**, 7523–7525.

64. Wei, C. H., Doherty, D. G. & Einstein, J. R. (1972) *Acta Cryst.* **B28**, 907–915.

65. Leung, Y. C. & Marsh, R. E. (1958) *Acta Cryst.* **11**, 17–31.

66. Goswami, K. N., Yadava, V. S. & Padmanabhan, V. M. (1977) *Acta Cryst.* **B33**, 1280–1283.

67. Fawcett, J. K., Camerman, N. & Camerman, A. (1975) *Acta Cryst.* **B31**, 658–665.

68. Ueki, T., Ashida, T., Kakudo, M., Sasada, Y. & Katsube, Y. (1969) *Acta Cryst.* **B25**, 1840–1849.

69. Ueki, T., Bando, S., Ashida, T. & Kakudo, M. (1971) *Acta Cryst.* **B27**, 2219–2231.

70. Rudko, A. D. & Low, B. W. (1975) *Acta Cryst.* **B31**, 713–725.

71. Kartha, G., Ambady, G. & Shankar, P. V. (1974) *Nature (London)* **247**, 204–205.

72. Konnert, J. & Karle, I. L. (1969) *J. Amer. Chem. Soc.* **91**, 4888–4892.

73. Meyer, W. L., Kuyper, L. F., Phelps, D. W. & Cordes, A. W. (1974) *J. Chem. Comm.* 339–340.

74. Groth, P. (1973) *Acta Chem. Scand.* **27**, 3419–3426.

75. Groth, P. (1974) *Acta Chem. Scand.* **A28**, 449–454.

76. Groth, P. (1977) *Acta Chem. Scand.* **A31**, 232–234.

77. Groth, P. (1975) *Acta Chem. Scand.* **A29**, 38–44.

78. Iitaka, Y., Nakamura, H., Takada, K. & Takita, T. (1974) *Acta Cryst.* **B30**, 2817–2825.

79. Groth, P. (1973) *Acta Chem. Scand.* **27**, 3217–3226.

80. Karle, I. L. (1974) *J. Amer. Chem. Soc.* **96**, 4000–4006.

81. Karle, I. L. (1974) *Biochemistry* **13**, 2155–2162.

82. Petcher, T. J., Weber, H. P. & Ruegger, A. (1976) *Helv. Chim. Acta* **59**, 1480–1488.

83. Flippen, J. L. & Karle, I. L. (1976) *Biopolymers* **15**, 1081–1092.

84. Karle, I. L. & Karle, J. (1963) *Acta Cryst.* **16**, 969–975.

85. Karle, I. L., Gibson, J. W. & Karle, J. (1970) *J. Amer. Chem. Soc.* **92**, 3755–3760.

86. Lecomte, C., Aubry, A. & Protas, J. (1974) *Acta Cryst.* **B30**, 1992–1996.

87. Zalkin, A., Forrester, J. D. & Templeton, D. H. (1966) *J. Amer. Chem. Soc.* 88, 1810–1814.

88. Lecomte, C., Aubry, A. & Protas, J. (1974) *Acta Cryst.* **B30**, 2343–2348.

89. Karle, I. L. (1975) *J. Amer. Chem. Soc.* 97, 4379–4386.

90. Duax, W. L., Hauptman, H., Weeks, C. M. & Norton, D. A. (1972) *Science* **176**, 911-913.

91. Neupert-Laves, K. & Dobler, M. (1975) *Helv. Chim. Acta* **58**, 432–443.

92. Toniolo, C. (1977) *Bioorg. Chem.* in press and references therein.

93. Brown, J. N. & Teller, R. G. (1976) *J. Amer. Chem. Soc.* **98**, 7565–7569.

94. Kopple, K. D., Schamper, T. J. & Go, A. (1974) *J. Amer. Chem. Soc.* **96**, 2597–2605.

95. Deber, C. M. (1974) *Macromolecules* 7, 47–51.

96. Cody, V., Duax, W. L. & Hauptman, H. (1973) *Int. J. Pept. Protein Res.* **5**, 297–308.

97. Balasubramanian, R., Lakshminarayanan, A. V., Sabesan, M. N., Tegoni, G., Venkatesan, K. & Ramachandran, G. N. (1971) *Int. J. Pept. Protein Res.* **3**, 25–33.

3 → 1 AND 4 → INTRAMOLECULAR HYDROGEN BONDS IN *CYCLO*(Gly-Pro-Gly-D-Ala-Pro) (CRYSTAL STRUCTURE ANALYSIS)

ISABELLA L. KARLE, *Laboratory for the Structure of Matter, Naval Research Laboratory Washington, D.C. 20375*

Complementary studies on the conformation of *cyclo* (Gly-Pro-Gly-D-Ala-Pro) have been performed in solution by Pease and Watson[1] using nuclear magnetic resonance and circular dichroism and in the crystalline state by X-ray diffraction analysis. The conformation derived from both studies is quite similar. Cyclic peptides are constrained to contain bends in the backbone and offer good models for establishing the various types of intramolecular hydrogen bonds and for determining the torsional angles and molecular dimensions. Crystal structure analysis is particularly well-suited for this task since the coordinates of the atoms can usually be determined to an accuracy near 0.002–0.003 Å and all other geometric quantities are derived from them with equivalent accuracy. The present study is the first example of a cyclic peptide containing both a 3 → 1 and a 4 → 1 hydrogen bond.

Crystals of *cyclo* (Gly-Pro-Gly-D-Ala-Pro) were grown from MeOH/ether solution by Pease and Watson[1]. They were stable in the dry state and X-ray intensity data were collected from one crystal on an automatic four-circle diffractometer with Cu K_α radiation to a scattering angle of $2\theta = 126°$. The space group is the noncentrosymmetric orthorhombic $P2_12_12_1$ with cell dimensions a = 10.254(2) A, b = 21.320(5) A, c = 8.565(1) A and there are four molecules in the unit cell. The structure of the crystal was derived with the direct method[2] of phase determination using symbolic addition. Least squares refinement led to an agreement factor between calculated and observed structure factors of 7.2% for 1554 measured data.

A stereodiagram[3] of the molecule prepared from the experimentally determined coordinates is shown in Fig. 1. The striking feature of the structure is a pair of nearly parallel transannular C=O···HN bonds, reminiscent of cyclic hexapeptides such as *c*(Gly-D-Ala-D-Ala-Gly-Gly-Gly)[4] and *c*(L-Ala-L-Pro-D-Phe)$_2$.[5] The $N_4H \cdots O_1$ bond is of the Type II, 4 → 1 bond[6] first observed in ferrichrome-A[7]. A comparison of the conformational angles observed for the 4 → 1 bond in *c*(Gly-Pro-Gly-D-Ala-Pro) and those found in other peptide molecules in the crystalline state is shown in Table I. There exists a remarkable consistency in the angular values, despite differences in the length of the backbone and the variety of the side groups.

The other transannular hydrogen bond in the cyclic pentapeptide, $N_1H \cdots O_4$, is the recently observed 3 → 1 type. This type of bond also exists in dihydrochlamydocin[11], a naturally occurring cyclic tetrapeptide, and comparisons of

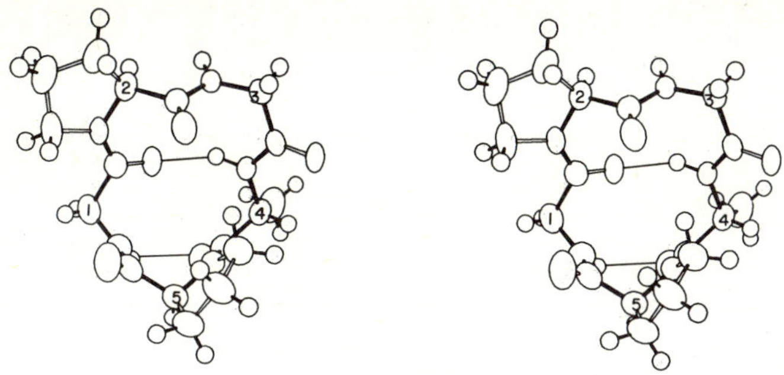

Fig. 1. Stereodiagram of *c*(Gly-Pro-Gly-D-Ala-Pro). The ellipsoids represent the thermal motion of the atoms at 50% probability, whereas the small spheres are the hydrogen atoms. The C^{α} atoms are labelled 1 to 5.

Table I. Conformational Angles $\Phi(N\text{-}C^{\alpha})$ and $\psi(C^{\alpha}\text{-}C')$ for 4 → 1 (Type II) Hydrogen Bonds.*

Peptide	Residue 2	Residue 3	Φ_2	ψ_2	Φ_3	ψ_3	N···O	Ref.
c-pentapeptide	L-Pro	Gly	-52°	126°	74°	12°	2.87 Å	This paper
Ferrichrome A	L-Ser	Gly	-57	132	82	-1	2.98	7
c-(L-Ala-L-Pro-D-Phe)$_2$	L-Pro	D-Phe	-60	122	78	9	3.20	5
linear tripeptide	L-Leu	Gly	-61	128	72		3.04	8
linear tripeptide	L-Pro	D-Lac	-62	140	91	-8	2.97	9
valinomycin	L-Val	D-Hyv	-63	129	96	-3	3.07	10
	L-Val	D-Hyv	-67	130	82	3	2.90	

*In Type II, 4 → 1 bonds residues 2 and 3 are either L,D or L, Gly.

conformational angles are shown in Table II. Again, the absolute values for ϕ, ψ and ω are entirely consistent in the three examples. The signs of the values are opposite in the present molecule from those in dihydrochlamydocin since the residues involved are of opposite hands. An important feature in the 3 → 1 type of bond is the *nonplanarity* of the peptide group containing the participating NH group. As can be seen in Table II, the ω values deviate from 180° by 18° to 24°. The other four ω values for this cyclic pentapeptide range from 174° to 178°.

Table II. Conformational Angles for 3 → 1 Hydrogen Bonds.

Peptide	Residue	Φ	ψ	ω	N···O	Ref.
c-pentapeptide	L-Pro	-86°	70°	-160°	2.92 Å	This paper
dihydrochlamydocin	D-Pro	+83	-73	+156	2.94	11
	Me_2 Gly	+72	-64	+162	2.82	

References

1. Pease, L. G. & Watson, C. (1977) in *Peptides: Proceedings of the 5th American Peptides Symposium*, Goodman, M. and Meienhofer, J., Eds., John Wiley & Sons, Inc., New York, 346–349.
2. Karle, J. & Karle, I. L. (1966) *Acta Cryst.* **21**, 849–859.
3. Johnson, C. K. (1965) ORTEP, ORNL-3794, Oak Ridge National Laboratory, Oak Ridge, Tenn.
4. Karle, I. L., Gibson, J. W. & Karle, J. (1970) *J. Amer. Chem. Soc.* **92**, 3755–3760.
5. Brown, J. N. & Teller, R. G. (1976) *J. Amer. Chem. Soc.* **98**, 7565–7569.
6. Venkatachalam, C. M. (1968) *Biopolymers* **6**, 1425–1436.
7. Zalkin, A., Forrester, J. D. & Templeton, D. H. (1966) *J. Amer. Chem. Soc.* **88**, 1810–1814.
8. Reed, L. L. & Johnson, P. L. (1973) *J. Amer. Chem. Soc.* **95**, 7523–7524.
9. Lecomte, C., Aubry, A. & Protas, J. (1974) *Acta Cryst.* **B30**, 2343–2348.
10. Karle, I. L. (1975) *J. Amer. Chem. Soc.* **97**, 4379–4386.
11. Flippen, J. L. & Karle, I. L. (1976) *Biopolymers* **15**, 1081–1092.

THE CRYSTAL AND MOLECULAR STRUCTURE OF A TETRAPEPTIDE, THE BENZYL ESTER OF Boc-L-Pro-Aib-L-Ala-Aib

G. D. SMITH, W. L. DUAX, *Medical Foundation of Buffalo, Inc., 73 High Street, Buffalo, New York 14203,* E. W. CZERWINSKI, N. E. KENDRICK, G. R. MARSHALL and F. S. MATHEWS, *Department of Physiology and Biophysics, Washington University, School of Medicine, St. Louis, Missouri 63110*

The title tetrapeptide (Figure 1) comprises residues 2 through 5 of the linear polypeptide alamethicin,[1] an antibiotic which affects ion transport through natural and synthetic membranes and contains the unusual amino acid, aminoisobutyric acid (Aib).

This compound crystallizes in the monoclinic space group $P2_1$ with cell dimensions a = 11.147 Å, b = 13.877 Å, c = 10.417 Å, β = 100.9° and Z = 2. The structure was finally solved through the use of the direct methods program QTAN[2]. The tetrapeptide has been refined by full-matrix least-squares, treating the thermal vibration of all non-hydrogen atoms anisotropically, to a residual of 0.10 for 2496 observed data.

An ORTEP drawing of the structure is shown in Figure 2. Bond distances and angles are unexceptional. The conformation of the first three residues (Pro-Aib-Ala) approximates that of a single turn of a 3_{10}-helix. Two weak intramolecular hydrogen bonds are present in this single turn of the helix and occur between the amino group of Ala^3 and the carbonyl oxygen of Boc(N-O distance, 2.97 Å) as well as between the amino group of Aib^4 and the proline carbonyl oxygen (N-O distance, 3.09 Å). In addition, there is a single intermolecular hydrogen bond between the amino group of Aib^2 and the carbonyl oxygen of Aib^4 of a translationally related molecule (N-O distance, 2.88 Å). Thus, every potential hydrogen bond donor is involved in a hydrogen bond.

All peptide linkages deviate by less than 8° from *trans*. The (ϕ, ψ) torsion angles[3] are Pro^1 (-56°, -35°), Aib^2 (-52°, -38°), Ala^3 (-93°, -13°) and Aib^4 (48°, 42°), respectively. These values are plotted in Figure 3 along with theoretical values for a 3_{10}- and an α-helix. While the torsion angles of the first two

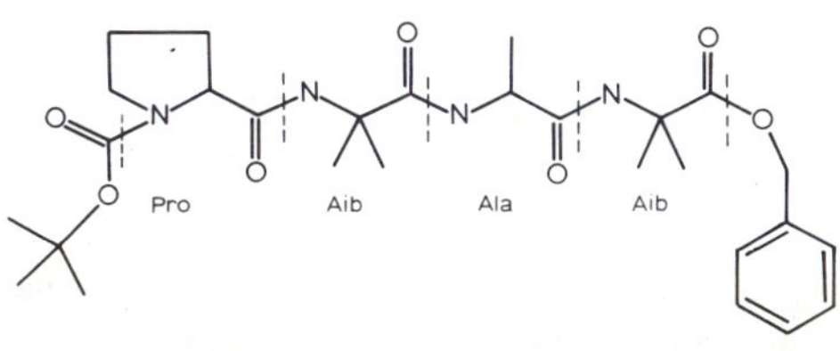

Fig. 1. Boc-L-Pro^1-Aib^2-L-Ala^3-Aib^4.

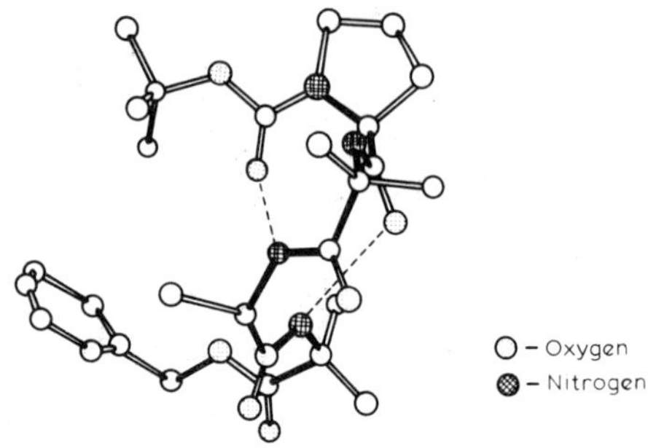

Fig. 2. Structure of the benzyl ester of Boc-L-Pro-Aib-L-Ala-Aib.

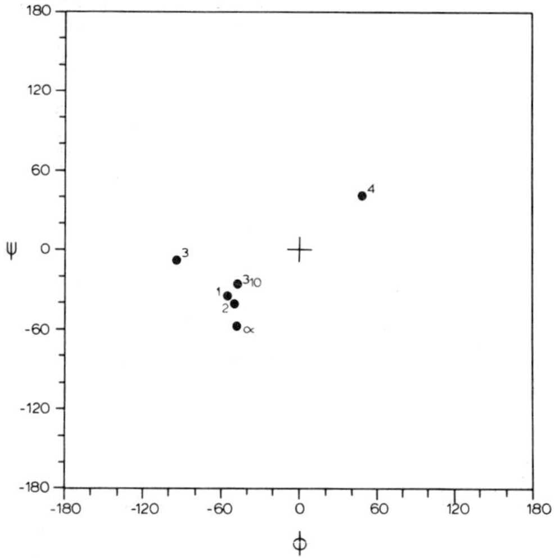

Fig. 3. ϕ, ψ plot for (1), L-Pro^1; (2) Aib^2; (3) L-Ala^3; (4) Aib^4 as well as the expected values for a 3_{10} helix and an α helix.

residues and to some extent the third residue closely approximate the accepted values for a 3_{10}- or an α-helix, the torsion angles of the fourth residue are considerably removed from this area of the graph. However, this should not be surprising since there are no intramolecular hydrogen bonds to maintain the helical conformation in this area.

Both empirical[4] and potential energy calculations[5,6] show that the allowable conformations which an Aib residue may adopt favor that of the right- and left-handed α-helix. The ϕ and ψ torsion angles of the two Aib residues obtained from this study are nearly equal in magnitude, opposite in sign from each other and deviate by less than 9° from the theoretically predicted values. The relationship between the two residues can be seen from the Newman projections of the ϕ and ψ torsion angles, illustrated in Figure 4. Since these two Aib residues have different environments, in terms of the groups to which they are covalently

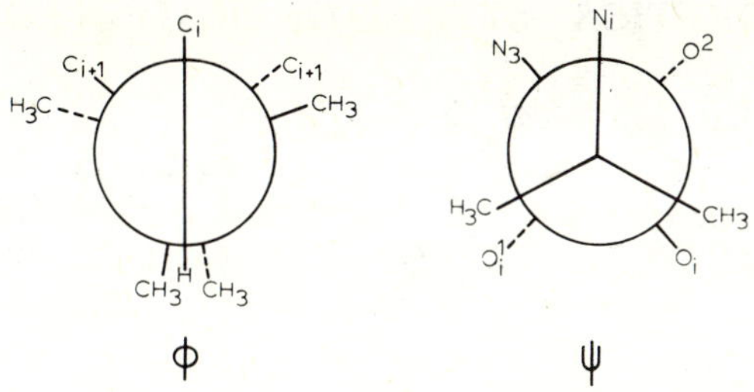

Fig. 4. Newman projection about the ϕ and ψ torsion angles of Aib^2 (solid lines) and Aib^4 (broken lines).

bonded and also the number and arrangement of the hydrogen bonds, it seems very unlikely that the observed relationship is fortuitous. Therefore, we conclude that the observed magnitudes of ϕ and ψ describe highly preferred conformations for the Aib residue.

The conformation of the pyrrolidine ring can best be described as nearly midway between the half-chair and envelope forms of a five-membered ring. The asymmetry parameters[7] were calculated to be 7.9 for a mirror plane passing through the β-carbon atom and 8.9 for a two-fold axis passing through the nitrogen atom. Further confirmation is provided by the fact that the γ-carbon atom is displaced -0.16 Å from the plane defined by N, C^{α} and C^{δ} while the β-carbon atom is displaced +0.47 Å from this same plane.

Research supported by Grant No. GM-19684 from the National Institute of General Medical Sciences, DHEW (GDS and WLD) and Grant No. PCM 74-03059 from the National Science Foundation (EWC, NEK, GRM and FSM).

References

1. Martin, D. R. & Williams, R. J. P. (1976) *Biochem. J.* **153**, 181–190.

2. Computer program written by D. A. Langs, Medical Foundation of Buffalo, Inc., Buffalo, New York.

3. Kendrew, J. C., Klyne, W., Lifson, S., Miyazawa, T., Nemethy, G., Phillips, D. C., Ramachandran, G. N. & Scheraga, H. A. (1970) *J. Mol. Biol.* **52**, 1–17.

4. Johnson, M. E. & Wu, T. T. (1976) *J. Theor. Biol.* **60**, 183–195.

5. Marshall, G. R. & Bosshard, H. E. (1972) *Circ. Res., Suppl. II* **30** & **31**, 143–150.

6. Burgess, A. W. & Leach, S. J. (1973) *Biopolymers* **12**, 2599.

7. Duax, W. L., Weeks, C. M. & Rohrer, D. C. (1976). In *Topics in Stereochemistry,* vol. 9, Eliel, E. L. & Allinger, N., Eds., Wiley-Interscience, New York, pp. 271–383.

THEORETICAL STUDIES OF PROTEIN FOLDING

A. T. HAGLER, *Chemical Physics Department, Weizmann Institute of Science, Rehovot, Israel,* B. HONIG, *Physical Chemistry Department, Hebrew University, Jerusalem, Israel,* and R. SHARON, *Chemical Physics Department, Weizmann Institute of Science, Rehovot, Israel*

Introduction

There has been considerable research devoted to the computer simulation of the dynamics of polypeptide chains. Interest has focused particularly on the folding problem and related questions pertaining to the factors that determine the three-dimensional conformation of native proteins. Recently a number of attempts have been made to fold a complete protein with computer simulation.[1-5] Pancreatic trypsin inhibitor (PTI) has been chosen for studies of this type, since it is a small protein, and its folding has been studied extensively experimentally,[6] The first claims of a successful folding simulation were made by Levitt and Warshel.[1,2] These workers used a highly simplified model of a polypeptide chain in which the backbone was represented in terms of only three pseudo torsional potentials (ascribed as Ala, Gly and Pro) while each side chain was assigned a spherical potential centered at a single point. An extended starting conformation was assumed which, following a series of minimizations and subsequent randomizations to escape local minima, folded into a conformation which in some ways, resembled the native protein.

We carried out a study aimed at defining the factors that lead to the formation of tertiary structure.[7] Our initial strategy has been to perform energy minimizations on polypeptide sequences consisting of only alanines and glycines. These two amino acids were chosen since they are extremely simple yet contain a great deal of structural information. Alanine, which is the smallest amino acid containing a β-carbon, tends to form secondary structures while glycine allows a chain to fold back on itself. By studying the conformations obtained from these two residues alone we can determine the effect that the presence or absence of a β-carbon in a particular position along the chain has on the computed conformation. In this way we can not only study the role of glycines in protein folding, but also perform control experiments on other folding simulations such as those of Levitt and Warshel[1,2] in which the results were attributed primarily to the use of "time averaged forces" between all 20 types of amino-acid residues. Our approach requires that we introduce no *a priori* assumptions and for this reason no modifications to the potential functions[8,9] were made.

Folding Simulations for Pancreatic Trypsin Inhibitor (PTI)

In our first set of folding simulations, we represented all glycines, asparagines and aspartic acids as glycines. (These are the amino acids which were ascribed a glycine backbone potential in the work of Levitt and Warshel[1,2].) All other residues were taken to be alanines. It should be noted[10] that by representing the sequence in this way, glycine residues are present in each of the three major turns of PTI but are absent from the extended regions.

The results of the first folding simulation we attempted are shown in Fig. 1b. It is clear that the conformation of Fig. 1b superficially resembles native PTI. A structure resembling an anti-parallel β-sheet has formed between residues 16 and 36 and the chain turns and folds back on itself at approximately the right places. The r.m.s. deviation from the native structure we obtain is 6.1 Å which is well within the range used to define successful folding simulations in previous studies.

The result shows that a polypeptide of 58 amino acids consisting of just alanines and glycines can be folded on a computer to a conformation containing many of the elements of tertiary structure normally associated with proteins. Moreover, the computed structure bears a resemblance to PTI. It is as close or closer to the native structure of PTI as has been achieved heretofore, both in terms of topology and r.m.s. deviation.

The general similarity obtained is related entirely to the definition of the *backbone* in terms of two types of residues, one which favors turns in the chain at the correct place and the other which favors extended regions. It appears that the "time-averaged" potentials[1,2] that have been introduced to represent *side chain* interactions, have little, or nothing, to do with the quality of the final result. Our results demonstrate that it is possible to obtain a structure which resembles the native conformation, from a sequence which contains essentially no relation to the true primary structure and therefore *no* information relevant to its actual *in vivo* folding. Clearly, using overly permissive criteria as to the success of a folding simulation can lead to incorrect conclusions as to the validity of a particular representation of protein structure.[1,2]

A close comparison of the various computed structures to native PTI however, reveals significant differences which may be more significant than the artifactual similarities. In particular, neither the threading of the N-terminal sequence through the 30-51 covalent loop nor the 180° twist in the β-sheet is reproduced. In fact, none of the published folding simulations have yielded these features so that all attempts to fold PTI on a computer have failed to predict key elements of its tertiary structure. Moreover, none of the computed structures would fold correctly even if more accurate potentials were introduced following the folding step, since a major rearrangement is required in order to thread the chain correctly. Since a large number of folding simulations have been attempted, the failure to obtain even a single positive result with regard to these key features may imply that a very fundamental error is inherent in these studies. It is

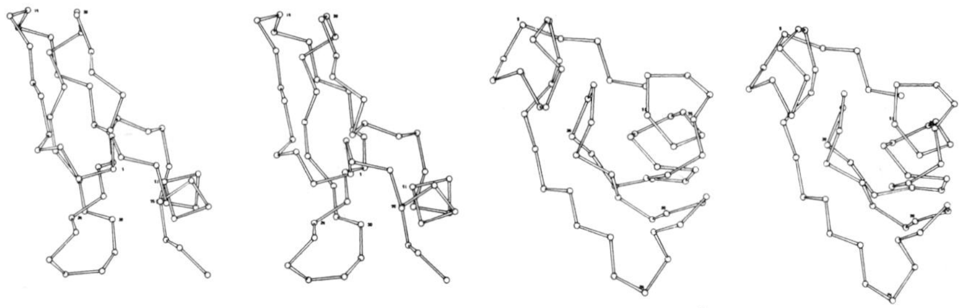

a b

Fig. 1. Stereo figures of PTI represented by connecting the C^{α} atoms of each residue (a) experimental structure[11] (b) results of folding simulation of Ala-Gly sequence with glycine residues placed in positions occupied by Asp, Asn or Gly in native protein (see text).

possible that quite specific directed interactions are operating throughout the folding pathway and that a significant fraction of the information present in the primary sequence is used in defining this pathway. While it is difficult to draw general conclusions, it is clear that crude representations of early steps in folding have biased current simulations to *incorrect* local minima.

The Role of Glycine Residues

Our calculations (see Table I) demonstrate the role of glycine residues in stabilizing folded structures. This results from their ability to assume conformations that are forbidden to other residues and which facilitate the formation of turns. The computational result that a polypeptide sequence tends to form turns at glycine positions is the first theoretical demonstration of this characteristic behavior.

In another set of experiments we used a starting conformation corresponding to that of an α-helix. As can be seen in Table I, when all residues were taken to

Table I. Conformational Energies of Various Ala-Gly Sequences

Initial Structure	Gly Position in PTI Sequence	Initial Energy	Final Energy
Extended	Asp Asn Gly	333	195
Extended	-	330	202
Helix	Asp Asn Gly	3207	52
Helix	-	3727	30
Folded	Asp Asn Gly	-	59
Folded	-	343	113

be alanine, a stable conformation of 30 kcal was obtained following minimization. This helical structure is clearly destabilized by the presence of glycine residues which raise its energy to 52 kcal, close to that of the folded conformation. Thus, the role of glycine as a helix-breaker found in statistical studies is also reflected by these calculations.

References

1. Levitt, M. & Warshel, A. (1975) *Nature* **253**, 694–698.
2. Levitt, M. (1976) *J. Mol. Biol.* **104**, 59–107.
3. Burgess, A. & Scheraga, H. A. (1975) *Proc. Nat. Acad. Sci. USA* **72**, 1221–1225.
4. Tanaka, S. & Scheraga, H. A. (1975) *Proc. Nat. Acad. Sci. USA* **72**, 3802–3806.
5. Kuntz, I. D., Crippen, G. M., Kollman, P. A. & Kimelman, D. (1976) *J. Mol. Biol.* **106**, 983–994.
6. Creighton, T. (1975) *J. Mol. Biol.* **95**, 167–199 and references therein.
7. Hagler, A. T. & Honig, B. (1977) submitted for publication.
8. Hagler, A. T., Huler, E. & Lifson, S. (1974) *J. Amer. Chem. Soc.* **96**, 5319–5327.
9. Hagler, A. T. & Lifson, S. (1974) *J. Amer. Chem. Soc.* **96**, 5327–5335.
10. Honig, B., Ray, A. & Levinthal, C. (1976) *Proc. Nat. Acad. Sci. USA* **73**, 1974–1978.
11. Huber, R., Kukla, D., Ruhlman, A. & Steigemann, W. (1971) Cold Spring Harbor Symp. Quant. Biol. **36**, 141–148.

PREDICTION OF PROTEIN SECONDARY STRUCTURE

PETER Y. CHOU and GERALD D. FASMAN,
Graduate Department of Biochemistry
Brandeis University
Waltham, Massachusetts 02154

The protein conformational predictive method of Chou and Fasman[1] based on a statistical analysis of 15 proteins has been refined using the X-ray atomic coordinates and φ, ψ dihedral angles from 29 proteins.[2] The total number of residues (N=4741) is approximately double that of the previous survey, with 38% in helices, 20% in β-sheets, and 30% in β-turns. Tetrapeptides whose C_i^α-C_{i+3}^α distances were below 7 Å, and not in helical regions were characterized as β-turns. A detailed analysis of 459 β-turns classified according to 11 bend types was presented earlier,[3] and it was shown that reverse chain folding is stabilized by anti-parallel β-sheets, as well as helix-helix and α-β interactions.

When the frequency of residues in the α-helical, β-sheets, and β-turn regions are divided by their respective average frequency, their conformational parameters are obtained: $P_\alpha = f_\alpha/\langle f_\alpha \rangle$, $P_\beta = f_\beta/\langle f_\beta \rangle$, and $P_t = f_t/\langle f_t \rangle$. These conformational potentials are shown in Table I, arranged in hierarchical order along with their assignments as formers, indifferent, and breakers of α- and β-regions as defined earlier[1]. Although most values remained fairly constant, Met showed the most dramatic change, and is now assigned as a strong α-former and a weak β-former. Other changes in assignments include Asn, Asp, and His. It should be noticed that all five charged residues are unfavorable for β-sheet formation with $P_\beta < 1.00$, while three of the charged residues Asp, His, and Arg are helical indifferent with $P_\alpha \approx 1.00$. On the other hand, helix breaking residues (Pro, Gly, Asn) are strong β-turn formers with $P_t > 1.50$, while β-formers occur infrequently in bend regions.

The positional preferences of amino acids in β-turns can be seen in Table I. The predominance of Pro in the 2nd rather than the 3rd position, Trp in the 4th but not in the 2nd position, Cys and His in the 1st but not in the 2nd position are no doubt due to stereochemical considerations which make certain residues more energetically stable at specific position of the β-turn.

To enhance further the predictive accuracy based on the P_α and P_β values derived from single residues, dipeptide and tripeptide conformational parameters, as well as helix and β-sheet boundary frequencies, were derived from 29 proteins and utilized in a computer predictive algorithm[2]. The conformational profile of thioredoxin using single residue information (P_α and P_β values of Table I) is shown in Fig. 1C. As can be seen, the helical and β-sheet peaks are in excellent agreement with the α and β regions observed from X-ray (Fig. 1B). A notable exception is the short helix 59-63 which contains two α-breakers (Asn

TABLE I
Conformational Parameters for a α-Helical, β-Sheet and β-Turn Residues in 29 Proteins

P_α			P_β			P_t		f_i		f_{i+1}		f_{i+2}		f_{i+3}	
Glu	1.51	H_α	Val	1.70	H_β	Asn	1.56	Asn	0.161	Pro	0.301	Asn	0.191	Trp	0.167
Met	1.45		Ile	1.60		Gly	1.56	Cys	0.149	Ser	0.139	Gly	0.190	Gly	0.152
Ala	1.42		Tyr	1.47		Pro	1.52	Asp	0.147	Lys	0.115	Asp	0.179	Cys	0.128
Leu	1.21		Phe	1.38	h_β	Asp	1.46	His	0.140	Asp	0.110	Ser	0.125	Tyr	0.125
Lys	1.16	h_α	Trp	1.37		Ser	1.43	Ser	0.120	Thr	0.108	Cys	0.117	Ser	0.106
Phe	1.13		Leu	1.30		Cys	1.19	Pro	0.102	Arg	0.106	Tyr	0.114	Gln	0.098
Gln	1.11		Cys	1.19		Tyr	1.14	Gly	0.102	Gln	0.098	Arg	0.099	Lys	0.095
Trp	1.08		Thr	1.19		Lys	1.01	Thr	0.086	Gly	0.085	His	0.093	Asn	0.091
Ile	1.08		Gln	1.10		Gln	0.98	Tyr	0.082	Asn	0.083	Glu	0.077	Arg	0.085
Val	1.06		Met	1.05		Thr	0.96	Trp	0.077	Met	0.082	Lys	0.072	Asp	0.081
Asp	1.01	I_α	Arg	0.93	i_β	Trp	0.96	Gln	0.074	Ala	0.076	Thr	0.065	Thr	0.079
His	1.00		Asn	0.89		Arg	0.95	Arg	0.070	Tyr	0.065	Phe	0.065	Leu	0.070
Arg	0.98	i_α	His	0.87		His	0.95	Met	0.068	Glu	0.060	Trp	0.064	Pro	0.068
Thr	0.83		Ala	0.83		Glu	0.74	Val	0.062	Cys	0.053	Gln	0.037	Phe	0.065
Ser	0.77		Ser	0.75	b_β	Ala	0.66	Leu	0.061	Val	0.048	Leu	0.036	Glu	0.064
Cys	0.70		Gly	0.75		Met	0.60	Ala	0.060	His	0.047	Ala	0.035	Ala	0.058
Tyr	0.69	b_α	Lys	0.74		Phe	0.60	Phe	0.059	Phe	0.041	Pro	0.034	Ile	0.056
Asn	0.67		Pro	0.55	B_β	Leu	0.59	Glu	0.056	Ile	0.034	Val	0.028	Met	0.055
Pro	0.57	B_α	Asp	0.54		Val	0.50	Lys	0.055	Leu	0.025	Met	0.014	His	0.054
Gly	0.57		Glu	0.37		Ile	0.47	Ile	0.043	Trp	0.013	Ile	0.013	Val	0.053

P_α, P_β, P_t are conformational parameters of helical, β-sheet and β-turns. f_i, f_{i+1}, f_{i+2}, f_{i+3} are bend frequencies in the four positions of the β-turn. H_α, H_β, etc., as defined previously (1).

59 and 63) in addition to α-breakers at 64-65 (Pro Gly). The predicted conformation of thioredoxin (34% helix, 24% β-sheet), shown in Fig. 2, agrees well with the 39% helix and 28% β-sheet found from X-ray studies[4]. Three of the β-turns 8-11, 49-52, and 67-70 were located exactly from the bend probability peaks (Fig. 1D), while two bends, 32-35 and 68-71, were localized within ± 1 residue.

The quality of prediction[1] for thioredoxin is Q_α = 77%, Q_β = 89%, and Q_t = 83% based on single residue information. However, with the aid of computer analysis, region 86-91 in thioredoxin may be predicted as β-sheet in agreement with X-ray, resulting in Q_α = 85% and Q_β = 94%. Thus, improvements of prediction are obtained by the consideration of dipeptide, tripeptide, and conformational boundary analysis[2]. Likewise the predictive accuracy for superoxide dismutase (Q_α = 47%, Q_β = 76%, Q_t = 74%) and triose phosphate isomerase (Q_α = 85%, Q_β = 81%, Q_t = 60%) obtained from the predictive rules[2] compare favorably with several other computer predictive algorithms currently in use[5].

The 3-dimensional structure of proinsulin has been proposed based on a predicted β-turn at 15-18 flanked by two helices in the C-peptide[6]. Using the bend frequencies from 29 proteins (Table I), it was shown for ten mammalian species that a high β-turn potential exists in the 12-17 region, and more importantly none outside of it[2]. While no biological role has been designated for the C-peptide, the β-turn conservation predicted in the proinsulins is probably essential for directing the proper folding of the C-peptide helices. This conformation possibly masks the receptor binding region of the hormone, thus making the precursor, proinsulin, inactive.

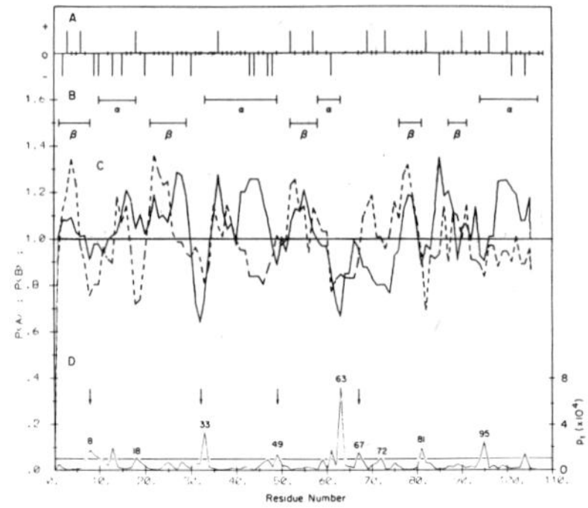

Fig. 1. Predicted conformational profile of thioredoxin. (A) Distribution plot of charged (+ / – = upward/downward bars) and hydrophobic residues (small bars on zero baseline). (B) The α and β regions from X-ray data. (C) The average P_α (–) and P_β (- - -) potential of residues i to i+3. (D) β-turn probability profile with arrows indicating X-ray results.

The importance of the β-turn in chain-folding may also be seen in the predicted structure of 7 homologous proteinase inhibitors[2]. Although only 5 invariant residues are present in addition to the 6 Cys residues (of approximately 60 residues), the β-turns are conserved, with the onset of chain reversal at residues 12, 25, 37, and 41, as shown by X-ray studies for pancreatic trypsin inhibitor. While the function of a protein depends on its unique three-dimensional topology, one may still gain significant information from secondary structural prediction. Empirical predictive algorithms can provide a useful starting conformation for energy minimization procedures, thus limiting the search for the native tertiary structure. It may also be helpful for the interpretation of X-ray diffraction data at low-resolution as was done in the case of tobacco mosaic virus[7].

The conformational parameters, P_α, P_β, and P_t (Table I) provide an expedient way of detecting regions in proteins with potential for conformational changes due to mutations or changes in solvent conditions. Applications of these parameters have suggested an $\alpha \rightarrow \beta$ transformation in the 52-57 region of *lac* repressor mutant[8] AP46 and in the β1-6 region of sickle cell hemoglobin[9]. An interesting $\beta \rightarrow \alpha$ transformation in region β1-8 of 4-Zn insulin in 6% NaCl was observed by X-ray studies[10]. This is not surprising since the β1-7 region also has α-potential, but was predicted[1] as β-sheet due to its higher β-potential $\langle P_\beta \rangle$=1.15 > $\langle P_\alpha \rangle$=1.07.

Another application of secondary structural predictions is the recognition of structural domains in homologous sequences [e.g., glutamate dehydrogenases[11], growth hormones[12], immunoglobulins[13]]. Finally, the predictive method may suggest the rational design of synthetic analogs for experimental testing to see whether conservation or changes in conformation will produce alteration or retention of hormonal or enzymatic activity [e.g., ribonuclease S-peptide[14], proinsulin C-peptide[15], secretin[16]]. Hence, the relative simplicity and accuracy

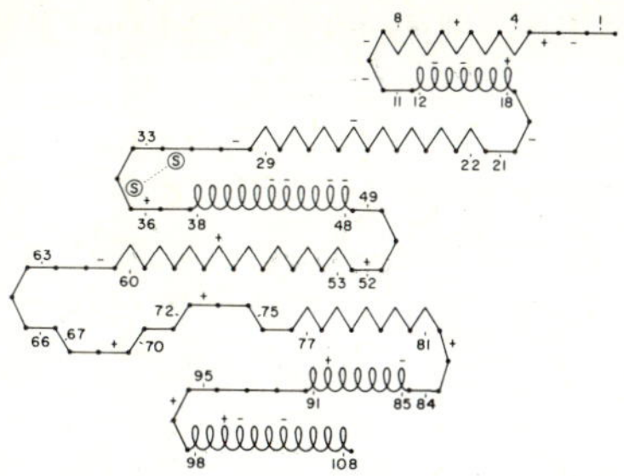

Fig. 2. Schematic diagram of the predicted secondary structure of thioredoxin. Residues are represented as helical (ℓ), β-sheet (∧), and coil (–). β-turn tetrapeptides are denoted by chain reversals. Conformational boundary residues are numbered, and the positions of charged residues are shown.

of the Chou-Fasman method have resulted in its extensive use in predicting the conformation of biological macromolecules[2]. It is hoped that the careful and judicious application of the empirical predictive method will lead to further insights on protein folding.

This work was supported by Grants from the U.S. Public Health, NSF, and American Cancer Society. Publication No. 1151 from the Department of Biochemistry, Brandeis University, Waltham, MA 02154.

References

1. Chou, P. Y. & Fasman, G. D. (1974) *Biochemistry* **13**, 222–245.
2. Chou, P. Y. & Fasman, G. D. (1977) *Adv. Enzymol.* **45**, in press.
3. Chou, P. Y. & Fasman, G. D. (1977) *J. Mol. Biol.* in press.
4. Holmgren, A., Söderberg, B. O., Eklund, H. & Bränden, C. I. (1975) *Proc. Nat. Acad. Sci. USA* **72**, 2305–2309.
5. Lenstra, J. A. (1977) *Biochim. Biophys. Acta* **491**, 333–338.
6. Snell, C. R. & Smyth, D. G. (1975) *J. Biol. Chem.* **250**, 6291–6295.
7. Durham, A. C. H. & Butler, P. J. G. (1975) *Eur. J. Biochem.* **53**, 397–404.
8. Chou, P. Y., Adler, A. J. & Fasman, G. D. (1975) *J. Mol. Biol.* **96**, 29–45.
9. Chou, P. Y. (1974) *Biochem. Biophys. Res. Commun.* **61**, 87–94.
10. Bentley, G., Dodson, E., Dodson, G., Hodgkin, D. & Mercola, D. (1976) *Nature* **261**, 166–168.
11. Wooton, J. C. (1974) *Nature* **252**, 542–546.
12. Pena, C., Stewart, J. M., Paladini, A. C., Dellacha, J. M. & Santome J. A. (1975) in *Peptides: Chemistry, Structure, and Biology,* Walter R. and Meienhofer, J., Eds., Ann Arbor Science Publishers, pp. 523–528.
13. Low, T. L. K., Liu, Y. S. V. & Putnam, F. W. (1976) *Science* **191**, 390–392.
14. Dunn, B. M. & Chaiken, J. M. (1975) *J. Mol. Biol.* **95**, 497–511.
15. Vogt, H. P., Wollmer, A., Naithani, V. K. & Zahn, H. (1976) *Hoppe-Seyler's Z. Physiol. Chem.* **357**, 107–115.
16. Fink, M. L. & Bodanszky, M. J. (1976) *J. Amer. Chem. Soc.* **98**, 974–977.

PREDICTION OF CYCLIZATION TENDENCIES OF POLYPEPTIDES FROM THE MACROCYCLIZATION EQUILIBRIA.

M. MUTTER, *Institut für Organische Chemie der Universität, Auf der Morgenstelle 18, D-7400 Tübingen, Germany,*
and P. J. FLORY, *Department of Chemistry, Stanford University, Stanford, Ca. 94305, USA*

An elaborated form of the Jacobson-Stockmayer theory of cyclization equilibria is used to calculate the cyclization constants K_x for polypeptides with various conformational features. According to this theory[1], the propensity for cyclization, expressed in the equilibrium constant K_x, is calculated by the relation

$$K_x = 2W(0)\,\Gamma_0(1)/\sigma_{cx}N_A \tag{1}$$

where $W(r)$ is the probability density function of the chain vector r; when $r = 0$, $\Gamma_0(\gamma)$ is the probability distribution of $\gamma = \cos\Delta\theta$, $\Delta\theta$ being the angle between a hypothetical bond $(3x + 1)$ and bond 1 (Fig. 1); σ_{cx} corresponds to the symmetry number of the ring, N_A is Avogadro's number. In order to form a ring, (i) the dotted hypothetical bond $3x + 1$ in Fig. 1 must coalesce with bond 1 (joining atom 0 and 1) and (ii) the direction of the bond to be formed, $3x$, must yield acceptable bond angles at atoms $3x - 1$ and 0. Bond angles and bond lengths are assigned the values used previously[2], the peptide bond being in the

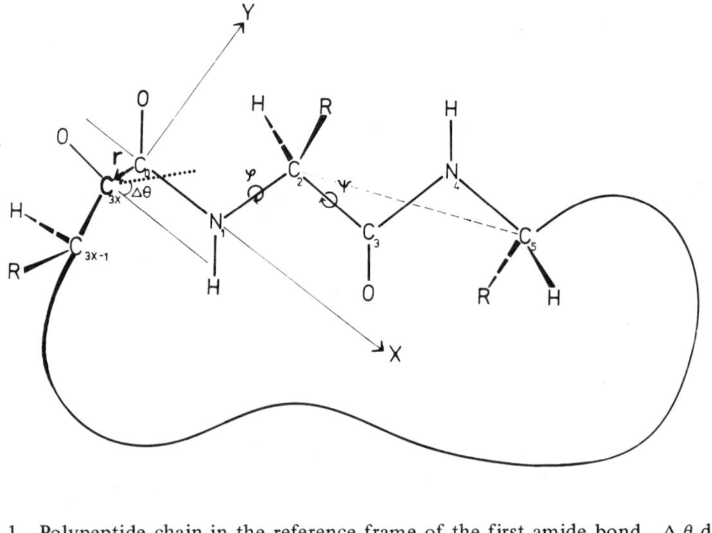

Fig. 1. Polypeptide chain in the reference frame of the first amide bond. $\Delta\theta$ denotes angle between the direction of the hypothetical $(3x + 1)^{th}$ bond and the first bond.

trans-planar configuration. Conformational energy calculations were performed using semiempirical potential functions[3]. In the case of X-Pro sequences, ECEPP[4] was used. The evaluation of $W(r)$ was obtained using three approximations[1]. First, Gaussian distribution was assumed; second, $W(0)$ was obtained in higher approximation by scalar Hermite expansion; third, $W(0)$ was estimated using a direct Monte Carlo procedure[1]. The angular correlation factor $2\,\Gamma_0(1)$ was evaluated according to the equation

$$2\,\Gamma_0(1) = \sum_{k=0}^{\infty} (2k+1)\,\langle P_k \rangle_{r=0} \qquad (2)$$

whereby $\langle P_k \rangle$ are averages of Legendre polynomials over all configurations of chains with $r = 0$[1]. The amino-acid sequences treated in this paper reflect the main structural features of naturally occurring and synthetic cyclic peptides, namely poly(L-Ala-D-Ala) (I), poly(D-Pro-L-Ala) (II), poly(Gly) (III), poly(L-Pro-L-Ala) (IV), poly(D-Ala-D-Ala-L-Ala-L-Ala) (V) and poly(L-Ala) (VI).

Results

Poly(Gly). K_x calculated on the basis of $W(0)$ in different approximations indicate that there is a departure from the Gaussian distribution throughout the investigated range. As a consequence, K_x is lowered by factors of ca. 1.7, 1.4 and 1.2 for $x = 6$, 10 and 12, respectively. The angular correlation factor, determined by $\langle P_1 \rangle_{r=0}$, is negative for all x ($2\,\Gamma_0(1) < 1$) and leads to a further depression of K_x. The relatively moderate deviations from the Gaussian distribution and a rapidly decreasing angular correlation with chain length reflect the high flexibility of this chain, as can be expected from the small values of the characteristic ratio, C_n, and the persistence vector, a (Table I).

Table I. Log K_x for the Peptide Sequences I – VI for Different Chain Lengths x, Calculated According to Eq (1). Numbers Labeled "+" Correspond to Estimated Maximal Values of K_x. In Columns 4 and 5, the Values for the Characteristic Ratio C_n and the Persistence Vector $a = \langle r \rangle$ are Given for $n = 20$.

sequence	6	10	20	C_{20}	a_{20}
I	-0.33	-1.60	-2.04	0.73	4.08
II	-1.61	-2.19	-2.55	1.41	5.9
III	-2.21	-2.20	-2.78	1.89	6.4
IV	-2.69+	-3.24+	-3.86+	3.26	10.7
V	-3.02+	-4.33+	-3.98+	3.62	12.0
VI	-3.95+	-4.19+	-4.67+	6.45	21.0

Poly(L-Ala). The values of K_x in Gaussian approximation are lower compared to poly(Gly) by factors of ca. 4 and 6 for $x = 10$ and 20, respectively, indicating a low density at $r = 0$; this is a direct consequence of the greater C_n and a (Table I). Furthermore, a strong deviation from the Gaussian distribution is indicated by the Monte Carlo approximation. Because of the extendedness of the chain, the moment method to evaluate $W(0)$ must fail in this case for short chain lengths. Given the values from the Monte Carlo estimate (Fig. 2), the depression of K_x due to departures from the Gaussian distribution amounts to factors of ca. 17, 8 and 3 for $x = 8$, 12 and 20, respectively. The Legendre polynomials for the evaluation of the angular correlation factor did not converge satisfactorily. Yet, a strong negative value of the leading term, $\langle P_1 \rangle_{r=0}$, for all x is indicated. The inclusion of the higher terms in $\langle P_k \rangle$ leads to a value of $2\,\Gamma_0(1)$ very close to zero. Whereas a quantitative evaluation of K_x is impossible, maximal values can be estimated (Table I).

Poly(D-Ala-L-Ala). Here, all three approximations of $W(0)$ gave identical results, i.e. the alternating D, L-chain obeys the Gaussian distribution for all $x \leqslant 6$. The K_x-values are considerably higher than those obtained for poly(Gly) (Fig. 2), in agreement with the very low values of C_n and a (Table I). The angular correlation assumes positive and negative values, depending on x.

Other sequences. The K_x-values of the sequences II, IV and V evaluated on the basis of $W(0)$ in higher approximation are plotted in Fig. 2. The low K_x-values of the (DDLL-) chain (V) could be expected from the large values of C_n and a (Table I). The insertion of L-Pro residues in a poly(L-Ala) chain (IV) leads to

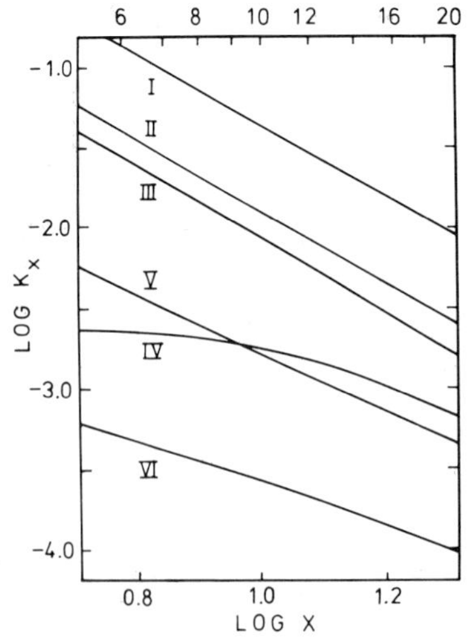

Fig. 2. Log K_x plotted against the number of residues x for the sequences I to VI (compare Table I) calculated from Equation (1) by neglect of angular correlations. $W(0)$ was estimated by Monte Carlo methods.

higher values of K_x compared to the poly(L-Ala) chain. The result for sequence II resembles that of sequence I.

Discussion

The present calculations show that the ring closure probabilities, expressed in the equilibrium constant K_x, span a very large scale for the various peptide sequences investigated here. The data of Table I lead to reliable predictions of the propensity for cyclization. The cyclization reaction should be most difficult for a poly(L-Ala) type chain which is in agreement with experimental findings[5]. At the opposite extreme, a large yield of cyclic products is predicted in the polymerization of D, L-Ala. This is confirmed by numerous successful syntheses of naturally occurring cyclic peptides[5,6]. Cyclic peptides containing Gly residues also are readily synthesized as expected from the present calculations. The general agreement between theoretical predictions and the experimental evidence indicates that K_x depends on both the density distribution $W(0)$ of the end-to-end distance r and the angular correlations.

One of the authors (M. M.) is grateful to the Deutsche Forschungsgemeinschaft for financial support.

References

1. Flory, P. J., Suter, U. W. & Mutter, M. (1976) *J. Amer. Chem. Soc.* 98, 5733–5739.

2. Flory, P. J. (1969) *Statistical Mechanics of Chain Molecules,* Interscience Publishers, New York, N. Y.

3. Brant, D. A. & Flory, P. J. (1965) *J. Amer. Chem. Soc.* 87, 2791–2800.

4. Scheraga, H. A., *Empirical Conformational Energy Program for Peptides* (ECEPP), Quantum Chemistry Program Exchange, QCPE Program Nr. 286.

5. Schröder, E. & Lübke, K. (1965) in *The Peptides* Vol. I, Academic Press, New York, N. Y., pp. 271–293.

6. Ovchinnikov, Yu. A. (1973) in *The Chemistry of Polypeptides,* Plenum Press, New York, N. Y., pp. 169–204.

ON THE USE OF CONFORMATIONAL ENERGY CALCULATIONS TO SHOW STRUCTURAL RELATIONSHIPS BETWEEN SMALL POLYPEPTIDES WHICH EXHIBIT OPIATE ACTIVITY AND THOSE WITH PITUITARY HORMONE-RELEASING ACTIVITY.

F. A. MOMANY, *Department of Chemistry, Memphis State University, Memphis, TN 38152,* C. Y. BOWERS, *Tulane Medical School, 1430 Tulane Ave., New Orleans, LA 70112,* J. K. CHANG, *Beckman Instruments, Palo Alto, CA 94304.*

Recently, experimental studies have shown that the small brain opiate acting pentapeptides, Met^5- and Leu^5-enkephalin, have hypophysiotropic activity.[1,2] A variety of analogs based on the enkephalin sequence have been studied and different analogs have demonstrated agonist or antagonist activity by a direct action on the pituitary *in vitro*.[1,2] In this paper, the conformational properties of some of these enkephalin analogs with luteinizing hormone (LH) and follicle-stimulating hormone (FSH) agonist or luteinizing hormone-releasing factor (LRF) antagonist activity are examined, and a mechanism for action at the receptor is proposed.

Conformational energy calculations have previously been carried out on the native enkephalins,[3,4] and also on native LRF.[5,6] The enkephalin conformations found by these calculations depend upon subtle changes in the end-groups and upon D- and L-amino acid substitutions in the enkephalin sequence.[4] Conformer C of Ref. 4 was described as being the most probable opiate-active conformation; now the lowest energy enkephalin conformer,[3,4] (Conformer IA of Ref. 4) is postulated to be that conformer responsible for LH/FSH activity. This postulate is further enhanced by comparing the lowest energy enkephalin conformer (see Figure 1) with the calculated structure of LRF[5,6] (see Figure 2). The orientation and relative position of the aromatic rings in these distinctly different molecules compare favorably. In both molecules the Tyr ring lies below the backbone structure, and has one ring face exposed to solvent or receptor surface. The Phe^4 ring of Met^5-enkephalin can easily take up a position equivalent to the His^2 ring of LRF (dotted lines of Figure 1), and the positively charged N-terminus of Met^5-enkephalin is oriented in a region of space close to that taken by the positively charged Arg^8 side chain of LRF. The location of the C-terminus of Met^5-enkephalin is close to that found for the <Glu^1 residue of LRF.

It has been shown *in vitro*[1,2] that Met^5-, Leu^5-, [D-Phe^2]-Met^5- and [D-Leu^2]-Met^5-enkephalin-amides in dosages of 3-100 μg/ml medium released LH and FSH

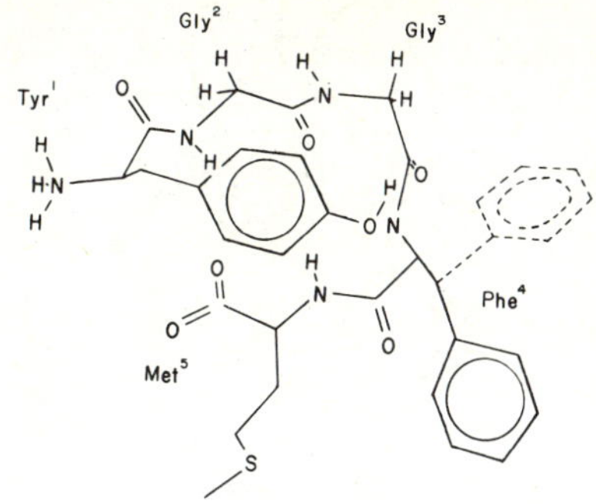

Fig. 1. Low energy conformer of Met^5-enkephalin.

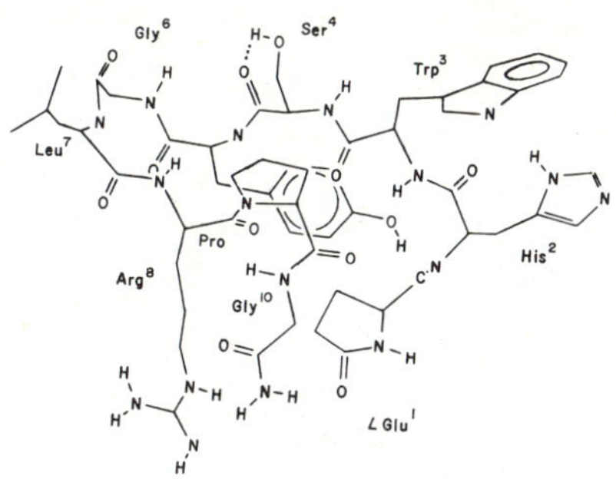

Fig. 2. Low energy conformer of luteinizing hormone-releasing factor (LRF).

and that Met^5-enkephalin, Leu^5-enkephalin, [DOPA^1]-, [Phe^1]-, [Trp^1]-Met^5-enkephalin, all inhibited the LRF response when the carboxyl terminus was free. Conformational energy calculations have shown[4] that upon amidation of the C-terminus, the population of the low energy conformers (A and C of Ref. 4) become nearly equal, thus a sufficient population of Met^5-enkephalin-NH_2 retains the conformation shown in Figure 1 to act as an agonist.

Using the structural similarities between Met^5-enkephalin and LRF, we have predicted some potentially potent analogs. The first pentapeptide sequence predicted to enhance LH release is:

$$^{+}H_3N - Tyr^1 - Ala^2 - D\text{-}Trp^3 - Phe^4 - Met^5 - NH_2 \qquad \text{(I)}$$

This sequence meets the criteria that the Gly^3 residue in the lowest-energy enkephalin structure (see Figure 1) will allow incorporation of a D-residue. Further, LRF shows a Trp residue in the equivalent region (see Figure 2). Sequence I was investigated by conformational energy calculations, and the lowest energy conformation is shown in Figure 3. Clearly, this structure mimics the conformation of Met^5-enkephalin and select regions of LRF very closely. Synthesis and testing of the above analog are being carried out. Preliminary *in*

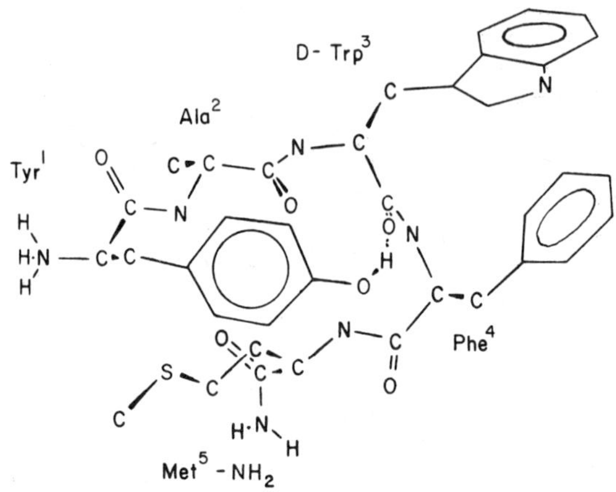

Fig. 3. Low energy conformer of [Ala^2, D-Trp^3]-Met^5-enkephalin-NH$_2$

vitro tests have at present proven inconclusive, and a report on the final results must await further study. The results of activity tests on this molecule, as well as other predicted analogs, such as:

$$^{+}H_3N - Tyr^1 - Ser^2 - D\text{-}Trp^3 - His^4 - Met^5 - NH_2 \qquad \text{(II)}$$

will be presented in complete detail elsewhere.[7] Testing of the C-terminal carboxylated analog (I) for potent antagonist activity is in progress.

A possible mechanism for the action of LRF and the active analogs studied here can now be proposed. A schematic diagram of this mechanism is shown in Figure 4. The distance, R, in Figure 4 appears to be relatively constant for all the calculated agonist and antagonist[7] conformers. Further, for every agonist, there is an amide or <Glu group situated between the charged group and the ring. We propose that the phosphate of a phosphorylated serine is moved by the action of the positive charge, exposing an area normally protected. The <Glu or amide group then moves into the exposed region, probably forming a hydrogen-bond (using the carbonyl group of the substrate). By activating this site in this way the system is turned on. The primary role of the His or Phe ring is to maintain a specific distance to the -NH_3^+ or Arg$^+$ group. This may be accomplished by fitting the ring into a specific 'pocket'. This distance criteria may explain why some enkephalin analogs with Phe^4 act as agonists, even though LRF requires the smaller His ring for *potent* agonist activity. The antagonist response of the enkephalin analogs with a free carboxyl terminus can now be explained since the COO^- group would situate itself in the same position that the phosphate had previously occupied, and thus turn off the mechanism in a similar manner as does the phosphate. Some inhibitors appear to have an aliphatic face in this central position,[7] and by the mechanism proposed here, simply do not turn the

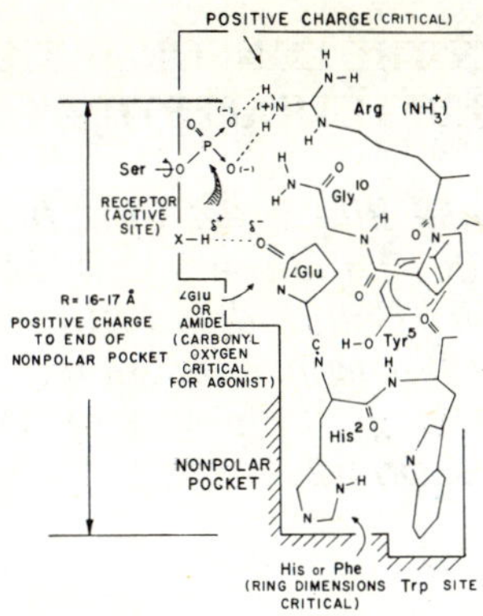

Fig. 4. Schematic mechanism for LH/FSH activity.

system on. Substitution of a carboxyl group in the 1-position in LRF does not result in an inhibitor analog. This may be due to a conformational change in the molecule, which prohibits binding at the active site.

We wish to thank the Memphis State University Faculty Research Fund for a summer grant, and the MSU Computing Center for extensive use of the computing facilities (FAM); NIH-AM18494 for support (CYB) and Dr. Albert Parlow for the NIAMDD program for the RIA reagents.

References

1. Bowers, C. Y., Fong, B. T. W. & Chang, J. K., Abstract #212, 1977 FASEB Meeting, Chicago, ILL.

2. Bowers, C. Y., Chang, J. K., Momany, F. A., Lam, Y., Knudsen, R. & Folkers, K., Abstract #000, *Endrocinology* 77 Meeting, London, England.

3. Isogai, Y., Nemethy, G. & Scheraga, H. A. (1977) *Proc. Nat. Acad. Sci. USA* 74, 414–418.

4. Momany, F. A. (1977) *Biochem. Biophys. Res. Comm.* **75**, 1098–1103.

5. Momany, F. A. (1976) *J. Amer. Chem. Soc.* 98, 2990–2996.

6. Momany, F. A. (1976) *J. Amer. Chem. Soc.* 98, 2996–3000.

7. Momany, F. A. (1977) *J. Med. Chem.* submitted.

STUDIES ON THE CONFORMATION OF CYCLIC TETRAPEPTIDES

G. MANJULA, *Molecular Biophysics Unit, Indian Institute of Science, Bangalore-560012, India* and C. RAMAKRISHNAN, *Molecular Biophysics Unit, Indian Institute of Science, Bangalore and Department of Biophysics and Theoretical Biology, University of Chicago, Chicago, Illinois 60637*

In an earlier paper,[1] the conformational studies on cyclic tetrapeptides with all the four peptide units in *trans* configuration (and non-planar) have been reported. At that time, no information regarding the nature of the peptide units in a cyclic tetrapeptide was available from any optical or X-ray studies. Subsequently, the crystal structure and nuclear magnetic resonance studies on a number of cyclic tetrapeptides have been reported.[2-10] In many of these examples,[2-8] the backbone of the cyclic tetrapeptide has been composed of alternating *cis* and *trans* peptide units. Another common feature in these cases is that one or more of the peptide units have methyl group at the N-atom (sarcosyl residue or *N*-methylated amino acid). Except in the case of *c*(Gly-Pro)$_2$, the cyclic ring has some symmetry, either strict or approximate, for its backbone. Thus, it would be appropriate to study the conformation of symmetrical cyclic tetra peptides with alternating *cis* and *trans* peptide units with a symmetry in the ring. Such a study has been carried out with two-fold and inversion symmetries and the results are presented in this paper. The symbols *t* and *c* will be used to denote the *trans* and *cis* peptide units respectively.

Method

The method used to form cyclic tetrapeptides with a two-fold symmetry is the same as that given by Ramakrishnan and Sarathy[1] except that the starting pair of peptide units is composed of linked *c* and *t* units. The inversion symmetric structures are generated by taking the midpoint of the virtual bond joining the terminal α-carbon atoms of the linked peptide units. Obviously, the symmetry condition imposes restrictions on the ranges of the parameters ϕ and ψ at the α-carbon atoms. In the present case, the geometry and the stereochemistry are applied together to arrive at conformations that are permissible.

Results

The geometrically permitted regions of the (ϕ-ψ) plane for a two-fold symmetry (designated by *d*) and the stereochemically allowed regions (permitted by extreme limits) are shown as superposed in Fig. 1(a) and 1(b). It can be seen

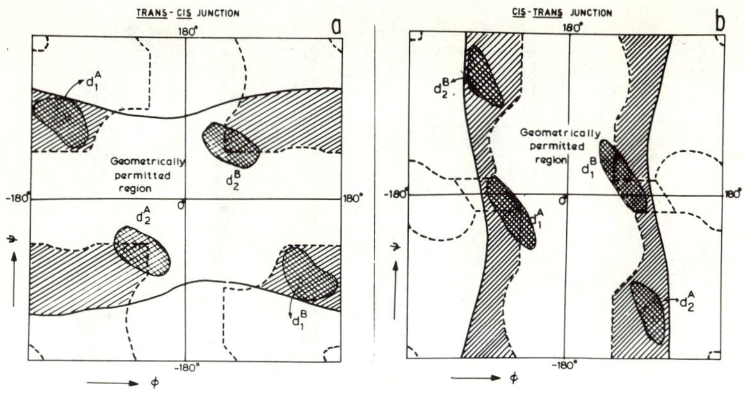

Fig. 1. Superposition of the region (————————) geometrically permitted to form a cyclic tetrapeptide of two-fold symmetric type on the corresponding contact map. (- - - - - - - - - -) outer limit for glycyl residue. The resultant permitted regions are shown as striped and the low energy regions are shown as cross-hatched. (a) at *t-c* junction; (b) at *c-t* junction.

that the superposition results in two distinct common regions and hence two distinct types of two-fold symmetric conformations, designated by d_1 and d_2. A similar procedure for the inversion symmetry (designated by i) also results in two distinct regions of conformations, designated as i_1 and i_2. The peptide units are taken to be planar. Table I gives the ϕ,ψ parameters and the energy values corresponding to the minimum energy conformation for glycyl residues.

Since many of the observed cyclic tetrapeptides have *N*-Me peptide units and glycyl residues, the effect of introducing methyl groups at nitrogen atoms is studied with a glycyl side group. The details can be found elsewhere[11] and the main results are: (a) methyl substitution at one or both of the *c* units is possible for all the four types; (b) simultaneous substitution at both the *t* units is possible

Table I. The (ϕ,ψ) and Energy Values (for Glycyl Residues) for the Minimum Energy Conformations Corresponding to Different Symmetrical Types of Cyclic Tetrapeptides

Symmetry and type	c-t junction ϕ(°)	ψ(°)	t-c junction ϕ(°)	ψ(°)	Energy kcal/mole per res.
d_1	-70	-20	-135	85	0.07
d_2	100	-125	-50	-60	0.27
i_1	85	-155	-135	65	-1.27
i_2	-70	-35	-35	-80	1.16

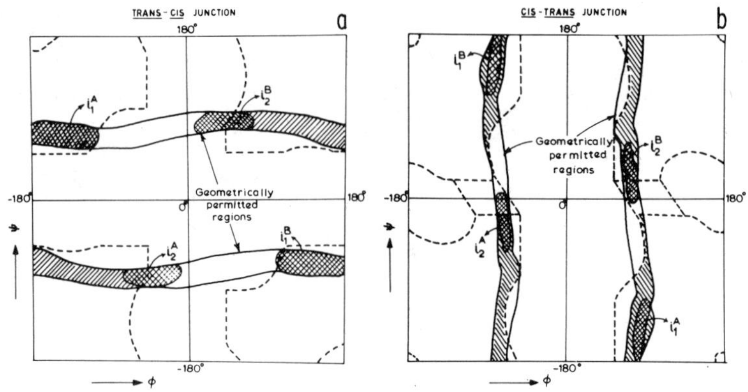

Fig. 2. Same as Figure 1 for inversion symmetric type.

for i_1 type only. i.e., cyclotetrasarcosyl can be and is of this type; (c) substitution at any of the *t*-units is not possible for types d_1 and i_2; (d) substitution at only one of the two *t* units is possible for type d_2. The conformations of cyclic tetrapeptides having mixed glycyl and sarcosyl residues, so far observed[4,8] are of i_1-type.

A similar type of study has also been made with analyl residues (L and D) at the α-carbon atoms and with peptide units, both as normal and *N*-methylated. The results cannot be generalized within narrow limits, but enable one to obtain the symmetry types that are possible for any given sequence of isomers. For example, a sequence of all L or all D amino acids can assume only the *d*-type of conformation (the symmetry being applied to backbone only). When the compound has mixed L and D residues, either *d* or *i* types become allowed, but not both. Further details will be published elsewhere. The simultaneous presence of the methyl group at the nitrogen atom and nonglycyl residue at the α-carbon atom imposes more restrictions on the types that any sequence would take. For instance, if all the residues are of the same isomeric type, then the methyl group cannot be accommodated at either *c* or *t* units. It would be worthwhile to study the conformation of such a compound, if it could be synthesized. In the case of dihydrotentoxin,[7] for the observed sequence of isomers, only *t* units and not *c* units can take in methyl groups, according to our study; this agrees with observations. A comparative study of the *c-t-c-t* sequence and all *t* sequence for *c*(Gly)$_4$ and *c*(Sar)$_4$ indicates a preference of the latter sequence with S_4 symmetry for *c*(Gly)$_4$ and the former sequence with *i* symmetry for *c*(Sar)$_4$; this is in agreement with observations.[4,9]

Contribution No. 00000 from Molecular Biophysics Unit, Indian Institute of Science, Bangalore, India.

The authors would like to thank Professor G. N. Ramachandran for his interest and suggestions. The work has been partly supported by a grant from Department of Science and Technology, India and United States Public Health Service Grant AM-11493 in Chicago.

References

1. Ramakrishnan, C. & Sarathy, K. P. (1968) *Biochim. Biophys. Acta* **168**, 402–410.
2. Konnert, J. & Karle, I. L. (1969) *J. Amer. Chem. Soc.* **91**, 4888–4892.
3. Dale, J. & Titlestad, K. (1969) *Chem. Commun.* 656–659.
4. Groth, P. (1970) *Acta Chem. Scand.* **24**, 780–790.
5. Dale, J. & Titlestad, K. (1970) *Chem. Commun.* 1403–1404.
6. Deber, C. M., Fossel, E. T. & Blout, E. R. (1974) *J. Amer. Chem. Soc.* **96**, 4015–4017.
7. Meyer, W. L., Kuyper, L. F., Lewis, R. B., Templeton, G. E. & Woodhead, S. H. (1974) *Biochem. Biophys. Res. Commun.* **56**, 234–240.
8. Declercq, J. P., Germain, G., Van Meerssche, M., Debaerdemaeker, T., Dale, J. & Titlestad, K. (1975) *Bull. Soc. Chem. Belg.* **84**, 275–287.
9. Grathwohl, C., Tum-kyi, A., Bundi, A., Schwyzer, R. & Wuthrich, K. (1975) *Helv. Chim. Acta* **58**, 415–423.
10. Flippen, J. L. & Karle, I. L. (1976) *Biopolymers* **15**, 1081–1092.
11. Manjula, G. M. (1977) Ph.D. Thesis. Indian Institute of Science.

SOLVENT ACCESSIBILITIES IN PEPTIDES

P. K. PONNUSWAMY and P. MANAVALAN, *Department of Physics, Autonomous Postgraduate Center, Tiruchirapalli 620 020 Tamilnadu, India*

Introduction

We have made a 'hardsphere-like' theoretical study on a few dipeptides and a pentapeptide to determine the probable backbone and side-chain conformations which are preferred for solvent interaction. We calculated the accessibility of a water molecule to these model solute systems and compiled solvent accessibility maps which were then used to interpret the observed conformations of amino-acid residues in protein crystals, taking into consideration the (ϕ, ψ) potential energy maps already available.

Methods

A water molecule is rolled along the envelope of the van der Waals surface of the solute molecule and the accessible surface area and a related parameter S, the static accessibility for each of the atoms are computed[1]. The number S, in turn, indicates the possible extent of the solvent interaction with that atom of the solute molecule. We considered the glycyl, alanyl and seryl dipeptides of the form CH_3-$C'O_{(1)}$-$NH_{(1)}$-$C^{\alpha}HR$-$C'O_{(2)}$-$NH_{(2)}$-CH_3 and the pentapeptide CH_3CO-Ala^1-Ala^2-Ala^3-Ala^4-$NHCH_3$. The accessibility S was computed as a function of the dihedral angles ϕ, ψ and χ.

Results and Discussion

Fig. 1 depicts the accessibility contours for the constituent atoms of the glycyl dipeptide and a water molecule at a given cross section. Fig. 2 depicts the accessibility (ϕ, ψ) map for the atom O_1 of alanyl dipeptide.

From Fig. 2 it is seen that the atom O_1 is highly accessible to a water molecule at the locations of the maximum accessibility (X) and the probable range for solvent accessibility is given by the area enclosed by the X-5 contour. Accessibility maps were thus computed for the atoms H_1, O_1, H_2 and O_2 which are the sites for solvent interaction. The accessibility potentiality of the entire dipeptide system was represented by superposition maps for the glycyl and alanyl dipeptides[2] given in Figs. 3 and 4, respectively.

These maps were constructed by superposing the O_1, H_1, O_2 and H_2 maps and shading the areas which fell within the isoaccessibility contour of X-5 so as to indicate the selective solvent accessibility of these four atoms. These maps thus

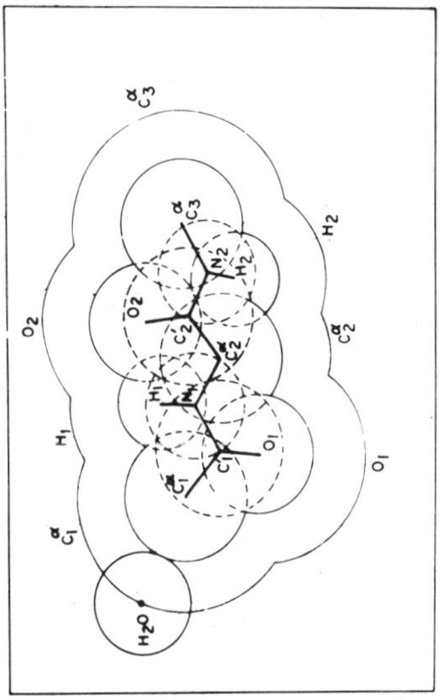

Fig. 1. van der Waals and accessibility contours of a cross section of glycyl dipeptide.

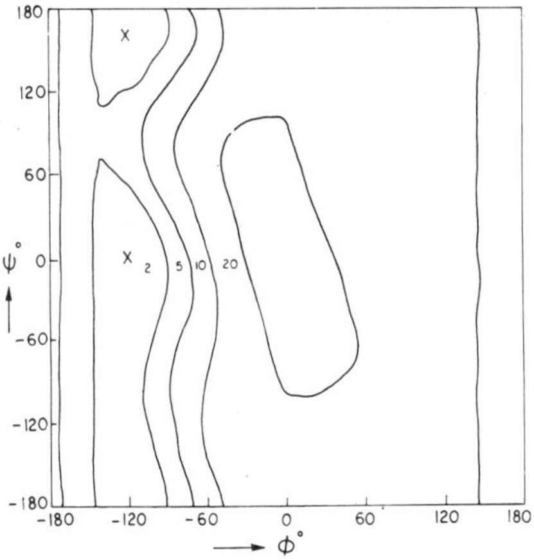

Fig. 2. Accessibility map for the atom O_1 in alanyl dipeptide.

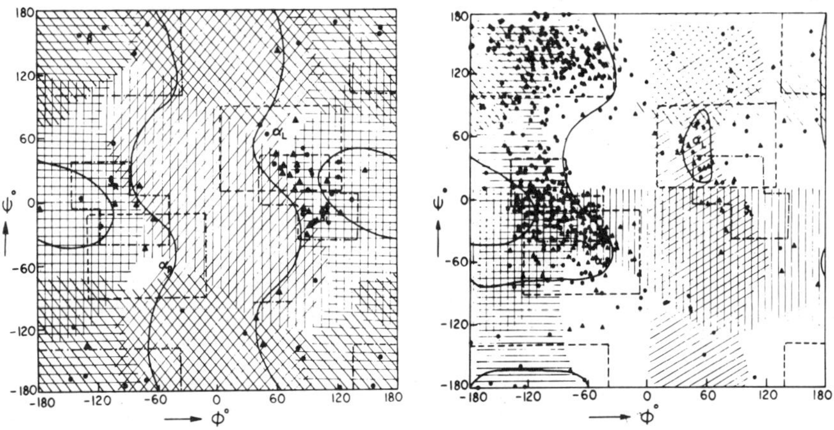

Figs. 3 & 4. Superposition maps for glycyl and alanyl dipeptides. … represent the best accessible areas for the atoms O_1, H_1, O_2 and H_2, respectively. The observed ϕ, ψ values of the middle two glycyl (Fig. 3) and nonglycyl (Fig. 4) residues of bend regions (▲) and residues in irregular regions (●) in ten proteins are plotted. Boundaries for conformational states α_R, α_L, β, ζ_R and ζ_L are marked.

help to identify areas where more than one atom could accommodate solvent molecules. The glycyl superposition map indicates that the atoms O_1, H_1 and O_2 simultaneously have higher solvent accessibilities near the ζ_R and ζ_L regions. The alanyl map indicates that the atoms O_1 and H_2 have higher accessibilities in the β-region, whereas O_1 and O_2 have higher values in the ζ_R and α_R regions. A part of α_L-region allows atoms H_1 and H_2 and a part of ζ_L region allows atoms H_1 and O_2 to have higher accessibilities. These results show that different atoms contribute individually and collectively to the total solvent accessibility of the glycyl and alanyl dipeptides depending on the backbone conformation: distorted α_R-helical or extended structural parts and also a conformation in the ζ_R region are favored for solvent interaction. This is indicated by the good agreement found when the glycyl and nonglycyl residues are plotted on the respective glycyl and alanyl superposition maps.

For seryl dipeptide the O^{γ} atom approaches different backbone atoms selectively to overlap areas for specific χ_1 values, and hence the accessibilities of the corresponding atoms are drastically decreased. From a model and the S vs χ_1 graphs, the probable hydrogen bonding situations with backbone/side chain/water/external group in the seryl dipeptide have been examined. The results suggest that the conformation with $\chi_1 \cong 60^\circ$ is favorable for internal backbone-side chain hydrogen bonding and that with $\chi_1 \cong -60^\circ$ is the most preferred conformation for hydrogen bonding interaction via a solvent molecule or an external solute group. Another interesting result is that the backbone conformations ζ_R and α_L are better than the β and α_R conformations when solvent accessibility is considered. These results are in good agreement with the observations made on the seryl residues in protein crystals.

For the penta-alanine, solvent accessibilities were computed for the ordered conformations β, C_7, α_R and α_L as well as for nine types of bend conformations[3]. The results indicate that the second and third residues for type I bend are more accessible to water molecules than the other ordered structures. Contrary to the other ordered structures, the accessibility values of the atoms H_2, O_2 and O_3 are almost equal to those found for the alanyl dipeptide, indicating that the addition of residues on either side in proper conformations does not affect the solvent accessibility of the middle two residues. This makes the type I bend structure the most probable conformation for solvent interaction. Accessibilities computed for other types of bends indicate that the bend types IV, V and VII are of open structures compared to types I, II and III; hence they may not promote hairpin-type chain reversals in proteins. Of the latter three types, only type I bend has both higher accessibilities for the middle two residues and lower values for the end residues.

References

1. Lee, B. & Richards, F. M. (1971) *J. Mol. Biol.* **55**, 379–400.
2. Manavalan, P., Ponnuswamy, P. K. & Srinivasan, A. R. (1977) *Biochem. J.* In press.
3. Lewis, P. N., Momany, F. A. & Scheraga, H. A. (1973) *Biochim. Biophys. Acta.* **303**, 211–229.

THE NON-PLANAR STRUCTURE OF SELF-ASSOCIATED *N*-METHYLACETAMIDE

F. FILLAUX, M. H. BARON, and C. de LOZE, *Laboratoire de Spectrochimie Infrarouge et Raman, 2 rue Henri Dunant, 94320, Thiais, France*

For Physicochemists studying peptides, polypeptides or proteins, an important point is whether the peptide group is planar or not. Experimental data show that significant deviation from planarity may be observed[1] (characterized by three parameters θ_N, θ_C and $\Delta\omega$).[2]

N-methylacetamide ($CH_3CONHCH_3$) is a good model of the peptide group from structural and vibrational points of view. A non-planar structure of the matrix-isolated monomer has been suggested on the basis of infrared spectra.[3] Since the peptide group is often hydrogen bonded to neighboring groups or molecules, a spectral study has been undertaken on self-associated $CH_3CONHCH_3$ (liquid and crystal).

Above 200 cm^{-1} the vibration spectra are well-known and interpreted assuming a planar structure[4-6] on the basis of X-ray data.[7] Not much data is available on the low frequency region.[8] We shall discuss results obtained in this region by Raman and neutron inelastic scattering (NIS) and show that a new interpretation of the infrared γ NH region is necessary. This leads to conclusions on the structure of *N*-methylacetamide.

Neutron Inelastic Scattering (NIS) and Raman Spectra

The NIS technique is especially relevant for proton motions of large amplitude. It is therefore a good method for identifying the torsions of the CH_3 groups. The Raman spectra are applicable for internal modes to which correspond strong polarizability variations of the molecule (including the deformations of bonds rich in π electrons) while the torsions of the CH_3 groups are generally too weak to be observed.

With both techniques the spectra have been recorded below 200 cm^{-1} at the temperature of liquid helium. Four isotopic derivatives ($CH_3CONHCH_3$, $CH_3CONDCH_3$, $CH_3CONHCD_3$, $CD_3CONHCH_3$) have been used for NIS, ten for Raman (the same as above, plus $CH_3CO^{15}NHCH_3$, $CD_3CONHCD_3$ and all the ND compounds), (manuscripts in preparation). No significant differences have been observed between the NH and ND compounds. The frequencies above 110 cm^{-1} are gathered in Table I for NH samples (below 110 cm^{-1} are located the external modes).

Table I. NIS and Raman Frequencies (cm^{-1}) of *N*-Methylacetamide

$CH_3CONHCH_3$		$CD_3CONHCH_3$		$CH_3CONHCD_3$		$CD_3CONHCD_3$	$CD_3CO^{15}NHCH_3$	Assignment
NIS	Raman	NIS	Raman	NIS	Raman	Raman	Raman	
5 K	35 K	5 K	35 K	5 K	35 K	35 K	35 K	
164 s	-			164 s				τ (C) CH_3
	164 s		168 s	153 m sh	145 s 140 w sh	146 st 142 sh	167 s	Extra Bands (τ C N)
153 s		154 s		-				
126 m	127 s	126 w	136 s 130 s	-	114 w	120 w	131 s 127 s	Extra Bands (τ C N)

The CH_3 torsions give two very strong NIS bands and are not observed in Raman spectra. The corresponding potential functions have been calculated.

$$(N)\text{-}CH_3 \quad V = (1.76/2)\,(1 \pm \cos\phi)\text{kcal/mole}$$

$$(C)\text{-}CH_3 \quad V = (2.02/2)\,(1 \pm \cos\psi)\text{kcal/mole.}$$

An interesting result concerns the "extra bands" observed in Raman and NIS. They undergo important isotopic effects upon substitutions (N)-CH_3 (N)-CD_3: they moreover shift upwards in frequency when the (C)-CH_3 is deuterated or when ^{14}N is replaced by ^{15}N. These bands can therefore not be due to external modes nor to CH_3 torsions. We assign them to an internal torsion mode around the C–N bond.

The existence of two (sometimes split) bands at a 30 cm^{-1} distance may be interpreted through a potential function with double minimum. The calculations of isotopic effects are consistent with the experimental results.

This model implies that the (N)-CH_3 group may occupy two symmetrical positions (relative to the mid-plane of the molecule) 0.27 Å apart and separated by a potential barrier of 0.6 kcal/mole. Calculations are in progress to determine precisely the normal coordinate in order to evaluate the corresponding $\Delta\omega$ angle.

The existence of this internal τ C–N mode is in contradiction with previous assignments of this mode at 220 cm^{-1} (amide VII). Besides, since the 200 cm^{-1} is also present, there are now more than (3 N - 6) internal vibrations. This difficulty will be solved through reinterpreting the infrared 400–800 cm^{-1} region.

Infrared Spectra

In this region according to Miyazawa[5] are expected a skeletal deformation around 430 cm^{-1}, the amide IV (δ CO 630 cm^{-1}), V (γ NH 725 cm^{-1}) and VI (γ CO 600 cm^{-1}) bands. In the ND compound the amide IV′ is little perturbed, the amide V shifts to the amide V′ (γ ND 505 cm^{-1}) and the amide VI disappears. This last phenomenon is unexpected and it is assumed that the amide IV′

and VI′ overlap. Using low temperature it is possible to obtain sharp bands and to analyze precisely the isotopic shifts (Fig. 1): no amide VI′ band is identified around 630 cm^{-1}.

These results become understandable assuming that the amide VI band is actually a γ NH absorption. The amide VI′ band of the liquid is then probably not distinguished from the amide V′. At low temperature these two bands shift to 536 and 568 cm^{-1}.

This new assignment brings the number of observed internal modes back to the (3 N - 6) expected.

The presence of two γ NH bands may be interpreted assuming two NH bands out of the CNC$^{\alpha}$ plane which may occupy two positions on each side of this plane (Fig. 2). Taking into account the results obtained for the τ C–N and

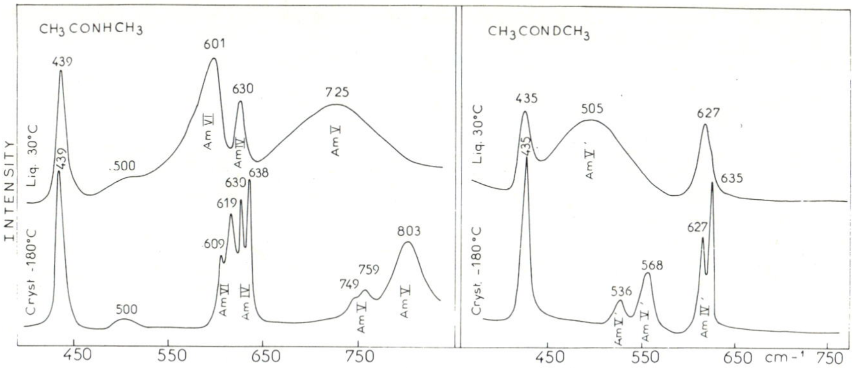

Fig. 1. Infrared spectra of *N*-methylacetamide.

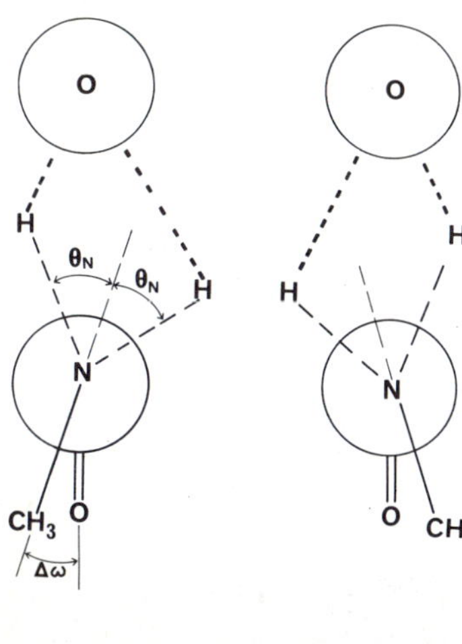

Fig. 2. Two possible positions of the hydrogen bonds in *N*-methylacetamide.

leading to an angle $\Delta\omega \neq 0$ there are two distinct hydrogen bonds which can give two absorptions (Fig. 2). The existence of two symmetrical forms ($\pm\ \Delta\omega$) leads to four positions for the NH proton, which are symmetrical two by two. The spectroscopic consequences of such a system are under investigation.

Conclusion

The evidence of a low frequency internal mode, through comparing Raman and NIS data and analyzing isotopic shifts, leads to a better assignment of the amide V–VI infrared region showing that there are two deformation modes. A new model of non-planar amide group is proposed with non-linear hydrogen bonds.

A study of the other spectral regions is in progress. The similarities between the spectra of *N*-methylacetamide and of polypeptides suggest that these results may be extended to the structure of the peptide group.

References

1. Ramachandran, G.N. & Kolaskar, A.S. (1973) *Biochim. Biophys. Acta* **303**, 385–388.
2. Ramachandran, G. N., Laksmminarayanan, A. V. & Kolaskar, A. S. (1973) *Biochim. Biophys. Acta* **303**, 8–13.
3. Fillaux, F. & de Lozé, C. (1976) *J. Chim. Phys.* **73**, 1010–1017.
4. Miyazawa, T., Shimanouchi, S. & Mizushima, S. (1958) *J. Chem. Phys.* **29**, 611–616.
5. Miyazawa, T. (1960) *J. Mol. Spectr.* **4**, 155–167.
6. Schneider, B., Horeni, A., Pivcova, H. & Honzl, J. (1965) *Coll. Czechos. Chem. Comm.* **30**, 2196–2214.
7. Katz, J. L. & Post, B. (1960) *Acta Cryst.* **13**, 624–628.
8. Itoh, K. & Shimanouchi, T. (1967) *Biopolymers* **5**, 921–930.

SOLUTION STRUCTURES OF PEPTIDES

V. T. IVANOV, *Shemyakin Institute of Bioorganic Chemistry, USSR Academy of Sciences, Moscow, USSR*

The major goal of investigators in the field of peptide research is the determination of the mode of action of their physiologically active representatives on the molecular level. Until recently, approaches to this problem were confined to synthesis of analogs of peptides and using them to set up structure-activity correlations. While one could thereby find portions of the molecule responsible for the activity, only a hazy picture of the action mechanism could be formed, because no account had been taken of the spatial relations of the active groups in the molecule, i.e., of its stereochemistry.

The early stages of such work were predominated by X-ray analysis which, among other things, yielded the geometric parameters of the peptide bond valence angles and bond lengths of the C^{α} atoms (see review Refs. 1 and 2). More recently, physicochemical studies of the peptides in solution have become predominant, embracing practically every known type of biologically active members of this class. Such studies, when applied to model peptides, allow one to follow systematically the effects of primary structure on the conformation of peptides.

Although X-ray studies of crystalline specimens far excel that of solution studies in resolution and accuracy, many essential questions can be answered only by studies in solution. A large part of the potential surface of each amino-acid residue of a peptide is energetically allowed. Even a simple compound such as AcAlaNHMe exists in solution simultaneously in many forms. With larger peptides, the combinations of such forms become practically unlimited. Their solutions usually contain a number of energetically close, interconverting forms, the equilibrium being shifted in one way or another depending upon the conditions – such as the solvent species and its ionic content, other substances interacting with the peptide, temperature, peptide concentration, etc.

The complexity of conformational studies and the limited capacities of separate physicochemical techniques have resulted in the use of several methods, each of which gives independent and complementary information on the peptide structure. Such methods include CD, ORD, IR and nmr studies. Karplus' type assumptions of dihedral angles from nmr spectra have also been useful.[3,4] The findings can serve as the basis for theoretical conformational analysis. The correctness of the resulting structures can sometimes be checked by comparison of calculated values and experimental measurements. When no unequivocal structure can be deduced, one must examine other factors contributing to structure such as intramolecular contacts between sequentially widely separated groupings with the aid of techniques such as the pH-dependencies of nmr and CD spectra, radiationless energy transfer determined from fluorescence data, EPR determination of interspin distances between two inserted iminoxyl radicals, etc. Naturally, the reliability of the conclusions could be considerably increased by simultaneous X-ray analysis of the substance in its crystalline form. For more details, see Refs. 5, 6.

The ring structure of a cyclic peptide limits its conformational mobility, simplifying the picture and considerably facilitating its study. As a result, a number of generalities have been made concerning the spatial structures of peptides, including the effects upon them of factors such as the size of the side chains, the configurations of the residues, the presence of *N*-methylamide and ester groups, the ring size etc. Cyclic penta- and hexapeptides exhibit a strong tendency to form $4 \rightarrow 1$ intramolecular hydrogen bonds (so-called β-turns) and show a high probability of *cis*-amide bond formation when the structure includes a proline or *N*-methylamide amino function.

Cyclopeptides have allowed us to develop the multiple-technique approach and to compare X-ray structures with those in solution. Cyclopeptides yield identical structures for rigid molecules in solution with those from X-ray analysis. Conformationally mobile substances generally do not show the same structures by X-ray and solution measurements, but they shed light on the conformational rearrangements accompanying changes in the solution conditions. This can be seen in the case of rigid *cyclo*[-(Val-MeIle-D-Hyiv)$_2$-] which formed a typical pleated sheet structure under all solution conditions.[7] Noteworthy are the similar values for the ϕ, ψ parameters found for the solutions and obtained a year later[8] for the crystal (Table I).

Another cyclic hexapeptide *cyclo*[-(Val-Sar)$_3$-] which cannot form a pleated sheet, assumes an entirely different set of conformations. In nonpolar media, it is preferentially in a symmetric form with three $3 \rightarrow 1$ hydrogen bonds closing 7-membered rings. Enhancement of the solvent polarity leads to conformational rearrangement and the appearance of a series of new structures with one or two

Table I. Conformational Parameters of *cyclo* [-(Val-MeIle-D-Hyiv)$_2$-]

Angle	Condition	Val	MeIle	D-Hyiv
φ°	Solution	-130±10	50±10	90±40
	Crystal	-130.1, -127.2	53.5, 57.9	104.3, 99.1
ψ°	Solution	130±30	30±30	0±10
	Crystal	104.5, 107.6	40.0, 37.5	-15.6, -9.6
ω°	Solution	~180	~180	~180
	Crystal	173.2, 169.7	168.8, 170.8	166.2, 170.1

cis-N-methylamide bonds. Their content reaches 77% in acetonitrile and 95% in dimethylsulfoxide; the conformation proposed possesses a single β-turn with the *cis*-amide bonds in the remaining part of molecule.[9] The addition of 0.5 mole of sodium thiocyanate to the acetonitrile solution of this cyclic compound shifts the equilibrium again in favor of the all *trans* form. The complex which results with Na^+ is of a "sandwich" type structure shown in Fig. 1a. The addition of more salt shifts the equilibrium towards an equimolar complex, with retention of the backbone conformation, but with the carboxyl ligands drawn somewhat into the interior (Fig. 1b). Judging from the change in the vicinal $^3J_{(H\text{-}NC^{\alpha}\text{-}H)}$ constants, the angle of twist thereby is in excess of 10°.

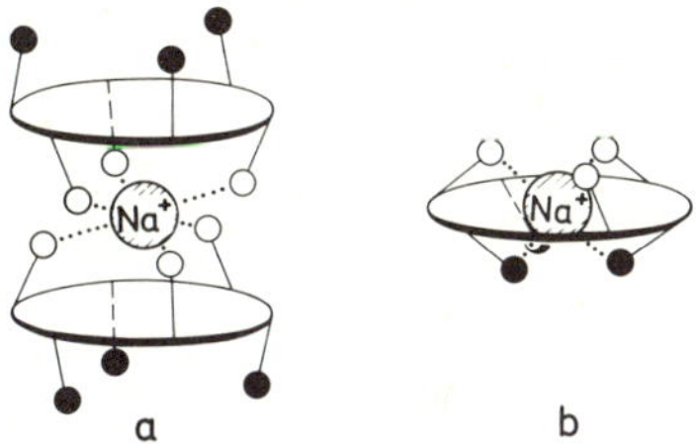

Fig. 1. Conformation of *cyclo*[-(Val-Sar)$_3$-] Na^+ complexes with (a) 2:1 and (b) 1:1 macrocycle: ion ratios (schematically) • – amide carbonyl oxygens ○ – methylamide carbonyl oxygens.

Naturally occurring cyclopeptides tend to form specific spatial structures ideally adapted to their biological function. Striking examples of such compounds are the membrane-active peptides of valinomycin and gramicidin S groups. Valinomycin is the classic representative of the group of ionophores, i.e. substances binding metal ions and transporting them across a membrane.

We investigated the structure of valinomycin and its alkali metal complexes in solution as far back as in 1969 and proposed a series of interconverting spatial forms for the free antibiotic and a rigid, "bracelet" conformation for its complexes with K^+, Rb^+ and Cs^+.[10-12] Subsequent studies were carried out by a variety of methods and procedures including IR,[13-15] Raman[16,17] and CD[13,15,18,19] spectroscopies, conformational calculations,[18-23] low angle X-ray scattering,[24] ultrasonic absorption,[13] and various types of nmr spectroscopy.[13,14,18-20,25-34] These investigations completely confirmed the basic conclusions of this initial work and added a number of novel details to the results.

In the complexed form (Fig. 2) there is a fused ring system of six β-turns in which all six NH groups are hydrogen bonded to the hydroxy acid (amide) carbonyls. The inward pointing amino acid (ester) carbonyls form a molecular cavity the size of which corresponds to the size of the non-hydrated K^+, Rb^+ or Cs^+ ions, but which is too big for the Na^+ ion. The ions situated within the center of the cavity and held there by ion-dipole interaction with the carbonyl oxygens are screened from the solvent by a system of H-bonds and the peripheral alkyl

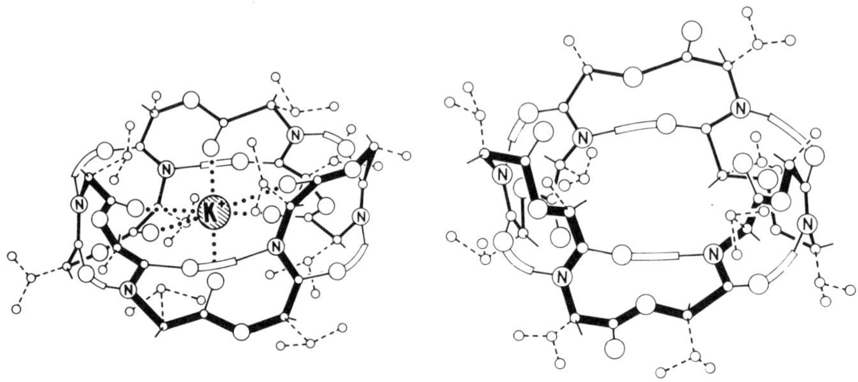

Fig. 2. Conformation of valinomycin complexed with K^+ (on the left) and in nonpolar solvents (on the right) (Form A).

groups. The torsion angles ϕ and ψ proposed for the K^+ complex in solution[12] later turned out to be very close to those found for the crystal by X-ray analysis[35] (Table II). The established structure made clear the reason for the exceptional K/Na complexing selectivity of valinomycin and also shed light on a number of its other properties, as well as on the structure-activity relationship in the extensive class of its analogs.[5]

Table II. Conformational Parameters of the Different Spatial Forms of Valinomycin

Conformational type	References	Angle	D-Val	L-Lac	L-Val	D-Hyiv
Complexes with K^+, Rb^+ and Cs^+ in solution	12	φ°	55	-70	-60	75
		ψ°	-110	-20	135	15
Crystalline valinomycin ($0.5KJ_3+KJ_5$) complex	35	φ°	59; 57; 58	-73;-66;-73	-58;-60;-58	79; 86; 79
		ψ°	-133;-129;-131	-12;-25;-16	131;133;133	3; -5; 8
Form A (non polar solvents)	36	φ°	90	-120	-80	120
		ψ°	-90	0	90	0
Monoclinic crystal, Modification B_1	37, 38	φ°	67; 54; 105	-71;-100;-165	-110;-67;-71	147;81;98
		ψ°	-136;-133;-68	-11;13;31	80;132;130	-6;3;-7
Monoclinic crystal, Modification B_2	37, 38	φ°	67; 64; 108	-71;-97;-166	-108;-66;-59	147;78;98
		ψ°	-136;-134;-68	-9; 7; 22	78;132;129	-10;7;-5
Triclinic crystal	37, 38	φ°	65; 106; 65	-98;-160-77	-102;-64;-63	-145;80;98
		ψ°	-135;-71;-134	10; 21;-7	74;131;128	-8; 2; -4
Triclinic crystal, Modification I	39	φ°	63; 60; 108	-74;-98;-164	-108;-67;-163	146;82;96
		ψ°	-134;-135;-69	-6; 14; 23	78;130;129	-11;3;-3
Triclinic crystal, Modification II	39	φ°	68; 63; 104	-75;-96;-162	-110;-65;-68	150;77;99
		ψ°	-134;-134;-71	-11; 6; 27	78;132;130	-12;8;-8

In order to understand the behavior of valinomycin in membranes, and also the dynamics of its complexing behavior, it is necessary to know the conformation in the free state as well as in the complexed form. In nonpolar media, there exists the same system of H bonds as in the complex (which we define as form A) (see Fig. 2), but the IR, CD and pmr spectra reveal a number of differences. Thus, one of the β-turns in the D-Val-Lac region is destabilized and the weakened H-bond migrates with high frequency (on the nmr time scale) about the ring.[15] In solvents of medium polarity only three β-turns are retained in the Val-D-Hyiv regions, resulting in a relatively planar (so-called propeller) conformation as shown in Fig. 3. In polar solvents, especially on heating, all intramolecular hydrogen bonds are disrupted in favor of NH solvation by the solvent. Apparently intermediate forms with 5, 4, 2 and 1 H-bonds also exist in the solution.[13]

The structures represented in Figs. 2b and 3 were deduced comparatively recently from measurements of the heteronuclear $^3J_{(^{13}C,H)}$ constants. Considering the novelty and potential of such an approach, we shall discuss the results in more detail. In the first studies that had led to proposal of form A with its six β-turns, it was pointed out that in this conformation the depsipeptide chain could fold in two ways A_1 and A_2 of opposite chirality.[10-12,25,26] On the basis of the $^3J_{(H\text{-}NC^{\alpha}\text{-}H)}$ constants, preference was given to form A_1 with *cis* orientation of the protons in the N-NC$^{\alpha}$-H fragments (ϕ for D-Val $\sim 60^\circ$ and for Val $\sim -40^\circ$). Form A_2 with *gauche* oriented protons, also assumed by the complex was rejected, because it was associated with the ϕ angles of $\sim -80^\circ$ and $\sim 90^\circ$, respectively, then considered to be incompatible with the formation of β-turns. Such an indirect approach to the structural determination was not entirely

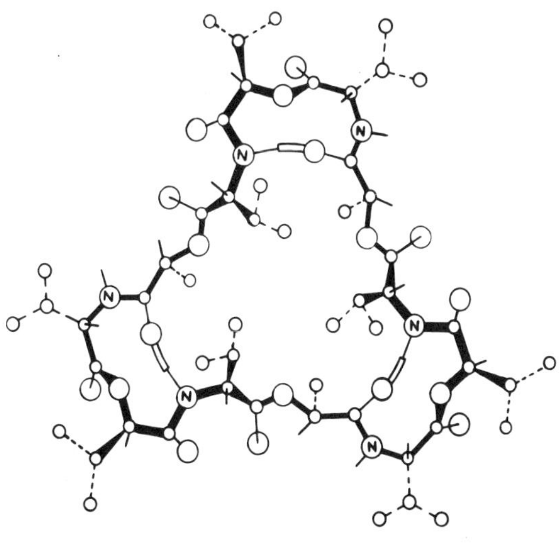

Fig. 3. Propeller conformation of valinomycin.

satisfactory – the more so when subsequent computational[40-42] and crystallographic[43-45] studies showed the actual possibility of such angles in a β-turn. Gavrilov in our laboratory measured several heteronuclear vicinal couplings (Fig. 4); for their interpretation, we made use of the $^3J_{(^{13}C'\text{-}NC^{\alpha}\text{-}H)}$ recently deduced stereochemical dependence.[46,47] As one can see from Fig. 5, form A_2

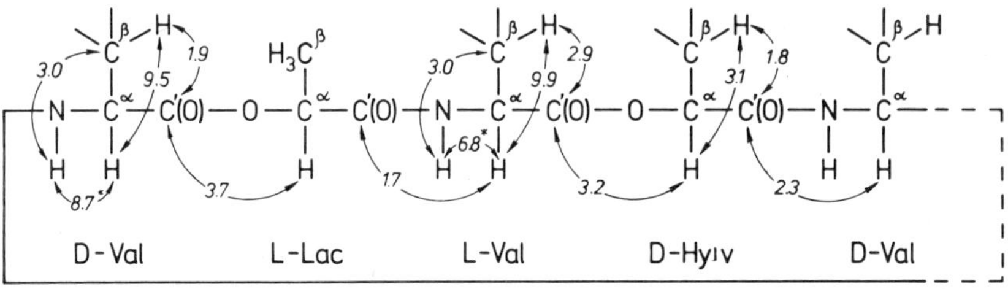

Fig. 4. Vicinal coupling constants found for valinomycin in chloroform.

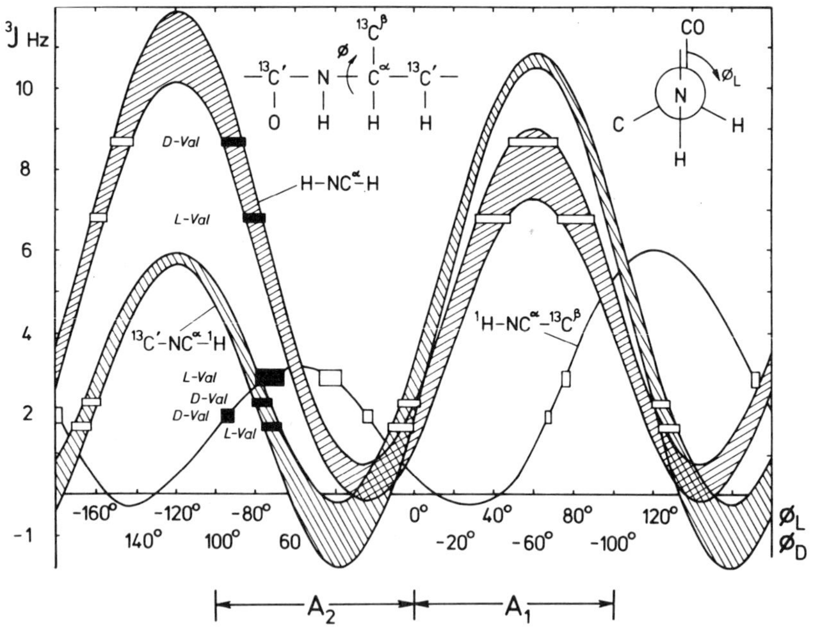

Fig. 5. Stereochemical dependences of vicinal spin-spin coupling constants describing internal rotation about the N-C$^{\alpha}$ bond. Rectangles refer to experimentally found couplings and dihedral angles related to them. The height of the rectangle corresponds to the experimental error of the spectral measurements, its width – to the range of possible dihedral angles. Black rectangles designate the self-consistent set of angles and coupling constants.

turned out to be the preferable one and, on the contrary, form A_1 must be rejected. A third constant, namely $^3J_{(^{13}C'\text{-}C^{\alpha}N\text{-}H)}$ characterizing the substituent orientation about the N-C^{α} bond (see Figs. 4 and 5) can be used to determine the angle ϕ if needed. Its stereochemical dependence, deduced theoretically,[46] requires experimental verification.

The conformation proposed has ϕ and ψ parameters (Table II) close to those of the complexed form, differing mainly in the ester carbonyl orientation; these carbonyls point inwards in the complex but more outwards in the uncomplexed valinomycin. Such conformation is ideally adapted for "sandwiching" of the metal ion. Our proposal that such sandwich complexes are actually formed[48,49] has not as yet been experimentally confirmed, although the existence of 2:1 macrocycle:cation complexes in the case of certain valinomycin analogs is beyond doubt.[50]

It is of interest to compare the findings for the solution and crystal structures of valinomycin. In Duax's[37,38] and Karle's[39] laboratories two crystalline modifications of valinomycin have been analyzed. Five different conformers were discerned in these modifications, all with very close ϕ and ψ angles (Table II); they lack an axis of symmetry, but do have a pseudo-symmetric center. As in form A_1, they have six intramolecular hydrogen bonds, but of these only four belong to the usual $4 \rightarrow 1$ type; the two remaining ones formed by ester carbonyls belong to the $5 \rightarrow 1$ type and closed 13-membered rings. The torsional angles in form A and in the crystal differ at most by 50° (Table II), the mean difference being much less. Thus, the crystalline conformation may be regarded as a distorted A_2 form, in which some of the ester carbonyls are twisted outward owing to electrostatic interactions. The energy barrier of transition between these two forms is apparently not large.

The participation of two ester carbonyls in hydrogen bonding can be clearly seen from the fact that the crystalline specimens subjected to the X-ray analysis exhibit splitting of the 1760 cm^{-1} bands in the IR[15] and Raman[16,17] spectra. Naturally no such splitting is apparent in the case of form A with its almost equivalent ester groupings in nonpolar solvents. Despite this, some authors[34] are inclined to assume that the predominant solution form is of the type shown in the crystal. That this is not so became quite obvious after Pletnev and Galitskii in our laboratory were able to decipher the crystalline structure of an analog of valinomycin, *cyclo* [-(D-Val-L-Hyiv-L-Val-D-Hyiv)$_3$-], differing from the parent compound in having α-hydroxyisovaleryl isopropyl instead of a lactyl methyl side chain. This depsipeptide forms a regular A_2 structure and its IR spectra, in both the crystalline and solution forms, closely resemble the IR spectra of valinomycin in nonpolar solvents.

Gramicidin S is one of the first biologically important peptides to have been subjected to conformational study. X-ray investigations[51,52] were unable to define the three-dimensional coordinates but were continued, bringing in ever new spatial models. In all, over 12 such models were proposed, each contradicting the other.[6] The composite application of the solution spectral methods[53,54] confirmed one of the structures to be a pleated sheet which had been proposed by Hodgkin and Oughton from X-ray studies[52] and by Schwyzer on the basis of chemical data.[55-57]

For this structure, several sets of ϕ, ψ parameters were proposed (Table III). The theoretically computed ϕ angles are in partial disagreement with the experimental $^3J_{(H\text{-}NC\alpha\text{-}H)}$ constants. For instance, a constant of 10.2 Hz (Val) corresponds to an angle of $-142° \leqslant \phi \leqslant -98°$,[3] but not to the range $-91° \leqslant \phi \leqslant -82°$ given in Table III. The CD curves of gramicidin S retain their shape almost unchanged over a wide range of solvent polarities, evidence of conformational rigidity.[53] A characteristic feature of the conformation is the presence of four strong hydrogen bonds between the valyl and leucyl CO and NH groups. The position of the β-turns is fixed by the unusual fragments Pro-D-Phe; in any other position, they are either impossible or sterically less advantageous.[60]

Among the gramicidin S analogs, a very close correlation is observed between the stability of the pleated sheet structure and the antimicrobial activity. Thus, in Izumiya's laboratory, it was shown that with change of the side chains, the antimicrobial activity diminishes, and simultaneously, the exceptionally strong negative Cotton effect of the n-π^* transition, characteristic of the "pleated sheet" structure[61] weakens. We have synthesized the following four gramicidin S analogs with retention of all or the majority of the functional and hydrophobic groups of the antibiotic, but with modification of its characteristic steric relationships and have subjected them to IR, CD and pmr analysis.[60]

cyclo [-(Val-Orn-Leu-D-Phe-Gly)$_2$-]	[Gly^5,Gly^{10}]-gramicidin S
cyclo [-(Pro-D-Phe-Leu-Orn-Val)$_2$-]	retro-gramicidin S
cyclo [-(Gly-D-Phe-Leu-Orn-Val)$_2$-]	[Gly^5,Gly^{10}]-retro-gramicidin S
cyclo [-(Val-Orn-Leu-Phe-Pro)$_2$-]	All-L-gramicidin S

Table III. Torsional Angles Proposed for Gramicidin S, (I,II) Based on Theoretical Analysis and (III) from Experimental Data

No	Parameter	Val	Orn	Leu	D-Phe	Pro	References
I	$\varphi°$	-82	-132	-143	58	-68	58
	$\psi°$	137	152	82	-116	-31	
II	$\varphi°$	-87; -91	-138; -148	-138; -159	64; 71	-75	59
	$\psi°$	122;129	126; 130	103; 116	-139;-140	-15;-18	
III	$\varphi°$	-120	-110	-120	55	-60	53
	$\psi°$	120	110	110	110	-40	

The first of these analogs obtained by 2 prolyl to 2 glycyl residue substitutions retains the general features of the gramicidin S chain, but differs by an augmented conformational mobility in the β-turn region. Two conformations exist in equilibrium: the original gramicidin S structure and a structure in which the D-Phe carbonyls assume the opposite orientation (cf. Fig. 6). The pleated sheet was not detected in the second or third analog in polar solvents, but apparently it is present in nonpolar solvents; the fourth analog showed no evidence of specific structures under any of the conditions investigated. Parallel to the structural destabilization in this series, a diminution of the antimicrobial activity was observed (see Table IV). Similar behavior was displayed by the gramicidin S depsianalogs obtained by successive replacement of two or four amide bonds by ester bonds thus depriving the antibiotic of two or four intramolecular hydrogen bonds:[63]

cyclo[-(Hyiv-Orn-Leu-D-Phe-Pro)$_2$-] [Hyiv^1, Hyiv^6]-gramicidin S

cyclo[-(Hyiv-Orn-Hyic-D-Phe-Pro)$_2$-] [Hyiv^1, Hyic^3, Hyiv^5, Hyic^{10}]-gramicidin S

Hyiv = HOCH(CHMe$_2$)COOH: Hyic = HOCH(CH$_2$CHMe$_2$)COOH

We proposed that a prerequisite for the conformation-activity correlation of the antibiotic is spatial orientation of the protonated ornithine groups and that such spatial disposition is the result of the specific structure of the peptide backbone. To test the validity of this proposal, spin labels were introduced into the ornithine side chains and the distance between them was estimated from ESR spectra. For gramicidin S itself, the following results were obtained (Fig. 7)[64] in ethanol at room temperature. The spin labels collide at a high rate, indicating the presence of conformation A (see Fig. 7). The collision rate greatly diminishes in chloroform, owing to the appearance of C conformers (Fig. 7). The average

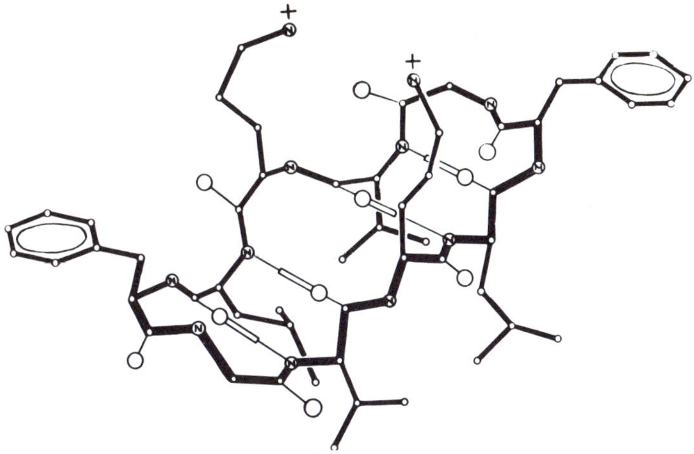

Fig. 6. Conformation of [Gly^5, Gly^{10}]-gramicidin S differing from gramicidin S in the type of β-turn.

Table IV. Antimicrobial Activity and ESR Data on Gramicidin S (GrS) Derivatives[6,62]

Compound	Minimal growth inhibiting concentration (γ/ml)				Mean distance between free radicals in the spin labelled derivatives (Å, -196°, EtOH)
	Staph. aureus 209P	Bac. subtilis	E. coli	Mycob. phlei	
GrS	0.51	0.5-1	2	2-4	12.5
Enantio-GrS	0.5-1	0.5-1	2	2-4	12.5
[Gly^5,Gly^{10}]-GrS	18	9-18	9-18	12-18	15
Enantio-[Gly^5,Gly^{10}]-GrS	18	9-18	9-18	12-18	15
"All L" GrS	60-75	9-18	18	12-25	16.5
Retro-GrS	60-75	9-18	18	18-25	18
Retro-[Gly^5,Gly^{10}]-GrS	100	12-18	9-18	37-60	18
[Hyiv^1,Hyiv^6]-GrS	9	2	18	12	15
[Hyiv^1,Hyic^5,Hyiv^6,Hyic^{10}]-GrS	>100	>100	>100	>100	>30

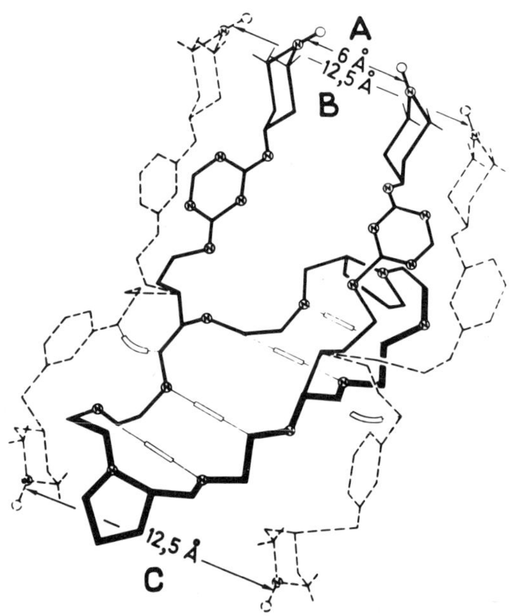

Fig. 7. Spin-labeled gramicidin S with different orientations of Orn side chains.

distance in both solvents, as measured by the dipole-dipole broadening in the spectra is ~ 12.5 Å at -196°C. With alcohol as a solvent conformation B is observed which corresponds to a spacing of 8–10 Å between the ornithine amino groups. Analysis of the data on the analogs (Table IV) showed that as the distance between the free radicals and consequently, between the ornithine amino groups increases, the antimicrobial activity, as expected, decreases.

We then noted that the distance between the positively charged amino groups of gramicidin S approximately corresponds to that between two negatively charged phosphate groups in a lipid monolayer (7-9 Å).[6] These facts suggest that an important stage in the antibiotic-target (biological membrane) interaction is the simultaneous binding of two neighboring phospholipid molecules (stage *a* in Fig. 8). Interaction with the protein constituents of biomembranes seems less probable, if one takes into account that the enantiomeric pairs given in Table IV display the same antimicrobial potency.

Subsequent studies on lipid mono- and bilayers[65] revealed a highly surface active lipid-gramicidin complex with a 2:1 molar ratio forms at the water-lipid interface. The complex inserts itself into the membrane and, judging from the area per complexed molecule, gramicidin S undergoes spatial reorganization, assuming a more compact form than the pleated sheet structure, possibly that of a mixed α,β form (stage *b* in Fig. 8). The lipid regions then become surrounded by lipid-gramicidin complexes in a mosaic pattern (stage *c*) and the membrane, thus divided into separate blocks, passes into colloidal solution (stage *d*) as already had been observed in the interaction of gramicidin S with phospholipid liposomes.[66] We regard this process as a possible mechanism of action of the antibiotic in biological systems.

The conformational study of linear peptides in solution has met with more modest success. If one disregards regular oligopeptides and short tri-tetrapeptides, the only molecule whose conformational equilibrium has been described in detail is bradykinin.[67-72] The difficulties of solution studies are enhanced in the linear peptides by their aforementioned conformational lability. The possible diminution of such mobility by lowering the solution temperature is of interest.

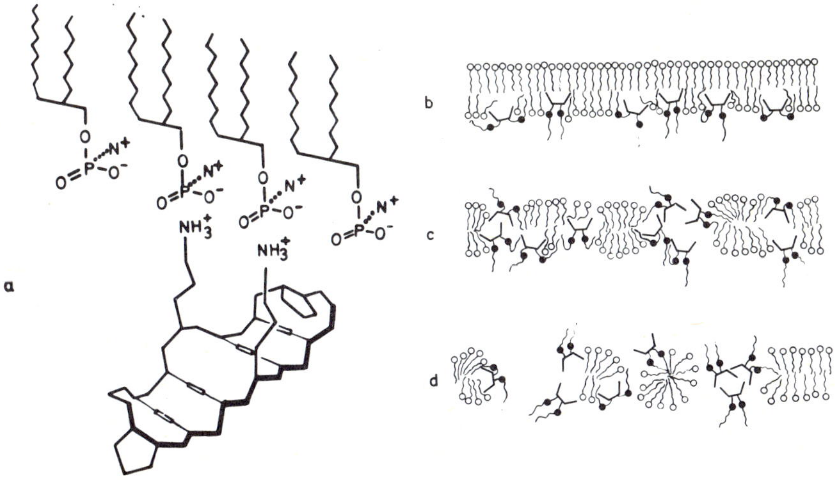

Fig. 8. Consecutive stages of the gramicidin S interaction with a lecithin membrane.

When studying the CD spectra of bradykinin, we noted that in all solvents the positive dichroic band at 220 nm increased in intensity on cooling (Fig. 10). We ascribed this band at first to the aromatic $A_{lg} \rightarrow B_{lu}$ transition, considering the increased intensity as due to diminished rotation of the phenyl group.[67,68] However, later we discovered references in the literature to similar temperature dependences in many other peptides – oxytocin and its analogs,[73,74] [Val^5]-angiotensin II and [Ile^5,Ala^8]-angiotensin II,[75] desmosin-containing elastin fragments,[76] β-melanocyte stimulating hormone, and a number of β-lipotropin fragments,[77-78] and even in polypeptides devoid of aromatic residues such as poly(L-Glu), poly(L-Asp) and poly(L-Lys) with ionized side chains,[79] (Ala-Gly-Gly)$_n$,[80] (Ala-Ala-Gly)$_n$[81] and poly(L-proline) II.[79]

In order to define the extent of this phenomenon and ascertain its source, we investigated the CD curves of the simple peptide model AcAlaNHMe, whose n–π* band also displayed strong temperature dependence (Fig. 9).[72] An earlier detailed study had shown that at room temperature the set of conformers of this

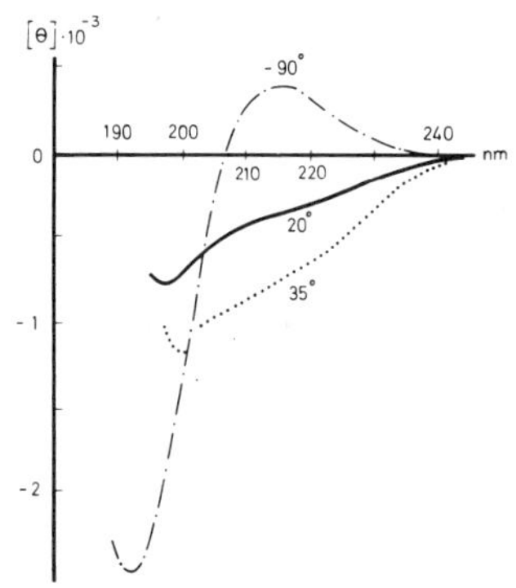

Fig. 9. CD spectra of AcAlaNHMe in ethanol.

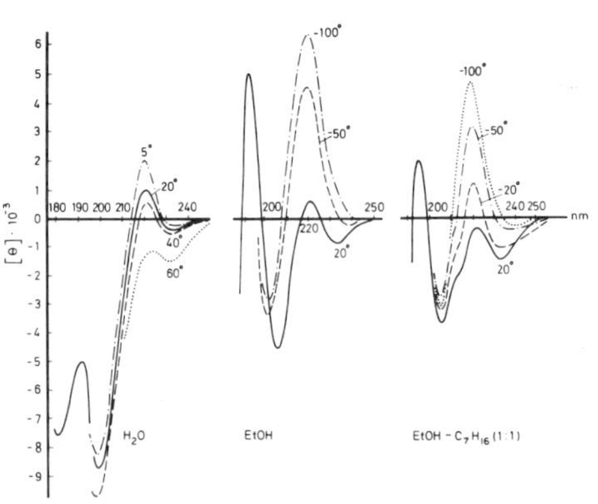

Fig. 10. CD spectra of bradykinin.

compound is "spread" over the potential surface, whereas on cooling one should expect its "concentration" near the minimum for extended structures.[82] It is to such type of conformational "concentrating" that we ascribe the temperature dependence of the CD curves in the other aforementioned cases, including that of bradykinin. This not only provides a novel interpretation of the bradykinin CD curves (assignment of the effect at the 220 nm band to the peptide rather than aromatic chromophores), but also indicates marked simplification of the conformational equilibria of linear peptides when their solutions are cooled to -50° – -100°C.

References

1. Marsh, R. E. & Donohue, J. (1967) *Adv. Prot. Chem.* **22**, 235–256.

2. Corey, R. B., Marsh, R. E. (1968) in *Progress in the Chemistry of Organic Natural Products*, vol. 26, Zechmeister, L., Ed., Springer-Verlag, Wien-New York, pp 1–47.

3. Bystrov, V. F., Ivanov, V. T., Portnova, S. L., Balashova, T. A. & Ovchinnikov, Yu. A. (1973) *Tetrahedron* **29**, 873–877.

4. Kopple, K. D., Wiley, G. R. & Tauke, R. (1973) *Biopolymers* **12**, 627–636.

5. Ovchinnikov, Yu. A., Ivanov, V. T. & Shkrob, A. M. (1974) *Membrane Active Complexones*, North-Holland/American Elsevier, Amsterdam.

6. Ovchinnikov, Yu. A. & Ivanov, V. T. (1975) *Tetrahedron* **31**, 2177–2209.

7. Ivanov, V. T., Lavrinovich, I. A., Portnova, S. L., Spassov, S. L., Meshcheryakova, E. A., Arkhipova, S. F., Senyavina, L. B. & Ovchinnikov, Yu. A. (1975) *Bioorgan. Khim.* **1**, 25–32.

8. Shishova, T. G., Andrianov, V. I., Simonov, V. I., Ivanov, V. T. & Ovchinnikov, Yu. A. (1976) *Bioorgan. Khim.* **2**, 1597–1605.

9. Ivanov, V. T., Lavrinovich, I. A., Portnova, S. L., Lapshin, V. V., Kostetsky, P. V. & Ovchinnikov, Yu. A. (1974) *Irv. Akad. Nauk SSSR. Ser. Khim.*, 2320–2328.

10. Ivanov, V. T., Laine, I. A., Abdullaev, N. D., Senyavina, L. B., Popov, E. M., Ovchinnikov, Yu. A. & Shemyakin, M. M. (1969) *Biochem. Biophys. Res. Commun.* **34**, 803–811.

11. Shemyakin, M. M., Ovchinnikov, Yu. A., Ivanov, V. T., Antonov, V. K., Vinogradova, E. I., Shkrob, A. M., Malenkov, G. G., Evstratov, A. V., Ryabova, I. D., Laine, I. A. & Malnik, E. I. (1969) *J. Membrane Biol.* **1**, 402–410.

12. Ivanov, V. T., Laine, I. A., Abdullaev, N. D., Pletnev, V. Z., Lipkind, G. M., Arkhipova, S. F., Senyavina, L. B., Meshcheryakova, E. A., Popov, E. M., Bystrov, V. F. & Ovchinnikov, Yu. A. (1971) *Khim. Prirodn. Soed.*, 221–227.

13. Grell, E., Funk, T. & Eggers, F. (1975) in *Membranes*, vol. 3, Eisenman, G., Marcel Dekker, New York, pp 2–126.

14. Von Dreele, P. H. & Stenhouse, I. A. (1974) *J. Amer. Chem. Soc.* **96**, 7546–7549.

15. Ovchinnikov, Yu. A. & Ivanov, V. T. (1974) *Tetrahedron* **30**, 1871–1890.

16. Asher, I. M., Rothschild, K. Y. & Stanley, H. E. (1974) *J. Mol. Biol.* **89**, 205–222.

17. Asher, I. M., Rothschild, K. Y., Anastassakis, E. & Stanley, H. E. (1977) *J. Amer. Chem. Soc.* **99**, 2032–2039.

18. Patel, D. J. & Tonelli, A. E. (1973) *Biochemistry* **12**, 486–496.

19. Davies, D. G. & Tosteson, D. C. (1975) *Biochemistry* **14**, 3962–3969.

20. Mayers, D. F. & Urry, D. W. (1972) *J. Amer. Chem. Soc.* **94**, 77–81.

21. Max, N. L. (1973) *Biopolymers* **12**, 1565–1574.

22. Sundaram, K. (1974) *Int. J. Quant. Chem.* **8**, 565–583.

23. Maigret, B. & Pullman, B. (1975) *Theor. Chim. Acta* **37**, 17–36.

24. Krigbaum, W. R., Kuegler, F. R. & Oelschlager, H. (1972) *Biochemistry* **11**, 4548–4551.

25. Ohnishi, M. & Urry, D. W. (1969) *Biochem. Biophys. Res. Commun.* **36**, 194–202.

26. Ohnishi, M. & Urry, D. W. (1970) *Science* **168**, 1091–1092.

27. Bystrov, V. F., Ivanov, V. T., Koz'min, S. A., Mikhaleva, I. I., Khalilulina, K. Kh., Ovchinnikov, Yu. A., Fedin, E. I. & Petrovskii, P. V. (1972) *FEBS Lett.* **21**, 34–38.

28. Ohnishi, M., Fedarko, M. C., Baldeschwieler, Y. D. & Johnson, L. F. (1972) *Biochem. Biophys. Res. Commun.* **46**, 312–320.

29. Patel, D. J. (1973) *Biochemistry* **12**, 496–501.

30. Urry, D. W., Kumar, N. G. (1974) *Biochemistry* **13**, 1829–1831.

31. Servis, K. L. & Patel, D. J. (1975) *Tetrahedron* **31**, 1359–1362.

32. Pitner, T. P., Walter, R. & Glickson, J. D. (1976) *Biochem. Biophys. Res. Commun.* **70**, 746–751.

33. Glickson, J. D., Gordon, S. L., Pitner, T. P., Agresti, D. G. & Walter, R. (1976) *Biochemistry* **15**, 5721–5729.

34. Davies, D. B. & Khaled, M. A. (1976) *J. Chem. Soc. Perkin II*, 1327–1334.

35. Neupert-Laves, K. & Dobler, M. (1975) *Helv. Chim. Acta* **58**, 432–442.

36. Bystrov, V. F., Gavrilov, Yu. D., Ivanov, V. T. & Ovchinnikov, Yu. A. (1977) *Eur. J. Biochem.*, in press.

37. Duax, W. L., Hauptman, H., Weeks, C. M. & Norton, D. A. (1972) *Science* **176**, 911–913.

38. Smith, G. D., Duax, W. L., Langs, D. A., De Titta, G. T., Edmond, J. W., Roher, D. C. & Weeks, C. M. (1975) *J. Amer. Chem. Soc.* **97**, 7242–7248.

39. Karle, I. L. (1975) *J. Amer. Chem. Soc.* **97**, 4379–4386.

40. Chandrasekaran, R., Lakshminarayanan, A. V., Pandya, V. V. & Ramachandran, G. N. (1973) *Biochim. Biophys. Acta* **303**, 14–27.

41. Lewis, P. N., Momany, F. A. & Scheraga, H. A. (1973) *Biochim. Biophys. Acta* **303**, 211–229.

42. Maigret, B. & Pullman, B. (1974) *Theor. Chim. Acta* **35**, 113–128.

43. Burgess, A. W., Ponnuswamy, P. K. & Scheraga, H. A. (1974) *Israel J. Chem.* **12**, 239–286.

44. Karle, I. L. (1974) *Biochemistry* **13**, 2155–2162.

45. Karle, I. L. (1974) *J. Amer. Chem. Soc.* **96**, 4000–4006.

46. Solkan, V. N. & Bystrov, V. F. (1974) *Izv. Akad. Nauk SSSR. Ser. Khim.*, 1308–1313.

47. Bystrov, V. F., Gavrilov, Yu. D. & Solkan, V. N. (1975) *J. Magn. Res.* **19**, 123–129.

48. Ivanov, V. T., Fonina, L. A., Uvarova, N. N., Kox'min, S. A., Karapatnikskaya, T. B., Chekhlayeva, N. M., Balashova, T. A. & Ovchinnikov, Yu. A. (1975) in *Peptide Chemistry, Structure and Biology*, Walter, R. & Meienhofer, J., Eds., Ann Arbor Sci. Publ., Ann Arbor, pp 195–201.

49. Ivanov, V. T. (1975) *Ann. N. Y. Acad. Sci.* **264**, 221–242.

50. Fonina, L. A., Savelov, I. S., Avotina, G. Ya., Ivanov, V. T. & Ovchinnikov, Yu. A. (1976) in *Peptides 1976*, Loffet, A., Ed., Editions de l'Universite de Bruxelles, Bruxelles, pp 635–640.

51. Schmidt, G. M., Hodgkin, D. C. & Oughton, B. M. (1957) *Biochem. J.* **65**, 744–750.

52. Hodgkin, D. C. & Oughton, B. M. (1957) *Biochem. J.* **65**, 752–756.

53. Ovchinnikov, Yu. A., Ivanov, V. T., Bystrov, V. F., Miroshnikov, A. I., Shepel, E. N., Abdullaev, N. D., Efremov, E. S. & Senyavina, L. B. (1970) *Biochem. Biophys. Res. Commun.* **39**, 217–225.

54. Stern, A., Gibbons, W. A. & Craig, L. C. (1968) *Proc. Nat. Acad. Sci. USA* **61**, 734–741.

55. Schwyzer, R., Sieber, P. & Gorup, B. (1958) *Chimia* **12**, 90–91.

56. Schwyzer, R. (1958) in *CIBA Foundation Symposium on Amino Acids and Peptides with Antimetabolic Activity*, pp 171–184.

57. Schwyzer, R. (1959) *Record Chem. Progr.* **20**, 147–167.

58. De Santis, P. & Liquori, A. M. (1971) *Biopolymers* **10**, 699–710.

59. Dygert, M., Go, N. & Scheraga, H. A. (1975) *Macromolecules* **8**, 750–761.

60. Miroshnikov, A. I., Snezhkova, L. G., Sychov, S. V., Chervin, I. I., Senyavina, L. B., Ivanov, V. T. & Ovchinnikov, Yu. A. (1977) *Bioorgan. Khim.* **3**, 180–191.

61. Kato, T., Waki, M., Matsuura, S. & Izumiya, N. (1970) *J. Biochem. (Japan)* **68**, 751–753.

62. Snezhkova, L. G., Shepel, E. N., Ryabova, I. D., Miroshnikov, A. I., Ivanov, B. T. & Ovchinnikov, Yu. A. (1975) *Bioorgan. Khim.* **1**, 347–358.

63. Zhyze, A. L., Kogan, G. A., Krit, N. A., Andronova, T. M., Filatova, M. P., Senyavina, L. B., Meshcheryakova, E. A., Ryabova, I. D., Ravdel, G. A. & Shukina, L. A. (1974) *Mol. Biol. (USSR)* **8**, 84–90.

64. Ivanov, V. T., Miroshnikov, A. I., Snezhkova, L. G., Ovchinnikov, Yu. A., Kulikov, A. V. & Likhtenshtein, G. I. (1973) *Khim. Prirodn. Soed.*, 91–98.

65. Semenov, S. N., Melnik, E. I., Snezhkova, L. G., Miroshnikov, A. I. & Ivanov, V. T. (1977) *Bioorgan. Khim.* **3**, in press.

66. Pache, W., Chapman, D. (1972) *Biochim. Biophys. Acta* **255**, 348–357.

67. Ivanov, V. T., Filatova, M. P., Reissman, Z., Reutova, T. O., Efremov, E. S., Pashkov, V. S., Galaktionov, S. G., Grigoryan, G. L. & Ovchinnikov, Yu. A. (1975) in *Peptides: Chemistry, Structure and Biology*, Walter, R. & Meienhofer, J., Eds., Ann Arbor Science Publ., Ann Arbor, p. 151–157.

68. Ivanov, V. T., Filatova, M. P., Reissman, Z., Reutova, T. O., Kogan, G. A., Efremov, E. S., Pashkov, V. S., Galaktionov, S. G., Grigoryan, G. L. & Bystrov, V. F. (1977) *Bioorgan. Khim.*, in press.

69. Efremov, E. S., Filatova, M. P., Reutova, T. O., Stepanova, L. N., Reissman, Z. & Ivanov, V. T. (1977) *Bioorgan. Khim.* **3**, in press.

70. Filatova, M. P., Reissman, Z., Reutova, T. O., Ivanov, V. T., Grigoryan, G. L., Shapiro, A. M. & Rosantsev, E. G. (1977) *Bioorgan. Khim.* **3**, in press.

71. Galaktionov, S. G., Sherman, S. A., Shenderovich, M. D., Nikiforovich, G. V. & Leonova, V. I. (1977) *Bioorgan. Khim.* **3**, in press.

72. Ivanov, V. T., Filatova, M. P., Reissman, Z., Reutova, T. O. & Chekhlyaeva, N. M. (1977) *Bioorgan. Khim.*, in press.

73. Urry, D. W., Quadrifoglio, F., Walter, R. & Schwartz, I. L. (1968) *Proc. Nat. Acad. Sci. USA* **60**, 967–974.

74. Fric, I., Kodicek, M., Jost, K. & Blaha, K. (1974) *Collection* **39**, 1271–1289.

75. Fermandjian, S., Morgat, Y. L., Fromageot, P. (1971) *Eur. J. Biochem.* **24**, 252–258.

76. Foster, J. A., Bruenger, E., Rubin, L., Imberman, M., Kagan, H., Mecham, R. & Franzbau, C. (1976) *Biopolymers* **15**, 833–841.

77. Makarov, A. A., Esipova, N. G., Pankov, Yu. A., Lobachev, V. M. & Grishkovskii, B. A. (1976) *Biochem. Biophys. Res. Commun.* **67**, 1378–1383.

78. Makarov, A. A., Esipova, N. G., Pankov, Yu. A., Lobachev, V. M. & Grishkovskii, B. A. (1976) *Biophisika* **21**, 233–238.

79. Tiffany, M. L. & Krimm, S. (1972) *Biopolymers* **11**, 2309–2316.

80. Rippon, W. B. & Walton, A. G. (1971) *Biopolymers* **10**, 1207–1212.

81. Doyle, B. B., Traub, W., Lorenzi, G. P., Brown, F. R. & Blout, E. R. (1970) *J. Mol. Biol.* **51**, 47–59.

82. Ivanov, V. T., Kostetsky, P. V., Meshcheryakova, E. A., Efremov, E. S., Popov, E. M. & Ovchinnikov, Yu. A. (1973) *Khim. Prirodn. Soed.*, 363–378.

AN APPROACH TO THE UNEQUIVOCAL DETERMINATION OF PEPTIDE CONFORMATIONS

DAVID COWBURN, ALAN J. FISCHMAN, DAVID H. LIVE, WILLIAM C. AGOSTA, *The Rockefeller University, New York, New York, 10021,* and HERMAN R. WYSSBROD, *The Rockefeller University* and *Department of Physiology and Biophysics, Mount Sinai Medical and Graduate Schools of the City University of New York, New York, New York, 10029*

Substantial efforts have been devoted over the last few years to the study of the relationships between the conformations of peptide hormones and their biological activities. Such studies have often built upon observations of the effects that altering the primary structure have on the biological activities. Although conformational studies have shed further light on the primary structure-activity relationship, it seems fair to say that the number of major new insights resulting from conformational studies has been somewhat limited. It is possible that one of the reasons for this relatively limited success is that time-dependent features of the peptide conformation, in which the structure assumes several stereoisomeric conformers, have been considered by only a few investigators.[1]

In attempting to understand these time-dependent or dynamic conformations more fully, we have concentrated our attention on determining the distribution of rotamers about individual torsion angles, using as our principal technique the measurement of vicinal homo- and heteronuclear coupling constants in a series of specifically designed and synthesized isotopic isomers of oxytocin. This peptide hormone is considered to have a relatively flexible conformation in aqueous solution, but the number of conformers contributing to the overall dynamic conformation is probably small.[2,3]

The isotopic isomers synthesized so far (isomers *1*-*5*, Fig. 1) contain substitutions of deuterium necessary for the simplification of coupled spin systems or for the removal of proton resonances at similar frequencies or for stereospecific assignment, and also enrichments of ^{13}C and ^{15}N at specific sites for the direct observation of vicinal heteronuclear coupling constants by ^{1}H nmr. As an example, by comparison with studies of the free amino acid[4] and measurement from isomers *1, 4, 5* the relative distribution among rotamers about the C^{α}-C^{β} bond (χ^1) of the leucyl residue can be precisely described as [0.73 (χ^1 = -60°, state I); 0.22 (χ^1 = 180°, state II); 0.04 ($\chi^1 \doteq$ 60°, state III)] in oxytocin, and (0.65, I; 0.27, II; 0.08, III) in the free tripeptide tail (L-prolyl-L-leucyl-glycyl amide). Thus the presence or absence of the oxytocin ring perturbs the rotation of this side chain only modestly (less than 0.5 kcal/mole). This suggests that if the tripeptide tail does interact non-covalently with the ring in oxytocin, then the leucyl side chain is probably not involved.

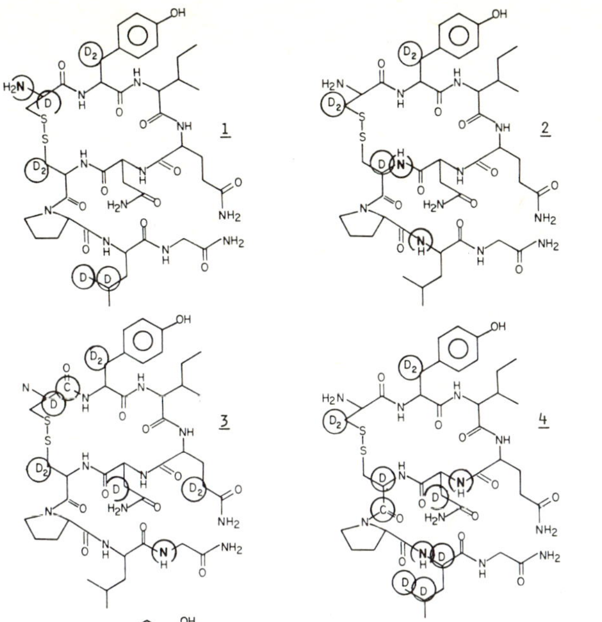

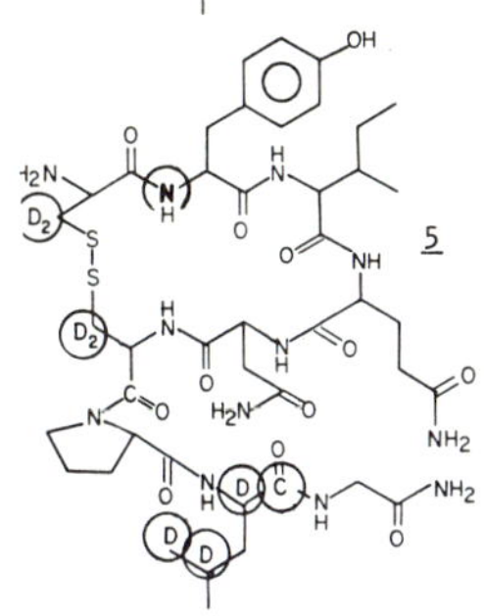

Fig. 1. Synthetic isotopic isomers of oxytocin incorporating, at the circled positions, deuterium (>99%), ^{15}N (>95%), and ^{13}C (>90%). Peptides were made by the solid-phase method on benzhydrylamine resin.

In Fig. 2, the 220-MHz spectra of the β-proton resonances of the half-cystyls 1 and 6 and tyrosyl 2 are illustrated in unsubstituted oxytocin (A) and the five isomers (I-*1* - I-*5*) along with a simulated spectrum of unsubstituted oxytocin in line broadened (B) and stick form (C). This figure demonstrates the simplifications available from the synthesis of isotopic isomers. It should be noted that extraction of the smaller vicinal coupling constants often requires line-fitting or resolution-enhancement techniques. The combination of ^{1}H-^{1}H, ^{15}N-^{1}H, and ^{13}C-^{1}H vicinal coupling constants so obtained for the two half-cystyl residues are not consistent with any set of rotamers at χ^1 of -60°, 60° and 180°. The constants are close to those expected for fixed torsion angles ($\chi_1^1 = -120°$; $\chi_6^1 = +120°$), eclipsed conformations which had been suggested by previous nmr studies of unsubstituted oxytocin.[2,5] However, it is probable that substantial libration occurs, given the magnitude of the temperature dependencies of coupling constants[2] and line widths.

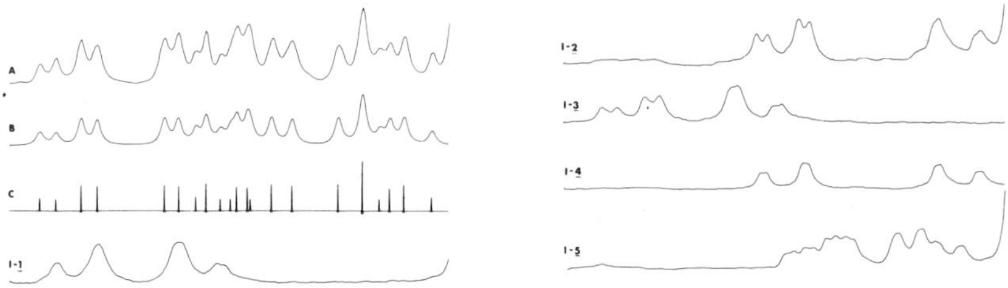

Fig. 2. ^{1}H nmr spectra (220 MHz) between 2.885 and 3.584 ppm downfield from TSP, at 19°C, of compounds of Fig. 1 and unsubstituted oxytocin. See text.

Investigations of the torsion angles of the backbone are continuing; determinations about the angle ψ still present major problems.

The isomers described contain several ^{15}N enrichments, and have been used, with other methods to assign the ^{15}N chemical shifts of oxytocin in water at pH 4.0. These are N'_1, 17.0; N'_2, 102.6; N'_3, 98.7; N'_4, 98.7; $(N_4^{\epsilon 2}/N_5^{\delta 2})$, 90.8/91.0; N'_5, 95.3; N'_6, 98.7; N'_7, 116.1; N'_8, 101.2; N'_9, 89.3; and N_{amide}, 85.8. All chemical shifts are in ppm downfield from $^{15}NH_4^+$ (5 *M* NH_4NO_3 in 2 *M* HNO_3).[6] All these values are within a few ppm of those expected from simple model compounds.

This general approach has been discussed by others;[7] based on our experimental investigations, it is our opinion that it represents an exhaustively rigorous approach to the problem of peptide conformation in solution. Although somewhat laborious, it is the most straightforward method presently available.

References

1. For summaries of previous work see Feeney, J. (1975) *Proc. R. Soc. Lond. A* **345**, 61–72 and *Peptides: Chemistry, Structure, Biology, Proceedings of the Fourth American Peptide Symposium*, Walter, R. & Meienhofer, J., Eds., Ann Arbor Science, Ann Arbor, Mich., pp. 61–291.

2. Wyssbrod, H. R., Ballardin, A., Schwartz, I. L., Walter, R., Van Binst, G., Gibbons, W. A., Agosta, W. C., Field, F. H. & Cowburn, D. (1977) *J. Am. Chem. Soc.* **99**, 5273–5276, & references therein.

3. Glickson, J. D. (1975) in *Peptides: Chemistry, Structure, Biology, Proceedings of the Fourth American Symposium*, Walter, R. & Meienhofer, J., Eds., Ann Arbor Science, Ann Arbor, Mich., pp. 787–802, & references therein.

4. Fischman, A. J., Wyssbrod, H. R., Agosta, Wm. C. & Cowburn, D., *J. Amer. Chem. Soc.*, in press.

5. Boicelli, C. A., Bradbury, A. F. & Feeney, J. (1977) *J. Chem. Soc.* **Perkin II**, 477–482.

6. Hawkes, G. E., Randall, E. W. & Bradley, C. H. (1976) *Nature* **257**, 767–772.

7. For a review see Bystrov, V. F. (1976) *Prog. NMR Spectroscopy* **10**, 41–81.

DEVELOPMENT OF NMR AND FLUORESCENCE METHODS FOR DETERMINING PEPTIDE CONFORMATIONS IN SOLUTION

JERRY D. GLICKSON, ROBERT E. LENKINSKI, N. RAMA KRISHNA, DAVID G. AGRESTI, *Departments of Biochemistry, Chemistry, Physics, and Medicine, and the Comprehensive Cancer Center, University of Alabama in Birmingham, Birmingham, Alabama 35294,* and RODERICH WALTER, *Department of Physiology and Biophysics, University of Illinois Medical Center, Chicago, Illinois 60680*

We have adapted a number of techniques that have proven useful in studies of other molecules to the investigation of the solution conformation of peptides.

Paramagnetic Metal Probes. A generalized formalism based on the Marquardt[1] or Simplex[2] algorithm has been developed for the analysis of nmr shift and relaxation data.[3] This method was employed to fit data on hen-egg-white lysozyme complexes with lanthanides[4] and with transition metals to the following equations

$$\delta_d = K_1 (3 \cos^2\theta - 1)/r^3 + K_2 (\sin^2\theta \cos 2\phi)/r^3 \tag{1}$$

$$\Delta\nu^M_{1/2} = 1/\pi T_{2M} = cf(\tau)/r^6 \tag{2}$$

where δ_d is the electron-nuclear dipolar contribution to the chemical shift, K_1 and K_2 are constants related to the elements of the magnetic susceptibility tensor of the metal, (r,θ,ϕ) are the spherical polar coordinates of a given hydrogen in a coordinate system in which the metal is at the origin and the susceptibility tensor is diagonalized, $\Delta\nu^M_{1/2}$ is the increment in the proton nmr linewidth associated with binding of a metal such as Gd^{3+}, T_{2M} is the corresponding transverse relaxation time, c is a constant which contains electronic terms of the metal ion and $f(\tau)$ is a function of the correlation times of the complex. Statistical hypothesis tests employing the Hamilton R-factor ratio method[5] led to rejection of the commonly assumed "axial model" (i.e. $K_2 = 0$ or $K_2 \ll K_1$) with a confidence level in excess of 97.5%. When the geometry of the complex has been independently determined (as in the case of protein-metal complexes for which X-ray data is available), this formalism can be employed to compare the validity of two or more sets of nmr spectral assignments. Alternatively, if the spectral assignments have been independently determined (as in the case of many peptides), the formalism can be employed to compare the validity of two or more models for the preferred solution conformation. Conformational averaging and internal rotation can be incorporated in the analysis.

Fluorescence Studies of Metal Complexes. Fluorescence spectroscopy was employed to identify peptides capable of forming complexes suitable for nmr spectral analysis. Angiotensin II (λ_{exc}=265 nm, λ_{em}=302 nm) forms complexes with both Ca^{2+} ($K_{diss} \sim 10^{-4}$ M^{-1}) and its paramagnetic fluorescent analog, Tb^{3+} (K_{diss} = 1.3 ± 0.2 × 10^{-4} M^{-1}). Energy transfer from Tb^{3+} (enhancement of tyrosine fluorescence by a factor of 1.7 ± 0.1) implicates the phenol group as a probable ligand. Since [Asn^1, Val^5]-angiotensin II does not form complexes with Ca^{2+} or Tb^{3+}, the Asp^1 carboxyl is also a likely ligand, as is the C-terminal carboxyl. Similar studies have demonstrated that Tb^{3+} forms complexes with the glycopeptide antibiotic bleomycin, and with a peptide fragment of thymopoietin, which retains the activity of the parent hormone.

Intramolecular Nuclear Overhauser Effect (NOE). We have recently demonstrated that the intramolecular NOE of peptides originates from a predominantly dipolar mechanism; we have derived a set of equations suitable for the analysis of NOE data of peptides and other complex biomolecules which do not satisfy the "extreme-narrowing" limit.[6] The conformation obtained using NOE data for the K^+-valinomycin complex in chloroform is (with slight modifications) consistent with the conformation determined by X-ray crystallography.[7] Quantitative evaluation of NOE data on valinomycin in dimethyl sulfoxide is now in progress. Our previous semi-quantitative interpretation of these data[6] indicated that the III-1 structure of Patel and Tonelli[8] was preferred in this solvent.

Quantitative Analysis of Solvent Saturation Experiments. It has previously been demonstrated that the extent of solvent exposure of NH and CH hydrogens is reflected in changes in the intensities of their corresponding resonances when the solvent is saturated. Quantitation of these effects can be accomplished by combining the transfer of saturation measurements (for NH protons) and intermolecular NOE measurements (for CH hydrogens) with T_1 experiments performed in the presence of solvent saturation. For NH hydrogens this yields the rate of proton exchange with the solvent.[9,10] For CH hydrogens the theory of Krishna and Gordon[11] yields the intermolecular dipolar relaxation rate. The above rates serve as indices of solvent exposure of NH and CH hydrogens, respectively. Performing such an analysis on *cyclo*(Gly-Phe) and *cyclo*(Gly-Tyr) in dimethyl sulfoxide, we obtain values of 0.298 sec^{-1} and 0.460 sec^{-1} for the "average" intermolecular dipolar relaxation rates of the aromatic CH protons of these diketopiperazines, respectively. The greater relaxation rate for the tyrosyl hydrogens may reflect formation of a hydrogen bonded complex between the phenol and dimethyl sulfoxide.[12]

Fluorescence Measurements of Tyrosine Side Chain Exposure. Cowgill[13] has reported that the extent of solvent exposure of the aromatic side chain of tyrosine is related to its fluorescence quantum yield. Table I shows that in aqueous

Table I. Tyrosine Quantum Yields in Water of Peptide Hormones

Compound	Relative Quantum Yield[a]	Compound	Relative Quantum Yield[a]
oxytocin	0.26	[Phe^{4}]-oxytocin	0.26
oxypressin	0.19	[Pro^{3}, Gly^{4}]-oxytocin	0.35
1-penicillamine oxytocin	0.35	[Lys^{8}]-vasopressin	0.26
[Asp^{5}]-oxytocin	0.38	angiotensin II	0.36
[Glu^{4}]-oxytocin	0.21	[Asn^{1}, Val^{5}]-angiotensin II	0.35
[Phe^{3}, Phe^{4}, Met^{8}]-oxytocin	0.21		

a Ratio of the tyrosyl quantum yield to that of the free amino acid.

solution relative quantum yields in excess of 0.3, characteristic of partially shielded tyrosines, are observed for oxytocin derivatives for which stabilization of the folded conformation has been reported – e.g. 1-penicillamine oxytocin[14] and [Pro^{3}, Gly^{4}]-oxytocin[15] – and for angiotensin, for which nmr solvent saturation experiments indicate shielding of the *meta*-hydrogens from the solvent.[16] In agreement with solvent saturation studies, oxytocin and [Lys^{8}]-vasopressin exhibit lower quantum yields consistent with a more exposed structure. These observations suggest that the fluorescence quantum yield may serve as a useful measure of tyrosyl side-chain exposure in conformational studies of peptides.

We have also monitored the fluorescence intensity of tyrosine as a function of solvent composition in H_2O/D_2O mixtures. H_2O quenches tyrosyl fluorescence; D_2O does not. The data fit the Stern-Volmer equation

$$F_o/F = 1 + K_p\,[H_2O] \tag{3}$$

where the reciprocal of the fractional change in fluorescence, F_o/F, is linearly related to the molar concentration of the quencher, $[H_2O]$, and $K_p = k_3\tau_f$, where τ_f is the fluorescence lifetime (2×10^{-9} sec) and k_3 the bimolecular collision rate (2.5×10^{7} lmol^{-1} sec^{-1}). The collisional rates for various peptides are being measured to ascertain if, as expected, k_3 is related to the extent of solvent exposure of the tyrosyl side chain.

This research was supported by USPHS grants CA-13148 and AM-18399. The authors acknowledge the use of NMR Facilities for Biomolecular Research located at Carnegie-Mellon University (NIH Grant No. RR-00292) and at the Francis Bitter National Magnet Laboratory of the Massachusetts Institute of Technology (NSF Contract No. C670 and NIH Grant No. RR-00995).

References

1. Marquardt, D. W. (1963) *J. Soc. Ind. Appl. Math.* **11**, 431–41.
2. O'Neall, R. (1971) *J. Roy. Stat. Soc. C* **20**, 338–45.
3. Agresti, D. G., Lenkinski, R. E. & Glickson, J. D. (1977) *Biochim. Biophys. Acta,* **76,** 711–19.
4. Campbell, I. D., Dobson, C. M. & Williams, R. J. P. (1975) *Proc. Roy. Soc. Lond. A* **345,** 41–59.
5. Hamilton, W. C. (1965) *Acta Cryst.* **18,** 502–510.

6. Glickson, J. D., Gordon, S. L., Pitner, T. P., Agresti, D. G. & Walter, R. (1976) *Biochemistry* **15**, 5721–5729.

7. Neupert-Laves, K. & Dobler, M. (1975) *Helv. Chim. Acta* **58**, 432–442.

8. Patel, D. J. & Tonelli, A. E. (1973) *Biochemistry* **12**, 486–501.

9. Waelder, S., Lee, L. & Redfield, A. G. (1975) *J. Amer. Chem. Soc.* **97**, 2927–2928.

10. Bleich, H. E. & Glasel, J. A. (1975) *J. Amer. Chem. Soc.* **97**, 6585–6586.

11. Krishna, N. R. & Gordon, S. L. (1973) *J. Chem. Phys.* **58**, 5687–5696.

12. Deslauriers, R., Grzonka, Z., Schaumburg, K., Shiba, T. & Walter, R. (1975) *J. Amer. Chem. Soc.* **97**, 5093–5100.

13. Cowgill, R. W. (1976) in *Biochemical Fluorescence*, vol. 2, Chan, R. F. & Edelhoch, H., Eds., M. Dekker, New York, 441–486.

14. Meraldi, J.-P., Yamamoto, D., Hruby, V. J. & Brewster, A. I. R. (1975) in *Peptides: Chemistry, Structure and Biology*, Walter, R. & Meienhofer, J., Eds., Ann Arbor Sci. Publ., Ann Arbor, MI, pp. 803–814.

15. Wyssbrod, H. R., Ballardin, A., Gibbons, W. A., Roy, J., Schwartz, I. L. & Walter, R. (1975) in *Peptides: Chemistry, Structure and Biology*, Walter, R. & Meienhofer, J., Eds., Ann Arbor Sci. Publ., Ann Arbor, MI, pp. 815–822.

16. Pitner, T. P., Glickson, J. D., Rowan, R., Dadok, J. & Bothner-By, A. A. (1975) in *Peptides: Chemistry, Structure and Biology*, Walter, R. & Meienhofer, J., Eds., Ann Arbor Sci. Publ., Ann Arbor, MI, pp. 159–164.

MULTIPLE CONFORMATIONS OF THE ZWITTERIONIC AND CATIONIC FORMS OF ENKEPHALINS AND OTHER PEPTIDES

CLAUDE R. JONES, JOSEPH B. ALPER, MEI-CHANG KUO, and WILLIAM A. GIBBONS, *Enzyme Institute and Department of Biochemistry, College of Agricultural and Life Sciences, University of Wisconsin, Madison, Wisconsin 53706.*

Conformational studies can probe structural requirements of analgesic activity by aiding the identification of common opiate-enkephalin structural features. Our preliminary investigation[1] in DMSO showed that the Met^5 amide proton had a very small chemical shift temperature dependence: this, plus $^3J_{NHCH}$ and $^3J_{CHCH}$ values suggested a model for the Met^5-enkephalin zwitterionic form with the following features: (a) Met^5 amide proton hydrogen bonded to Gly^2 carbonyl, (b) the COO^- and NH_3^+ termini in close proximity, (c) a β-turn of residues 2-5 as the major backbone conformation, (d) extensive C^α-C^β internal rotation and thus multiple, rapidly interconverting molecular conformations.

To investigate the salt bridge as a source of conformational stability[1] we examined the cationic form (COOH) of the peptide which lacks the interaction of charged terminii. We have also investigated the D-Ala^2 derivative because of suggestions[2,3] that its enhanced activity might be due to a different folded conformation with different factors contributing to its stability.

Results and Discussion

Spectra of the zwitterionic and cationic forms of Met^5- and [D-Ala^2]-Met^5-enkephalin appear very similar but for both peptides the amide resonances of Phe^4 and Met^5 are reversed in going from the zwitterionic to cationic forms. This comparison resolves the apparent discrepancies in the various published reports[1,4,5] of enkephalin conformational parameters, some authors studied the cationic[5] and others the zwitterionic forms[1,4]. Furthermore, the values of $^3J_{NHCH}$, which are sensitive to the backbone angles, ϕ, and its average are different for the two forms (Table I). Similarly, the $\Delta\delta/\Delta T$ values for the Met^5 amide proton indicate a solvent shielded environment for the zwitterionic form but a solvent exposed one for the cationic form. The principal conformations or extent of conformational averaging of the cationic form differs from the zwitterionic form. This seems to confirm the importance of the salt bridge to the stability of the physiological, zwitterionic form.

The nmr parameters in Table I of [D-Ala^2]-Met^5- and Met^5-enkephalin zwitterions are essentially the same; the parameters of the two cations are also identical. Thus, no experimental justification is found for any enhanced activity of the D-Ala^2 analog due to stabilization of a conformation different than that

Table I. Conformational Data of the Zwitterionic (Z) and Cationic (C) Forms of Met^5- and [D-Ala^2]-Met^5-enkephalin

Parameter	Peptide	Residue			
		2	3	4	5
$^3J_{NHCH}$	Met (Z)			8.3	7.2
in Hz	Met (C)	5.2, 5.9	5.4, 6.0	8.4	7.9
	D-Ala (Z)	6.5	4.9, 6.5	8.6	7.0
	D-Ala (C)	7.5	5.9, 5.9	8.5	7.8
$\Delta\delta/\Delta T$	Met (Z)	7.8	3.1	7.5	0.9
in ppb/°C	Met (C)	4.1	4.3	5.7	6.3
	D-Ala (Z)	6.2	2.7	7.1	-0.3
	D-Ala (C)	4.0	3.7	5.3	6.0

of Met^5-enkephalin[2,3]. Other possible explanations of the enhanced activity are: first, our measurements failed to detect a statistically small but biologically important conformational species; second, the conformation of the two peptides is the same but the extra methyl group confers new receptor-peptide properties; third, the enhanced activity is due to the increased persistence observed for this analog[2].

All three C^α-C^β rotamers are significantly populated[6] for each residue studied (Table II) in both Met^5- and [D-Ala^2]-Met^5-enkephalin; thus for a given backbone there are at least 27 molecular conformations, each differing in at least one side-chain conformation. The five most probable molecular conformations are in Table II; even the most favored accounts for only 13% of the instantaneous conformations. Multiple tertiary conformations, each with a small cumulative statistical weight, are a common property of peptides as shown in Table II by the results obtained for gramicidin S[7] and oxytocins[8].

Most of the 22-minimum-energy conformation which have been reported[3,9,10] are for non-physiological forms of the peptides and thus cannot be directly compared to the nmr parameters. Despite lack of quantitative agreement, the calculations do predict several folded enkephalins with similar energies. This leads to conformational averaging and can explain the lack of exact agreement in Table III between the observed nmr parameters and those predicted[11] from idealized models[12] of several of the possible turns which could give a solvent shielded Met^5 amide proton. Although multiple conformations exist for the zwitterionic forms of these peptides, they cannot be regarded as random or random coil peptides; instead the time-averaged conformation seems to most resemble a 2-5-β-I turn. We cannot yet exclude other regular or distorted conformations, β-II turns, fortuitous averaging of several quite different conformations nor minor contributors to a time-averaged conformation which is mainly a

Table II. Tertiary Structure Data for Met^5- and [D-Ala^2]-Met^5-enkephalin Cationic (C) and Zwitterionic (Z) Forms.

Peptide	Residue	Coupling Constants (Hz)			Populations (%)		
		$^2J_{CHCH}$	$^3J_{CHCH}$		tg^+	tg^-	gg
Met (Z)	Tyr	-13.7	6.2	7.5	44	33	23
	Phe	-13.9	4.0	10.1	68	13	19
	Met	-13.3	4.5	7.3	43	17	40
Met (C)	Tyr	-14.0	5.9	8.5	50	26	24
	Phe	-13.9	4.2	8.4	53	15	32
D-Ala (Z)	Tyr	-13.6	7.0	7.9	48	40	12
	Phe	-13.9	4.0	10.6	76	13	11
D-Ala (C)	Phe	-13.6	3.7	9.9	67	10	23

Cumulative Statistical Weights (S. W.) of Tertiary Conformations

Met^5-enkephalin (Z)				Gramicidin S				
Tyr	Phe	Met	S. W.	Val	Orn	Leu	Phe	S. W.
180	180	180	13%	180	-60	180	180	13%
180	180	+60	12%	180	-60	-60	180	12%
-60	180	180	9%	+60	-60	180	180	8%
-60	180	+60	9%	+60	-60	-60	180	7%
+60	180	180	7%	180	180	180	180	5%

Table III. Comparison of Observed Backbone Coupling Constants (in Hz) and those Predicted[11] for Folded Conformations[12]

Residue	Observed	Predicted for Turn Type					
		I	I'	II	II'	III	III'
Gly	4.9, 6.5	3.7, 5.9	5.9, 3.7	3.7, 5.9	5.9, 3.7	3.7, 5.9	5.9, 3.7
Phe	8.3 to 8.6	9.0	3.9	4.9	7.3	3.6	5.7

β-I turn. In fact, the use of the β-I turn model as a working hypothesis would, from Table III, seem to require some distortion from the idealized angles or contributions from minor conformers. Concentration effects on enkephalin spectra exist[13] but cannot explain spectral discrepancies[1,4,5]. $^3J_{NHCH}$ values, at best, are consistent with a small amount of gramicidin S dimer; Urry and co-workers[13] agree with the original β-turn model[1,4].

References

1. Jones, C. R., Gibbons, W. A. & Garsky, V. (1976) *Nature* **262**, 779–782.

2. Coy, D. H., Kastin, A. J., Schally, A. V., Morin, O., Caron, N. G., Labrie, F., Walker, J. M., Fertel, R., Bernston, G. G. & Sandman, C. A. (1977) *Biochem. Biophys. Res. Commun.* **73**, 632–638.

3. Momany, F. A. (1977) *Biochem. Biophys. Res. Commun.* **75**, 1098–1103.

4. Garbay-Jaureguiberry, C., Roques, B. P., Oberlin, R., Anteunis, M. & Lala, A. K. (1976) *Biochem. Biophys. Res. Commun.* **71**, 558–565.

5. Bleich, H. E., Cutnell, J. D., Day, A. R., Freer, R. J., Glasel, J. A. & McKelvy, J. F. (1976) *Proc. Nat. Acad. Sci. USA* **73**, 2589–2593.

6. Pachler, K. G. R. (1964) *Spectrochim. Acta* **20**, 581. (Packler's method alone does not allow unambiguous assignment of χ_1 = 180 and –60 rotamers.)

7. Kuo, M., Jones, C. R. & Gibbons, W. A., in preparation for Biochemistry.

8. Nicholls, L. J. F., Ford, J. J., Jones, C. R., Manning, M. & Gibbons, W. A. (1977) *Peptides: Proceedings of the 5th American Peptide Symposium,* Goodman, M. & Meienhofer, J., Eds., John Wiley & Sons, Inc., New York, pp. 165-167.

9. De Coen, J. L., Humblet, C. & Koch, M. H. J. (1977) *FEBS Lett* **73**, 38–42.

10. Isogai, Y., Nemethy, G. & Scheraga, H. A. (1977) *Proc. Nat. Acad. Sci. USA* **74**, 414-418.

11. Cung, M. T., Marraud, M. & Neel, J. (1974) *Macromolecules* **7**, 606–613.

12. Venkatachalam, C. M. (1968) *Biopolymers* **6**, 1425–1436.

13. Khaled, M. H., Long, M. M., Thompson, W. D., Bradley, R. J., Brown, G. B. & Urry, D. W. (1977) *Biochem. Biophys. Res. Commun.* **76**, 224–231.

CONDENSATION PRODUCTS OF Met-Val-Gly-Pro-Asn-Gly AND THEIR CONFORMATIONS

KENNETH D. KOPPLE, MEOW-CHAN FENG and ANITA GO,
Department of Chemistry, Illinois Institute of Technology, Chicago, Illinois 60616

Judging from the frequency of occurrence of proline in the *i+1* and asparagine in the *i+2* positions of protein β-turns, and the frequency of methionine and valine in extended chains,[1,2] the repeating hexapeptide sequence Met-Val-Gly-Pro-Asn-Gly might be expected to fold in solution to a cross-β backbone[3] in which there is a turn at Pro-Asn and an extended region centered on Met-Val. We have allowed an 0.5 *M* solution of H-Met-Val-Gly-Pro-Asn-Gly-ONp in dimethylformamide to self-condense, and have examined the products.

The reaction product was extracted with 0.1 *M* ammonium bicarbonate and with ethanol. The polymer remaining insoluble contained 20% of the hexapeptide units. Fractionation of the soluble portion on BioGels P-2, P-4 and P-6 afforded two pure crystalline, ninhydrin-negative peptides, identified by proton magnetic resonance spectra and elution volumes as cyclic monomer (8%) and cyclic dimer (30%). The remainder of the hexapeptide units appeared as soluble oligomers, ranging up to heptamer according to elution volumes and end-group analysis by dinitrophenylation.[4,5] Proton magnetic resonance spectra of the soluble oligomers showed that they had undergone extensive deamidation of the asparagine side chain upon storage in ammonium bicarbonate buffer. This was confirmed by amino-acid analysis, and they were not further investigated.

The water-insoluble polymer, which did have the proper amino-acid analysis and a consistent set of N-H proton resonances, was soluble in fluorinated alcohols and in formic acid, but was insoluble in dimethyl sulfoxide, dimethylformamide and common organic solvents. Exhaustive treatment of water-swollen polymer with fluorodinitrobenzene produced a dinitrophenyl derivative; from its infrared spectrum an average chain length of 7–8 hexapeptide units was estimated.[4,5] Solutions of the polymer in fluorinated alcohols exhibited circular dichroism (CD) suggesting a mixture of random coil and β-conformation,[6-9] and addition of water up to 50% did not change the CD spectrum significantly. Films of the polymer also showed the CD of a mixture of random coil and β-conformation. However, on exposure to water vapor for a few hours, their CD spectra became characteristic of β-structure alone, although the estimated ellipticities were about one half those normally observed. Circular dichroism data are given in Table I.

Films cast on AgCl disks for infrared spectra were oriented by rolling a glass rod through the evaporating solution. The amide I and N-H stretching absorptions exhibited weak perpendicular dichroism. On exposure of the film to water vapor, the dichroism increased and the amide I band was resolved into two, the

Table I. Circular Dichroism of Met-Val-Gly-Pro-Asn-Gly Derivatives

Derivative	State	Extrema, λ, nm ($10^{-3}[\theta]_R$, deg-cm^2/dmol)		
cyclic monomer	in TFE	<192 (> +5)	210sh(-10)	222 (-13.3)
cyclic dimer	in TFE	<192 (< -16)		220 (+2.4)
polymer	in HFP[a]	194 (-10.4)		223 (-3.3)
polymer	film[b]	198 (-2.2)		222 (-1.5)
polymer	film(H_2O)[b,c]	195 (+7.5)	214 (-3.3)	

a) Dichroism same at 0.2 and 1.2 mg/ml.

b) Polymer film from HFP on quartz, average thickness 0.094 mg/cm^2, cast so that CD spectra are unaffected by rotation of the film about the optical axis. $[\theta]_R$ calculated on the basis of 0.094 mg/cm^2 for comparison with solution values.

c) Same film, after exposure to 100% relative humidity at room temperature for 10 hr. Linear dichroism is still absent. The circular dichroism does not increase on further exposure to water vapor.

stronger at 1640 cm^{-1} with a dichroic ratio of 1.5, and the weaker at 1665 cm^{-1} with a ratio of 1.2. The N-H stretching absorption at 3300 cm^{-1} also had a dichroic ratio of 1.5.

In its more ordered form the polymer appears from these observations largely to consist of extended residues joined by hydrogen bonds perpendicular to the peptide chains. The intended self-associated folded chain is probably excluded by the perpendicular dichroism of the C=0 and N-H stretching absorptions, and by the insolubility of the polymer in a highly ordered form. Although the presence of proline residues precludes a fully extended β-sheet, one possible structure is an in-phase parallel sheet with a 12-residue repeat, in which the Gly-Pro sequence has the dihedral angles (φ,ψ) Gly, -60°, 180°, Pro, -50°, 120°, and the other residues all have the -120°, 120° backbone angles of the parallel β-structure (Figure 1). When the peptide chains associate in this manner the asparagine amide groups can form hydrogen bonded chains transverse to the backbone.

A more interesting result of the experiment is the formation of a 30% yield of cyclic dimer in the concentrated polymerization mixture. This suggests that the

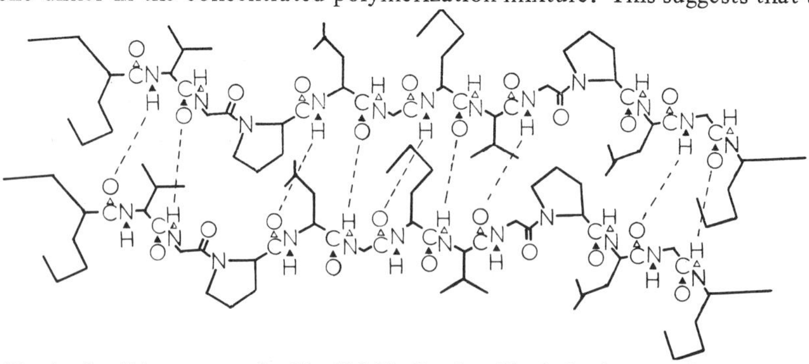

Fig. 1. Possible structure for Met-Val-Gly-Pro-Asn-Gly derivatives: an in-phase parallel sheet.

cyclic dimer represents a conformation close to one preferred by the open H-(Met-Val-Gly-Pro-Asn-Gly)$_2$-ONp. For the cyclic dodecapeptide a likely conformation has been derived from nuclear magnetic resonance studies which will be described elsewhere in detail.[10] The data indicate a structure with C_2 symmetry on the nmr time scale. Key observations are of the relative exchange rates of the peptide (N-H) protons and of the exposure of these protons to the line-broadening effects of a nitroxyl cosolute, which combine to suggest that the methionine and both glycine peptide protons are shielded from the solvent, that the valine N-H is exposed, and that the methionine carbonyl is buried. The same conformation is retained in water, hexafluoro-2-propanol and dimethyl sulfoxide. The large chemical shift non-equivalences (0.6 - 0.8 ppm) of the enantiotopic α-protons of both glycine residues suggest that the backbone conformation is not subject to much internal motion.

In the proposed form of the cyclic dimer (Figure 2) all residues are in backbone conformations allowed for the L-series except the glycine following asparagine. This cyclic structure can be reached readily from a segment of the intended cross-β-backbone, in which an extended Gly-Met-Val-Gly sequence occurs between two Pro-Asn turns. The principal changes required are rotations at the glycine of the Asn-Gly-Met sequence. Formation of the cyclic dimer therefore may be a consequence of the conformational freedom introduced by a glycine unit; had this glycine been an L-series residue instead, polymerization might have competed more effectively.

That the polymer apparently does not adopt the cross-β-conformation may also stem from the absence of side chains on two residues in the repeating unit; not only is the flexibility of an individual chain thereby increased, making less probably any single structure such as the intended one, but reduced side chain interference facilitates interchain association at sufficiently high concentration. The particular sheet structure proposed for the ordered film would not be possible if the glycine before proline were replaced by an L-series residue.

This work was supported by National Institutes of Health Grant GM 14069.

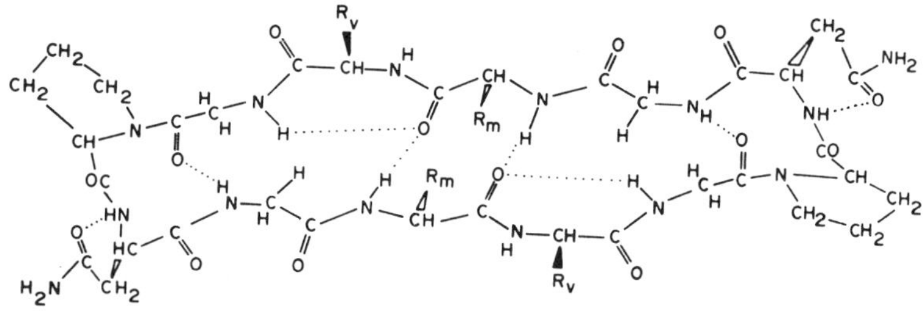

Fig. 2. Proposed form for Met-Val-Gly-Pro-Asn-Gly derivatives: a cyclic dimer.

References

1. Chou, P. Y. & Fasman, G. D. (1974) *Biochemistry* **13**, 222–245.
2. Chou, P. Y., Adler, A. J., & Fasman, G. D. (1975) *J. Mol. Biol.* **96**, 29–45.
3. Geddes, A. J., Parker, K. D., Atkins, E. D. T. & Beighton, E. (1968) *J. Mol. Biol.* **32**, 343–358.
4. Scheidt, U. & Restle, H. (1954) *Z. Naturforsch.* **9b**, 182.
5. Okamoto, S. & Hamamoto, M. (1964) *Agr. Biol. Chem. (Tokyo)* **38**, 55.
6. Greenfield, N. & Fasman, G. D. (1969) *Biochemistry* 8, 4108–4116.
7. Chen, Y. H., Yang, J. T. & Chau, K. H. (1974) *Biochemistry* **13**, 3350–3359.
8. Balcerski, J. S., Pysh, E. S., Bonora, G. M. & Toniolo, C. (1976) *J. Amer. Chem. Soc.* **98**, 3470–3473.
9. Rinaudo, M. & Domard, A. (1976) *J. Amer. Chem. Soc.* **98**, 6360–6364.
10. Kopple, K. D. & Go, A. (1977) *J. Amer. Chem. Soc.*, in press.

CONFORMATION OF BRADYKININ IN AQUEOUS SOLUTION. ELECTROMETRIC TITRATION OF THE HORMONE AND RELATED PEPTIDES

A. C. M. PAIVA and L. JULIANO, *Department of Biophysics and Physiology, Escola Paulista de Medicina, C. P. 20.388, 04023 São Paulo, S. P., Brazil*

The solution conformation of the linear nonapeptide hormone bradykinin is the object of some controversy. The first CD studies led to the proposal of random conformation in aqueous solution,[1,2] but Cann et al.[3] interpreted the CD spectra as indicating a partly ordered structure due to a 3 → 1 hydrogen bond bridging the Pro^7 residue. On the other hand, a combination of a variety of spectral methods and theoretical conformational analysis led Ivanov et al.[4] to propose a folded structure with a 3 → 1 hydrogen bond across Gly^4, in which the C-terminal carboxylate group forms a strong ionic bond with the Arg^1 side-chain guanido group.

More recent studies indicated that, although the bradykinin molecule may occur in a number of possible conformations in aqueous solution, it tends to increased order in a more hydrophobic environment.[5] Further CD studies[6] with β-homoprolyl^7- and β-homophenylalanyl^8-bradykinin were interpreted as favoring the model of Ivanov et al.[4] over that of Cann et al.[3] Marlborough and Ryan[6] further propose that the folded conformation may play a role in bradykinin activity, suggesting that the lower potency of β-homophenylalanine^8-bradykinin (compared to that of the β-homoproline^7 analog) might be due to perturbation of the proposed ionic interaction between the Arg^1 side-chain and the carboxylate group.

In order to investigate possible electrostatic interactions in aqueous solution, we have synthesized bradykinin and several related peptides (Table I), and determined their amino and carboxyl pK values. We have not attempted to titrate the guanido groups. The electrometric titrations were done as previously described,[7] in the presence of 0.15 *M* KCl. In the pH range 2.5–10.5, and in the temperature interval 10°–40°C, the titration of bradykinin was normal and reversible, as illustrated by the example of Figure 1. No sign of the reported[4] hydrolysis above pH 9 was found under our experimental conditions. The estimated errors of the pK values listed in Table I allow us to consider as significant (at the 95% confidence level) differences of 0.06 units for the case of carboxyl groups, and of 0.03 units for the amino groups.

The carboxyl pK in bradykinin (3.29) is the same as in the model compound acetyl-Arg (Table I), in contrast to a previous finding that this group could not be titrated by ^{13}C nmr, in the range of pH 2.5–5.[4] The pK value of the C-terminal

Table I. pK Values for the Amino and Carboxyl Groups of Bradykinin and Related Compounds in 0.15 *M* KCl at 25°C

Compound	Amino Acid Sequence	pK Carboxyl	pK Amino
1	Arg-Pro-Pro-Gly-Phe-Ser-Pro-Phe-Arg	3.29	7.41
2	Arg-Pro-Pro-Gly-Phe-Ser-Pro-Phe	3.35	7.43
3	Arg-Pro-Pro-Gly-Phe-Ser-Pro	3.35	7.49
4	Arg-Pro-Pro-Gly-Phe	3.10	7.55
5	Arg-Pro-Pro-Gly	3.35	7.67
6	Gly-Phe-Ser-Pro-Phe-Arg	3.35	7.90
7	Phe-Ser-Pro-Phe-Arg	3.33	7.19
8	Acetyl-Arg	3.30	—
9	Acetyl-Phe	3.39	—
10	Acetyl-Pro	3.36	—
11	Acetyl-Gly-Gly	3.55	—
12	Gly-Gly-NH_2	—	7.94

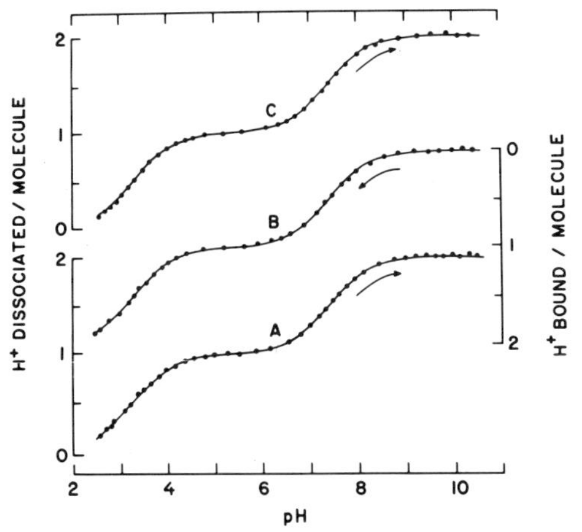

Fig. 1. Titration of a 1.5×10^{-3} *M* solution of bradykinin hydrochloride at 25°C, in 0.15 *M* KCl, with KOH (A), followed by titration with HCl (B), and then again with KOH (C). The points are experimental values and the lines were calculated from two Henderson-Hasselbalch terms with pK values 3.29 and 7.41.

Arg residue was not influenced by the removal of three or four N-terminal residues (Table I, compounds 6 and 7). Removal of one or two residues from the C-terminus also did not significantly alter the carboxyl pK, as is seen by comparison of compounds 2 with 9 and 3 with 10. However, removal of four (compounds 4 vs 9) and five (compounds 5 vs 11) C-terminal residues led to significant decreases in the carboxyl pK values. Thus, while there is a considerable effect of the amino group's positive charge on carboxyl ionization in the N-terminal tetrapeptide, the same is not seen in the C-terminal pentapeptide. This suggests the presence of some structure in the N-terminal lower homologs, which is also supported by the higher amino group pK values in compounds 4 and 5, as compared to that in bradykinin. Whether or not some structure would also occur at the N-terminus of the nonapeptide, cannot be inferred from our data. It is clear, however, that no electrostatic interaction between the N- and C-terminal ends of bradykinin occurs under our titration conditions. This contrasts with the significant effects observed in the case of the octapeptide angiotensin, under similar experimental conditions.[7] We believe that, although our bradykinin titrations were done at 0.15 ionic strength, a detectable lowering of the carboxyl pK should be expected if there were a strong ionic bond between that group and the Arg^1 side chain in aqueous solution. This ionic bond might be present in more hydrophobic media,[5] but it should not play a role in biological activity because the forces involved in receptor binding should be some orders of magnitude larger than those responsible for the proposed folded conformation of bradykinin.

References

1. Bodanszky, A., Ondetti, M. A., Rolofsky, C. A. & Bodanszky, M. (1971) *Experientia* **27**, 1269–1270.
2. Brady, A. H., Ryan, J. W. & Stewart, J. M. (1971) *Biochem. J.* **121**, 179–184.
3. Cann, J. R., Stewart, J. W. & Matsueda, G. (1973) *Biochemistry* **12**, 3780–3788.
4. Ivanov, V. T., Filatova, A., Reissman, T. O., Reutova, E., Efremov, E. S., Pashkov, V. S., Galaktinov, S. G., Grigoriam, G. L. & Ovchinnikov, Yu., in *Peptides: Chemistry, Structure and Biology*, Walter, R. & Meienhofer, J., Eds., Ann Arbor Science, Ann Arbor, pp. 151–157.
5. Marlborough, D. I., Ryan, J. W. & Felix, A. M. (1976) *Arch. Biochem. Biophys.* **176**, 582–590.
6. Marlborough, D. I. & Ryan, J. W. (1977) *Biochem. Biophys. Res. Comm.* **75**, 757–765.
7. Juliano, L. & Paiva, A. C. M. (1974) *Biochemistry* **13**, 2445–2450.

CONFORMATIONAL ANALYSIS OF TENTOXIN AND RELATED CYCLIC TETRAPEPTIDES. ISOLATION AND BIOASSAY OF TWO CONFORMERS OF D-MeAla TENTOXIN.

DANIEL H. RICH and PRADIP BHATNAGAR, *School of Pharmacy, University of Wisconsin, Madison, WI 53706*

Tentoxin, *cyclo*(*N*-methyl-L-alanyl-L-leucyl-*N*-methyl-(*Z*)-dehydrophenylalanyl-glycyl) (*1*), is a phytotoxin produced by *Alternaria tenuis.* (*N*-methyl (*Z*) dehydrophenylalanine = MePhe [(*Z*)Δ], where (*Z*) = zusammen.)[1] The toxicity of *1* is linked to the presence in susceptible species of a single tentoxin-binding-site on chloroplast coupling-factor-one (CF_1), a key protein involved in ATP synthesis.[2]

During our structure-activity studies, we obtained evidence that the conformation proposed[3] for *1* could not be correct. Synthetic [L-Pro^1]-tentoxin (*2*),[4] which can not adopt the proposed[3] conformation because of the L-Pro ring system, was found to have the same biological activity (Figure 1a,b) and CD and nmr parameters as *1* (Table I).

In *2* (Table I) the chemical shift of the Gly outer (3.55 ppm) and inner (5.2) protons is similar to those in the 8-position of *cyclo*$(Sar)_4$.[5] The $^3J_{NH\text{-}C\alpha H}$ coupling constant is 10.5 Hz which is consistent with a vicinal bond angle (θ) of 170 ± 20°.[6] The vinyl proton (7.74 ppm) is consistent with an *s-cis* CH=C–C=O configuration[7] rather than the proposed *s-trans* configuration for tentoxin.[3] By combining the dehydrophenylalanyl *s-cis* configuration with the glycine α-N-NH 170° bond angle with the conformationally restrictive proline residue,

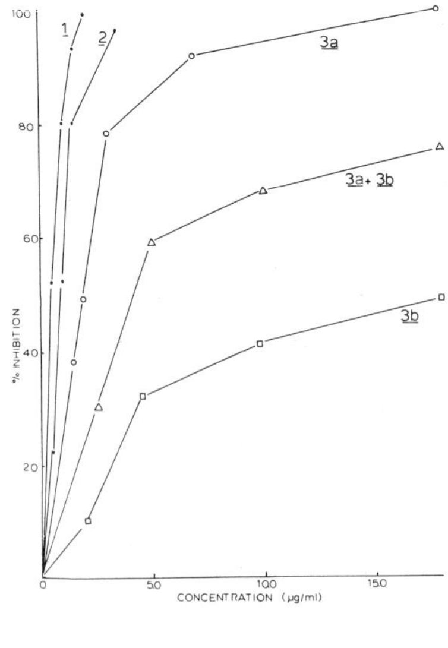

Fig. 1. Inhibition of coupled electron transport in lettuce chloroplasts by tentoxin analogs. Tentoxin, *1*; [L-Pro^1]-tentoxin, *2*; [D-MeAla^1]-tentoxin, lower R_f conformer, *3a*; [D-MeAla^1]-tentoxin, higher R_f conformer, *3b*; *3a* + *3b* indicates an equilibrium mixture of *3a* and *3b.*

Table I. ^{1}H nmr and CD Parameters of Tentoxin Analogs.[a]

	1-Position	L-Leu	MePhe[(Z)Δ]	Gly	Ø CD in CH_3OH (Ø x 10^{-3} λ nm)
1 L-MeAla	αH 4.3 βH 1.52 NMe 2.77	αH 4.16 (8.9) δH 0.52, 0.61 NH 7.2 (8.9)	NMe 3.19 C_6H_5 7.27 βH 7.75	H_o 3.51 (2,15) H_i 5.2 (10,15) NH 7.97 (10,<2)	4.5 (315), 13 (305), 19 (284), 9 (265), -12 (250), -21 (245), -33 (240), -14 (235), -3 (230), 9 (225), -9 (220)
2 [L-Pro^1]	αH 4.49 βH 2.41 δH 3.54	αH 4.4 (7) δH 0.51, 0.63 NH 7.3 (7)	NMe 3.19 C_6H_5 7.41 βH 7.76	H_o 3.54 (1,14.8) H_i 5.13 (10.5, 14.8) NH 7.93 (10.5,1)	2 (320), 11 (305), 19 (281), 16 (270), 7 (260), -8 (250), -19 (245), -27 (235), -18 (230), -6 (227), 3 (226), 6 (225), 0 (220), -8 (218)
$3a_1$[b] [D-MeAla^1]	αH 4.67 βH 1.58 NMe 3.18[c]	αH 4.41 (7) δH 0.39[c], 0.55, 0.77 NH 8.6 (7)	NMe 3.33[c] C_6H_5 7.49[c] βH 7.26	H_o 3.68 (<2, 15.5) H_i 5.1 (10,15.5) NH 8.22 (10)	CD obtained as mixture of $3a_1$ and $3a_2$ at 27°. 2 (320), 5 (300), 8 (285), 12 (270), 15 (255), 0 (240), -9 (235)
$3a_2$[b] [D-MeAla^1]	αH 4.67 βH 1.58 NMe 3.03[c]	αH 4.41 δH 0.39[c], 0.55, 0.77 NH 8.57	NMe 3.27[c] C_6H_5 7.39[c] βH 7.77	H_o 3.5 (<1,18) H_i 4.22 (7.5, 18) NH 7.8 (7.5)	
3b [D-MeAla^1]	αH 4.48 βH 1.52 NMe 3.04	αH 4.17 (6) δH 0.53, 0.77 NH 6.0 (6)	NMe 3.18 C_6H_5 7.44 βH 7.73	H_o 3.7 (<1,16) H_i 5.19 (10,16) NH 6.94 (10,<1)	-3 (320), -10 (300), -15 (285), -9 (275), 0 (265), 9 (255), 18 (245), 0 (236), -9 (235)

[a]Chemical shifts at 30° C reported in ppm from internal TMS. Coupling constants in Hz. Only vicinal NH-CαH and ring geminal coupling constants reported.

[b]Data obtained at -30° C in deuteriochloroform.

[c]Assignment of δ protons to individual conformer is ambiguous.

the major features of the [L-Pro^1]-tentoxin conformation are defined (Figure 2). (^{13}C nmr established a *cis* Gly-Pro bond and UV data established a *cis* Leu-MePhe[(*Z*)Δ] bond.) There is rapid rotation about the ψ_{Pro},ϕ_{Leu} angles. Because of the close spectral similarity between *2* and *1*, tentoxin must also have this conformation. The torsion angles for *1* and *2* are: ϕ_1, -80; ψ_1, -10. ϕ_2, -120; ψ_2, +70. ϕ_3, -90; ψ_3, -20. ϕ_4, -130; ψ_4, +70°.

[D-MeAla^1]-tentoxin (*3*) and D-Pro^1-tentoxin (*4*) were synthesized and found to exist in multiple conformations. Two *conformers* (*3a, 3b*) of compound *3* (not diastereomers) were isolated by preparative TLC at 4°C. The energy of activation for equilibration between conformers is 23 ± 1 kcal.

Low temperature nmr studies (-30°) showed that the slow moving conformer *3a* is a mixture (1:1) of two conformers, $3a_1$ and $3a_2$ (Table I). The nmr data show that in conformer $3a_1$ θ_1,θ_2 for GlyNH-CH_1H_2 is about 180°, and 0° while in $3a_2$, θ_1,θ_2 for GlyNH-CH_1H_2 is 90° and either ± 30° or ± 140°. Two types of vinyl protons are evident (7.74, 7.26 ppm) and these can be assigned to the *s-cis* ($3a_2$) and *s-trans* ($3a_1$) configurations respectively. The *s-trans* conformer $3a_1$ is closely related to the *cyclo*(Sar)$_4$ conformation.[6] Conformations $3a_1$ and $3a_2$ differ by a rotation about the ψ_{Phe},ϕ_{Gly} bonds. The activation barrier (E_a) for this rotation is 13 kcal/mole.

Analysis of conformer 3b. The glycine chemical shifts and coupling constants in *3b* (Table I) indicate Gly θ_1,θ_2 is 170° and -70°, and the vinyl proton (7.73 ppm) is consistent with an *s-cis* configuration. Together these indicate that in *3b* the vinyl proton, the Gly inner proton and the MePhe[Δ] carbonyl[4] are approximately coplanar (± 30°) as was found for *1*. However there are significant

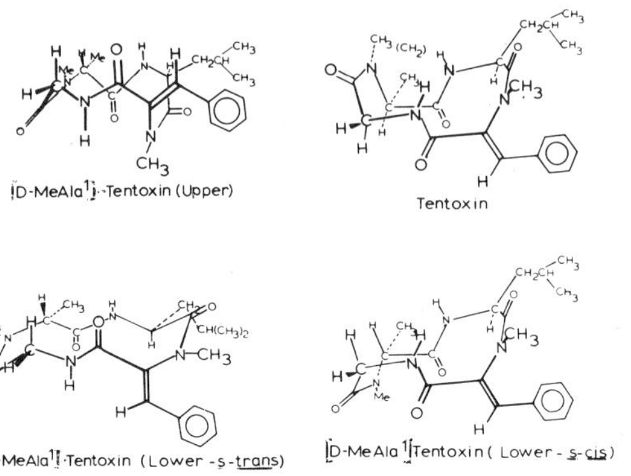

Fig. 2. Solution conformations of tentoxin analogs: upper right, *1* and *2*; upper left, *3b*, lower left, *3a*$_1$; lower right, *3a*$_2$.

differences in the spectral properties of *1* and *3b*. In *3b* the Leu NH (6 ppm) is upfield which suggests it is in a shielding environment. Furthermore the CD spectra for *1* and *3b* are very different. In *1*, the CD is positive at 284 nm (Φ = +19,000) whereas in *3b* it is negative (Φ = –15,000). All other tentoxin analogs, *2* and *3a* have positive elipticities at 280 nm. To explain these differences the conformation shown in Figure 2 is proposed for *3b*.

Effect of conformation on biological activity. The inhibition of coupled electron transport[1] by the tentoxin conformers is shown in Figure 1. Inhibition was measured at 25°C and data collected in the first 2–3 min before significant equilibration *3a* ⇌ *3b* occurs ($T_{1/2}$ = 20 min). It is apparent that the biological activities of the two conformers *3a* and *3b* differ particularly at the lower concentrations, and that the activity of *3a* approaches that of *1*. The ID_{50} of *3a* is 2 μg/ml vs. 1 μg/ml for *1*. The *s-cis* conformer *3a*$_2$ is essentially isosteric with *1* except for the projection of the Gly-MeAla amide bond. The data establish that in this system, different conformers of the same molecule have different biological activities.

We thank National Institutes of General Medical Sciences (GM 19311) for research support and Prof. R. D. Durbin, Dr. J. A. Steele and T. Uchytil for biological data.

References

1. IUPAC Nomenclature Commission (1970) *J. Org. Chem.* **35**, 2849–2867.

2. Steele, J. A., Uchytil, T. F., Durbin, R. D., Bhatnagar, P. & Rich, D. H. (1976) *Proc. Nat. Acad. Sci. USA* **73**, 2245–2248.

3. Meyer, W. L., Templeton, G. E., Grable, C. I., Jones, R., Kuyper, L. F., Lewis, R. B., Sigel, C. W. & Woodhead, S. H. (1975) *J. Amer. Chem. Soc.* **97**, 3802–3809.

4. Rich, D. H., Bhatnagar, P., Mathiaparanam, P., Grant, J. A. & Tam, J. P. (1977) submitted for publication.

5. Dale, J. & Titlestad, K. (1970) *J. Chem. Soc. Chem. Commun.*, 1403–1404.

6. Bystrov, V. F., Ivanov, V. T., Portnova, S. L., Balashova, T. A. & Ovshinokov, Yu. A. (1973) *Tetrahedron* **29**, 873–877.

7. Kevill, D. N., Weiler, E. D. & Cromwell, N. H. (1964) *J. Org. Chem.* **29**, 1276–1278.

CONFORMATIONAL STUDIES OF CYCLIC PENTAPEPTIDES BY NMR SPECTROSCOPY

HORST KESSLER, YVES A. BARA, AXEL FRIEDRICH, WOLFGANG HEHLEIN, PETER KONDOR, MICHAEL MOLTER,
Institut für Organische Chemie, Laboratorium Niederrad, Universität Frankfurt a.M., Theodor-Stern-Kai 7, D-6000 Frankfurt am Main 70, W-Germany

Conformational Studies of Peptides in Solution are often complicated by a rapid equilibrium between different forms. For example *cyclo* [Gly_5] and *cyclo*-[Phe_5] both show equivalence of the five amino acids in their ^1H and ^{13}C nmr spectra in DMSO at room temperature.[1]

Although a symmetric conformation of cyclopentapeptides is allowed[2] we suppose that a rapid equilibration causes the apparent symmetry. The mean temperature coefficients of the NH proton chemical shifts support this assumption. Therefore the only molecules suitable for a more detailed investigation are those in which one conformation strongly dominates the others.

We synthesized all cyclic pentapeptides of the structure *cyclo* [Phe_nGly_{5-n}][3] and used the temperature coefficients of the NH-chemical shift, $\Delta\delta/T$, as a criterion for selection. We consider a peptide to populate mainly one conformation when showing extreme values for $\Delta\delta/T$ (large *and* small). The temperature coefficients (Figure 1) of *cyclo* [Phe_4-Gly], *cyclo* [Phe_2-Gly_3], and *cyclo* [Phe-Gly-Phe-Gly_2] are of a mean order of magnitude, but three of the six phenylalanine- and glycine-containing pentapeptides (*cyclo* [Phe_3-Gly_2], *cyclo* [Phe-Gly_4], and *cyclo* [Phe_2-Gly-Phe-Gly]) show relatively marked differences and were selected for detailed conformational studies.

Cyclo [Phe_3Gly_2] is discussed more in detail in Ref. 1. The assignment of the signals was achieved by the synthesis of derivatives with deuterated amino acids (α-deuterio-glycine, α-deuterio-L-phenylalanine). Their spectra are shown in Figure 2.

Phe^2-NH and Gly^5-NH exhibit small temperature coefficients in DMSO indicating intramolecular orientation. This is supported by the effects caused by $CDCl_3$ addition (high field shifts only for the NH signals of Phe^1, Phe^3, and Gly^4 above 50% $CDCl_3$). Assuming that all peptide bonds have the *trans* conformation, an intramolecular hydrogen bond of Phe^2-NH is only possible to the Gly^5-CO (3 → 1 hydrogen bond, γ-loop). Gly^5-NH is involved either in a γ-loop to the Phe^3-CO or a β-loop to the Phe^2-CO. The geminal coupling constant of Gly^4 ($^2J_{HCH}$ = 15.4 Hz) would support a β^{II}-loop. Dilution studies with trifluoroethanol and H/D exchange rates are in agreement with these conclusions.[1]

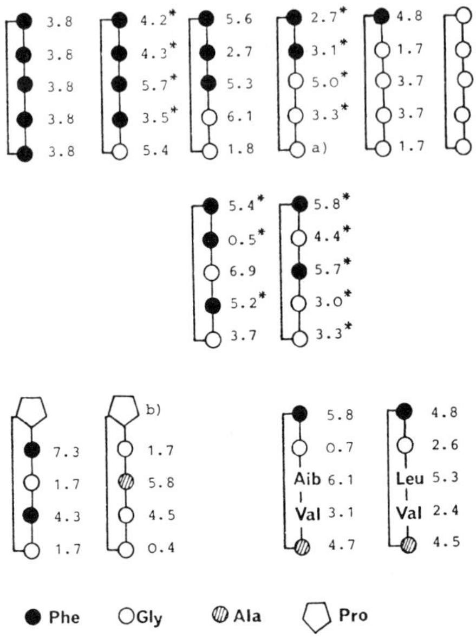

Fig. 1. Temperature coefficients of the NH chemical shift in cyclopentapeptides in DMSO at 90 MHz. Measuring range 50°–120°C. Values marked by an asterisk are not definitely assigned. Values in 10^{-3} ppm/K. a) Signal covered at 90 MHz. b) Results from Wüthrich and coworkers.[4]

For *cyclo* [Pro-Phe-Gly-Phe-Gly] we also recently suggested a β,γ-structure.[5] Such a structure may well be reasonable for the peptide *cyclo* [Pro-Gly-Ala-Gly$_2$],[4] where two small coefficients for Gly^2-NH and Gly^5-NH indicate intramolecular orientation. Cyclopentapeptides with larger side chains (*cyclo*-[Phe-Gly-Aib-Val-Ala] and *cyclo* [Phe-Gly-Leu-Val-Ala]), which we synthesized to improve the solubility in lipophilic media, again show two relatively small coefficients for Gly-NH and Val-NH (unpublished data).

It looks as if the cyclopentapeptides studied so far have a common structure with two intramolecular hydrogen bonds. In all of them, glycine seems to prefer a position where its NH-group forms a β- (or γ-) loop whereas the CO-group of the same glycine residue is always involved in a γ-loop.

All results indicate that γ-loops in smaller peptides are energetically less unfavored than usually assumed. In larger peptides the 180° folding of the backbone in a β-loop is ideal for further stabilization, whereas this is prevented by the 120° folding of a γ-loop. This geometric arrangement is probably responsible for the fact that γ-loops are rarely found in peptides and proteins.

This work was supported by the Deutsche Forschungsgemeinschaft and the Fonds der Chemischen Industrie.

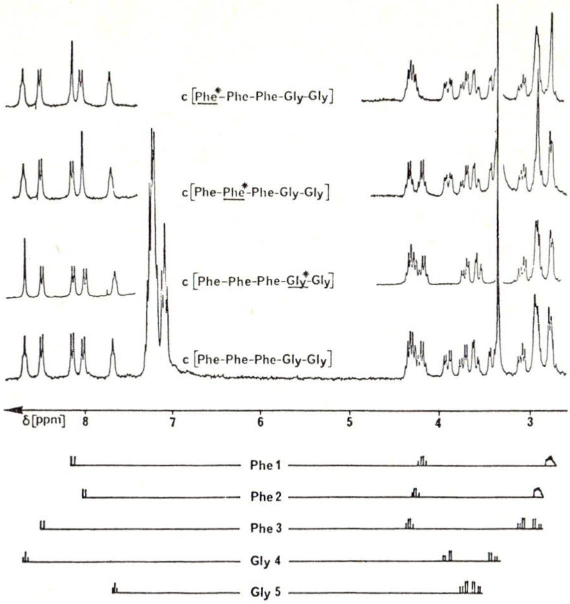

Fig. 2. 270 MHz-nmr spectra of *cyclo*[Phe_3-Gly_2] and some deuterated analogs in d_6-DMSO at 20°C.

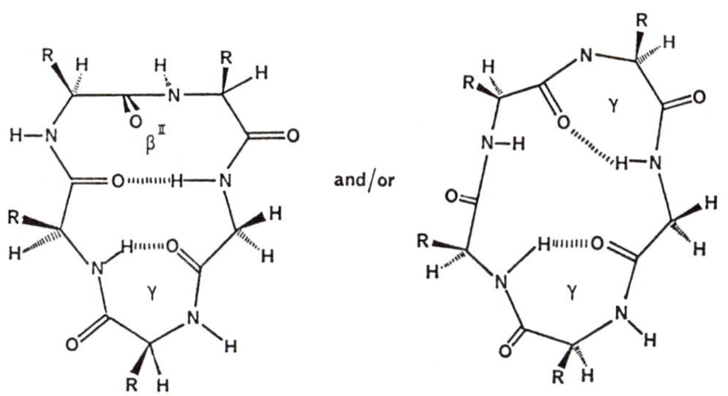

Fig. 3. Conformation of cyclopentapeptides in DMSO solution.

References

1. Bara, Y. A., Friedrich, A., Kessler, H. & Molter, M. (1977) *Chem. Ber.*, in press.

2. Ramakrishnan, C. & Sarathy, K. P. (1969) *Int. J. Prot. Res.* **1**, 63–71.

3. Bara, Y. A., Friedrich, A., Hehlein, W., Kessler, H., Kondor, P., Molter, M. & Veith, H.-J. (1977) *Chem. Ber.*

4. Meraldi, J. P., Schwyzer, R., Tun-kyi, A. & Wüthrich, K. (1972) *Helv. Chim. Acta* **55**, 1962–1973.

5. Demel, D. & Kessler, H. (1976) *Tetrahedron Lett.*, 2801–2804.

CONFORMATIONAL AND ION BINDING STUDIES ON A CYCLIC PENTAPEPTIDE: EVIDENCE FOR β AND γ TURNS IN SOLUTION

LILA G. PEASE and CHRISTOPHER WATSON, *Department of Chemistry, Amherst College, Amherst, Massachusetts 01002*

The cyclic pentapeptide *cyclo*(Gly-Pro-Gly-D-Ala-Pro) was designed from model building and sequence considerations so that a likely conformation might contain both a β-turn and a γ-turn, and hence two different types of transannular hydrogen bonds, one 1 ← 4 and the other 1 ← 3 (see Scheme 1). A similar conformation has been proposed for the pentapeptide, *cyclo*(Pro-Phe-Gly-Phe-Gly) in solution.[1] The simultaneous existence of both of these types of intramolecular hydrogen bonds in one peptide should provide an opportunity for comparison of their spectral properties and conformational details. In addition, since no other homodetic cyclic pentapeptides have been examined for ion binding capability, we have investigated the interaction of the present cyclic pentapeptide with cations.

Free Conformation

^{13}C nuclear magnetic resonance (nmr) spectra of *cyclo*(Gly-Pro-Gly-D-Ala-Pro) (synthesis to be reported elsewhere) in $CDCl_3$, CD_3CN, DMSO-d_6, and D_2O revealed that the peptide occurs in one major conformation which contains all-*trans* peptide bonds (to the exclusion of any other conformation, except in D_2O where 20% of the peptide adopts a one-*cis* conformer). Unusual chemical shifts are seen among the proline signals: in particular, one C^{α} and one C^{γ} resonance occur at exceptionally high field (see Figure 1). Although many factors

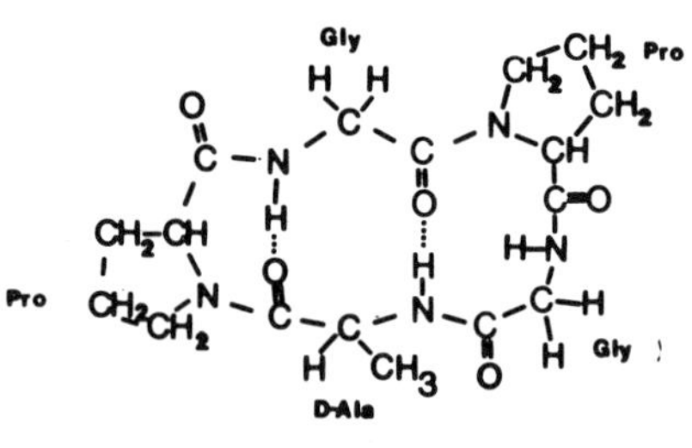

Scheme 1

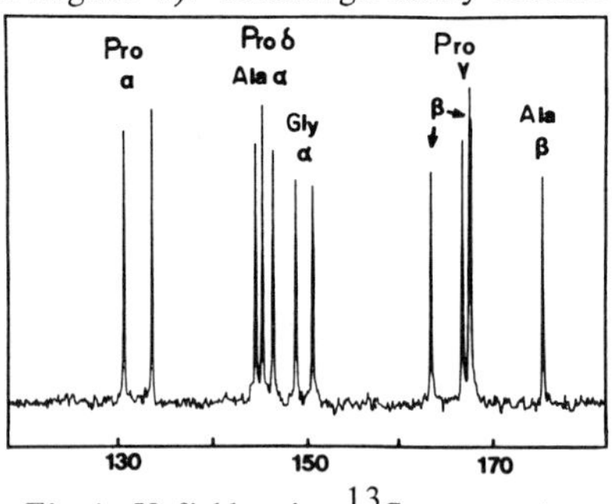

Fig. 1. Upfield region ^{13}C nmr spectrum (25.00 MHz) of *cyclo*(Gly-Pro-Gly-D-Ala-Pro) in $CDCl_3$. Concentration 100 mg/ml. Chemical shifts in this and Figure 3 are ppm upfield from external $^{13}CS_2$.

may contribute to ^{13}C chemical shifts, the dependence of the Pro C^{γ} resonance position on the proline ψ angle is well-established.[2-4] Here, as in *cyclo*(Pro-Gly)$_3$, the high field position signals a low *trans'* ψ angle, as expected for a γ turn with Pro in position 2 of the $1 \leftarrow 3$ hydrogen bond. Also consistent with the presence of a γ turn in the cyclic pentapeptide is a long wavelength minimum in the circular dichroism spectrum.[4]

^1H nmr experiments demonstrate the presence of *two* N-H's (those of the Gly preceding Pro and of D-Ala) which do not participate in intermolecular interactions, and are most likely intramolecularly hydrogen-bonded. The evidence includes: 1) low temperature coefficients of their chemical shifts in DMSO-d_6 and $CDCl_3$, 2) concentration independence of their resonances in $CDCl_3$, and 3) invariance of their chemical shifts in $CDCl_3$ solution as acetone (a hydrogen-bond acceptor) is added. $J_{N\alpha}$ values were the same (within experimental error) in $CDCl_3$, DMSO-d_6, and CD_3CN: D-Ala $J_{N\alpha}$ = 8.5 Hz; Gly (preceding Pro) $J_{N\alpha}$ = 9.5, 2.5 Hz; and Gly (preceding D-Ala) $J_{N\alpha}$ = 6.0, 7.0 Hz.

A conformation consistent with the nmr results and sterically reasonable (from molecular models) is described by the ϕ,ψ values shown in Table I. In Figure 2 is shown a Corey-Pauling Koltun (CPK) model of the proposed free conformation of *cyclo*(Gly-Pro-Gly-D-Ala-Pro). The spectral evidence suggests that the molecule is rigid, since markedly different chemical shifts are observed for the two prolyl and glycyl residues, and for the two H^{α}'s of one of the glycines ($\Delta\delta$ = 1.4 ppm).

Table I. ϕ and ψ Values for *Cyclo*(Gly-Pro-Gly-D-Ala-Pro)

	Gly	-	Pro	-	Gly	-	D-Ala	-	Pro
ϕ	120		-60		90		160		-80
ψ	180		120		0		-120		70

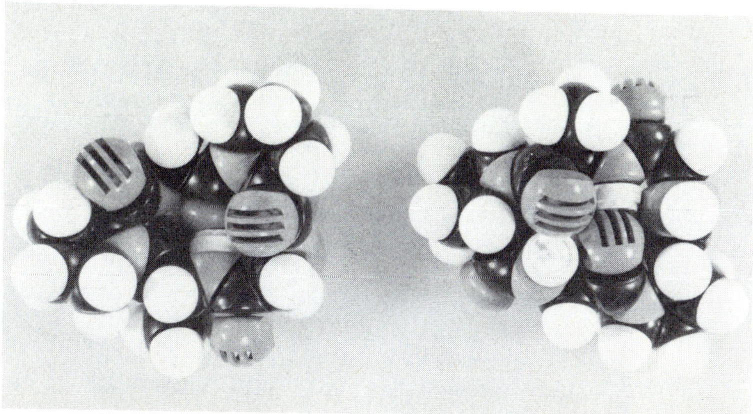

Fig. 2. CPK model of proposed solution conformation of *cyclo*(Gly-Pro-Gly-D-Ala-Pro). Note two hydrogen bonds.

The results of a crystal-structure determination on *cyclo*(Gly-Pro-Gly-D-Ala-Pro) are presented elsewhere in this volume.[6] The conformation observed in the crystal is very similar to that which is proposed in solution. It is also all-*trans*, and is stabilized by the same two intramolecular hydrogen bonds. The principal distinctions between the two conformations are the ϕ,ψ angles of the Gly preceding Pro and of the D-Ala, which are less extended in the crystal structure, and the presence of a non-planar Pro-Gly peptide bond (in the γ-turn), a feature which is not experimentally demonstrable in solution. The fact that the crystal and proposed solution conformations are so similar suggests strongly that the intramolecular interactions present in this cyclic peptide are energetically very important in stabilizing the observed conformation. The conclusion is also supported by the result that the all-*trans* β,γ-turn conformer is predominant in all solvents examined, regardless of their polarity or capability for hydrogen bonding.

Ion Binding

When perchlorate salts of Li^+, Mg^{++}, Mn^{++}, Ba^{++}, or Ca^{++} were added to acetonitrile solutions of the cyclic pentapeptide, large changes were observed in the circular dichroism spectrum. (Little or no binding was observed with Na^+, K^+, or NH_4^+.) Titrations of peptide solutions with cation revealed that at least two complexes were found, as *two* inflections occurred in a plot of the change in ellipticity as a function of log[cation].

^{13}C nmr spectra of peptide in CD_3CN were obtained in the presence of varying concentrations of $Mg(ClO_4)_2$. Particularly informative is the region containing the Pro C^β and C^γ signals (Figure 3). Two significant changes are observed: 1) A *new* set of resonances appears and initially grows as the cation concentration increases, then decreases. These resonances can be identified with a conformer containing one *cis* X-Pro bond. 2) *Shifts* are seen in the peaks originally associated with free peptide. In particular, the high field Pro C^β moves downfield as the cation is added, until it eventually crosses the other C^β resonance.

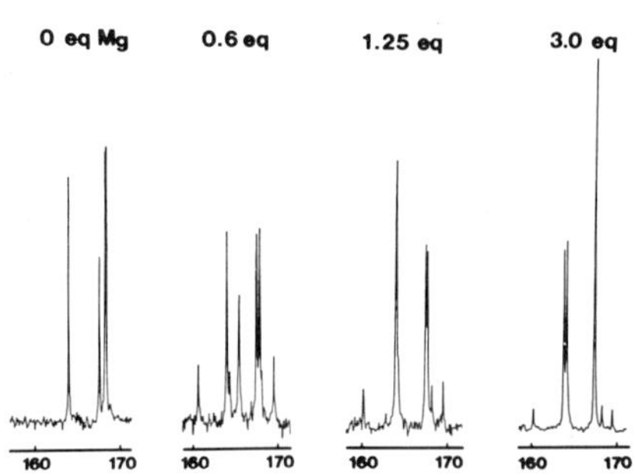

Fig. 3. Proline C^β and C^γ region of ^{13}C nmr spectra of *cyclo*(Gly-Pro-Gly-D-Ala-Pro) in CD_3CN at various $[Mg(ClO_4)_2]$. Peptide concentration 0.05 *M*.

These observations indicate that among the complexing species are both all-*trans* and one-*cis* structures, and that an all-*trans* species is predominant at high cation concentration. Model building suggests that the Gly-Pro bond is the more likely one to isomerize. Conformations can be found (among all-*trans* and one-*cis* structures) which have three carbonyls favorably oriented for ion binding. With the data presently available, it is not possible to define the stoichiometry or detailed conformations of the complexes formed.

Acknowledgment is made to the Donors of the Petroleum Research Fund, administered by the American Chemical Society, for the support of this research. We also thank the National Science Foundation and the Research Corporation for major support in the purchase of JEOL FX 60 and FX 100 nmr instruments. The capable technical assistance of Mr. Craig Leonardi is gratefully acknowledged.

References

1. Demel, D. & Kessler, H. (1976) *Tetrahedron Lett.* 2801–2804.
2. Madison, V., Atreyi, M., Deber, C. M. & Blout, E. R. (1974) *J. Amer. Chem. Soc.* **96**, 6725–6734.
3. Siemion, I. Z., Wieland, T. & Pook, K.-H. (1975) *Angew. Chem.* **14**, 702–703.
4. Deber, C. M., Madison, V. & Blout, E. R. (1976) *Acc. Chem. Res.* **9**, 106–113.
5. Bystrov, V. F., Portnova, S. L., Balashova, T. A., Koz'min, S. A., Gavrilov, Yu. D. & Afanas'ev, V. A. (1973) *Pure and Appl. Chem.* **36**, 19–34.
6. Karle, I. L. (1977) in *Peptides: Proceedings of the 5th American Peptide Symposium,* Goodman, M. & Meienhofer, J., Eds., John Wiley and Sons, Inc., New York, pp. 274–276.

ASYMMETRIC, HYDROGEN-BONDED PEPTIDE COMPLEXES

VINCENT MADISON, JERRY LASKY and BRUCE CURRIE
Department of Medicinal Chemistry, University of Illinois at the Medical Center, Chicago, Illinois 60680

As the bioactivity of a peptide is usually dependent on its conformation, reliable prediction of peptide conformation is essential to the design of compounds with a specific activity. Successful conformational predictions have depended on calibration of potential functions and parameters with extensive experimental data (for example, see Refs. 1 and 2).

In a number of cases specific peptide-solvent interactions have been shown to select the conformational state (for instance, in solutions of polyproline[3] or a cyclic hexapeptide[4]), but little experimental information is available on specific noncovalent interactions between peptides and other molecules.

The experiments described herein were focused on the geometry and the stability of hydrogen-bonded complexes between peptides and aromatic compounds. These data should prove useful in assessing the conformational consequences of intermolecular hydrogen-bonds and π-interactions for specific geometries in solution.

Experimental Procedures

The blocked dipeptide amide Cbz-D-Val-L-Pro-NH_2 was synthesized by the mixed-anhydride method as modified by Chang, et al.[5] The carbobenzoxy group was removed by catalytic hydrogenation and the cyclization effected by warming to 35°C while evaporating the solvent. The commercial, blocked dipeptide Cbz-L-Val-L-Pro-OMe was cyclized in an analogous manner. The cyclic dipeptides were crystallized from methanol/ether and recrystallized from chloroform. Their CD matched that which has been reported.[6]

CD measurements were made at 22°C, in spectral grade chloroform (Matheson, Coleman and Bell), which is stabilized with 0.03% pentene and dried over activated molecular sieves, on a Jasco J40A spectrometer which is equipped with a piezo-elastic modulator.

Results and Discussion

Neither the aromatic compounds nor the peptides utilized have CD in the aromatic region. However, in mixtures of a peptide and an aromatic compound in chloroform solution, induced CD arises in this region (Fig. 1), due to interactions between the aromatic and peptide chromophores. This induced CD is evidence for formation of specific asymmetric complexes.

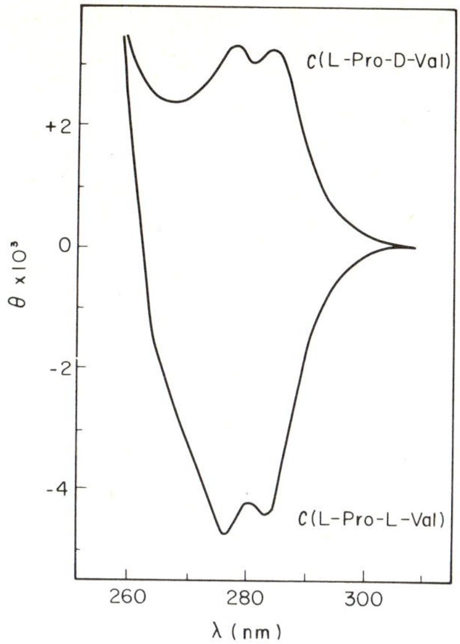

Fig. 1. Observed ellipticity for 0.2 *m* 2,6-dichlorobenzoic acid complexed with 0.3 *m* cyclic dipeptide in chloroform.

The geometry of the peptide-aromatic compound complex is dependent on the configuration of the amino acid in the cyclic dipeptide. Even though the peptide backbones of diastereomers such as *c*(L-Pro-L-Val) and *c*(L-Pro-D-Val) have similar conformations[7] and similar peptide CD[6], the induced CD band at 300 nm of pyridone complexes is positive for *c*(L-Pro-L-Val) and negative for *c*(L-Pro-D-Val). Further, for complexes of these peptides with 2,6-dichlorobenzoic acid, the induced CD is positive for the D-Val compound but negative for the L-Val compound (Fig. 1). Binding of either compound to *c*(L-Pro-L-Pro) is weaker, indicating that the -CONH- group is the primary binding site in the valine peptides.

A reasonable model for the complex between 2,6-dichlorobenzoic acid and a cyclic dipeptide contains two hydrogen bonds in a planar, six-membered ring, with the benzene ring perpendicular to this plane (Fig. 2). In this model the electric transition moments of the secondary amide and benzene chromophores (μ_1 and μ_2) are perpendicular and the interaction energy between the two is zero (Fig. 2), thus no CD would arise in the 280 nm aromatic band due to these two chromophores.[8] (A small CD band could result from interaction of the tertiary amide and benzene chromophores.) The favored side-chain rotamer of *c*(L-Pro-L-Val) places a methyl group above the peptide N-H.[7] This methyl group could displace the acid carbonyl oxygen downwards as indicated in Fig. 2. The resulting tilt would produce a negative interaction energy and a negative CD band as is observed (Fig. 1). Similarly, in *c*(L-Pro-D-Val) a methyl group is placed below the N-H, which could displace the oxygen upwards, thus reversing the sign of the interaction energy and reversing the sign of the CD band (Figs. 1 and 2). For pyridone complexes, a similar reversal of the aromatic CD bands is predicted for displacements from a planar, hydrogen-bonded ring.

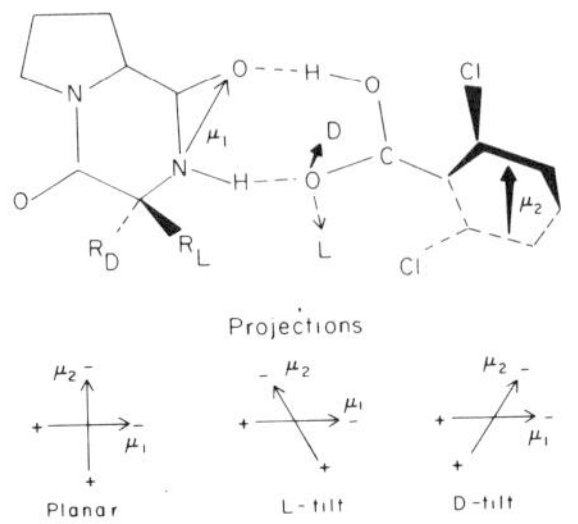

Fig. 2. Peptide-dichlorobenzoic acid complex. Transition moment projections are shown for three geometries (see text).

Avoiding self-association, a binding constant of about 100m^{-1} and a molar ellipticity of −4000° at 300 nm were determined by the CD for the pyridone-*c*(L-Pro-D-Val) complex. Further data are necessary to see if the demonstrated differences in complex geometry will yield significant differences in stability.

Conclusions

Observations of induced CD bands indicate that 2-pyridone and 2,6-dichlorobenzoic acid form specific, asymmetric complexes with cyclic dipeptides. A plausible model for these complexes contains two hydrogen bonds in a six-membered ring. The observed sign changes for the CD of complexes of either aromatic compound with *c*(L-Pro-L-Val) vs. *c*(L-Pro-D-Val) can be explained by opposite deflections from a planar ring. This ring is a null plane for the induced CD so that the CD is exquisitely sensitive to small deviations in hydrogen-bond geometry.

We are grateful to the National Institutes of Health (GM22514) and the Physical Chemistry Division of Lilly Research Laboratories for financial support.

References

1. Warshel, A., Levitt, M. & Lifson, S. (1970) *J. Mol. Spectrosc.* **33**, 84–99.
2. Hagler, A. T., Huler, E. & Lifson, S. (1974) *J. Amer. Chem. Soc.* **96**, 5319–5335.
3. Strassmair, H., Engel, J. & Zundel, G. (1969) *Biopolymers* **8**, 237–246.
4. Madison, V., Atreyi, M., Deber, C. M. & Blout, E. R. (1974) *J. Amer. Chem. Soc.* **96**, 6725–6734.
5. Chang, J. K., Sievertsson, H., Currie, B. L., Bogentoft, C., Folkers, K. & Bowers, C. Y. (1972) *J. Med. Chem.* **15**, 623–627.
6. Madison, V., Young, P. E. & Blout, E. R. (1976) *J. Amer. Chem. Soc.* **98**, 5358–5364.
7. Young, P. E., Madison, V. & Blout, E. R. (1976) *J. Amer. Chem. Soc.* **98**, 5365–5371.
8. Schellman, J. A. (1968) *Acct. Chem. Res.* **1**, 144–151.

ANALYSIS OF THE ABSORPTION SPECTRUM OF ACTINOMYCIN D AND THE FORMATION OF ITS COMPLEXES WITH DEOXYNUCLEOTIDES

HENRY E. AUER, *Department of Biochemistry, University of Rochester School of Medicine, Rochester, N.Y. 14642,*
BARBARA E. PAWLOWSKI-KONOPNICKI, YU-CHIH CHEN CHIAO and THOMAS R. KRUGH, *Department of Chemistry, University of Rochester, Rochester, N. Y. 14627*

The chromophore of the antibiotic actinomycin D (AMD) is the 2-amino-3-oxo-4,6-dimethylphenoxazine (herein termed phenoxazone) ring, which is comprised of a benzenoid portion and a quinoid portion at opposite ends of the ring system. Bonded to each of these (at the 1 and 9 positions) are two identical cyclic pentapeptides. This antibiotic forms a long-lived intercalated complex with DNA with an apparent requirement for guanine as one of the bases at the binding site. The structure of a cocrystalline complex of AMD:deoxyguanosine (1:2) shows that the guanine rings are stacked on opposite faces of the phenoxazone ring over the benzenoid and quinoid portions, respectively.[1] Hydrogen bonds between the cyclic peptides and the 2-amino groups of the guanine residues are observed, which evidently contribute to the selectivity for this residue.

These observations have stimulated us to examine model systems comprised of AMD and simple deoxynucleotides, using primarily circular dichroism (CD) spectroscopy. In this report we present an analysis of the spectral patterns which are obtained with AMD monomers, AMD dimers and complexes formed between AMD and certain deoxynucleotides. In addition, we describe titrations of AMD with nucleotides, which were monitored using absorption and CD spectroscopy. From the results it emerges that the visible absorption band of AMD is considerably more complex than may have been supposed and that certain wavelengths can discriminate between the binding processes occurring at the two sites.

Results

Analysis of the Visible Absorption Band of Actinomycin D

2-Aminodeoxyadenosine (n^2dAdo) has the same hydrogen bonding potential as does dG for interaction with AMD, and forms a complex of high affinity with the drug.[2] CD spectra of the n^2dAdo:AMD complex at two stages of formation are shown in Fig. 1. Of the three prominent bands, that at 490 nm appears only at low nucleotide:drug ratios, and is not apparent when lower AMD concentrations are used. For these reasons this band is assigned to a complex of high affinity formed between n^2dAdo and AMD dimers. The remaining two bands,

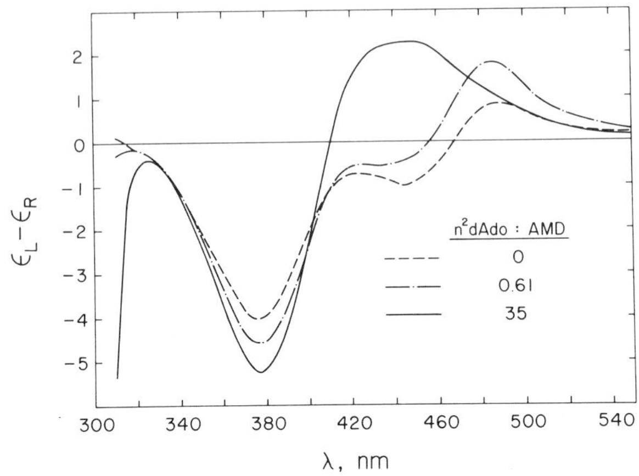

Fig. 1. CD spectra of 1.0×10^{-4} *M* AMD in the presence of added 2-aminodeoxyadenosine with the molar ratios indicated, at 0°C, 5*mM* potassium phosphate pH 7.0.

positive at about 450 nm and negative at about 375 nm are observed at saturation (Fig. 1) as well as in the spectrum of the complex with dGMP (Table I). Table I summarizes features of the optical spectra of AMD observed in this wavelength region under a variety of conditions, including the absorption spectrum which reveals no indication of the spectral complexity apparent in the CD spectra. It is seen from Table I that the CD bands at 370 nm, 440 nm and 490 nm respond in very different and characteristic ways to changes in the ambient conditions. As a consequence we may conclude that the absorption envelope manifested by AMD is in fact comprised of three electronic transitions centered, respectively, approximately at the wavelengths mentioned.

Table I. Spectral Features of Actinomycin D.[a]

State or Complex	370 nm	440 nm	490 nm
Absorption spectrum[b]	Unresolved tail	Abs. Max.	Unresolved tail
CD, Monomer[c]	(-)	(-) weak	Absent
CD, Dimer, high temp[c]	(-)	(-) 452 nm	Absent
CD, Dimer, low temp[c]	(-)	(-) 441 nm	(+)
CD, n^2dAdo:Dimer			(+)
CD, n^2dAdo or dGMP	(-)	(+)	(+)[d]
CD, pdG-dC	(-) strong	Absent	(-) 470 nm[e]
CD, poly(dG-dC)·poly(dG-dC) or DNA	(-) strong	(-) sh	(-) 465 nm[e]

a) Signs of CD extrema are shown as (+), positive and (-), negative; sh indicates shoulder. b) Ref. 3. c) Ref. 4. d) Unresolved positive tail extends to 550 nm. e) Zero intensity observed above 500 nm.

Wavelength Dependence of Nucleotide Titration Profiles

Titrations of AMD with certain deoxynucleotides were monitored in both absorption and CD modes at several wavelengths of interest. 380 nm and 465 nm were chosen as representative wavelengths for the transitions identified at 370 nm and 490 nm, respectively. The results with dGMP are shown in Fig. 2.

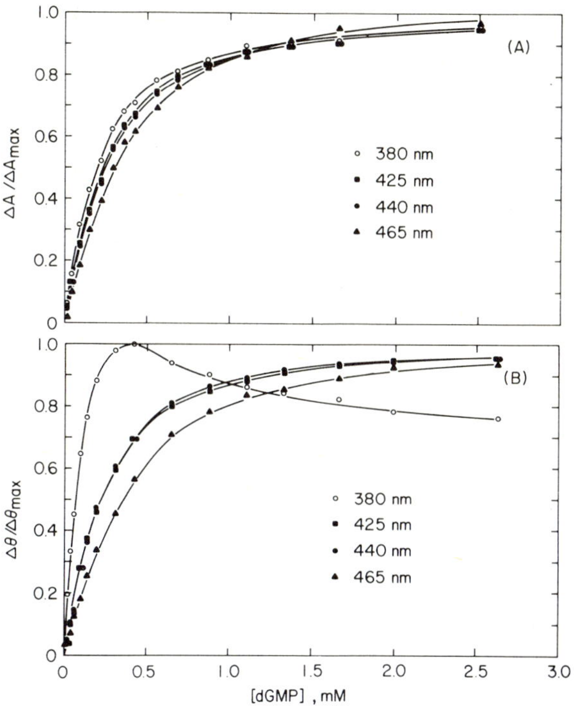

Fig. 2. Absorption (A) and CD (B) titration profiles of 1.0×10^{-4} *M* AMD with added dGMP, presented as the fraction of the maximal change observed at each wavelength. The actual titrations were carried to at least 7 *mM* dGMP. The changes are measured with respect to free AMD and are all positive values except for 380 nm, which represents an intensification of a negative extremum.

The binding isotherms at the various wavelengths do not superimpose. This can arise only if a) there exist two or more electronic transitions among the wavelengths chosen, b) they reflect processes localized on different portions of the phenoxazone chromophore, and c) these binding processes have distinguishable binding affinities. The sequence of binding strengths obtained in both absorption and CD titrations is 380 nm>425 nm≃440 nm>465 nm. (The biphasic nature of the CD titration profile at 380 nm was unaltered when the AMD concentration was changed. We conclude that it is due to interaction between the strong and weak binding processes manifested at this wavelength.) While titration of AMD with dGMP has been carried out before using nmr spectroscopy[2] the distinction of two binding processes with different intensities of complex formation could not be achieved, due to the high ratio of AMD to free dGMP required in that work.

dAMP binds much more weakly to AMD than does dGMP. In CD titrations with dAMP it is again observed that the 380 nm profile titrates with high affinity and the 465 nm profile characterizes a weak binding process. Comparison with the analogous nmr experiment[2], in which two binding processes were clearly manifested, permits identification of the 370 nm transition with processes occurring preponderantly at the benzenoid binding site and of the 490 nm transition with binding primarily at the quinoid site. The 440 nm transition apparently reflects significant contributions from both the benzenoid and quinoid portions of the phenoxazone ring.

Conclusion

The analysis presented here has successfully resolved the visible absorption spectrum of actinomycin D into a minimum of three unique electronic transitions. Complex formation at each of the two nucleotide binding sites of the drug is detectable with a high degree of spectroscopic selectivity by two of these bands. These findings open the way for more detailed study of the interactions of AMD with nucleotides and nucleic acids.

Supported by the American Cancer Society, the U.S. Public Health Service and the U.S. Energy Research and Development Administration.

References

1. Jain, S. C. & Sobell, H. M. (1972) *J. Mol. Biol.* **68**, 1–20.
2. Krugh, T. R. & Chen, Y. C. (1975) *Biochemistry* **14**, 4912–4922.
3. Ziffer, H., Yamaoka, K. & Mauser, A. B. (1968) *Biochemistry* **7**, 996–1001.
4. Auer, H. E., Pawlowski-Konopnicki, B. E. & Krugh, T. R. (1977) *FEBS Lett.* **73**, 167–170.

CARBON-13 ENRICHED HUMAN ENCEPHALITOGENIC (EAE) NONAPEPTIDE: SYNTHESIS AND CONFORMATIONAL STUDIES

C. M. DEBER, P. D. ADAWADKAR, and J. TOM-KUN,
Research Institute, Hospital for Sick Children, and Department of Biochemistry, University of Toronto, Toronto, Ontario, Canada

Experimental allergic encephalomyelitis (EAE), an autoimmune disease related to multiple sclerosis, can be induced in laboratory animals through intradermal injection of myelin basic protein, a principal protein component of the membranous myelin sheath protecting the nerves.[1] When it was discovered that a relatively small tryptic fragment of this protein, subsequently designated "EAE nonapeptide," was nearly equally active on a mole/mole basis in disease induction,[2] interest focused on the relationships between structure and function of this peptide, and the aspects of conformation responsible for its potent biological activity. The structure of EAE nonapeptide, Phe-Ser-Trp-Gly-Ala-Glu-Gly-Gln-Lys, derived from bovine myelin basic protein is given in Figure 1; the human analog is identical, except that the C-terminal Lys is replaced by Arg.

We now report the total synthesis, characterization, and preliminary measurements of spin-lattice relaxation times (T_1's), of human EAE nonapeptide, which has been enriched with carbon-13 both in backbone and side-chain atoms in a total of five positions: both Gly α-carbons, both Gly carbonyl carbons, and the Ala β-carbon (methyl group). Carbon-13 enrichment was carried out with the anticipation that the resulting enhanced resonances could be readily detected in nmr samples containing low, physiologically relevant levels of nonapeptide (ca. 1 mg/ml), and hence useful for relaxation and linewidth measurements under such conditions.

Human EAE nonapeptide was synthesized using solid phase peptide synthesis, starting with Boc-Arg(NO_2) – 1% cross-linked resin, and employing Boc-amino acids, DCC coupling, TFA deprotection, with minor modifications of standard procedures. The crude peptide product, cleaved from the resin with HF, was chromatographed on Bio-Gel P2, and subsequently on carboxymethyl-cellulose (CM-52). (In a second preparation, using Aoc-Arg(ϵ-Tos) – resin as starting material, only a single chromatography on Bio-Gel P2 was required.) In each synthesis, a major component (ca. 20%) of the crude product was shown to be EAE nonapeptide on the basis of the following observations: (1) correct amino-acid analysis; (2) single spot on TLC in several systems, with behavior similar to authentic bovine EAE nonapeptide; (3) appropriate carbon-13 spectrum; and (4) biological activity (i.e., the synthetic material displayed the capacity to induce EAE disease in guinea pigs). Details of the synthesis and characterization of this peptide will be presented elsewhere (manuscript in preparation).

Measurements of T_1 spin-lattice relaxation parameters for various carbon atoms in a peptide can be used to construct a dynamic picture of the motional properties of the peptide in solution,[3] and help to identify regions of specific conformational features, or relative immobilization of certain regions of the peptide chain, which might be induced by the binding or incorporation into a liposome or cell membrane. Such conformational events may occur upon interaction of EAE nonapeptide with liposomes or with lymphocytes, whose sensitization by EAE nonapeptide has been implicated in the initial steps of induction of EAE disease.[4,5] [In a related investigation, we have recently been able to observe the carbon-13 enriched resonance of methylated (at methionine S-CH_3) myelin basic protein incorporated (30% by weight) into a lipid vesicle (50/50 w/w phosphatidylcholine/phosphatidic acid). Preliminary results based on an increase observed in resonance linewidth (from 1 to 15 Hz vs. protein in free solution), and a significant drop in T_1, suggest substantial "immobilization" of the protein in the membrane (Deber, C. M., Moscarello, M. A., and Wood, D. D., in prep.).]

During the progress of synthesis of the human analog, T_1's were obtained on a sample of bovine EAE nonapeptide (see Figure 1). Overlap of several α-carbon resonances prevented unambiguous chemical shift assignments and measurement of their individual T_1's. Thus, the value of 99 millisec represents the "NT_1 value" of a spectral envelope of six α-carbon resonances which appeared to recover full amplitude at comparable rates. However, such data do not preclude the possibility of increased molecular motion near peptide chain ends, as often manifested by increasing T_1's toward terminal residues. The present results are comparable to the findings of Deslauriers et al.[6] for T_1 values of peptide hormones of molecular weight similar to EAE nonapeptide, including luteinizing hormone-releasing factor and several others.[7]

The carbon-13 spectrum of synthetic, carbon-13 enriched human EAE nonapeptide, depicting only the enriched resonances, is presented in Figure 2. Natural abundance atoms can be visualized clearly only after ca. 100,000 scans under these spectral conditions (some may be noted as bumps in the baseline at the spectrum amplitude in Figure 2), but clear spectra of the enriched resonances can be obtained in a few minutes with less than 1,000 transients. Spectra of

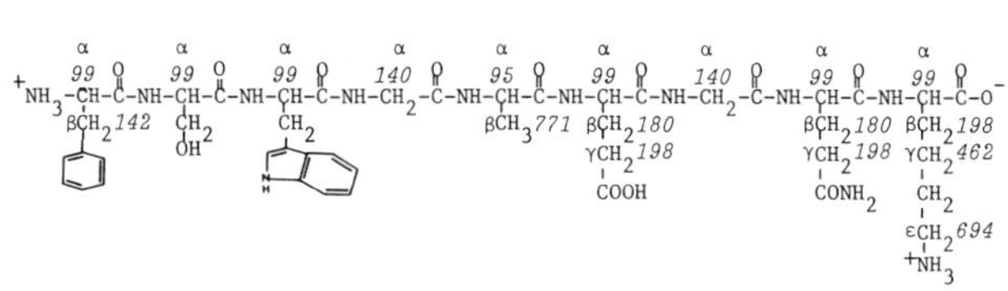

Fig. 1. Structure of bovine EAE nonapeptide. In the human analog, Arg replaces Lys at the C-terminus. Numbers in italics are values of NT_1 spin-lattice relaxation times (where N = number of directly attached protons) for natural abundance bovine EAE nonapeptide, given in milliseconds. T_1 data were recorded on an XL-100 spectrometer, 25 MHz, 50 mg/ml in D_2O.

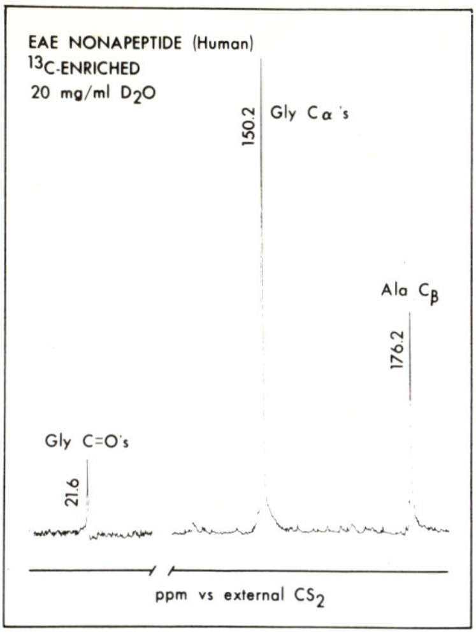

Fig. 2. Carbon-13 nmr spectrum (25 MHz) of synthetic human EAE nonapeptide (Phe-Ser-Trp-Gly^4-Ala-Glu-Gly^7-Gln-Arg), displaying the C-13 enriched resonances. Gly^4 C$^\alpha$ has been enriched 60%, and Gly^7 C$^\alpha$ has been enriched 85%. Ala C$^\beta$ is enriched 60%. Gly carbonyls are enriched 15–20%. See text for further discussion of this spectrum.

enriched resonances of 1 mg/ml solutions of this peptide are readily obtainable in a few hours. Values of NT_1 determined for enriched resonances of human EAE nonapeptide are given in Table I. These results demonstrate (a) that NT_1 values for corresponding resonances in bovine and human EAE nonapeptide are comparable, and (b) that placing the peptide at pH = 1 does not appear to have induced any conformational and/or stoichiometric changes. Despite the range of values in Table I for both resonances, they may likely be treated as essentially the same within experimental error. It is further noted that although there are two Gly residues in the molecule (which have been enriched with differing percentages of carbon-13 to facilitate assignments), in no spectra studied thus far have separate resonances been observed for the two Gly α-carbons. Separate resonances might be expected in instances where differences in local magnetic environments are not completely averaged, as might be the case in a region of ordered peptide structure.

The present results suggest a conformation for bovine and human EAE nonapeptide devoid of specific secondary or tertiary structure under the spectral conditions employed, but provide essential control data for forthcoming nmr studies of interactions of EAE nonapeptide with biological membranes.

Table I. NT_1 Values of Bovine and Human EAE Nonapeptide

NT_1 Values of Bovine and Human EAE Nonapeptide

Source	pH	NT_1 (millisec)^d Gly C^α	Ala C^β
Human[a]	7	156	825
Human[b]	7	176	831
Human[a]	1	160	777
Bovine[c]	7	140	771

[a,b] Two separate syntheses of ^{13}C-enriched human nonapeptide.
[c] Sample studied in Figure 1. [d] Estimated uncertainty in NT_1's: $\pm$ 10%.

We thank the Medical Research Council of Canada (Grant # MA-5810) and the Multiple Sclerosis Society of Canada for their generous support of this research. Synthetic bovine EAE nonapeptide was the kind gift of Dr. Ruth Nutt and her colleagues at Merck, Sharp, and Dohme, West Point, Pa.

References

1. Eylar, E. H. (1972) *Ann. N. Y. Acad. Sci.* **195**, 481–91.

2. Westall, F. C., Robinson, A. B., Caccam, J., Jackson, J. & Eylar, E. H. (1971) *Nature* **229**, 22–24.

3. For a review, see Gurd, F. R. N. & Keim, P. (1973) *Methods Enzymol.* **27**, 836–911.

4. Kies, M. W. (1973) in *Biology of Brain Dysfunction*, Vol. 2, Gaull, G. E., Ed., Plenum Press, New York, pp. 185–274.

5. Stone, S. H. (1961) *Science* **134**, 619–620.

6. Deslauriers, R., Levy, G. C., McGregor, W. H., Sarantakis, D. & Smith, I. C. P. (1975) *Biochemistry* **14**, 4335–4343.

7. For a review, see Deslauriers, R. & Smith, I. C. P. (1976) *Topics in Carbon-13 NMR Spectroscopy*, Vol. 2, Levy, G. C., Ed., Wiley-Interscience, New York, pp. 1–80.

MEASUREMENT OF THE BROWNIAN MOTION OF OLIGOPEPTIDE CHAIN ENDS RELATIVE TO EACH OTHER

ELISHA HAAS, EPHRAIM KATCHALSKI-KATZIR and IZCHAK Z. STEINBERG, *The Weizmann Institute of Science, Rehovot, Israel*

Linear flexible oligomer and polymer molecules have an enormous number of conformations when in the random coil state. Due to Brownian motion of the chain segments, the conformation of each molecule in solution changes perpetually and the separation between the ends of the molecules varies ceaselessly. The distribution of end-to-end distances of an ensemble of molecules at equilibrium and its change with time upon the disturbance of the equilibrium reflect the variability of the chain conformations and the dynamics of the interconversion of conformations, respectively. We have used nonradiative transfer of excitation energy[1] between chromophores attached to the ends of oligopeptide molecules as a tool for the experimental estimation of the distribution of end-to-end distances of the labeled peptide chains and of the Brownian motion of the chain ends relative to each other.

A series of oligopeptide derivatives, dansyl-[N^5-(2-hydroxyethyl-)-L-glutamine]$_n$-*N*-hydroxyethyl-L-β-naphthylalanine (abbreviated DG_nN) was prepared. The repeating units were N^5-(2-hydroxyethyl-)-L-glutamine residues, n varying from 4 to 9. The β-methyl-naphthalene chromophore served as the energy donor, whereas the dansyl chromophore served as the energy acceptor. The donor fluorescence was found to decay monoexponentially in the absence of acceptor. However, when the donor was attached to a flexible oligopeptide containing an acceptor, its fluorescence decay kinetics deviated markedly from monoexponential behavior.[2] The deviation was interpreted in terms of the spread in distances between donor-acceptor pairs on the various oligomer molecules, resulting in variability in efficiency of energy transfer between the pairs. The distribution function, $N(r)$, of end-to-end distances was evaluated for the above oligopeptides from the fluorescence decay of the donor, $I(t)$, obtained in a highly viscous (glycerol) solution,[2] by use of nonlinear least-squares procedures[3] for the analysis of the experimental data.

When the above peptides were dissolved in mixtures of glycerol and trifluoroethanol, the average lifetime decreased markedly upon increasing the relative content of trifluoroethanol in the mixture, and hence decreasing the viscosity. Under the same conditions the fluorescence lifetime of the donor in a peptide lacking the acceptor changed only slightly (from 61.0 nsec in glycerol to 63.5 nsec in trifluoroethanol). The decrease in average donor lifetime upon the decrease in viscosity in peptides containing both donor and acceptor is due to the perturbation of the equilibrium end-to-end distribution of distances of the ensemble of excited molecules by the energy transfer process, which is more favorable for short than for long distances. The relative amounts of excited oligomer molecules of a long end-to-end distance within the ensemble of excited molecules increases during the lifetime of the donor excited state. However, as a result of the Brownian motion of the ends of the oligomer, the excited molecules with a relatively long end-to-end distance assume, on the average, a shorter end-to-end distance. The overall drift toward the equilibrium distribution results in a net diffusional "flux" of chain ends in each molecule towards one another, thus facilitating the energy-transfer process. Assuming a Fick equation for the above drift, the total change with time in the concentration of excited oligomer molecules of end-to-end distance r, can be represented by Eq. (1):

$$\frac{\partial \bar{N}^*(r,t)}{\partial t} = -\frac{1}{\tau}\{1+(\frac{R_0}{r})^6\}\bar{N}^*(r,t) + \frac{1}{N_0(r)}\frac{\partial}{\partial r}\{N_0(r)D\frac{\partial \bar{N}^*(r,t)}{\partial r}\} \quad (1)$$

where $N_0(r)$ is the equilibrium end-to-end distance distribution function obtained in a viscous solution in which Brownian motion is negligible;[2] $N^*(r,t)$ is the end-to-end distribution function in the ensemble of excited molecules at time t after excitation; $\bar{N}^*(r,t)=N^*(r,t)/N_0(r)$, τ is the donor fluorescence lifetime in the absence of an acceptor; R_0 is the Förster constant,[1] and D is the intramolecular diffusion coefficient of the chain ends. Equation (1) can be used only when R_0 has a single value shared by all the donor-acceptor pairs in the sample. Detailed discussion of the experimental implications of this condition is given elsewhere.[4,5] Since τ, R_0, and $N_0(r)$ are known from independent experiments and calculations, Eq. (1) can be solved numerically using a postulated value for D, and the appropriate boundary conditions. A computed decay curve of the donor fluorescence, $I_c(t)$, is derived by means of Eq. (2):

$$I_c(t) = k\int_a^{r_{max}} N^*(r,t)\,dr = k\int_a^{r_{max}} N_0(r)\bar{N}^*(r,t)\,dr \quad (2)$$

In Eq. (2), k denotes a constant characteristic of the experimental set up: a and r_{max} are, respectively, the approximate distance of closest approach of the naphthalene and dansyl groups, and the end-to-end distance of the fully stretched molecule. The computed $I_c(t)$ was compared with the experimental donor fluorescence decay curve, $I(t)$, (after convolution with the experimental excitation pulse profile).[3] Using an algorithm for a nonlinear least-squares estimation of parameters, a search was made for that value of D which, when used with Eqs. (1) and (2), yields a calculated donor fluorescence decay curve, $I_c(t)$, which fits best to the experimental decay curve.

The intramolecular diffusion coefficients of the chain ends relative to one another were calculated for the oligopeptides with n=4,5,8,9 in solvent mixtures of glycerol-trifluoroethanol of various viscosities. The results are summarized in

Fig. 1. The values of D obtained in solutions of viscosities ranging between 1 cp and 200 cp fall in the range of 5×10^{-9}–$10^{-7}cm^2/sec$, increasing systematically as the viscosity of the solvent is decreased (Fig. 1). For example, D is $0.8\times10^{-7}cm^2/sec$ for a peptide containing eight N^5-2-hydroxyethyl-L-glutamine residues (n=8) dissolved in a 7:3 (w/w) mixture of trifluoroethanol/glycerol, possessing a viscosity of 7.6 cp. This value is approximately an order of magnitude lower than that of the diffusion coefficients expected for small molecules comparable in size to the chromophores attached to the ends of the oligopeptides. It is thus evident that the "internal friction" of the oligomers, due to restrictions in change, in molecular conformation reduces by about an order of magnitude the diffusion rate of the chain ends in the above example.

Extrapolation of the data in Fig. 1 to zero viscosity yields a value for D that is probably well below $10^{-6}cm^2/sec$, although it cannot be obtained accurately. This shows that for all of the oligopeptides studied the internal friction is finite and appreciable. Furthermore, the internal friction is higher for the shorter chains, indicating that the shorter chains restrict the flexibility of motion of their ends more than the longer chains do.

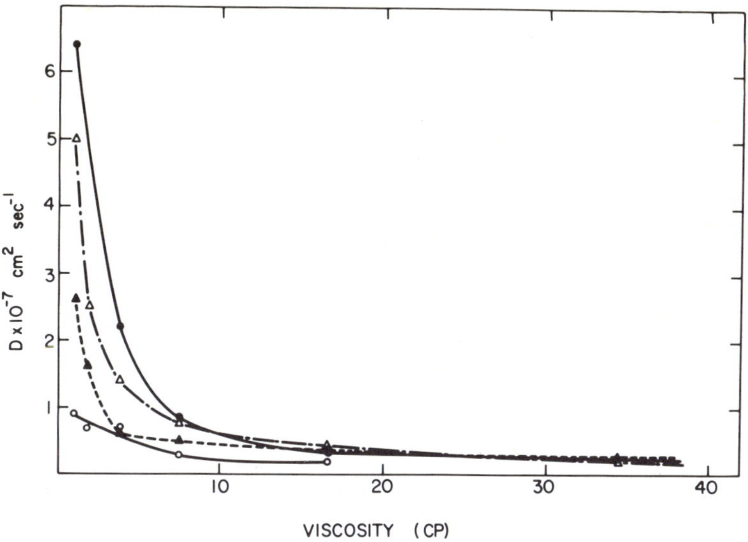

Fig. 1. Diffusion coefficients of one molecular end relative to the other of DG_nN (n=4,5,8,9) as a function of solvent viscosity. ○–○ DG_4N, ▲–▲ DG_5N, △–△ DG_8N, ●–● DG_9N. Temperature, 25°C.

References

1. Förster, Th. (1948) *Ann. Phys. Leipzig* **2**, 55–75.

2. Haas, E., Wilcheck, M., Katchalski-Katzir, E. & Steinberg, I. Z. (1975) *Proc. Nat. Acad. Sci. USA* **72**, 1807–1811.

3. Grinvald, A. & Steinberg, I. Z. (1974) *Anal. Biochem.* **59**, 583–598.

4. Haas, E., Katchalski-Katzir, E. & Steinberg, I. Z. (1977) *Biopolymers*, in press.

5. Steinberg, I. Z., Haas, E. & Katchalski-Katzir, E., in preparation.

CONFORMATIONAL ANALYSIS OF ADRENOCORTICO-TROPIC HORMONE RELATED TO NON-RADIATIVE ENERGY TRANSFER

M. LECLERC, *Laboratoire de Chimie Générale I, Faculté des Sciences, Université Libre de Bruxelles, B-1050 Bruxelles, Belgium,* S. PREMILAT, *Laboratoire de Biophysique, Centre du ler Cycle, Université de Nancy I, F-54037 Cedex, France,* and A. ENGLERT, *Laboratoire de Chimie Générale I, Faculté des Sciences, Université Libre de Bruxelles, B-1050 Bruxelles, Belgium*

The dimensions of small molecules of biological interest, such as peptide hormones, can be derived from the efficiency of electronic excitation energy transfer by the Förster mechanism[1], if suitably chosen luminophores are attached to distant sites of the molecule.[2] Recently it has been shown that the methods of theoretical conformational analysis based on energy calculations from semiempirical potential functions can be used for the interpretation of energy transfer data. By deriving all relevant experimental properties from a theoretical model of the molecules studied[3], this method provides a complete description of molecular dimensions, compatible with the experimentally determined properties. Here we describe a theoretical model for adrenocorticotropic hormone which is used to calculate properties related to non-radiative energy transfer. These theoretical results are compared with experimental data taken from the literature.[4-6] These data refer to the efficiency of energy transfer from tyrosine to tryptophan in the segments 2-9 and 9-23 in the fully active fragment ACTH (1-24)[4] (Fig. 1) and from Trp^9 to N^ϵ-dansyllysine^{21} in the same but dansylated fragment[5,6] determined in water and various other solvents.

```
       (2)                         (9)        (12)
(H)-Ser-Tyr-Ser-Met-Glu-His-Phe-Arg-Trp-Gly-Lys-Pro-
Val-Gly-Lys-Lys-Arg-Arg-Pro-Val-Lys-Val-Tyr-Pro(OH)
(13)                        (21)     (23)
```

Fig. 1. The primary sequency of human ACTH (1–24). Amino acids considered as hydrophobic are underlined.

These studies, as well as a number of other investigations by various experimental methods (dialysis[7], CD[8], hydrogen exchange[9]), indicate that the behavior of ACTH in solution "is best explained by a mobile equilibrium involving many different shapes".[7] It seems therefore appropriate to use a statistical approach to describe the various conformation dependent properties of the molecules. Thus, a set of chains have been generated by the Monte Carlo method.

The conformational energy is computed taking into account the interactions between all the atoms in the chain, as described previously by one of us.[10] Chains are constructed from standard geometrical parameters[11] but all the amino-acid side chains (except for Gly,Pro,Tyr,Trp,Phe and Lys(Dns)) have been replaced by a suitable composite atom.[12] The geometrical parameters of the dansyl luminophore have been obtained from X-ray data on related compounds.[13,14] The S-N bond has been represented as a double-bond[14] in a *trans* conformation with respect to C-N and S-C bonds. The values for the angle of rotation around the S-C bond, have been taken equal to ±120° because of steric constraints. The direction of the transition moments for all the luminophores has been determined by quantum mechanical calculations.[15]

The solvent effect is roughly taken into account by eliminating the attractive part of the van der Waals potential for interactions involving a hydrophilic residue.[10] Otherwise all the parameters in the potential functions are taken from Scheraga.[11,16] (The dielectric constant is equal to 3.5). The full charges of side chains which are ionized at neutral pH have been decreased by half, in order to represent the average screening effect of water and counter-ions. The charge distribution on the dansyl group has been computed by an INDO quantum-mechanical program.

The calculated mean dimensions of the various segments of the chain are shown in Table I.

Table I. Various Moments of the Distribution $f(r)$ in Tyr^2-Trp^9, (a); Trp^9-Tyr^{23}, (b) and Trp^9-Lys(Dns)^{21}, (c).

	$\langle r \rangle$	$\langle r^2 \rangle^{1/2}$	$\langle r^{-6} \rangle^{-1/6}$
(a)	12.0	13.5	6.0
(b)	24.3	26.7	8.0
(c)	21.7	24.0	6.5

The distribution function $f(r)$ of the distances r between Trp^9 and Tyr^{23} is shown in Figure 2.

The calculated transfer efficiencies are compared with the experimental values in Table II. The agreement is satisfactory if the averaging regime is considered as partly static ($\langle T \rangle_{s'}$) or purely static ($\langle T \rangle_s$). In both these regimes, the lifetimes of molecular conformations are long compared to the transfer time, but in the former regime the rotational motion of the luminophores is short on the same time scale. If however the regime were dynamic ($\langle T \rangle_d$), which implies that all the possible conformations would be adopted successively by the molecule during the transfer time, there would be no agreement between experimental and calculated properties. It can be argued that in molecules with a considerable number of internal rotation angles the regime is always nearly static. It is indeed

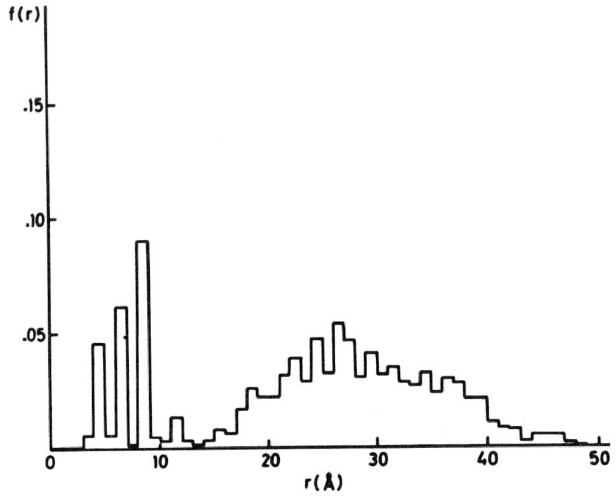

Fig. 2. The distribution function $f(r)$ of the distance r between Trp^9 and Tyr^{23}.

Table II. A Comparison of the Experimental and Calculated Transfer Efficiencies

Luminophores implied in the transfer	Experimental efficiencies		Calculated efficiencies		
	$<R_o>$	$<T>$	$<T>_{s'}$	$<T>_s$	$<T>_d$
Tyr^2-Trp^9	9.8	0.5±0.15	0.47	0.36	0.83
Tyr^{23}-Trp^9	11.0	0.15±0.10	0.20	0.18	0.84
Trp^9-$Lys(Dns)^{21}$	19.4	0.45	0.44	0.33	0.98

reasonable to assume that only a very small fraction of the total number of molecules undergoes a conformational change which leads to a noticeable variation of interluminophore separation during the transfer time of electronic excitation energy.

It should be added that for the dansylated segment 9–21 the efficiency computed without charges on the side chains agrees well with the experimental value for the protected (uncharged) peptide in water: $<T>_{exp} = 0.64$ and $<T>_{s'}$ (calculated) = 0.57. In conclusion it appears that energy transfer data on ACTH can be interpreted satisfactorily by a model of statistical chain.

Most of the calculations were performed on the IBM 370/168 computer at Orsay, France during the Workshop on Models for Protein Dynamics, May–July 1976. We acknowledge Dr. B. Maigret for help in the quantum-mechanical calculations. One of us (M. L.) is "Aspirant au Fonds National de la Recherche Scientifique".

References

1. Förster, Th. (1948) *Ann. Phys. (Leipzig)* **2,** 55–75.
2. Stryer, L. & Haugland, R. P. (1967) *Proc. Nat. Acad. Sci. USA* **58,** 719–736.
3. Leclerc, M., Prémilat, S., Guillard, R., Renneboog-Squilbin, C. & Englert, A. (1977) *Biopolymers* **16,** 531–544.
4. Eisinger, J. (1969) *Biochemistry* **8,** 3902–3907.
5. Schiller, P. W. (1972) *Proc. Nat. Acad. Sci. USA* **69,** 975–979.
6. Schiller, P. W. & Schwyzer, R. (1973) in *Peptides,* Nesvadba, H., Ed., North-Holland Publishing Company, Amsterdam, pp. 354–366.
7. Craig, L. C., Fisher, J. D. & King, T. P. (1965) *Biochemistry* **4,** 311–318.
8. Holladay, L. A. & Puett, D. (1976) *Biopolymers* **15,** 43–59.
9. Li, C. H. (1956) *Adv. Prot. Chem.* **11,** 101–190.
10. Prémilat, S. & Maigret, B. (1977) *J. Chem. Phys.* **66,** 3418–3425.
11. Scheraga, H. A. (1968) *Adv. Phys. Org. Chem.* **6,** 103–184.
12. Pletnev, V. Z., Popov, E. M. & Kadymova, F. A. (1974) *Theoret. Chim. Acta* **35,** 93–96.
13. Cruickshank, D. W. J. (1957) *Acta Cryst.* **10,** 504–508.
14. O'Connell, A. M. & Maslen, E. N. (1967) *Acta Cryst.* **22,** 134–145.
15. Oth, J. (1977) manuscript in preparation.
16. Poland, D. & Scheraga, H. A. (1967) *Biochemistry* **6,** 3791–3800.

CONFORMATIONAL ASPECTS OF POLYPEPTIDE INTERACTIONS WITH IONS

E. PEGGION, A. COSANI, M. PALUMBO, and M. TERBOJEVICH, *Biopolymer Research Center, Institute of Organic Chemistry, University of Padova, Via Marzolo, 1 – 35100 Padova – Italy.*

Introduction

During the last decade an impressive amount of work has been carried out using synthetic peptides as model systems for more complex enzyme and protein molecules. Different research groups have given different interpretations of the concept of "model compound". So, for instance, detailed knowledge of the various secondary structures of peptide chains has been obtained using synthetic, high molecular weight homopolypeptides or copolypeptides.[1] Such models have been also used to gain information on interactions among aromatic chromophores in the side-chains.[2] On the other hand, some workers focused their attention on cyclic peptides as model systems. In a recent paper Blout and coworkers[3] stressed the reasons for such an interest. A suitable model compound must be simple from a structural point of view and must also be of biological relevance. The molecule should possess few conformational states, which should be easily characterizable by experimental and theoretical conformational analysis techniques.[3] Undoubtly synthetic cyclic peptides meet the above criteria, especially if we consider that many natural products are indeed cyclic compounds.

A third class of model peptides extensively investigated by various authors includes sequential polypeptides.[4] Many conformational details have been obtained, especially in the field of collagen-like structures, for sequential polypeptides containing the glycine residue at every third position.[5] Finally, conformational studies on short, linear oligopeptides have given new information on the tendency of various amino acid residues to form parallel and anti-parallel β-structures.[6] From all these studies, using a combination of different techniques we have gained substantial knowledge of the conformation and structure of model compounds. This leads to a better understanding of the conformation of more complicated protein molecules.

If we extend the investigations to the conformational aspects of the interactions with ions of polypeptide model systems, we may improve our knowledge of the structure and stereochemistry of metallo-enzymes. This is undoubtedly a broad field of application of the above mentioned classes of synthetic polypeptides. Interaction between proteins and metal ions might occur in several ways, which are sometimes extremely complicated in character. In some cases protein molecules and metal ions form complexes of well-defined stoichiometry, and the presence of the appropriate ion is essential for biological activity. In other cases, protein systems provide the transport of metal ions through dynamic equilibria between ions and multiple binding sites in the protein moiety. Investigations on the conformational aspects of interactions with ions with appropriate model systems could provide a powerful tool to answer some fundamental questions such as: the stereochemical requirements of the binding site which provide the "frame" ligand for the proper ion, the factors which determine coordination of the right ion to the right enzyme, and therefore the way of action of ion-containing proteins.

A substantial amount of work has been carried out on interactions between metal ions and simple amino acids and peptides. The structural aspects[7] and the optical rotatory properties[8] of metal ion complexes of small peptides have been recently reviewed in an exhaustive way by Martin and coworkers.[7,8] In some cases oligopeptide sequences reproducing the amino acid sequence of natural products have been shown to interact specifically with a given ion as strongly as the native protein.[9]

On the other hand, high-molecular-weight synthetic polypeptides have been shown to form metal complexes which in some cases exhibit stereospecific catalytic activity analogous to ion promoted enzyme reactions.[10] The conformational aspects of high-molecular-weight polypeptide interactions with ions have been investigated in detail in an attempt to understand the reasons for catalytic stereospecific activity. In this paper we will review recent studies on interactions between metal ions and poly(α-amino acid) systems, with particular emphasis on those investigations dealing with the stereochemistry of complex formation and its effects on the conformation of the polypeptide chain.

Interaction of Poly(L-histidine) $(\text{L-His})_n$ with Cu(II) ions

Histidine residues have been recognized as important binding sites for metal ions in a number of natural products[11,12] and in synthetic peptides.[7,8,11] In particular, histidine residues have been suggested as ligands for copper ions in a number of proteins, including copper containing oxidases[12] and the blue and non-blue copper proteins.[13]

Pecht et al.[14] investigated the interaction between $(\text{L-His})_n$ and cupric ions. They first reported that the polymer forms copper complexes which exhibit oxidase activity about two orders of magnitude higher than that of Cu(II) aquo-complex.[14] A careful characterization of the catalytic polymeric complex of $(\text{L-His})_n$ was subsequently carried out by Levitzki et al.[15] in order to ascertain the way of binding of the metal ions to the polymer chain. The characterization was carried out using several techniques which include potentiometric titration, visible and ultraviolet absorption spectroscopy, and circular dichroism (CD). From the potentiometric titration curve of the polymer in the presence of Cu(II) ions, the authors were able to detect two kinds of complexes. Complex I starts to form at pH 3, and its formation is essentially complete at pH 5. At this pH value *four* protons are titrated per mole of bound copper. The spectral properties in the visible absorption region suggested that at least one of these protons comes from a peptide nitrogen. This hypothesis was further substantiated by the CD properties both in the visible absorption and in the ultraviolet absorption region. Two oppositely signed bands have been observed in the region of copper *d-d* transition with a rather high rotational strength. Since the polymer is in the random coil conformation until pH 5, optical activity must arise from asymmetric induction from the chiral centers of the single amino acid residues. With coordination of imidazole nitrogens only at the square-planar coordination positions of copper, a rather small induced optical activity should be expected, owing to the distance of the chiral α-carbon atom from the complex. When at least one deprotonated peptide nitrogen is coordinated at the square-planar coordination positions of Cu(II) ions the asymmetric α-carbon atom is much closer and enhancement of rotational strength is therefore expected. The structure proposed by Levitzki et al.[15] for complex I of $(\text{L-His})_n$ involves three imidazole nitrogens of residues 1, 2, and 4, and one peptide nitrogen of the residues 3, in the classical square-planar geometry. Actually, with histidine residues a structure is possible in which one imidazole nitrogen and the adjacent deprotonated peptide nitrogen occupy two of square-planar coordination sites of Cu(II) ions, forming a stable hexa-atomic chelate ring containing the asymmetric α-carbon atom (Figure 1). This way of coordination of Cu(II) ions has been shown to occur frequently in histidine-containing peptides.[11] At strongly alkaline pH a biuret type complex of $(\text{L-His})_n$ was observed,[15] also characterized by strong optical activity. In analogy to observations in oligopeptides[11] Levitzki et al.[15] proposed a structure in which four consecutive deprotonated peptide nitrogens occupy the square-planar positions of copper, with possible coordination of imidazole nitrogens at the apical positions. Space-filling models showed that this structure can be repeated indefinitely, forming a rigid helix.

It is well established that $(\text{L-His})_n$ undergoes a disorder-order conformational transition between pH 5 and 6, the ordered form being a β-pleated sheet.[2] In order to determine to what extent complex formation affects the conformational change, we have recently re-investigated the interactions between Cu(II) ions and $(\text{L-His})_n$. Between pH 3 and 5, the formation of complex I, described by Levitzki et al.[15] was observed. The visible and ultraviolet absorption and CD spectra were in quantitative agreement with those reported by the above mentioned authors. However, on increasing the pH value above 5, the formation of a new, previously undetected complex, was revealed, with completely different CD properties (Figure 2). The spectrum of the complex, which was named complex II, is very weak with two positive bands at 640 nm and 485 nm, and a negative one at ~540 nm. The well-defined isodichroic point at ~595 nm suggests the presence of a two-component equilibrium system with additive contributions to

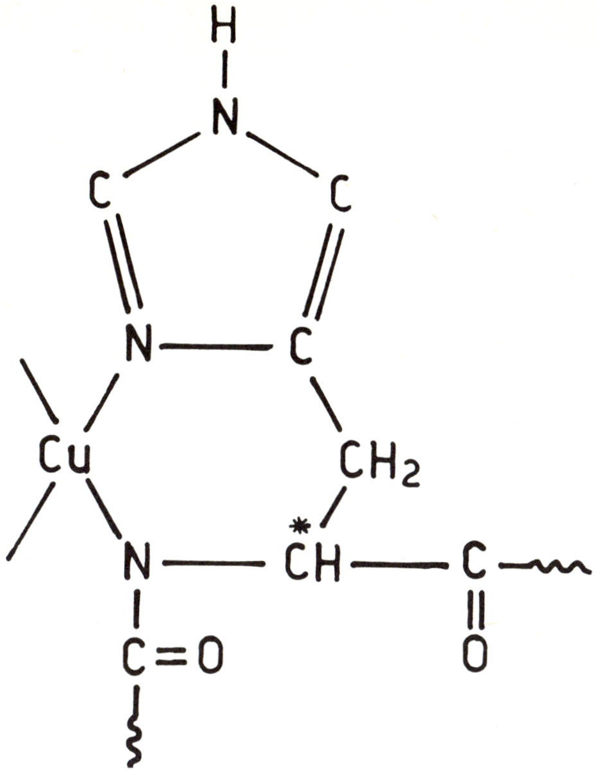

Fig. 1. A possible structure of histidine peptides coordinated to Cu(II) ions.

the optical activity. At strongly alkaline pH values the usual biuret-type complex, already described by Levitzki et al.[15] was observed. Since the formation of complex II occurs within the pH range where the polymer undergoes a coil-to-β-form transition, its structure must be related in some way to the conformation of the polypeptide chain. Actually we have shown that the stability of complex II of $(L\text{-His})_n$ with Cu(II) ions depends upon at least three factors, namely pH, conformation of the polypeptide chain and Cu/peptide molar ratio. If cupric ions are added to a polymer solution at constant pH 4.5, complex II and not complex I is formed immediately after the addition of metal ions. However, the CD pattern is time-dependent and it slowly transforms into the spectrum of complex I. At Cu/peptide molar ratios not exceeding 0.1, on increasing the pH of the solution above 6 or more, again the CD pattern typical of complex II is formed, but, in the far UV we also observe the simultaneous formation of CD pattern of $(L\text{-His})_n$ β-structure (Figure 3). We therefore conclude that, under the above described experimental conditions, the conformational changes of the polymer also modify the way of binding of cupric ions. The visible absorption

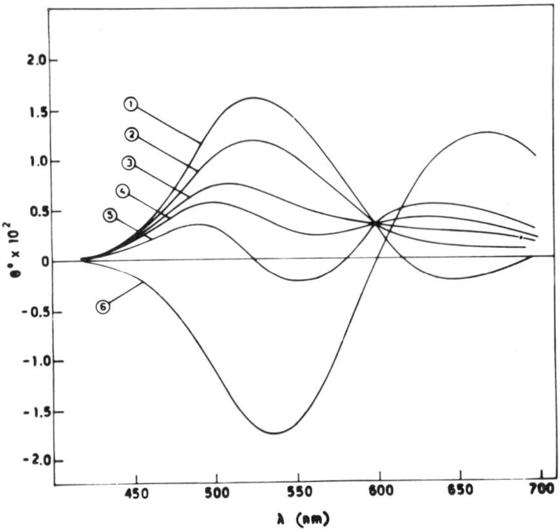

Fig. 2. CD properties of the Cu(II)-(L-His)$_n$ system in the visible region. Cu/His = 0.12; 1) pH 6.05; 2) pH 7.07; 3) pH 7.60; 4) pH 7.88; 5) pH 8.0; 6) pH > 12.

characteristics and the very weak rotatory strength of complex II do suggest coordination of imidazole nitrogens only to the square-planar coordination positions of copper. In other words, the formation of the β-pleated sheet causes substitutions of deprotonated peptide nitrogens with side-chain imidazole nitrogens at the coordination sites of copper. This is an example of the effect of peptide conformation on the way of binding of a metal ion. Actually, the above results have been obtained using Cu/peptide molar ratios not higher than 0.05. If more cupric ions are added to the solution of complex II at pH 6, again the formation of the CD spectrum of complex I is observed, and the β-structure tends to disappear. At Cu/peptide ratios of the order of 0.4, the formation of complex II can be detected at pH values of the order of 7.0. It seems therefore that increasing amounts of Cu(II) ions shift the coil-β transition and the consequent formation of complex II to higher pH values. The structure of complex I appears to be incompatible with the β-form of (L-His)$_n$. We must point out that at Cu/peptide molar ratios 0.4 or higher, the β-structure of (L-His)$_n$ can be detected with difficulty from the CD pattern in the far UV because of partial overlapping of intense charge-transfer bands of residual complex I with imidazole and peptide transitions.[16] In order to promote β-structure formation, we should further increase the pH, but this also enhances the coordination capacity of the peptide nitrogen ligand. At alkaline pH values the tendency of the peptide chain to form the β-structure is overcome, and the biuret complex is formed in

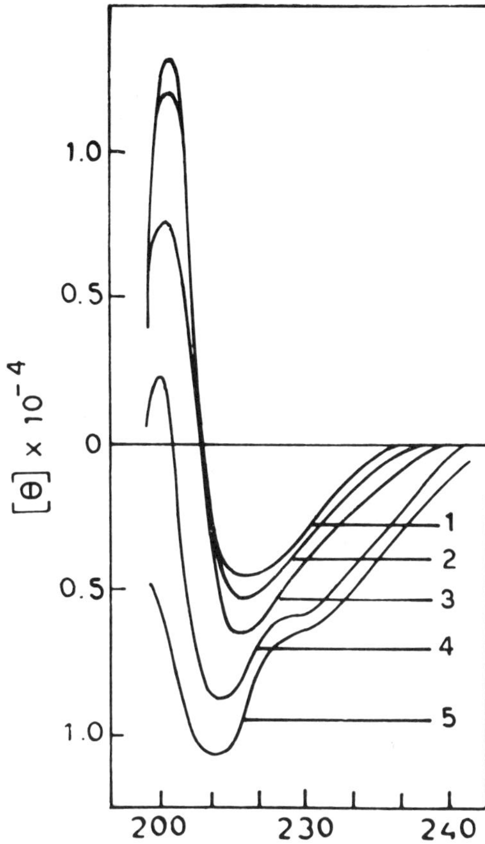

Fig. 3. CD properties of the Cu(II)-(L-His)$_n$ system in the peptide absorption region. Cu/His = 0.12; 1) pH 7.04; 2) pH 6.56; 3) pH 6.06; 4) pH 5.54; 5) pH 4.47.

which four consecutive deprotonated peptide nitrogens are coordinated to Cu(II) atoms in the classical square geometry. As previously mentioned, Levitzki et al.[15] proposed for the biuret complex a structure which implies conformational rigidity. It would be of interest to investigate the implications of such a structure on the hydrodynamic behavior of the complex solution.

As a conclusion, from all the work carried out on the interactions between cupric ions and (L-His)$_n$, it appears that both pH and macromolecular conformation affect the way of binding of metal ions to the polymer. At pH $\leqslant 5$ the presence of imidazole nitrogens favors deprotonation of an adjacent peptide nitrogen, consistent with observations in copper complexes of histidine-containing oligopeptides.[11] At pH $\geqslant 6$, the tendency of coordination of peptide nitrogens depends upon the conformation of the polypeptide chain, and it is prevented by the β-structure formation. At higher pH values, the competition between protons and Cu(II) ions toward peptide nitrogen ligands is shifted in favor of the metal, with consequent disruption of the β-structure.

Interaction of Cu(II) ions with Poly(α-amino acids) Containing Side-chain Amino Groups

Hatano and coworkers[17-20] first reported the catalytic properties of $(\text{L-Lys})_n$-Cu(II) complexes, and a detailed analysis was carried out in the attempt to correlate catalytic activity with conformational properties of the polymeric complex. The authors found that an initial complex is formed at pH 8.5 and that four protons are released per mole of bound copper. From the visible absorption spectrum and from the vanishingly small optical rotation of the complex, the authors concluded that only side-chain amino groups are coordinated to copper ions. At pH > 10 a second complex was observed, whose remarkable optical activity and spectral characteristics are typical of biuret-type complexes. Nozawa and Hatano[19] proposed for the high pH complex a structure in which the polymer is in the α-helical conformation, and four amino nitrogens are coordinated to square planar-positions of copper, with peptide nitrogen coordination at apical positions (Figure 4). This second complex was found to accelerate the oxidation of D-DOPA more than that of L-DOPA. The selectivity ratio (i.e. the ratio between the rates of oxidation of the D-isomer and the L-isomer) was found to be strongly dependent upon temperature. However, for experiments carried out at constant temperature and at various Cu/Lys ratios no apparent correlation was observed between helix content and selectivity ratio.

The problem of interaction between $(\text{L-Lys})_n$ and copper ions has been independently investigated by Garnier and Tosi[21] and by our group.[22,23] We have extended our studies to $(\text{L-Orn})_n$ and to the polymer of diaminobutyric acid, $(\text{L-A}_2\text{bu})_n$, with the specific purpose of establishing the relationship between complex structure and polymer conformation. Again we have observed that two kinds of complexes are formed: complex I stable at pH $\geqslant 8.5$ and complex II, stable at pH > 10.5. However there is one significant difference between our results and those of Nozawa and Hatano.[19] Until pH 8.5 two (and not four) protons are released per mole of bound copper in the case of $(\text{L-Lys})_n$ and $(\text{L-Orn})_n$. The same results have been obtained by Garnier and Tosi[21] on the Cu(II) complex of $(\text{L-Lys})_n$. Only with $(\text{L-A}_2\text{bu})_n$ four protons have been titrated up to pH 8.5. At least one of these protons arises from a peptide nitrogen. The visible absorption spectra and the very small rotational strength of type I complexes of $(\text{L-Lys})_n$ and $(\text{L-Orn})_n$ are consistent with the data reported by Hatano and coworkers[17-20], and imply coordination of side-chain amino nitrogens only to Cu(II) ions. Complex I of $(\text{A}_2\text{bu})_n$ exhibits entirely different features. The visible absorption spectrum is blue-shifted down to 585 nm; furthermore, the CD properties reveal a high rotational strength in the

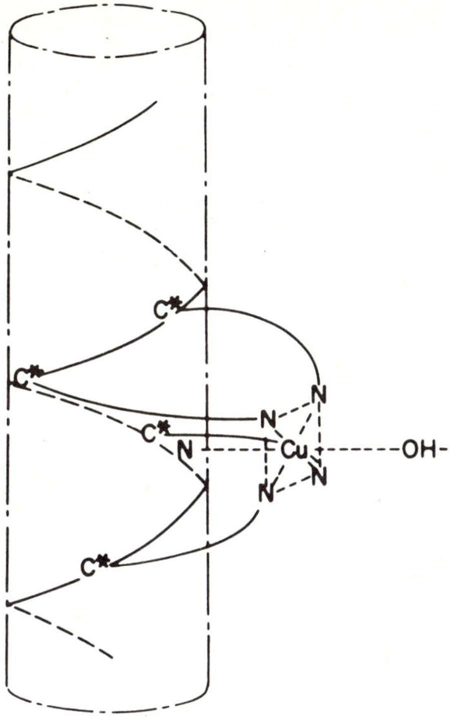

Fig. 4. A schematic structure of the $(\text{L-Lys})_n$-Cu(II) complex (from Ref. 19).

visible and in the near ultraviolet absorption region (Figure 5). Both these results suggest coordination of at least one deprotonated peptide nitrogen to the square-planar coordination positions of copper.[15] Since the polypeptide chain of $(\text{L-A}_2\text{bu})_n$ is in the random coil conformation in the entire range of examined pH, optical activity of complex I of $(\text{L-A}_2\text{bu})_n$ must be ascribed to asymmetric induction from the chiral centers of single amino-acid residues. With $(\text{L-A}_2\text{bu})_n$ a structure is possible in which the asymmetric α-carbon atom is part of a stable hexa-atomic chelate ring which contains a side-chain amino group, an asymmetric α-carbon atom, and the adjacent, deprotonated peptide nitrogen coordinated at two of the square-planar positions of Cu(II). Such a structure justifies the high optical activity observed in complex I of $(\text{L-A}_2\text{bu})_n$. From Figure 1 the analogies with the copper complex of $(\text{L-His})_n$ formed at pH 5 (complex I) are clearly evident. The structural similarities between the two complexes are reflected in their CD properties which are shown for comparison in Figure 6. This suggests a very similar way of binding of copper ions in the two cases and leads to the conclusion that a diaminobutyric acid residue could be equivalent to an histidine residue as ligand for Cu(II).

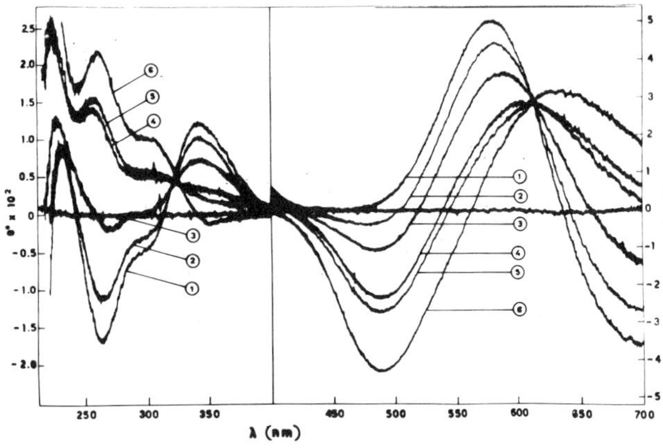

Fig. 5. CD spectra of the Cu(II)-(L-A_2bu)$_n$ system at different pHs. Cu/A_2bu = 0.25. 1) pH 8.29; 2) pH 9.31; 3) pH 10.18; 4) pH 11.67; 5) pH 12.09; 6) pH > 13 (from Ref. 22).

At pH 8.5 type II complexes of the three polymers are formed in which deprotonated peptide nitrogens are involved. The CD spectra are different from one another, implying different geometries of the three complexes and, perhaps different apical interactions of the side-chain amino groups. Turning now to the conformational aspects of type II complexes, the main question is whether their structure is compatible with the α-helical confirmation of the polymeric backbone. This aspect concerns only (L-Lys)$_n$ and (L-Orn)$_n$, since the diaminobutyric acid polymer is unable to assume an ordered conformation even at pH 14.[24] The following experiment has been therefore performed. To a polymer solution of (L-Lys)$_n$ kept at constant pH 11.0 (where the polymer is completely in the α-helical form at 25°C) increasing amounts of Cu(II) ions were added. The CD patterns recorded after each addition of metal ions clearly indicate that the helix content decreases with the extent of complex II formation. This is shown in Figure 7, where the helix content appears to decrease linearly with the extent of complex formation, and drops to zero at Cu/peptide molar ratios of the order of 0.15. Substantially identical results have been obtained with complex II of (L-Orn)$_n$, and have also been independently confirmed by Garnier and Tosi (personal communication). The conclusion therefore follows that the α-helical structure is incompatible with type II complex formation, and previously proposed models based on a helical arrangement of the peptide backbone are therefore inconsistent.[19] This conclusion is in line with the results reported by other investigators on various systems including proteins.[25] In a recent paper Gray and coworkers[26] observed that a short section of α-helix

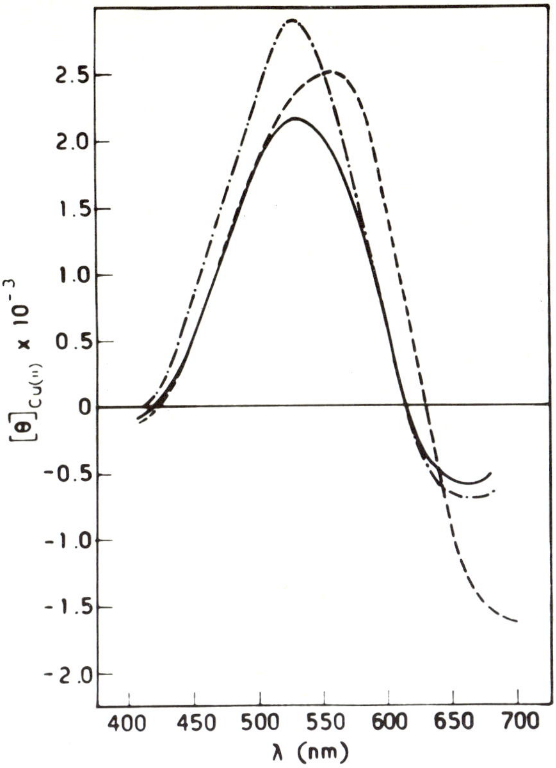

Fig. 6. CD spectra of the low-pH cupric complexes of $(L\text{-His})_n$ and of $(L\text{-}A_2bu)_n$ in the visible region. ——— $(L\text{-His})_n$ - Cu(II) (from Ref. 15) -·-·-·-· $(L\text{-His})_n$ - Cu(II) (this work) ------- $(L\text{-}A_2bu)_n$ - Cu(II) (from Ref. 22)

in the apoprotein apoplastocyanine is strongly perturbed upon Cu(II) incorporation, which probably occurs through coordination at one deprotonated peptide nitrogen.

The particular conformational properties of $(L\text{-Lys})_n$, $(L\text{-Orn})_n$ and $(L\text{-}A_2bu)_n$ in water/methanol mixtures gave also the possibility of checking the effect of type I complex formation on the α-helical structure of the polypeptide backbone. We have shown that the α-helix of $(L\text{-Lys})_n$ and $(L\text{-Orn})_n$, stable in water/methanol mixtures containing 90% methanol, is not perturbed by formation of complex I in which side-chain groups only are involved. On the contrary, in the case of $(L\text{-}A_2bu)_n$, the formation of complex I, where at least one peptide nitrogen is involved, causes collapse of the helix structure.

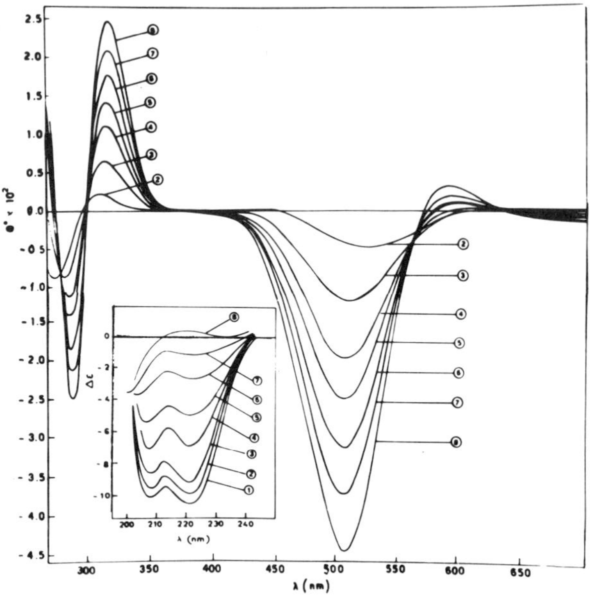

Fig. 7. CD spectra of $(L\text{-}Lys)_n$ - Cu(II) at pH 11 and 25°C. 1) Cu/Lys = 0; 2) Cu/Lys = 0.016; 3) Cu/Lys = 0.033); 4) Cu/Lys = 0.057; 5) Cu/Lys = 0.082; 6) Cu/Lys = 0.106; 7) Cu/Lys = 0.131; 8) Cu/Lys = 0.164 (from Ref. 23).

Finally we mention some results of investigations on the interaction of copper ions with $(L\text{-}Lys)_n$ in the β-structure. Potentiometric titrations, visible and ultraviolet absorption and CD measurements carried out at 50°C again showed the formation of complexes I and II identical to those observed at room temperature.[27] Addition of Cu(II) ions to a $(L\text{-}Lys)_n$ solution kept at constant pH 10.8 and at 50°C (where the polymer is completely in the β-form) causes the progressive disruption of the β-structure (Figure 8).

From all these results we can draw the general conclusion that whenever peptide nitrogens are coordinated to square-planar positions of Cu(II), neither the α-helix nor the β-structure can be formed. Furthermore, in $(L\text{-}A_2bu)_n$ the presence of an amino group spaced by two methylene residues from the asymmetric α-carbon makes possible deprotonation of an adjacent peptide nitrogen even at physiological pH values. This behavior is analogous to that observed in histidine-containing oligopeptides and confirms the importance of deprotonated amide nitrogens as binding sites for metal ions even in biological systems.

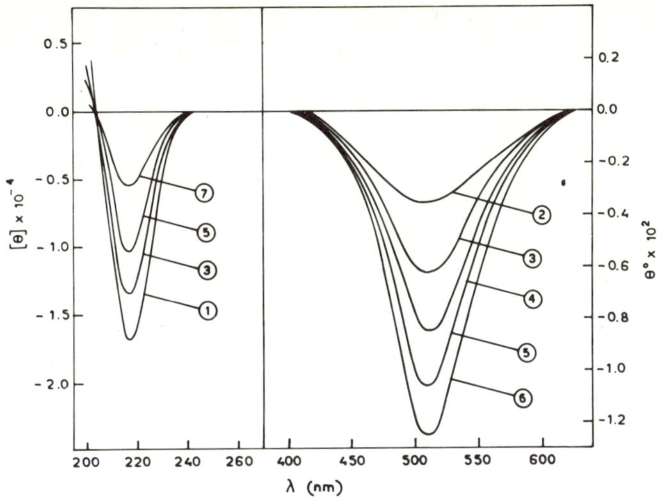

Fig. 8. CD spectra of (L-Lys)$_n$ -Cu(II) at pH 10.3 and at 50°C. 1) Cu/Lys = 0; 2) Cu/Lys = 0.015; 3) Cu/Lys = 0.029; 4) Cu/Lys = 0.044; 5) Cu/Lys = 0.058; 6) Cu/Lys = 0.073; 7) Cu/Lys = 0.109 (from Ref. 27).

Stereospecific Interaction of Metal Ions with Polypeptides

In the last part of this review we will report some examples of stereospecific complex formation induced by asymmetric conformations of the polypeptide matrix. This problem is related to the very fundamental question of what are the stereochemical requirements for the coordination of the proper ion in the proper way to a peptide system.

De Santis and coworkers[28,29] studied the complex formation between Cu(II) ions and the Schiff base resulting from reaction of salicylaldehyde with the cyclic peptide gramicidin S. On the basis of spectroscopic data, X-ray analysis, and conformational analysis of model compounds these authors have been able to show that, depending upon the solvent, there are two possible orientations of the two diastereotopic faces of the square, *trans*-planar, metal chelate chromophore toward the cyclic peptide asymmetric matrix (Figure 9). The R form is stabilized in solvents containing hydroxyl groups (ethanol, trifluoroethanol etc.), while the S form is stable in trimethylphosphate, chloroform, and pyridine. The authors extended their investigations to copper chelates of salycylaldehyde derivatives of (L-Lys)$_n$, (L-Orn)$_n$ and (L-A$_2$bu)$_n$, namely to copper complexes of *bis*-(*o*-hydroxybenzylidene amine) derivatives of lysine, ornithine and diaminobutyric acid polymers.[30] These modified polypeptides can be considered simple models for pyridoxal enzymes. In fact it has been shown that derivatives of

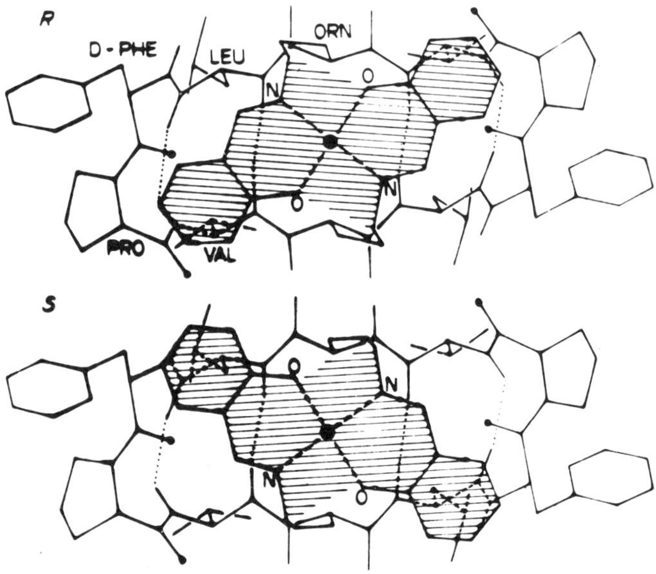

Fig. 9. The two structures of the gramicidin-*bis*(*o*-hydroxybenzylidene amine) –Cu(II) complex (from Ref. 30).

pyridoxal and salycylaldehyde in the presence of metal ions can be used as catalysts for the same kind of reactions catalyzed by pyridoxal enzymes.[31] The hypothesis has been made that catalytic activity depends upon reactivity of metal chelates of the prosthetic group.[31] De Santis and coworkers[30] observed a quasi-perfect inversion of the CD properties of copper chelates of $(L\text{-}Lys)_n$ and $(L\text{-}Orn)_n$ in the absorption region of the complexes, without appreciable alteration of the right-handed, α-helical form of the polypeptide backbone (Figure 10). The results have been interpreted by assuming that the asymmetric conformation of the polypeptide matrix induces stereospecificity in the formation of the complex, in such a way that the ornithine polymer "binds" one diastereotopic face of the metal chelate side-chain, while the lysine polymer binds the opposite one. In particular, the $(L\text{-}Orn)_n$ complex exhibits CD properties very similar to the S form of the corresponding complex of gramicidin S in chloroform, and the $(L\text{-}Lys)_n$ complex exhibits CD characteristics very similar to the R form of the gramicidin S complex. On the basis of these results the authors, using normal methods of conformational analysis, constructed the following models in which every copper ion bridges together one pair of Schiff base side-chains (Figure 11). In the case of the $(L\text{-}Orn)_n$ complex, each Cu(II) can bridge the side-chain chromophores in 1–5 positions, while, in the case of the $(L\text{-}Lys)_n$ complex,

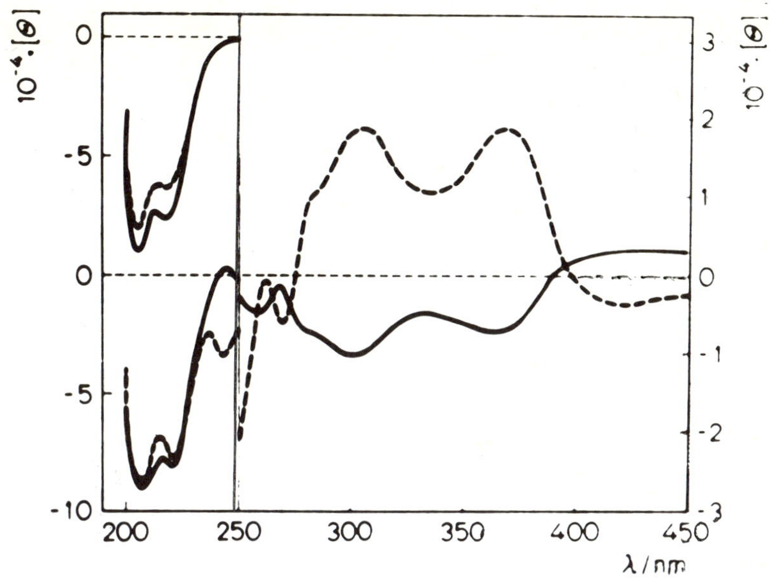

Fig. 10. CD spectra of the Cu(II) complexes of the *o*-hydroxybenzaldehyde Schiff bases with (L-Lys)$_n$ (——) and (L-Orn)$_n$ (-----). The spectra of the upper-left side refer to (L-Lys)$_n$ and (L-Orn)$_n$ under the same experimental conditions (from Ref. 30).

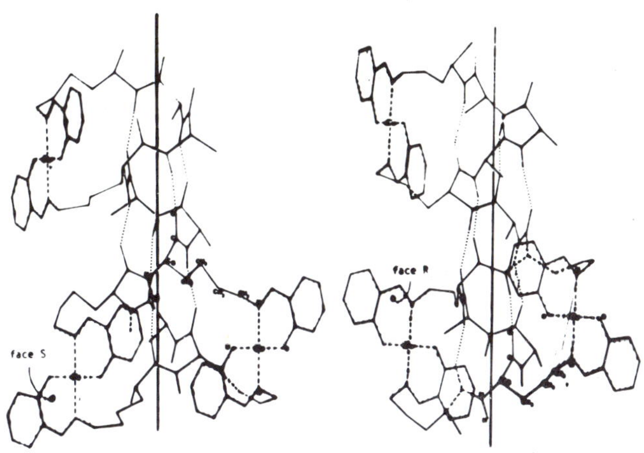

Fig. 11. Structure of the Cu(II) complexes of the *o*-hydroxybenzaldehyde Schiff bases with (L-Lys)$_n$ (left) and (L-Orn)$_n$ (right) (from Ref. 32).

bridging occurs between side-chains in positions 1-4. Analogous results have been obtained using Schiff bases derivatives of $(L\text{-Lys})_n$ and $(L\text{-Orn})_n$ with pyridoxal.[32] Also in this case in fact a stereospecific binding of the square-planar Cu(II) complex of *bis*-pyridoxalidene imine group to the α-helical matrix was observed, with opposite configurations for the complexes of $(L\text{-Lys})_n$ and $(L\text{-Orn})_n$.

In all the examined cases conformational analysis shows that the optimum stoichiometry should be one copper ion per pair of pyridoxalenimine side-chains, but a detailed, direct determination of the stoichiometry of the complexes has not been carried out by the authors.[30]

A further example of stereospecific complex formation induced by asymmetry of the polypeptide backbone has been recently reported by our group.[33] We have described the synthesis and the conformational properties of acetoacetyl derivatives of $(L\text{-Lys})_n$, $(L\text{-Orn})_n$ and $(L\text{-}A_2bu)_n$. These modified polymers, which contain the chelating β-ketoamide group in the side-chains have been shown to assume the right-handed α-helical form in aqueous solution in a wide range of pH. Interaction of these polymers with metal ions has been investigated using several techniques, which include classical methods for determination of stoichiometry and stability constants of the complexes, visible and ultraviolet absorption spectroscopy, and CD. It has been found in particular that the three examined polymers bind strongly Fe(III) ions at pH≃2.5, forming complexes with stoichiometry 1:2, i.e. each ferric ion bridges one pair of side-chains. From the absorption spectra (summarized in Table I) we concluded that only oxygen atoms of the side-chain β-ketoamide groups are involved in the complexes. This behavior is different from that of the model compound *n*-hexyl-acetoacetamide, which, under identical conditions, binds only one ferric ion with coordination of the amide nitrogen.

Table I. Maximum Absorption Wavelengths and Corresponding Molar Extinction Coefficients of the Ferric Complexes of $[Lys(Acac)]_n$, $[Orn(Acac)]_n$, and $[A_2bu(Acac)]_n$[a]

Ferric Complex of	Metal-Ligand Ratio	$\lambda_{max}(\varepsilon)$	$\lambda_{max}(\varepsilon)$
model compound	1:1	258(12,100)	536(780)
$[Lys(Acac)]_n$	1:2	257(20,900)	478(1,450)
$[Orn(Acac)]_n$	1:2	257(19,200)	478(1,380)
$[A_2bu(Acac)]_n$	1:2	257(20,700)	478(1,430)

[a]The spectral data relative to the model ferric complex are reported for comparison. The ε values are always calculated per mole of complex formed, i.e., per Fe^{3+} ion.

Because of solubility reasons we used very low Fe(III)/peptide ratios, so that, in our experiments no more than 5% of side-chain chromophores were usually involved in complex formation. Under these conditions there is no effect of complex formation on the α-helical structure of the polymeric backbone. Conversely, we have found an effect of the polymer conformation on the stereochemistry of the ferric complex formation of β-ketoamide sidechains. There is in fact evidence for stereospecific complex formation induced by the asymmetric structure of the polypeptide matrix. The evidence is based on the CD properties of the complexes in the charge-transfer transition region. In all three cases there are two strong oppositely signed bands at 250 nm and 270 nm of almost identical rotatory strength (Figure 12), the cross position being coincident with the ultraviolet absorption maximum. This pattern strongly suggests an exciton splitting

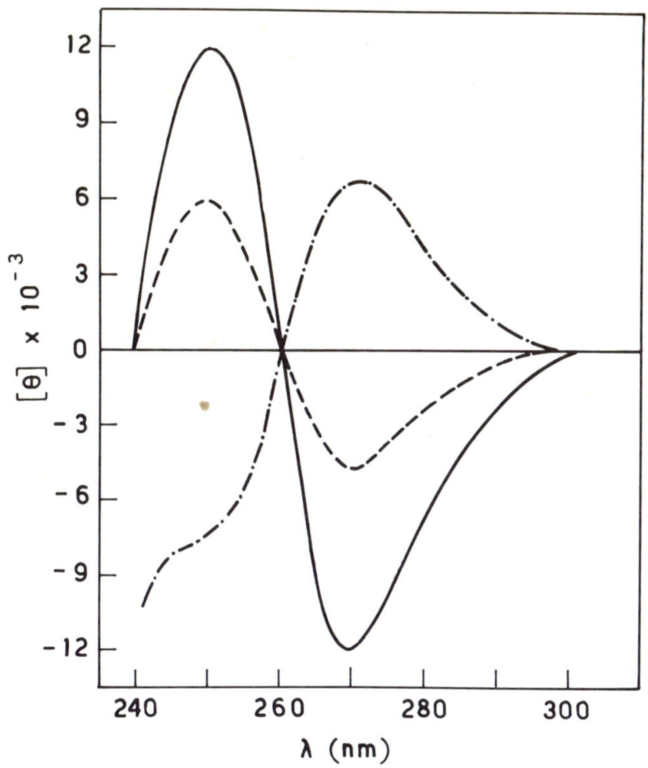

Fig. 12. CD spectra of the ferric complexes of $\{$L-Lys(Acac)$\}_n$ (----), $\{$L-Orn(Acac)$\}_n$ (-•-•-•-), and $\{$L-A$_2$bu(Acac)$\}_n$ (——) in the near UV region (from Ref. 33)

of the $\pi \rightarrow \pi^+$ transition of the β-ketoamide group due to degenerate coupling of the $\pi \rightarrow \pi^+$ transition moments of two ligands bound to the same metal ion. It is well known that *bis*- and *tris*-chelates of bidentate ligands (such as acetylacetone) with metal ions in an octahedral coordination can be intrinsically dissymmetric with consequent very high rotational strength. For these complexes both theoretical calculations and experimental results[34] showed that the ligand $\pi \rightarrow \pi^+$ transitions are split into two oppositely signed bands giving a conservative spectrum. From the sense of the splitting it was also possible to determine the absolute configuration of the complexes.[34] The CD results obtained on the ferric complexes of the three polymers strongly suggest the formation of intrinsically dissymmetric complexes. The predominant formation of one diastereoisomeric form over the other is induced by the conformation of the polypeptide matrix. Here again we observed that the absolute configuration of the ferric complex of the acetoacetyl derivative of $(\text{L-Orn})_n$ is just the opposite of those of $(\text{L-Lys})_n$ and $(\text{L-A}_2\text{bu})_n$ derivatives. This result is in line with the findings of De Santis[32] on copper complexes of salicylaldehyde and pyridoxal derivatives of the same polymers. Finally, we observed that the rotational strength of the polymeric complexes increases as the distance of the side-chain groups from the helical backbone diminishes. This result suggests a more effective asymmetric induction of the helical peptide matrix on the stereospecific formation of chiral complexes, when the distance between the complex and the peptide backbone is decreased.

The above reported examples show how in model systems a particular peptide conformation might induce steric restrictions on the way of binding of metal ions, leading to stereospecificity in the complex formation process.

Conclusions

There are several implications in the work described in the present review. The first concerns the field of catalysis. The advantages of using catalytically active species anchored to polymeric matrices are well-known. However an overwhelming majority of investigations up to date have involved cross-linked polystyrene as the polymeric support, with consequent poor information about the effects of particular, asymmetric polymer structures on the catalytic phenomenon. When a catalytic active species is anchored to an asymmetric polypeptide matrix, we might obtain chiral catalytic systems, which in principle can display stereospecificity in analogy to enzyme promoted reactions.

A second aspect of this work is strictly correlated with the first and concerns the use of these systems as models for metallo-enzymes. As already stressed in the Introduction to the present paper, useful information can be obtained on the way of coordination of metal ions to polypeptides, which can lead to the understanding of the control mechanism determining coordination of metal ions in biological systems. However, detailed knowledge of the structure of ions containing proteins at the binding site is very difficult to obtain. Besides the presence of appropriate ligand groups in a well-defined space arrangement, the whole structure of the molecule is designed to produce optimum conditions for specific biological activity. It is because of this complexity that the model system approach to this problem, including that described in the present work, is still very rudimentary. A significant advance in this area requires intensive research in peptide synthesis and conformation and in coordination chemistry.

References

1. For a comprehensive review see (1967) *Poly-α-Amino Acids.* Fasman, G. D., Ed., Marcel Dekker Inc., New York.

2. For a review, see Peggion, E. Cosani, A., Terbojevich, M. & Palumbo, M. (1977) in *Optically Active Polymers,* E. Selegny Ed., pp. 0000–0000.

3. Deber, C. M., Madison, U. & Blout, E. R. (1975) *Acc. Chem. Res.* **9**, 106–113.

4. Mattice, W. L., & Mandelkern, L. (1971) *Biochemistry* **10**, 1926–1933; 1934–1942.

5. For a review, see Traub, W., (1974) in *Peptides, Polypeptides and Proteins,* Blout, E. R., Bovey, F. A., Goodman, M., & Lotan, N., Eds., Wiley Interscience, New York, p. 146, and references therein quoted.

6. Goodman, M., Toniolo, C., & Naider, F. (1974) in *Peptides, Polypeptides and Proteins,* Blout, E. R., Bovey, F. A., Goodman, M. & Lotan, N., Eds., Wiley Interscience, New York, p. 308.

7. Martin, R. B., Petit-Ramel, M. M. & Sharff, J. P. (1973) in *Metal Ions in Biological Systems,* Vol. 2, Sigel, H., Ed., Marcel Dekker Inc., New York, p. 17.

8. Martin, R. B. (1974) in *Metal Ions in Biological Systems,* Vol. 1, Sigel, H., Ed., Marcel Dekker Inc., New York, p. 129.

9. For example, Camerman, N., Camerman, A. & Sarkar, B. (1976) *Can. J. Chem.* **54**, 1309–1316; Kruck, T. P. A., Show - Jy, L. & Sarkar, B. (1976) *Can. J. Chem.* **54**, 1300–1308.

10. Hatano, M. & Nozawa, T. (1976) in *Metal Ions in Biological Systems,* Vol. 5, Sigel, H., Ed., Marcel Dekker Inc., p. 277.

11. Sigel, H. (1970) *Acc. Chem. Res.* **3**, 201–208.

12. Lontie, R. & Vanquickenborne, L. (1974) in *Metal Ions in Biological Systems* Vol. 3, Sigel, H., Ed., Marcel Dekker Inc., p. 183.

13. Felsenfeld, G. (1960) *Arch. Biochem. Biophys.* 87, 246–251.

14. Pecht, I., Levitzki, A. & Anbar, M. (1967) *J. Amer. Chem. Soc.* **80**, 1587–1591.

15. Levitzki, A., Pecht, I. & Berger, A. (1972) *J. Amer. Chem. Soc.* **94**, 6844–6849.

16. Tsangaris, J. M., Wen Chang, J. & Martin, R. B. (1969) *J. Amer. Chem. Soc.* **91**, 726–731.

17. Hatano, M., Yoneyama, M., Ito, I., Nozawa, T. & Nakay, M. (1969) *J. Amer. Chem. Soc.* **92**, 2165–2166.

18. Hatano, M., Nozawa, T., Ikeda, S. & Yamamoto, T. (1971) *Makromol. Chem.* **141**, 1–9; 11–19.

19. Nozawa, T. & Hatano, M. (1971) *Makromol. Chem.* **141**, 21–29.

20. Nozawa, T. & Hatano, M. (1971) *Makromol. Chem.* **141**, 31–41.

21. Garnier, A. & Tosi, L. (1977) *Biochem. Biophys. Res. Commun.* **74**, 1280–1283.

22. Palumbo, M., Cosani, A., Terbojevich, M. & Peggion, E. (1977) *J. Amer. Chem. Soc.* **99**, 939–941.

23. Palumbo, M., Cosani, A., Terbojevich & M. Peggion, E. (1977) *Macromolecules* **10**, in press.

24. Gourke, M. J. & Gibbs, J. H. (1971) *Biopolymers* **10**, 7957.

25. Osterberg, R. (1974) in *Metal Ions in Biological Systems,* Vol. 3, Sigel, H., Ed., Marcel Dekker Inc., pp. 45–84.

26. Solomon, E. J., Hare, J. W. & Gray, H. B. (1976) *Proc. Nat. Acad. Sci. USA* **73,** 1389–1393.

27. Palumbo, M., Cosani, A., Terbojevich, M. & Peggion, E. (1977) *Biopolymers* **16,** 109–119.

28. Camiletti, G., De Santis, P. & Rizzo, R. (1970) *Chem. Comm.,* 1073.

29. De Santis, P., D'Ilario, L., Lamanna, G., Morosetti, S. & Savino, M. (1973) *Biopolymers* **12,** 423–433.

30. Dentini, M., De Santis, P., Savino, M. & Verdini, S. (1974) *Makromol. Chem.* **175,** 327–337.

31. Snell, E. E., Braunstein, A. E., Severin, E. S. & Torehinsky, Yu. M. (1968) *Pyridoxal Catalysis: Enzyme and Model Systems,* Interscience, New York.

32. De Santis, P. (1976) *La Chimica e Industria* **2,** 25–29.

33. Palumbo, M., Cosani, A., Terbojevich, M., Bacchion, G. & Peggion, E. (1976) *Biopolymers* **15,** 2241–2253; 2254–2262.

34. Mason, S. F. (1973), in *Optical Rotatory Dispersion and Circular Dichroism,* Ciardelli Ed., Heyden and Sons Ltd, New York, p. 196; Bosnich, B., *ibidem,* p. 254.

PREFERENTIAL ADSORPTION AND CONFORMATIONAL TRANSITION OF POLY[N^5-(3-HYDROXYPROPYL)-L-GLUTAMINE] IN WATER/2-CHLOROETHANOL MIXTURES.

CLAUDE LOUCHEUX, MICHEL MORCELLET and CLAUDE FEYEREISEN, *Laboratoire de Chimie Macromoléculaire, Université des Sciences et Techniques de Lille*
Boîte Postale 36, 59650 Villeneuve d'Ascq, France

It is well known that the conformational transition in globular proteins in water/organic solvent mixtures is accompanied by changes in preferential and absolute adsorption[1]. Very few studies have been carried out on synthetic polypeptides[2-6]. We wish to report a study on poly[N^5-(3-hydroxypropyl)-L-glutamine], $[Gln(CH_2)_3OH]_n$ in water/2-chloroethanol mixtures, in which conformational properties, preferential adsorption and partial specific volume of $[Gln(CH_2)_3OH]_n$ have been investigated.

Experimental details (circular dichroïsm, measurement of the preferential adsorption parameter $(\partial g_3/\partial g_2)_m$ can be found elsewhere[6]. The partial specific values $\bar{v}^o$ of $[Gln(CH_2)_3OH]_n$ has been determined by densitometry in the usual manner.

Conformational Transition and Preferential Adsorption to $[Gln(CH_2)_3OH]_n$ in Water/2-Chloroethanol Mixtures.

The conformational transition in $[Gln(CH_2)_3OH]_n$ was followed at 25°C by the variation of the residue ellipticity at 222 nm when 2-chloroethanol is added to an aqueous solution of the polypeptide (M_r = 253,000). $[\theta]_{222}$ increases rapidly with increasing 2-chloroethanol concentration, the transition midpoint being located at a volume concentration of 20% 2-chloroethanol. In 90% by volume of 2-chloroethanol, the maximum helicity is attained (f_N = 0.78). This indicates good helicogenic properties of the organic solvent. Table I shows the variation of preferential adsorption of 2-chloroethanol in the different solvent mixtures. These data show that a preferential solvation of 2-chloroethanol takes place in the 5-20% organic solvent concentration range. It vanishes rapidly before preferential hydration occurs, the cross-over point being located at the conformational transition midpoint. Comparing the variation of preferential adsorption and the variation of the helical content f_N (Table I), one notices that the variations parallel each other. Moreover, the inversion of preferential solvation by 2-chloroethanol into preferential hydration can be interpreted by the following:

At low 2-chloroethanol concentration, the activity of water is sufficiently weakened by the presence of the organic solvent so that the mixed solvent acts

Table I. Preferential Adsorption to $[Gln(CH_2)_3OH]_n$ and Conformational Transition in Water/2-Chloroethanol Mixtures at 25°C

2-chloroethanol concentration vol %	helical content f_N *	preferential adsorption (in g of organic solvent per g of polypeptide)
5	0.22	+ 1.14
10	0.26	+ 0.33
20	0.46	+ 0.08
30	0.57	- 0.43
60	0.73	- 0.46

* as determined by circular dichroïsm measurements (see Ref. 6)

as a poorer hydrogen bonding agent, thus favoring the intramolecular hydrogen bonds and enhancing the helical conformation.

When increasing the 2-chloroethanol content, the coil-to-helix transition is achieved. At the same time, the preferential adsorption of 2-chloroethanol changes into a preferential hydration. This means that the water content of the solvation layer decreases more slowly than the water content of the bulk solvent. The remaining water is adsorbed to the hydrophilic sites of $[Gln(CH_2)_3OH]_n$: the side chain amide bond and the terminal hydroxyl group. This is consistent with Inoue and Izumi's observations[4] for $[Gln(CH_2)_3OH]_n$ in water/dioxane mixtures: when the dioxane concentration is increased, preferential hydration increases too.

The fact that a complete helicity is not obtained may be explained by the remaining water adsorbed on the polypeptide.

Partial Specific Volume of $[Gln(CH_2)_3OH]_n$ in Water/ 2-Chloroethanol Mixtures.

Figure 1 shows the variation of the partial specific volume of $[Gln(CH_2)_3OH]_n$, $\bar{v}^o$, as a function of 2-chloroethanol concentration, which follows a S-shaped curve. Such a S-shaped curve is usually attributed as a consequence of the conformational transition and has been noticed in the variation of the refractive index increment[7,8] of $[Gln(OBzl)]_n$ in mixed solvents, of the density increment of $[Gln(CH_2)_3OH]_n$ in water/2 chloroethanol mixtures[6]; pH induced volume changes show similar variations[9].

Nevertheless, the partial specific volume after dialysis of the solution against the solvent mixture, $\bar{v}^{o*}$, shows a quite different variation (see Fig. 1): $\bar{v}^{o*}$ is a linear function of 2-chloroethanol concentration. By dialysis, the preferential adsorption phenomenon is suppressed. Therefore we suggest that the S-shaped curve is due to the preferential adsorption to the polypeptide in the mixed

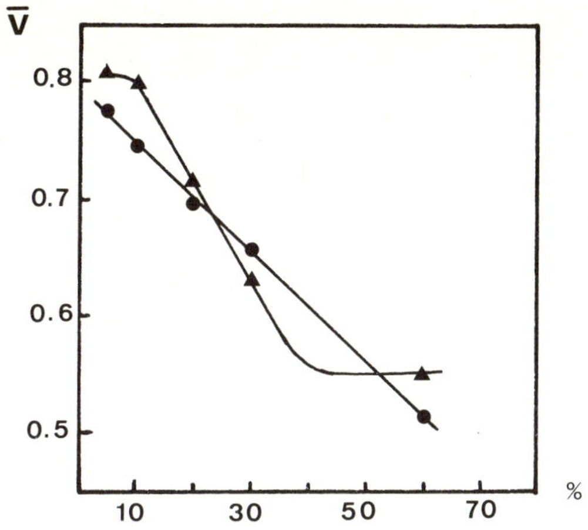

Fig. 1. Variation partial specific volume of $[Gln(CH_2)_3OH]_n$ in ml/g before (▲) and after (●) dialysis, at 25°C, with 2 chloroethanol concentration.

solvents. In fact, this feature is not surprising, since it is well known that $\bar{\nu}^o$ depends on solvation[9]. The linear decrease of the partial specific volume after dialysis, $\bar{\nu}^{o*}$, when the preferential adsorption phenomenon is suppressed, indicates that helical and randomly coiled residues contribute to $\bar{\nu}^{o*}$ independently of each other, and that the observed contraction has its origin in the conformational transition. The S-shaped curve shown by $\bar{\nu}^o$ appears then as a feature due to preferential adsorption: the inversion of preferential adsorption at 20% in volume 2-chloroethanol induces a rapid change in the variation of $\bar{\nu}^o$.

References

1. Inoue, H. & Timasheff, S. N. (1972) *Biopolymers* **11**, 737–743.
2. Strazielle, C., Dufour, C. & Marchal, E. (1970) *C. R. Acad. Sci., Ser.* C 271, 720–727.
3. Morcellet, M. & Loucheux C. (1975) *Polymer* **16**, 401–405.
4. Inoue, H. & Izumi, T. (1976) *Biopolymers* **15**, 797–812.
5. Kohiyama, J. Mori, T., Yamamoto, K. & Iijima, T. (1977) *J. Chem. Soc. Farad. Trans. I* 73, 203–212.
6. Feyereisen, C., Morcellet, M. & Loucheux, C. (1977) *Macromolecules* **10**, 485–488.
7. De Groot, K., Feyen, J., de Visser, A. C., van de Ridder, G. & Bantjes, A. (1971) *Kolloid Z. Z. Polym* **246**, 578–579.
8. Katime, I. & Roig, A. (1974) *An. Quim.* **70**, 853–855.
9. Noguchi, H. & Yang, J. T. (1963) *Biopolymers* **1**, 359–370.

CIRCULAR DICHROISM OF POLY(γ-[2-(9-CARBAZOLYL)-ETHYL]-L-GLUTAMATE) IN LIQUID CRYSTALLINE AND SOLID STATES

MASAHIRO HATANO, HIROYUKI NOMORI and MASAO YOSHIKAWA, *Chemical Research Institute of Non-aqueous Solutions, Tohoku University, Katahira 2-1-1, Sendai, 980, Japan*

It is well known that a bulky aromatic side chain substituent of a polymer plays an important role for the main-chain conformation of the polymer. The conformation of polyglutamates and polyaspartates is strongly influenced by the perturbation caused by their side chain substituents.[1] On the other hand, poly(-*N*-vinylcarbazole) is compelled to a transition at 306°C as a result of the steric hindrance between the carbazolyl groups.[2]

We will present in this paper an interesting result on the drastic change of the induced circular dichroism of the carbazolyl group (Cz) attached to the polypeptide main chain, with the variation of states of the polypeptide.

Recently, we synthesized poly(γ-[2-(9-carbazolyl)ethyl]-L-glutamate), $[Glu(OCH_2CH_2Cz)]_n$ and poly(β-[2-(9-carbazolyl)ethyl]-L-aspartate), $[Asp(OCH_2CH_2Cz)]_n$ from the corresponding NCA monomers, respectively.[3]

In the series of poly(D-glutamate esters) with aromatic side chains,[1,4-6] we clarified various side-chain effects on the interaction energy among the aromatic side chains, the conformations of the peptide main chain and side chain, and their helix stability. From the results of their reactivity in the polymer-ester interchange reaction and the induced circular dichroism (ICD) in the aromatic chromophores, we propose four types of conformations of the side chain aromatic groups; namely, freely rotating conformation, the restricted conformation by steric hindrance among aromatic groups, the restricted conformation by the interaction with neighboring side-chain groups, and the restricted conformation by the interaction with the amide group in the main peptide chain. The ICD of the aromatic side-chain group provides information whether the aromatic side chain rotates freely or not. Poly(γ-benzyl-L-glutamate) $[Glu(OBzl)]_n$ and poly$[Glu(OCH_2CH_2Cz)^8\ Glu(OMe)^{92}]$ in tetrachloroethane exhibit no optical activity with respect to the side-chain chromophores. This demonstrates the free rotation of benzyl and *N*-carbazolylethyl groups in the polypeptides. However, the *N*-carbazolylethyl group is optically active with respect to $[Glu(OCH_2CH_2Cz)]_n$, and the group must be restricted in terms of free rotation by the large steric repulsion among the carbazolyl groups in $[Glu(OCH_2CH_2Cz)]_n$.

The observed ICD bands will be assigned on the basis of molecular orbital calculation[7] and magnetic circular dichroism (MCD).[8]

Experimental

$[Glu(OCH_2CH_2Cz)]_n$ and $[Asp(OCH_2CH_2Cz)]_n$ were synthesized through the N-carboxyanhydrides (NCAs) of their parent amino-acid esters, which were obtained through esterification of L-glutamic acid and L-aspartic acid with 2-(9-carbazolyl)ethanol.

Circular dichroism (CD) and optical rotatory dispersion (ORD) spectra were measured by a JASCO J-20A dichrograph and a J-15 spectropolarimeter, respectively. A 12.2-kG electromagnet was used for measurement of the MCD spectra. Spectroscopic assignments were given for the observed MCD spectra of $[Glu(OCH_2CH_2Cz)]_n$, on the basis of the PPP calculations[8] and of the MCD data for monomeric carbazolyl unit.

Results and Discussion

The CD spectrum of $[Glu(OCH_2CH_2Cz)]_n$ at low concentration in dichloromethane shows one peak at 221 nm, and the infrared absorption spectrum and the value of Moffitt's b_o (-677) strongly suggest that $[Glu(OCH_2CH_2Cz)]_n$ exists in the right-handed α-helical conformation in dilute dichloromethane solution. Furthermore, the polypeptide shows CD even in the absorption region of the side-chain carbazolyl chromophore at 230–350 nm. The CD profile in this region coincides well with that of absorption spectrum of $[Glu(OCH_2CH_2Cz)]_n$ both in shape and in extrema wavelengths (Fig. 1). On the other hand, the CD spectrum of $[Glu(OCH_2CH_2Cz)]_n$ in the solid state indicates different ICD bands from those in dilute solution in the absorption region of the side-chain chromophore, and the same conformation as in the dilute solution (Fig. 1). On

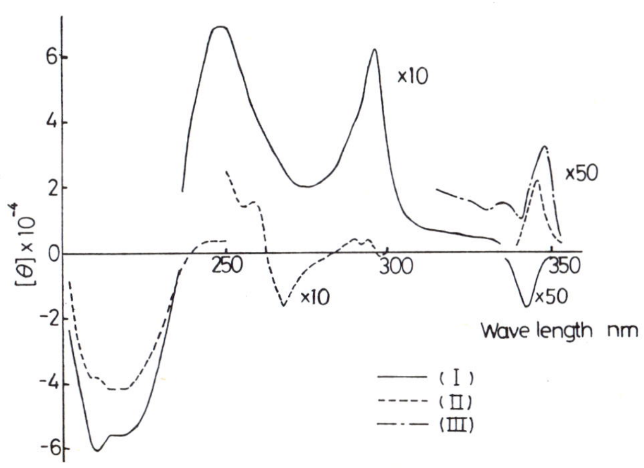

Fig. 1. CD spectra of $[Glu(OCH_2CH_2Cz)]_n$ in solid state (I), dilute solution (II), and liquid crystalline state (III).

heat treatment of the $[Glu(OCH_2CH_2Cz)]_n$ film, the CD bands increased gradually in the CD magnitudes. In the $[Glu(OCH_2CH_2Cz)]_n$ film including acridine orange (AO) as a dye probe, the ICD observed in the absorption region also increased in magnitude after thermal treatment at 120°C for 2 hours (Fig. 2). In the solid state, the CD signs and magnitudes depend strongly on the wavelengths corresponding to the electronic transitions of the carbazolyl group. For example, the magnitudes of CD bands at 250 and 290 nm were intensified by the annealing of $[Glu(OCH_2CH_2Cz)]_n$ film.

An observation on the $[Glu(OCH_2CH_2Cz)]_n$ film cast from dichloromethane under a polarized microscope indicates no appearance of the liquid crystalline phase. These observations indicate that the helix content of $[Glu(OCH_2CH_2Cz)]_n$ in the solid state is substantially similar to that in dilute solution, and remains almost constant throughout the annealing. But, the side chain of $[Glu(OCH_2CH_2Cz)]_n$ in the solid state is held more rigidly than in solution, and the orientation of the side chain around the helix axis is enhanced by annealing as well as in thermal transition of poly(*N*-vinylcarbazole) at 302°C.

Poly(γ-methyl-D-glutamate)[9] and poly(γ-benzyl-L-glutamate)[9,10] can form a type of lyotropic cholesteric mesophase, when dissolved in dichloroethane or dichloromethane at higher concentrations than 0.70 *M* of the glutamyl residues. In former case, the supramolecular helical structure of the molecules retains in solid state. $[Glu(OCH_2CH_2Cz)]_n$ was dissolved in a dichloroethane solution of poly(γ-methyl-D-glutamate), $[Glu(OMe)]_n$, at the molar concentration ratio of 1:50 on the basis of the glutamyl residues, and a film was cast from the mixed solution. Fig. 3 shows the CD spectrum of the film, exhibiting a cholesteric pitch band around 750 nm and the well-resolved CD bands in the 250–350 nm region. This suggests that the carbazolyl group of $[Glu(OCH_2CH_2Cz)]_n$ intercalated in

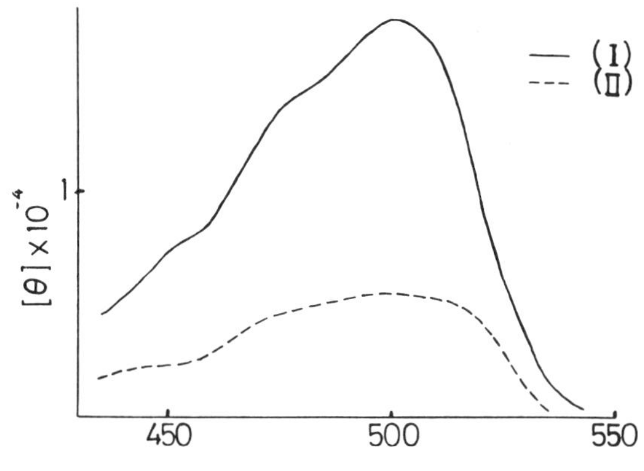

Fig. 2. ICD of AO buried in $[Glu(OCH_2CH_2Cz)]_n$ film. (I), after annealing at 120°C for 2 hr; (II), before the annealing.

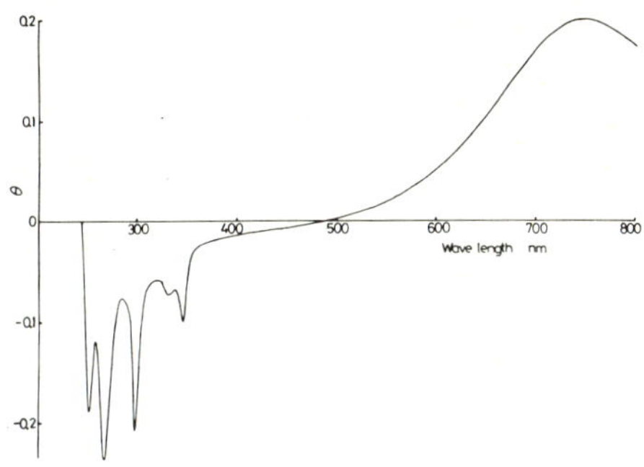

Fig. 3. CD spectrum of the film casted from a mixed solution of $[Glu(OCH_2CH_2Cz)]_n$ and $[Glu(OMe)]_n$ in dichloroethane.

the cholesteric mesophase of $[Glu(OMe)]_n$ was perturbed strongly by the dissymmetric field due to the supramolecular helical structure of $[Glu(OMe)]_n$. In the concentrated solution of $[Glu(OCH_2CH_2Cz)]_n$ in dichloroethane, the CD bands with opposite signs to those of the $[Glu(OCH_2CH_2Cz)]_n$ intercalated in the cholesteric mesophase of $[Glu(OMe)]_n$ can be observed in the 300–360 nm region. This shows that the carbazolyl group is perturbed by the dissymmetric field of different sense from that in the $[Glu(OMe)]_n$ liquid crystal. But, the CD band of the $A_1 \leftarrow A_1$ transition of the carbazolyl group has the same sign as that of the $B_2 \leftarrow A_1$ transition (see Table I) in both cases. *N*-Ethylcarbazole in the cholesteric phase of $[Glu(OBzl)]_n$ solution in tetrachloroethane exhibited the similar CD profile having the same signs in both transitions of $A_1 \leftarrow A_1$ and $B_2 \leftarrow A_1$. This is not case in Saeva's result[11] where the CD curve for *N*-ethylcarbazole in the thermotropic liquid crystalline phase consists of two regions with opposite signs. As Sackman and Voss[12] and Saeva[11] indicated, the CD sign depends on the polarization character of the corresponding electronic transition of the dye only when the helical arrangement of the dyes along the cholesteric

Table I. Calculated and Observed Transition Energies

Sym.	Calculated ΔE (eV)	(f)	Observed ΔE (eV)	(nm)
$A_1 \leftarrow A_1$	3.79	(0.006)	3.57	(347)
$B_2 \leftarrow A_1$	4.34	(0.166)	4.20	(295)
$B_2 \leftarrow A_1$	4.43	(0.409)	4.68	(265)
$B_2 \leftarrow A_1$	5.14	(1.380)	5.17	(240)
$A_1 \leftarrow A_1$	5.28	(0.012)	5.32	(233)
$A_1 \leftarrow A_1$	5.33	(0.314)	———	

helical axis is realized. Our results show such a helical arrangement of carbazolyl side chains can be formed (Fig. 1) in the $[Glu(OCH_2CH_2Cz)]_n$ film, whereas $[Glu(OCH_2CH_2Cz)]_n$ in the liquid crystalline phase does not form the helical arrangement of side-chain chromophores.

References

1. Konishi, Y. & Hatano, M. (1976) *Polym. Lett.* **14**, 351–359.
2. Crystal, R. G. (1971) *Macromolecules* **4**, 380–384.
3. Nomori, H., Yoshikawa, M. & Hatano, M. (1976) *Makromol. Chem.* **177**, 3077–3081.
4. Konishi, Y. & Hatano, M. (1976) *Polym. Lett.* **14**, 219–223.
5. Konishi, Y. & Hatano, M. (1976) *Polym. Lett.* **14**, 303–305.
6. Konishi, Y. & Hatano, M. (1976) *J. Polym. Sci., Polym. Chem. Ed.* **14**, 2329–2334.
7. Tajiri, A., Tanikawa, K. & Hatano, M. to be submitted in *Makromol. Chem.*
8. Kaito, A., Tajiri, A. & Hatano, M. (1976) *J. Amer. Chem. Soc.* **98**, 384–388; 7932–7938.
9. Tsuchihashi, N., Nomori, H. & Hatano, M. (1975) *Bull. Chem. Soc. Japan* **48**, 29–32.
10. Saeva, F. D. & Olin, G. R. (1973) *J. Amer. Chem. Soc.* **95**, 7882–7884.
11. Saeva, F. D. (1972) *J. Amer. Chem. Soc.* **94**, 5135–5136.
12. Sackmann, E. & Voss, J. (1972) *Chem. Phys. Lett.* **14**, 528–532.

CONFORMATIONAL ASPECTS OF PROTECTED CO-OLIGOPEPTIDES WITH ALTERNATING L-ISOLEUCINE AND D-ALLOISOLEUCINE RESIDUES

TULLIO PAGANETTI, VINCENZO RIZZO, ANTONIO BAICI, PETER NEUENSCHWANDER and GIAN PAOLO LORENZI,
Technisch-Chemisches Laboratorium, ETH,
8092 Zurich, Switzerland

Co-oligo- and copolypeptide chains formed by alternating L-isoleucine (A) and D-alloisoleucine (B) residues have chiral lateral groups all of the same configuration.

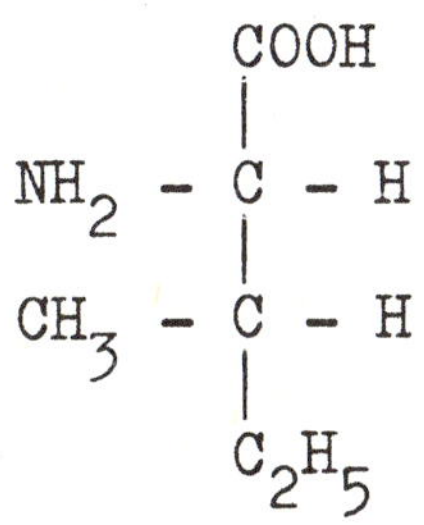

A: H-L-Ile-OH

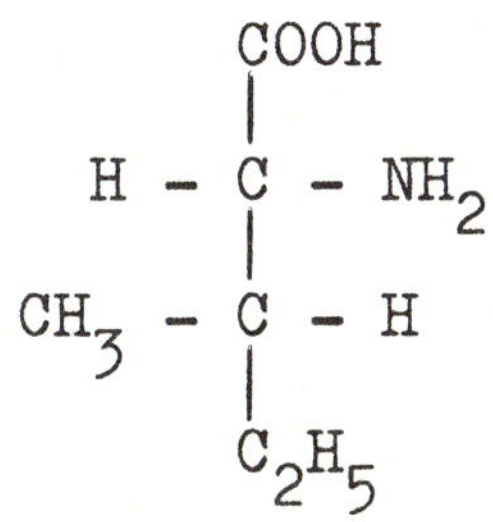

B: H-D-aIle-OH

With the aim of assessing the influence this might have on the conformation of the L,D-peptide chains, we have begun to study the members of the series Boc-(D-aIle)$_m$-(L-Ile-D-aIle)$_n$-OMe with m = 0 and n = 1, 2, 3 and 4 (II, IV, VI and VIII) and with m = 1 and n = 1, 2 and 3 (III, V and VII). These co-oligopeptides have been synthesized in a stepwise fashion by standard racemization-free methods from commercial amino acids of very high optical purity. VII and VIII possess a remarkable solubility in cyclohexane, which is shared only by II among the other co-oligopeptides. In this nonpolar solvent, as indicated[1] by a number of complementary techniques, VII and VIII favor β-helical conformations[2] with a specific sense of twist. This communication presents and discusses in terms of chain-length dependence nmr, CD and UV data for all synthesized co-oligopeptides in trifluoroethanol solution.

Results and Discussion

The NH-region above 6.5 ppm of the 90-MHz FT nmr spectra of the co-oligopeptides in trifluoroethanol is reported in Fig. 1. The adjacent region at higher field is seriously disturbed by the strong signal of the OH-proton of the

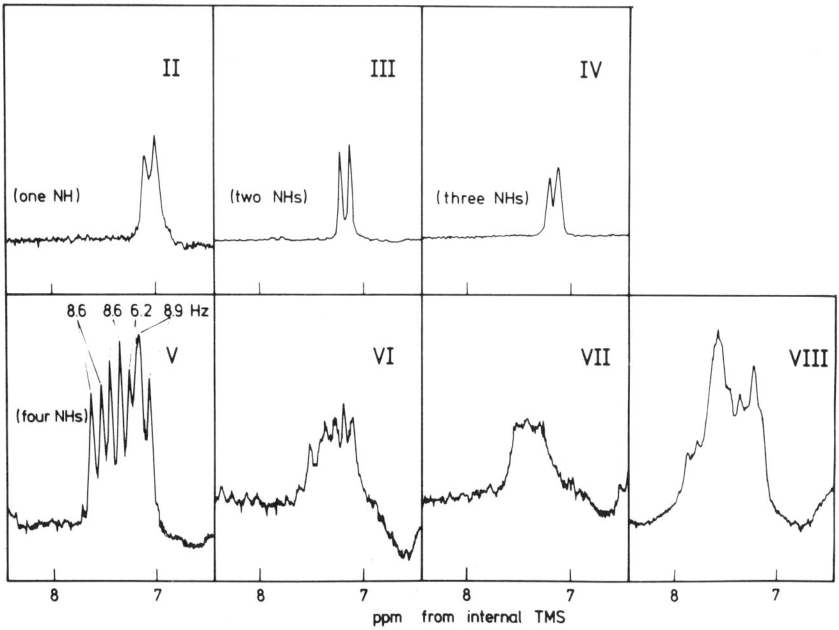

Fig. 1. Partial 90 MHz FT ^{1}H nmr spectra of the co-oligopeptides II-VIII in trifluoroethanol solution (Concn, 0.02–0.08 g/ml; t, 25°C; external lock).

solvent and it is not shown. Thus in the reported partial spectra of II, III, IV and V, one NH-proton – probably the urethane one – is not represented, and in those of VI, VII and VIII one or more protons may be missing. A single doublet for two NHs of III and three NHs of IV (Fig. 1) suggests a conformational equilibrium with many rapidly interconverting conformers. In V, on the contrary, the relatively well separated doublets for the different NHs hint at a conformational equilibrium with a restricted number of conformations. Though poorly resolved, the partial spectra of VI, VII and VIII also point to an analogous situation.

The CD spectra of the co-oligopeptides in trifluoroethanol solution are shown in Fig. 2. No substantial concentration effect on the CD was noticed. In the case of IV and V, the highest concentrations investigated were in the range of those used for nmr measurements (Fig. 1). The profile of the CD spectra of V, VI and VII is similar to that observed[1] for VII and VIII in cyclohexane. This similarity suggests the prevalence of a helical conformation of the β-type in the conformational equilibrium of these co-oligopeptides in trifluoroethanol. For V, this prevalence is supported by the high values of the vicinal coupling constants $^{3}J_{NH\text{-}CH}$ which characterize its nmr spectrum (that of the NH-proton not shown in Fig. 1, at 6.02 ppm, is 9.3 Hz), which are in the range of those expected for β-helices. Understandably, this type of conformation appears at V, since V, as a

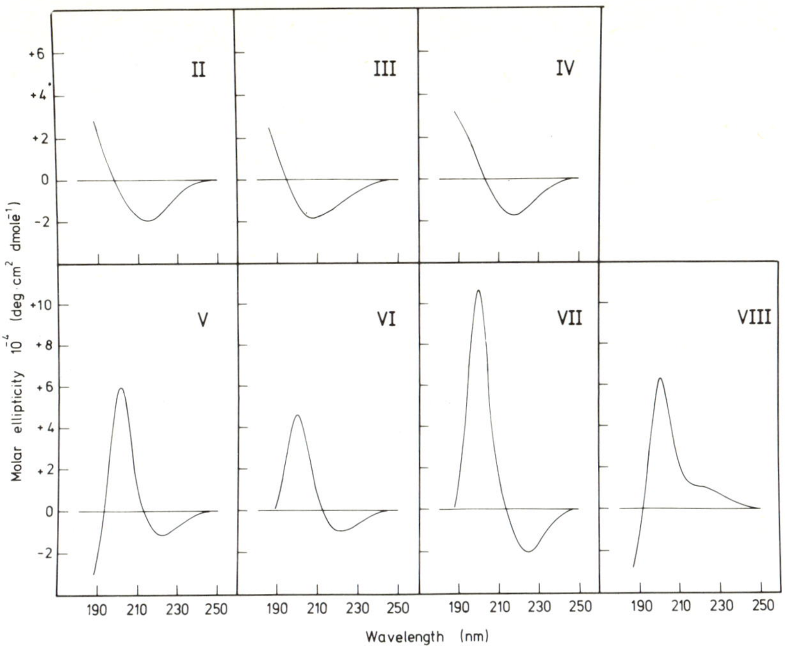

Fig. 2. CD spectra of the co-oligopeptides II-VIII in trifluoroethanol solution (Concn, 0.2 mg/ml; t, 25°C).

pentapeptide, has the minimum number of residues required for one turn of a β-helix with about 4.4 residues per turn, with one normal interturn hydrogen bond.

The sign and relative position of the bands in the CD spectra of V, VI and VII are the same as those calculated preliminarily by Bayley[3] for left-handed β-helices with 4.4 residues per turn. The profile of the CD spectrum of VIII in trifluoroethanol is different from that of the oligomers V, VI and VII in the same solvent and suggests the presence of another ordered conformation. Conformational energy calculations by Hesselink and Scheraga[4] have shown that α-helical structures may be energetically favorable for L,D-sequences. One cannot exclude the possibility that the CD spectrum of VIII may reflect a significant presence of the left-handed α-helical conformation in the conformational equilibrium. Consistent with this possibility, there is some indication of a weak hypochromism in the UV absorption band arising from the $\pi - \pi^*$ transition setting in at VIII (Fig. 3). While further study is needed to clarify these aspects, recent CD measurements[5] have hinted at the prevalence of an α-helical conformation in the conformational equilibrium of Boc-(L-Val-D-Val)$_4$-OMe in trifluoroethanol. With respect to VIII this should have the opposite handedness, since the ellipticity bands have different signs. The differences in the CD spectra of the two octapeptides indicate that the configuration of the asymmetric carbon atoms of the lateral chains may

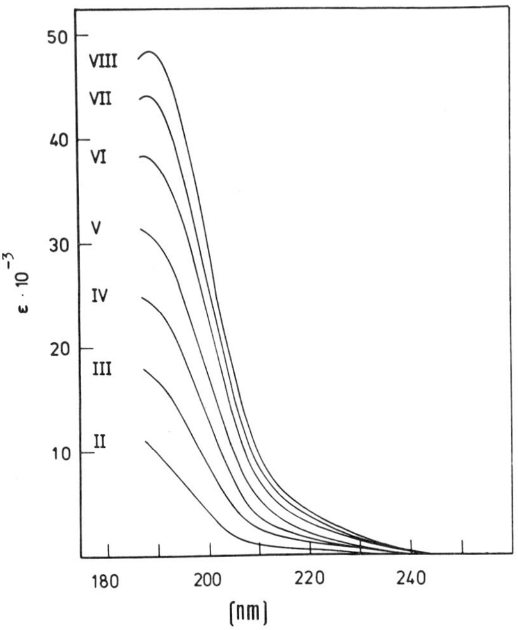

Fig. 3. UV spectra of the co-oligopeptides II-VIII in trifluoroethanol solution (Concn, 0.2 mg/ml; t, 25°C).

be an important factor in determining the conformational equilibrium of VIII in trifluoroethanol.

Conclusions

In trifluoroethanol VII assumes preferentially a β-helical conformation similar to that which it favors in cyclohexane. However, this conformation is destabilized in trifluoroethanol, since the intensity of the dichroic bands of VII in this solvent is substantially lower than in cyclohexane. V and VI also favor a similar β-conformation in trifluoroethanol. For VIII the conformational equilibrium is different in trifluoroethanol from what it is in cyclohexane. It is possible that in trifluoroethanol the left-handed α-helical conformation contributes significantly to the conformational equilibrium.

References

1. Lorenzi, G. P. & Paganetti, T. (1977) *J. Amer. Chem. Soc.* **99**, 1282–1283.
2. De Santis, P., Morosetti, S. & Rizzo, R. (1974) *Macromolecules* **7**, 52–58.
3. Bayley, P. M. (1973) *Prog. Biophys. Mol. Biol.* **27**, 3–76.
4. Hesselink, F. T. & Scheraga, H. A. (1972) *Macromolecules* **5**, 455–463.
5. Tomasic, L., unpublished results.

STRUCTURE OF γ-BRANCHED HOMO-OLIGOPEPTIDES

M. PALUMBO, G. M. BONORA, C. TONIOLO, E. PEGGION, *Biopolymer Research Center of C. N. R., Institute of Organic Chemistry, University of Padova, 35100 Padova, Italy;* and
E. STEVENS, *Department of Chemistry, Brown University, Providence, Rhode Island 02912*

The present study is a part of our continuing investigation of conformational properties of a series of linear homo-oligopeptides (from dimer to heptamer) derived from α-amino acid residues containing hydrocarbon side-chains.[1] We have already shown that the position of side-chain branching is of paramount importance in directing oligopeptide conformation. Relative to γ-position branching at the β-position: i) markedly increases the stability of corresponding intermolecular β-forms,[2] ii) seems to favor the unusual parallel arrangement of peptide chains within the pleated sheet structure,[3-6] and iii) decreases the propensity to form intramolecularly H-bonded folded structures.[3] In this communication we wish to compare the conformational preferences of the three γ-branched homo-oligopeptide series having the general formula *t*-Boc-(L-X)$_n$-OMe (n=2-7; X= Leu, Cha, and Phe). For this stereochemical analysis we have employed IR absorption and CD to 150 nm.

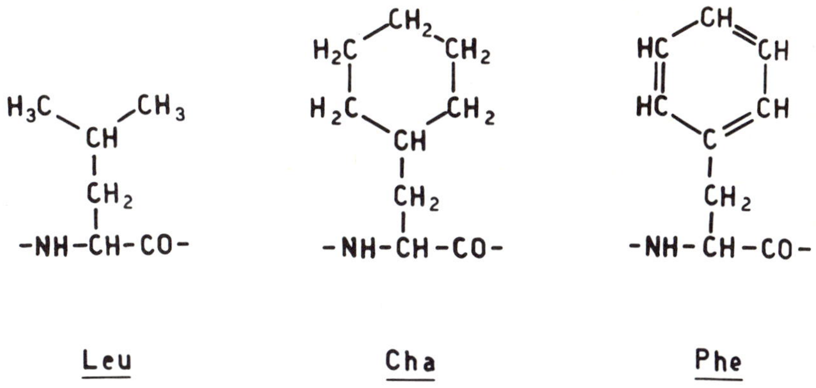

From the IR absorption data in the solid state it appears that the higher oligomers of the three series tend to adopt a β-conformation (the corresponding amide I band is found at 1633-1641 cm^{-1}), most probably of the intermolecular type.[3,5,6] The peculiar stereochemical requirements of the cyclohexyl moiety of Cha peptides apparently decrease their tendency to form the aforementioned secondary structure. In fact, the amide I transition associated with the random-coil structure (at 1655-1660 cm^{-1})[3] still remains clearly visible at the hexamer and heptamer levels in that series. From an analysis of the three homo-heptamer methyl ester hydrochlorides the 1690 cm^{-1} band, indicative of the presence of chains in the antiparallel arrangement,[3,5,6] occurs only in the Leu and Cha peptides. (The *t*-butyloxycarbonyl blocking group was removed to avoid superposition of bands from the urethane and amide groups.)[3,5,6] In contrast, we did not obtain evidence of the occurrence of this band in the homo-heptamer methyl ester hydrochloride derived from Phe, which does exhibit β-structure formation. On the basis of the examination of molecular models, we tentatively suggest that side-chain-side-chain interactions are the major factor responsible for the onset of the postulated unusual parallel β-form in Phe oligopeptides[7-9]. Only in this chain arrangement are the phenyl rings of adjacent chains (lying alternatively on opposite sides of the pleated sheet within the same chain) always parallel and equidistant. The closest contact of the carbon atoms of the phenyl rings is approximately 3.5 Å.

The vacuum-ultraviolet CD spectra of films of *t*-Boc-(L-X)$_7$-OMe (X= Leu, Cha, and Phe) are shown in Figure 1. The molar ellipticities were calculated assuming no difference in molar ellipticity between film and solution near 200 nm.[10,11] The Leu heptamer curve is typical of a β-sheet which contains a mixture of parallel and antiparallel oriented chains.[4,12] The Phe heptamer curve, with a cross-over near 195 nm, is typical of the parallel β-sheet previously observed in the case of *t*-Boc-(L-Val)$_7$-OMe.[4] Apparently the aromatic side-chain transitions are not optically active enough to perturb this general pattern. The Cha heptamer spectrum displays features characteristic of disordered chains. All of these conclusions are consistent with those based on the IR absorption data reported here and with earlier vacuum-ultraviolet CD measurements.[4,12]

The onset of intramolecularly H-bonded folded forms was investigated in solvents of low polarity (e.g., $CDCl_3$) using the IR dilution technique.[13] On the basis of our findings and literature data,[14] we assume that at the lowest concentrations examined (below $5 \times 10^{-4} M$) the H-bonded N-H absorption is due almost entirely to intramolecularly H-bonded folded species. The di- and tripeptides from Leu, Cha and Phe possess low, although not negligible, amounts of these conformers. A striking difference between the aliphatic (Leu and Cha) and aromatic (Phe) peptides stands out at the tetramer level. Phe peptides clearly exhibit a lower extent of intramolecularly C=0 . . . H–N folded forms (the

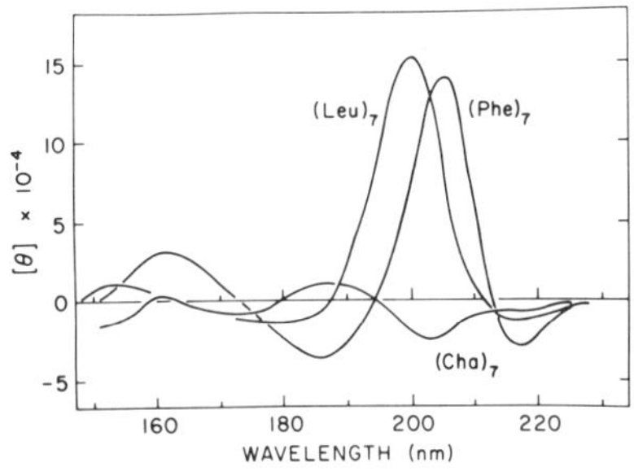

Fig. 1. Vacuum-ultraviolet circular dichroism spectra of *t*-Boc-(L-X)$_7$-OMe(X=Leu, Cha, and Phe). The films were cast from F_3EtOH solutions (conc. ≃ 2mg/ml). Scale for *t*-Boc-(L-Cha)$_7$-OMe spectrum is reduced by a factor of 4.

corresponding band appears at about 3345 cm^{-1}) than those derived from Leu and Cha. We propose that competition between the aromatic ring and the amide carbonyl in the intramolecular interaction with the N-H groups would be the major factor for the reduced occurrence of these conformers in Phe peptides.[15] The large (20 cm^{-1}) shift to higher frequency of the band associated with the "free" N-H vibration at 3415 cm^{-1} of Phe higher oligomers upon hydrogenation of the aromatic rings strongly supports our interpretation.

In solvents of high polarity (1, 1, 1, 3, 3, 3-hexafluoroisopropanol [F_6 Pr^i OH] and 1, 1, 1-trifluoroethanol [F_3EtOH]) ordered secondary conformations are absent in all Leu and Cha peptides, as shown by CD analysis. Addition of water to F_3EtOH solutions induces the formation of associated structures in the Leu homo-heptamer and Cha homo-hexamer. Dilution disrupts the aggregated forms. The stability of the associated structures in F_3EtOH/water mixtures is higher in Cha peptides (Figure 2), probably due to stronger hydrophobic interactions.

In conclusion, the position of branching being equal, steric requirements, electronic properties, and hydrophobic character of the amino-acid side chains all have relevant effects in determining the nature and stability of oligopeptide conformations. In this connection it is of interest that the α-helical structures of Cha and Phe homopolymers exhibit different conformational stabilities. This result was explained in terms of different contributions from non-covalent bonds in the side chains.[16]

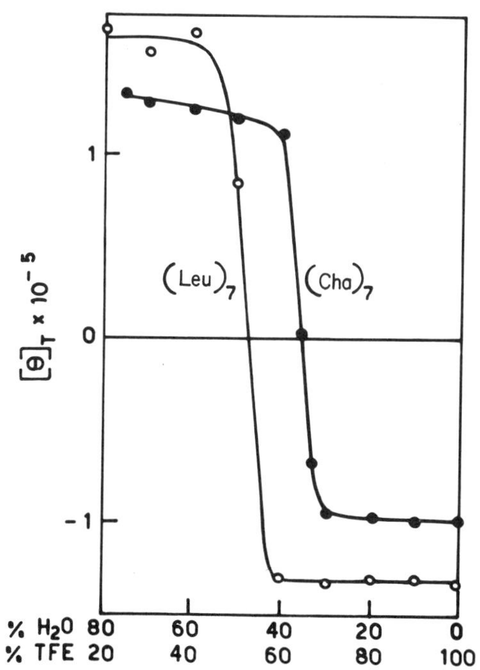

Fig. 2. Total molar ellipticity values at 200 nm of *t*-Boc-(L-X)$_7$-OMe (X=Leu and Cha) *versus* solvent composition (F_3EtOH/water) (conc. $3 \times 10^{-4} M$).

This study was supported by grants BMS 73-01799 from NSF, GM-22347 from NIH, and 1099 from NATO.

References

1. For the previous paper in this series see Bonora, G. M. & Toniolo, C. (1977) *Gazz. Chim. Ital.* **107**, in press.

2. Toniolo, C. & Bonora, G. M. (1975) in *Peptides: Chemistry, Structure, and Biology*, Walter, R. and Meienhofer, J., Eds., Ann Arbor Science, Ann Arbor, pp. 145–150.

3. Palumbo, M., Da Rin, S., Bonora, G. M. & Toniolo, C. (1976) *Makromol. Chem.* **177**, 1477–1492.

4. Balcerski, J. S., Pysh, E. S., Bonora, G. M. & Toniolo, C. (1976) *J. Amer. Chem. Soc.* **98**, 3470–3474.

5. Toniolo, C., Bonora, G. M., Palumbo, M. & Pysh, E. S. (1976) in *Peptides 1976*, Loffet, A., Ed., Presses Univ. de Bruxelles, Bruxelles, pp. 597–600.

6. Toniolo, C. & Palumbo, M. (1977) *Biopolymers* **16**, 219–224.

7. Wei, C. H., Doherty, D. G. & Einstein, J. R. (1972) *Acta Cryst.* **B28**, 907–915.

8. Marsh, R. E. & Glusker, J. P. (1961) *Acta Cryst.* **14**, 1110–1116.

9. Harada, Y. & Iitaka, Y. (1977) *Acta Cryst.* **B33**, 247–249.

10. Bonora, G. M., Maglione, A. & Toniolo, C. (1974) *Polymer* **15**, 767–770.

11. Toniolo, C. & Bonora, G. M. (1974) *Gazz. Chim. Ital.* **104**, 843–848.

12. Kelly, M. M., Pysh, E. S., Bonora, G. M. & Toniolo, C. (1977) *J. Amer. Chem. Soc.* **99**, 0000–0000.

13. Mizushima, S., Shimanouchi, T., Tsuboi, M. & Souda, R. (1952) *J. Amer. Chem. Soc.* **74**, 270–271.

14. Shields, J. E., McDowell, S. T., Pavlos, J. & Gray, G. R. (1968) *J. Amer. Chem. Soc.* **90**, 3549–3556.

15. Néel, J. (1971) in *Proc. IXth IUPAC Macromol. Microsymp.*, Sedlacek, B., Ed., Butterworths, London, pp. 201–225.

16. Peggion, E., Strasorier, L. & Cosani, A. (1970) *J. Amer. Chem. Soc.* **92**, 381–386.

CONFORMATIONAL STUDIES ON POLYOXYETHYLENE-BOUND PEPTIDES

M. MUTTER, H. MUTTER and E. BAYER, *Institut für Organische Chemie der Universität, Auf der Morgenstelle 18, D-7400 Tübingen, Germany*

The liquid-phase method for peptide synthesis[1] has currently been used for the effective synthesis of model peptides in relation with conformational studies[2,3]. The use of a soluble macromolecular protecting group i.e. polyoxyethylene $(CH_2CH_2O)_n$, allows the investigation of conformational properties of peptides in a great variety of organic solvents, including water. Thus, even peptide sequences which are insoluble using low molecular weight protecting groups become accessible for conformational studies in solution. For the direct investigation of conformational properties of peptides bound to $(CH_2CH_2O)_n$ i) the $(CH_2CH_2O)_n$ group must allow the application of all currently used methods such as nmr, CD, IR or Raman spectroscopy and ii) the conformational behavior of the peptide must not be influenced by the C-terminal macromolecular protecting group. In order to establish these two prerequisites for the general use of $(CH_2CH_2O)_n$-peptides in conformational studies, we synthesized several homooligopeptides with strong tendencies to adopt ordered structures and compared the conformation under various conditions with their low molecular-weight ester analogs.

Results and Discussion

Spectroscopic properties of $(CH_2CH_2O)_n$

Polyoxyethylene shows no Cotton effect and is optically transparent up to the far UV-region; consequently, the measurement of the CD in each step of the synthesis has become a routine method in the liquid-phase procedure for recording any conformational changes with growing peptide chain[2]. The solubilizing effect of $(CH_2CH_2O)_n$ upon the peptide chain in solvents suitable for CD-investigations such as F_3 EtOH, F_3 AcOH or H_2O is very strong. In the ^{1}H-nmr-spectra, $(CH_2CH_2O)_n$ shows only one singlet for the CH_2-group at δ = 3.66 ppm, thus not interfering with the signals of the peptide which are sensitive toward conformational changes. Fig. 1 shows the ^{1}H-nmr spectra of Boc(Ala)$_8$-O$(CH_2CH_2O)_n$ (M_r 6,000) in DMSO with the characteristic double peak[4,5] for the α-C-H-protons at 4.26 ppm and 4.58 ppm. Also, IR- and Raman methods were applied directly to the $(CH_2CH_2O)_n$-peptide in the solid state as well as in solution. The characteristic IR-absorption bands of $(CH_2CH_2O)_n$ do not interfere with the regions of interest for conformational studies of peptides; moreover,

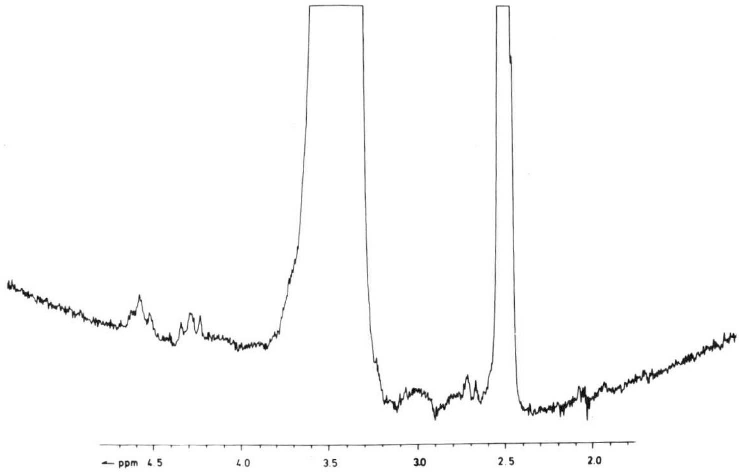

Fig. 1. ^{1}H-nmr-spectra of Boc(L-Ala)$_8$-O(CH$_2$CH$_2$O)$_n$ (M$_r$ 6 × 10^3) in DMSO at 25°C (conc. 6 mg peptide/ml; recorded by a WH 90, Fa. Bruker, 163 scans).

$(CH_2CH_2O)_n$ is Raman-inactive. Thus, all currently used investigation methods can be applied directly to $(CH_2CH_2O)_n$ bound peptides.

The influence of the $(CH_2CH_2O)_n$-group.

First, oligomers of (L-Glu)$_n$ were synthesized by the stepwise procedure of the liquid-phase-method using $(CH_2CH_2O)_n$, M$_r$ 20,000. For the detection of any influence of the $(CH_2CH_2O)_n$-chain on the conformation of the peptide, the CD was measured after each step under various conditions and compared with the data obtained for the free oligomers. The CD spectra of the $(CH_2CH_2O)_n$-peptides in H_2O, pH 3.9 are shown in Fig. 2. The formation of an α-helical structure starts at n=7 and amounts to about 60% for n=20. Identical CD-spectra were obtained when the $(CH_2CH_2O)_n$-group was cleaved from a sample and measured under the same conditions. The addition of $(CH_2CH_2O)_n$ to the free peptide also had no influence upon the CD-spectra. The helix-coil transition induced by the continuous neutralization of the COOH-side chains did not display any detectable difference in the conformational behavior between the free and the $(CH_2CH_2O)_n$-bound peptide. The fully neutralized oligomer with n=20 shows the expected random-coil conformation (curve 20*). Also, the solvent-dependent properties were not changed by the $(CH_2CH_2O)_n$-group. Both series showed higher ellipticities in the helix-promoting solvent F_3EtOH compared to water; the addition of increasing amounts of F_3 AcOH resulted in the disruption of the helical structure and displayed no difference in the stability of the secondary structure between $(CH_2CH_2O)_n$-bound and free oligomers. Whereas the $(CH_2CH_2O)_n$-group did not interfere with the formation of an α-helical

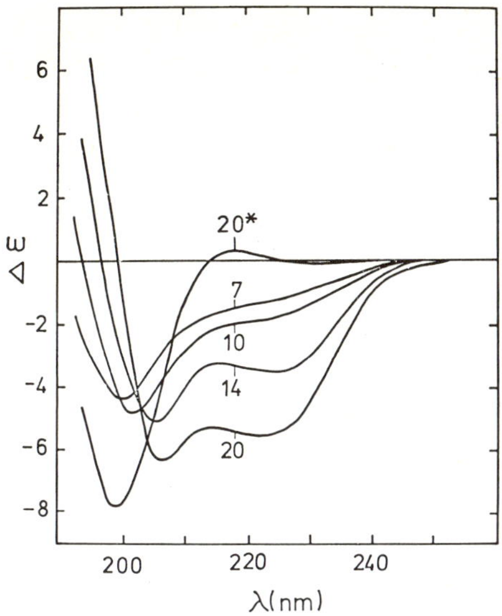

Fig. 2. CD-spectra of Boc(L-Glu)$_n$-$(CH_2CH_2O)_n$ (M_r 2×10^4) in H_2O, pH 3.9 at 25°C (conc. 0.2 mg peptide/ml); recorded by a dicrograph JASCO J-20.

structure, the influence upon the intermolecular aggregation of peptide chains, as realized in a β-conformation, seemed to be more critical. To delineate this effect, we synthesized several hydrophobic oligomers such as (Ile)$_n$, (Val)$_n$ and (Ala)$_n$ using $(CH_2CH_2O)_n$ of various molecular weights and studied the CD in F_3 EtOH and in H_2O. In an earlier study, we found a pure β-conformation for (L-Ala)$_{10}$-$O(CH_2CH_2O)_n$ in H_2O[2]. Figure 3 shows the CD spectra of Boc(L-Ile)$_n$-$O(CH_2CH_2O)_n$, Boc(L-Val)$_8$-$O(CH_2CH_2O)_n$ and Boc(L-Ala)$_8$$O(CH_2CH_2O)_n$ in F_3 EtOH.

Whereas the lower oligomers of (Ile)$_n$ have more or less random conformations, the higher oligomers, starting at a chain length of n=7 adopt β-conformation. Even concentrations as low as 0.02 mg peptide/ml did not result in a significant disruption of the β-sheets for the octapeptide of Ile. The

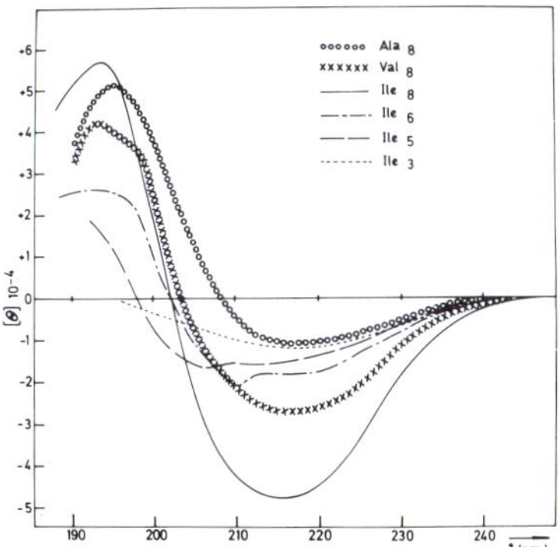

Fig. 3. CD-spectra of Boc(L-Ile)$_n$-$O(CH_2CH_2O)_n$ (M_r 10^4), Boc(L-Val)$_8$-$O(CH_2CH_2O)_n$ (M_r 10^4) and Boc(L-Ala)$_8$-$O(CH_2CH_2O)_n$ (M_r 6×10^3) in F_3 EtOH (conc. $5 \times 10^{-4}M$).

characteristic CD-data for the different oligomer-series are in agreement with those of the low molecular weight analogs[6], indicating that the $(CH_2CH_2O)_n$-group also has no influence upon the formation of β-conformations. We attribute this finding to the low density and the high flexibility of the $(CH_2CH_2O)_n$-coil under the experimental conditions. We conclude that conformational studies on $(CH_2CH_2O)_n$-bound peptides are also relevant for the corresponding free peptides. The present investigations confirm the results of conformational studies on sequential $(CH_2CH_2O)_n$-peptides, where no influence of the $(CH_2CH_2O)_n$-group could be detected[2,3]. Thus, the liquid-phase-method for peptide synthesis offers a new tool for conformational studies of peptides.

We are grateful to Miss J. Brun for very helpful technical assistance in the CD-investigations.

References

1. Mutter, M. & Bayer, E. (1974) *Angew. Chem.* 88, 101–102.
2. Mutter, M. & Bayer, E. (1974) *Angew. Chem., Int. Ed. Engl.* **13**, 88–89.
3. Mutter, M., Uhmann, R. & Bayer, E. (1975) *Justus Liebig's Ann. Chem.*, 901–915.
4. Mutter, M., Mutter, H., Uhmann, R. & Bayer, E. (1975) *Biopolymers* **15**, 917–927.
5. Mayr, W., Oekonomopulos, R. & Jung, G., *Biopolymers*, submitted.
6. Goodman, M., Toniolo, C. & Naider, F. (1974) in *Peptides, Polypeptides and Proteins*, Blout, E. R., Bovey, F. A., Goodman, M. & Lotan, N., Eds., Wiley Interscience, New York, pp. 308–319.
7. Paolillo, L., Temussi, P. A., Bradbury, E. M., Cary, P. D., Crane-Robinson, C. & Hartman, P. G. (1974) in *Peptides, Polypeptides and Proteins,* Blout, E. R., Bovey, F. A., Goodman, M. & Lotan, N., Eds., Wiley Interscience, New York, pp. 177–189.
8. Albert, K., Bayer, E., Mutter, H. and Mutter, M., in preparation.
9. Toniolo, C., Bonora, G. M. and Fontana, A. (1974), *Int. J. Pept Protein Res.*, 6,371–380.

SYNTHESES AND PROPERTIES OF POLYPEPTIDES AND CROSS-LINKED, INSOLUBLE, PEPTIDE MATRICES OF ELASTIN.

R. S. RAPAKA, KOUJI OKAMOTO, M. M. LONG and D. W. URRY,
Laboratory of Molecular Biophysics, Department of Biochemistry and the Cardiovascular Research and Training Center, University of Alabama Medical Center, Birmingham, Alabama 35294.

The molecular structure of elastin has been investigated by several laboratories. The models suggested are the corpuscular model[1], the modified Partridge model[2], the "Oiled-Coiled" model[3], the random net-work[4] and the fibrillar model[5]. The fibrillar model was based on data obtained from the electron microscopy[6,7] and solution studies on the repeating peptide sequences of tropoelastin[8], -Val^1-Pro^2-Gly^3-Gly^4-, -Val^1-Pro^2-Gly^3-Val^4-Gly^5- and -Ala^1-Pro^2-Gly^3-Val^4-Gly^5-Val^6-. The fibrillar model consists of a sequential arrangement of β-spiral segments followed by α-helical segments. It has been further proposed that the β-spiral portion consists of an elastomeric polypentapeptide (PPP) region followed by the less mobile polyhexapeptide (PHP) segment which aligns and interlocks the chains to facilitate cross-linking of the lysyl residues of the α-helical segments[9]. Conformational studies show that the β-spirals of the PHP are less dynamic compared to the PPP due to an H-bonded ring between Val^6 N-H of unit i and Val^6 C-O of unit i+1 and also due to intermolecular interlocking of hydrophobic ridges.

Recently the cross-linked PPP was synthesized[10]. It confirmed the functional role of the pentapeptide repeats because it exhibited strikingly similar elastomeric properties of aortic elastin and calcifying properties of α-elastin coacervates. It was also shown by electron microscopy to have a filamentous structure[9,11,12]. To elucidate the functional role of the hexapeptide, the synthesis of the hybrid cross-linked peptide, cross-linking between the PPP and the PHP,[13] and of the cross-linked PHP were undertaken. The elastomeric, coacervating and calcifying properties were then investigated.

The synthesis of Boc-Val-Ala-Pro-Gly-OH[13],F_3AcOH·H-Val-Ala-Pro-Gly-Val-Gly-ONp[13] and H-(Val-Pro-Gly-ϕ-Gly)$_n$-Val-OMe were reported earlier[10]. The synthetic route for these peptides (Schemes 1 and 2) is similar to that of the cross-linked PPP[10]. Cross-linking was achieved by primary amide bond formation between the ϵ-amino group of the lysyl residue (ϕ^4) of one copolymer [either H-(Val^1-Pro^2-Gly^3-ϕ^4-Gly^5)$_n$-Val-OMe or H-(Val^6-Ala^1-Pro^2-Gly^3-ϕ^4-Gly^5)$_n$-Val-OMe] and the γ-carboxyl of the glutamyl (ϕ'^4) of another copolymer, during a temperature elicited phase separation with flow orientation. The synthesis of the cross-linked polymers proceeded smoothly with good yields. Initial observations show that the hybrid cross-linked polymer is a poor elastomer and the cross-linked PHP is non-elastomeric.

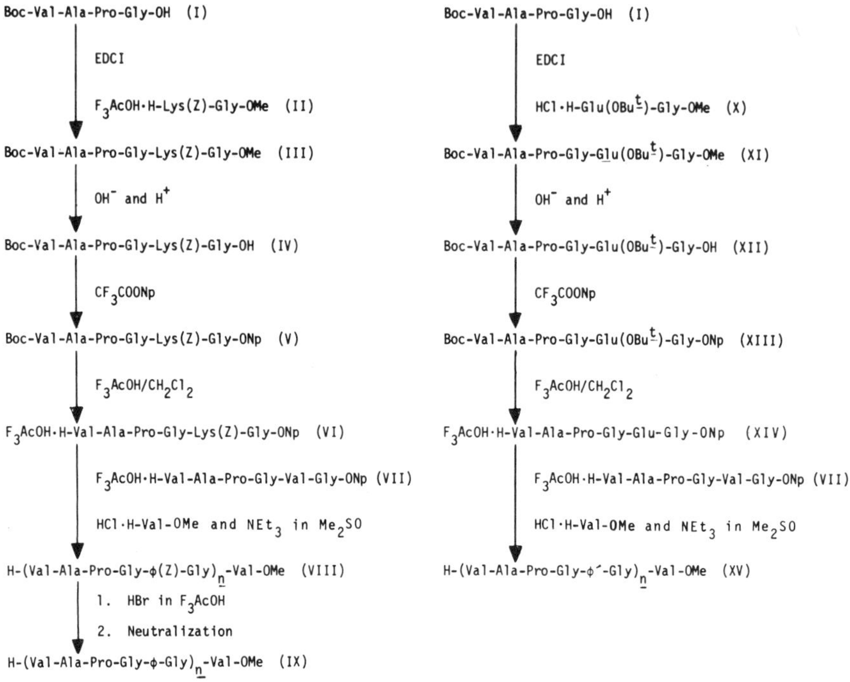

Scheme 1. Syntheses of H-(Val-Ala-Pro-Gly-ϕ-Gly)$_n$-Val-OMe and H-(Val-Ala-Pro-Gly-ϕ'-Gly)$_n$-Val-OMe (where ϕ is Val or Lys and where ϕ' is Val or Glu).

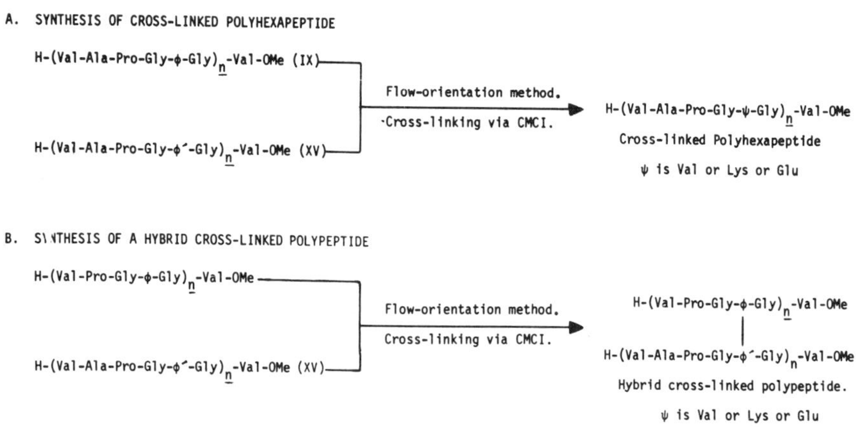

Scheme 2. Synthesis of Cross-Linked Peptides.

Properties of the Peptides

Figure 1A is a scanning electron micrograph of the high polymer of Val^1-Pro^2-Gly^3-ϕ^4-Gly^5 taken with a JEOL JSM-U3 instrument at 25 kV accelerating voltage. The fibers were drawn from a water saturated mass of non-crosslinked high polymer, dried and coated with aluminum. During formation the fibers were cohesive, weakly retractive and stable in air. In contrast, the high polymer of the hexamer -Val^6-Ala^1-Pro^2-Gly^3-ϕ^4-Gly^5- (Figure 1B), formed sheets which tore when pulled. The pentamer and hexamer differ only by one amino-acid residue in primary structure yet these micrographs show the microscopic forms of PPP and PHP to be dramatically dissimilar.

The cross-linked PHP (X-PHP) does not initiate calcification as readily as the cross-linked PPP (X-PPP). Both the X-PPP and X-PHP were incubated with a sterile calcifying medium of 1 ml of bovine serum augmented with 3.0 *mM* exogenous $CaCl_2$ and KH_2PO_4 for 72 hours at 37°C. The results of this calcification experiment are shown in Figure 2. Figures 2A, 2B and 2C are the SEM micrograph, Ca X-ray map, and X-ray spectrum of the X-PPP, respectively. The spectrum (Fig. 2C) shows the presence of calcium at 3.7 keV, phosphorus at 2.0 keV and the coating aluminum at 1.4 keV. As previously reported[12] calcification occurs throughout the bulk of the material. The X-PHP initiated calcification only in selected areas on the peptide matrix (Figs. 2D–2F). The spectrum (Fig. 2F) taken in spot mode, is of a calcium- and phosphorus-enriched area indicated by the arrow in Fig. 2E. The phosphorus map is spatially similar to the calcium map. These data demonstrate that under the conditions of this calcification experiment, the X-PHP is not as readily calcified as the X-PPP.

It is known that polypentapeptide coacervates[9]. Substituted polypentapeptides $(V^1P^2G^3\phi^4G^5)_n$ and $(VPG\phi'G)_n$ (where ϕ is Val or Lys and ϕ' is Val or Glu) behave similarly. Polyhexapeptide $(V^6A^1P^2G^3V^4G^5)_n$ does not coacervate but rather precipitates *irreversibly*. X^4-substituted polyhexapeptides undergo *reversible* precipitation. This is due to the disruption of the periodic hydrophobic ridges present in $(V^6A^1P^2G^3V^4G^5)_n$. In X-4 substituted analogs, where X is an

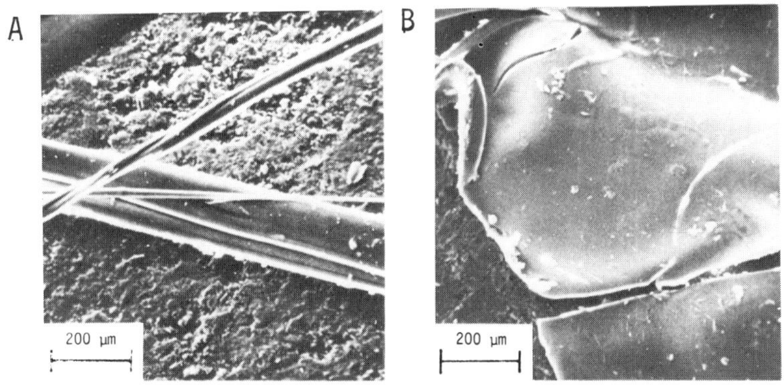

Fig. 1. SEM micrographs of (Val-Pro-Gly-ϕ-Gly)$_n$ (A), and (Val-Ala-Pro-Gly-ϕ-Gly)$_n$ (B) precursors of the synthetic cross-linked matrices.

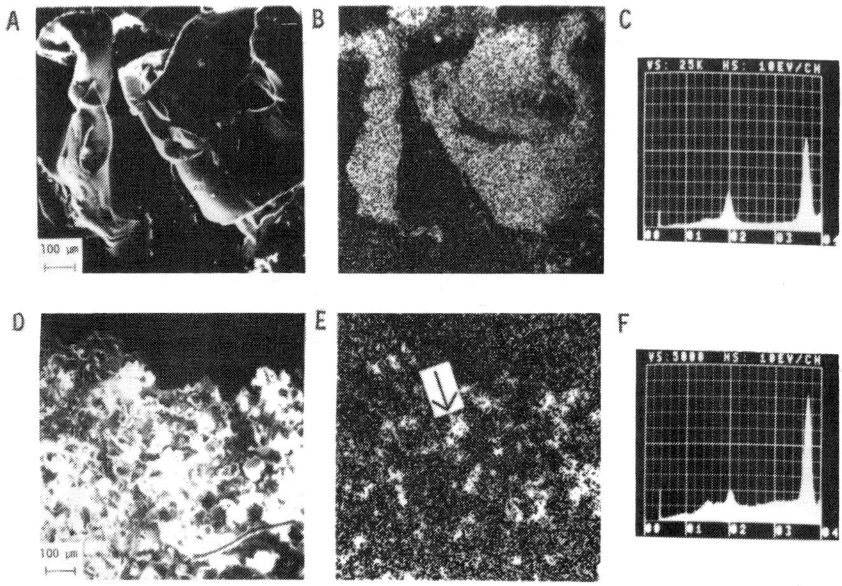

Fig. 2. SEM micrograph, Ca X-ray map, and X-ray spectrum (A, B, C) of the cross-linked pentamer high polymer after calcification. Parts D, E and F are the corresponding figures for the cross-linked hexamer high polymer after incubation with a calcifying medium. A, B, D, E have the same magnification.

ionic side chain, reversible precipitation is seen only at a pH where ionization is suppressed.

This work was supported in part by the National Institutes of Health, Grant No. HL-11310.

References

1. Partridge, S. M. (1966) *Fed. Proc., Fed. Amer. Soc. Exp. Biol.,* **25**, 1023–1029.
2. Weis-Fogh, T. & Anderson, S. O. (1970) *Nature,* **227**, 718–721.
3. Gray, W. R., Sandberg, L. B. & Foster, J. A. (1973) *Nature,* **246**, 461–466.
4. Hoeve, C. A. J. & Flory, P. J. (1974) *Biopolymers,* **13**, 677–686.
5. Urry, D. W. (1974) *Adv. Exp. Med. Biol.* **93**, 211–243.
6. Cox, B. A., Starcher, B. C. & Urry, D. W. (1974) *J. Biol. Chem.,* **249**, 997–998.
7. Gotte, L., Giro, M. G., Volpin, D. & Horne, R. W. (1974) *J. Ultrastruct. Res.* **46**, 23–33.
8. Urry, D. W. & Long, M. M. (1976) *CRC Crit. Rev. Biochem.* **4**, 1–45.
9. Urry, D. W. (1977) *Perspect. Biol. Med.* (in press).
10. Okamoto, K. & Urry, D. W. (1976) *Biopolymers,* **15**, 2337–2351.
11. Urry, D. W., Okamoto, K., Harris, R. D., Hendrix, C. F. & Long, M. M. (1976) *Biochemistry,* **15**, 4083–4089.
12. Urry, D. W., Long, M. M., Hendrix, C. F. & Okamoto, K. (1976) *Biochemistry* **15**, 4089–4094.
13. Okamoto, K., Rapaka, R. S. & Urry, D. W. (1977) *Biopolymers* (in press).

SYNTHETIC SEQUENTIAL POLYHEPTAPEPTIDES AS MODELS FOR THE TWO-STRANDED α-HELICAL COILED-COIL STRUCTURE OF TROPOMYOSIN

R. S. HODGES, S. A. ST-PIERRE, and F. -S. TJOENG,
Department of Biochemistry and Medical Research Council Group on Protein Structure and Function University of Alberta, Edmonton, Alberta, Canada

Tropomyosin contains two α-helical chains of 33,000 daltons which are in register, parallel and arranged in a coiled-coil structure. Tropomyosin is known to be intimately involved in the calcium regulated system of muscle contraction and relaxation. Thus the amino-acid sequence of tropomyosin contains the sequences specific for; the coiled-coil structure, bringing the chains into register, the binding sites for troponin and actin, and the possible conformational change during the contraction process.[1] To investigate these structure-function relationships our approach was to synthesize model polyheptapeptides of known molecular weight, and test for the formation and stability of the two-stranded coiled-coil structure.

The chemical synthesis of the polyheptapeptides was achieved by solution synthesis of the protected heptapeptides followed by polymerization to a sufficiently high molecular weight to form the coiled-coil.[2] The chromatographic profile of the deprotected polypeptide on an 8 *M* urea column showed one broad peak due to the polydispersity of polypeptide obtained after the active ester polymerization.[3] This presentation deals with conformational studies on five synthetic polyheptapeptides using circular dichroism and molecular-weight analysis by column chromatography.

Results and Discussion

The formation of the two-stranded coiled-coil in the synthetic polymers was indicated by a high helical content in benign media and the doubling of the molecular weight in benign media relative to denaturing media. Unlike tropomyosin, our synthetic models do not have a sequence specificity to maintain the two chains in register, as a result, polypeptide chains of equal size can stagger giving a large distribution of molecular weights for the associated chains.[3] This phenomenon became apparent on comparing the narrow molecular weight distribution in denaturant to the broad distribution obtained in benign medium. Additional evidence for this staggering phenomenon came from comparative measurements of CM-tropomyosin and the polyheptapeptides in a solution containing hexafluoroisopropanol. Since the tropomyosin chains are in register and assumed to be completely helical, we would not expect any significant increase in helicity upon the addition of a solvent that induces helicity in a single-chained, potentially helical polypeptide. No such increase is seen with CM-tropomyosin, as shown in Table I. In contrast, the polyheptapeptides [Leu-Glu-Ser-Leu-Glu-Ser-Lys]$_n$ and [Leu-Glu-Ala-Leu-Glu-Ala-Lys]$_n$ both of molecular weight 9,500 in denaturant showed increases in ellipticities of -6,650 to -17,000 and -15,700 to -21,900, respectively. This suggested the existence of permanent regions of two-stranded coiled-coil structure in the polyheptapeptides interrupted by non-helical, single-stranded regions of varying lengths. These regions that do not contain the coiled-coil structure would result in lower molar ellipticity values for the polyheptapeptides relative to tropomyosin (Table I). Fraction 5 of the peptide [Leu-Glu-Ser-Leu-Glu-Ser-Lys]$_n$ contained only 20% α-helix relative to tropomyosin in benign media, however fractionation of fraction 6 on BioGel P-200 in benign buffer gave material with helical content of 75% (Table I). Fractions 3,5, 9 and 13 of the polypeptide [Leu-Glu-Ser(Ac)-Leu-Glu-Ser(Ac)-Lys]$_n$ from the fractionation in 8 *M* urea showed that the α-helical content decreased as the chain length of the polypeptide decreased (Table I).

Table I. Molecular Parameters of Synthetic Polyheptapeptides Compared to CM-tropomyosin

Polypeptide	Fraction No.[a]	% TFE or HFIP[b] (v/v)	$[\theta]_{222}$ (deg cm dmole^{-1})	% Helix	Molecular Weight in Denaturant
[L-G-$\underset{Ac}{S}$-L-G-$\underset{Ac}{S}$-K]$_n$	3	0	-16,300	48	11,000
	5	0	-11,800	35	9,500
	9	0	-9,000	27	8,000
	13	0	-7,300	22	5,000
[L-G-S-L-G-S-K]$_n$	5	0	-6,650	20	9,500
		80	-17,000	50	
	6[d]	0	-25,500	75	9,000
[L-G-A-L-G-A-K]$_n$	1	0	-15,700	46	9,500
		80	-21,900	65	
CM-tropomyosin		0	-34,000	100	33,000
		90	-30,900	91	
[A-G-S-A-G-S-K]$_n$	4	0	-2,000	6	9,500
[A-G-S-L-G-S-K]$_n$	4	0	-2,400	7	9,500
[L-G-S-L-G-S-K]$_n$	5	0	-6,650	20	9,500
[A-G-S-A-G-S-K]$_n$	4	80	-9,000	27	9,500
[A-G-S-L-G-S-K]$_n$	4	80	-12,000	35	9,500
[L-G-S-L-G-S-K]$_n$	5	80	-17,000	50	9,500

[a] The deprotected polypeptides were fractionated on a 2.6cm x 100cm Sephadex G-100 column equilibrated with 8M urea, 0.2M KCl, 0.05M Tris/HCl buffer at pH 6.0.

[b] Hexafluoroisopropanol, HFIP; trifluoroethanol, TFE.

[c] A 1.1M KCl/0.05M phosphate buffer at pH 7.0 was used for all benign measurements.

[d] Fraction 6 of this polypeptide was further fractionated on a 1.6 x 100cm BioGel P-200 column equilibrated with benign buffer.

Initially we wanted to test the hypothesis of Hodges et al.[4] that the hydrophobes in positions 2 and 5 of the sequence of the polyheptapeptide, $[X\text{-}N\text{-}X\text{-}X\text{-}N\text{-}X\text{-}X]_n$, where N is a non-polar residue, were responsible for the formation and stabilization of the coiled-coil structure. Therefore, in order to maintain a sequence representative of tropomyosin in our heptapeptide, with hydrophobic residues only at positions 2 and 5, the hydrophilic residue serine was placed in positions 4 and 7 although serine is not considered a good helix-forming residue. It was postulated that the leucine residues at positions 2 and 5 would provide strong hydrophobic interactions to form and stabilize the coiled-coil and overcome the destabilizing effect of the serine residues.[3] As expected, a considerable increase in ellipticity was observed when comparing the acetylated serine polymer to the deacetylated polymer; a further increase occurred when the serine residues were replaced with alanine which is considered a good helix-forming residue (Table I).

The CD spectra of three polyheptapeptides of the same molecular weight differing only in the hydrophobic residues at position 2 and 5 are shown in Fig. 1. In benign buffer, with or without trifluoroethanol, comparison of the three peptides showed that the ellipticities increase as the size of the hydrophobic side chain increased (Fig. 1 and Table I). In the presence of the helix-inducing solvent this increase in α-helix with the change in hydrophobic residues is more pronounced. However, we do not know at this time whether the increase in α-helix

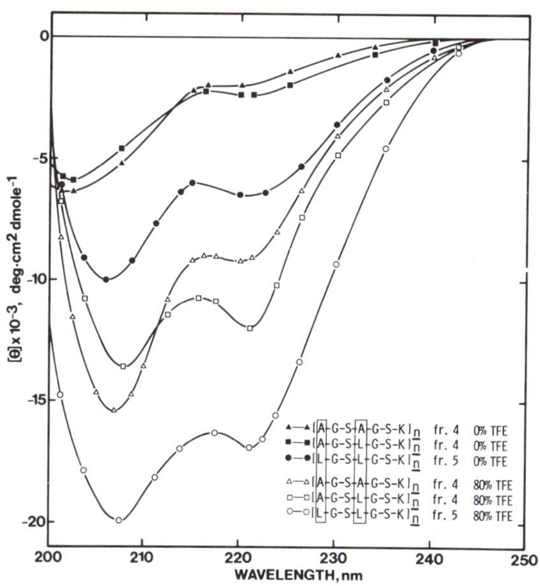

Fig. 1. Circular dichroism spectra of three synthetic polyheptapeptides in benign buffer.

with the helix inducing solvents is an increase only in single chain helix or whether we may be forming the two-stranded, coiled-coil structure.

The formation of very stable coiled-coil regions in our sequential polypeptides was confirmed by comparison with CM-tropomyosin on temperature denaturation. It was seen that 75% of the original helicity of our peptide was preserved even at 70°C, at which temperature CM-tropomyosin was completely denatured.[3] Similarly, denaturation studies show that in 4 *M* urea 95% of the original helicity remains in fraction 7 of the polyheptapeptide $[\text{Leu-Glu-Ser-Leu-Glu-Ser-Lys}]_n$ whereas 50% remains in CM-tropomyosin. This model polypeptide has both increased thermal and denaturant stability relative to tropomyosin, which can be due only to the presence of leucine residues in the hydrophobic positions. In tropomyosin the hydrophobic positions are occupied by a variety of non-polar residues, many of which are alanine, and this may decrease the stability of tropomyosin.

Work is continuing in the synthesis of model polypeptides with the chains in register so that a more detailed study can be carried out on the properties of the coiled-coil and that an attempt be made to understand the possible conformational changes that occur in tropomyosin during the contraction process.

References

1. Smillie, L. B. (1976) *PAABS Revista* **5**, 183–263.
2. St.-Pierre, S. A. & Hodges, R. S. (1977) *Can. J. Biochem.* (in press).
3. St.-Pierre, S. A. & Hodges, R. S. (1976) *Biochem. Biophys. Res. Commun.* **72**, 581–587.
4. Hodges, R. S., Sodek, J., Smillie, L. B. & Jurasek, L. (1972) *Cold Spring Harbor Symp. Quart. Biol.* **37**, 299–310.

NMR AND CIRCULAR DICHROISM STUDIES OF A REPEATING PEPTIDE OF A PROTEIN IN HUMAN PAROTID SALIVA H-Gly-Pro_4-OH

MARIE-H. LOUCHEUX-LEFEBVRE, *Institut de Recherches sur le Cancer de Lille, B.P. 3567, 59020 Lille Cédex* and
NICOLE HELBECQUE, *Laboratoire Associé n° 04-0268 du CNRS, Place de Verdun, 59045 Lille Cédex, France*

Previous studies have demonstrated the presence of proline-rich proteins and glycoproteins in human parotid saliva. (See the first seven references listed in Ref. 1.) The interest of these components is related to their high affinity to hydroxyapatite minerals. The low molecular weight unglycosylated proline-rich proteins are metabolic precursors during the biosynthesis of the glycoprotein. Therefore they would represent the polypeptide moiety of the proline-rich glycoprotein.

Several investigations dealing with the determination of the primary structure have shown the existence of the Gly-Pro_4 repeating sequence (P. Degand, personal communication). Therefore the synthesis of the pentapeptide H-Gly-Pro_4-OH and its polycondensation were carried out. In this work ^{1}H and ^{13}C nmr and CD studies of the pentapeptide are presented. The corresponding polymer was also studied by the CD technique and the results obtained were compared with those previously observed for proline-rich proteins.[1]

Nmr Studies

^{13}C nmr spectra were obtained using a Perkin-Elmer R 32 pulse Fourier transform spectrometer at 22.63 MHz. The proton-decoupled ^{13}C spectrum of H-Gly-Pro_4-OH in D_2O (zwitterionic form) was recorded. The assignments of different lines were obtained from a series of off-resonance spectra and are given in Table I. For comparison, the chemical shifts previously observed for the residue Pro in a *trans* conformation about either the Gly-Pro bond in a Gly-Pro dipeptide[2,3] or the Pro-Pro bond in polyproline II, $(Pro)_n$II,[3] are indicated. The conclusions of this study are: i) only the *trans* conformation is observed; ii) the carbon resonances of the C-terminal proline are fairly separated from those of other proline residues; iii) it is impossible to assign the carbon resonances due to each inside proline residue.

Pmr spectra were obtained on a Varian 300 MHz Spectrometer. The continuous wave ^{1}H spectra of H-Gly-Pro_4-OH was recorded at different pDs. As expected, they exhibit a very high complexity and only the resonances due to the α protons can be used to obtain new information (Fig. 1). For the zwitterionic form one observes, from up- to-downfield the α protons of Gly (residue 1), C-terminal proline (residue 5) and the three inside prolines (residues 2, 3 and 4) overlapping with the HOD peak. When the pD decreases, the peak of residue 5 is

Table I. Assignment of ^{13}C Peaks. Solvent: D_2O

Resonance Assignment[a]	H-Gly-L-Pro_4-OH[b]	H-Gly-L-Pro-OH (Ref. 2) trans	H-Gly-L-Pro-OH (Ref. 3) trans	$(Pro)_n$II (Ref. 3)
C^{α} 1	41.1			
C^{α} 5	62.5	62.8	62.9	
C^{α} 2 3 4	59.5			59.4
C^{β} 5	29.9	30.3	30.5	
C^{β} 2 3 4	28.8			28.8
C^{β} 2 3 4	28.6			
C^{γ} 5	24.9	25.0	25.2	
C^{γ} 2 3 4	25.2			25.5
C^{δ} 5	47.6	47.5	47.6	
C^{δ} 2 3 4	48.1			48.5
C^{δ} 2 3 4	48.3			

[a]in ppm relative to external TMS.

[b]H-Gly-L-Pro-L-Pro-L-Pro-L-Pro-OH

1 2 3 4 5

shifted downfield. Moreover, the peaks corresponding to residues 1 and 5 remain the same in shape (a singlet and a quartet, respectively) but those peaks corresponding to residues 2, 3 and 4 vary. Fig. 2 shows the extended spectra which correspond to them at pD 6.5 and 3.6. The following conclusions may be drawn from these results: in the zwitterionic form, all the Pro residues of the pentapeptide adopt a *trans* conformation. When the pD decreases, the Pro residues 2 and 5 remain in a *trans* conformation but the peak of Pro residues 3 and 4 reveals the presence of *cis* and *trans* isomers.

Circular Dichroism Study

The CD spectra were recorded with a Jobin-Yvon R. J. Mark III dichrograph under a stream of high purety dry N_2, with a cell thickness of 0.01 cm, in 0.01 *M* phosphate buffer solutions at pD 6.45.

H-Gly-Pro_4-OH and its polymer exhibit similar CD spectra, characterized by a strong negative band at 202 nm and reminiscent of a $(Pro)_n$II helical structure spectrum (Fig. 3). As a comparison the CD spectrum of Proline-rich proteins[1] was also drawn in Fig. 3. This latter spectrum was fitted with a calculated spectrum obtained by the linear combination of $(Pro)_n$II,[4] type I class B β-turn,[5] and unordered[6] spectra. It corresponds to 64% of $(Pro)_n$II, 29% of β-turn and 7% of unordered structures.[1]

The variation of the ellipticity at 202 nm, $[\theta]_{202}$, when $CaCl_2$ is added is shown in Fig. 4. As in the case of proline-rich proteins[1] the results are in good agreement with the decline of an ordered structure when $CaCl_2$ is added.

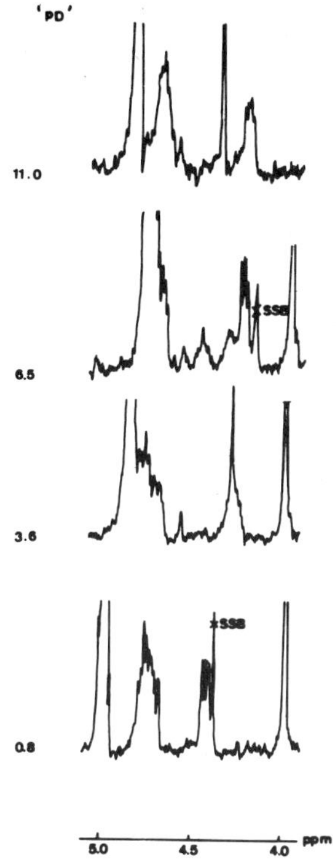

Fig. 1. Pmr spectra of H-Gly-L-Pro_4-OH in D_2O at different pD. $C^{\alpha}H$ resonances.

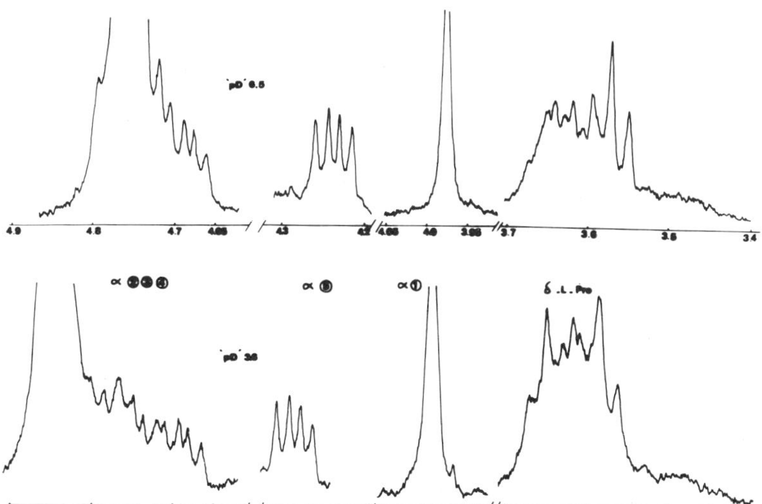

Fig. 2. Extended Pmr spectra at pD 6.5 (zwitterionic form) and pD 3.6, in D_2O.

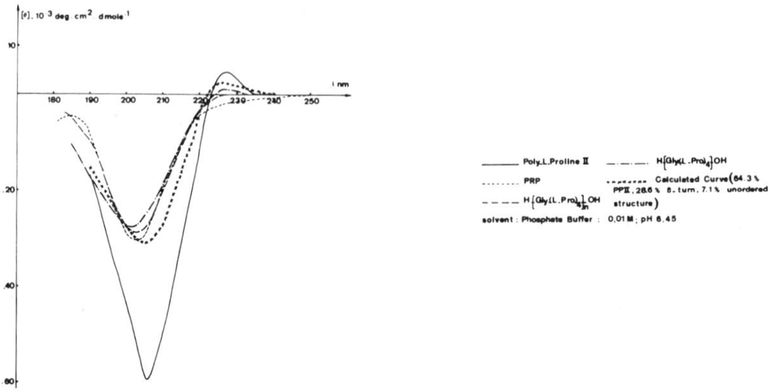

Fig. 3. Circular dichroism spectra. PRP = proline-rich proteins; PP II = polyproline II.

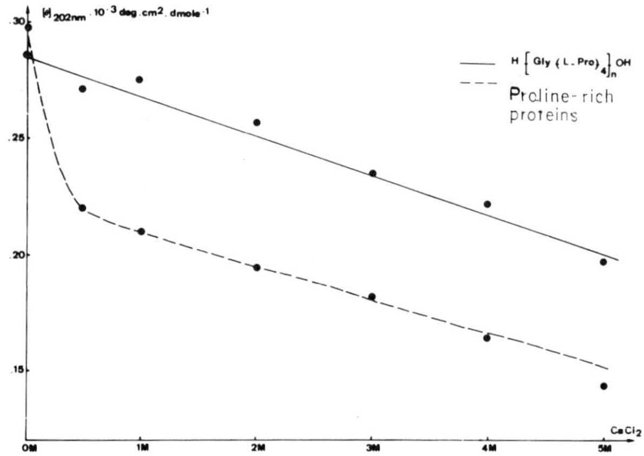

Fig. 4. Variation of $[\theta]_{202}$ versus the concentration of $CaCl_2$. In 0.01 M phosphate buffer solution, at pD 6.45.

We are indebted to Pr M. Anteunis, University of Gent (Belgium) where the ^1H nmr studies were performed.

References

1. Aubert, J. P., Boersma, A., Loucheux-Lefebvre, M. H., Degand, P. & Biserte, G. (1975) *FEBS Lett.* **56**, 263–267.

2. Evans, C. A. & Rabenstein, D. L. (1974) *J. Amer. Chem. Soc.* **96**, 7312–7317.

3. Dorman, D. E. & Bovey, F. A. (1973) *J. Org. Chem.* **38**, 2379–2383.

4. Ronish, E. W. & Krimm, S. (1974) *Biopolymers* **13**, 1635–1651.

5. Woody, R. W. (1974) in *Peptides, Polypeptides and Proteins*, Blout, E. R., Bovey, F. A., Goodman, M. & Lotan, N., Eds., Wiley Interscience, New York, 338–350.

6. Chen, J. H., Yang, J. T. & Martinez, H. M. (1972) *Biochemistry* **11**, 4120–4131.

BACTERIAL CELL WALL PEPTIDOGLYCAN: A SEMI-CRYSTALLINE MATRIX OR A RANDOM POLYMER? AN *IN VIVO* ^{15}N NMR STUDY

AVIVA LAPIDOT and CHARLES S. IRVING,
Department of Isotope Research, Weizmann Institute of Science, Rehovot, Israel

The peptidoglycan layer of the bacterial cell wall serves to maintain the shape of the cell and protects the fragile cytoplasmic membrane from high osmotic pressures associated with hypertonic environmental conditions. Very little is known about the physicochemical origin of the shape and mechanical properties of the peptidoglycan layer.[1] At present there is no indication of whether the conformation and mobility of the peptidoglycan subunits are determined by intramolecular, noncovalent bonding interactions (like polypeptides in solution), or by constraints on the peptidoglycan layer originating from the dimensions of the cell or forces acting on cell wall. Using nmr to measure the effect of various denaturing conditions on the motional freedom of peptidoglycan subunits, we determined what contributions hydrogen bonding, electrostatic interactions, and covalent cross-linking make to the motional properties of peptidoglycan.

Recent ^{15}N nmr studies of intact cells have shown that cell-wall resonances can be conveniently studied *in vivo* by ^{15}N nmr spectroscopy of highly ^{15}N-enriched gram-positive bacteria, such as *Bacillus licheniformis*.[2] Changes in the relative motional properties of nitrogen groups in a macromolecule can be conveniently detected in proton-decoupled ^{15}N nmr spectra by changes in the relative intensities of the resonance. To a large extent resonance intensities are determined by the ^{15}N-(^{1}H) nuclear Overhauser factor, which is a sensitive function of molecular correlation time.[2] An increase in the relative intensity of a resonance in the absence of a change in the chemical composition of the sample indicates either an increase in tumbling rate of the macromolecule, an increase in segmental motion or oscillations in a subunit, a decrease in cooperativity between subunit motions, or a narrowing of the distribution of correlation times among the subunits.[3]

The proton-decoupled ^{15}N nmr spectrum of peptidoglycan disaccharide oligopeptides obtained from the lysozyme digest of purified *B. licheniformis* ATCC 9945 bacterial cell walls freed of teichoic and teichuronic acids, displayed amide resonances at 245.5, 250.5, 251.3, 253.4, 256.0, 263.4, 267.5 and 268.8 ppm. Additional resonances at 252.6 and 254.0 ppm were observed in the proton-decoupled ^{15}N nmr spectrum of peptidoglycan oligopeptides and disaccharide oligopeptides obtained from the autolysates of autolyzed cell walls (Fig. 1A). The 253.4 ppm resonance has been assigned to the acetamide nitrogens of *N*-acetylmuramic acid and *N*-acetylglucosamine residues. From among this group

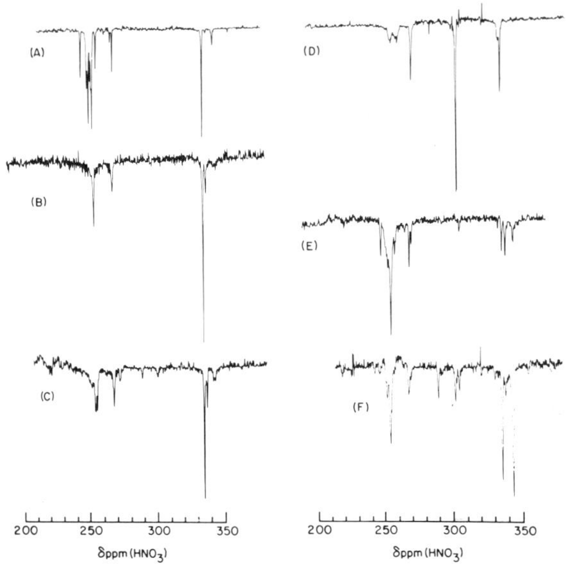

Fig. 1. The proton-decoupled ^{15}N nmr (9.12 Hz) spectra of A) normal cell walls of ^{15}N enriched *B. licheniformis* ATCC 9945 autolyzed for 1 hr, 37°C, pH 8.0, B) intact cells in H_2O, C) intact cells in H_2O at pH 2.0, D) peptidoglycan cell wall fraction in 6 *M* guanidine hydrochloride (27°C) prepared from cell walls whose teichoic and teichuronic acids were removed by trichloroacetic acid hydrolysis (37°C, 24 hours) and suspended in 6 *M* guanidine HCl, then boiled in 4% sodium dodecyl sulfate for 5 min and returned to 6 *M* guanidine hydrochloride, E) spheroplasts in 2 *M* sucrose prepared by treating intact cells with lysozyme (2 mg/25 ml, pH 8.0, 1 hr, 25°C), F) partially autolyzed intact cells. Spectral conditions were as in Ref. 2.

of peptidoglycan resonances, the proton-decoupled ^{15}N nmr spectrum of highly ^{15}N-enriched intact *B. licheniformis* cells at 27°C (Fig. 1B) display only the 253.4 and 267.5 ppm peptidoglycan resonances, in addition to a 335.6 ppm resonance originating from a peptidoglycan amino group, a 337.0 ppm resonance originating from the D-Alanine amino group of cell wall teichoic acid, and a 343.0 ppm resonance originating from protein lysine-N_ω nitrogens. Assignments were made on the basis of isotope labelling experiments, cell fraction experiments, and comparison of chemical shifts to those of standard compounds. Only two peptidoglycan resonances were observed in the native cell wall *in vivo* indicating (1) the polymer is not in a flexible extended conformation and (2) that segmental motions or oscillations must exist within the peptidoglycan subunits.

Lowering the environmental pH of the cells to pH 2.0 led to the appearance of the 250.5 and 252.6 ppm peptidoglycan resonances and caused the 253.4 ppm resonance to decrease in intensity and shift to 254.6 ppm (Fig. 1C). More dramatic changes in the spectrum would be expected for the acid denaturation of a charged, folded polypeptide in solution. The limited changes in the motion of peptidoglycan caused by alteration of the charge distribution in peptidoglycan is consistent with those expected for amphoteric polyelectrolyte gels.[4]

Measurements of the effects of hydrogen-bond-breaking reagents (such as urea or guanidine hydrochloride) on peptidoglycan particles freed of teichoic and teichuronic acids has been proposed as a method for determining the relative importance of hydrogen bonding in peptidoglycan.[5] The proton-decoupled ^{15}N nmr spectrum of *B. licheniformis* cell walls, which have been freed of teichoic and teichuronic acids, are similar to the whole cell and cell wall spectra—with the exception of the appearance of a new resonance at 248.8 ppm, the decrease in intensity of the 253.4 ppm resonance, and the disappearance of the teichoic acid D-alanine amino resonance at 337.0 ppm. The ^{15}N spectrum of *B. licheniformis* cell walls, free of teichoic and teichuronic acids, displayed in 6 *M* guanidine hydrochloride a number of unexpected resonances originating from denatured proteins (201.7, 204.1, His-$N_{\epsilon,\ell}$; backbone peptides 256.4; 262.2, 264.8, Gln-N_ϵ or Trp-N_ϵ; 291.2, Arg-N_ϵ; 304.4, Arg-$N_{\omega,\omega'}$; 341.9, Lys-N_ω). When these proteins were removed by boiling the cell wall in 4% sodium dodecyl sulfate for 5 minutes, the ^{15}N spectrum (Fig. 1D) of peptidoglycan particles in 6 *M* guanidine hydrochloride was identical to that of untreated particles in distilled water. The ^{15}N spectrum of peptidoglycan particles at 60°C in 6 *M* guanidine hydrochloride remained quantitatively unchanged. The resistance of cell wall peptidoglycan to thermal and guanidine hydrochloride denaturation is remarkable and demonstrates the small contribution that hydrogen bonding makes to the motional properties of peptidoglycan. In this respect, peptidoglycan differs markedly from a folded polypeptide in solution.

Limited lysozyme hydrolysis of peptidoglycan β-glycosidic bonds in intact, but osmotically protected *B. licheniformis* cells, resulted in a ^{15}N whole cell spectrum (Fig. 1E) that displayed relatively intense peptidoglycan resonances at 245.5, 250.5, 251.3, 256.0 and 268.8 ppm. Autolytic cleavage of the *N*-acetylmuramic acid-L-Ala peptide bond of peptidoglycan in the whole cells exposed to anaerobic shock led to the similar appearance of peptidoglycan amide resonances at 245.5, 253.5, 267.5 and 268.8 ppm in the whole cell ^{15}N nmr spectrum (Fig. 1F). The effect of breaking the peptidoglycan cross-links appeared reversible, since cross-linking of the free amino groups of partially hydrolyzed peptidoglycan of whole cells with glutaraldehyde yielded an ^{15}N spectrum similar to that of a normal, intact cell.

Conclusions

The motions available to the peptidoglycan subunits do not appear to depend on non-covalent bonding interactions. A possible explanation of our results is that the morphology, growth of cell walls, and/or the osmotic forces acting on the cell wall produce a mechanical strain within the peptidoglycan matrix that forces the subunits into either a fully extended, taut conformation or a compressed, folded conformation. Hydrolytic cleavage at a relatively small number of sites might be sufficient to release the strain on the cell wall and allow the subunits to assume more flexible conformations. These results point out some of the features of polypeptides in cell organelles in which the mechanical properties of the cell organelle, rather than intra-molecular non-covalent bonding interactions, determine its conformation and motional properties.

References

1. Ghuysen, J. M. & Shockman, G. D. (1973) in *Bacterial Membranes and Walls,* Leive, L., Ed., Marcel Dekker, Inc., New York, pp. 37–130.
2. Lapidot, A. & Irving, C. S. (1977) *Proc. Nat. Acad. Sci. USA* **74**, 1988–1992.
3. Schaefer, J. (1976) *Macromolecules* **6**, 882–888.
4. Muroga, Y., Noda, I. & Nagawawa, M. (1969) *J. Phys. Chem.* **73**, 667–672.
5. Kelemen, M. V. & Rogers, H. J. (1971) *Proc. Nat. Acad. Sci. USA* **68**, 992–996.

BETA-HELICAL CONFORMATIONS OF PEPTIDES AND PROTEINS

STEPHEN J. KENNEDY, HENRY R. BESCH, JR., AUGUST M. WATANABE, ROGER W. ROESKE, *Depts. of Biophysics, Pharmacology, Medicine and Biochemistry, Indiana University School of Medicine, Indianapolis, IN 46202*, and ALAN R. FREEMAN, *Dept. of Physiology, Temple University School of Medicine, Philadelphia, PA 19140*

The proposal of the α-helix and β-pleated sheet as allowed conformations of polypeptide chains[1] was based on the application of a set of rigid physicochemical restrictions which have become known as the Pauling-Corey postulates of protein structure. One of these, the assumption of equivalence, excludes from consideration all conformations in which individual amino-acid residues are not identically oriented. Recently, however, a new helical conformation unique to polypeptide chains containing alternating L- and D-amino acid residues has been proposed by Urry[2] and independently by Ramachandran and Chandrasekharan.[3] This structure, the $\beta^{6}_{3,3}$-helix, contains residues which are non-equivalent: two points are required to describe the conformation on a Ramachandran plot. The non-equivalence of residues in the β^{6}-helix suggests that other regular configurations might exist which heretofore have been excluded by the simplifying assumption of equivalence.

Theory

In order to generalize the structural theme of the β^{6}-helix, it is necessary to consider all possible residue conformations capable of participating in a parallel hydrogen bonding scheme. The four local conformations in which the carbonyl groups of adjacent peptide bonds are oriented either parallel or antiparallel to each other satisfy this requirement. These conformations are labeled β, β_D, δ_1, and δ_2 in Figure 1 and the central part of Figure 2. The conformations β and β_D correspond to parallel β-pleated sheets composed of L- or D-amino acids, respectively. The conformations δ_1 and δ_2 are roughly equivalent to the right- and left-handed α-helices. As shown in Figure 1, the conformations β and δ_1 are fully allowed for L-amino-acid residues. The conformations β_D and δ_2 are the mirror images of β and δ_1 and are allowed only for D-residues. These four residue conformations can be combined in a variety of ways to generate regular helical structures which satisfy all of the Pauling-Corey postulates except equivalence. Several of the many possible β-helices, including the previously described $\beta^{6}_{3,3}$- and $\beta^{6}_{2,4}$-helices[2,3] are shown in Figure 2. The $\beta^{13}_{4,9}$- and $\beta^{15}_{5,10}$-helices do not require the presence of D-amino acids and should be allowed conformations of proteins.

Because the β-helices are large in diameter, they can accommodate ions and water molecules in their interiors. The larger β-helices have amino-acid side chains located inside the helix, and these groups can supply ligands for ion coordination. In addition, the distinct interior and exterior regions of a given β-helix provide the potential for the generation of an amphipathic structure by amino-acid sequences with the appropriate pattern of polar and hydrophobic residues. Thus β-helices should be particularly applicable to the structure of membrane proteins. Because of the very direct relationship between amino-acid sequence and β-helical structure, it is to be expected that the three-dimensional structure of a β-helical membrane protein could be predicted from a knowledge of its amino-acid sequence. Conversely, for a given β-helix, amino-acid sequences can be chosen which should form the helix in a lipid bilayer environment.

A Synthetic β^{12}-helical Ion Channel

The peptides HCO-(Leu-Ser-Leu-Gly)$_6$-OH and H-(Leu-Ser-Leu-Gly)$_{12}$-OH were designed and synthesized in an attempt to produce a β^{12}-helical ion channel. The synthesis was accomplished by the solid-phase method utilizing repeated couplings of the tetrapeptide Boc-Leu-Ser(OBz)-Leu-Gly-OH with DCC and HOBt. Both of these peptides have been shown to produce channels in an artificial lipid bilayer system.[4] The unit conductance (Figure 3) and ion selectivity of the channels are consistent with the β^{12}-helical structure and ion coordination by serine hydroxyl groups inside the channel.

A β^{13}-helical Model of Glycophorin

Examination of the amino-acid sequences of membrane proteins known to span the bilayer could provide examples of β-helical structure. The only integral membrane protein which has been sequenced to date is glycophorin, the major glycoprotein of human erythrocyte membranes, and a β^{13}-helical model has been developed[5] for the central (membrane penetrating) segment of this protein. The arrangement of residues 55-107 of glycophorin into four turns of β^{13}-helix is shown in Figure 4. The β^{13}-helix places more hydrophobic residues in the membrane interior than the α-helix proposed[6] for residues 73-95, and simultaneously allows all thirteen charged residues within its sequence to be exposed to water, either at the ends or in the interior of the helix.

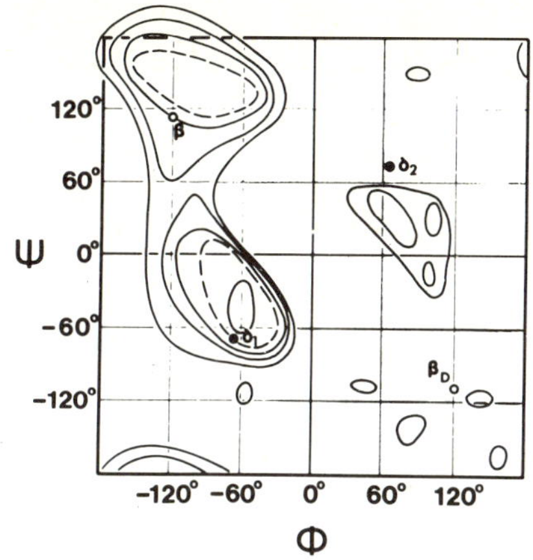

Fig. 1. Ramachandran plot of the four residue conformations allowed to participate in β-helices.

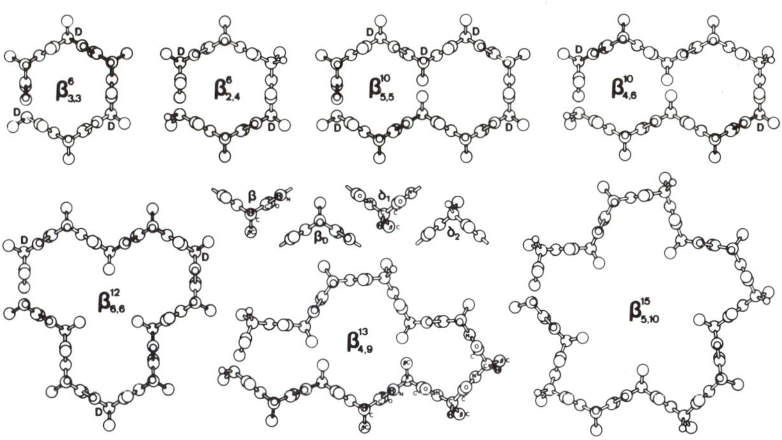

Figure 2. Single turns of some β-helices.

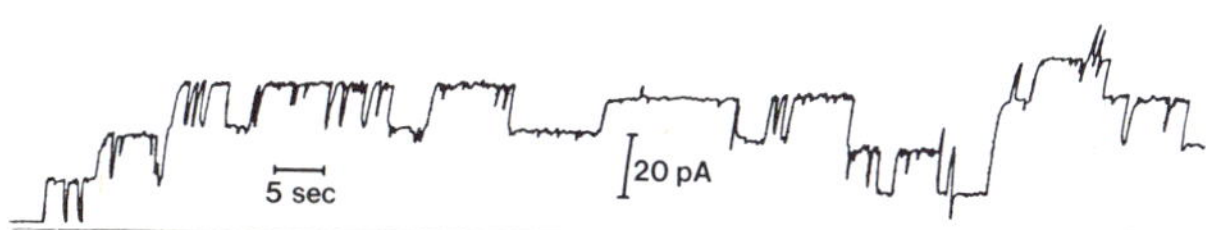

Fig. 3. Single channel current fluctuations in an artificial lipid bilayer membrane obtained in the presence of HCO-(Leu-Ser-Leu-Gly)$_6$-OH.

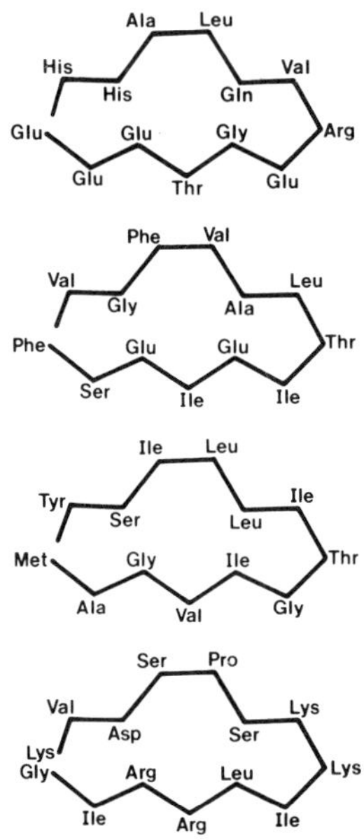

Fig. 4. A β^{13}-helical model of the central segment of glycophorin. The sequence begins with Glu 55 at upper left, and generates a helix with a hydrophobic exterior. Charged residues appear only at the ends or in the interior of the helix.

References

1. Pauling, L. & Corey, R. B. (1951) *Proc. Nat. Acad. Sci. USA* **37**, 235–240.

2. Urry, D. W. (1972) *Proc. Nat. Acad. Sci. USA* **69**, 1610–1614.

3. Ramachandran, G. N. & Chandrasekharan, R. (1972) *Indian J. Biochem. Biophys.* **9**, 1–11.

4. Kennedy, S. J., Roeske, R., Freeman, A. R., Watanabe, A. M. & Besch, H. R., Jr. (1977) *Science* **196**, 1341–1342.

5. Kennedy, S. J., Watanabe, A. M. & Besch, H. R., Jr. (1977) Submitted.

6. Segrest, J. P., Kahane, I., Jackson, R. L. & Marchesi, V. T. (1973) *Arch. Biochem. Biophys.* **155**, 167–183.

THE STUDY OF PROTEIN CONFORMATION USING CLEAVED PEPTIDES: APPLICATION TO HISTONES H3 AND H4.

C. CRANE-ROBINSON, H. HAYASHI, P. D. CARY, T. MOSS, E. M. BRADBURY, *Portsmouth Polytechnic, Portsmouth, U.K.*, L. BÖHM, *University of Cape Town, South Africa,* G. BRIAND, P. SAUTIERE, *Institute Cancer, Lille, France* D. KRIEGER, G. VIDALI, *Rockefeller University, New York,* P. N. LEWIS and J. TOM-KUN, *University of Toronto, Canada.*

The basic unit of chromatin, the nucleosome, consists of a 'core particle' of ~140 b.p. of DNA and 2 molecules each of histones H3, H4, H2A and H2B, linked by ~60 b.p. of DNA to the next nucleosome. The diffraction pattern of chromatin and many of the fragments of nuclease digestion can be obtained from a reconstituted complex of DNA and the histones H3 and H4 alone[1,2] but not from a DNA/H2A/H2B reconstitute. H3 and H4 are therefore the main organizers of DNA in the nucleosome. They both have highly asymmetric and highly conserved amino-acid sequences, the N-terminal 20-30 residues being very basic and low in apolar residues, while the remainder has a composition like that of globular proteins. The probable existence of two domains suggests a structural independence and indicates that cleaved peptides and complexes prepared therefrom might take up the native conformation.

H3 and H4 form a tetrameric complex in solution having a frictional ratio of 1.99 in 50 *mM* acetate/bisulphite, pH5 and 1.74 in 25 *mM* phosphate, pH7.[3] A high value of (f/f_0) indicates either an asymmetric or an expanded structure and the ionic strength dependence of (f/f_0) fits the behavior of an expanded polyelectrolyte. The nmr spectrum of $(H3/H4)_2$ shows several peak perturbations characteristic of globular proteins. It was concluded[3] that the complex has a globular core with free N-terminal 'tails'. To verify this model and to define the regions of both chains critical for complexation, we mixed several peptide combinations and used the perturbations in the aromatic region of the nmr spectrum as the criterion of complex formation. The peptides used in the study were; from H4, 38–102 (chymotryptic digestion) and 1–84 (cyanogen bromide treatment); from H3, 1–90 and 1–120 (cyanogen bromide treatment) and 42–135 (N-bromosuccinimide treatment).

Figure 1 shows the nmr spectra of 3 out of a total of 7 peptide combinations tested. The peaks at positions 1 and 2 are characteristic of the native complex (top spectrum) and it is seen that loss of 41 and 37 residues from the N-terminal of H3 and H4, respectively does not inhibit complexation (3rd spectrum). Loss of residues 91–135 from H3 prevents complexation (bottom spectrum) but loss

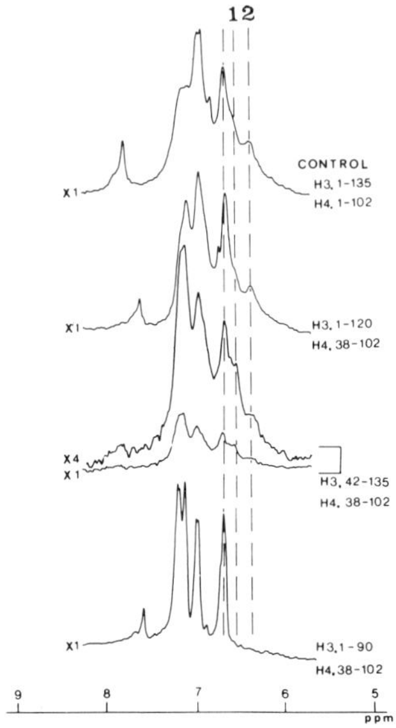

Fig. 1. Aromatic region nmr spectrum of peptide mixtures.

of residues 121-135 does not (2nd spectrum). Figure 2 summarizes the results, and it follows that it is the C- rather than the N-terminal portions of H3 and H4 that are important for complexation. Removal of 36% and 30% from the N-termini of H4 and H3 respectively does not inhibit complexation but removal of 18% and 33% from their C-termini does. Loss of 16 residues from the C-terminus of H3 does not, however, prevent complexation. The results are fully consistent with the $(H3/H4)_2$ complex having free N-terminal 'tails' and demonstrate that regions critical for complexation are: H3, (42-120) and H4, (38-102).

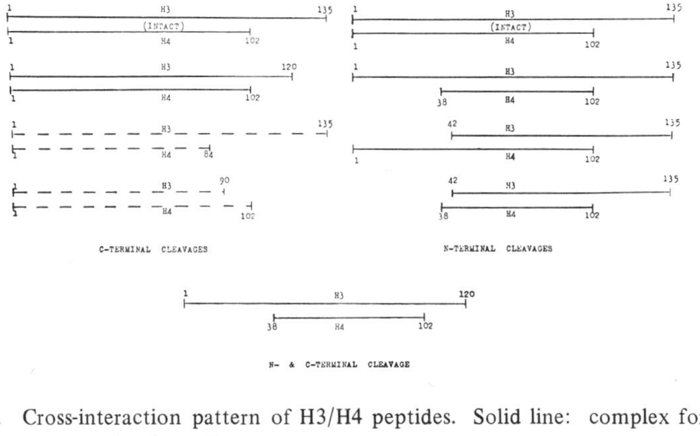

Fig. 2. Cross-interaction pattern of H3/H4 peptides. Solid line: complex formation, broken line: no complex formation.

Cleaved peptides have also been used to locate secondary structure in H4. On increase of ionic strength, H4 self-aggregates with the rapid formation of 25% α-helix and a slower formation of 30% β-structure.[4] In the multimeric state of H4 ~30 N-terminal residues remain mobile in solution.[5] The experimental approach was as follows: peptides were studied as random coils (pH3.5, no salt) and in 0.2 *M* NaCl, pH3.5 (conditions for a structured state). Nmr spectra were used to monitor self-aggregation and to define the part of the chain included in the aggregate. The IR amide I spectra (Figure 3) were used to detect β-structures using the intensity of the 1620 cm^{-1} band.[6] CD ellipticities at 222 nm (Figure 4)

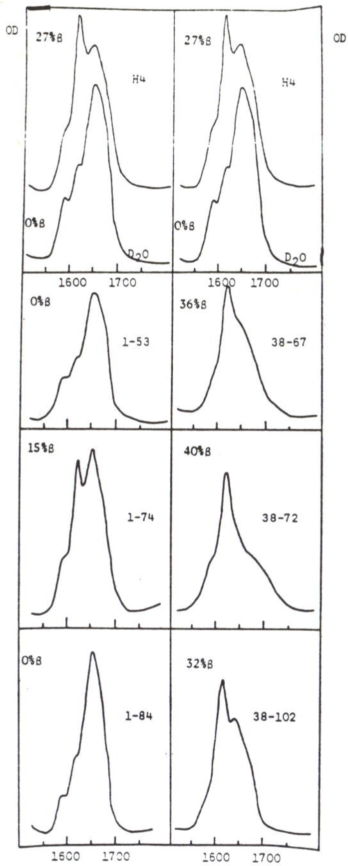

Fig. 3. Amide I bands of H4 and 6 peptides in 0.2 *M* NaCl/D_20, pD3.5 and H4 in D_20, pD3.5. Doublet centered at 1610 cm^{-1} due to arginyl side chains.

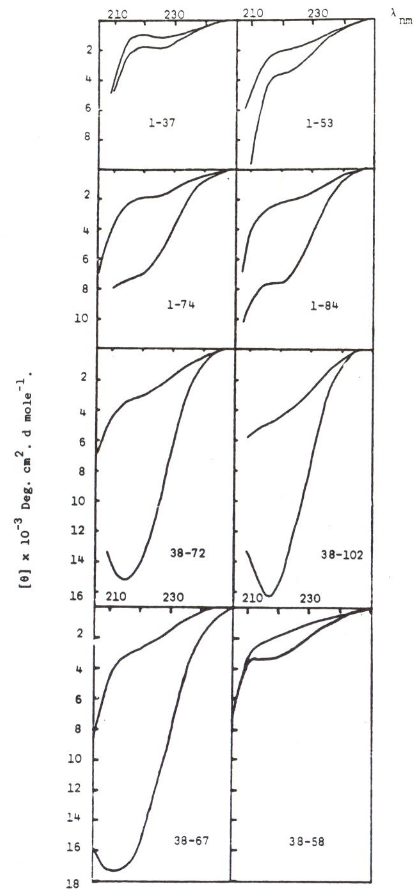

Fig. 4. CD spectra of 8 peptides from H4 in H_20 (upper spectra) and 0.2 *M* NaCl pH3.5.

were used to obtain the helicity (0% helix = -1000° and 100% helix = -30,000°) after first subtracting the IR-determined β-contribution (100% β = -3360°). The peptides were prepared as follows: 1-23, 25-67 and 69-102 by acetic acid hydrolysis and 38-58 and 38-67 by chymotryptic cleavage of 25-67. 1-37, 38-102 and 38-72 were products of chymotryptic cleavage, while 1-53 and 1-74 were products of *Staphylococcus aureus* V8 protease action (cleavage at Glu), both on intact H4. Nmr spectra of H4 (1-102) and peptides 1-74 and 38-102 all show dipolar broadening due to aggregation. In the aggregated state of intact H4, one of two histidines remains mobile while the remaining aromatics are not observable. The nmr spectra of 1-74 and 38-102 show that this is His 18 and not His 75. The conclusions to be drawn from the results in Table I are as follows: a) Residues

Table I. The Content of α-helix and β-structure Forming Amino-acid Residues in Histone H4 and its Different Fragments in 0.2 *M* NaCl.

Peptide	*Number of Residues α helix in 0.2 M NaCl*	*Number of Residues β structure in 0.2 M NaCl*	*Aggregation?*
Intact H4	25	27	Yes
1-23	0	0	No
1-37	0	0	No
1-53	<5	0	No
1-74	13	11	Yes
1-84	18	0	Yes
25-67	12	9	Yes
38-58	<3	0	No
38-67	14	11	Yes
38-72	14	14	Yes
38-102	27	21	Yes
69-102	0	0	No

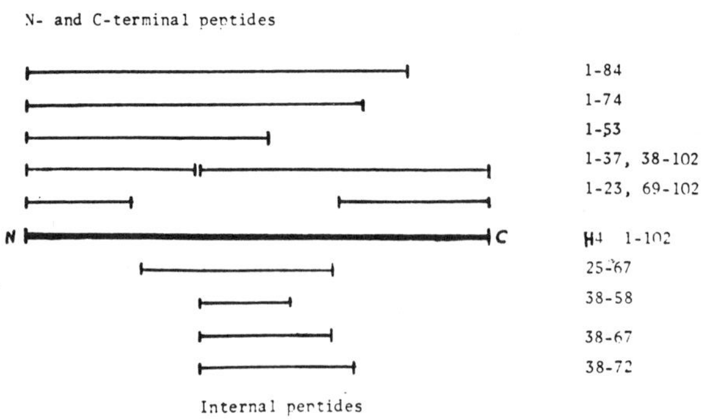

1-37 contain no helix:38–102 and 1–102 have the same helicity as do 38–72 and 1–74 (both pairs differ by the first 37 residues) and 1–37 has no helix. b) Since peptides 25–67, 38–67, 38–72 and 1–74 all self-associate and form ~13 helical residues, there is a 13-residue helix between 38 and 67. c) Since peptides 1–53 and 38–58 have no helicity, this helix is at the C-terminal end of 38–67. Since all predictive schemes[7,8] agree that helix before 49 is unlikely, we assign the helix to residues 55–67. A recent prediction by Fasman et al.[9] concludes that residues 57–67 are helical. d) The 13 helical residues in 1–74 are ~1/2 that of intact H4. A second helix of similar length therefore exists. Since 1–84 contains more than 13 helical residues but less than the 25 of intact H4, the second helix lies in the region of residue 84. e) β structure forms in the C-terminal residues of H4 since 1–84 has no β structure while intact H4 and 38–102 both form the same amount.

References

1. Camerini-Otero, R. D., Sollner-Webb, B. & Felsenfeld, G. (1976) *Cell* **8**, 333-347.
2. Moss, T., Crane-Robinson, C. & Bradbury, E. M. (1977) *Nucl. Acids Res.* **4**, 2477–2485.
3. Moss, T., Cary, P. D., Crane-Robinson, C. & Bradbury, E. M. (1976) *Biochemistry* **15**, 2261-2267.
4. Lewis, P. N., Bradbury, E. M. & Crane-Robinson, C. (1975) *Biochemistry* **14**, 3391–3400.
5. Bradbury, E. M. & Rattle, H. W. E. (1972) *Eur. J. Biochem.* **27**, 270–281.
6. Chirgadze, Yu. N., Shestopalov, B. V. & Venyaminov, S. Yu. (1973) *Biopolymers* **12**, 1337–1351.
7. Vorobev, V. I., Birstein, T. M., Aleksanyan, V. I. & Zalenskii, A. O. (1972) *Sov. Mol. Biol.* **6**, 346–352.
8. Lewis, P. N. & Bradbury, E. M. (1974) *Biochim. Biophys. Acta.* **336**, 153-164.
9. Fasman, G. D., Chou, P. V. & Adler, A. J. (1976) *Biophysical J.* **16**, 1201-1238.

CONFORMATIONAL STUDIES ON THE CYANOGEN BROMIDE FRAGMENTS OF THERMOLYSIN

ANGELO FONTANA and CLAUDIO VITA, *Institute of Organic Chemistry, Biopolymer Research Centre, C.N.R., University of Padova, I-35100 Padova, Italy*

Conformational studies on peptide fragments of well-characterized proteins have been carried out in several laboratories in order to elucidate the mechanism by which proteins assume their characteristic conformations. It has been shown that in aqueous solution protein fragments contain little secondary structure.[1,2] However, since in some instances the entire amino acid sequence of a protein is not required for native-like structure, perhaps there is a critical size necessary to obtain the medium- and long-range interactions that stabilize the conformation resulting from short-range interactions. In order to investigate this question for thermostable proteins, we studied fragments of thermolysin, an endoprotease for which the amino-acid sequence[3] and three-dimensional structure[4] have been elucidated. This single polypeptide chain protein of 316 amino acids is folded into two distinct halves, the covalent connection being performed by a segment of α-helix. The protein was cleaved with cyanogen bromide at the two methionine residues and the peptide fragments 1-120, 121-205 and 206-316 were isolated in pure form following a procedure similar to that described by Titani et al.[5] The occurrence of secondary structure in the isolated fragments was investigated by far-ultraviolet circular dichroism (CD) measurements.

Fragment 1-120 was found to be insoluble in aqueous buffers at neutral pH. At pH 9.6, where this fragment could be dissolved, the CD spectrum in the region 250–200 nm was indicative of an essentially disordered conformation.

Fragment 121-205 contains all the amino acids involved in the binding of the functional zinc ion and of three of the four calcium ions in the native enzyme. CD studies on this fragment revealed that the secondary structure of the peptide is dramatically affected by calcium ions. As shown in Figure 1, in the absence of these ions the CD spectrum below 250 nm is indicative of little regular structure, whereas the addition of calcium causes a marked decrease in the magnitude of ellipticity. The resulting CD spectrum, with minima near 208 and 222 nm, is typical of a polypeptide with a substantial portion of the peptide bonds in an α-helical conformation.[6] The profile of $[\theta]$ at 220 nm *versus* calcium concentration indicates a sharp conformational transition occurring at 1-10 mM concentration of the ion, followed by a linear increase up to 0.9 M concentration. Calcium ions play a specific role in the conformation of the peptide fragment at neutral pH, since 0.2 M NaCl was found to have no significant effect on the secondary structure of the peptide.

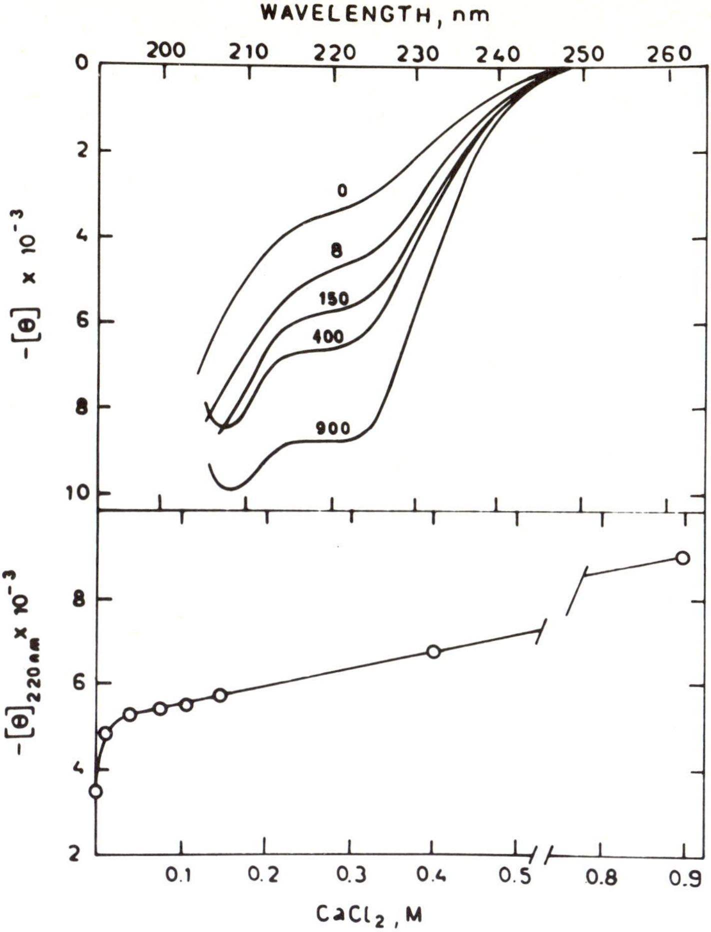

Fig. 1. (*top*) Far-ultraviolet CD spectra (mean residue ellipticity) of fragment 121-205 of thermolysin in 20 *mM* Tris/HCl buffer, pH 7.2, in the presence of different concentrations of $CaCl_2$. Numbers near the curves indicate *mM* concentration of $CaCl_2$ in Tris buffer. The peptide concentration was 0.15 mg/ml and spectra were measured in a fused quartz cell of 0.1 cm path length. (*bottom*) Dependence of $[\theta]$ at 220 nm on $CaCl_2$ concentration. Experimental conditions were as in the upper part of the figure.

Fragment 206-316 in 20 *mM* Tris/HCl buffer, pH 7.2, at room temperature exhibits a far-ultraviolet CD spectrum with minima near 209 and 222 nm and a crossover point from negative to positive at 203 nm. The mean residue ellipticity at 22°C at both wavelengths was found to be $-13{,}600 \pm 300$ deg. cm^2. $dmol^{-1}$ (Figure 2). At high temperature the peptide fragment undergoes a large conformational alteration and exhibits a change in the form of the CD spectrum which is that to be expected for extensive unfolding.

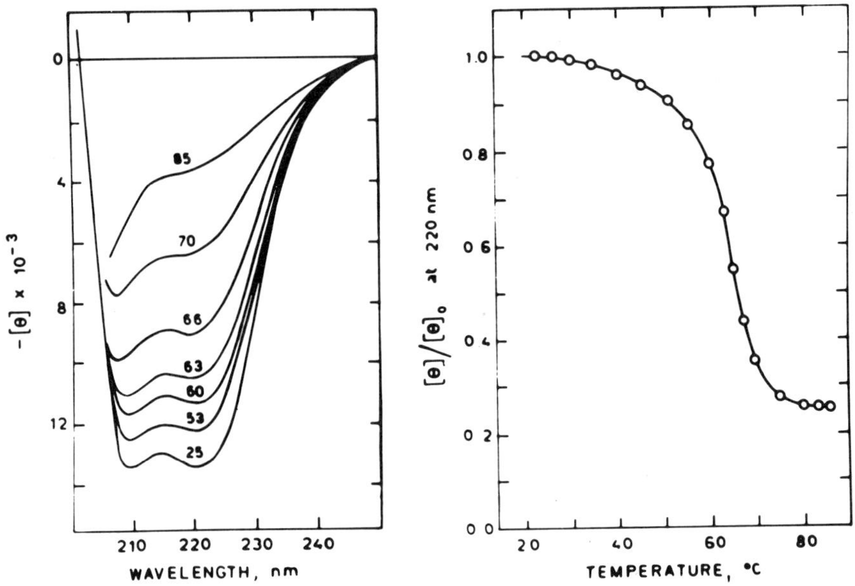

Fig. 2. (*left*) Far-ultraviolet CD spectra of fragment 206-316 of thermolysin at different temperatures in 20 *mM* Tris/HCl buffer, pH 7.2. Numbers near the curves indicate temperature in degrees centigrade. The peptide concentration was 0.2 mg/ml. (*right*) Temperature dependence of $[\theta]/[\theta]_0$ at 220 nm of fragment 206-316, where $[\theta]_0$ is the mean residue ellipticity at 22°C.

Figure 2 (right) shows the temperature dependence of the mean residue ellipticity $[\theta]$ at 220 nm normalized at 22°C. A cooperative thermal transition is seen in the range 60–70°C. A melting temperature (t_m) of 65° ± 0.5°C has been obtained in three different experiments. The thermal denaturation of the fragment appears also fully reversible, since the same variation of ellipticity with temperature is obtained on heating or cooling. In addition, the CD spectrum at room temperature of a heated solution of the fragment is indistinguishable from that obtained with an unheated sample.

The conformation of fragment 206-316 also was shown to be rather stable to the denaturing action of urea and guanidine hydrochloride, as monitored by the determination of the mean residue ellipticity value at 220 nm. The conformational transitions found, which are fully reversible upon removal of the denaturant, have midpoints at 5 *M* for urea and 2 *M* for guanidine. These concentrations are remarkable considering that fragment 206-316 consists of a polypeptide chain with no disulfide bridges, metal ligands or cofactors which would be expected to stabilize polypeptide structures.

Using the formula given by Greenfield and Fasman[6] to quantify the CD data, 33% of amino-acid residues of fragment 206-316 are in an α-helix secondary structure, on the basis of the mean residue molar ellipticity at 208 nm. Following the procedure of Chen et al.,[77] which takes into consideration the chain-length dependence of helical segments, the percentage of α-helix was found to be 45%. In the native enzyme, the corresponding sequence has been found 50% helical by X-ray crystallography.[4] These numerical data would indicate that fragment 206-316 appears to retain a similar percentage of α-helical secondary structure as in the native protein.

These results indicate that fragments 121-205 and 206-316 maintain significant levels of secondary structure in aqueous solution. The conformational transition occurring in fragment 121-205 in the presence of calcium ions is best interpreted in terms of binding of these ions to the negatively charged polypeptide, which contains 15 carboxylate groups, with the elimination of the steric repulsion between the negative charges and consequent α-helix induction. In the native enzyme calcium would induce the formation and stabilize the helical structure of that region corresponding to the fragment. The conformational properties of fragment 206-316, which corresponds approximately to one domain of thermolysin, appear to support the proposal of Wetlaufer,[8] that distinct compact regions, or domains, in globular proteins would fold independently from the remainder of the protein molecule.

References

1. Wetlaufer, D. B. & Ristow, S. (1973) *Ann. Rev. Biochem.* **42**, 135–138.
2. Anfinsen, C. B. & Scheraga, H. A. (1975) *Adv. Protein Chem.* **29**, 205–300.
3. Titani, K., Hermondson, M. A., Ericsson, L. H., Walsh, K. A. & Neurath, H. (1972) *Nature (London) New Biol.* **238**, 35–37.
4. Matthews, B. W., Weaver, L. H. & Kester, W. R. (1974) *J. Biol. Chem.* **241**, 8030–8044.
5. Titani, K., Hermondson, M. A., Ericsson, L. H., Walsh, K. A. & Neurath, H. (1972) *Biochemistry* **11**, 2427–2435.
6. Greenfield, N. & Fasman, G. (1969) *Biochemistry* **8**, 4108–4116.
7. Chen, Y. H., Yang, J. T. & Chau, R. H. (1972) *Biochemistry* **13**, 3350–3359.
8. Wetlaufer, D. B. (1973) *Proc. Nat. Acad. Sci. USA* **70**, 697–701.

SOLUTION SYNTHESIS OF COMPLEX PEPTIDES BY THE MAXIMUM PROTECTION PROCEDURE

SHUMPEI SAKAKIBARA, *Peptide Institute, Protein Research Foundation, Minoh-shi, Osaka 562, Japan*

One of the attractive features of solution methods of peptide synthesis is the possibility of detecting impurities in each reaction product. Theoretically, such detectable impurities can be removed by combinations of known separation techniques. If a homogeneous product can be isolated after each reaction, it should be possible to synthesize any peptide with a high degree of purity. In order to minimize the need for the separation of impurities, all possibility of the formation of side products in each reaction should be eliminated. Ideally, all the functional groups attached to the side chains in every component should be protected in solution procedures, as is customary in solid phase peptide synthesis. All protective groups attached to the final product should be removable by the technique which will be used for the final deprotection. Currently, procedures being applied to the final deprotection reaction are: (1) trifluoroacetic acid treatment at room temperature for 0.5~1 hour; (2) HF treatment at 0°C for 1 hour; (3) sodium treatment in liquid ammonia for 10 to 15 seconds, repetitively over 15-30 minutes; or (4) catalytic hydrogenolysis in the presence of palladium. Among these, we chose the HF procedure for the following reasons: (1) HF is a powerful solvent for all the amino-acid derivatives; thus, the final stage of deprotection should not be complicated by solubility problems; (2) Boc-amino acids which are currently used for solid phase synthesis are being manufactured in quantity and at very high purity; side-chain protective groups attached to those Boc-amino acids are known to be highly stable under conditions used for removal of N^{α}-Boc-groups and to be removable in HF at 0°C; and (3) because it is possible to avoid, in most cases, the unfavorable side reactions which sometimes result from the use of this HF procedure. With these facts in mind, we have developed a procedure for peptide synthesis which we call, "Solution Synthesis of Peptides by the Maximum Protection Procedure."

During repeated synthesis of bradykinin at the Peptide Institute following the procedure outlined in Fig. 1, the formation of a faster moving impurity was observed on paper electrophoresis at pH 4.8. This impurity was isolated by gel filtration with Sephadex G-25 and 0.2 *N* acetic acid. Amino-acid analysis and elemental analysis of the isolated material suggested that its structure was:

```
        Arg-Pro-Pro-Gly-┐
                        │
Arg-Pro-Pro-Gly-Phe-Ser-Pro-Phe-Arg
```

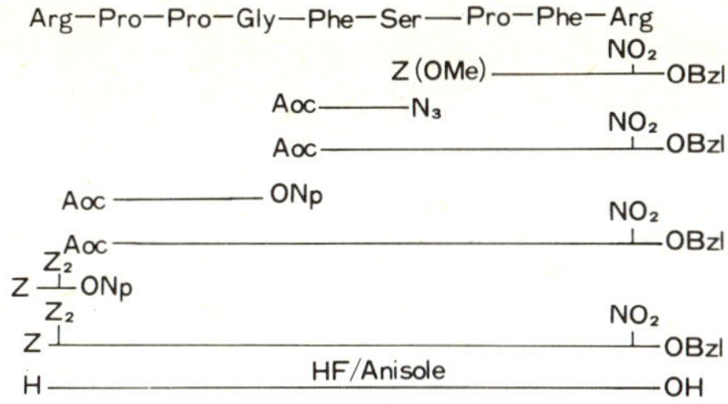

Fig. 1. A previous route for the synthesis of bradykinin.

This material was almost inactive in the uterus contraction test, but biological activity was found when it was treated with 1.5 *N N*-methylmorpholine for 5 hours at room temperature. Formation of bradykinin after the treatment was also confirmed by isolating the main product by gel filtration on Sephadex G-15. This product was identical to standard bradykinin in paper electrophoresis, bioassay, and amino-acid analysis.[1]

Recently, Bodanszky et al.[2] pointed out the occurrence of similar side reactions with unprotected Ser or Asp residues as follows: (1) the hydroxyl group in the Ser-residue is active enough to form the *O*-acyl derivatives with active esters of Boc-amino acids especially when imidazole or a His residue is present in the reaction mixture. (2) An unprotected Asp residue in a protected peptide fragment is converted to a kind of mixed anhydride with a Boc-amino acid *p*-nitrophenyl ester, which is further converted to a β-aspartyl peptide with an amino component or to a succinimide-type derivative within the molecule.[3] Our observation described above and these reported facts led us to conclude that the protection of all the functional groups is necessary even when the active ester procedure is applied to the solution synthesis of peptides.

There are a number of strategies which can be used to synthesize complex peptides by solution procedures. However, from the standpoint of isolating homogeneous materials, it is important to minimize the application of stepwise elongation with Boc-amino acids to obtain larger peptide fragments, since current analytical techniques cannot provide sufficient information on the purity of every synthetic intermediate. This is particularly true when all the functional groups are protected by similar hydrophobic protective groups.

The azide procedure is attractive when protected peptide esters can be converted to hydrazides by treatment with hydrazine. However, accumulated evidence indicates that the risk of racemization of the activated amino-acid residue by the azide procedure is not appreciably lower than with other activation procedures.[4] Further, removal of side products which may arise from undesirable decomposition of excess azide is not always completely successful. Thus, we decided to use the carbodiimide procedure because of the simplicity of its application and its higher reactivity.[5,6]

To use the carbodiimide procedure for fragment condensation, every carboxyl component must be prepared as a protected peptide with a free carboxyl group at the C-terminus. If Asp(OBzl), Asn, Glu(OBzl), and/or Gln residues are located in a peptide fragment, the C-terminal ester group cannot be removed by simple saponification. In such cases, a protected peptide acid can be synthesized by stepwise elongation of free amino acids or peptides with Boc-amino acid active esters. During the synthesis of such a C-terminal-free peptide, unexpected racemization was observed at the C-terminal amino-acid residue when pentachlorophenyl or *p*-nitrophenyl esters of Boc-amino acids were coupled with free peptides or amino acids.[7] As shown in Table I, the role of 1-hydroxybenzotriazole (HOBt) as the catalyst for the racemization reaction is evident. The mechanism of the racemization may be explained as mixed anhydride formation between the protected peptides and active esters, as has been suggested by Natarajan and Bodanszky.[3] However, no catalytic activity of imidazole was observed.

Table I. Racemization of the C-Terminal Amino Acid Residue during Coupling Reaction of Boc-Lys[Z(Cl)]-OX(1.2 eq.) with Various Peptides or Amino Acids in DMF at Room Temperature for 40 hours.

Amino Component	Active Esters -OX	TEA	Additives	D/L+D[a] %
Gly-Phe	-ONp	2 eq.	HOBt(0.25 eq.)	13.3
	-OPcp	2	HOBt(0.25 eq.)	40.7
	-OSu	2	HOBt(0.25 eq.)	< 2.0
	-OSu	1	Im(1 eq.)	< 2.0
	-OSu	1	Im(1 eq.)+HOBt(0.25 eq.)	4.8
	-ONp	1	Im(1 eq.)	3.2
	-ONp	1	Im(1 eq.)+HOBt(0.25 eq.)	9.1
	-ONp	2	——	< 2.0
	-OPcp	2	——	< 2.0
	-OSu	2	——	< 2.0
	-ONp	2	HONb(0.25 eq.)	< 2.0
Gly-Leu	-ONp	2	HOBt(0.25 eq.)	16.5
	-OPcp	2	HOBt(0.25 eq.)	41.4
	-OSu	2	HOBt(0.25 eq.)	< 2.0
Phe	-ONp	2	HOBt(0.25 eq.)	22.7
Pro	-ONp	2	HOBt(0.25 eq.)	3.8
	-ONp	2	——	2.7
Asn-Phe-Phe	-ONp	0	HOBt(0.25 eq.)	19.9
	-ONp	0	HOSu(0.25 eq.)	< 2.0
	-ONp	0		2.2

a These values were determined by a modified Manning and Moore procedure[7] after acid hydrolysis with 6 *N* HCl at 110°C. for 22 hours. These values include some additional racemization (1.5~1.8%) acquired during acid hydrolysis.
Np, *p*-Nitrophenyl; Pcp, Pentachlorophenyl; Su 1-Hydroxysuccinimide; HONb, *N*-Hydroxy-5-norbornene-2,3-dicarboximide; Z(Cl),2-Cl-Benzyloxycarbonyl.

DCC is widely used with HOBt or HOSu for coupling a protected peptide with a protected amino component.[5,6] This reagent, however, is known to form inactive urea-type derivatives with carboxyl components. In order to facilitate the removal of this side product from the reaction mixture, we used water-soluble carbodiimide, 1-ethyl-3-(3-dimethylaminopropyl)-carbodiimide, together with a slight excess of HOBt. The extent of racemization following such a convenient coupling procedure has been measured previously; it is comparable to that resulting from DCC/HOBt or the azide procedures.[8]

It is important when using the HF procedure to minimize the modification of amino-acid residues by carbonium cations which are formed in the reaction mixture. This is also important in solid phase peptide synthesis when HF is used in the final step. Many peptides have been removed from the resin and deprotected by the HF procedure and the purity of the products has been studied by many workers. However, problems concerning the formation of succinimide-type derivatives at Asp-Gly, Asp-Ser, or Asp-His bonds and the alkylation of Tyr and Trp residues by benzyl or *t*-butyl cations remain unsolved.[9-11] Several alternative solutions have been proposed to avoid the former problem; that is, application of alternate protective groups which are stable in HF but removable under other mild conditions.[12,13] Alkylation of Trp or Tyr residue in HF is unavoidable, but this problem can be overcome by using a combination of appropriate scavengers in the reaction mixture followed by strict purification of the final product. Modification of Trp residue occurs even when trifluoroacetic acid is used in the final deprotection procedure. In 1972, Wünsch et al.[14] noted the formation of substantial amounts of side products when the protected human-gastrin analog

$$<\text{Glu-Gly-Pro-Trp-Leu-Glu(OBu}^t\text{)-Glu(OBu}^t\text{)-Glu(OBu}^t\text{)-Glu(OBu}^t\text{)-}$$
$$\text{Glu(OBu}^t\text{)-Ala-Tyr(Bu}^t\text{)-Gly-Trp-Leu-Asp(OBu}^t\text{)-Phe-NH}_2$$

was treated with trifluoroacetic acid (TFA) at room temperature for one hour. In addition to the main product, a considerable amount of side product was isolated by countercurrent distribution or partition chromatography. It was thought that the compound had a modified Trp-residue in the molecule. After repeated model reactions with Trp, the structure of the modified Trp-residue was recently determined to be 1-*t*-butyl-tryprophan.[14] They also suggested that the yield of the alkylated compound was higher in HF than in TFA. We dissolved Boc-Trp in TFA for one hour at room temperature and distilled off excess TFA under reduced pressure. The residue was charged on the top of a column of Diaion HP-20 (Equivalent to Amberlite XAD-II), and then eluted with aqueous methanol of different concentrations; intact Trp was washed out with 20% MeOH, and a modified compound was eluted with 50% MeOH. The isolated material appeared to be homogeneous on cellulose TLC and gave a single peak on amino-acid analysis as shown in Fig. 2. Recovery of the modified material from the mixtures of Boc-Trp with TFA or HF under different conditions is shown in Table II.

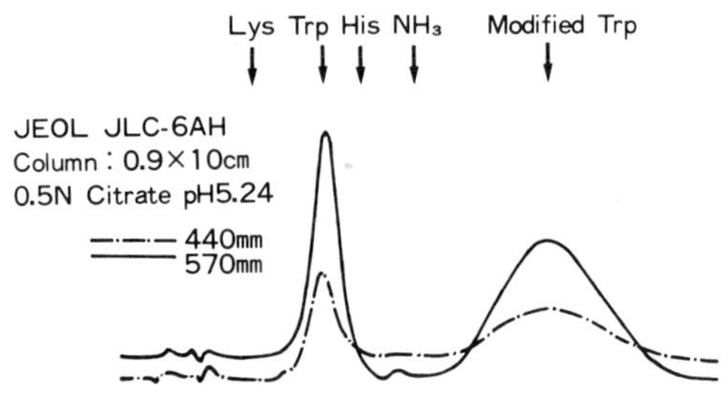

Fig. 2. Amino-acid analysis of modified Trp.

Table II. Effect of Scavengers on the Alkylation of the Trp-Residue during Acidolysis of Boc-Trp.

Reagent	Scavengers(eq.)[a]	Recovery(%) Trp	Trp(Bu^t)
TFA at 0°C/10min r.t./50min	D(30) + E(10)	95	3.5
	D(100) + E(10)	97	1.2
	A(30) + E(10)	92	6.8
	A(100) + E(10)	93	6.0
	E(10)	92	5.4
	E(100)	93	4.0
	none	67	14.9
HF at 0°C/1 hr.	D(10) + E(10)	85	<1
	A(10) + E(10)	86	<1
	E(10)	88	<1
	none	57	19.0

a. A, Anisole; D, Dimethylsulfide; E, Ethanedithiol

As can be seen in the table, anisole is a weaker scavenger than dimethylsulfide in TFA-reactions and addition of an unusually large amount of dimethylsulfide effectively prevented modification of Trp during cleavage of the Boc-group in TFA. Lower recovery of Trp from the HF-reaction mixtures may be explained by the formation of some insoluble materials, but the amount of alkylated derivatives detected was negligible. These results suggest that the use of HF for treating protected peptides with Trp does not prevent recovery of the final product. As can be expected from the above mentioned facts, such an alkylated derivative is more lipophilic than the desired product. It can be removed readily from the main product by means of countercurrent distribution, high performance liquid chromatography, column chromatography on Sephadex LH-20, or by partition chromatography on Sephadex G-25.

The final and largest problem in the solution synthesis of peptides by the maximum protection procedure is the low solubility of each intermediate in neutral organic solvents. Since the solubility of peptides is sequence-dependent, it could not be predicted prior to actual synthesis. Although we have not yet accumulated enough data in our laboratory, we anticipate that the solubility of protected peptides is not as low as originally expected. The following list of peptides has been synthesized by the maximum protection procedure in our laboratory without any serious solubility problems:

```
          NO2                  Bzl         NO2
Z(OMe)-Arg-Pro-Pro-Gly-Phe-Ser-Pro-Phe-Arg-OBzl

    Z   Tos                     Bzl         NO2
Boc-Lys-Arg-Pro-Pro-Gly-Phe-Ser-Pro-Phe-Arg-OBzl

      Z   Tos                     Bzl         NO2
Z-Met-Lys-Arg-Pro-Pro-Gly-Phe-Ser-Pro-Phe-Arg-OBzl

                        Tos     Z(Cl)
Z-Pyr-Gly-Leu-Pro-Pro-Arg-Pro-Lys-Ile-Pro-Pro-OBzl

                    OBzl    Bzl     OBzl
Boc-Trp-Ala-Gly-Gly-Asp-Ala-Ser-Gly-Glu-OBzl

              Bzl Bzl         Tos
Z-Pyr-His-Trp-Ser-Tyr-Gly-Leu-Arg-Pro-Gly-NH2

        Bz(Cl2) OBzl  Z(Cl)   Tos Tos     Bzl(Cl2)
Z-Pyr-Leu-Tyr-Glu-Asn-Lys-Pro-Arg-Arg-Pro-Tyr-Ile-Leu-OBzl

     Bzl(OMe) Z(Cl)                  Z(Cl) Bzl     Bzl Bzl Bzl(OMe)
Z-Ala-Gly-Cys-Lys-Asn-Phe-Phe-Trp-Lys-Thr-Phe-Thr-Ser-Cys-OBzl

    Tos     Z(Cl)
Aoc-Arg-Pro-Lys-Pro-Gln-Gln-Phe-Phe-Gly-Leu-Met-NH2

                            Z(Cl)                   Bzl(Cl)
Boc-Trp-His-Trp-Leu-Gln-Leu-Lys-Pro-Gly-Gln-Pro-Met-Tyr-OBzl

┌──────────────────────────┐
Bzl         Bzl Bzl                    Z(Cl)   Bzl     OBzl
└Ser-Asn-Leu-Ser-Thr-Asu-Val-Leu-Gly-Lys-Leu-Ser-Gln-Glu-Leu-His-

     Z(Cl)       Bzl Bzl(Cl2) Tos Bzl OBzl
     Lys-Leu-Gln-Thr-Tyr-Pro-Arg-Thr-Asp-Val-Gly-Ala-Gly-Thr-Pro-NH2
```

Applying the techniques discussed above, the synthesis of [Asu^{1,7}]-human-calcitonin was carried out recently in our laboratory (Asu, aminosuberic acid). The sequence and the protective groups attached to the side chains are indicated below.

```
┌──────────────────────────┐
            Bzl Bzl                     Bzl Bzl(Cl2) Bzl OBzl        Z(Cl)
└Gly-Asn-Leu-Ser-Thr-Asu-Met-Leu-Gly-Thr-Tyr-Thr-Gln-Asp-Phe-Asn-Lys-
                                     10                  15

         Bzl                 Bzl
Phe-His-Thr-Phe-Pro-Gln-Thr-Ala-Ile-Gly-Val-Gly-Ala-Pro-NH2
    20                  25                  30
```

Fragment (1-10) was synthesized as reported previously in a synthesis of [Asu^{1,7}]-eel calcitonin (Fig. 3).[15] The azide coupling of the cyclic part to Met-Leu-Gly was effective when a twofold excess of the tripeptide was used. The unbound tripeptide was easily removed by washing with water. The most complex problem in the synthesis of the linear part (11-32) was encountered during the purification of the fragment (24-32), which was hardly soluble in DMF (Fig. 4). Once the suspension was heated, it turned to a pasty mass and it took an extremely long time to filter out the solvent. All the condensation reactions were monitored by thin-layer chromatography and the purity of each product was insured by amino-acid analysis.

Removal of excess starting materials from the reaction product gradually became easier as fragment condensation was repeated and the chain length increased, since those fully protected fragments were almost insoluble in the usual organic solvents. After removal of the Boc-group with TFA, however, the TFA salts were generally soluble in a mixture of DMF and *N*-methylpyrolidone.

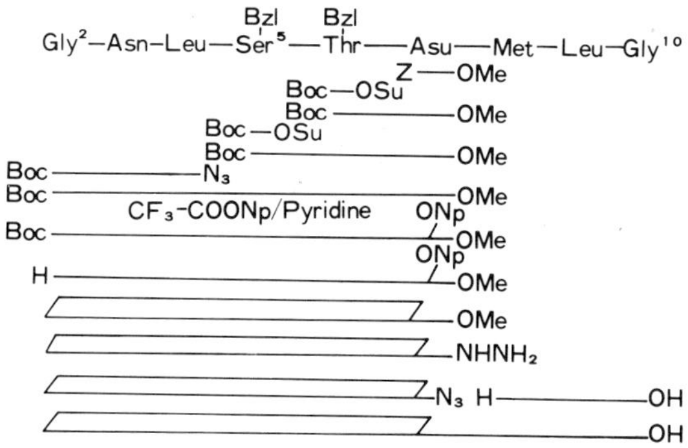

Fig. 3. Synthesis of [Asu^{1,7}]-human-calcitonin (1-10).

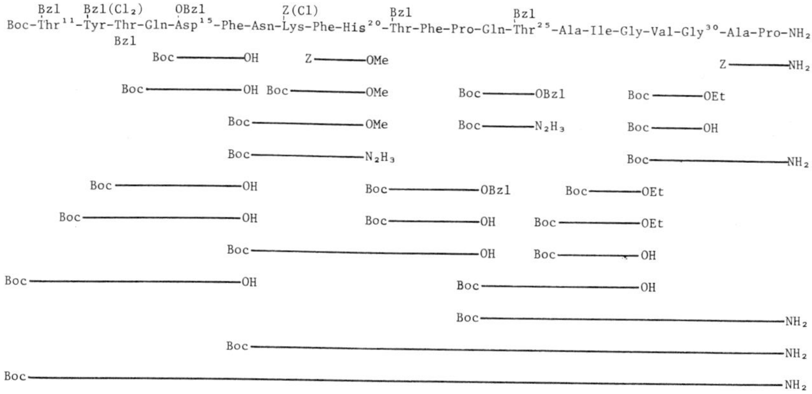

Fig. 4. Synthesis of [Asu^{1,7}]-human-calcitonin (11-32).

The TFA salt was neutralized with TEA in DMF and the free base was precipitated by the additions of water. The N-terminal deblocked peptide was then subjected to the next coupling reaction after drying in a desiccator. Fortunately, the protected peptides with a free amino group at the N-terminus were generally soluble in a mixture of DMF and *N*-methylmorpholine.

In order to confirm the purity of major intermediates, the protected fragments (17-32) and (11-32) were treated with HF in the presence of anisole, and each product was subjected to CM-cellulose column chromatography. As shown in Fig. 5 and in Table III, the major peaks were judged to be the desired peptides, and contamination by side products was confirmed to be negligible.

After final coupling between fragment (1-10) and the linear fragment (11-32) (Fig. 6), the product (1.5 g) was treated with HF (20 ml) at -5°C for 1 hour in

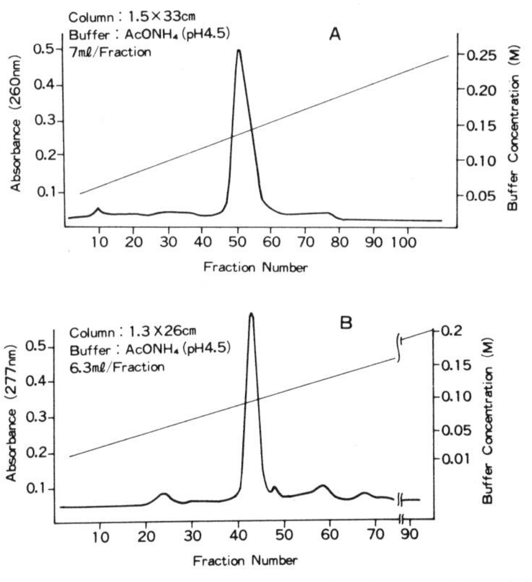

Fig. 5. CM-cellulose column chromatography of HF-treated h-calcitonin fragments: A, (17-32); B, (11-32).

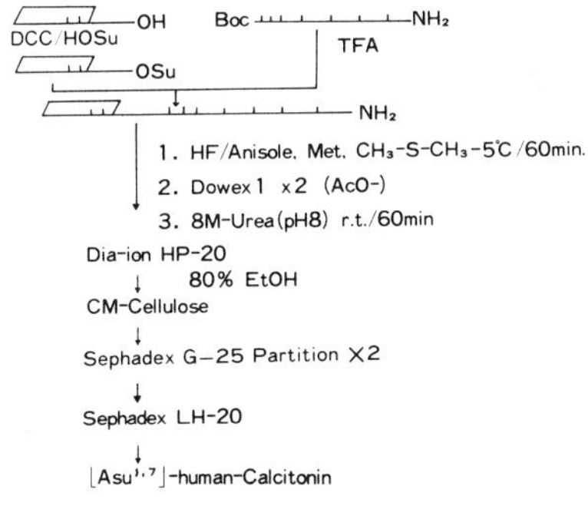

Fig. 6. Synthesis of fully protected [Asu^{1,7}]-human-calcitonin and its deprotection with HF.

the presence of anisole (3 ml), Met (1.5 g), and dimethylsulfide (1 ml). Excess HF was then removed under reduced pressure below 0°C. The residue was dissolved in 20% acetic acid and the solution was passed through a column of Dowex 1 × 2 (AcO^-) to eliminate F-ions. The column was washed with the same solvent and the crude product in the effluent was absorbed on Dia-ion HP-20 to concentrate the peptide on the column. It was then eluted with 80% ethanol and lyophilized (740 mg). The crude material was dissolved in 8 *M* urea solution of pH 8.0 at room temperature for 30 minutes to reverse all the *N* to *O* shift of peptide bonds at the Ser and Thr residues.

The treated peptide was absorbed on Dia-ion HP-20, excess urea was washed out with water, and the treated peptide was eluted with 80% ethanol and lyophilized. The crude product was purified by ion exchange chromatography on CM-cellulose (189 mg), partition chromatography on Sephadex G-25 for two times (132 mg), and final gel filtration on Sephadex LH-20 (102 mg); $[\alpha]_D^{26}$ -69.6° (c 0.72, 1 *M* AcOH) (Fig. 7).

As judged from the chromatograms and from amino-acid analysis (Table III), the final product was homogeneous and showed 390 MRC U/mg of hypocalcemic activity when porcine MRC standard was used as a reference.

As can be seen in Table III, no racemization of the Phe residue at position 16 was found by amino-acid analysis after aminopeptidase M (AP-M) digestion with an isolated free peptide fragment (11-32). However, after L-amino acid oxidase digestion of the acid hydrolyzate, 15% of Phe remained undigestible; if all the

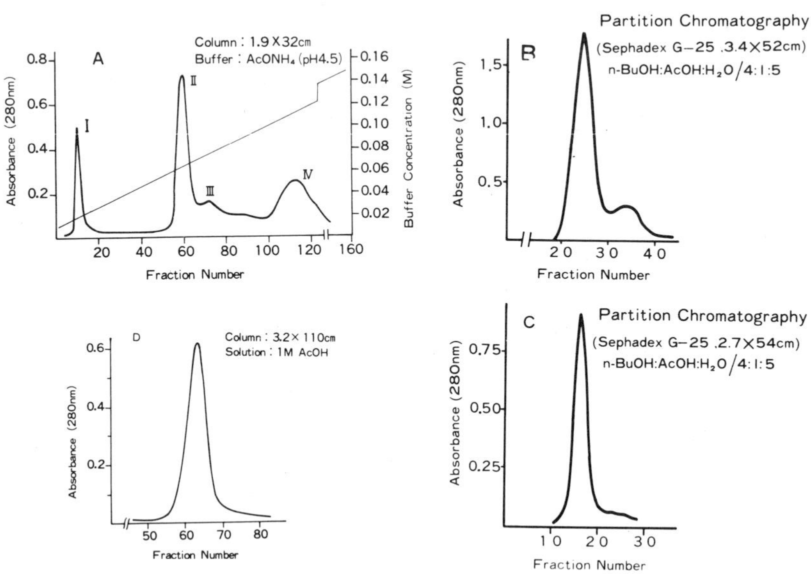

Fig. 7. Purification of synthetic[$Asu^{1,7}$]-human-calcitonin. A, CM-cellulose chromatography; B,C, partition chromatography on Sephadex G-25; D, gel filtration on Sephadex LH-20.

Table III. Amino Acid Analysis of [Asu^{1,7}]-human-calcitonin and Its Fragments, (11-32) and (17-32)

	(17-32)	(11-32)		(1-32)
	HCl	HCl	AP-M	HCl
Lys	1.02 (1)	0.91 (1)	1.03 (1)	1.07 (1)
His	1.02 (1)	0.84 (1)	0.91 (1)	1.05 (1)
Asp	0.85 (1)	1.94 (2)	0.73 (1)	2.98 (3)
Asn			ND (1)	
Thr	2.00 (2)	3.00 (4)	ND (4)	4.62 (5)
Ser				1.00 (1)
Gln			ND (2)	
Glu	1.17 (1)	2.20 (2)		2.16 (2)
Pro	1.68 (2)	2.14 (2)	1.94 (2)	2.10 (2)
Gly	2.00 (2)	2.00 (2)	2.06 (2)	4.00 (4)
Ala	2.06 (2)	1.98 (2)	1.94 (2)	2.10 (2)
Val	1.06 (1)	0.96 (1)	1.05 (1)	1.01 (1)
Met				0.95 (1)
Ile	1.07 (1)	1.00 (1)	1.05 (1)	1.10 (1)
Leu				1.96 (2)
Tyr		0.98 (1)	1.01 (1)	0.94 (1)
Phe	2.16 (2)	3.06 (3)	3.00 (3)	3.14 (3)
Asu				0.97 (1)

Thr^{11}-Tyr-Thr-Gln-Asp-Phe-Asn-Lys-Phe-His-Thr-Phe-Pro-Gln-Thr-Pro-Gln-Thr-Ala-Ile-Gly-Val-Gly-Ala-Pro^{32}-NH_2,

remaining Phe was attributed to the racemization of Phe residue at position 16 during the activation reaction and a control value for the racemization during acid hydrolysis (1.5% per one Phe residue) is taken into consideration, the extent of racemization during activation of Phe at position 16 was considered to be 10.5% in a stage of the fragment synthesis. However, the same test with the purified final product indicated the presence of only 4% of D-Phe in the acid hydrolyzate. If a control value of 1.5% each is subtracted for three Phe residues from the observed value of 4%, it is reasonably judged that 10.5% of the racemized product at position 16 had been removed completely during the final purification procedures. Thus, it must be emphasized that each coupling reaction of a peptide fragment containing an optically active amino acid residue at the C-terminus should be followed by a determination of the extent of racemization even though the coupling procedure had been proved to be racemization-free by standard test systems. However, if a small extent of racemization is unavoidable, partition chromatography appears to be very effective in yielding a product of high purity.

The most advantageous property of the Asu-derivatives is their stability during storage; a previously synthesized [Asu^{1,7}]-eel-calcitonin[15] was stored over two months at 65°C, and no deterioration in the biological activity was observed during this period (Fig. 8). On the other hand the biological activity of natural eel-calcitonin was reduced to one-third of the original level within one month. Stability of dilute solutions of the [Asu^{1,7}]-eel-calcitonin at higher pH (40 U/ml) was also better than with natural-type eel-calcitonin even though it was dissolved in a dilute gelatin solution (Fig. 9).

Immuno-cross-reactivity of the [Asu^{1,7}]-human-calcitonin against antisera to human-calcitonin was measured by Dr. Abe of the National Institute of Cancer Research in Tokyo. The precipitation curve was shown to be parallel to that of human-calcitonin (Fig. 10).

Immunization of rabbits with [Asu^{1,7}]-human-calcitonin is now being developed by Professor Kumahara, of Osaka University, and his group. They succeeded in obtaining an antibody which was useful in a dilution of 1:2500. As can be seen in Fig. 11, 100 pg of calcitonin was measurable with this antisera, which was shown to cross-react with human-calcitonin. Since the synthetic analog has no disulfide bond in the molecule, no complications were encountered in ^{125}I-labeling of this material. This system was therefore considered to be more practical for the radioimmunoassay of human-calcitonin than systems which contain disulfide-peptides.

In conclusion, the maximum protection procedure in solution synthesis was demonstrated to be useful for the synthesis of complex peptides such as calcitonin. For the synthesis of much more complex peptides, further experimental work must be conducted in which the length of the peptides under study is gradually increased.

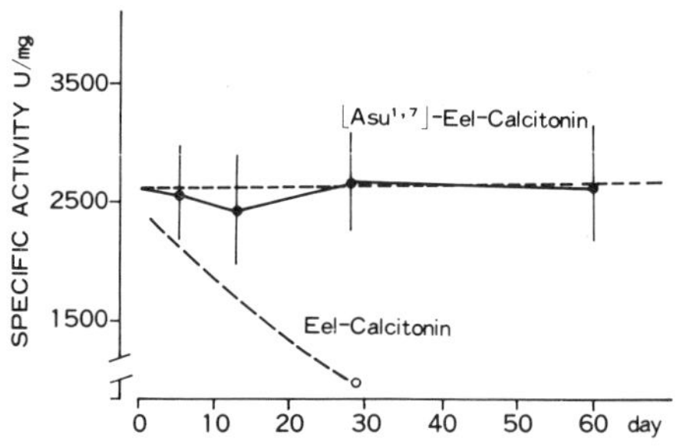

Fig. 8. Stability of powdered eel-calcitonins during storage at 65°C.

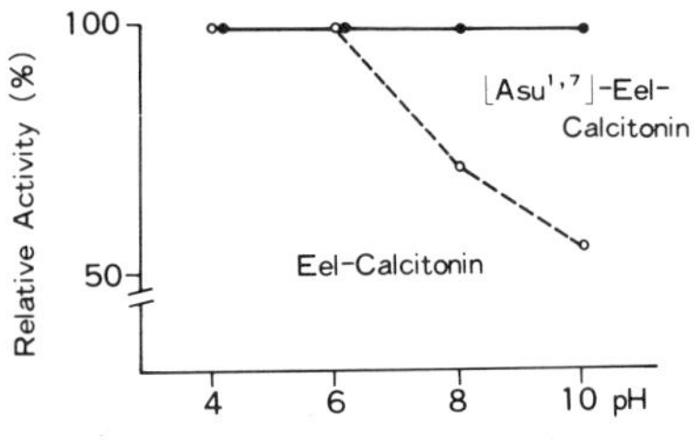

Fig. 9. Activity change of eel- and [Asu^{1,7}]-eel-calcitonin solutions at various pHs during incubation at 65°C for 4 hrs.

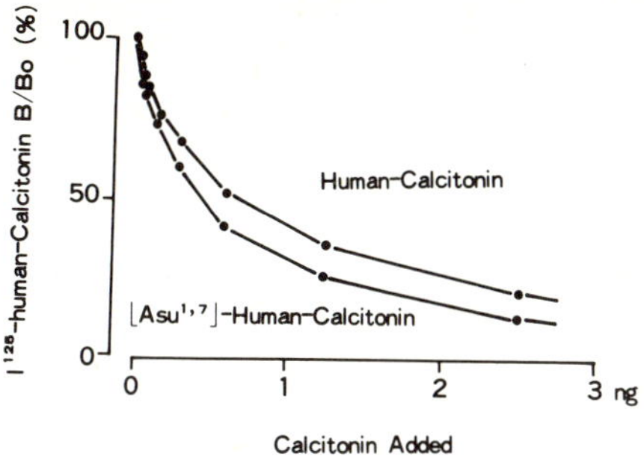

Fig. 10. Cross-reaction of [Asu^{1,7}]-human-calcitonin with anti-human calcitonin.

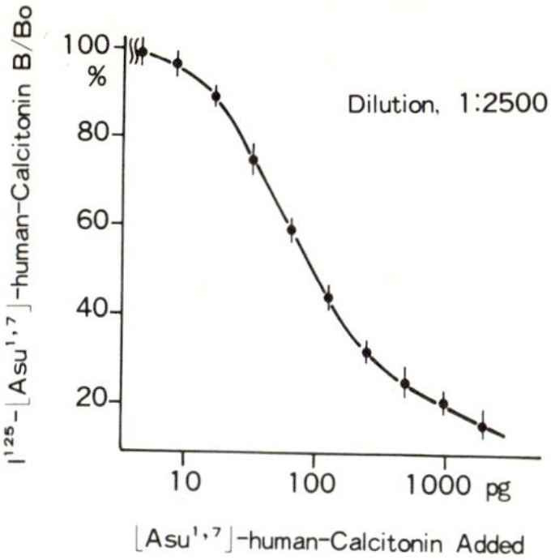

Fig. 11. The standard curve in radioimmunoassay for [Asu^{1,7}]-human-calcitonin using an antiserum in a dilution of 1:2500.

References

1. Tsubaki, T., Kimura, T. & Sakakibara, S. (1972) *Proceedings of the 9th Symposium on Peptide Chemistry*, Yanaihara, N., Ed., Protein Research Foundation, Minoh, Osaka, p. 84–89.

2. Bodanszky, M., Fink, M. L., Klausner, Y. S., Natarajan, S., Takemoto, K., Yiotakis, A. E. & Bodanszky, A. (1977) *J. Org. Chem.* **42**, 149–152.

3. Natarajan, S. & Bodanszky, M. (1976) *J. Org. Chem.* **41**, 1269–1272.

4. Finn, F. M. & Hofmann, K. (1976) *The Proteins*, vol. II, Neurath, H. & Hill, R. L., Ed., Academic Press, New York, p. 128–131.

5. Weygand, F., Hoffmann, D. & Wünsch, E. (1966) *Z. Naturforsch B.* **21**, 426–428.

6. König, W. & Geiger, R. (1970) *Chem. Ber.* **103**, 788–798.

7. Mihara, S., Takaya, T., Morikawa, T., Emura, J. & Sakakibara, S. (1977) *Peptide Chemistry 1976*, Nakajima, T., Ed., Protein Research Foundation, Minoh, Osaka, p. 32–35.

8. Emura, J., Morikawa, T., Takaya, T. & Sakakibara, S. (1977) *Peptide Chemistry 1976*, Nakajima, T., Ed., Protein Research Foundation, Minoh, Osaka, p. 36–40.

9. Baba, T., Sugiyama, H. & Seto, S. (1973) *Chem. Pharm. Bull.* **21**, 207–209.

10. Erickson, B. W. & Merrifield, R. B. (1973) *J. Amer. Chem. Soc.* **95**, 3750–3756.

11. Wünsch, E., Jaeger, E., Kisfaludy, L. & Löw, M. (1977) *Angew. Chem. Int. Ed. Engl.* **16**, 317–318.

12. Yang, C. C. & Merrifield, R. B. (1976) *J. Org. Chem.* **41**, 1032–1041.

13. Suzuki, K., Nitta, K. & Sakai, Y. (1976) *Chem. Pharm. Bull.* **24**, 3025–3033.

14. Wünsch, E., Jaeger, E., Deffner, M. & Scharf, R. (1972) *Hoppe-Seyler's Z. Physiol. Chem.* **353**, 1716–1720.

15. Morikawa, T., Munekata, E., Sakakibara, S., Noda, T. & Otani, M. (1976) *Experientia* **32**, 1104–1106.

SPECIAL FEATURES OF LARGE-SCALE PEPTIDE SYNTHESIS

MAX FEURER, *CIBA-GEIGY Limited, Basle, Switzerland*

Possibilities and Problems of Peptide Manufacture

Behind the spectacular success of laboratory peptide syntheses the progress in technology has remained somewhat hidden, and I am glad to have the opportunity to present to you this aspect of peptide chemistry. In principle it appears possible today to synthesize peptides up to about 50 amino acids in the *laboratory*, following methods compiled in the monumental volumes of Houben-Weyl.[1]

Is this statement also true for chemical manufacture? I would answer this question positively; but actually only relatively few synthetic peptides are produced industrially. Looking for an explanation we must realize that chemistry is not the only determining step for the therapeutic utilization of an active compound. The hurdles of biological tests, the detailed evaluation of biologists, pharmacologists and clinical investigators constitute a filter through which only a few biologically active peptides have passed. Another reason arises from the fact that only a few companies will take the risk of building up the appropriate infrastructure for peptide production and analytical control, accept the strict requirements by health authorities, and pay the costs of introduction, just to find out that the market is too small for this peptide hormone. From this point of view, the possibilities of peptide synthesis are only one piece in a complicated mosaic. Nevertheless, within the last 20 years about a dozen industrially synthesized peptide hormones reached the market (Table I). On the whole, the list reflects the scientific sequence of the discoveries, isolations and laboratory syntheses in the field of peptide chemistry. After this general survey I should like to present two examples in more detail, namely: angiotensin II, the compound we started with; and corticotropin-(1-24), for which we used a general concept of synthesis.

Starting Production: (Asn^1, Val^5)-angiotensin II

Since there is practically no literature on peptide production and patents are generally not too communicative, it might be interesting to show you the original scheme of synthesis when we started production in 1958 (Scheme 1).

From the present point of view of peptide chemistry this scheme may appear to be rather daring. But in spite of chemical deficiencies we learned a lot technically. A small production unit was set up and we had the courage to abandon glass vessels in favor of ordinary steel and glass-lined reaction kettles for the production of the intermediates on a relatively large scale. In addition, we developed and installed a Craig distribution apparatus suitable for the industrial

Table I. Industrial Synthetic Peptide Hormones

Peptide	No.of amino acids	Trade Name	Producer	Intro-duction
Oxytocin	9	SYNTOCINON	SANDOZ	1956
[Asn^1,Val^5]-angiotensin II	8	HYPERTENSIN	CIBA	1959
Lys-Vasopressin	9	VASOPRESSIN DIAPID (USA)	SANDOZ	1961 1970
Corticotropin-(1-24)	24	SYNACTHEN	CIBA	1967
Pentagastrin	5	PEPTAVLON	ICI	1969
Desamino-oxytocin	9	SANDOPART	SANDOZ	1971
Desamino-D-Arg-vasopressin [DDAVP]	9	MINIRIN	FERRING	1973
TRH	3	TRF TIREGAN THYPINONE	ROCHE HOECHST ABBOTT	1974
Salmon-calcitonin	32	CALCITONIN CALCIMAR	SANDOZ ARMOUR	1974 1975
LH-RH	10	RELISORM L	ABBOTT	1975
Human-calcitonin	32	CIBACALCIN	CIBA-GEIGY	1977

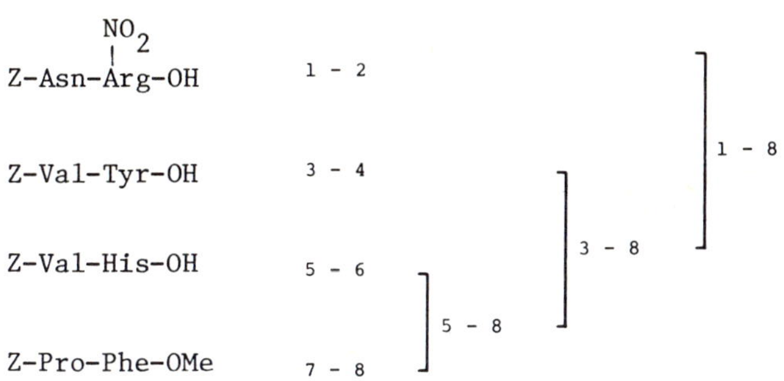

Scheme 1. Original scheme for the synthesis of [Asn^1, Val^5]-angiotensin II, 1958.

purification of the final product. With these innovations we started the industrial production of peptide hormones. As peptide chemistry progressed, the synthetic scheme was obviously improved, and many extras were added, but I particularly wanted to demonstrate how simply one can start peptide production.

Introduction of Concepts: Corticotropin-(1-24)

When we decided, in the early sixties, to attempt the synthesis of corticotropin-(1-24), we did not want to repeat our mistakes made with angiotensinamide, and we started thinking about concepts. Let us look at the concepts used in the original scheme for the synthesis of corticotropin-(1-24), (Scheme 2).

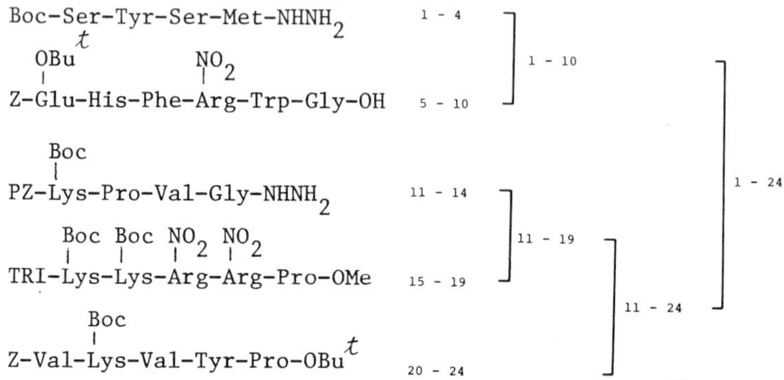

Scheme 2. Synthesis of corticotropin-(1-24), R. Schwyzer.[2]

The concepts consisted mainly of:

- Fragment condensation of medium sized peptide sequences. This approach permits the concomitant work on several fragments by various members of a group in research, development and production. The fragments were so chosen as to avoid racemization.
- Protection of the side chains by substituents which can be easily removed in the final steps of the synthesis.

Even though this scheme permitted the preparation of small amounts of corticotropin-(1-24), we realized that concepts, as such, are not sufficient for a technical synthesis. We encountered insurmountable obstacles when the first 100 grams of the active compound had to be prepared in the laboratory. Choice of fragments and the great number of different protecting groups caused complications. As a consequence a revised scheme had to be elaborated, (Scheme 3).

This revised scheme proved to be suitable not only for enlarged laboratory preparations but also for industrial manufacture. I should like to emphasize that simple and sound chemistry is an absolutely essential basis for technology.

Evaluation of Purity

The aim of the industrial synthesis of peptides consists in the preparation of pharmaceuticals with a similar degree of purity as other synthetic medicinals. The responsibility of the manufacturer to the patient and the requirements by health authorities have to be fulfilled in every respect. The application of ingenious purification techniques at the very end of a synthesis might be regarded as sufficient to obtain a pure, active compound. According to our experience, purification of the intermediates should be an integral part of the whole synthesis. In industrial manufacture it has proven to be essential to check each starting

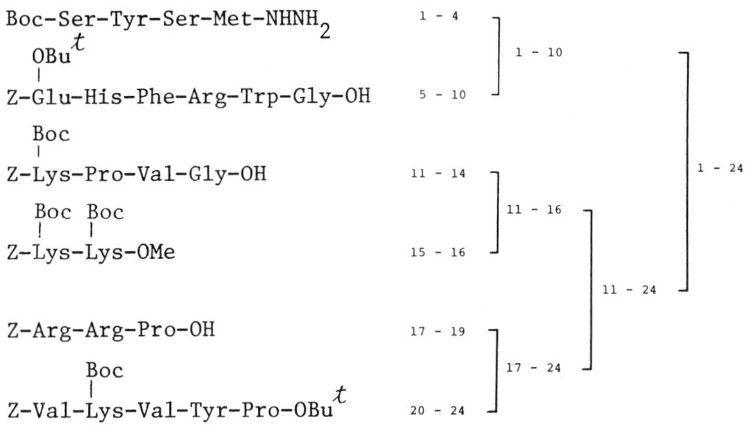

Scheme 3. Revised synthesis of corticotropin-(1-24).

material and isolated intermediate with regard to identity and purity before it is used in subsequent reaction steps. Only in this way can we prove the correct sequence and achieve a constant quality of the final peptide hormone.

Analytical Methods

Characterization and determination of content of an active compound like corticotropin-(1-24) demand highly developed analytical methods. The peptide contains 8 basic and 2 carboxylic groups and is isolated as an acetate with about 6 molecules of acetic acid. Consequently the substance contains about 10% acetic acid and in addition about 10% water which is taken up by the hygroscopic lyophilizate on equilibration with moist nitrogen. The content of the three components is determined for each batch by accurate analytical methods, i.e. for: peptide – high pressure liquid chromatography; for acetic acid – gas chromatography; and for water – Karl Fischer titration.

Other analytical tests for identity and purity are: amino-acid analysis, enzymatic degradation for testing the optical purity, electrophoresis and thin layer chromatography (TLC), UV absorption at different wavelengths, optical rotation, determination of the methionine(4)-sulfoxide by titration, and absence of microorganisms.

Dosage

The proper application of precise physico-chemical analytical methods to each batch of an industrial synthetic peptide permits the quantitative determination of the content and the evaluation of purity far more accurately than biological tests with their inherent large limits of error. As a consequence, the therapeutic dosage of a carefully manufactured peptide hormone is no longer dependent on the standardization by International Units, but can be safely expressed in weight, usually in milligrams.

Genealogy

According to the regulations of the health authorities the manufacturer has to present all analytical data of all intermediates and starting materials used in each step. For the production of a peptide with 20-30 amino acids this may concern hundreds or thousands of data following a genealogic system covering the whole synthesis. Instead of compiling all data by hand we set up a computer program which is able to arrange all the given necessary data for each batch of the final product by following the path of synthesis back to the starting materials, the amino acids.

Organization of Peptide Synthesis

I believe that the successful industrial manufacture is highly dependent on an efficient organization of all activities. This framework should be considered as a unit of science and technology with a continuous flow of information back and forth between the individual sections. This approach ensures the constant adaption of technical processes to new techniques which are either published in the literature or elaborated within the sections. As an example, a general scheme of how this problem can be solved is shown in Scheme 4.

Research	Rapid synthesis of small amounts of pure peptide - full protection of amino acids - fragment condensation - chemically safe coupling methods - frequent use of special purifications	Analytics
Development	Adaption of synthesis for industrial production - best suited fragments - simple coupling methods and purification - minimal protection Pilot scale up	
Production	Pilot scale up - technical adaption of procedures - trouble shooting - equipment - personnel - safety, GMP, ecology - starting materials - costs, planning, investment	

Scheme 4. Peptide synthesis organization.

Research concentrates its efforts on the synthesis of new peptides in small quantities for preliminary pharmacological tests. The methods used are specifically directed to a rapid and chemically safe synthesis. In order to take no risks, all functional groups in the side chains of amino acids are usually protected. For special purifications, such as counter-current distribution, small automatic units should be available. The freedom from development work of our research team has contributed to the rapid succession of new syntheses such as porcine tyrocalcitonin, human calcitonin, human parathyroid hormone, human corticotropin and human insulin.

Development is, according to our experience, as indispensable in peptide chemistry as it is in any other field of pharmaceuticals. The members of this group should be well acquainted with peptide chemistry, but also know exactly the requirements of production. The tasks of development consist essentially in 1) the elaboration of the most efficient scheme for the technical manufacture of a peptide; 2) the search for the best-suited fragments which can be purified by conventional means and are easily coupled with each other on a large scale; and 3) the reduction of protecting groups to a safe minimum in order to eliminate as many chemical steps as possible. These activities involve mainly laboratory work. Considering the great number of steps in a peptide synthesis this basic work requires a much longer period of time and more manpower than customary in the development of synthetic pharmaceuticals.

The next step consists in the scale-up, which is usually carried out jointly by development and production, because the size of pilot and production batches do not differ significantly. Sometimes this undertaking by production shows some difficulties, as the English Nobel-prize winner J. Cornforth formulated: *It is no good offering an elegant, difficult and expensive process to an industrial chemist, whose ideal is something to be carried out in a disused bath tub by a one-armed man who cannot read, the product being collected continuously through the drainhole in 100% purity and yield.*[3]

Production is mainly responsible for the technical and financial aspects of peptide syntheses and the chemical trouble-shooting for processes in production. For equipment we generally use stainless or glass-lined steel kettles fitted with normal stirrers, vapor columns, condensers and receivers (Fig. 1). A heating-cooling system allows regulation of the temperature in the kettle continuously between -20° and +70°C. For hydrogenations special high speed stirrers are used in order to shorten the reaction time. Special equipment is required for handling the last steps, especially the final purification. We prefer Craig distribution apparati for routine use, since they have proven to be versatile even on an industrial scale (Fig. 2). This method permits not only a very efficient separation of by-products but also a close control of any stage of the purification.

The training of the staff in production takes more time than usual and the standard of skill has to be higher because of the small number of batches of each

Fig. 1. Typical 200-gallon reactors used in production-scale peptide synthesis.

Fig. 2. Large-scale Craig countercurrent distribution apparatus with 80 tubes for peptide production.

step. The same crew has to carry out dozens of different reactions at short intervals. The necessary close supervision makes it clear that the number of foremen and technicians is higher than in other manufacturing units. Safety and ecology require no special attention, the quantity of each batch being very small compared to the usual technical order of magnitude. But even in small amounts hydrazine- or azide-containing sewage may turn out to be highly toxic for fish. In consequence, we destroy these compounds before they reach the sewage system.

Costs, Planning, Investment

Costs

After one has surmounted the numerous chemical difficulties of a 50–100 step synthesis of a new, active peptide, one is ready to start development. The first question of management will be: "How much does the active compound cost?" Because this simple question cannot be answered as such, you should return: "How much do we need for the next years?" Depending on the production scale, the same compound may cost up to 5 times more if production remains very small (Fig. 3).

For an initial approximation of costs it is sufficient to calculate with an average yield. It should not be forgotten, with a yield of 75%, only 10% of the starting material remains after 8 subsequent steps. To determine labor costs, a model reaction can be carried out which shows how much time is needed for preparations, the reaction and the isolation of the compound, as well as the cleaning of the equipment. In a second approximation, specific yields and adjusted labor costs can be taken in account. The latter might easily amount to the half of the real total costs, the raw materials probably 10–20%.

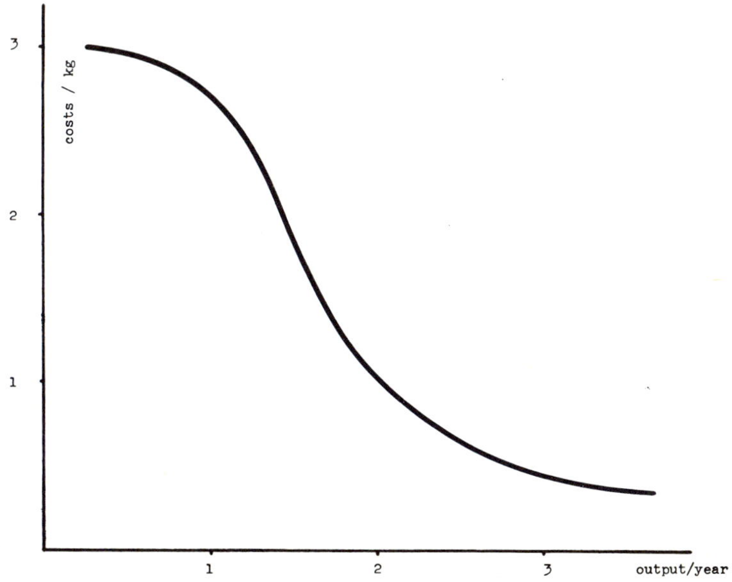

Fig. 3. The correlation of costs versus output.

Planning

The initial planning of a fragment peptide synthesis can be compared with the setting-up of a railroad time table, the trains running with unknown speed. But at the junction points, they should arrive at the right time in the right composition, to be connected and pursue their course to the next junction point.

Instead of planning the synthesis as a whole, it is advisable to determine in advance some key and stock intermediates. The chosen compounds should be stable, crystalline if possible, and have reliable analytical criteria of purity. Those key intermediates also serve as a buffer for unforeseen needs, not indicated in the budget. This may happen frequently because marketing prefers to order over short periods. If the amount key intermediates are too small, the active compound cannot be manufactured in time. On the other hand, if the key intermediates are overstocked, the company loses money on the uselessly invested capital.

Investment

To determine the necessary capacity for a one-step reaction, such as a cementation furnace, we can use the equation: capacity × time = output. An extreme contrary to this method should be applied to find the lay-out of a peptide productic unit. In the same vessel a variety of reactions must be carried out and dozens of different reactants processed, following each other in short intervals. For the manufacture of our peptide hormones, groups of three kettles between 20 and 200 gallons combined with one centrifuge proved to be most useful. The minimal number of such groups depends on: the budget of the active compound; the most time-consuming path of the synthesis; the desired stock of key intermediates; the yield of each step; the average duration of one batch; and the term for the final product.

Obviously the set-up of a new peptide production unit is a very time-consuming affair. After the planning together with the engineers, the hurdle of approval by the management is sometimes not taken in account. The installation of the equipment takes months; then it must be checked in detail by simulated runs. Finally production can begin. All this time your company does not see one gram of the active compound.

At the first glance an investment for peptide manufacture seems to be a very courageous enterprise – and that is a fact. But without any risk, there is nothing to win, and today, there are enough examples proving full success. Somebody *does* earn the money to enable our research team to continue its fine work on the active compounds of tomorrow.

References

1. Houben-Weyl (1974) Methoden der organischen Chemie Band 15, *Synthese von Peptiden*, Georg Thieme Verlag Stuttgart.

2. Schwyzer, R. & Kappeler, H. (1963) *Helv.* **46**, 1550–1572.

3. Cornforth, J. (1976) *Nachr. Chem. Techn.* **24**, 34.

REDUCTION OF METHIONINE SULFOXIDE TO METHIONINE IN PEPTIDES AND PROTEINS

RICHARD A. HOUGHTEN and CHOH HAO LI, *Hormone Research Laboratory, University of California, San Francisco, California 94143*

The oxidation of methionine to methionine sulfoxide in peptides and proteins occurs frequently, both by design[1], and as an undesired side reaction during oxidative cleavage at tryptophan[2], isolation from natural sources[3], electrophoresis[4], and solid phase synthesis[5]. While the intentional introduction of methionine sulfoxide can be readily accomplished[1], the reduction back to methionine has not been adequately investigated[6–8]. We have now investigated the reduction of methionine sulfoxide in peptides and proteins with specific reducing agents (mercaptoacetic acid, mercaptosuccinic acid, *N*-methyl mercaptoacetamide, mercaptoethanol, sodium dithionite, sodium *meta*-bisulfite, sodium thiosulfate, dithiothreitol, 2- and 3-mercaptopropionic acid), and under various conditions: concentration of reducing agent (2–100%), temperature (4°–37°C), pH (2–9), solvent (H_2O–100% acetic acid), and the presence of denaturing agents (8 *M* urea and 6 *M* guanidine hydrochloride).

The relative ability of a 0.75 *M* solution of each of the reducing agents examined to convert methionine sulfoxide back to methionine at pH 3.5 was as follows: mercaptosuccinic acid, 100%; mercaptoacetic acid, 90%; *N*-methyl mercaptoacetamide, 80%; 2-mercaptopropionic acid, 75%; dithiothreitol, 55%; $Na_2S_2O_4$, 40%; 3-mercaptopropionic acid, 35%; mercaptoethanol, 30%; $Na_2S_2O_3$, 30%; and $Na_2S_2O_5$, 20%. At pH 8.5 the relative rates were as follows: *N*-methyl mercaptoacetamide, 100%; dithiothreitol, 75%; mercaptoacetic acid, 70%; mercaptosuccinic acid, 60%; 2-mercaptopropionic acid, 60%; 3-mercaptopropionic acid, 50%; mercaptoethanol, 50%; $Na_2S_2O_4$, none; $Na_2S_2O_5$, none. When the reduction was carried out with the commonly used mercaptoacetic acid, mercaptoethanol, or dithiothreitol in increasing amounts of acetic acid (Table I), the rate of reduction was found to decrease steadily with increase in acetic acid concentration. Variation of the pH between 2 and 9 of the reaction solution in the reduction of the free amino acid and a model heptapeptide containing methionine sulfoxide significantly effected only reactions with mercaptocarboxylic acids. These had a decrease in rate with higher base concentrations. With increasing concentrations of reducing agents, an increase in rate of reduction was seen until the reducing agent reached a concentration of about 50% (v/v) after which the rate dropped rapidly. In the presence of 8 *M* urea or 6 *M* guanidine hydrochloride, the reduction rate was reduced about 20% in all cases.

While mercaptosuccinic acid and mercaptoacetic acid had the highest rates of reduction, they both suffered from the serious disadvantage of giving varying amounts of acetylated products when a methionine sulfoxide containing

Table I. Reduction of L-Methionine-d, 1-Sulfoxide

	Mercaptoacetic acid							Mercaptoethanol							Dithiothreitol			
% HOAc	100	75	50	25	10	5	H_2O	100	75	50	25	10	5	H_2O	75	25	5	H_2O
Reaction time (hours) 6	0	4	15	25	35	55	50	0	0	3	5	7	10	12	3	10	20	20
17	3	15	35	60	70	80	75	1	3	6	15	22	25	30	5	20	35	45
40	15	50	80	90	98	100	100	2	5	15	30	50	60	62	15	70	80	85
64	20	65	85	95	100	100	100	5	10	25	40	55	65	70	25	70	94	94
98	25	70	95	97	100	100	100	7	12	40	65	78	82	80	30	82	98	98
165	35	88	97	100	100	100	100	10	35	55	80	90	92	90	45	92	99	98

At 37°C; reducing agents at 0.725 M; methionine sulfoxide = 1.0 mg/ml; determined by the ready separation of methionine sulfoxide and methionine by paper electrophoresis at pH 2.1; ninhydrin visualization.

heptapeptide was reduced. This acetylated by-product varied between 3 and 45%, and was highest at basic pH's and lowest at pH 7.0. Since this occurs because of thioglycolides formed from the reducing agents, we prepared *N*-methyl mercaptoacetamide which lacked the possibility of thioglycolide formation. It was found that, while only slightly less reactive than mercaptoacetic acid and mercaptosuccinic acid, this reducing agent gave no acetylation by-product.

Using *N*-methyl mercaptoacetamide at a concentration of 7–28% (0.72–2.88 *M*) at 37°C, under acidic, neutral, or basic conditions, with and without denaturant present, we were able to reduce quantitatively methionine sulfoxide back to methionine in 12–18 hr in two peptides (the model heptapeptide and α-MSH) and several proteins (reduced and alkylated human and bovine growth hormone[9], and ovine prolactin) having all of their methionines oxidized to sulfoxide (Table II). Thus, we conclude that the reagent of choice for the reduction of methionine sulfoxide back to methionine in peptides and proteins is *N*-methyl mercaptoacetamide.

Table II. Reduction of 7/7-Met (O)-Prolactin

Reaction time (hrs)	MA-5%	MMA-7%	ME-5%	DTT-10%	MA-20%	MMA-26%	ME-20%
5	79	73	38	56	90	97	54
11	98	92	57	82	100	100	83
24	100	100	81	92	100	100	96

Concentration, 5.0 mg/ml; reaction temperature, 37°C; unreduced methionine sulfoxide was determined as methionine sulfone by amino acid analysis after dialysis against 50% HOAc, lyophilization, CNBr treatment in 70% HCO_2H, lyophilization x 2, treatment with 70% HCO_3H at room temperature for 1 hr, lyophilization x 2 and hydrolysis in 6N HCl. MA, mercaptoacetic acid; MMA, N-methyl mercaptoacetamide; ME, mercaptoethanol; DTT, dithiothreitol.

We thank Mr. Kenway Hoey for technical assistance and the entire Hormone Research Laboratory for their patience during the occasional olfactory assaults which occurred during this work. This study was supported in part by NIH Grants 1F32CA-05770-01, GM-2907, and AM-6097.

References

1. Houghten, R. A. & Li, C. H. (1976) *Biochem. Biophys. Acta* **439**, 240–249.
2. Omenn, G. S., Fontana, A. & Anfinsen, C. B. (1970) *J. Biol. Chem.* **245**, 1895–1902.
3. Lo, T., Dixon, J. S. & Li, C. H. (1961) *Biochem. Biophys. Acta* **53**, 584–586.
4. Jacobs, S. (1973) *Analyst* **98**, 25–33.
5. Norris, K., Halstrom, J. & Brunfeldt, K. (1971) *Acta Chem. Scand.* **25**, 945–954.
6. Dedman, M. L., Farmer, T. H. & Morris, C. (1956) *Biochem. J.* **66**, 166–177.
7. Iselin, B. (1961) *Helv. Chim. Acta* **44**, 61–78.
8. Polzhofer, K. P. & Ney, K. H. (1971) *Tetrahedron* **27**, 1997–2001.
9. Houghten, R. A. & Li, C. H. (1977) *Int. J. Pept. Protein Res.* in press.

FURTHER STUDIES ON RACEMIZATION IN PEPTIDE SYNTHESIS

J. KOVACS, Y. HSIEH, K. Y. HUI and S. E. KIM, *Department of Chemistry, St. John's University, New York, New York 11439*

It was reported at the Fourth American Peptide Symposium[1] that Z-Cys(Bzl)-ONp (I) racemizes about as fast as the dipeptide Z-Gly-Cys(Bzl)-ONp (II) with NEt_3 but couples 8 times slower with H-Val-OMe in THF. The k_c/k_r values for I and II are 2.66 and 14.6, respectively, which indicate that Cys dipeptide active ester probably will give optically purer products in coupling than the corresponding Cys active esters. Similar behavior was observed for the OPcp esters. The IR spectra did not indicate the presence of any detectable oxazolone during racemization. It was concluded that II probably racemizes mainly through α-hydrogen abstraction.

Preliminary studies on the mechanism of racemization of II will be reported here. Kemp and Rebek[2] demonstrated that the kinetic isotope effect can be used as a means of distinguishing enolization from oxazolone formation under certain conditions. Active ester I, deuterated on the α-carbon was prepared from Ac-DL-Cys(Bzl)-OH (III). Compound III was converted to oxazolone IV with DCC and IV was deuterated with $AcOD/NEt_3$. Deuterated Ac-DL-Cys(Bzl)-OH (V) crystallizes slowly on acidification with CF_3COOD with concomitant formation of acetic anhydride (gas chromatography). The free acid V precipitates out of acetic acid solution of oxazolone IV and acetic anhydride can be detected by IR. Racemic Ac-(α ^{2}H)Cys(Bzl)-OH was resolved with acylase I in the presence of Co^{++} ion in good yield. Optically pure VI was converted to VII which was practically 100% deuterated as determined by mass spectrometry using *m/e* 332 ion (Scheme 1).

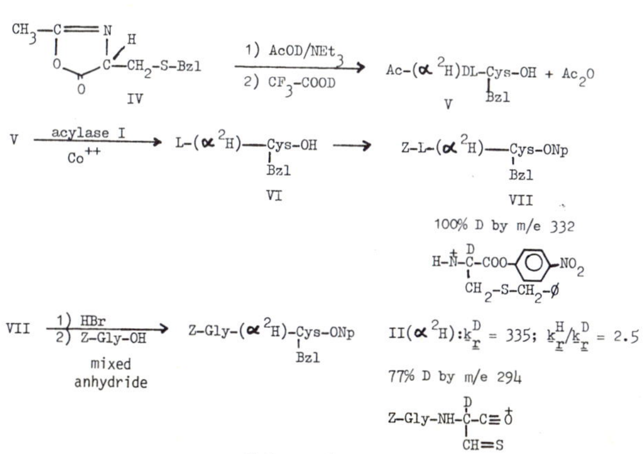

Scheme 1

Racemization of VII in THF with 7 eq. of NEt_3 gave an isotope effect of k_H/k_D = 7.5. Deuterium content of the racemic ester (94% racemized) was only 64%. The only proton source is the amide hydrogen and this value therefore indicates exchange with the α-deuterium.

The deuterated dipeptide II (α ^{2}H) prepared from VII by the backing-off procedure[4] showed 77% deuterium content by mass spectrum using *m/e* 294 ion. Racemization of II (α ^{2}H) was carried out under the same conditions as the undeuterated dipeptide II (in THF with 7 eq. of NEt_3) and an isotope effect of k_H/k_D = 2.5 was observed. This indicates that proton abstraction is the rate-determining step and that the racemization seems to proceed by α-hydrogen abstraction or enolization is parallel with oxazolone formation. Further study is in progress for additional information.

After the anomalous behavior of Cys (Refs. 3, 5, 6) and Cys dipeptide[1] active ester derivatives was observed, it was of interest to test the effect of the side chain of Met on k_r and k_c. Here the sulfur is attached to the γ-carbon. Table I shows the experimental and calculated[1] k_r and k_c values for Z-Met and Z-Gly-Met active esters. Data indicate that Met behaves more like simple amino acids: the k_c/k_r values are high, indicating "good behavior," that is lower degree of racemization can be expected during coupling.

It is known that some of the carboxyl activated histidine derivatives racemize unusually fast[7-12] during coupling. Preliminary rate studies were initiated to provide quantitative data concerning the effect of side-chain protecting groups on racemization and coupling. Table II shows the k_r, k_c and k_c/k_r values for Z-His(Bzl)-OPcp and Z-His(Z)-OPcp. The benzyl protected imidazole histidine active ester has the highest racemization rate among all amino acids studied (k_r = 626) and the lowest k_c/k_r value (14). This indicates that the extent of racemization which can be expected during coupling is higher than that for other amino acids. Even without NEt_3, a fast autoracemization takes place in THF solution as shown in Table II. These experiments indicate intermolecular-imidazole-side-chain-catalyzed racemization rather than intramolecular racemization.

Table I. Experimental and Calculated Racemization and Coupling Rate Constants for Benzyloxycarbonylmethionine and Benzyloxycarbonylglycylmethionine Active Esters

Compound	$k_r \times 10^6 M^{-1} sec^{-1}$ a)b)		$k_c \times 10^2 M^{-1} sec^{-1}$ c)		k_c/k_r	
Z-Met-OPfp	56.35	(58.53)[d)]	16.79	(10.00)	2980	(1724)
Z-Gly-Met-OPfp	4710	(6146)	61	(85)	130	(138)
Z-Met-OPcp	2.80	(3.90)	0.36	(0.27)	1286	(698)
Z-Gly-Met-OPcp	312.50	(409.7)	1.84	(2.30)	59	(57)
Z-Met-ONp	8.65	(6.24)	0.076	(0.06)	88	(97)
Z-Gly-Met-ONp	307.5	(655.6)	0.55	(0.506)	18	(7.8)

a) With NEt_3 in THF. b) Concentration of active esters was 0.05 M. All k_r values are average of two experiments with 7 & 35 eq. of NEt_3. c) With Val-OMe in THF. Concentration of active ester and Val-OMe was 0.13 M. d) Predicted values in parentheses.[1]

Table II. Racemization and Coupling Rate Constants of His Active Ester Derivatives in THF

Compound	Ester Conc.	NEt_3 Conc.	$k_r x10^6$ sec^{-1}	$k_r x10^6$ $M^{-1}sec^{-1}$	$k_c x10^2$ $M^{-1}sec^{-1}$ b)	k_c/k_r
Z-His(Bzl)-OPcp	0.15	No	60[a]	387[a]		
	0.05	No	29[a]	580[a]		
	0.05	0.35	225[a]	626[a]	0.9[c]	14.5
Z-His(Z)-OPcp	0.05	0.35		36	0.23[c]	64

a) The average of 3 experiments. b) With Val-OMe, the ester conc. & Val-OMe conc. was 0.13 M. c) The average of 2 experiments.

If the imidazole protecting group is an electron withdrawing group (in our experiment, carbobenzoxy), then the racemization rate decreases 17 times while the coupling rate decreases only 4 times, which raises the k_c/k_r value by 4.4 times. This value (64) is close to that of Z-Cys(Bzl)-OPcp (41), but still much lower than those for all other amino acids.[1] We conclude, in accordance with other investigators,[7-13] that for the protection of the imidazole side chain of histidine an electron withdrawing group should be used preferentially in order to decrease racemization or autoracemization. The order of ease of racemization of amino-acid active esters investigated under the same conditions is as follows: His > Cys > Ser > Asp > Phe > Met > Glu > Ala > Trp.

Racemization of the penultimate amino acid on activation of the C-terminal carboxyl group was observed by Bergman and Zervas[13] and investigated later by Weygand et al.[14] and by Taschner and coworkers.[15] Kinetic studies of the racemization of Z-Cys(Bzl)-Gly-OPcp in DMF and 7 eq. of NEt_3 gave a k_r value of 43 × $10^{-6}M^{-1}$ sec^{-1}. The racemization of Z-Phe-Gly-OPcp under the same conditions gave k_r value of 78.5 × $10^{-6}M^{-1}$ sec^{-1}. These rates are significant when compared with Z-Cys(Bzl)-Gly-OEt, which shows no detectable racemization with NEt_3.[6] Here the oxazolone is responsible for racemization. These and previous results indicate that coupling fragments at the glycine residue might not be completely free of racemization.

We wish to thank Dr. C. Iden of S.U.N.Y. at Stony Brook and Mr. Rocco Alessandro of the Department of Chemistry, St. John's University for measurement and evaluation of the high resolution spectra and Mrs. Christina Loskofsky for her assistance in infrared spectroscopy.

References

1. Kovacs, J., Cover, R., Jham, G., Hsieh, Y. & Kalas, T. (1975) *Peptides: Chemistry, Structure and Biology, Proceedings of the Fourth American Peptide Symposium*, Walter, R. & Meienhofer, J., Eds., Ann Arbor Sci. Publ., Ann Arbor, MI, pp. 317-323.

2. Kemp, D. S. & Rebek, F. (1970) *J. Amer. Chem. Soc.* **92**, 5792-5793.

3. Kovacs, J., Cortegiano, H., Cover, R. E. & Mayers, G. L. (1971) *J. Amer. Chem. Soc.* **93**, 1541-1543.

4. Goodman, M. & Steuben, K. C. (1959) *J. Amer. Chem. Soc.* **81**, 3980-3983.

5. Kovacs, J., Mayers, G. L., Johnson, R. H., Cover, R. E. & Ghatak, U. R. (1970) *Chem. Commun.*, 53–54.

6. Kovacs, J., Mayers, G. L., Johnson, R. H., Cover, R. E. & Ghatak, U. R. (1970) *J. Org. Chem.* **35**, 1810–1815.

7. Merrifield, R. B. (1964) *Biochemistry* **3**, 1385–1390.

8. Windridge, G. C. & Jorgensen, E. C. (1970) *Intra-Sci. Chem. Rep.* **5**, 375–380.

9. Windridge, G. C. & Jorgensen, E. C. (1971) *J. Amer. Chem. Soc.* **93**, 6318–6319.

10. Syrier, J. L. M. & Beyerman, H. C. (1974) *Rec. Trav. Chim. Pays-Bas* **93**, 117–120.

11. Beyerman, H. C., Hirt, J., Kranenburg, P., Syrier, J. L. M. & Von Zon, A. (1974) *Rec. Trav. Chim. Pays-Bas* **93**, 256–257.

12. Veber, D. F. (1975) *Peptides: Chemistry, Structure and Biology, Proceedings of the Fourth American Peptide Symposium*, Walter, R. & Meienhofer, J., Eds., Ann Arbor Sci. Publ., Ann Arbor, MI, pp. 307–316.

13. Bergman, M. & Zervas, L. (1928) *Biochem. Z.* **203**, 280–292.

14. Weygand, F., Prox, A. & König, W. (1966) *Chem. Ber.* **99**, 1446–1450.

15. Dzieduszyka, M., Smulkowsky, M. & Taschner, E. (1973) *Peptides 1972, Proceedings of the Twelfth European Peptide Symposium*, Hanson, H. & Jakubke, H. D., Eds., North Holland Publ. Co., Amsterdam, pp. 103–107.

STUDIES ON PROTHROMBIN. SYNTHESIS OF PEPTIDES CONTAINING γ-CARBOXYGLUTAMIC ACID

RICHARD G. HISKEY and NORMAN T. BOGGS III,
The W. R. Kenan, Jr. Laboratories of Chemistry, Department of Chemistry, The University of North Carolina at Chapel Hill, 27514

In an earlier communication[1] we reported the preparation of various γ-carboxyglutamic acid (Gla) derivatives and developed routes for the chemical synthesis of peptides containing this interesting new amino acid. We now report our continuing studies on a convenient process for the separation and isolation of both enantiomers of a protected Gla intermediate VI that is suitable for the chemical synthesis of chiral peptides containing Gla. The synthesis of a useful model tripeptide V is also described.

Although the published synthetically useful methods of producing Gla derivatives have yielded only racemates,[1-4] one compound, (±)*N*-carbobenzyloxy-γ-di-*t*-butylglutamic acid-α-methyl ester [(±)I], has been useful in producing model systems and in developing coupling procedures.[1] Desiring a convenient model for Ca^{++} binding studies and pK_a measurements, the tripeptide IVb was synthesized as shown in Fig. 1. Selective cleavage of the *t*-butyl esters was accomplished at 0°C for 30 minutes in a 1:1 (v/v) mixture of methylene chloride and trifluoroacetic acid.

In preliminary studies on the classical resolution of (±)VI, a diastereomeric pair of salts was produced by adding solid L-tyrosine hydrazide (H-L-Tyr-N_2H_3, prepared by the method of Curtius[5]) to a warm solution of (±)VI[3] in methanol. Free acid was liberated from the tyrosine hydrazide salt by extraction into ether from 20% aqueous citric acid. After recrystallization of (±)VI•T from methanol, (L)-VI (configuration elucidated by acid hydrolysis to L-glutamic acid) was obtained only 80% optically pure, and the conversion efficiency from serine methyl

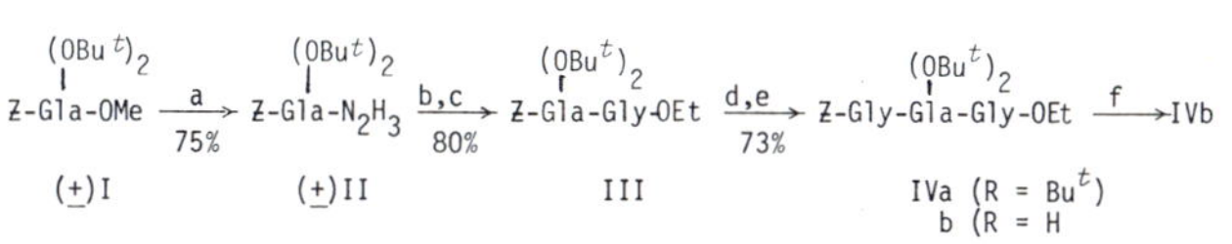

Fig. 1. a. $N_2H_4 \cdot H_2O$, CH_3OH. b. *i*-BuONO, THF, HCl @ –23%. c. H-Gly-Et, THF @ 0°. d. H_2, Pd.C, THF. e. *i*Butylcholoroformate, THF, ET_3N. f. TFA/CH_2Cl (1:1) @ 0°, 30 minutes.

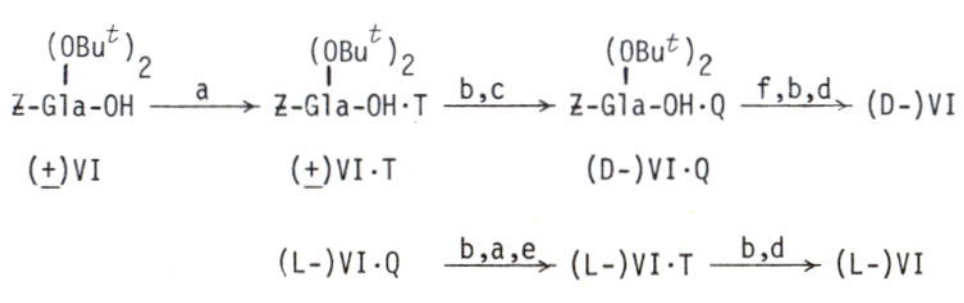

Fig. 2. a. CH_3OH, L-Tyrosine hydrazide. b. 20% Aq. citric acid/Et_2O. c. L-Quinine, EtOAc. d. Remove solvent. e. Recrystallization, CH_3OH. f. Recrystallized EtOAc.

ester hydrochloride was less than 5%. Although the resolution process was not a complete success, the insolubility of (±)VI•T in ether allowed the elimination of two purification steps in the synthesis of (±)VI from serine methyl ester hydrochloride. Preparation of (±)I followed, without purification, by base hydrolysis to (±)VI yielded a crude oil. Treatment of a cooled solution of the oil and H-L-Tyr•N_2H_3 in a minimum amount of methanol with an excess of dry ether precipitated the (±)VI•T salt in 59% overall yield for the three transformations.

A more effective resolution procedure involved using the quinine salt procedure of Marki and Schwyzer[6] to obtain D-VI, and the H-L-Tyr-N_2H_3 procedure to obtain L-VI from the oily quinine salt of this enantiomer. The process provides 15% D(-)VI and 18% L(+)-VI these percentages do not include partially resolved fractions. Thus, a convenient and efficient route to synthetic quantities of the Gla derivative L(+)- or D(-)-II from serine methyl ester hydrochloride is now available.

To assay optical purity, the scheme in Fig. 3 was utilized. The data verifies the assignment of absolute configuration previously suggested[3] and indicates that the optical purity of IV is sufficient for use as a synthetic intermediate. Table II contains a selection of rotation data obtained during resolution.

Finally, optically pure D(±)-II and D(±)-IX were prepared as indicated in Fig. 4. Table I presents the properties of racemic and resolved Gla derivatives.

Table I. Racemic and Resolved Gla Derivatives

	Mp.°C	$[\alpha]_D$ temp (degrees) This Study	Schwyzer and Coworkers[a]
(+)V·T	125-130	$+22.66^{22}$(c=2.0,CH_3OH)	
D-VI·Q	139-141	-68.6^{26}(c=1.0, $CHCl_3$)	-72.4(c=1, $CHCl_3$)
L-VI·T	149-150	$+22.2^{22}$(c=1.0, CH_3OH)	
D-VI	87	$+11.6^{22}$(c=2.26, CH_3OH)	
		-9.7^{22}(c=0.89, $CHCl_3$)	-11.3(c=1.1, $CHCl_3$)
L-VI	80-84	-10.3^{21}(c=2.0, CH_3OH)	-10.9(c=1.1, CH_3OH)
(±)VII	148.5-150	-	+12.3(c=1.1, $CHCl_3$)
D-VII	163-164[b]	-5.7^{26}(c=1.0, CH_3OH)	-5.7(c=1, CH_3OH)
L-VII		$+5.8^{24}$(c=1.0, CH_3OH)	+5.6(c=1, CH_3OH)
(±)I	oil	-[c]	
D-I	oil	$+1.25^{21}$(c=1.2, $CHCl_3$)	
(±)II	108.5-110		
D-II	63.5-64.5	-7.2^{24}(c=1.0, $CHCl_3$)	
D, D-IX	oil	$+11.6^{23}$(c=1, CH_3OH)	
		-9.3^{21}(c=1, $CHCl_3$)	
L, L-IX	-	-	-11.7(c=1.2, CH_3OH)

a. Resolution and rotation data of Schwyzer and coworkers[3,6,7]. b. Variable, a second measurement being 146-148d. c. Rotations observed from 0.5-2.0 degrees. However, hydrolysis yielded racemic glutamic acid.

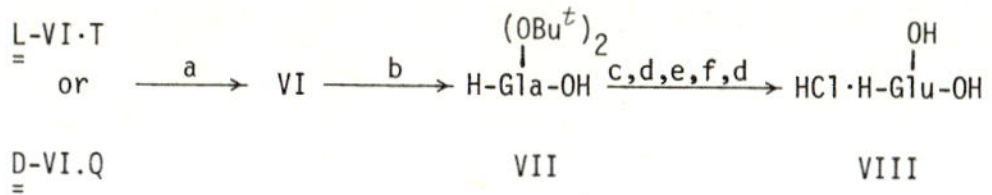

Fig. 3. a. 20% aq. citric acid/Et_2O. b. H_2, Pd. C, CH_3OH. c. 6 *N* HCl, 110°C, 3 hr. d. Dissolve in H_2O, lyophilize X3. e. Extraction X3 of aq. solution with Et_2O. f. Recrystallization from 6 *N* HCl.

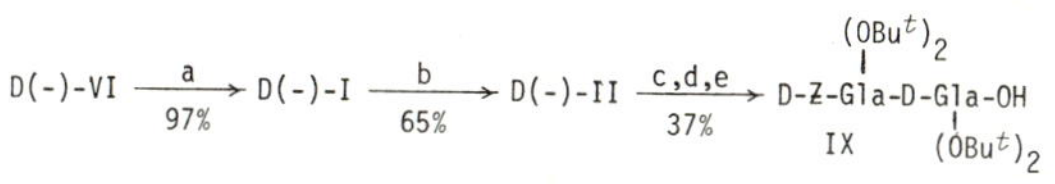

Fig. 4. a. CH_2N_2, Et_2O. b. $N_2H_4 \cdot H_2O$, CH_3OH. c. DMF, THF/HCl, *n*-BuONO, -23°C. d. Et_3N. e. D(-)-VII slurried in DMF, 0°C.

Table II. Resolution Rotation Data[a] $[\alpha]_D$ Degrees

Solvent	Salt $CHCl_3$	VI $CHCl_3$	VI CH_3OH	VII CH_3OH	VIII H_2O	% Optical Purity[m]
(D-)VI·Q	-67.5(24)	-8.6(24)	-	-5.9(24)	-20.5(22)	100±2
	-68.6(23)	-9.7[b]	+11.6[c]	-5.4[f]	-19.6[g]	98±2[n]
	-67.6(26)	-8.4(22)[d]	+9.8(22)[e]	-5.1[f]	-	
Solvent:	CH_3OH	$CHCl_3$	CH_3OH	CH_3OH	H_2O	
(L-)VI·T	+22.3(24)	-	-10.3(21)[h]	+5.8(24)	+19.3	95±2
					+19.6[k]	95±2[n]
	+22.0(24)	+9.4(26)[i]	-8.5(26)[j]	+5.8(22)	+19.6(22)[l]	98±2[n]

a. $[\alpha]_D$, Temperature indicated in brackets, concentrations at 1.00 unless otherwise indicated. *b*. c=0.86. *c*. c=26. *d*. c=1.20. *e*. c=2.55. *f*. Not dried over P_2O_5. *g*. c=0.84. *h*. c=2.0. *i*. c=2.49. *j*. c=1.05. *k*. c=0.965. *l*. c=0.92. *m*. Defined as $(\alpha_{obs}/\alpha_{std})X100$, where α_{std} is the rotation of authentic l-glutamic acid hydrochloride (+20.3±0.4, c=1.00, H_2O). *n*. Corrected for concentration variation of the standard.

References

1. Boggs III, N. T., Gawley, R. E., Koehler, K. A. & Hiskey, R. G. (1975) *J. Org. Chem.* **40**, 2850–2851.
2. Morris, H. R., Thompson, M. R. & Dell, A. (1975) *Biochem. Biophys. Res. Commun.* **62**, 856–861.
3. Marki, W. & Schwyzer, R. (1975) *Helv. Chim. Acta.* **58**, 1471–1477.
4. Weinstein, B., Watrin, K. G., Loie, H. E. & Martin, J. C. (1976) *J. Org. Chem.* **41**, 3634–3635.
5. Curtius, T. (1917) *J. Prakt. Chem.* **95**, 349–360.
6. Marki, W. & Schwyzer, R. (1976) *Helv. Chim. Acta.* **59**, 1591–1592.
7. Marki, W., Oppliger, M., Thanei, P. & Schwyzer, R. (1977) *Helv. Chim. Acta.* **60**, 798–806.

IMPROVED SYNTHESIS OF FOLATE CONJUGATES

CHARLES N. C. DREY, *School of Chemistry, Robert Gordon's Institute of Technology, Aberdeen,* and GARETH P. PRIESTLEY, *Department of Chemistry and Polymer Technology, Polytechnic of the South Bank, London.*

Synthesis of folate conjugates have been previously detailed.[1-10] These suffer severally from a number of practical difficulties such as low yields, protracted isolation and purification procedures and removal of protecting groups. Collective experience suggested that acidolytic or reductive methods of cleavage of protecting groups was a necessary requirement and that C-terminal elongation of peptide residues may well provide significant advantages in handling glutamic acid peptides. It would thus be possible to isolate and characterize intermediate peptide acids or their quarternary ammonium salts at each stage of peptide synthesis.

In a preliminary model synthesis incorporating the picolyl ester,[11] starting materials were prepared from *N*-benzyloxycarbonyl-γ-*t*-butyl-L-glutamate,[12,13] namely *N*-benzyloxycarbonyl-α-4-picolyl-L-glutamate and α-4-nitrobenzyl-γ-*t*-butyl-L-glutamate. Coupling of the latter using the carbonic mixed anhydride (CMA) procedure[14] afforded the fully protected dipeptide (*1*) in 81% (see Scheme 1). Treatment of (*1*) with trifluoroacetic acid furnished the free γ-carboxyl dipeptide (*2*) in 86%. At this point the peptide could be elongated as necessary through a peptide acid and α-4-nitrobenzyl-γ-*t*-butyl-L-glutamate. To assess the route the peptide was 'stopped' by coupling (*2*) and α,γ-di-4-nitrobenzyl-L-glutamate with the CMA method[14] to yield the protected tripeptide (*3*); subsequent deprotection yielded the free aminopeptide (*4*) dihydrobromide in good yield.

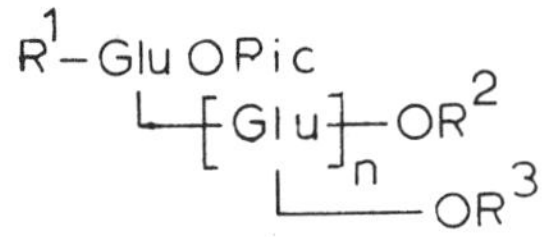

(*1*) R^1 = Z, R^2 = pNO$_2$Bzl,* R^3 = Bu^t, n = 1

(*2*) R^1 = Z, R^2 = pNO$_2$Bzl, R^3 = H, n = 1

(*3*) R^1 = Z, R^2 = R^3 = pNO$_2$Bzl, n = 2

(*4*) R^1 = H, R^2 = R^3 = pNO$_2$Bzl, n = 2

(*5*) R^1 = N^{10}-trifluoroacetylpteroyl, R^2 = R^3 = pNO$_2$Bzl, n = 2

* (pNO$_2$Bzl = 4-nitrobenzyl)

Scheme 1

We found that a critical step in the preparation of pure conjugates was the rigorous purification of pteroic acid.[15] Derivatization to yield *N*[10]-trifluoroacetylpteroic acid[3-7] then proceeded smoothly in high yield. Coupling of the free amino peptide (*4*) with the latter using the CMA method[14] in dimethylformamide[7], gave after washing and isolation the fully protected conjugate (*5*) in 55% yield.

Modification of the synthesis pioneered by Meienhofer and co-workers[7] incorporating methodology from the above route led to a successful combination as described below.

C-terminal peptide synthesis in conjunction with block coupling of intermediates (Scheme 2), furnished fully protected tri-(*6*), penta-(*7*) and hepta-L-glutamates (*8*).[7] The key intermediate *N*-benzyloxycarbonyl-α-*t*-butyl-L-glutamyl-α-*t*-butyl-L-glutamate[16] was prepared by salt coupling[10,16-18] using the CMA method.[7] Extension to the protected tripeptide (*6*)[7] then gave entry to a 2+3 and 2+5 'blocking', yielding the complete series of peptides.[7] Amino deprotection and coupling to *N*[10]-trifluoroacetylpteroic acid[3-7] furnished analytically pure conjugates in 69, 69, and 80% yield for the tri-(*9*), penta-(*10*), and hepta-pteroylglutamates[7] (*11*), respectively. Additional simplification of the route gave access to the partially protected tripeptide (*12*) *via* two successive C-terminal, salt coupling reactions.[10,16-18] Removal of the *N*-protecting group giving (*13*), was followed by coupling to *N*[10]-trifluoroacetylpteroic acid[3-7] yielding the pteroylglutamyl acid (*14*) in 71% yield. Complete deprotection[7] led directly without further purification to pure pteroyl triglutamate (*15*) in 70% recovery.[19]

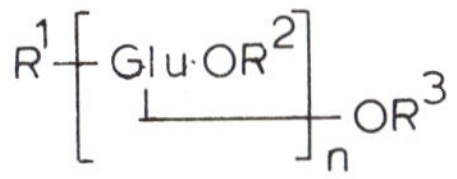

(*6*) R^1 = Z, $R^2 = R^3$ = Bu^t, $n = 3$

(*7*) R^1 = Z, $R^2 = R^3$ = Bu^t, $n = 5$

(*8*) R^1 = Z, $R^2 = R^3$ = Bu^t, $n = 7$

(*9*) R^1 = N^{10}-trifluoroacetylpteroyl, $R^2 = R^3$ = Bu^t, $n = 3$

(*10*) R^1 = N^{10}-trifluoroacetylpteroyl, $R^2 = R^3$ = Bu^t, $n = 5$

(*11*) R^1 = N^{10}-trifluoroacetylpteroyl, $R^2 = R^3$ = Bu^t, $n = 7$

(*12*) R^1 = Z, R^2 = Bu^t, R^3 = H, $n = 3$

(*13*) R^1 = H, R^2 = Bu^t, R^3 = H, $n = 3$

(*14*) R^1 = N^{10}-trifluoroacetylpteroyl, R^2 = Bu^t, R^3 = H, $n = 3$

(*15*) R^1 = pteroyl, $R^2 = R^3$ = H, $n = 3$

Scheme 2

Pteroylglutamates may conveniently be prepared by C-terminal peptide coupling reactions in conjunction with rigorously purified pteroic acid. Simplification of the method by a minimal-protection salt-coupling strategy led to the rapid preparation of pteroyl tri-γ-L-glutamic acid in high yield.

One of us (G. P. P.) wishes to thank the Wellcome Trust for a grant, and in addition we thank Dr. S. W. Wilkinson for helpful discussions.

References

1. Boothe, J. H., Mowat, J. H., Hutchings, B. L., Angier, R. B., Waller, C. W., Stokstad, E. L. R., Semb, J., Gazzola, A. L. & SubbaRow, Y. (1948) *J. Amer. Chem. Soc.* **70**, 1099-1102.
2. Boothe, J. H., Semb, J., Waller, C. W., Angier, R. B., Mowat, J. H., Hutchings, B. L., Stokstad, E. L. R. & SubbaRow, Y. (1949) *J. Amer. Chem. Soc.* 71, 2304-2308.
3. Krumdiek, C. L. & Baugh, C. M. (1969) *Biochemistry* 8, 1568-1572.
4. Nair, M. G. & Baugh, C. M. (1973) *Biochemistry* 12, 3923-3927.
5. Baugh, C. M., Stevens, J. C. & Krumdieck, C. L. (1970) *Biochim. Biophys. Acta* **212**, 116-125.
6. Kisliuk, R. L., Gaumont, Y. & Baugh, C. M. (1974) *J. Biol. Chem.* **249**, 4100-4103.
7. Godwin, H. A., Rosenberg, I. H., Ferenz, C. R., Jacobs, P. M. & Meienhofer, J. (1972) *J. Biol. Chem.* 247, 2266-2271.
8. Bieri, J. H. & Viscontini, M. (1973), *Helv. Chim. Acta* 56, 2905-2911.
9. Khalifa, E., Bieri, J. H. & Viscontini, M. (1973) *Helv. Chim. Acta* 56, 2911-2919.
10. Sengupta, P. K., Bieri, J. H. & Viscontini, M. (1975) *Helv. Chim. Acta,* 58, 1374-1379.
11. Camble, R., Garner, R. & Young, G. T. (1969) *J. Chem. Soc.* (C), 1911-1916.
12. Itoh, M. (1969), *Chem. Pharm. Bull. (Japan)* 17, 1679-1686.
13. Klieger, E. & Gibian, H. (1962) *Annalen* 655, 195-210.
14. Anderson, G. W., Zimmerman, J. E. & Callahan, F. M. (1967) *J. Amer. Chem. Soc.* 89, 5012-5017.
15. Houlihan, C. M., Boyle, P. H. & Scott, J. H. (1972) *Analyt. Biochem.* 46, 1-6.
16. Hollosi, M., Kajtar, M. & Bruckner, V. (1969) *Acta Chem. Acad. Sci. Hung.* 62, 305-319.
17. Sachs, H. & Brand, E. (1954) *J. Amer. Chem. Soc.* 76, 1811-1814.
18. Burkhardt, H. & Mitchell, H. K. (1961) *Arch. Biochem. Biophys.* 94, 32-34.
19. Drey, C. N. C. & Priestley, G. P. (1977) *J. Chem. Soc. Chem. Commun.* 144-145.

AN IMPROVED LEUCINE ENKEPHALIN SYNTHESIS

GERALD SIGLER, WILLIAM FULLER and RAYMOND VAVREK, *Bio-Organic Department, Calbiochem, 10933 North Torrey Pines Road, La Jolla, California 92037*

Since the discovery and structure elucidation in 1975 of the endogenous opiate peptides, leucine- and methionine-enkephalin,[1] the demand for research quantities of these substances has risen sharply. As a result, we became interested in developing a rapid solution method for their synthesis which could be adapted to large scale production.

In order to minimize the number and the complexity of the protected amino-acid derivatives which would be employed as starting materials, a strategy of minimum protection was chosen. With leucine-enkephalin as the prototype, two alternate routes were devised and carried out (Schemes 1 & 2). The key features of these syntheses are: 1) the use of salt couplings to obviate the need for ester protection, 2) the synthesis of the C-terminal sequence via the *N*-carboxyanhydride of phenylalanine, 3) the application of a mixed anhydride fragment coupling to obtain the *N*-protected pentapeptide, and 4) the incorporation of tyrosine with the phenolic-OH unprotected.

The dipeptide, Boc-Tyr-Gly, was prepared by coupling Boc-Tyr-OSu with sodium glycinate in a mixed-solvent system of THF/aqueous $NaHCO_3$. The coupling reaction proceeded smoothly on a 0.3 molar scale; over 85% yield of the dipeptide was obtained on acidification of the concentrated reaction mixture.

Crystalline phenylalanine-*N*-carboxyanhydride (Phe-NCA) was prepared in large quantities by the procedure of Fuller, et al.[2] The coupling of Phe-NCA to Leu was carried out in a biphasic solvent system of acetonitrile/0.2 molar Na_2CO_3 (60:40) at -10°C.[3,4] As was found with the synthesis of other simple oligopeptides,[4] these conditions gave negligible amounts of polymerized by-products of the type $(Phe)_x$-Leu. The major side product of the reaction was

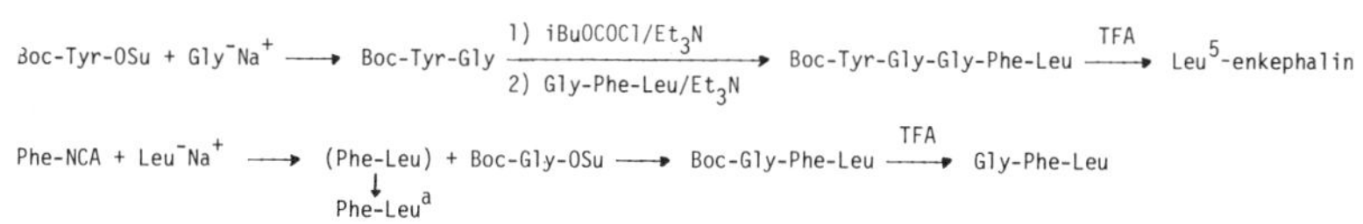

Scheme 1. 2 + 3 fragment coupling.

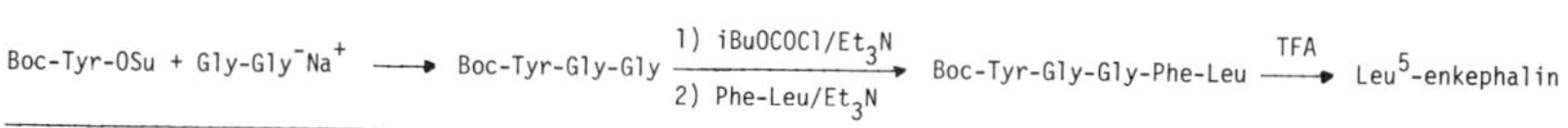

a Isolated for use in Scheme 2

Scheme 2. 3 + 2 fragment coupling.

free Phe, resulting from the hydrolysis of the unreacted Phe-NCA, which due to technical difficulties, could not be removed from the stable emulsion formed during the reaction. After adjusting the reaction mixture to pH 8.0 with HCl, Boc-Gly-OSu was added at room temperature. In spite of the rapid hydrolysis of the active ester in the reaction medium, which gave Boc-Gly as the major contaminant in the product, it was possible to obtain an overall 55% yield of Boc-Gly-Phe-Leu by the selective crystallization of the protected tripeptide from ethyl acetate/hexane on a 10 mmolar scale. However, an attempt to scale up the reaction to 0.2 molar resulted in a much lower yield of the tripeptide. The Boc-tripeptide was deprotected with TFA, neutralized, and the tripeptide was crystallized from 50% 2-propanol.

For the 2 + 3 fragment coupling, the mixed anhydride of Boc-Tyr-Gly with isobutylchloroformate was reacted with the triethylammonium salt of Gly-Phe-Leu in a mixture of THF and DMF. The Boc-pentapeptide was separated from the excess of Boc-Tyr-Gly by several precipitations from methanol with water. Attempts to purify further the Boc-Leu^5-enkephalin by crystallization were largely unsuccessful. However, after deprotection with TFA, a five-gram sample of the leucine-enkephalin salt was easily purified by preparative high-pressure liquid chromatography (Prep LC/Waters Associates) on silica gel columns with a chloroform/methanol/water system. The yield of purified leucine enkephalin acetate, based on the amount of Gly-Phe-Leu used in the fragment coupling, was 40%.

Because of the difficulties in scaling-up the preparation of the tripeptide Gly-Phe-Leu by the *in situ* NCA procedure, an alternate synthesis of leucine-enkephalin was pursued in which the salt coupling with Boc-Gly-OSu was eliminated (Scheme 2). Using essentially the same procedure as for the synthesis of Boc-Tyr-Gly, Boc-Tyr-OSu was coupled to the sodium salt of Gly-Gly to give a 70% yield of pure Boc-Tyr-Gly-Gly as an oil. The dipeptide, Phe-Leu, was isolated in 45% yield from the reaction of sodium leucinate with Phe-NCA on a 0.2 mole scale. The mixed anhydride of Boc-Tyr-Gly-Gly was then coupled with the triethylammonium salt of Phe-Leu in a mixture of THF and DMF. The crude product contained large amounts of Boc-Tyr-Gly-Gly and several minor Pauly-positive side products which moved faster on TLC than the Boc-pentapeptide. Preparative-high-pressure-liquid chromatography on 20 grams of the crude material with $CHCl_3$-MeOH (90:10) was successful in removing Boc-Tyr-Gly-Gly from the mixture, but only partially separated the other minor protected impurities.

In conclusion, two alternate syntheses of leucine-enkephalin have been carried out using salt couplings only. However, neither route has been optimized on a large scale.

The main advantages of our approach were found to be: 1) the ease of preparation of the acylating components, 2) the rapidity of the coupling methods and the generally facile work-up procedures, and 3) the lack of any deprotection steps other than removal of the Boc-group.

On the other hand, the major difficulties with the method which complicated scale-up procedures were: 1) The low yields in the *in situ* reaction of Boc-Gly-OSu with the salt of the C-terminal dipeptide due to hydrolysis of the active ester, 2) the lack of crystallinity of the two intermediates Boc-Tyr-Gly-Gly and Boc-Tyr-Gly-Gly-Phe-Leu, and 3) a mixture of tyrosine-containing side-products obtained from the mixed anhydride salt coupling which were difficult to separate from the desired Boc-pentapeptide. However, deprotection of the crude Boc-Leu^5-enkephalin gave a mixture of products which was easily purified by high pressure liquid chromatography.

References

1. Hughes, J., Smith, T. W., Kosterlitz, H. W., Fothergill, L. A., Morgan, B. A. & Morris, H. R. (1975) *Nature* **258**, 577–579.

2. Fuller, W. D., Verlander, M. S. & Goodman, M. (1976) *Biopolymers* **15**, 1869–1871.

3. Katakai, R., Oya, M., Toda, F., Uno, K. & Iwakura, Y. (1973) *Macromolecules* **6**, 827–831.

4. Iwakura, Y., Uno, K., Oya, M. & Katakai, R. (1970) *Biopolymers* **9**, 1419–1427.

PREPARATION AND CONFORMATIONAL ANALYSIS OF GALLINE AND CLUPEINES

F. MARCHIORI, G. BORIN, B. FILIPPI, G. M. BONORA, and C. TONIOLO, *Biopolymer Research Center, C. N. R., Institute of Organic Chemistry, University of Padova, 35100 Padova, Italy.*

Protamines are present in animal mature sperm cell nuclei as basic proteins of relatively low molecular weight (4,000-9,000) associated with DNA.[1] They contain in a linear polypeptide chain 30 – 70 amino acid residues, 70% of which are arginines that are inserted in the sequence as clusters of 2-7 residues separated by blocks of 2-4 neutral amino acids. The significance of this unique structural feature is currently not understood, although it is reasonable to believe that it is related to their property of forming specific complexes with DNA where various types of interaction other than electrostatic are operative. We intend to prepare protamines, their fragments and selectively modified analogs thereof by synthesis or chromatographic purification of commercial products followed by chemical or enzymatic cleavage, and to investigate their conformational preferences and the stability of their complexes with nucleotides and DNA using various spectroscopic techniques. We first focused our attention on galline (from domestic fowl)[2] and clupeines (from herring).[1] In this communication we describe the synthesis of the octadecapeptide C-terminal fragment of galline (G-I) and the preliminary results of a conformational analysis of clupeines.

For protamines, the major synthetic problem arises from the presence of long sequences of arginyl residues. To overcome this difficulty, our synthetic approach is based on the substitution of all arginines by ornithines which can be transformed into arginines by amidination. To make this approach reliable, the amidination reaction was studied using ornithyl peptides of different length prepared as an intermediate for the synthesis of the C-terminal sequence 48-65 of galline,[2] previously reported as a single molecular species:[3] H-Ser-Gly-Gly-Val-Arg_4-Tyr-Gly-Ser-Arg_6-Tyr-OH.

The entire sequence was synthesized following the schemes shown in Figures 1 and 2. The protected peptides A, B and C were transformed into the corresponding hydrazides and successively coupled by the azide procedure starting from fragment D. All compounds were obtained in pure form and fully characterized. The peptide intermediates were deblocked by hydrogenolysis followed by treatment with 90% trifluoroacetic acid and amidinated using 3,5-dimethylpyrazole-1-carboxamidine nitrate at pH 9.3 (48 h at 37°C).[4] The reaction mixture was purified on a G-10 Sephadex column equilibrated and eluted with 5% acetic acid. The Sakaguchi and ninhydrin positive fractions were collected and lyophilized. The amino-acid composition of the products, both in the acid hydrolyzates and in the aminopeptidase M digests, show the complete

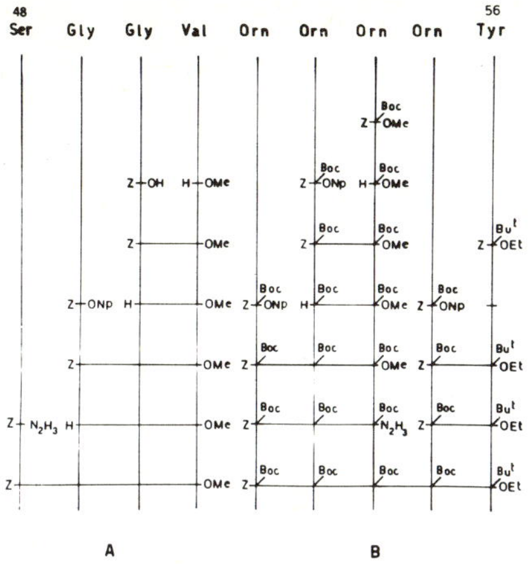

Fig. 1. Scheme for the synthesis of subfragments A (sequence 48-51) and B (sequence 52-56).

transformation of ornithine into arginine residues. When ornithine is present as the N-terminal residue, the reaction is selective for the δ-amino group of the side chain; when glycine is present as the N-terminal residue, the reaction occurs to some extent at the α-amino group as well. The results obtained show that the proposed approach is very promising for the synthesis of protamines.

Because of the still unclarified possible significance of heterogeneity in clupeines, the three major components, YI, YII, and Z (Scheme 1), were separated by chromatography on a column of CM-52 carboxymethylcellulose. The homogeneity of the various fractions was checked by analytical rechromatography

YI: H-Ala·AAAA·Ser·Ser·Ser·A·Pro·Ile·AAAA·Pro·AAA·Thr·Thr·AAAA·Ala·Gly·AAAA-OH (residues 1–31; 1, 5, 10, 15, 20, 25, 31 marked)

YII: H-Pro·AAA·Thr·AA·Ala·Ser·A·Pro·Val·AAAA·Pro·AA·Val·Ser·AAAA·Ala·AAAA-OH (residues 1–30; 1, 30 marked)

Z: H-Ala·AAAA·Ser·AA·Ala·Ser·A·Pro·Val·AAAA·Pro·AA·Val·Ser·AAAA·Ala·AAAA-OH (residues 1–31; 1, 5, 10, 15, 20, 25, 31 marked)

Scheme 1. The primary sequence[1] of clupeines YI, YII, and Z. Arginine is represented by A.

on the same column, amino-acid analysis, and determination of N-terminal amino acid.

A CD investigation of the three chromatographically pure clupeines was carried out as a function of pH, solvent polarity, temperature, added salts, and structure-disrupting agents. In 2-chloroethanol the three proteins assume the right-handed α-helical form in considerable amount (40 – 50%). Addition of water to the alcoholic solution destroys this ordered structure. The differences in the extent of formation and the stability of the α-helical conformation among the three clupeines appear to be minor. The CD spectra of clupeine Z at neutral

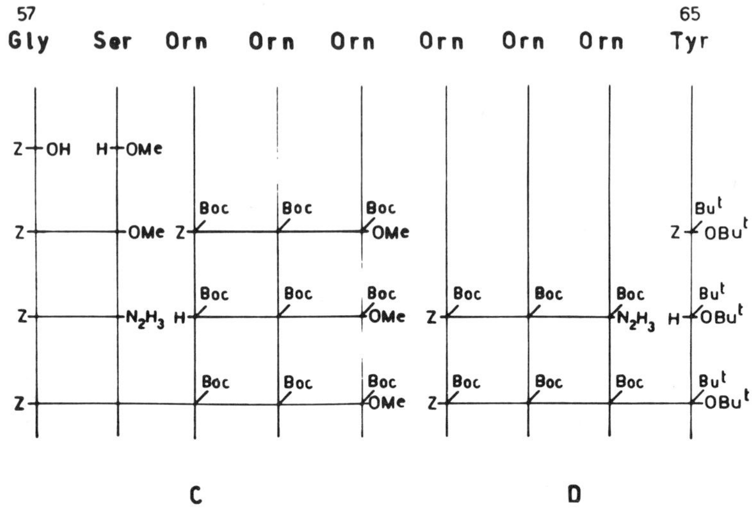

Fig. 2. Scheme for the synthesis of subfragments C (sequence 57-61) and D (sequence 62-65).

pH and in 2-chloroethanol are shown in Figure 3. In the 0.1 -4.0 $\times$ $10^{-1}M$ range, in contrast to components YI and Z, the YII component does not seem to be

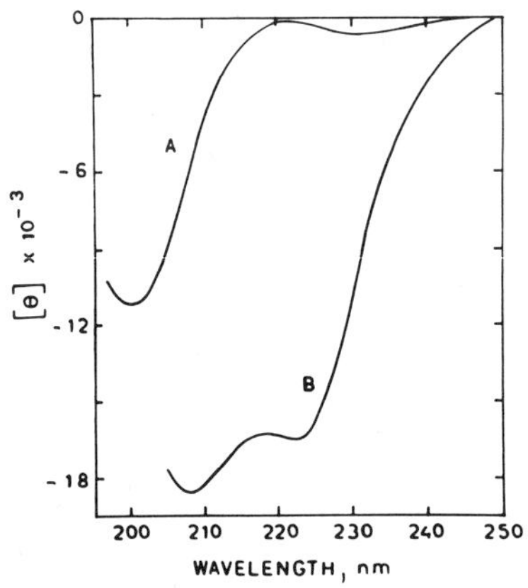

Fig. 3. Circular dichroism spectra of clupeine Z in water (pH 6.0) (A) and in 2-chloroethanol (B) at 23°C.

sensitive to the addition of phosphate mono- and di-anions. The possible implications of this finding to the interaction of clupeines with nucleotides and DNA are under investigation.

References

1. Ando, T., Yamasaki, M. & Suzuki, K. (1973) *Protamines: Isolation, Characterization, Structure and Function,* Molecular Biology, Biochemistry, and Biophysics Series, no. 12, Springer, Verlag, Berlin.
2. Nakano, M., Tobita, T. & Ando, T. (1976) *Int. J. Pept. Protein Res.* 8, 565–578.
3. Nakano, M., Tobita, T. & Ando, T. (1975) *Int. J. Pept. Protein Res.* 7, 31–46.
4. Ariely, S., Wilchek, M. & Patchornik, A. (1966) *Biopolymers* 4 91–96.

SYNTHESIS OF GLYCOTRIPEPTIDES CONTAINING L-SERINE AND 2-ACETAMIDO-2-DEOXY-β-D-GLUCOPYRANOSYL RESIDUES

HARI G. GARG and ROGER W. JEANLOZ, *Departments of Biological Chemistry and Medicine, Harvard Medical School, and Massachusetts General Hospital, Boston, Massachusetts 02114*

Synthetic model compounds containing 2-acetamido-2-deoxy-D-glucopyranosyl L-asparagine and L-serine residues are of interest for elucidating the significance of the L-serine (or L-threonine) residue located one amino-acid residue away, in the direction of the C-terminal, from the L-asparagine residue involved in the *N*-(L-aspart-4-oyl)-2-deoxy-β-D-glucopyranosylamine carbohydrate-protein linkage of *N*-glycoproteins,[1] and for the study of the base-catalyzed β-elimination of carbohydrate chains in *O*-glycoproteins, such as submaximillary mucins, blood group substances, and proteoglycans.[2]

Compound *1* was obtained by condensation of 2-acetamido-3,4,6-tri-*O*-acetyl-*N*-[*N*-(benzyloxycarbonyl)-L-aspart-4-oyl]-2-deoxy-β-D-glucopyranosylamine[3] with glycyl-L-serine methyl ester [obtained by treatment of the *N*-benzyloxycarbonyl derivative[4] with 30% hydrogen bromide in acetic acid, followed by saponification with triethylamine] in the presence of Woodward reagent K under the conditions described earlier.[5]

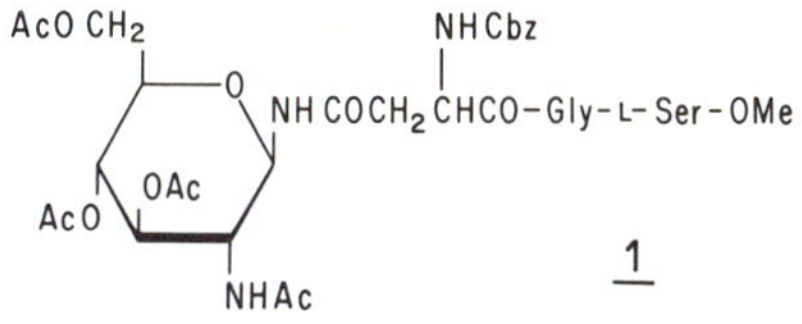

1

Compound *3* was obtained from *O*-(2-acetamido-3,4,6-tri-*O*-acetyl-2-deoxy-β-D-glucopyranosylamine)-*N*-(benzyloxycarbonyl)-L-serine methyl ester,[6] by removal of the benzyloxycarbonyl group, condensation with *N*-(benzyloxycarbonyl)-glycine *p*-nitrophenyl ester,[7] to give *2*, removal of the benzyloxycarbonyl group, and condensation with *N*-(benzyloxycarbonyl)-L-asparagine *p*-nitrophenyl ester.[8]

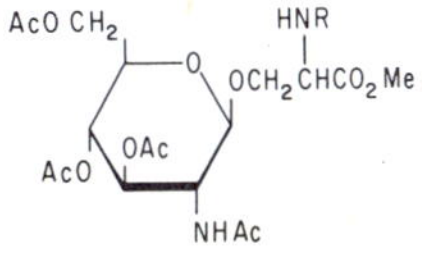

2 R = Cbz-Gly
3 R = Cbz-L-Asn-Gly

Finally *7, 8, 9,* obtained by catalytic hydrogenation of the *N*-benzyloxycarbonyl derivatives[9] *4, 5,* and *6*, respectively, were *O*-deacetylated with triethylamine to give *10, 11,* and *12*, respectively, see Scheme 1. The physical properties of the new glycopeptides are reported in Table I.

Table I. Physical Characteristics of the Glycopeptides[a]

Compound	M.p. (°)[b]	$[\alpha]_D^{21}$ [c]	t.l.c. (R_f)[d]
2-Acetamido-3,4,6-tri-*O*-acetyl-*N*-[*N*-(benzyloxycarbonyl)-L-aspart-1-oyl-(glycyl-L-serine methyl ester)-4-oyl]-2-deoxy-β-D-glucopyranosylamine (1)	225--228 (MeOH)	+57.7 (1.1, C)	0.65 (A)
O-(2-Acetamido-3,4,6-tri-*O*-acetyl-2-deoxy-β-D-glucopyranosyl)-L-serine derivative of			
N-(Benzyloxycarbonyl)-glycyl-L-serine methyl ester (2)	168--169 (EtOH)	+31.5 (1.7, C)	0.64 (B)
N-(Benzyloxycarbonyl)-L-asparaginyl-glycyl-L-serine methyl ester (3)	188--190 (dec, EtOH)	-11.8 (0.25, F)	0.38 (A)
L-Seryl-L-alanyl-L-alanine methyl ester (7)	192--193 ($MeOH/Et_2O$)	-21.0 (1.1, F)	0.72 (C)
L-Seryl-L-alanyl-glycine ethyl ester (8)	173-174 (dec) ($EtOH/Et_2O$)	-14.5 (0.6, F)	0.70 (C)
L-Seryl-glycyl-L-alanine methyl ester (9)	94--96 (2-PrOH)	-21.3 (1.7, F)	0.56 (C)
O-(2-Acetamido-2-deoxy-β-D-glycopyranosyl)-L-serine derivative of			
L-Seryl-L-alanyl-L-alanine methyl ester (10)	195--196 (dec, EtOH)	-32.0 (1.1, F)	0.16 (D)
L-Seryl-L-alanyl-glycine ethyl ester (11)	173--174 ($EtOH/H_2O$)	-25.2 (1.3, F)	0.12 (D)
L-Seryl-glycyl-L-alanine methyl ester (12)	145--146 (2-PrOH)	-26.6 (1.7, F)	0.19 (C)

[a] For methods, see Ref. 9.

[b] In parentheses, solvent of crystallization.

[c] In parentheses, concentration (*c*) and solvent: C, $CHCl_3$; F, $HCONMe_2$.

[d] In parentheses, solvent (v/v): A (14:1, $CHCl_3/MeOH$); B (19:1, $CHCl_3/EtOH$); C (7:3, $CHCl_3/MeOH$); D (1:1, $CHCl_3/MeOH$).

Treatment of *4, 5, 6, 7, 8,* and *9* with 2 *M* sodium borohydride in a 0.5 *M* sodium hydroxide solution for 20 h at 40°C almost completely cleaved glycopeptides *4, 5,* and *6*, whereas *7, 8,* and *9* were unaffected.

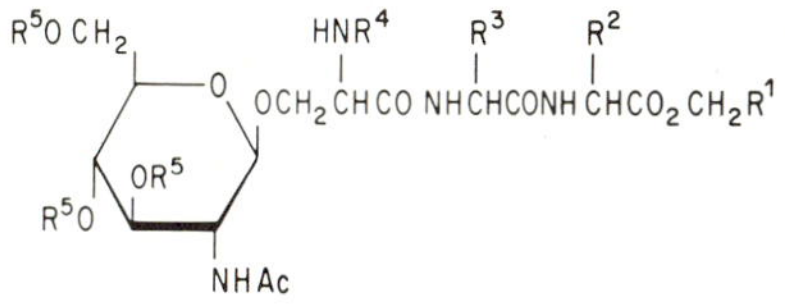

	R¹	R²	R³	R⁴	R⁵
4	H	Me	Me	Cbz	Ac
5	Me	H	Me	Cbz	Ac
6	H	Me	H	Cbz	Ac
7	H	Me	Me	H	Ac
8	Me	H	Me	H	Ac
9	H	Me	H	H	Ac
10	H	Me	Me	H	H
11	Me	H	Me	H	H
12	H	Me	H	H	H

Scheme 1.

Attempts to transfer the glycosyl residue from the L-seryl residue of *3* to the L-asparagine residue to give *1*, under the conditions (treatment with stannic chloride) described for O → N migration of the glycosyl residue in pyrimidine nucleoside derivatives,[10] were unsuccessful.

This is publication No. 732 of the Robert W. Lovett Memorial Group for the Study of Diseases Causing Deformities, Harvard Medical School at Massachusetts General Hospital, Boston MA 02114. The work was supported by a USPHS Research Grant (AM-03564).

References

1. Marshall, R. D. & Neuberger, A. (1972) in *Glycoproteins,* Gottschalk, A., Ed., Elsevier, Amsterdam, pp. 453–470.
2. Gottschalk, A. (1972) in *Glycoproteins,* Gottschalk, A., Ed., Elsevier, Amsterdam, pp. 470–476.
3. Garg, H. G. & Jeanloz, R. W. (1972) *Carbohydr. Res.* **23**, 437–439.
4. Harris, J. I. & Fruton, J. S. (1951) *J. Biol. Chem.* **191**, 143–151.
5. Garg, H. G. & Jeanloz, R. W. (1974) *Carbohydr. Res.* **32**, 37–46.
6. Garg, H. G. & Jeanloz, R. W. (1976) *Carbohydr. Res.* **49**, 482–488.
7. Bodanszky, M. & Du Vigneaud, V. (1962) *Biochem. Prep.* **9**, 110–112.
8. Bodanszky, M. & Du Vigneaud, V. (1959) *J. Amer. Chem. Soc.* **81**, 5688–5691.
9. Garg, H. G. & Jeanloz, R. W. (1975) in *Peptides: Chemistry, Structure and Biology,* Walter, R. & Meienhofer, J., Eds., Ann Arbor Science, Ann Arbor, Michigan, pp. 379–384.
10. Thacker, D. & Ulbricht, T. L. V. (1968) *J. Chem. Soc., C,* 333–337.

PRODUCTION OF SEMISYNTHETIC MYOGLOBIN BY STEPWISE FRAGMENT CONDENSATION.

FRANK R. N. GURD, WILLIAM H. GARNER, RICHARD D. DIMARCHI and CHI-CHIN WANG, *Department of Chemistry, Indiana University, Bloomington, IN 47401*

The specific incorporation of alternative amino acids or isotopically enriched amino acids into proteins is required for the detailed examination of structure-function relationships. Semisynthetic approaches depend on cleavage of the naturally occurring parent protein to obtain peptide segments that can be altered or substituted with isotopically enriched amino acids and then used for coupling to obtain a reconstructed protein incorporating both the natural peptide segment, or segments, and the artificially introduced residues. In general the methods require reversible intermediate protection of some functional groups, cleavage of the polypeptide chain at appropriate places, activation and coupling of the isolated fragments with modified natural fragments or with synthetic peptides, and deprotection[1,2].

This progress report deals with semisynthetic studies on myoglobin in which the whole protein or a defined fragment of it is isolated with a free terminal α-amino group and with all ε-amino groups protected by acetimidation[3]. The free α-amino group is used as the site of coupling with a carboxyl-activated, suitably protected amino acid or peptide to produce the desired sequence in a protected form which can then be reconverted to the simple polypeptide form by deprotection. Two applications are described here. The first is the specific coupling of a glycine residue enriched in ^{13}C at the amino terminus of the intact protein, an application which tests all stages except cleavage. The second application involves specific chemical cleavage of the amino-protected myoglobin to yield the large fragment comprising residues 15–153, followed by recoupling in stages to rebuild the segment containing the first 14 amino acids.

Preparation of Amino-Protected Protein. Acetimidation of 20g purified sperm whale myoglobin dissolved in water (30 mg/ml) was accomplished with 120g methylacetimidate at 15°C for 1 hour at pH 9.8[3]. The initial separation of the protein reaction products was achieved on DEAE A-50 Sephadex at pH 8.9 and 4°C. Each fraction was purified on Sephadex CM C-50 pH 6.5. The first component was acetimidated at all 19 lysine groups as well as the α-amino group and is suitable for the cleavage treatment to prepare fragment 15–153. The second component could be separated into two fractions, one of which bears the blocking groups on all 19 lysine residues but has the α-amino group free. This material is suitable for the preparation of the glycine adduct described below. The acetimidation procedure has the advantage that the charge state of the amino groups is retained which appears responsible for the retention of water solubility. The acetimidation procedure yields more free α-amino product at higher pH (10.5) but at the expense of deamidation as shown by isolation of later chromatographic products followed by titration, electrophoresis and isolation of peptide fragments. The purity of the lysine-protected form with the α-amino group free was assured by coupling with tetrahydrophthalic anhydride, chromatography, and removal of the tetrahydrophthaloyl group at pH 6.0 in 30 hours at 22°C.

Preparation and Properties of the N^{α}-Glycylmyoglobin. A 100-fold excess of *N*-trifluoroacetyl ^{13}C-glycine *N*-hydroxysuccinimide ester was mixed with the myoglobin derivative containing the free α-amino group, in water at pH 6.8 at 15°C for 4 hours. The protected adduct was isolated from the unreacted protein by Sephadex C-50 chromatography, and the protective groups were removed in concentrated ammonium hydroxide/acetic acid (15:1, v/v) for 28 hours at 16°C at an apparent pH of 11.5, with only minimal deamidation. The final product was isolated chromatographically, shown to be homogeneous electrophoretically, and the presence of the glycine amino-terminal residue confirmed by automated Edman degradation. Overall yield was 3%. The cyano- or aquo-ferrimyoglobin form was studied by physical techniques. Absorbance spectra, circular dichroism spectra and isoionic point were not distinguishable from those of the unmodified protein controls[5].

In separate experiments both carbons of the glycine adduct were enriched with respect to ^{13}C. The product enriched in C^{α} is illustrated in Fig. 1, with the enriched site resonance identified by the numeral 2. From the pH dependence

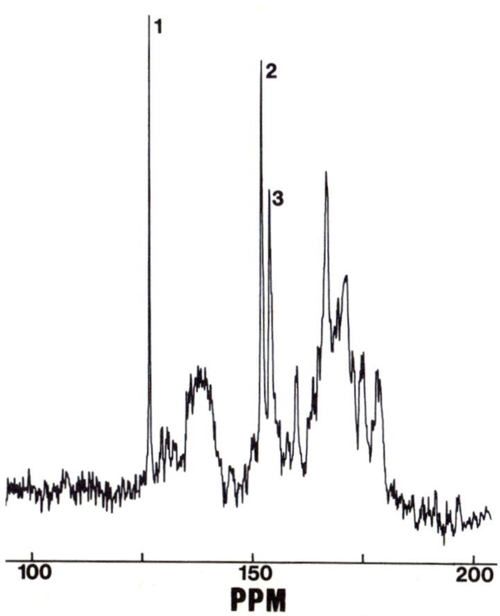

Fig. 1. Proton-decoupled ^{13}C spectrum at 67.9 MHz of N^{α}-[2-^{13}C]-glycylmyoglobin. The single adduct peak at pH 7.7 is marked 2 with the ε carbons of lysine numbered 3. Chemical shifts are expressed upfield of external CS_2, with internal dioxane taken as 126.3 ppm (peak 1).

of the chemical shift the pK of the α-amino group was found to be 7.75 to 7.80. T_1 values were determined at three magnetic field strengths. NT_1 values were fit to a stochastic diffusion model for internal rotational motion in one degree of freedom of a group attached to a relatively rigid myoglobin molecule reorienting as a sphere with an overall correlation time, τ_R, of 17 nanoseconds. The internal correlation time values, τ_G, are listed in Table I. They show definite drift with field strength but unequivocally demonstrate that there is a high degree of rotational freedom, especially in comparison with the mass of the C^α resonances in myoglobin[6]. The trend towards shorter internal correlation times with higher pH has also been observed in peptides, in particular a pentapeptide with amino-terminal glycine whose rotational motions have been analyzed by a combination of ^{13}C and ^{2}H relaxation measurements[7].

Replacement of N-Terminal Sequence 1–14. Des-(1–14)-myoglobin was prepared from the completely amino-blocked form described above, following the removal of the heme, by treatment with a 20-fold molar excess of 2-(2-nitrophenylsulphenyl)-3-bromo-3′-methyl indolenine in 67% acetic acid at 4°C for 24 hr[8] with the inclusion of a 50-fold molar excess of phenol. Addition of 2% (v/v) β-mercaptoethanol stopped the reaction and reduced any methionine sulfoxide. The yield was 55-65%. The product was purified by gel filtration and characterized. The rebuilding[4] was done by coupling Trp^{14} as a single residue, followed by sequence 6–13 and finally 1–5 through the azides. Secondary protective groups were required during solid-phase synthesis and coupling. The overall yield in the reconstruction steps was near 3%. The properties of the reconstituted material are compared with the virgin protein in Table II.

Future Objectives. Although quite compact, myoglobin exhibits marked internal mobility.[5] The N-terminal portion appears to be relatively sensitive to denaturation[9], and hence substitutions in the region of the molecule dealt with here should produce information about stability and the functional importance of internal motions. The two approaches described are being extended by stepwise degradation of the N-terminal sequence followed by rebuilding, and by coupling of sequences corresponding to residues 1–13 as the synthetic tridecapeptide,

Table I. Nmr Relaxation Times, T_1, and Internal Rotational Correlation Times, τ_G, for N^α-Glycylmyoglobin

	14.1kG		23.5kG		63.5kG	
pH	NT_1 ms	τ_G ps	NT_1 ms	τ_G ps	NT_1 ms	τ_G ps
7.0	96	124	154	92	343	55
9.5	133	67	249	43	440	38

Table II. Spectral Properties of Semisynthetic Myoglobin Derivatives

	Absorbance Ratio A_{409}/A_{280}	[θ] ($deg \cdot cm^2/dmole$) 208 nm	222 nm
Virgin Aquoferrimyoglobin	5.35	-21,800	-24,900
N^{α}-Glycyl Myoglobin	5.26	-21,022	-24,011
Des-(1-14)-Myoglobin	1.47	-13,908	-10,418
Reconstituted Myoglobin	5.36	-20,500	-20,200

appropriately protected, to the large fragment 14-153 obtained by extension of the isolated fragment 15-153.

Supported by PHS Grants HL-05556 and -14680. This is the 85th paper in a series, following Ref. 7.

References

1. Rees, A. R. & Offord, R. E. (1976) *Biochem. J.* **149**, 487–493.
2. Dun, B. M., DiBello, C., Kirk, K. L., Cohen, L. A. & Chaiken, I. M. (1974) *J. Biol. Chem.* **249**, 6295–6301.
3. Garner, W. H. & Gurd, F. R. N. (1975) *Biochem. Biophys. Res. Commun.* **63**, 262–268.
4. Wang, C. C., Garner, W. H. & Gurd, F. R. N. (1977) *Fed. Proc.* **36**, 890.
5. Jones, W. C., Jr., Rothgeb, T. M. & Gurd, F. R. N. (1986) *J. Biol. Chem.* **251**, 7452–7460.
6. Visscher, R. B. & Gurd, F. R. N. (1975) *J. Biol. Chem.* **250**, 2238–2242.
7. Neireiter, G. W., Garner, W. H., Buchak, B., Clouse, A. O., Addleman, R. E. & Gurd, F. R. N. (1977) *J. Biol. Chem.* (submitted).
8. Fontana, A. (1972) *Methods Enzymol.* **25**, 419–423.
9. Friend, S. H., Hanania, G. I. H. & Gurd, R. S. (1977) *Fed. Proc.* **36**, 890.

THE PRESENT STATUS OF PEPTIDE SYNTHESIS BY FOUR-COMPONENT CONDENSATION AND RELATED CHEMISTRY

IVAR UGI, GERHARD EBERLE, HEINER ECKERT, INGER LAGERLUND, DIETER MARQUARDING, GISELHER SKORNA, REINHARD URBAN, LORENZ WACKERLE, and HASSO v. ZYCHLINSKI, *Organisch-Chemisches Institut der Technischen Universität, 8 München 2, Arcisstr. 21, Germany*

Presently, Scheme 1 represents the most effective way to synthesize peptide fragments by stereoselective four-component condensations[1-3] (4CC). Here (*1*) is an amino acid or peptide residue (P_1) with an N-terminal blocking group B_N, and a free carboxyl group, and (*4*) is an α-isocyano acid derivative, mostly an isocyano derivative of a dipeptide residue (P_2) with a C-terminal blocking group[4,5] B_C. Under suitable conditions the chiral amine component (*2*) has very strong asymmetric inducing power, and affords generally in this 4CC a highly preferential synthesis of the *n*-isomers (*n* = R,S or S,R; *p* = R,R or S,S) of (*5*). In favorable cases an isomer ratio *n:p* = 200:1 has been achieved.[6-8] In the presence of (*6*) the acidolysis of (*5*) yields the peptide derivatives (*7*) and (*8*) from which the chiral template (*2*) can be resynthesized. Since the acidolysis of (*5n*) proceeds much slower than that of (*5p*), it is possible to obtain (*5n*) in a very high degree of isomer purity through partial acidolysis of the 4CC products (*5n*) and (*5p*), and isomerically pure (*7*) is obtained after total acidolysis. Here the exponential stereoselectivity effect of independent parallel reactions[9] is utilized. Thus, >99.98% pure *N*-formyl-tetravaline-methylester has been prepared by 4CC and selective cleavage according to Scheme 1.[6,7]

It has been observed in several cases that the 4CC products (*5*) are considerably more soluble in organic solvents (e.g. *n*-pentane, dichloromethane, methanol) than the corresponding peptide derivatives (*7*). This may serve as a basis for circumventing solubility problems in peptide syntheses.

The chiral amine component (*2*) of Scheme 1 is best synthesized according to Scheme 2. The conversion of optically active (*11*) into (*2*) is less convenient via

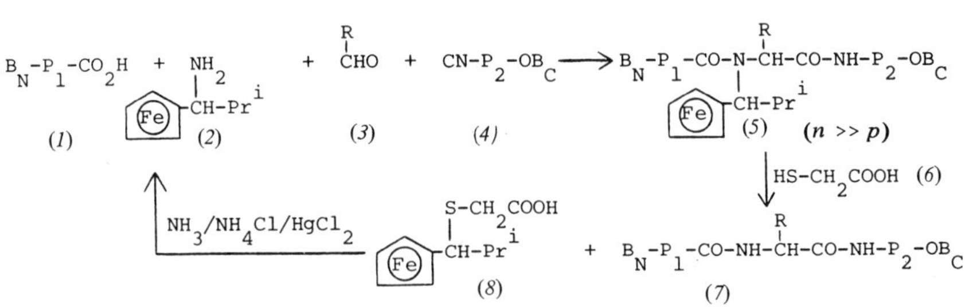

Scheme 1

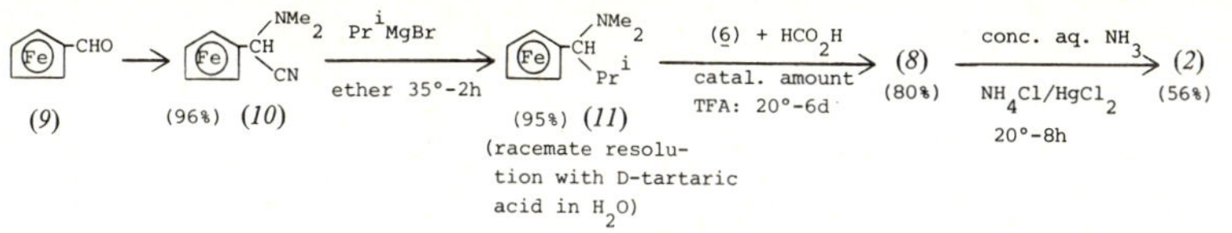

Scheme 2

the azide which is obtained by treating (*11*) with methyl iodide in the presence of sodium azide (monoglyme/water; 65h at 40°C; 65%), and subsequently reduced to (*2*) with $Na_2[AlH_2(OCH_2CH_2OCH_3)_2]$ (benzene; 2h at 30°C, 81%).[10]

The combination of peptide fragments (*12*) and (*13*) according to Scheme 3[8,11] has been improved since the last report.[8] In particular, the acidolysis (*16*) → (*17*) can now be carried out in one step, and in almost quantitative yield, instead of the less effective two steps which were needed in our previous procedure.[8] The one-step cleavage is achieved by dissolving the 4CC product (*15*) in five to ten parts of trifluoroethanol at 0°C and adding within one half hour the same amount of TFA. As a rule this step can be carried out in the presence of carbonium ion scavengers like 1- or 2-methylindole or -anisole, or (*6*), in one to two hours (according to TLC). The reaction mixture is subsequently poured into anhydrous diethyl ether, and the ether insoluble peptide derivative (*16*) is washed with diethyl ether until free of (*18*) and TFA. Contaminating traces of (*18*) can be removed from (*17*) by distribution between water and ethyl acetate. This procedure was first developed for the synthesis of H-Gly-Ala-Leu-OH,[12] and later used in the synthesis of a variety of peptides, including a heptapeptide from (*12*) with three amino acid units, and (*13*) with four. The fact that (*19*), (*20*), (*21*), and (*15*) undergo almost exclusively the 4CC in methanol,[13] and the Passerini reaction[14] in dichloromethane, practically without competition from the 4CC (see Scheme 4) is significant for the combination of peptide fragments by 4CC,[13] as well as the attachment of *N*-protected amino acids to isocyanomethyl polystyrene by the Passerini reaction with dichloro-acetaldehyde (*21*). Since (*5*) and (*16*) are cleaved by acidolysis, peptide syntheses by 4CC according to Schemes 1 and 3 sometimes require protective groups which are not acid labile. Therefore, a new protective group technique

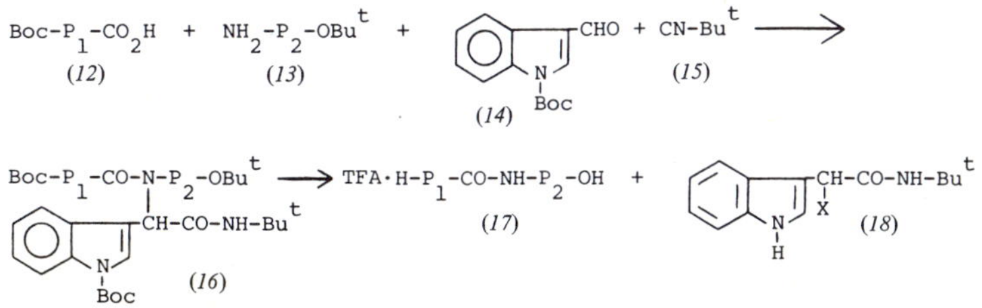

Scheme 3

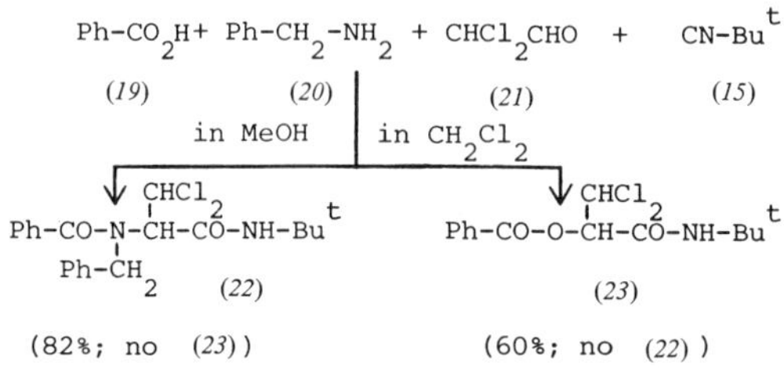

Scheme 4

was developed which is based upon the rapid cleavage under mild conditions of β-halogenated alkoxy compounds by cobalt-I-phthalocyanine anion[13,15-18] (*24*). The latter reagent also converts (*22*) into *N*-benzyl-benzamide,[13] and is useful for the cleavage of resin bound peptide derivatives (*28*) → (*29*) + (*30*) according to Scheme 5, as a part of a solid phase peptide synthesis.[19] Isocyanomethyl polystyrene (*25*) is prepared from aminomethyl polystyrene. This is made from a copolymer of 2% divinyl benzene and styrene by reacting with *N*-chloromethyl-phthalimide in the presence of sulfuric acid or trifluoromethane sulfonic acid,[20] and subsequent dephthaloylation with hydrazine[20] or by shaking with 27% aqueous methylamine and dichloromethane at 20°C for 48h (conversion 100%).

Formylation of the amino resin is achieved by shaking with $Ac\text{-}O_2CH$ (72h at 20°C, 100%). The formyl aminomethyl resin is phosgenated (0°C, vigorously stirred) in the presence of water, dichloromethane and triethylamine (40-60%, depending upon the degree of functionalization of the resin) to yield (*25*) with 0.08, 0.25, and 0.50 isocyanomethyl groups per styrene unit. Phosgenation in the absence of the aqueous phase affords less than 15% conversion to the isocyanide, according to thiocyanate titration.[21,22]

We acknowledge the financial support of our work on 4CC by the Deutsche Forschungsgemeinschaft and our stereochemical studies by Stiftung Volkswagenwerk.

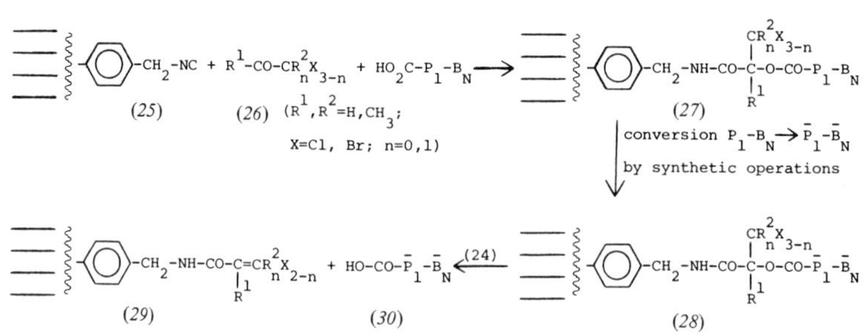

Scheme 5

References

1. Ugi, I. (1962) *Angew. Chem.* **74**, 9–22; *Int. Ed.* **1**, 8–21.

2. Gokel, G., Lüdke, G. & Ugi, I. (1971) in *Isonitrile Chemistry*, Ugi, I., Ed., Academic Press Inc., New York, ch. 8.

3. Gokel, G., Hoffmann, P., Kleimann, H., Klusacek, H., Lüdke, G., Marquarding, D. & Ugi, I. (1977) in *Isonitrile Chemistry*, Ugi, I., Ed., Academic Press, Inc., New York, ch. 9.

4. Urban, R., Marquarding, D., Seidel, P., Weinelt, A. & Ugi, I. (1977) *Chem. Ber.* (in press).

5. Skorna, G. & Ugi, I. (1977) *Angew. Chem.* 89, 267–268; *Int. Ed.* **16**, 259–260.

6. Urban, R., Eberle, G., Marquarding, D., Rehn, D., Rehn, H. & Ugi, I. (1976) *Angew. Chem.* **88**, 644–646; *Int. Ed.* **15**, 627–628.

7. Urban, R. & Ugi, I. (1975) *Angew. Chem.* 87, 67–69; *Int. Ed.* **14**, 61–62.

8. Ugi, I., Aigner, H., Beijer, B., Ben-Efraim, D., Burghard, H., Bukall, P., Eberle, G., Eckert, H., Marquarding, D., Rehn, D., Urban, R., Wackerle, L. & von Zychlinsky, H. (1976) in *"Peptides 1976,"* Loffet, A., Ed., Editions de L'Universite, Bruxelles, pp. 159–180.

9. Brandt, J., Jochum, C., Ugi, I. & Jochum, P. (1977) *Tetrahedron* **33**, 1353.

10. Marquarding, D., Burghard, H., Ugi, I., Urban, R. & Klusacek, H. (1977) *J. Chem. Res.* (S), pp. 82–83; (M), p. 0915.

11. Wackerle, L. & Ugi, I. (1975) *Synthesis*, 598–599.

12. Izumiya, N. & Muraoka, M. (1969) *J. Amer. Chem. Soc.* **91**, 2391–2392.

13. Lagerlund, I., Eckert, H. & Ugi, I. (1977) *Tetrahedron* (in press).

14. Marquarding, D., Gokel, G., Hoffmann, P. & Ugi, I. (1971) in *Isonitrile Chemistry*, Ugi, I., Ed., Academic Press Inc., New York, ch. 7.

15. Eckert, H. & Ugi, I. (1975) *Angew. Chem.* 87, 847; *Int. Ed.* **14**, 825–826.

16. Eckert, H. & Ugi, I. (1976) *J. Organomet. Chem.* **118**, C55, C59.

17. Eckert, H. & Ugi, I. (1976) *Angew. Chem.* 88, 717–718; *Int. Ed.* **15**, 681.

18. Eckert, H. (1977) *Synthesis*, 332–334.

19. Stewart, J. M. & Young, J. D. (1969) in *Solid Phase and Peptide Synthesis*, W. H. Freeman & Co., San Francisco.

20. Mitchell, A. R., Kent, S. B. H., Ericson, B. W. & Merrifield, R. B. (1976) *Tetrahedron Lett.*, 3795–3798.

21. Arora, A. S., v. Hinrichs, E. & Ugi, I. (1974) *Z. Anal. Chem.* **269**, 124.

22. Skorna, G. & Ugi, I., *Chem. Ber.* (submitted).

SOME RECENT DEVELOPMENTS IN SOLID PHASE PEPTIDE SYNTHESIS

R. B. MERRIFIELD, GEORGE BARANY, WESLEY L. COSAND, MARTIN ENGELHARD, and SVETLANA MOJSOV,
The Rockefeller University, New York, New York 10021

Since the Fourth American Peptide Symposium, there has been continuing progress in the development and application of the solid phase approach to peptide synthesis. Each of the main steps in the method, as illustrated in Fig. 1, has received some attention, and modifications in the overall strategy have been studied. The result has been to improve the quality of individual syntheses and to broaden the range of procedures that can be applied to new synthetic problems.

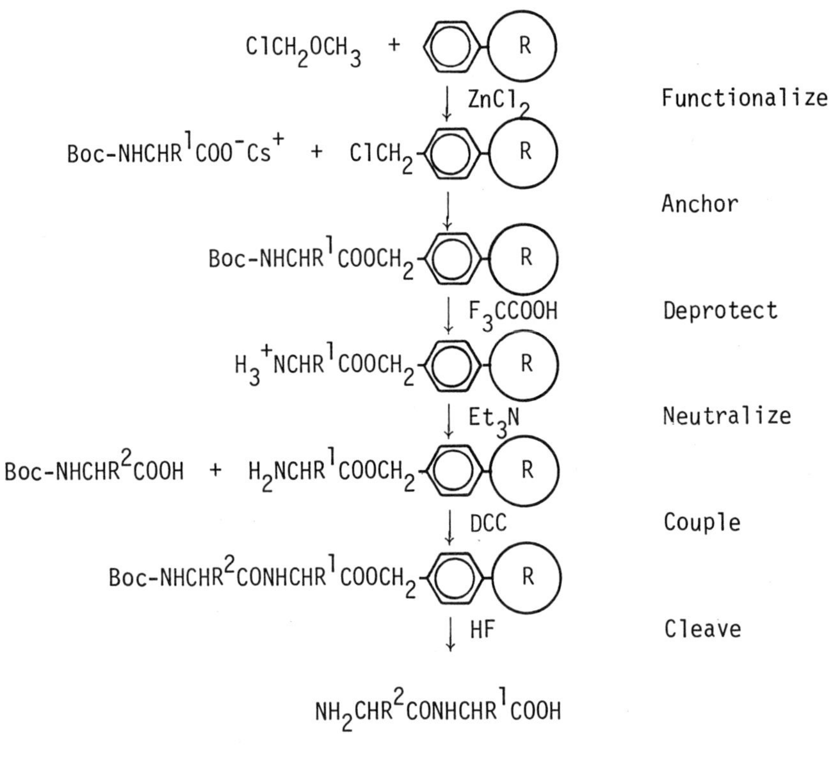

Fig. 1. A typical scheme for solid phase peptide synthesis.

The Support

Most solid phase syntheses have continued to use the well-tested and readily available styrene-divinylbenzene copolymer beads.[1] The 1% crosslinked resin is now standard, but the more open 0.5% crosslinked gel has clear advantages when handled in a centrifugal reaction vessel.[2] Styrene that has been radiation-grafted onto a Teflon core gives results comparable to those from the standard beads. A useful feature of the Teflon beads is that they can be applied in conjunction with standard beads for simultaneous syntheses of analogs in the same vessel, with subsequent separation by density difference at the end of the synthesis.[3] Theoretical arguments have been presented for the superiority of polyacrylamide supports over polystyrene supports and quite good results have been obtained with the former.[4] Further comparative data are needed to establish the relative merits of the two materials. Soluble supports of the polyethylene glycol type continue to be of value for liquid phase syntheses,[5] and the tactical advantages of the liquid-phase–solid-phase approach justify a continuing effort in the exploration of this promising technique.[6]

Attachment to the Support

Recent independent studies in our laboratory[7] and that of Sparrow[8] have led to new linkages between the peptide chain and the solid support that have demonstrated advantages. We sought to avoid the gradual acidolytic loss of peptide chains that has been shown to occur during a long stepwise synthesis using the conventional benzyl ester linkage. Therefore, we designed and synthesized a support that was predicted to possess increased acid stability. A phenylacetamidomethyl (Pam) group was inserted between the polystyrene matrix and the

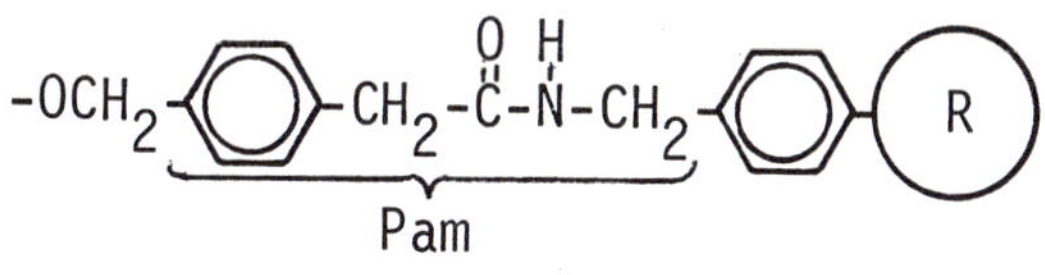

(1)

peptide.[7] In 50% TFA/CH_2Cl_2 the observed rate of cleavage of a peptide from this resin was only 1% as fast as that from the usual support. This improved stability also avoids the generation of hydroxymethyl groups, due to loss of peptide chains, and the accompanying side reactions which then ensue. Sparrow[8] has inserted a long extender arm, which also contains an acetamido linkage, between the peptide and the polystyrene resin. The improved yield of synthetic

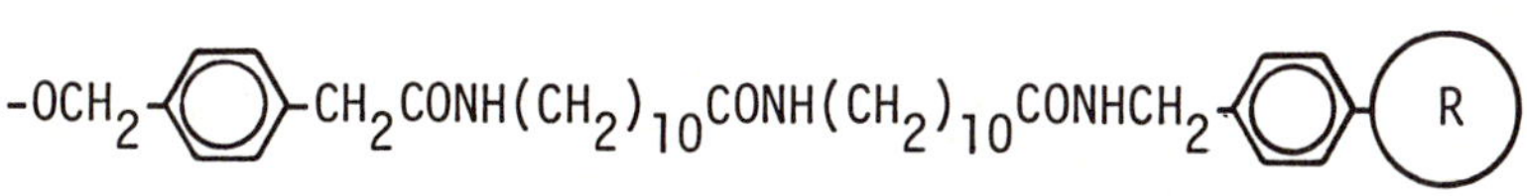

(2)

57-residue apoliprotein CI made on this support[9] was attributed to the flexibility of the arm and to the removal of the peptide chain from the vicinity of the polystyrene backbone. Whether the improved syntheses with the new linkages are due to their increased stability or to their role as extenders or both is not yet certain.

To facilitate the preparation of the Pam resin, a new synthesis of aminomethyl-copoly(styrene-divinylbenzene) was developed, which involved a direct one-step amidomethylation.[10]

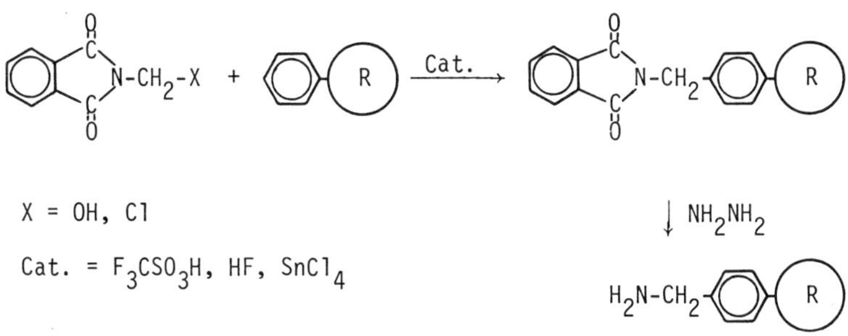

(3)

The preparation is easily controlled and circumvents the need for chloromethylation of the polymer.

Linkages to the resin that can be cleaved by other mechanisms have been examined. For example peptides esterified to 2-hydroxyethylsulfonylmethyl-polystyrenes are stable to acid but can be cleaved by β-elimination with methoxide ion.[11] Photolabile linkages also continue to receive attention.[12-14]

Side Chain Protecting Groups

The proper choice of side chain protecting groups is an integral part of peptide synthesis but cannot be covered in detail here. A problem of interest to us has been the acid rearrangements[15] of tyrosine alkyl ethers to stable *C*-alkyl derivatives. It was shown that deprotection of *O*-benzyltyrosine, or of peptides containing this residue, by means of 10%/anisole/HF could give rise to 10–70% of 3-benzyltyrosine *via* an intramolecular route. *O*-(2,6-dichlorobenzyl)-tyrosine was shown to be much more acid stable and to reduce the rearrangement product several fold, but not to zero. Three new tyrosine ethers have now been synthesized.[16]

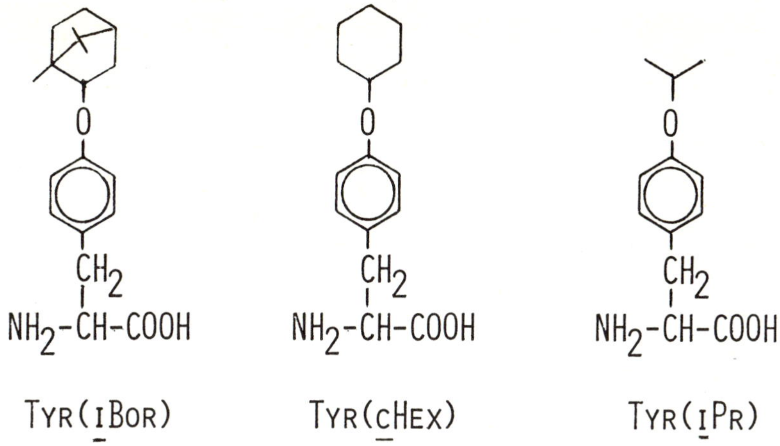

The isobornyl and cyclohexyl derivatives were prepared in good yield by the reaction of *N*-trifluoroacetyltyrosine methyl ester with boron trifluoride and either camphene or cyclohexene. The isopropyl derivative was obtained in low yield from isopropylbromide and the copper complex of tyrosine. The rate of loss of the protecting groups in 50% TFA/CH_2Cl_2 is shown in Fig. 2. Table I lists the calculated first-order rate constants and loss per cycle, together with the extent of rearrangement to the 3-alkyl derivatives occurring in HF/anisole.

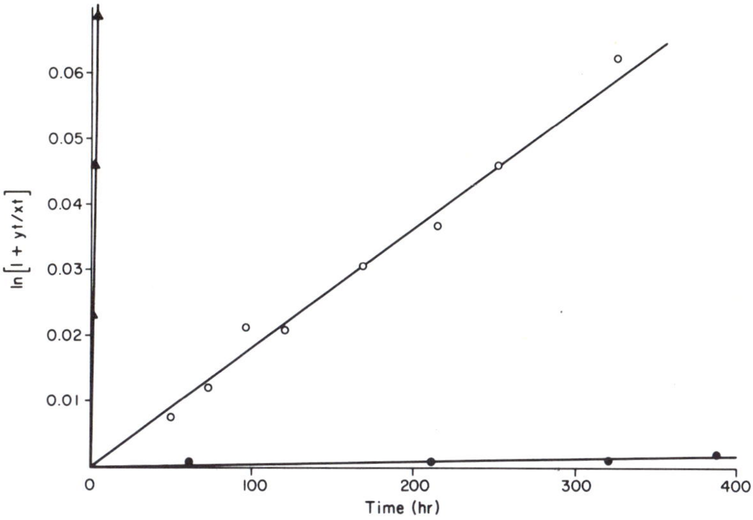

Fig. 2. Plot for the apparent first-order loss of protecting groups from tyrosine in 50% TFA/CH_2Cl_2 at 20°C. (○) Tyr(cHex), (●) Tyr(iPr), (▲) Tyr(Bzl) (from Ref. 15).

Table I. Acid Stability of Protected Tyrosine

Compound	k_1 (sec^{-1})	Loss/cycle (50% TFA) (%)	3-Alkyl-Tyr (HF-Anisole) (%)
Tyr(iBor)	2×10^{-2}	100	0
Tyr(Bzl)	6.4×10^{-6}	0.76	15
Tyr(cHex)	5.1×10^{-8}	0.006	0.5
Tyr(iPr)	1.5×10^{-9}	0.00017	3.5
Tyr(2,6-Cl_2Bzl)	1.2×10^{-9}	0.00014	5

For solid phase synthesis of tyrosine peptides the *O*-cyclohexyl derivative has the best combination of properties. Its acid stability is satisfactory for a multi-step synthesis and, although the extent of rearrangement is not zero, it is acceptably low for many purposes. Boc-Tyr(cHex) was used for the solid phase synthesis of angiotensin II, and the unfractionated, cleaved product was shown to contain only 0.3% of the 3-cHex-Tyr isomer. In this instance the side-product was readily removed chromatographically, and the pure angiotensin II was fully active in the pressor assay. Trudelle and Spach[17] observed a side-product when Tyr(Bzl) was deprotected in HF or HBr/TFA, but detected none when HBr/HOAc was the reagent. This finding was confirmed by Bell et al.[18] The rearrangement can also be avoided by protection of tyrosine with benzyloxycarbonyl[15] or 2,4-dinitrophenyl[19] groups, which can be removed by acid or by thiols, respectively. Both groups, however, are slightly sensitive to amine nucleophiles.

Orthogonal Protection

A frequent criticism of the standard solid phase methodology has been that selectivity in the removal of the α-amino, α-carboxyl, and side-chain protecting groups has been achieved merely by relying on different degrees of sensitivity to acidic reagents. Thus, the rates are different, but the mechanisms of the reactions are similar. Fine tuning has resulted in practical, but obviously not ideal, schemes. In a number of instances, additional selectivity has been achieved by choosing groups that are removable by other chemical mechanisms, but most laboratories still rely on quantitative rate differences such as illustrated in Fig. 3.

The ideal synthesis would depend on a set of completely independent reactions, and such a scheme of protection has been called *orthogonal.*[20] This term

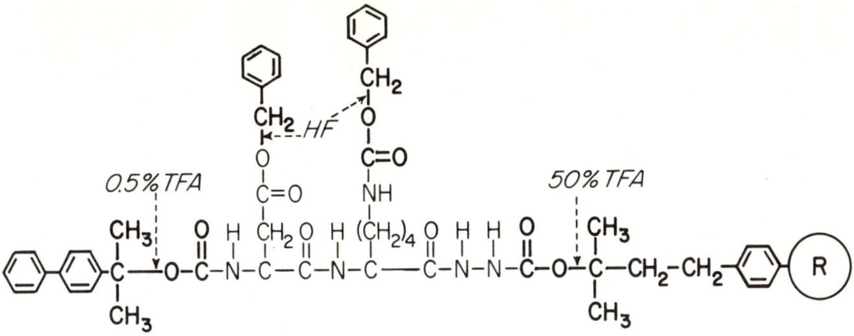

Relative stability to TFA: 1 — 1,000,000 — 3,000

Fig. 3. A scheme for the solid phase synthesis of protected peptides based on differential sensitivity to acid.

means that each class of protecting groups is cleaved by a specific mechanism and that all other groups are stable to reaction by that mechanism. Therefore any group can be removed in any order without effect on any other group. This, of course, is a very old idea, which has been commonly applied in simple cases, but its incorporation into a general scheme for a multistep solid phase synthesis has not been achieved.

A scheme which can be envisioned as meeting the requirements of orthogonal protection is illustrated in Fig. 4. This makes use of acid-labile side-chain protection, a photolabile carboxyl group to anchor the peptide to the resin, and a new thiol-sensitive α-amino protecting group. Thus, there are three kinds of reactions and three classes of protecting groups, each of which is sensitive to one of the reactions and stable to the others.

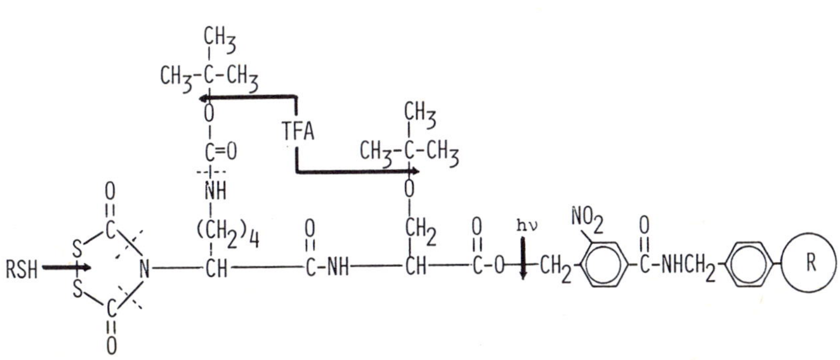

Fig. 4. An orthogonal protection scheme.

The new thiol-sensitive amino protecting group,[21] dithiasuccinoyl (Dts), was developed specifically for this purpose and appears to meet the requirements. Several Dts-amino acids were prepared by the reactions shown in Fig. 5.

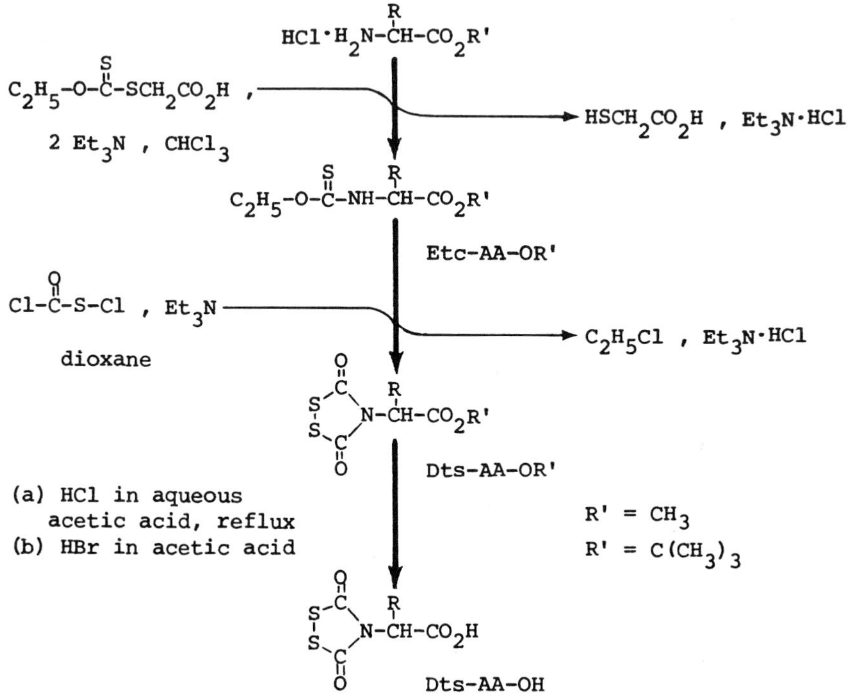

Fig. 5. The synthesis of Dts-amino acids.

The structure of these stable, readily crystallizable derivatives was established by elemental analysis, infrared and ultraviolet spectra, proton and carbon-13 nuclear magnetic resonance, electron impact and chemical ionization mass spectrometry (molecular ion and ion fragmentation patterns, with accurate mass measurements on key ions) and chemical transformations. The 1,2,4-dithiazolidine-3,5-dione ring is readily opened upon mild reduction with a thiol, liberating two moles of carbonyl sulfide and generating the free amine. Nucleophilic attack at the carbonyl by the thiol to give a thiourethane does not seem to occur. Dts-compounds are resistant to strong acids, including HBr in acetic acid and hot 6 *N* HCl; and to weak bases such as aqueous sodium bicarbonate and α-amino groups of peptides. Cleavage at the carbonyl occurs with strong bases, yielding the free amine with aqueous NaOH and the urea with benzylamine.

Solid Phase-Fragment Synthesis

Although the stepwise solid phase method is rapid and convenient, it has potential limitations that are well recognized. One of these, the increasing difficulty of separating the desired peptide from potential deletion peptides as the length of the peptide increases, can be minimized by coupling fragments rather than single amino acids. The advantage of the solid phase-fragment approach was recognized early by Weygand and Ragnarsson[22] and subsequently by others. Recently, this approach has gained in popularity and it is possible to make comparisons of the two techniques in a few instances. The fragment synthesis of bovine pancreatic trypsin inhibitor by Yajima et al.[23] is especially noteworthy because of the high yields of the couplings between fragments as large as 19 residues and the resin-bound peptide chain. The yields for the four peptide couplings ranged from 72 to 100% and the yield from the starting peptide fragments, which were synthesized by conventional solution methods, to the final product was 7.5%. The overall yield of protein based on starting amino acid derivatives is not clear, but we estimate it to be about 0.2%. Because of the large differences in size between the 58-residue product and any termination or deletion peptides resulting from incomplete reactions, the synthetic trypsin inhibitor could be readily purified by gel filtration. After further purification by affinity chromatography, the product had a specific activity of 82%. Yajima et al.[23] concluded that such a fragment approach was better than simple stepwise solid phase synthesis.

Recently, a stepwise solid phase synthesis of the same trypsin inhibitor was achieved by Tan and Kaiser.[24] It was purified by ion-exchange and affinity chromatography and was indistinguishable from natural trypsin inhibitor by polyacrylamide gel electrophoresis, amino-acid analysis, peptide maps, and circular dichroism spectra. Trypsin was inhibited stoichiometrically and, moreover, the dissociation constant for the trypsin-synthetic trypsin inhibitor complex was found to be $8 \times 10^{-14}\,M$, in very good agreement with the value of $6 \times 10^{-14}\,M$ reported for the native inhibitor. From these two results it can be concluded that the product was 100% active. The fully active product was obtained in 2.9% yield from the first amino acid on the resin, and the overall yield, based on starting amino acids, was 1.2%. One can conclude that a molecule of this complexity can be made by either approach, but, while the stepwise method was better in this instance, it would be unwarranted to generalize about relative merits at this time.

A solid phase-fragment synthesis of ACTH-(1-24) by chain elongation at the carboxyl end has recently been reported.[25] Six short protected peptide fragments were synthesized in solution by the oxidation-reduction method and then coupled through their freed amino groups to the growing carboxyl-activated peptide chain that was anchored to a methyl chloroformylated resin. Fully active hormone was obtained in 25% overall yield based on the first dipeptide-resin.

In 1967, Wünsch[26] announced his elegant synthesis of crystalline glucagon by classical fragment methods. The material was pure by numerous criteria and fully active in several assays. This has remained a standard by which other syntheses are measured. In 1975, the Protein Synthesis Group of the Shanghai Institute of Biochemistry reported a new synthesis of glucagon using a fragment synthesis on a solid support.[27] They followed the basic Wünsch scheme rather closely. However, the Nps group was replaced by Bpoc, $Arg(Z)_2$ was replaced by $Arg(Adoc)_2$, DCC/HOBt was used instead of DCC/HOSu, and fragment IV was three residues shorter. The first three residues at the carboxyl terminus were added stepwise to a pellicular poly(styrene-divinylbenzene) resin, which then served as the starting point for the coupling of the remaining fragments which had been synthesized by solution methods. The couplings were reported to be essentially quantitative. The completed protected peptide-resin was cleaved and fully deprotected in HF/anisole. Two peaks were obtained by gel filtration, and the glucagon fraction was purified further by DEAE-cellulose chromatography. The yield from the first amino acyl-resin was about 17%. The final preparation (3.4 mg) was homogeneous, indistinguishable from natural glucagon by polyacrylamide electrophoresis, and gave typical rhombic dodecahedral crystals. It was essentially fully active in the rabbit blood sugar assay.

We have just completed a stepwise solid phase synthesis of glucagon (Mojsov and Merrifield), according to the sequence of Bromer et al.,[28] using the protecting groups shown in Fig. 6. The support was the alkoxybenzyl alcohol

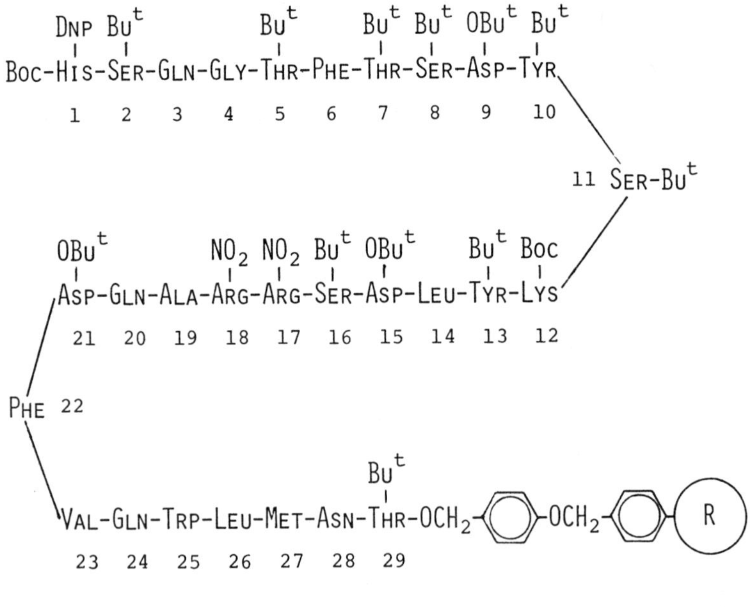

Fig. 6. The structure of synthetic protected glucagon.

resin of Wang.[29] The α-amino protecting group was Bpoc in all but the last step where Boc was used. The couplings were with DCC (3-fold, double coupling). The Dnp group was removed from histidine by thiophenol and the remaining side-chain groups, except Arg(NO_2), were removed during the cleavage step with 50% TFA/CH_2Cl_2 (45 min, 25°C). The mild acid procedure was used to minimize the possibility of $\alpha \rightarrow \beta$ rearrangement of the Asp^{15}-Ser^{16} sequence. The nitro groups could then be removed by HF/anisole or by catalytic hydrogenation. The crude free peptide was purified by gel filtration and by DEAE-cellulose chromatography in 6 *M* urea, Tris buffer pH 8.5, with a NaCl gradient from 0 to 0.4 *M*. The main peak (2.4% yield from the first amino acyl-resin) eluted at the position of native glucagon. The isolated product was readily crystallized from water at pH 8.5. Full characterization is still underway, but preliminary data indicate good purity and close identity with natural glucagon. The hyperglycemia assay showed essentially full activity compared with native glucagon and the product is also active in an adenyl cyclase assay with isolated liver membranes. We conclude that glucagon can be synthesized by classical methods, by the solid phase-fragment method, or by the stepwise solid phase method. The purified products appear to be chromatographically homogeneous, fully active, and indistinguishable from native glucagon, and it has been possible to obtain crystalline hormone from all three syntheses. We stress that our results are new and incomplete and further comparisons are not possible at this time.

Use of An Internal Structural Control for Inactive Synthetic Peptide Analogs

The interpretation of data from synthetic peptide analogs that have very low activity must be made with caution.[30,31] The lack of activity may be due to the intended new structure itself or to some unrecognized difficulty in the synthesis. We think an internal structural control can be very useful in such instances. If the inactive product can be converted into a fully active peptide that is indistinguishable from the naturally occurring parent molecule, we have in effect an internal control for the correctness of the structure of the analog.

Examples of this idea include those involving the chemical conversion of inactive protected peptides into active free peptides, the enzymatic formation of active peptides from synthetic precursors of low activity, or the conversion of inactive protected peptides into active free peptides, the enzymatic formation of active peptides from synthetic precursors of low activity, or the conversion of one amino-acid residue into another. A convenient transformation of the latter type is the conversion of an ornithine residue into an arginine residue by guanidination.[32] This route has been applied several times to avoid the difficulties associated with the direct incorporation of arginine into peptides. For example, Bodanszky et al.[33] synthesized active ACTH-(6-10) via the inactive Orn^8 analog.

We have tested the value of an internal structural control in a study of the role of arginine in the neurotoxin, apamin.[34] This octadecapeptide from bee venom has been isolated and purified,[35] and its structure determined.[36-38] Its synthesis has been described by Van Rietschoten et al.[39] and Sandberg and Ragnarsson.[40] Chemical data suggested that the two arginines at positions 13 and 14 might be required for activity,[41] and we designed a synthetic approach to the question in which these positions would contain N^{δ}-trifluoroacetylornithine in place of arginine. The protecting groups could be removed to give the di-ornithine derivative and it could then be guanidinated to the di-arginine derivative. The latter would also contain homoarginine (Har) at position 4, but that was acceptable because it was already known that [Har^4]-apamin is fully active.[41] If either of the two analogs were to prove inactive but could be converted into active [Har^4]-apamin, we would have the desired internal control and more reliable conclusions about these replacements of arginine could be drawn.

The peptides were synthesized by the stepwise solid phase method[1,31,42] using the protecting groups shown in Fig. 7

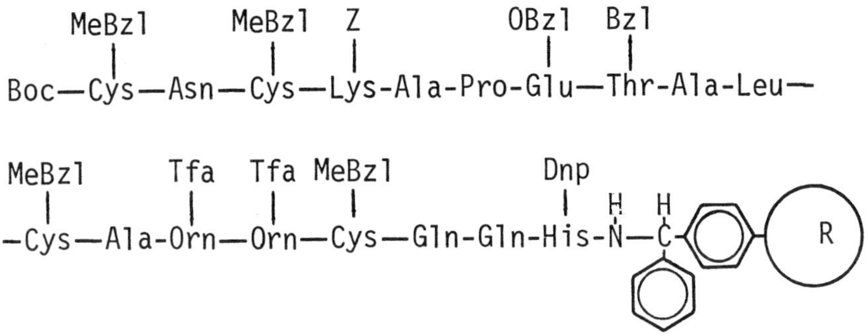

Fig. 7. Protected [Orn^{13,14}]-apamin-benzhydrylamine resin.

The support was a benzhydrylamine resin, coupling was with DCC alone except with Boc-Asn-OH and Boc-Gln-OH where hydroxybenzotriazole was added. The Dnp group was removed from His with 2.5% thiophenol in DMF, and the cleavage and deprotection of side chains was with 10% anisole/HF, for 30 min, at 0°C. The [Orn(Tfa)^{13,14}]-apamin and [Orn^{13,14}]-apamin were each reduced and then oxidized in air and purified chromatographically. The [Orn^{13,14}]-apamin was guanidinated with *O*-methylisourea and purified by Sephadex G-25 and carboxymethylcellulose chromatography. The work-up of the three peptides is outlined in Fig. 8 and the elution profile for [Har^4]-apamin is shown in Fig. 9.

The coupling, cleavage, deprotection, and guanidination steps all proceeded in good yields. The largest losses were at the oxidation steps and during the several chromatographic steps, giving an overall yield of 14%. The synthetic

Protected Peptide-Resin	152 μmol
↓ 1, Thiophenol 2, HF	↓ 91%
[Orn(Tfa)^{13,14}]-peptide	138 μmol
↓ 1, Piperidine 2, DTT 3, G-25	↓ 88%
[Orn^{13,14}]-peptide$(SH)_4$	122 μmol
↓ Oxidize	
[Orn^{13,14}]-peptide$(-S-S-)_2$	↓ 32%
↓ 1, CM-52 2, G-25	
[Orn^{13,14}]-apamin	39.2 μmol
↓ 1, O-Methylisourea 2, G-25 3, CM-52 4, G-25	↓ 53%
[Har^4]-apamin	20.9 μmol

Fig. 8. Work-up of synthetic apamin analogs

peptides gave good amino acid analyses and were homogeneous by ion exchange chromatography and polyacrylamide gel electrophoresis (Fig. 10).

Neither the [Orn(Tfa)^{13,14}]-apamin nor [Orn^{13,14}]-apamin produced any symptoms of neurotoxicity whatever upon intravenous injection into mice at the highest levels that could be tested. Therefore, they were below 5% and 1% active respectively. As expected, the synthetic [Har^4]-apamin was fully active compared with native apamin and [Har^4]-apamin derived from it. The data are summarized in Table II.

These results provide the necessary internal control and strengthen the conclusion that analogs of the correct amino-acid sequence were synthesized. We can then conclude that the observed inactivity of the ornithine derivative was due to the intended structure and can infer that at least one of the arginines at

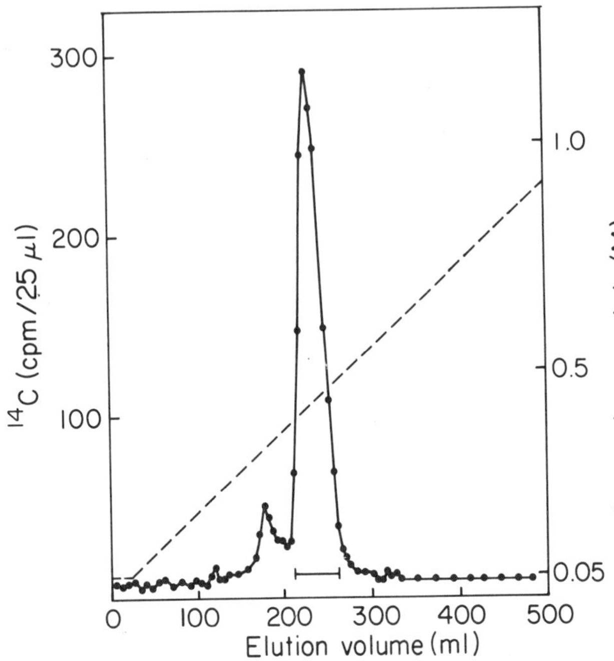

Fig. 9. Chromatographic purification of [Har^{4}]-apamin on CM-cellulose.

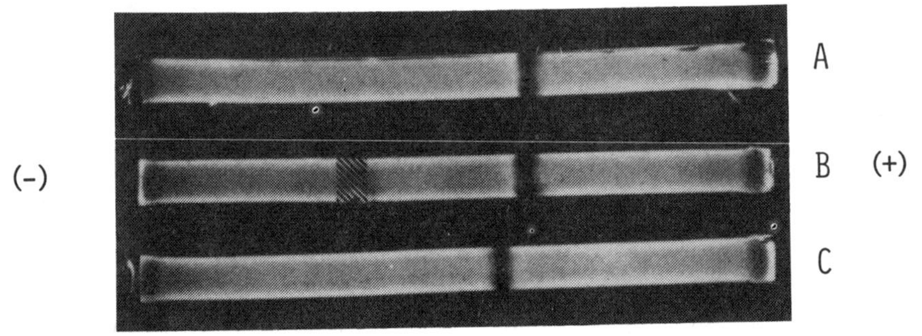

Fig. 10. Gel electrophoresis of apamin analogs. Gels were 30% polyacrylamide – 0.3% *N,N'*-methylenebisacrylamide, run in 0.9 *M* acetic acid, 5 hr, 150 v. Stained with amido black. A, natural apamin; B, synthetic [Har^{4}]-apamin; C, synthetic [Orn^{13,14}]-apamin. The cross-hatched band marks the position of the running dye, methyl green.

positions 13 and 14 plays an important role in the action of apamin. The level at which arginine functions is not established, but it may provide electrostatic, steric, or conformational properties that are necessary for proper folding, stability, transport, or action at the receptor site.

Table II. Neurotoxic Activity of Natural Apamin and Synthetic Apamin Analogs

Compound	LD_{50} [a] (nmol)	Relative activity
Natural apamin	47	1.00
Natural [Har^4]-apamin		1.00 [c]
Synthetic [Har^4]-apamin	44	1.07
Synthetic [$Orn^{13,14}$]-apamin	>1200 [b]	No activity (<1%)
Synthetic [$Orn(Tfa)^{13,14}$]-apamin	>280 [b]	No activity (<5%)

[a] Calculated from data at six concentrations on groups of six mice each.

[b] No observable symptoms at this dose.

[c] Taken from Vincent et al.[41]

References

1. Merrifield, R. B. (1963) *J. Amer. Chem. Soc.* **85**, 2149–2154.
2. Birr, C. & Lochinger, W. (1971) *Synthesis*, 319–321.
3. Tregear, G. W. (1972) in *Chemistry and Biology of Peptides*, Meienhofer, J., Ed., Ann Arbor Sci. Publ., Michigan, pp. 175–178.
4. Atherton, E. & Sheppard, R. C. (1975) in *Peptides 1974*, Wolman, Y., Ed., Wiley, New York, pp. 123–128.
5. Bayer, E. & Mutter, M. (1972) *Nature* **237**, 512–513.
6. Frank, H. & Hagenmaier, H. (1975) *Experientia* **31**, 131–133.
7. Mitchell, A. R., Erickson, B. W., Ryabtsev, M. N., Hodges, R. S. & Merrifield, R. B. (1976) *J. Amer. Chem. Soc.* **98**, 7357–7362.
8. Sparrow, J. T. (1976) *J. Org. Chem.* **41**, 1350–1353.
9. Sigler, G. F., Soutar, A. K., Smith, L. C., Gotto, A. M., Jr. & Sparrow, J. T. (1976) *Proc. Nat. Acad. Sci. USA* **73**, 1422–1426.
10. Mitchell, A. R., Kent, S. B. H., Erickson, B. W. & Merrifield, R. B. (1976) *Tetrahedron Lett.*, 3795–3798.
11. Tesser, G. I., Buis, J. T., Wolters, E. T. M. & Bothe-Helmes, E. G. (1976) *Tetrahedron* **32**, 1069–1072.
12. Rich, D. H. & Gurwara, S. K. (1975) *J. Amer. Chem. Soc.* **97**, 1575–1579.
13. Rich, D. H. & Gurwara, S. K. (1975) *Tetrahedron Lett.*, 301–304.
14. Wang, S. S. (1976) *J. Org. Chem.* **41**, 3258–3261.
15. Erickson, B. W. & Merrifield, R. B. (1973) *J. Amer. Chem. Soc.* **95**, 3750–3756.

16. Engelhard, M. & Merrifield, R. B. (1977) submitted.

17. Trudelle, Y. & Spach, G. (1972) *Tetrahedron Lett.*, 3475–3478.

18. Bell, J. R., Jones, J. H. & Webb, T. C. (1975) *Int. J. Pept. Protein Res.* 7, 237–244.

19. Fridkin, M., Hazum, E., Tauber-Finkelstein & Shaltiel, S. (1977) *Arch. Biochem. Biophys.* **178**, 517–526.

20. Barany, G. & Merrifield, R. B. (1977) submitted.

21. Barany, G. & Merrifield, R. B. (1977) *Fed. Proc.* **36**, 864.

22. Weygand, F. & Ragnarsson, U. (1966) *Z. Naturforsch.* **21 b**, 1141–1144.

23. Yajima, H., Kiso, Y., Okada, Y. & Watanabe, H. (1974) *J. Chem. Soc., Chem. Commun.*, 106–107.

24. Tan, N. H. & Kaiser, E. T. (1976) *J. Org. Chem.* **41**, 2787–2793.

25. Maruyama, H., Matsueda, R., Kitazawa, E., Takahagi, H. & Mukaiyama, T. (1976) *Bull. Chem. Soc. Jap.* **49**, 2259–2267.

26. Wünsch, E. (1967) *Z. Naturforsch.* **22 b**, 1269–1276.

27. Protein Synthesis Group, Shanghai Institute of Biochemistry (1975) *Scientia Sinica* **18**, 745–768.

28. Bromer, W. W., Sinn, L. G., Staub, A. & Behrens, O. K. (1956) *J. Amer. Chem. Soc.* 78, 3858–3860.

29. Wang, S. S. (1973) *J. Amer. Chem. Soc.* **95**, 1328–1333.

30. Rudinger, J. (1971) in *Drug Design*, Medicinal Chemistry Monograph Series, vol. II, Ariens, E. J., Ed., Academic Press, New York, pp. 319–419.

31. Merrifield, R. B. (1969) *Adv. Enzymol.* **32**, 221–296.

32. Fruton, J. S. (1949) *Adv. Protein Chem.* **5**, 1–82.

33. Bodanszky, M., Ondetti, M. A., Burkhimer, C. A. & Thomas, P. L. (1964) *J. Amer. Chem. Soc.* **86**, 4452–4459.

34. Cosand, W. L. & Merrifield, R. B. (1977) *Proc. Nat. Acad. Sci. USA* **74**, 2771–2775.

35. Habermann, E. & Reiz, K. G. (1965) *Biochem. Z.* **341**, 451–466.

36. Haux, P., Sawerthal, H. & Habermann, E. (1967) *Hoppe-Seyler's Z. Physiol. Chem.* **348**, 737–738.

37. Shipolini, R., Bradbury, A. F., Callewaert, G. L. & Vernon, C. (1967) *Chem. Commun.*, 679–680.

38. Callewaert, G. L., Shipolini, R. A. & Vernon, C. A. (1968) *FEBS Lett.* **1**, 111–113.

39. Van Rietschoten, J., Granier, C., Rochat, H., Lissitsky, S. & Miranda, F. (1975) *Eur. J. Biochem.* **56**, 35–40.

40. Sandberg, B. E. B. & Ragnarsson, U. (1975) *Int. J. Pept. Protein Res.* 7, 503–504.

41. Vincent, J. P., Schweitz, H. & Lazdunski, M. (1975) *Biochemistry* **14**, 2521–2525.

42. Erickson, B. W. & Merrifield, R. B. (1976) in *The Proteins*, 3rd ed., vol. 2, Neurath, H. and Hill, R. L., Eds., Academic Press, New York, pp. 257–527.

Note added in proof: During the discussion period for this paper we learned that the term 'orthogonal' has been used in a similar way in the laboratory of Dr. D. S. Kemp.

SOLID PHASE SYNTHESIS ON POLYAMIDE SUPPORTS: β-ENDORPHIN

E. ATHERTON and R. C. SHEPPARD, *MRC Laboratory of Molecular Biology, Hills Road, Cambridge CB2 2QH, England.*

Some years ago we advanced a simple model for the solid phase synthesis system.[1] It regards the peptide-polymer assembly as analogous to the simple graft copolymers well known and understood in solution chemistry. Consideration of the solvation and precipitation properties of such copolymers led to the conclusion that steric interference would be minimized only if *both* the backbone and pendant polymer chains were freely solvated by the same solvent media. For peptide chains this would require polar organic solvents of the type known to solvate and hence destabilize secondary structures in synthetic polypeptides (e.g. dimethylformamide). For simultaneous solvation of the polymer backbone, the latter should be structurally related to the peptide. This leads directly to the concept of polyamide supports for solid-phase peptide synthesis and the maximum use of polar organic solvents for coupling and deprotection reactions.

Initial experiments showed that the strong internal hydrogen bonding of primary polyamides was not disrupted by aprotic organic solvents, such resins usually being freely permeated only by water.[2] Tertiary amide polymers, however, swell well in both aqueous and polar organic media. We have concentrated attention on polydimethylacrylamide resins prepared by direct polymerization,[3,4] and now record our recent experiences using a resin of this type. Elsewhere we have described related studies on oligonucleotide synthesis[5,6] and protein degradation.[7]

The preparation of cross-linked polydimethylacrylamide by persulphate-initiated emulsion copolymerization of mixtures of dimethylacrylamide, *N,N'*-bisacryloylethylene diamine, and *N*-acryloyl-*N'*-*t*-butyloxycarbonyl-β-alanylhexamethylene diamine has already been reported.[3] This last named monomer provides a protected functionality in the resin, as well as an internal reference amino acid and a spacer molecule. Additional spacer or reference amino acids can be added if desired before acylation of the terminal amino group by any one of a number of reversible linkage agents. Thus far we have concentrated on the simple benzyl alcohols (I) and (II). (See Scheme 1.) These may be usefully introduced together with the first amino acid of the sequence proper using activated ester derivatives, e.g. (III) under reaction conditions similar to those used for subsequent amino acids. Esters of (I) are cleaved by liquid HF in the usual way. Esters of (II) are remarkably resistant to HF but are rapidly cleaved by methanolic ammonia, hydrazine, and hydroxide ion.

(I) (II) (III)

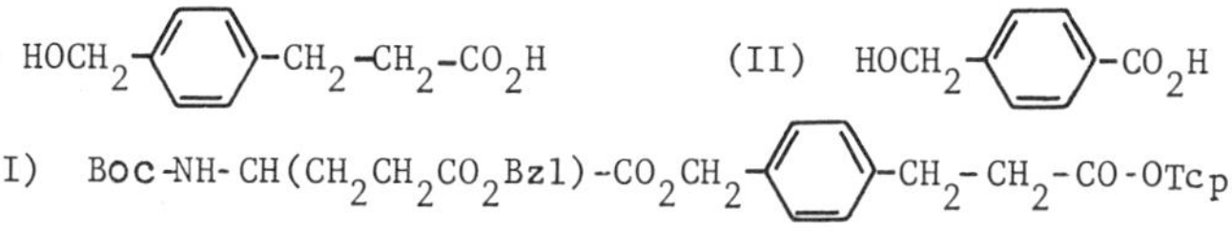

(IV) Tyr-Gly-Gly-Phe-Met-Thr-Ser-Glu-Lys-Ser-Gln-Thr-Pro-Leu-Val-Thr-Leu-Phe-Lys-Asn-Ala-Ile-Ile-Lys-Asn-Ala-Tyr-Lys-Lys-Gly-Glu

Scheme 1

The synthesis of peptides containing up to 17 residues has already been reported.[2-4] The method is illustrated here by the more substantial 31 residue sequence of human β-endorphin (lipotropin C fragment) (IV), a peptide with powerful opiate-like activity.[8,9] The synthesis was initiated with the active ester (III) and the following 30 residues added using HCl/AcOH for deprotection steps, 10% diisopropylethylamine in DMF for neutralization, and preformed symmetrical amino-acid anhydrides in DMF for acylation.[3] The assembly was complete in ten days. Ninhydrin tests were used extensively for checking completeness of coupling; they were invariably negative within 10-20 minutes of introduction of the anhydride and no repeated acylation or deprotection steps were carried out. Samples were removed for amino-acid analysis at intervals chosen to give a complete picture of apparent incorporation levels (Table I), and samples were additionally removed at about ten residue intervals for complete solid-phase Edman degradation.[7] This technique established that no detectable rearrangement[10] of Boc-glycine anhydride had occurred with resultant over incorporation at residues 2 and 3. It also confirmed efficient incorporation of Ile[23] for which only a poor amino-acid analysis is obtained.

The completed peptide was cleaved from the resin with HF/anisole and freed from high molecular weight aggregates on Sephadex G-50. Further purification was achieved by repeated chromatography on carboxymethyl cellulose. The final product gave an excellent amino-acid analysis (Table I), a single ninhydrin reacting spot on thin-layer chromatography and electrophoresis, and was judged by Dr. J. F. W. Deakin of the National Institute for Medical Research, London, to be nearly equally active with natural ovine β-endorphin in the rat-tail flick test for analgesic properties.

More recent bioassays by Dr. G. E. Metcalf of Reckitt and Colman, Hull, have given much lower potencies. This appears to be associated with post-synthesis oxidation of the methionine residue at position 5 within the enkephalin sequence. Hydrolysis of the peptide resin now shows (5 months after synthesis) a high proportion of methionine sulphone at this position.

The overall yield in this synthesis was 10%. This is much less than the 30% obtained in the superb solid phase synthesis recently reported by Li et al.[11] It seems probable that the major factor responsible for this difference is the efficiency of the HF-cleavage step, for we observe substantial losses and aggregation at this stage,[12] in contrast to the results reported by Li et al.[11] We are encouraged by the evident efficiency of the chain assembly process, and current work is centered on alternatives to the HF-cleaved benzyl ester linkage.

Table I. Amino-Acid Analysis Data for Resin-Bound and Free Synthetic β-Endorphin. All ratios are referred to glutamic acid = 1, 2, or 3 as appropriate and are uncorrected for destruction or incomplete release. The two isoleucine residues are present in an -Ile-Ile- sequence which is incompletely hydrolyzed under the standard conditions (6 *N*, HCl, 110°C, 18 h).

Step	3	7	9	17	19	22	27	31			
Residues	31-29	31-25	31-23	31-15	31-13	31-10	31-5	31-1	HF-G50	CM 52-urea G50	2nd CM52
Glu (Bzl)	1.00	1.00	1.00	1.00	1.00	2.00	2.00	3.00	3.00	3.00	3.00
Gly	1.09	1.08	1.06	1.03	1.06		1.12	3.25	3.19	2.92	2.89
Lys (Z-Cl_2)	1.07	2.19	3.20	4.24	4.13	3.92	4.91	5.25	5.08	4.58	4.75
Tyr (Bzl-Cl_2)		0.96	0.98	0.99	1.00	0.97	1.00	2.06	2.05	1.88	2.02
Ala		1.03	1.01	1.98	1.97	1.95	2.02	2.08	2.15	1.94	2.00
Asn (Xn)		0.96	0.96	1.90	1.89	1.89	1.93	1.96	2.06	1.94	1.95
Ile			0.83	1.33	1.48	1.52	1.55	1.51	1.19	1.30	1.43
Phe				0.92	0.94	0.90	0.90	1.97	1.99	1.87	1.97
Leu				0.97	1.94	1.84	1.95	1.94	2.04	1.92	1.95
Thr (Bzl)				1.01	1.01	1.75	2.83	2.80	2.81	2.73	2.72
Val				0.92	0.94	0.85	0.92	0.95	0.98	1.00	0.99
Pro					0.71	0.93	0.75	1.05	1.07	0.97	0.99
Ser (Bzl)						0.95	1.76	1.81	1.79	1.67	1.80
Met							1.14	1.15	0.92	0.93	0.94

We thank Mario Caviezel and Hilary Over for invaluable assistance, and Dr. J. F. W. Deakin and Dr. G. E. Metcalf for the bioassay.

References

1. Sheppard, R. C. (1973) in *Peptides 1971 Proc. 11th European Peptide Symposium,* Nesvadba, H., Ed. North Holland, Amsterdam, pp. 111–125.

2. Atherton, E. & Sheppard, R. C. (1975) in *Peptides 1974 Proc. 13th European Peptide Symposium,* Wolman, Y., Ed., Wiley, New York, pp. 123–127.

3. Atherton, E., Clive, D. L. J., & Sheppard, R. C. (1975) *J. Amer. Chem. Soc.* **97,** 6584-6585.

4. Atherton, E., Clive, D. L. J., East, D. A. & Sheppard, R. C. (1976) in *Peptides 1976 Proc. 14th European Peptide Symposium,* Loffet, A., Ed., Edition de l'Universite de Bruxelles, pp. 291–297.

5. Gait, M. J. & Sheppard, R. C. (1976) *J. Amer. Chem. Soc.* **98,** 8514–8516.

6. Gait, M. J. & Sheppard, R. C. (1977) *Nucleic Acids Res.* **4,** 1135–1158.

7. Atherton, E., Bridgen, J. & Sheppard, R. C. (1975) *FEBS Lett* **64,** 173–175.

8. Bradbury, A. F., Smythe, D. G., Snell, C. R., Birdsall, N. J. M. & Hulme, E. C. (1976) *Nature* **260,** 793–795.

9. Loh, H. H., Tseng, L. F., Wei, E. & Li, C. H. (1976) *Proc. Nat. Acad. Sci. USA* **73,** 2895–2898.

10. Merrifield, R. B., Mitchell, A. R. & Clarke, J. E. (1974) *J. Org. Chem.* **39,** 660–668.

11. Li, C. H., Yamashiro, D., Tseng, L-F. & Loh, H. H. (1977) *J. Med. Chem.* **20,** 325-328.

12. Cf. Tregear, G. W. (1975), in *Peptides 1974 Proc. 13th European Peptide Symposium,* Wolman, Y., Ed., Wiley, New York, pp. 177-189.

INTERCHAIN REACTIONS (CYCLO-OLIGOMERIZATIONS) DURING THE CYCLIZATION OF RESIN-BOUND PEPTIDES

M. ROTHE, A. SANDER, W. FISCHER, W. MÄSTLE, and B. NELSON, *Lehrstuhl Organische Chemie II, University of Ulm, 79 Ulm, Germany*

During the cyclization of activated linear peptides, bifunctional linear molecules are commonly used which, in principle, may yield a series of linear and cyclic oligopeptides and polypeptides formed simultaneously by intermolecular reactions. Even if the ring-closure is carried out at high dilution, cyclo-oligomerizations take place depending on the structure and concentration of the linear starting peptide, the nature of the solvent, temperature etc.[1]

A new approach to the preparation of cyclic peptides which should provide for the formation even of strained rings has been developed by Fridkin et al.[2] By linking the linear peptide to an insoluble polymer at only low degree of substitution, a situation approaching infinite dilution should be obtained by isolation of the individual molecules from each other within the polymer with no possibility of interaction.[2,3]

Nevertheless, severe difficulties have been encountered in the utilization of this method.[4] Only few examples of cyclization on polymer supports have been reported so far, viz. the preparation of some cyclic dipeptides (in rather low yields!), tetrapeptides, and a hexapeptide.[2,5] Therefore, we reinvestigated the problem using the linear subunit Val-Orn(Pht)-Leu-D-Phe-Pro of the cyclic decapeptide gramicidin S for the ring-closure on solid supports. Earlier studies of our group[6] and others[7] resulted in the detection of interchain reactions during the usual Merrifield method, i.e. the occurrence of intermolecular aminolysis of the benzyl ester link between the peptide chain and the resin. Thus, we expected a considerable amount of interchain reactions (cyclo-oligomerizations) during the cyclization of the peptides linked to the resin by an activated ester bond. These reactions should be detected by the formation of gramicidin S consisting of 2 identical pentapeptide units.

The synthesis of the pentapeptide and the cyclization reactions were performed according to the safety-catch method used by Flanigan and Marshall.[5] (See Scheme 1.)

At first, the Boc-protected pentapeptide mentioned above was synthesized on a cross-linked (4-hydroxy)phenylthiomethyl polystyrene. Subsequently it was activated by oxidation of the thioether group to the sulfone by *m*-chloroperbenzoic acid. After deprotection with TFA/CH_2Cl_2 and neutralization of the protonated α-amino group by an excess of triethylamine the polymer-bound activated pentapeptide sulfonylphenyl ester was cyclized in different solvents (DMF, CH_2Cl_2, benzene, pyridine) at room temperature and at 80°C.

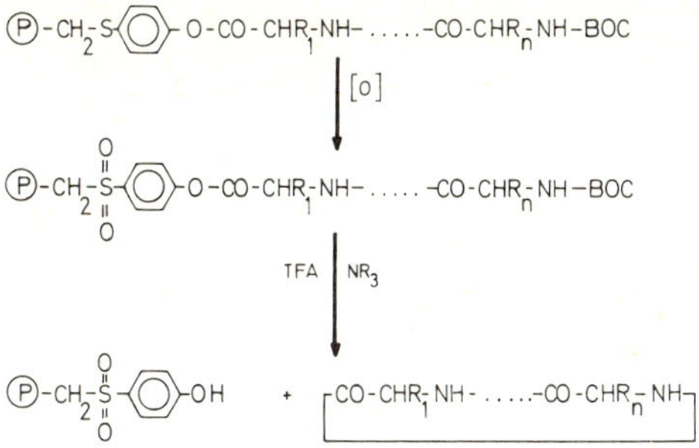

Scheme 1. Cyclization of resin-bound peptides.

Ring-closure occurred during the first hour of the reaction. In all the cyclizations studied the cyclodecapeptide, diphthalyl gramicidin S, formed by cyclodimerization was obtained as the main product. In addition, different amounts of higher ring oligomers were formed depending on the solvent used, viz. the cyclic trimer (triphthalyl sesquigramicidin S) and the 60-membered cyclic tetramer (tetraphthalyl digramicidin S). They were separated from each other and from still higher cyclopeptides (n>4, M_r 2,800) by gel permeation chromatography (Fig. 1). Suprisingly, the formation of the "monomeric" cyclopentapeptide (phthalyl semigramicidin S) was not observed at all at room temperature. Actually, only interchain reactions had taken place. At 80°C very low yields (up to 3%) of this compound were obtained. The cyclization products were identified by amino acid analysis and IR spectra. Mass spectra were obtained up to the dimer (m/e = 1,401). The ring sizes of the higher oligomers were calculated from a straight line obtained by plotting the elution volume versus log M_r.

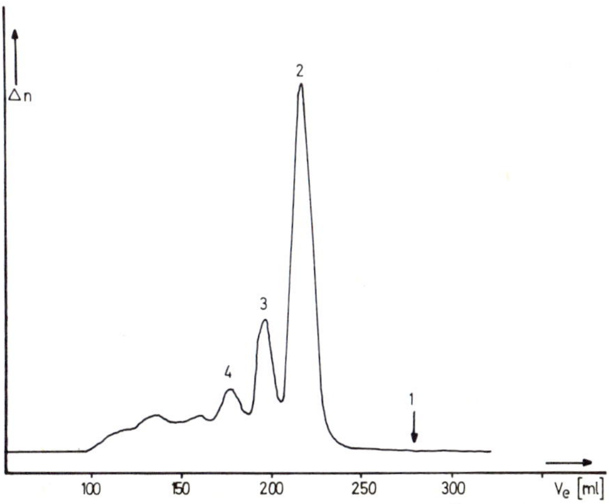

Fig. 1. Cyclization of resin-bound gramicidin S – pentapeptide. Gel permeation chromatography of c[Val-Orn(Pht)-Leu-D-Phe-Pro]$_n$(n=1,2,3,4); Sephadex LH20/MeOH.

As mentioned above, cyclodimerization proved to be the main reaction and the degree of cyclo-oligomerization is strongly solvent-dependent. In DMF and CH_2Cl_2 only 25-35% of gramicidin S and 20-30% of higher rings were obtained. Thus, the formation of the latter compounds is strongly favored in more polar solvents. In pyridine and benzene, practically no higher rings are formed whereas the yields of diphthalyl gramicidin S were the highest obtained (45%). In CH_2Cl_2 all the smaller rings (dimer and trimer) were obtained within the first 24 hours. A new cyclization of the same resin after repeated filtration and washing of the support resulted in the formation of exclusively very large rings (M_r>2800) after 10 days. As expected high molecular weight peptide chains and rings are formed particularly during the later stages of the reaction. Cyclizations at 80°C gave higher yields of the cyclopeptide mixture; cyclopentapeptide as well as the highest rings being formed. An influence of the cross-linking (1 and 2% DVB, resp.) of the resin on the relative ratio of the rings up to the tetramer could not be found but the content of the rings with n>4 increases with decreasing amount of cross-linking.

The steric isolation of the cyclizing peptide by attachment to a solid support and the exclusion of interchain reactions cannot be achieved in this way.[3] Instead, the expected cyclopentapeptide (which is in fact formed in solution) is only obtained in negligible amounts at elevated temperatures. Thus, the polymeric support causes steric hindrance of the cyclization to the "monomeric" ring, whereas interchain reactions with subsequent ring-closure are predominant. This may be explained by the formation of antiparallel pleated sheet structures of the oligomeric chains in which the chain ends come close together yielding the larger rings more easily.

As expected from these results, small rings are generally formed with more difficulties. We therefore cyclized linear tripeptides of *N*-alkyl amino acids on the solid support[8] which are known to give cyclotripeptides.[1] Ring-closure of Pro-Sar-Sar on the resin yielded about 50% of a cyclopeptide mixture (at 80°C even 66%) consisting of the complete series of cyclic oligotripeptides up to the 72-membered tetracosapeptide (n = 8) which again were separated by gel permeation chromatography. In pyridine, mainly the cyclotripeptide and cyclohexapeptide are formed, in DMF, however, the higher rings. Cyclization of triproline yields more than 80% of the 9-membered cyclotripeptide in dilute solution. In contrast, ring-closure on the resin gave only rather low yields (24% at 20°C, 45% at 80°C, DMF).

Interchain reactions are also responsible for the occurrence of cyclopeptides with failure sequences. During the preparation of the sequence Pro-Pro-Gly on the resin from Boc-Pro-Pro and glycyl polymer, the glycine residues that failed to couple are activated after the oxidation. During the cyclization they can react with the correct tripeptide, hexapeptide or nonapeptide sequence to give active esters of the sequence Gly-(Pro-Pro-Gly)$_n$ followed by ring-closure. Cyclopeptides *c*[Gly-(Pro-Pro-Gly)]$_n$ (n = 1-3) have been identified by mass spectrometry in addition to the expected ring oligomers (n = 2,3).

The so-far unknown cyclo-oligomerization of resin-bound peptides may explain the difficulties encountered with this method up to now. The occurrence of extensive interchain reactions shows that solid-phase synthesis is really no synthesis on a "solid" phase but takes place on highly solvated, flexible polymer chains exercising various conformations–more or less in solution.

We gratefully acknowledge financial support for this work by a grant from the DFG and the Fonds der Chemischen Industrie. This paper is dedicated to Prof. G. Manecke on the occasion of his 60th birthday.

References

1. Rothe, M., Theysohn, R., Mühlhausen, D., Eisenbeiss, F., & Schindler, W. (1972) in *Chemistry and Biology of Peptides,* Meienhofer, J., Ed., Ann Arbor Science Publishers, Ann Arbor, Mich., pp. 51–57.
2. Fridkin, M., Patchornik, A., & Katchalski, E. (1965) *J. Amer. Chem. Soc.* 87, 4646–4648.
3. Crowley, J. I. & Rapoport, H. (1976) *Acc. Chem. Res.* 9, 135–144.
4. Wieland, Th. (1972) in *Chemistry and Biology of Peptides,* Meienhofer, J., Ed., Ann Arbor Science Publishers, Ann Arbor, Mich., pp. 97–99.
5. Flanigan, E. & Marshall, G. R. (1970) *Tetrahedron Lett.* 2403–2406.
6. Rothe, M. & Mazánek, J. (1974) Justus Liebig's *Ann. Chem.* 439–459.
7. Beyerman, H. C., de Leer, E. W. & Van Vossen, W. (1972) *J. Chem. Soc. Chem. Commun.* 929.
8. Rothe, M. & Nelson, B. (1973) "Side Reactions during Solid-Phase Peptide Synthesis", Chemiedozententagung, Münster, Germany, Juli 23.

TOTAL SYNTHESIS ON GEL PHASE OF THE MAST-CELL-DEGRANULATING PEPTIDE BY FRAGMENT CONDENSATION.

CHRISTIAN BIRR, MARGOT WENGERT-MÜLLER and ANGELIKI BUKU. *Max-Planck-Institut für Medizinische Forschung, Abteilung Naturstoff-Chemie Jahnstr. 29, D-6900 Heidelberg 1, FRG.*

These investigations were initiated to develop an optimal route for the preparation of the mast-cell-degranulating (MCD) peptide, so that the structure-function relationships[1] of this strongly basic bee venom component, capable of counteracting rheumatic diseases[2], could be elucidated. The fully protected eicosadipeptide VII (Fig. 1) representing the structure of the MCD peptide[3] from bee venom was synthesized in the gel phase at the same time and subsequent[4] to the recently published conventional synthesis in solution[5].

The starting Ddz-amino acids were loaded onto the gel phase (<0.5% DVB-copolystyrene[6] by the cesium salt procedure[7], with complete conversion of the bromacetyl[8] anchor functions (0.5-1.1 meq/g). All operations on gel phase, including the release of fully protected peptide acids from the polymer by base hydrolysis (benzyltrimethylammonium hydroxide in MeOH/dioxane 1:3), were performed in the centrifugal reactor[9]. They were monitored continuously by photometry of circulating reaction liquids, based on the spectroscopic properties of the N^{α}-protecting Ddz-group[10] and its fission products, which also were utilized quantitatively for determinations of peptide incorporation on the polymer.

In preliminary syntheses on phenacetyl-type polymers we experienced conversion and blockage of parts of the initially loaded amino-acid residues during deprotection (5% TFA/CH_2Cl_2) and deprotonation (10% TEA in DMF/CH_2Cl_2 1:4). The side reaction was accompanied by a bright yellow coloration of the support and decreased incorporation of the second amino acid, due to a cyclol-type transformation of the C-terminal moiety (Fig. 2). Depending on diketopiperazine release (Fig. 3) on the dipeptide stage, a further decrease in the incorporation of the third amino acid was determined spectrophotometrically.

Both effects are drastically diminished in bulky sequences like Ddz-Ile-Lys(Z)-OCH_2CO-polystyrene and Ddz-Ile-Cys(Acm)-OCH_2CO-polystyrene; MCD-sequences (10-11) and (14-15). To circumvent completely these undesirable losses related to the use of phenacetyl binding sites on the polymer, which greatly facilitate the subsequent detachment of fragments via base hydrolysis, fully

Ddz-Ile-Lys(Z)-Cys(Acm)-Asn-Cys(SBu^t)-Lys(Z)-Arg(Tos)-His-Val-Ile-
-Lys(Z)-Pro-His-Ile-Cys(Acm)-Arg(Tos)-Lys(Z)-Ile-Cys(SBu^t)-Gly-
-Lys(Z)-Asp(OBzl)

Fig. 1. Fully protected MCD-peptide, VII. Ddz, α,α-dimethyl-3,5-dimethoxybenzyloxycarbonyl.

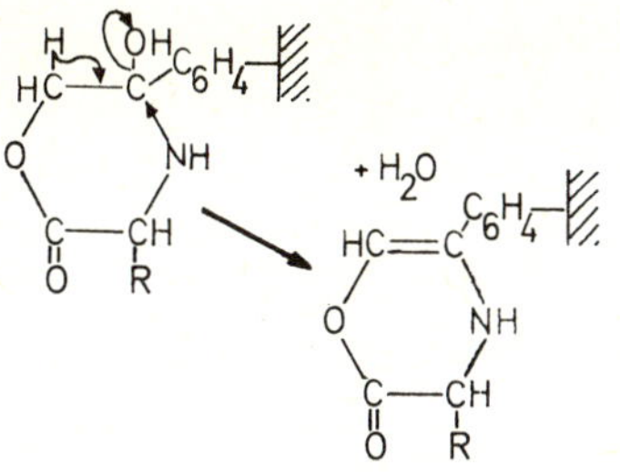

Fig. 2. Cyclol-type conversion.

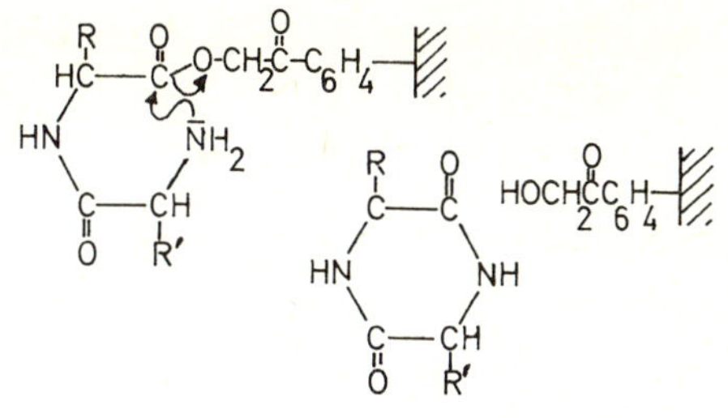

Fig. 3. Release of diketopiperazine.

protected tripeptide cesium salts (1.5-fold excess) can be attached onto the phenacetyl support in DMF with complete reaction of the bromo functions[11] at 40°C in 3 days. To suppress the formation of the cyclol and diketopiperazine on the first two stages of an actual synthesis, the deprotonation was shortened to two minutes. DCC was added prior to each Ddz-amino acid, both at least 3 times in a 3-fold excess calculated from the initial load on the polymer. 3-Nitrophthalic anhydride (0.1 molar/pyridine/10 min) was used as a blocker to suppress the formation of false sequences[12]. The urea derivative was rinsed out by CH_2Cl_2/MeOH (4:1). CH_2Cl_2 was the general solvent and washing medium. The syntheses and condensations of the following fragments are described in Fig. 4; yields are in brackets: I, 1-7 (82%); II, 8-11 (77%); III, 12-15 (50%); IV, 16-22 (84%); V, 1-11 (53%); VI, 12-22 (24%); VII, 1-22 (32%).

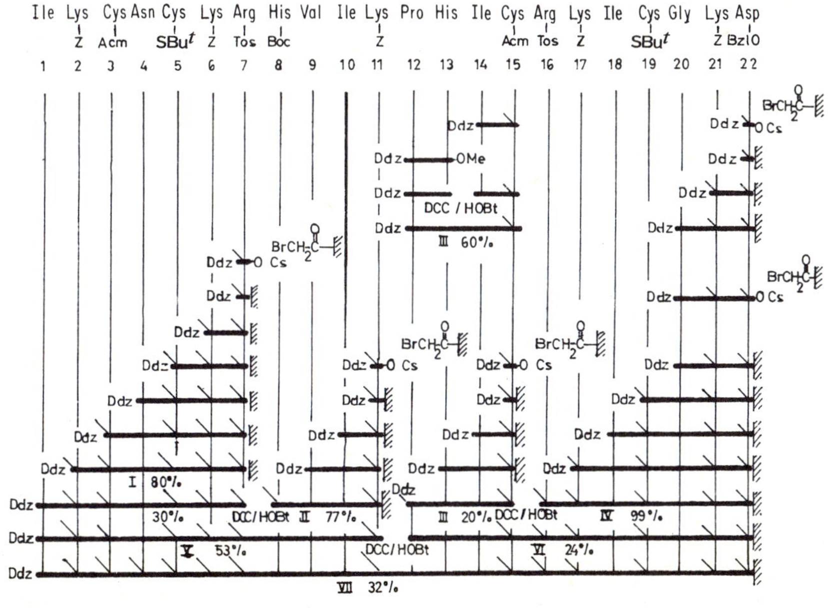

Fig. 4. Syntheses of MCD-peptide fragments.

After detachment, I was purified on Sephadex LH 20 with dioxane/MeOH (1:1) and on silica gel with $CHCl_3$/MeOH gradient; yield 34%, based on initial load on the polymer. The amino-acid analysis of the heptapeptide was in good agreement with theory.

Compound II remained on support for condensation with I (0.1 equiv., related to II on the polymer) by the DCC/HOBt procedure[13]. After 3 days complete incorporation of I was detected by Ddz-determinations in the reaction solution. V was released as described above and purified on silica gel with $ChCl_3$/MeOH gradient; yields 53% (126 mg); R_f ($CHCl_3$/MeOH/AcOH 85:10:5) 0.45. The amino-acid analysis showed the proper proportions.

Compound III was cleaved from the polymer. Solubility problems during work-up because of de-shielded histidine (final yield 20%, based on initial load on polymer) induced us to repeat the tetrapeptide synthesis from dipeptides in solution; yield of the crude product 60%. III was purified on silica gel with $ChCl_3$/MeOH gradient and crystallized from MeOH; mp. 170°C. The amino-acid analysis was in agreement with theory.

Compound IV was built up after re-attachment of purified Ddz-Gly-Lys(Z)-Asp(OBzl) cesium salt (MCD-sequence 20-22) onto bromoacetyl polystyrene. Starting from 0.55 meq/g tripeptide load on the polymer, the sequence was elongated stepwise up to position 16 with complete transformation of the individual Ddz-amino acids as determined photometrically.

Homogeneous IV on the polymer (4.5 gr; 0.21 meq/g) was elongated by 0.9 meq of pure III utilizing the DCC/HOBt method in DMF. After 3 days, 73% incorporation of III was measured by Ddz-determinations in the reaction solution. Remaining free amino functions from IV were blocked by 3-nitrophthalic anhydride. Yield of the MCD fragment VI (12-22) on the polymer was 24%, as determined after Ddz-deprotection.

The final fragment condensation was performed with 126 mg of V (1-11) onto VI on the polymer (3-fold excess) by the DCC/HOBt method in DMF. After 5 days completed incorporation of V was confirmed as described above, yielding 32% of the entirely protected MCD peptide. A quantitative Edman degradation on the polymer[14] is in progress. The product isolation and the bio-assay of the oxidized free eicosadipeptide for comparison with our former conventional preparation will be published elsewhere.

References

1. Breithaupt, H. & Habermann, E. (1968) *Naunyn-Schmiedebergs Arch. Pharmakol. Exp. Path.* **261**, 252–270.

2. Banks, B. E. C., Rumjanek, F. D., Sinclair, N. M. & Vernon, C. A. (1976) *Bull. Inst. Pasteur* **74**, 137–144.

3. Habermann, E. & Breithaupt, H. (1968) *Naunyn-Schmiedebergs Arch. Pharmakol. Exp. Path.* **260**, 127–128.

4. Wengert-Müller, M. (1977) Thesis, Heidelberg University, in preparation.

5. Buku, A., Altmann, R. & Birr. Chr. (1976) in *Peptides 1976* Loffet, A. Ed., Edition de l'Université de Bruxelles, pp. 153–157.

6. Birr, Chr. (1973) *Justus Liebigs Ann. Chem.* **1973**, 1652–1662.

7. Gisin, B. F. (1973) *Helv. Chim. Acta* **56**, 1476–1482.

8. Mizoguchi, T., Shigezane, K. & Takamura, N. (1970) *Chem. Pharm. Bull.* **18**, 1465–1474.

9. Birr, Chr. & Lochinger, W. (1971) *Synthesis* **1971**, 319-321.

10. Birr, Chr. (1975) in *Peptides 1974,* Wolman, Y. Ed., J. Wiley & Sons, New York-Toronto, pp. 117-122.

11. Wang, S. S., Gisin, B. F., Winter, D. P. Makofske, R., Kulesha, I. D., Tzougraki, C. & Meienhofer, J. (1977), *J. Org. Chem.* **42**, 1286–1290.

12. Penke, B. & Birr. Chr. (1974) *Justus Liebigs Ann. Chem.* **1974**, 1999-2002.

13. König, W. & Geiger, R. (1970) *Chem. Ber.* **103**, 788–798.

14. Birr, Chr. & Frank, R. (1975) *FEBS Lett* **55**, 65–67.

ALTERNATING LIQUID-SOLID PHASE SYNTHESIS OF LUTEINIZING HORMONE-RELEASING FACTOR

HARTMUT FRANK, HANSPAUL HAGENMAIER, ERNST BAYER, *Institut für Organische Chemie, D-7400 Tübingen, Auf der Morgenstelle 18, Fed. Rep. Germany,* and
DOMINIC M. DESIDERIO, *Institute for Lipid Research, Baylor College of Medicine, Houston, Texas 77030, USA.*

Synthesis of biologically active peptides for clinical purposes requires a strategy which (a) enables a side-product-free synthesis of every intermediate and (b) allows analytical procedures for control of reactions and purity of sequential peptides. Syntheses on polymeric supports[1,2] are appropriate for preparation of peptides for experiments in vitro but do not meet the above requirements.

Alternating liquid-solid phase peptide synthesis has been devised as a repetitive method for synthesis of pure protected peptides[3,4]. It allows the purification of each sequential peptide while retaining the simplicity of removal of reagents and by-products typical for support methods. Coupling of a sequential peptide with an excess of amino acid attached to a polymeric support through its temporary amino-protecting group affords clear distinction of reacted and unreacted species: the reaction product becomes attached to the polymer, while the reagents, side-products and, if the coupling reaction is incomplete, precursor stay in solution. In contrast to solid phase peptide synthesis, the progress of the coupling reaction can be controlled with high accuracy by photometric determination of amino groups in the filtrate with ninhydrin or fluorescamine[5].

The goal of this investigation is to study the potential of alternating liquid-solid phase peptide synthesis for the preparation of pure, biologically active peptides. Luteinizing hormone-releasing factor (LRF) with the sequence <Glu-His-Trp-Ser-Tyr-Gly-Arg-Pro-Gly-NH_2 has been selected as example. The peptide is synthesized in a stepwise manner employing amino acids protected at the α-amino group with the polymeric α-polystyryl benzyloxycarbonyl residue (Pboc) except for histidine, where the Boc residue is used.

In Table I the Pboc-amino acids, prepared as described elsewhere[6], are listed. The diphenylmethyl residue[7] (Ph_2Me) is employed to mask the C-terminal carboxamide group. Prolyl-glycyl-diphenylmethylamide trifluoroacetate as amino component is coupled with a four-fold excess of Pboc-Arg-(NO_2) preactivated for two hours with dicyclohexylcarbodiimide(DCC) and *N*-hydroxysuccinimide (HOSu) in DMF/CH_2Cl_2. During the first two hours the solution is kept neutral by gradual addition of *N*-methylmorpholine. The progress of the coupling reaction is determined quantitatively in the filtrate of the reaction mixture (Fig. 1). After two hours at 0°C the resin is rinsed alternately with DMF and EtOH and three times each with CH_2Cl_2, EtOH, dioxane, 10% AcOH in dioxane, 10%

AcOH in CH_2Cl_2, CH_2Cl_2, EtOH and ether. The Pboc-group is cleaved with 20% TFA in CH_2Cl_2 and the polymer is rinsed with DMF until the filtrate remains ninhydrin-negative. The solvents are evaporated, tripeptide trifluoroacetate precipitated with ether, redissolved in CH_2Cl_2/BuOH (3:1) and extracted with half-saturated, cold NaCl solution. During extraction the aqueous phase is adjusted to pH 4 with $NaHCO_3$. After drying, the organic phase is concentrated *in vacuo* and the oily residue triturated with ether. The precipitated tripeptide is tested by amino acid analysis and reversed-phase high performance liquid chromatography (HPLC). Elongation of the peptide chain up to the octapeptide is performed in a similar manner. After incorporation of the dichlorobenzyltyrosine residue separation from amino acid is effected by precipitation of the peptide diphenylmethylamide trifluoroacetate from DMF with cold half-saturated NaCl solution. A crucial question is whether the preceding sequential peptide can be retained so strongly on the resin by ionic forces or hydrogen bonding that it might still contaminate the elongated peptide chain.

Therefore the sequential peptides are subjected to reversed-phase HPLC. Due to the UV absorption of the C-terminal diphenylmethyl residue and of most protecting groups, amounts of less than 0.5 nmol are detected without prior derivatization. In Figure 2 the chromatogram of about 50 nmol of synthetic tripeptide is shown, before and after addition of 0.5 nmol of dipeptide. In all cases investigated, we could not detect the truncated peptide sequence. The amino-acid composition is determined after each elongation step (Table II). The completely assembled peptide is deprotected with hydrogen fluoride in the presence of anisole and dithioethane and purified by partition chromatography[8] on Sephadex LH-20 and column chromatography on Sephadex G-25 superfine. The overall yield is 26% based on dipeptide.

Purity of the synthetic peptide is tested by HPLC[9] yielding a single peak, thin-layer chromatography in seven different systems, amino-acid analysis and elemental analysis. The biological activity is 82% (± 50); in clinical application the material proved to be highly active in stimulating the release of luteinizing hormone. Racemization of amino-acid residues has been determined by gas chromatographic separation of *N*-pentafluoropropionyl-D,L-amino-acid isopropyl esters[11,12] after conversion of arginine into the more volatile ornithine: Pro 0.5%, Leu 0.5% (0.1%), Ser 15.1% (2.0) Glu 1.4% (0.1%) Tyr 0.3% (0.1%), Arg(Orn) 0.4% (0.1%) His 2.0% (n.d.), (racemization determined for L-amino acids treated in the same manner). High racemization of serine has been observed also in insulin A-chain synthesized in our laboratory by liquid phase method. This phenomenon is under investigation. The result of our investigation proves that alternating liquid-solid phase peptide synthesis can be employed for the synthesis of LRF. We consider that synthetic methods, employing the principle of phase transfer – not necessarily soluble/insoluble –, offer a great potential for the synthesis of pure, medium-sized peptide hormones in a stepwise manner similar in simplicity to the support-methods.

Table I. Amino-Acid Residues per Gram of Derivatized Polymer.

Arg(NO_2)	1.32 mmole
Leu	1.48 mmole
Gly	1.63 mmole
Tyr(Cl_2Bzl)	1.31 mmole
Ser(Bzl)	1.54 mmole
Trp	1.38 mmole

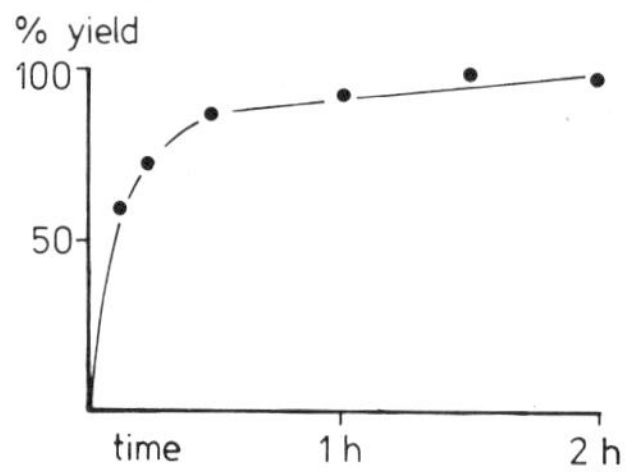

Fig. 1. Coupling of TFA-Pro-Gly-NH(Ph_2Me) and Pboc-Arg(NO_2) with DCC/HOSu.

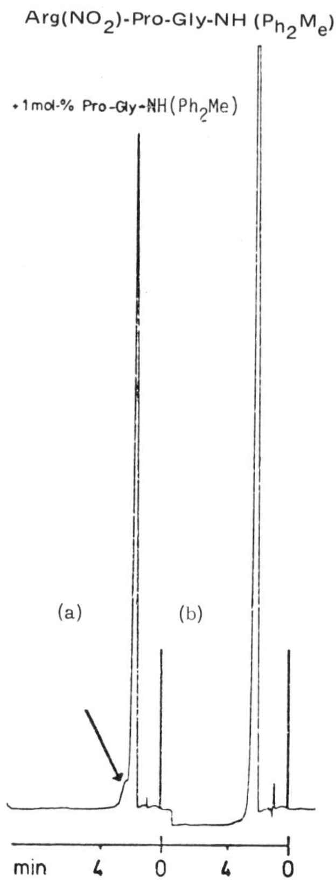

Fig. 2. Reversed phase HPLC of tripeptide derivative (a) with and (b) without addition of precursor. Conditions: Hibar RP-18 (Merck, Darmstadt), 5μ, 3 × 250 mm, 1.5 ml/min, UV-detection 254 nm. Elution: acetonitrile/1.0 *M* ammonium acetate pH 5.0, 25:75.

Table II. Amino Acid Composition of Sequential Peptide Derivates (Ph_2 Me) –
NH-Gly←Pro←Arg←Leu←Gly←Tyr←Ser←Trp←His←<Glu
(Arg: NO_2; Tyr: Cl_2Bzl; Ser: Bzl; His: Tos)

Gly	1.00	1.00	2.00	2.00	2.00	2.00	2.00	2.00
Pro	1.04	1.02	1.01	1.00	1.07	0.99	0.95	0.95
Arg	0.98	1.05	1.08	0.94	1.08	0.95	0.97	0.95
Leu		1.10	0.98	0.96	1.07	0.97	1.01	1.10
Tyr				1.01	1.10	1.01	1.04	1.01
Ser					0.95	0.76	0.89	0.55
Trp						n.d.	n.d.	n.d.
His							0.82	0.94
<Glu								0.97

References

1. Merrifield, R. B. (1963), *J. Amer. Chem. Soc.* **85**, 2149–2154.

2. Bayer, E. & Mutter, M. (1972) *Nature* **237**, 512–513.

3. Frank, H. & Hagenmaier, H. (1975) *Experientia* **31**, 131–132.

4. Frank, H., Meyer, H. & Hagenmaier, H. (1975) in *Peptides*, Walter, R. and Meienhofer, J., Eds., Ann Arbor Sci. Publ., Ann Arbor, Mich. pp. 439–445.

5. Hagenmaier, H. & Mutter, M. (1974) *Tetrahedron Lett*, 767–770.

6. Frank, H., Meyer, H. & Hagenmaier, H. (1977) *Chemiker-Ztg.* **101**, 188–193.

7. Geiger, R., König, W., Wissmann, H., Geisen, K. & Enzmann, F. (1971) *Biochem. Biophys. Res. Commun.* **45**, 767–773.

8. Burgus, R. & Rivier, J. (1976), in *Peptides 1976*, Loffet, A., Ed., Université Bruxelles, pp. 85–94.

9. Besch, P., Baylor College, Houston, Tx. personal communication.

10. Bayer, E., Gil-Av, E., König, W. A., Nakaparksin, S., Oro, J. & Parr, W. (1970), *J. Amer. Chem. Soc.* **92**, 1738–1740.

11. Frank, H., Nicholson, G. J. & Bayer, E. (1977), *J. Chrom. Sci.* **15**, 174–176.

POTASSIUM CYANIDE CATALYZED TRANSESTERIFICATION: A MILD PROCEDURE FOR REMOVING PEPTIDES FROM THE MERRIFIELD RESIN WITH THE PROTECTING GROUPS INTACT

GRAHAM MOORE and DENIS MCMASTER, *Division of Pharmacology & Therapeutics, Faculty of Medicine, University of Calgary, Calgary, Alberta T2N 1N4, Canada*

A problem which can be encountered when anhydrous hydrogen fluoride is used to remove peptides from a solid support with simultaneous cleavage of side chain protecting groups, is difficulty in separating the desired peptide from partial sequence peptides present in the reaction product. When faced with this situation, an alternative approach is to remove the peptide from the resin with all its protecting groups intact, and carry out purification steps at this stage of the synthesis, when smaller contaminating peptides may often be largely extractable into an appropriate organic solvent.

Transesterification. In the solid phase peptide synthesis procedure of Merrifield, a benzyl ester type bond is used to anchor peptides to the solid support.[1] Selective cleavage of this bond may be accomplished by transesterification in the presence of alcohol and a suitable catalyst. Methods have been devised for obtaining the methyl ester[2] and the readily hydrolyzable dimethylaminoethyl ester,[3] but strongly basic conditions, which may racemize the peptide, are required in order to obtain the benzyl ester.[2,3]

Potassium cyanide catalyzed transesterification. Potassium cyanide has been shown to be a mild catalyst of transesterification reactions involving unsaturated fatty acid esters.[4] We have examined the possible use of KCN as a catalyst in transesterification reactions involving peptides linked to the Merrifield resin and have found that, by using the appropriate alcohol, the methyl, ethyl or benzyl ester can be obtained. Boc-Gly-resin or Boc-Val-resin was stirred in alcohol in the presence of 1% KCN at room temperature and the time dependent release of the Boc-amino-acid ester monitored colorimetrically after acid deprotection and subsequent reaction with ninhydrin (Fig. 1). Based on steric considerations, it might be predicted that the reaction would proceed more readily with simple amino acids and aliphatic alcohols, i.e. Boc-Gly-resin → Boc-Gly-OMe, than with more sterically hindered reactants, i.e. Boc-Val resin → Boc-Val-OBzl. In general, the data in Fig. 1A, and in particular the observation that no transesterification took place in *t*-butyl alcohol, are in accord with this hypothesis. However, the initial rate of transesterification of Boc-Gly-resin in benzyl alcohol

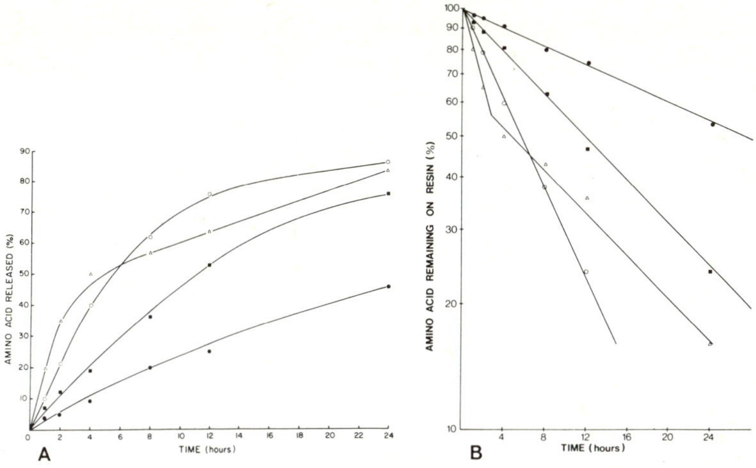

Fig. 1. Transesterification with potassium cyanide and alcohol. (A) Time dependent release of amino acid from the resin, (B) semilogarithmic plot for determination of first order rate constants. Boc-Gly-resin (200 mg; 0.63 mmole Gly/g) or Boc-Val-resin (400 mg, 0.40 mmole Val/g) was stirred in alcohol (10 ml) with KCN (100 mg) at room temperature. Aliquots (100 μl) were removed at each time interval, filtered and heated in 1*N* HCl at 95°C for 1 hour. After neutralization, the amount of amino acid present was determined by the colorimetric ninhydrin method of Rosen.[5] Percent release from the resin was determined by comparison to standard curves simulating 0–100% release. ○ Boc-Gly-resin/methanol, $k = 0.124$ hr^{-1}; △ Boc-Gly-resin/benzyl alcohol, k (initial) = 0.217 hr^{-1}, k (final) = 0.058 hr^{-1}; ■ Boc-Val-resin/methanol, k = 0.058 hr^{-1}; ● Boc-Val-resin/benzyl alcohol, k = 0.023 hr^{-1}.

was greater than all other reactions studied (Fig. 1A). When the data are plotted semilogarithmically, as in Fig. 1B, the formation of Boc-Gly-OBzl appears to be characterized by two first-order rate constants. Possibly adsorption of benzyl alcohol to the polymer expedites the initial phase of the reaction resulting in the observed kinetics.

For each reaction illustrated in Fig. 1, the Boc amino-acid esters were isolated from the reaction mixture and shown by thin-layer chromatography and proton-magnetic-resonance spectroscopy to be identical to authentic derivatives synthesized by classical solution methods.

Racemization. The possibility that the amino acid attached to the resin may racemize during the transesterification reaction was investigated. Z-Ala-Val resin was converted to Z-Ala-Val-OBzl by stirring the peptide-resin with 1% KCN in benzyl alcohol for 24 hrs. After isolation of the product and deprotection by hydrogenation, no detectable racemization (< 1%) was observed using an amino-acid analyzer as described by Manning and Moore.[6]

Peptides containing aspartic or glutamic acid. Side-chain-benzyl-ester-protecting groups of aspartyl and glutamyl residues are transesterified under these conditions and the possibility that these residues might also cyclize and undergo $\alpha \rightarrow \omega$ shift warranted investigation. When Z-Asp(Bzl)-Gly-Phe-Gly-resin was stirred with 1% KCN in benzyl alcohol for 24 hours, the product was found not to contain the succinimido derivative (infrared spectrum). After deprotection by hydrogenation, the tetrapeptide was digested with carboxypeptidase A and the appearance of products monitored by amino-acid analysis. Glycine, phenylalanine and α-aspartyl-glycine were released by the enzyme. However, an unidentified product with similar chromatographic properties to β-aspartyl-glycine [possibly β-Asp-(Gly-Phe)] was also present. This suggests that aspartyl and glutamyl residues should be protected with a non-transesterifiable protecting group, e.g. *t*-butyl ester.

Transesterification in 95% aqueous alcohol. Treatment of Boc-Gly-resin with 1% KCN in 95% aqueous methanol for 24 hours resulted in the formation of Boc-Gly-O^-K^+ as the only product. By monitoring the reaction by thin-layer chromatography, it was shown that the resin bond was first transesterified and subsequently saponified. Side-chain-benzyl-ester-protecting groups of aspartyl and glutamyl residues were transesterified but not saponified under these conditions. However, when either of the reactants was sterically hindered (e.g. Boc-Val-resin/95% methanol; Boc-Gly-resin/95% benzyl alcohol), the saponification step was not complete after 24 hours. This observation, together with the possibilities of (i) racemization and (ii) cyclization of acidic residues, renders this procedure, as a method for releasing peptides from the Merrifield resin with a free C-terminus, impractical for most purposes.

Synthesis of arginine-vasotocin. Arginine-vasotocin (AVT) was synthesized by a procedure involving KCN catalyzed transesterification with benzyl alcohol. After stirring Z-Cys(Bzl)-Tyr(Bzl)-Phe-Gln-Asn-Cys(Bzl)-Pro-Arg(Tos)-Gly-resin in benzyl alcohol containing 1% KCN for 48 hours, the fully protected peptide was obtained in 55% yield after trituration with methanol and recrystallization from acetic acid/methanol. After ammonolysis, deprotection (Na/NH_3), oxidation and two Sephadex G15 chromatography steps, the AVT had the same pharmacological potency (140 rat pressor units/mg) as a sample of AVT which had been obtained by liberating the peptide from the resin directly by ammonolysis, indicating that the transesterification procedure did not "damage" the peptide.

Influence of cation, solvent and crown ether on reaction rate. Several investigations were made to determine the possibility of increasing the rate of this transesterification reaction. These included efforts to (i) increase the solubility of the catalyst (potassium cyanide is marginally soluble in benzyl alcohol) and (ii) change the swelling characteristics and/or solvent interactions of the resin. The reaction of Boc-Val-resin in benzyl alcohol was used as a model in these studies since, from a practical viewpoint, an increase in the rate of this reaction was desirable (Fig. 1).

The replacement of potassium cyanide by cesium cyanide was found to have essentially no effect on the reaction rate even though gravimetric experiments indicated that the cesium salt was more soluble in benzyl alcohol than the potassium salt. Similarly, the inclusion of dioxane or dimethyl sulfoxide in the reaction mixture (equal volume ratio with benzyl alcohol) did not increase the rate of the reaction, and, in the case of dioxane, decreased the rate. In contrast, the substitution of potassium cyanide by potassium cyanide:dibenzo-18-crown-6-ether complex (prepared by dissolving molar equivalents of KCN and the crown ether in methanol followed by complete removal of the solvent) promoted the reaction (Fig. 2A), such that the first order rate constant was almost doubled (Fig. 2B). The use of a KCN:crown-ether complex would therefore appear to be an improvement to the general method.

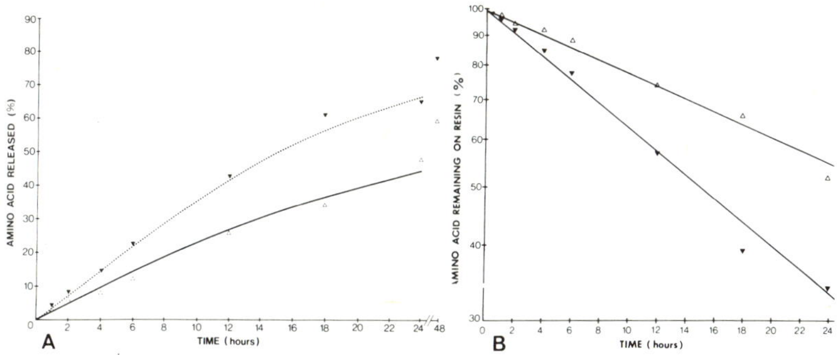

Fig. 2. Influence of crown ether on rate of transesterification. (A) Time dependent release of amino acid from resin; (B) semilogarithmic plot for determination of first order rate constants. Boc-Val-resin (400 mg; 0.40 mmole Val/g) was stirred in benzyl alcohol (10 ml) with KCN (100 mg) or KCN:dibenzo-18-crown-6-ether complex (120 mg) at room temperature. Rates of release were determined as described for Fig. 1. △ KCN, k = 0.025 hr^{-1}; ▾ KCN:crown ether, k = 0.046 hr^{-1}.

References

1. Merrifield, R. B. (1963) *J. Amer. Chem. Soc.* 85, 2149–2154.

2. Stewart, J. M. & Young, J. D. (1969) in *Solid Phase Peptide Synthesis*, W. H. Freeman and Co., San Francisco, pp. 45.

3. Barton, M. A., Lemieux, R. H. & Savoie, J. Y. (1973) *J. Amer. Chem. Soc.* **95**, 4501–4506.

4. Mori, K., Tominaga, M., Takigawa, T. & Matsui, M. (1973) *Synthesis* 790–791.

5. Rosen, H. (1957) *Arch. Biochem. Biophys.* 67, 10–15.

6. Manning, J. M. & Moore, S. (1968) *J. Biol. Chem.* **243**, 5591–5597.

CYSTEINE PROTECTION IN SOLID PHASE SYNTHESIS OF APAMIN

J. VAN RIETSCHOTEN, E. PEDROSO MULLER, and C. GRANIER, *Laboratoire de Biochimie Faculté de Médecine Secteur Nord Bd. P. Dramard, 13326 Marseille Cedex 3, France.*

Apamin[1] is the neurotoxic component of bee venom. It acts at the level of polysynaptic reflexes in the central nervous system.[2] The structure of this peptide (Fig. 1) has been determined[3-5] and the synthesis by solid-phase methodology reported.[6] *t*-Butylmercapto was used to protect the sulfhydryl function, but this group proved to be very unstable during the HF reaction. Reoxidation of the reduced peptide was optimal in 0.1 *M* Tris/HCl buffer at pH 8.0 at a concentration of about 10 μmoles of peptide per liter. Reoxidation took 48 hr by aeration in 0.1 *M* Tris or ammonium acetate. Under these conditions, synthetic apamin could be reduced and reoxidized with a yield of 70% after purification on CM-cellulose and gel filtration.

In this paper we report three protecting groups of cysteine tested for the synthesis of apamin: *p*-methoxybenzyl[7], ethylmercapto[8] and acetamidomethyl.[9] The three syntheses were performed in the same way and with the same side chain protections as for the first synthesis of apamin.[6] Reoxidation of the reduced peptide was also performed under identical conditions. Purification of the reoxidized peptide was accomplished as described[6] but with an additional equilibrium ion-exchange chromatography on CM-cellulose buffered with 0.17 *M* ammonium acetate at pH 6.0.

Cys-Asn-Cys-Lys-Ala-Pro-Glu-Thr-Ala-Leu-Cys-Ala-Arg-Arg-Cys-Gln-Gln-His-NH_2

Fig. 1. Structure of apamin.

Treatment of the peptide after assembly on the resin and before reoxidation was designed according to the protecting group of cysteine : ***p*-Methoxybenzyl-cysteine-containing apamin.** To minimize some of the undesirable side reactions of the HF treatment (1 hr, 0°C),[10,11] not only 10% anisole but also an excess of cysteine and ethylmethylthioether was added. After five washes with diethylether the resin was treated with AcOH (3 ml) and then H_2O (3 × 2 ml). The acidic solution containing the reduced peptide (yield of HF cleavage: 57%) was applied to a Biogel P_2 column equilibrated with 0.1 *M* acetic acid. The eluted peptide, detected by UV absorbance at 235 nm, was air-oxidized as described above and purified by ion-exchange chromatography and gel filtration. The purified peptide (7% of the crude peptide) was fully active with an LD_{50} of 60 μg for a 20 g mouse. **Ethylmercaptocysteine-containing apamin.** This peptide was treated as described previously[6] but without lyophilizing the partially deprotected crude peptide obtained after the HF reaction. The acidic mixture containing the crude peptide (yield of HF cleavage: 63%) was brought to pH 8.0 with conc. NH_4OH (some insoluble polymers formed). An equal volume of propanol was added, nitrogen bubbled through the mixture for 15 min and tributylphosphine added to cleave the *S*-protecting group and to reduce polymers. After 24 hr, the mixture was air-oxidized and the peptide purified as described above. The yield of fully active apamin (LD_{50} for a 20 g mouse: 61 μg) was 20% based on the crude peptide. **Acetamidomethylcysteine-containing apamin.** The HF reaction was performed in the usual way (1 hr, 0°C, 10% anisole) and the crude peptide lyophilized (66% yield). The tetra-*S*-Acm apamin was purified on CM-cellulose with a conductivity gradient from 0.11 *M* NH_4OAc pH 5.1 to 0.25 *M* NH_4OAc pH 6.3. The protected peptide eluted with approximately 0.14 *M* NH_4OAc (yield, 32% of the crude peptide). Amino-acid analyses of total enzymatic hydrolysates of several fractions of the eluate showed that only one contained close to 4 acetamidomethylcysteines and no cystine. All other fractions had a composition of 1.4 to 3.2 residues of Acm-cysteine and 0.5 to 2.5 residues of half-cystine. It is clear that some deprotection of Acm-cysteine had occurred. We were unable to determine the amount of deprotection of the peptide which remained on the resin, and it is unclear whether loss of the *S*-Acm group occurred during synthesis or during the HF reaction. The loss is estimated at 20% based on the relative amounts of polymeric peptide to monomeric peptide after a gel filtration of the crude peptide on Biogel P_4. The purified tetra-*S*-Acm apamin was treated with mercuric acetate in degassed dilute AcOH at pH 4.0 for 90 min as described by Veber et al.[12] The solution was then stirred for 20 hr in the presence of a large excess of mercaptoethanol. Desalting on a Biogel P_2 column in 0.1 *M* acetic acid was followed by air-oxidation. The desalting step is very important for complete removal of mercaptoethanol. Reoxidation in the presence of traces of this reagent yielded active apamin (LD_{50} for a 20 g mouse: 70 μg) but with a slightly different envenoming symptomatology, especially a retarded timing of the symptoms. We concluded without further evidence that some disulfide exchange occurred by reoxidation of the cysteines with mercaptoethanol. Purified apamin was recovered in 12% yield compared to crude peptide and its LD_{50} for a 20 g mouse was 58 μg.

The chemical purity and homogeneity of purified apamin were assessed by the symmetrical shape of the curve after equilibrium ion-exchange chromatography, by high voltage paper electrophoresis, and by amino acid analyses after acid hydrolysis and after total enzymatic digestion. The biological activity was tested by the quantitative toxicity assay measuring the LD_{50} for a 20 g mouse.

We found the same activity for the three peptides as for native apamin (see figures given above) and the envenoming symptomatology was identical for all toxins.

The *S-p*-methoxybenzyl protecting group provides the easiest manipulation of the peptide, but the final yield of apamin is the lowest. Using the two other protecting groups (ethylmercapto and acetamidomethyl) the yields are much better.

The correct disulfide bridging in syntheses of analogs is problematic and depends either on the right conformation of the reduced peptide in the reoxidation buffer or eventually on statistical reoxidation when the conformation has been altered. It would be most useful to have protecting groups of cysteine which are selectively removable to permit stepwise disulfide bridging as demonstrated in a recent classical synthesis of human insulin by Kamber et al.[13] But under the conditions used in solid phase synthesis no protecting group of cysteine seems to be completely stable and the ensuing difficulties increase rapidly with the number of disulfide bonds. It is however encouraging that a peptide as complex as apamin can be synthesized almost routinely in a good yield.

References

1. Habermann, E. & Reiz, K. G. (1965) *Biochem. Z.* **343,** 192–203.

2. Habermann, E. (1972) *Science* **177**, 314–322.

3. Haux, P., Sawerthal, H. & Habermann, E. (1967) *Hoppe Seyler's Z. Physiol. Chem.* **348,** 737–738.

4. Shipolini, R., Bradbury, A. F., Callewaert, G. L. & Vernon, C. A. (1967) *Chem. Commun.* 679–680.

5. Callewaert, G. L., Shipolini, R. & Vernon, C. A. (1968) *FEBS Lett.* **1,** 111–113.

6. Van Rietschoten, J., Granier, C., Rochat, H., Lissitzky, S. & Miranda, F. (1975) *Eur. J. Biochem.* **56,** 36–40.

7. Akabori, S., Sakakibara, S., Shimonishi, Y. and Nobuhara, Y. (1964) *Bull. Chem. Soc. Jap.* **37,** 433.

8. Inukai, N., Nakani, K. & Murakami, M. (1967) *Bull. Chem. Soc. Jap.* **40,** 2913–2918.

9. Veber, D. F., Milkowski, J. D., Denkewalter, R. G. & Hirschmann, R. (1968) *Tetrahedron Lett.* **26,** 3057.

10. Ivanov, V. T., Mikhaleva, I. I., Volpina, O. M., Myagkova, M. A. & Degin, V. I. (1976) in *Peptides 1976*, Loffet A., Ed., Editions de l'Université de Bruxelles, pp. 219–231.

11. Rocchi, R., Benassi, C. A., Tomatis, R., Ferroni, R. & Menegatti, E. (1976) *Int. J. Pept. Protein Res.* **8,** 167–175.

12. Veber, D. F., Milkowski, J. D., Varga, S. L., Denkewalter, R. G. & Hirschmann, R., (1972) *J. Amer. Chem. Soc.* **94,** 5456–5461.

13. Kamber, B., Hartmann, A., Jöhl, A., Mürki, F., Riniker, B., Rittel, W. & Sieber, P. (1975) in *Peptides: Chemistry, Structure, Biology.* Walter, R. and Meienhofer, J. Eds., Ann Arbor Science, Ann Arbor, pp. 477–485.

SYNTHESIS OF AN ANTIBODY FRAGMENT: CORRECTING LOSS OF PEPTIDE IN THE COURSE OF SYNTHESIS

JAMES BURTON, MICHAEL N. MARGOLIES, and EDGAR HABER. *Departments of Pathology, Surgery, and Medicine, Massachusetts General Hospital and Harvard Medical School, Boston, Massachusetts, 02114.*

The smallest antibody fragment known to have full antigen-binding activity is a noncovalent combination of the variable regions of the heavy and light chains[1]. Chemical synthesis of either domain and combination with the other, complementary variable region will yield a semisynthetic antibody fragment capable of binding antigen with the same strength and specificity as the intact immunoglobulin.

The homogeneous antibody chosen for the first synthesis, 3368, was isolated from the serum of a rabbit hyperimmunized with pneumococcal vaccine[2]. It binds to the cell wall polysaccharide, poly (β 1→3) cellobiuronic acid from type III pneumococcus. Synthesis of the 112-residue light-chain variable-region (V_L) was completed using standard solid phase techniques to yield 12 μmol of protected proteinyl-polymer. Eighty-two percent of the protein was lost from the support during synthesis (1.5% cycle^{-1}). After HF cleavage, dialysis of the product yielded a precipitate that was partly solubilized by reduction and carboxymethylation. The major fraction resulting from gel filtration of the soluble material had the expected amino-acid composition, but showed a diffuse pattern on SDS-gel electrophoresis[3]. The low yield of peptide that resulted from losses during chemical synthesis limited the number of purification steps that could be undertaken to isolate biologically active material from the crude, synthetic mixture. One objective of current research is to prevent these losses.

Two approaches designed to minimize loss of peptide from the support in the course of synthesis have been examined. First, a functional group more resistant to the reagent used for α-amino deprotection (50% TFA/CH_2Cl_2) was used to anchor the growing peptide to the polymer. Two syntheses of the model tetrapeptide B (Fig. 1) (residues 105–108 of 3368-V_L) were compared utilizing either the standard benzyl ester (BE) or the benzhydryl-amide (BHA) support. These contained 244 and 331 μmol peptide per gram of material, respectively. Treatment of both peptidyl-resins with 50% TFA/CH_2Cl_2 showed that peptide was lost from the BE-support 3.8 times faster than observed with the BHA-support (Fig. 2). Synthesis of 3368-V_L requires 57 hours of exposure to 50% TFA/CH_2Cl_2. Under these conditions 38% of the tetrapeptide would be lost from the BHA support while over 84% would be cleaved from the BE-support. The latter value correlates well with the value of 82% actually observed in the synthesis of

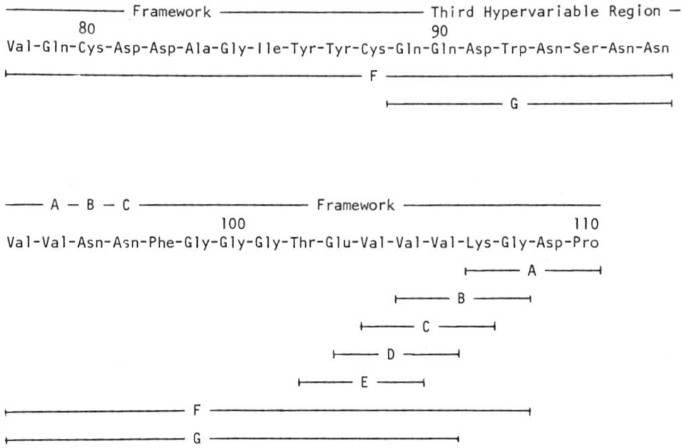

Fig. 1. Amino-acid sequence of antibody 3368 light-chain variable-region peptides.

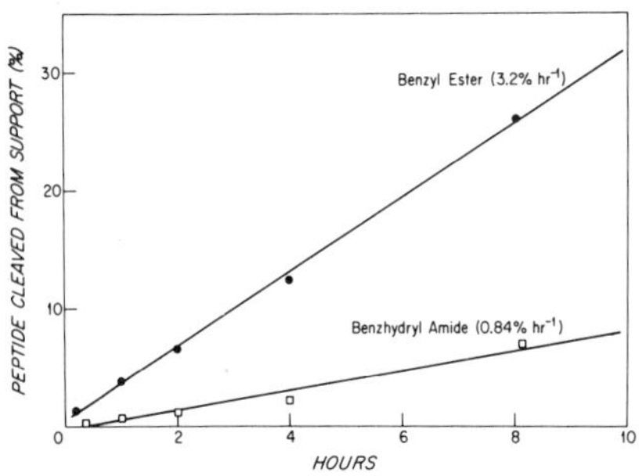

Fig. 2. Loss of Val-Val-Lys(2Cl-Z)-Gly from benzhydrylamide and benzyl ester supports in 50% TFA.

the entire 3368-V_L and demonstrates that the tetrapeptide is a useful model for predicting the behavior of the V_L during synthesis.

Three syntheses of 30+ residue peptides from the C-terminus of 3368-V_L were completed using the BHA support. After HF cleavage and gel filtration, automated Edman degradation showed all peptides to be heterogeneous. In the best synthesis, a 36-residue peptide (F), both preview and truncated sequences, were observed. The first step of Edman degradation showed in addition to the expected valine (82.5%): glutamic acid, 4.5%; glycine, 9.5%; alanine, 2.9%; tyrosine, 0.6%; and cysteine, trace. Because of the sequence heterogeneity of the peptides prepared on the BHA-resin, further attempts to synthesize V_L on the BHA-resin were abandoned in favor of the following approach.

Kamen and coworkers[4] reported that in their synthesis of lysozyme, use of HCl/dioxane rather than TFA/CH_2Cl_2 for α-amino deprotection resulted in the loss of less material from the support. An initial study showed that loss of the model tetrapeptide B from the BE-resin in 5.8 N HCl/dioxane was about two-thirds of that observed in 50% TFA/CH_2Cl_2. In subsequent studies, four additional tetrapeptide sequences, (A, C, D, and E), were evaluated as possible starting points for the synthesis of V_L fragments. The peptidyl resins used in the studies contain 250-300 μmol tetrapeptide per gram of support. The rate of loss of peptide varies between 2-5% hr^{-1} in 50% TFA/CH_2Cl_2. In HCl/dioxane, the rate of loss of peptide is less and steric factors appear to play a more significant role. Tetrapeptide E, is lost only one-seventh as rapidly as B (Fig. 3). If the synthesis of V_L is begun at Val^{105}, only 15% of the protein should be lost from the polymer in the course of synthesis.

X-ray crystallography of the structurally analogous V_L dimer REI[5] indicates that Lys^{107} is H-bonded to Ala^{13}. Other interactions of residues found between positions 105 and 110 and other parts of the antibody appear to be absent. Thus, on biological grounds, synthesis of the V_L can probably be initiated at any residue between 105 and 110.

A 20-residue fragment extending between Val^{106} and Gln^{89} was prepared on the standard BE-resin using HCl/dioxane for α-amino deprotection. Sequence analysis indicates that only the expected glutamine is present at the N-terminus.

In summary, stability of the chemical bond anchoring a peptide to the BHA resin is sufficient to permit synthesis of long sequences. Antibody sequences of intermediate length, however, are heterogeneous when prepared this way. Requisite stability may also be obtained by starting with a hindered C-terminal residue attached to the BE support and deprotecting with HCl/dioxane. Analysis of sequences prepared in the latter way indicates the peptides are homogenous at the 98+% level.

RABBIT V_L SEQUENCE

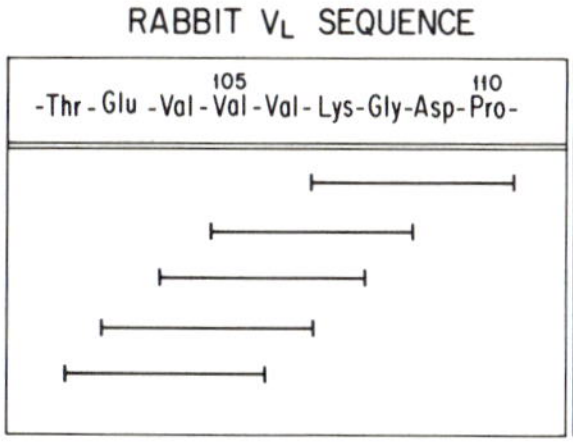

RATE OF LOSS (% hr^{-1})

	50% TFA/CH_2Cl_2	5.8N HCl/Dioxane	Ratio (TFA/HCl)
A	4.8	1.10	4.4
B	3.2	2.00	1.6
C	2.0	0.58	3.4
D	2.4	0.46	5.2
E	2.0	0.30	6.7

Fig. 3. Rate of loss of various antibody fragments from the support in 50% TFA/CH_2Cl_2 and 5.8 N HCl/dioxane.

References

1. Hochman, J., Inbar, D. & Givol, D. (1973) *Biochemistry* **12**, 1130–1135.

2. Margolies, M. N., Cannon, L. E., Strosberg, A. D. & Haber, E. (1975) *Proc. Nat. Acad. Sci. USA* 72, 2180–2184.

3. Burton, J., Rosemblatt, M. & Haber, E. (1977) in *Antibodies in Human Diagnosis and Therapy*, Haber, E. & Krause, R. M. Eds, Raven Press, New York, pp. 205–214.

4. Sharp, J. J., Robinson, A. B. & Kamen, M. D. (1973) *J. Amer. Chem. Soc.* **95**, 6097–6108.

5. Davies, D. R. & Padlan, E. A. (1977) in *Antibodies in Human Diagnosis and Therapy,* Haber, E. & Krause, R. M. Eds., Raven Press, New York, pp. 119–132.

TOTAL SYNTHESIS OF TAIWAN COBRA VENOM CARDIOTOXIN

KUNG-TSUNG WANG and CHI-HUEY WONG, *Institute of Biological Chemistry, Academia Sinica, Taipei, Taiwan, R.O.C.*

Taiwan cobra venom cardiotoxin (CTX III) contains 60 amino-acid residues with 4 disulfide bonds[1]. It is a good example for comparison of stepwise solid-phase synthesis and fragment solid-phase synthesis. A preliminary synthesis by stepwise method gave 75% of biological activity[2]. We resynthesized cardiotoxin by stepwise and fragment solid-phase methods.

The procedures of the stepwise synthesis were as described by Merrifield[3] with some modifications. For fragment solid-phase synthesis, cardiotoxin was divided into 16 fragments with terminal amino groups protected by either Boc- or Nps- groups. The other protecting groups were the same as in the stepwise synthesis.[2] The first three C-terminal amino-acid residues were coupled stepwise and then fragment coupling was started. The procedure was similar to the stepwise synthesis except that the coupling reactions were carried out by DCC-HOBt in DMF and reaction times were extended (2 days or more). The scheme for the fragment method is illustrated in Fig. 1.

The final peptide-resin at the end of synthesis was treated with liquid HF to liberate free peptides. After gel chromatography with Sephadex G-50, the folding of the deprotected peptide and disulfide bond formation was carried out by spontaneous air oxidation in 0.05 *M* phosphate buffer (pH 6.8) with protein concentration of 10^{-5} *M*. The crude cardiotoxin was then purified by Sephadex G-50 gel chromatography and repeated CMC column chromatography. The product obtained from the stepwise method and after 2nd CMC chromatography was further purified by chromatography on anti-cardiotoxin sera column.

The highly purified cardiotoxin from both methods were compared with natural cardiotoxin by disc gel electrophoresis, CD spectra (Fig. 2), amino-acid analysis, peptide maps of trypsin digest after performic acid oxidation, and biological properties (Table I). The results showed that physical and chemical properties were the same as natural cardiotoxin. The fragment solid phase method can give 100% active product with total yield of 0.5% while the stepwise method about 85% with overall yield of 0.1%.

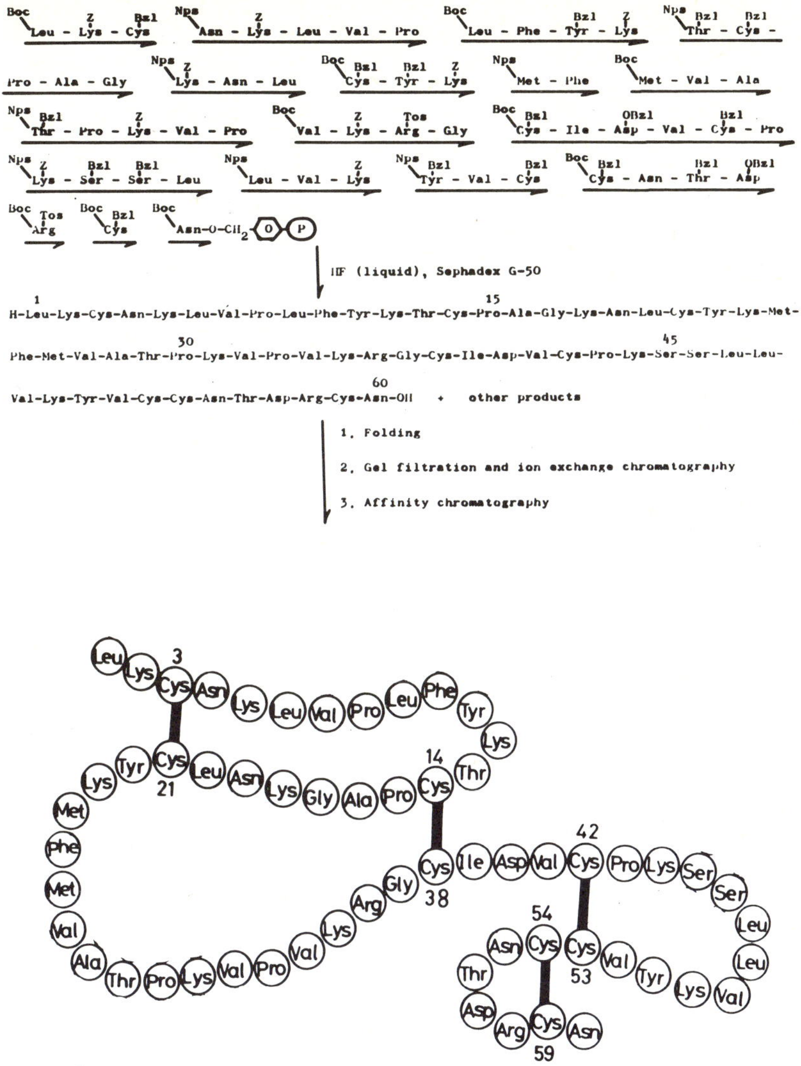

Fig. 1. The fragment solid phase synthesis of cobra venom cardiotoxin (CTX III).

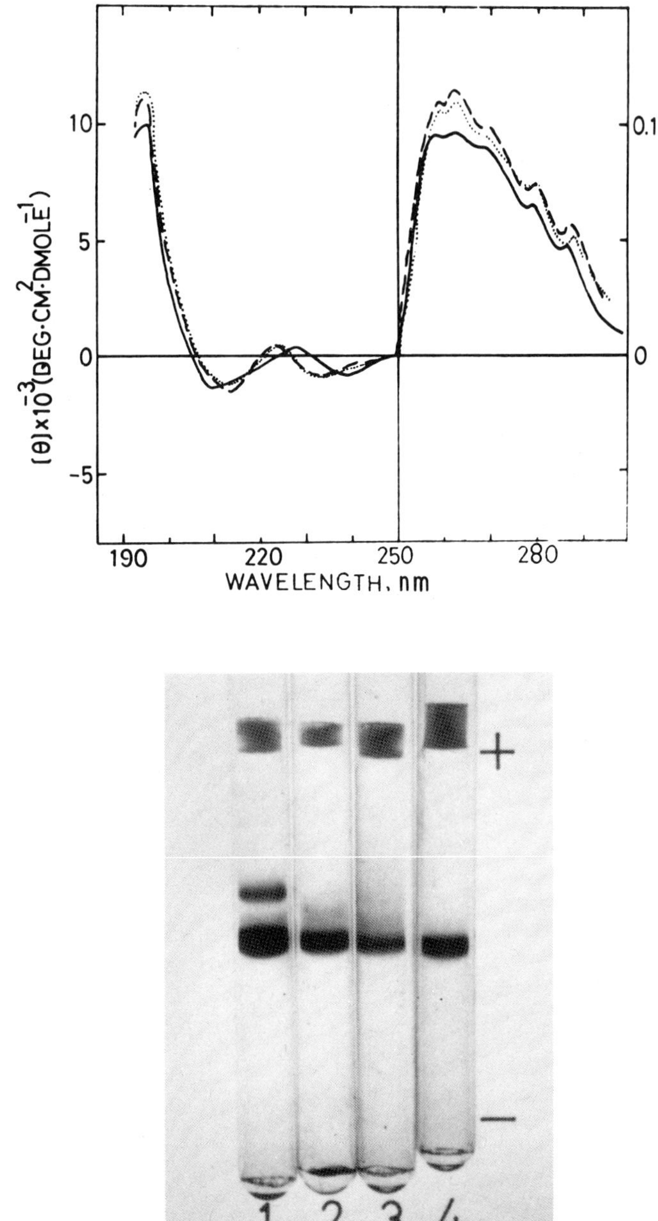

Fig. 2. (above) CD spectra of synthesized cardiotoxin. (–), stepwise method; (. . . .), fragment method; (– –), native cardiotoxin. (below) Disc gel electrophoresis of synthesized cardiotoxin. 1. stepwise method (2nd CMC) 2. stepwise method (2nd CMC and affinity chromatography) 3. fragment method (2nd CMC) 4. native cardiotoxin.

Table I. Biological Properties of Synthetic Cardiotoxin

Tests	Native CTX (III)	Synthesized CTX (III)	
		Stepwise	Fragment
Immunodiffusion[a]	+	+	+
LD_{50} (μ g/g)	2.5	3.3	2.3
Haemolysis[b]	100%	88%	90%
Muscle contraction[c]	+	+	+

a. The synthetic CTX (III) showed a cross reaction identical with that of CTX at a concentration of 0.2 mg/ml.
b. The values indicated were obtained from the average of two independent assays and native CTX (III) was taken as standard.
c. The three preparations have the same contraction patterns.

References

1. Hayashi, K. Takechi, M. Kaneda, N. & Sasaki, T. (1976) *FEBS Lett.* 66, 210–214.
2. Wong, C. H. & Wang, K. T. (1976) *J. Chinese Biochem. Soc.* 5, 1–6.
3. Merrifield, R. B. (1963) *J. Amer. Chem. Soc.* 85, 2149–2154.

REMOVAL OF N^{α}-BENZYLOXYCARBONYL GROUPS FROM SULFUR-CONTAINING PEPTIDES BY CATALYTIC HYDROGENOLYSIS IN LIQUID AMMONIA: STEPWISE SYNTHESIS OF SOMATOSTATIN

ARTHUR M. FELIX, MANUEL H. JIMENEZ, and JOHANNES MEIENHOFER, *Chemical Research Department, Hoffmann-La Roche Inc., Nutley, New Jersey 07110*

The ideal protecting group combination of N^{α}-benzyloxycarbonyl groups and side-chain protection by *t*-butyl derived groups[1,2] was recently extended to include sulfur-containing peptides which are known to resist catalytic hydrogenolysis in most solvents. When liquid ammonia, a powerful solvent for both protected and free peptides, was used as solvent for palladium catalyzed hydrogenation, quantitative removal of N^{α}-benzyloxycarbonyl groups was attained from *S*-benzylcysteine- and methionine-containing peptides.[3] These observations led to a novel synthesis of oxytocin *via* incremental chain elongation[4] in which catalytic hydrogenolysis in liquid ammonia was successfully utilized for cleavage of the N^{α}-benzyloxycarbonyl groups.

The presence, at every stage, of a sulfur-containing residue (cysteine) at the C-terminal creates a severe test of the novel catalytic hydrogenolysis in liquid ammonia when this procedure is employed in the stepwise synthesis of somatostatin.[5] For side-chain protection the *t*-butyl ester, *t*-butyl ether, and *N-t*-butyloxycarbonyl groups were used. The cysteine thiol functions were protected by the *S-t*-butyl[6,7] group which was also completely stable toward palladium-catalyzed hydrogenolysis in liquid ammonia. Peptide synthesis was carried out as shown in Scheme I by incremental chain elongation with N^{α}-benzyloxycarbonyl-L-amino acids. Coupling reactions were carried out by the *N*-hydroxysuccinimide ester method[8] for the preparation of compounds **10, 14, 16, 20, 22, 24, 26.** Compounds **2, 4, 6,** and **8** were prepared by the mixed-anhydride procedure with isobutylchloroformate[9] and *N*-methylmorpholine.[10] Incorporation of Z-Asn-OH and Z-Trp-OH was achieved by the *p*-nitrophenyl ester[11,12] and carbodiimide-hydroxybenzotriazole[13] methods, respectively.

Catalytic hydrogenolysis in liquid ammonia of the N^{α}-benzyloxycarbonyl groups proceeded in good yield (average yield 87% for the 12 reductions of the di- through tridecapeptide stages). The addition of dimethylacetamide (~0.06 ml/ml NH_3) and triethylamine (~4 equiv.) improved the rate and extent of Z-group cleavage in difficult cases and has therefore been incorporated into our standard recommended procedures.[14] Several of the intermediates were purified by high performance liquid chromatography on silica gel 60.[15,16] Nuclear magnetic resonance (100 MHz) revealed characteristic chemical shifts for the *t*-butyl

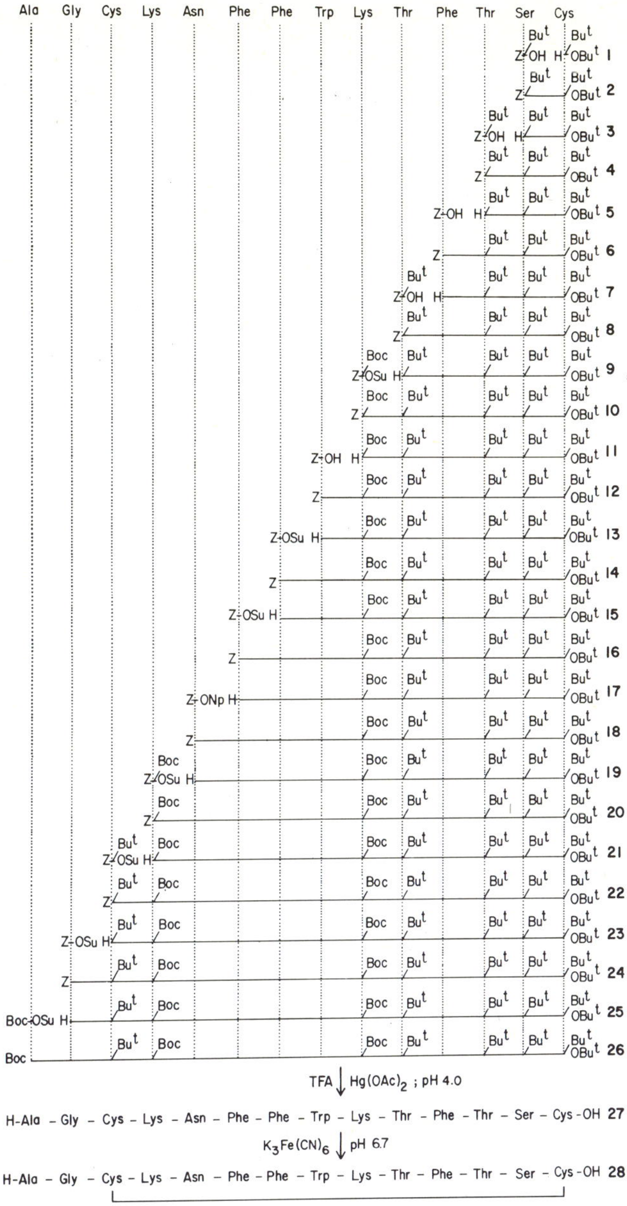

Scheme 1. Stepwise synthesis of somatostatin using catalytic hydrogenolysis in liquid ammonia for each N^{α}-Z-group cleavage.

derived protecting groups and permitted further characterization of all intermediates as exemplified in Figure 1 for the protected tetradecapeptide, **26**.

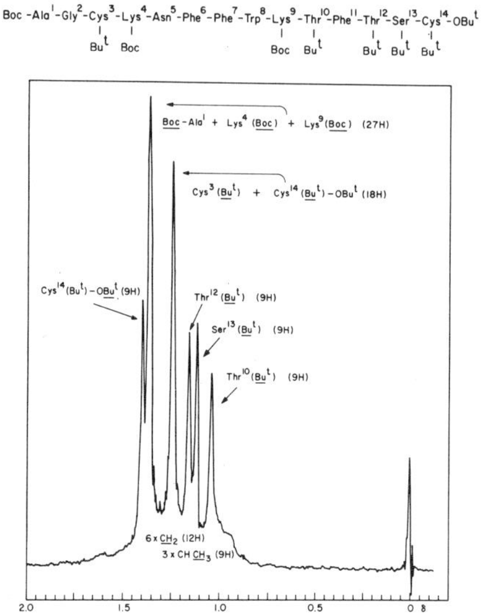

Fig. 1. The nmr spectrum (100 MHz) of the protected tetradecapeptide, **26**, in DMSO-d_6.

Deprotection of the *t*-butyl-derived groups was carried out by treatment with anhydrous trifluoroacetic acid. The removal of the *S*-*t*-butyl group under mild conditions was not previously established.[8,17,18] Consequently, this means of cysteine-thiol protection has not found general utility in peptide synthesis. We have now observed that the *S*-*t*-butyl group may be quantitatively cleaved under mild conditions by using mercuric acetate at pH 4 and 25°C for several hours. Independent studies by Fujino and Nishimura,[19] in which deprotection was achieved with mercuric trifluoroacetate, lend further support to the general utility of the *S*-*t*-butyl protecting group in peptide synthesis.

The product was purified by gel filtration and afforded chromatographically homogeneous dihydrosomatostatin, **27**. Oxidation[20] with $K_3Fe(CN)_6$ provided somatostatin, **28**, in 62% yield. Both products inhibited growth-hormone release *in vitro*.[21]

The authors thank Dr. T. Williams for the nmr spectra, Dr. T. Mowles for carrying out the growth hormone inhibition assays and Mr. T. Gabriel for assistance with the high performance liquid chromatography.

References

1. Schwyzer, R. (1966) *Naturwissenschaften* **53**, 189–197.

2. Finn, F. M. & Hofmann, K. (1976) in *The Peptides*, 3rd ed., Neurath, H. and Hill, R. L., Eds., Academic Press, New York, pp. 105–253.

3. Meienhofer, J. & Kuromizu, K. (1974) *Tetrahedron Lett.* **37**, 3259–3262.

4. Kuromizu, K. & Meienhofer, J. (1974) *J. Amer. Chem. Soc.* **96**, 4978–4981.

5. Brazeau, P., Vale, W., Burgus, R., Ling, M., Butcher, M., Rivier, J. & Guillemin, R. (1973) *Science* **179**, 77–79.

6. Chimiak, A. (1963) in *Peptides, Proceedings of the Fifth European Symposium,* Young, G. T., Ed., Pergamon Press, New York, p. 37.

7. Beyerman, H. C. (1963) in Peptides, Proceedings of the Fifth European Symposium, Young, G. T., Ed., Pergamon Press, New York, p. 53.

8. Anderson, G. W., Zimmerman, J. E. & Callahan, F. (1963) *J. Amer. Chem. Soc.* **85**, 3039.

9. Vaughan, Jr., J. R. & Osato, R. L. (1952) *J. Amer. Chem. Soc.* **74**, 676–678.

10. Anderson, G. W., Zimmerman, J. E. & Callahan, F. M. (1967) *J. Amer. Chem. Soc.* **89**, 5012-5017.

11. Bodanszky, M. & du Vigneaud, V. (1959) *J. Amer. Chem. Soc.* **81**, 5688–5691.

12. Bodanszky, M., Denning, Jr., G. S. & du Vigneaud, V. (1963) *Biochem. Prep.* **10**, 122-125.

13. König, W. & Geiger, R. (1970) *Chem. Ber.* **103**, 788–798.

14. Felix, A. M., Jimenez, M. H. & Meienhofer, J. (submitted) Org. Syn.

15. Gabriel, T. F., Michalewsky, J. & Meienhofer, J. (1976) *J. Chromatogr.* **129**, 287-293.

16. Gabriel, T. F., Jimenez, M. H., Felix, A. M., Michalewsky, J., & Meienhofer, J. (1977) *Int. J. Pept. Protein Res.* **9**, 129–136.

17. Callahan, F. M., Anderson, G. W., Paul, R. & Zimmerman, J. E. (1963) *J. Amer. Chem. Soc.* **85**, 201-207.

18. Kofod, H. (1975) in *Peptides (1974), Proceedings of the Thirteenth European Peptide symposium*, Wilman, Y., Ed., J. Wiley and Sons, New York, pp. 57–58.

19. Fujino, M. & Nishimura, O. (1976) *J. Chem. Soc. Chem. Commun.*, 998.

20. Rivier, J. E., Brown, M. R. & Vale, W. (1976) *J. Med. Chem.* **19**, 1010–1013.

21. Vale, W., Grant, G., Amoss, M., Blackwell, R. & Guillemin, R. (1972) *Endocrinology* **91**, 562-572.

PEPTIDE SYNTHESIS IN AQUEOUS SOLUTION WITH *o*-NITRO-*p*-SULFOPHENYL ESTERS

YAKIR S. KLAUSNER, TZVI H. MEIRI and EUGENIA SCHNEIDER, *Department of Organic Chemistry, The Hebrew University, Jerusalem, Israel.*

The natural medium of macromolecules such as enzymes, peptide hormones and nucleic acids is water. In organic solvents substantial changes of their conformation usually takes place. It is therefore advantageous to modify chemically and even to synthesize these molecules in water[1], a solvent which is also compatible with the methods of analysis and purification that must accompany each synthetic step.

Toward this end we explored the synthesis, water solubility and acylation properties in water, as well as in DMF of *o*-nitro-*p*-sulfophenyl esters of several *N*-protected amino acids. Attachment of a sulfonic acid residue in the form of sodium salt to the well-known *o*-nitrophenyl esters (ONo)[2,3] was anticipated to confer appreciable water solubility without considerably affecting the reactivity. Four methods for the preparation of the active esters were investigated: DCC[4], EEDQ[5], nitrophenyl-trifluoroacetate[6] and symmetrical anhydrides[7,8]. Of these, symmetrical anhydrides, employed in 2.5-3 fold excess proved to be the reagents of choice (Scheme 1). The symmetrical anhydrides were prepared in methylene-chloride[8] or tetrahydrofuran and added directly into a suspension of *o*-nitro-*p*-sulfophenol[9] sodium salt in pyridine. After removal of the methylenechloride *in vacuo* the suspension was stirred until practically complete dissolution of the phenol occurred (usually 4-16 hours). The active esters were isolated by precipitation with dry ether, followed by digestion with isopropanol or by column chromatography on silica gel.

Most of the *o*-nitro-*p*-sulfophenyl esters (ONs) so far tested have good water solubility (Table I) and are efficient acylating reagents in this solvent as well as in DMF. The well-known protected tetrapeptide[10] which comprises positions 11-14 in ACTH was synthesized in both solvents (Scheme 2).

In general, the peptide ester hydrochloride (Scheme 2) was dissolved in water and the active ester was added either in aqueous solution or as a solid. *N* NaOH was added and the pH was maintained at 8-8.5 with the aid of a pH meter.

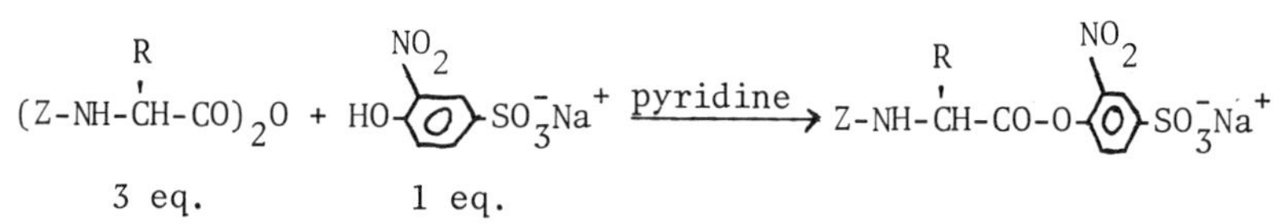

Scheme 1

Table I. Properties of Some Benzyloxycarbonylamino Acid *o*-Nitro-*p*-sulfophenyl Esters

Active ester	m p, °C[b]	$(\alpha)_D^{24}$ [c]	Yield, %[d]	Solubility in water		IR[g]	UV[h]: λ max(nm); ε	
Z-Val-ONs[a]	210-212(dec)	-17°	85	2[e]	0.04[f]	1770	253	5000
Z-Pro-ONs	105-109	-59°	72	2.5	0.05	1776	256	4600
Z-Lys(Boc)-ONs	160-162	-20°	76	not appreciably soluble		1774	255	4900
Z-Leu-ONs	198-200(dec)	-29°	94	12	0.27	1778	253	4600
Z-Phe-ONs	90- 95	-42°	64	10	0.19	1775	255	5100

[a] Suggested abbreviation for o-nitro-p-sulfophenyl esters (sodium salts): ONs. [b] The active esters do not possess sharp m p. The reported values refer to the starting of softening or the appearance of a yellow color. [c] C = 1; in dimethylformamide, containing 1% acetic acid. [d] Yield was calculated on the basis of o-nitro-p-sulfophenol sodium salt; purity was shown by t.l.c. and elemental analysis. [e] In percent. [f] In molar concentration. [g] Recorded in KBr, C=O active ester in cm^{-1}. [h] Recorded in ethanol.

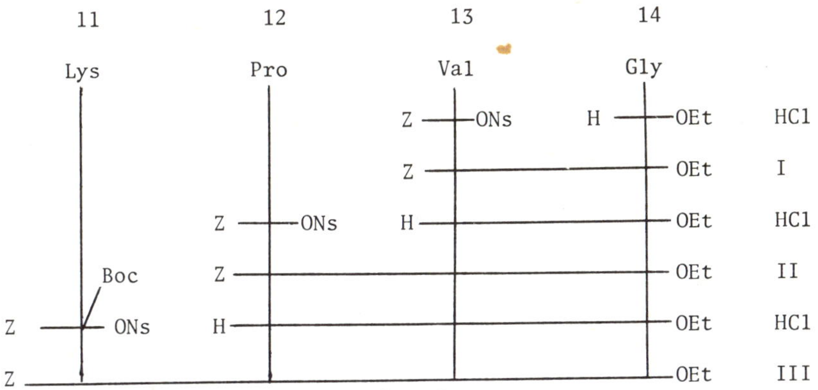

Scheme 2

Reaction times were usually short and good yields of protected peptides were usually obtained (Table II). In the case of the lysine derivative which is not appreciably soluble in water, the reaction was commenced with the active ester in suspension. Even in this case the ester was soon dissolved in water and the acylation was fast. It should be noted that the products are insoluble in water and were isolated by extraction with ethylacetate. Efficient acylation reactions took place also in DMF, and as expected, 1-hydroxybenzotriazole[11] was an excellent catalyst. The yields and physical characteristics of the tetrapeptide derivatives prepared in both solvents are given in Table II.

In a previous study it was shown[12] that the reactions between Z-Leu-ONp, Z-Leu-ONo and benzylamine in DMF proceeded with $t_{1/2}$ of 1.5 min. and 0.65 min, respectively. Under similar conditions the reaction between Z-Leu-ONs and benzylamine gave $t_{1/2}$ of 0.5 min. The reaction of Z-Leu-ONs with glutamic acid

Table II. Peptide Synthesis Using *o*-Nitro-*p*-Sulfophenyl Esters in Water or Dimethylformamide.

Peptide	Solvent	Yield, [a]%	Reaction time[b]
I	water	85	4
	DMF	89	2
II	water	88,5	2.5
	DMF	93	2
III	water	44	3
	DMF	69	5.5

[a] Purity was established by comparison of m p with the literature[10], t.l.c. and elemental analysis. 10% excess of the active ester was employed.

[b] In hours.

in water gave $t_{1/2}$ of 4.5 min and 2.1 min at pH 9 and pH 9.7 respectively. It is interesting to note that 1-hydroxybenzotriazole acts as a catalyst in aqueous solutions. Addition of 1 equivalent of HOBt to the above-mentioned reaction reduced the half reaction times to 2.2 min at pH 9 and to 1 min at pH 9.7.

References

1. Glass, J. D. Meyers, C., Schwartz, I. L. & Walter, R. (1975) in *Peptides 1974, Proc. of the 13th European Peptide Symp.*, Y. Wolman, Ed., Israel University Press, pp. 141–152.
2. Bodanszky, M. & Bath, R. J. (1969) *Chem Commun.* 1259–1260.
3. Bodanszky, M., Fink, M. L., Funk, K. W., Kondo, N., Yang-Lin, C. & Bodanszky, A. (1974) *J. Amer. Chem. Soc.* **96**, 2234–2240.
4. Sheehan, J. C. & Hess, G. P. (1955) *J. Amer. Chem. Soc.* **77**, 1067–1068.
5. Belleau, B. & Malek, G. (1968) *J. Amer. Chem. Soc.* **90**, 1651–1652.
6. Sakakibara, S. & Inukai, N. (1964) *Bull. Chem. Soc. Jap.* **37**, 1231–1232.
7. Muramatsu, I. & Hagitani, A. (1959) *Nippon Kagaku Zasshi,* **80**, 1497–1501.
8. Blake, J. & Li, C. H. (1975) *Int. J. Pept. Protein Res.* 7, 495–501.
9. Gnehm, R. & Knecht, O. (1906) *J. Prakt. Chem.* **73**, 519–537.
10. Schwyzer, R. & Rittel, W. (1961) *Helv. Chim. Acta* **44**, 159–169.
11. König, W. & Geiger, R. (1973) *Chem. Ber.* **106**, 3626–3635.
12. Bodanszky, M., Bath, R. J., Chang, A., Fink, M. L., Funk, K. W., Greenwald, S. M. & Klausner, Y. S. (1972) in *Chemistry and Biology of Peptides, Proc. of the 3rd American Peptide Symp.*, J. Meienhofer, Ed., Ann Arbor Science Publishers, pp. 203–207.

1,4-DINITRO-1,3-BUTADIENE-2,3-DICARBOXYLIC ANHYDRIDE, A NEW PEPTIDE COUPLING AGENT

BORIS WEINSTEIN, *Department of Chemistry BG-10*
University of Washington, Seattle, Washington 98195

Mixed anhydrides are useful as peptide coupling agents due to their ease of formation, quick reaction time, and facile work-up conditions. Symmetrical anhydride reagents are rather rare, but would be expected on several grounds to be better than the standard unsymmetrical ones in common use. To test these ideas, catechol on treatment with phosgene gave catechol carbonate, a known compound. Addition of methyl-L-alaninate, followed by N^{α}-benzyloxycarbonyl-L-phenylalanine, afforded in moderate yield the dipeptide. Application of the highly useful nuclear magnetic resonance procedure for the determination of racemization revealed the absence of any D-L isomer.

In order to improve the reagent, 1,4-dibromo-2,3-butanedione was refluxed with sodium nitrite in acetone to yield yellow 1,4-dinitro-2,3-butanedione. Alternately, a more efficient procedure involved the treatment of nitromethane with sodium hydride, followed by the addition of oxalyl chloride in benzene. Both preparations also furnished a small quantity of the mono isomer. Next, the dinitro-diketone was reacted with phosgene to give 1,4-dinitro-1,3-butadiene-2,3-dicarboxylic anhydride. This compound, in turn, was used to couple a variety of protected amino acids to afford dipeptides in high yield. The other product from this sequence is the starting dinitro-diketone, so the process is economical. Indeed, the latter material is water soluble and protected peptides can be isolated in remarkable purity.

The coupling agent has been used to make simple tri- and tetrapeptides, plus several interesting biological active peptides. After it has been applied to some other complex cases, then the results should indicate whether it can replace simpler anhydrides. Indeed, experience to date has shown that it could be the agent of choice with unprotected amino acids and can also provide a useful tool in solid phase procedures.

CYSTEIC ACID, A WATER SOLUBLE PROTECTING GROUP IN PEPTIDE SYNTHESIS

A. HUBBUCH, W. DANHO and H. ZAHN, *Deutsches Wollforschungsinstitut an der RWTH Aachen, 5100 Aachen, West Germany*

One of the main problems in the classical synthesis of peptides is the decreasing solubility of larger representatives of this class of compounds.[1] This creates cumbersome problems in coupling and purification. The convenient protecting groups used in peptide synthesis possess hydrophobic properties and enhance the solubility in organic solvents (Bpoc, Boc, Trt, Z). It would be of interest to investigate the influence of a hydrophilic protecting function on the solubility characteristics of peptides for the following reasons: a) Water soluble protecting groups are necessary, if the aim of peptide synthesis in aqueous solution[2] can be accomplished. b) Improved water solubility of protected synthetic peptides should provide advantages for the purification process.

There are, however, only few examples of water soluble protecting groups: The polar Msc-group[3] shows hydrophilic properties. Wieland[2] attempted the preparation of such a group, and recently Kunz reported the properties of the alkali-labile cationic Peoc-group[4]. Cysteic acid seems to be a suitable reagent for the following reasons: The sulfo group being the salt of a very strong acid, bears a negative charge within a wide pH range. It should be rather inert toward the common reagents used in peptide synthesis (especially DCC), and it can be easily removed by Edman degradation. Braunitzer et al.[5] showed the hydrophilic effect of sulfo groups, notably in counter-current distribution with polysulfonated isothiocyanates.

Cysteic acid peptides were described in analytical work done on wool[6] and they were products in the sequence analysis of proteins, such as insulin[7] and papain[8] for example. In this context a few examples of cysteic acid preparations are reported[6,9,10]. The authors employed performic acid[10] or bromine oxidation[6,9]. This method is not feasible for acid-labile protected peptides. Another restriction is the sensitivity of some amino acids towards oxidation. Therefore, two routes for the preparation of cysteic acid derivatives were developed. The first involves the oxidation of cystine derivatives and peptides, the second employs cysteic acid as starting material.

$$(\text{Y-}\overset{|}{\text{Cys}})_2 \qquad\qquad (\text{Y-}\overset{|}{\text{Cys}}\text{-NH-R}_1)_2$$
$$\downarrow \qquad\qquad\qquad\qquad \downarrow$$
$$\text{Y-Cys}(\text{O}_3\text{Na})\text{-OH} + \text{H}_2\text{N-R}_1 \longrightarrow \text{Y-Cys}(\text{O}_3\text{Na})\text{-NH-R}_1 \rightarrow\rightarrow \text{Y-Cys}(\text{O}_3\text{Na})\text{-NH-R}_2 \rightarrow \text{H}_2\text{N-R}_2$$
$$\uparrow$$
$$\text{Y-L} + \text{H-Cys}(\text{O}_3\text{Na})\text{-ONa}$$

The following cysteic acid derivatives, which are applicable to peptide synthesis were prepared:

Boc-Cys(O_3Na)-OH	Z-Cys(O_3Na)-OH	Pht-Cys(O_3K)-OH
Boc-Cys(O_3Na)-ONp	Z-Cys(O_3Na)-ONp	Pht-Cys(O_3K)-ONp
Boc-Cys(O_3Na)-ONSu		

Model peptides of the type Y-Cys(O_3Na)-Leu-OMe were synthesized via different procedures to investigate the best conditions for the introduction of cysteic acid in peptides (Table I). All procedures afforded homogenous material in yields ranging from 70-90% of theory. The identity of products prepared by these different procedures was proved by IR- and ^{1}H-nmr spectroscopy. It appears that methods involving mixed anhydride coupling of cysteic acid derivatives or catalytic oxidation with peroxide of the corresponding cysteine derivatives provide the best results.

In order to explore the influence of a N-terminal cysteic acid residue on the solubility characteristics of sparingly soluble (both in organic and aqueous solvents) peptides, fragments corresponding to various sections of the insulin molecule have been prepared as shown in Table II. It was found that Boc-Cys(O_3Na)-ONSu was the ideal coupling reagent. The reaction rate was high and no side

Table I. Various Procedures for the Introduction of Cysteic Acid in Peptides.

Y	DCCI/HOBT	Activ.ester	Mix.anhydr.	Performic acid oxidation	Catalytic oxidation with H_2O_2
Boc		+	+		+
Z		+	+	+	+
Pht	+	+	+		

Table II. Solubility Properties of Synthetic Model Cysteic Acid Peptides

	Cysteic Acid Peptides	Yield %	Solubility[4] DMF	Solubility[4] H_2O	K-values (CCD)
I	Boc-Cys(O_3Na)-Phe-Val-Asn-Gln-OH	31[1]	***	***	0.6[2]
II	Boc-Cys(O_3Na)-Phe-Val-Asn-Gln-His-Leu-Cys(Trt)-Gly-OH	44	**	-	4[3]
III	[Boc-Cys(O_3Na)-Phe-Val-Asn-Gln-His-Leu-Cys-Gly-OMe]$_2$	31[1]	**	-	-
IV	Boc-Cys(O_3Na)-Gly-Ile-Val-Glu(OBu^t)-OMe	59	***	*	-
V	Boc-Cys(O_3Na)-Ser(Bu^t)-Leu-Tyr(Bu^t)-Gln-OH	67	***	**	1.5[2]
VI	Boc-Cys(O_3Na)-Leu-Leu-Phe-OH	80	***	**	2[2]

[1] The low yields are due to high losses during column chromatography. No free base was detected by TLC.

[2] Sec.-butanol/water = 1:1; 0.3 g NH_4OAc/l

[3] DMF-system: methanol/chloroform/cyclohexane/DMF/water = 5:5:2:2:2

[4] *** very good, ** good, *poor, - insoluble

products could be observed. Purification was accomplished with Sephadex LH 20/MeOH or countercurrent distribution (CCD). In case of the synthesis of (V), the mixed anhydride coupling led to the formation of at least two side-products, in contrast to the result obtained for the preparation of Y-Cys(O_3Na)-Leu-OMe. Electrophoretic and nmr-data suggest the formation of a sulfonamide bond and sulfonation of the aromatic ring of tyrosine.

The effect of the anionic sulfo group on the behavior in CCD is clearly illustrated during the work up of the reaction mixture: the desired peptide derivative has a K-value of 1.5, whereas the side-products have K-values of 5.7.

The nonapeptide II exhibits a remarkable change in its solubility properties. While the N^{α}-Boc-protected octapeptide of insulin B-(1-8), Boc-Phe-Val-Asn-Gln-His-Leu-Cys(Trt)-Gly-OH, is poorly soluble in methanol or DMF and exhibits a distribution coefficient of K = 0.05 in the DMF-system for countercurrent distribution (see Table II), the corresponding cysteic acid derivative, II (see Table II) has good solubility in both methanol and DMF and a K-value of 4 in the DMF-system. Cysteic acid peptides show the desired change in the solubility characteristics and hence, offer advantages for the purification of hydrophobic and sparingly soluble peptide derivatives by counter current distribution.

We acknowledge gratefully the financial support of this work by Deutsche Forschungsgemeinschaft (AZ Za 5/29).

References

1. Lübke, K. & Klostermeyer, H. (1970) *Adv. in Enzymol.* **33**, 446–525. Wünsch, E., (1971) *Angew. Chem.* **83**, 773–812.
2. Wieland, Th. (1965) *Acta Chim. Hung.* **44**, 5–9.
3. Tesser, G. I. & Balvert-Geers, I. C. (1975) *Int. J. Pept Protein Res.* **7**, 295–305.
4. Kunz, H. (1976) *Chem. Ber.* **109**, 2670–2683.
5. Braunitzer, G., van der Walt, S. & Bless, B. (1975) *Hoppe-Seyler's Z. Physiol. Chem.* **356**, 1321–1323.
6. Consden, R., & Gordon, A. H. (1950) *Biochem. J.* **46**, 8–20.
7. Ryle, A. P., Sanger, F. F., Smith, L. F. & Kitai, R. (1955) *Biochem. J.* **60**, 541–556.
8. Light, A., Glaser, A. N. & Smith, E. L. (1960) *J. Biol. Chem.* **235**, 3159–3162.
9. White, J. (1933) *J. Biol. Chem.* **102**, 249–251.
10. Roberts, B. W. & du Vigneaud, V. (1953) *J. Biol. Chem.* **204**, 871–875.

THE 2-TRIMETHYLSILYLETHYL RESIDUE, A SELECTIVELY CLEAVABLE CARBOXYLIC ACID PROTECTING GROUP.

PETER SIEBER, RUDOLF H. ANDREATTA, KAREL EISLER, BRUNO KAMBER, BERNHARD RINIKER and HANS RINK,
Research Laboratories, Pharmaceutical Division, CIBA-GEIGY Ltd. Basel, Switzerland.

In peptide synthesis by way of the fragment condensation approach, there is still a great need for new carboxyl protecting groups. Carpino[1] found the 2-trimethylsilylethyloxycarbonyl residue suitable for use as an *N*-protecting group, being selectively cleavable with fluoride ions. Because this cleavage method seemed very promising, we examined the usefulness of the analogous ester function, the *2-trimethylsilylethyl* (Tmse) residue, $-CH_2-CH_2-Si(CH_3)_3$, as a carboxyl protecting group.

Methods of synthesis. The Tmse esters are prepared by one of the three methods shown below.

					Conditions
1	Z-AA-OH +	DCC +	HOTmse[2] +	pyridine	Acetonitrile
	1	1.1	1.2	3 mol	5 - 15 h 0°C
2	Z-AA-OH +	DCC +	HOBt +	HOTmse	DMF
	1	1.1	1	3 mol	1 d 25°C
3	Trt-AA-OH +	C_6H_{11}-N =	C-NH-C_6H_{11}[3]		Ethyl acetate
	1		OTmse	1 mol	1 d 50°C

2-Trimethylsilylethanol (HOTmse) is synthesized from readily available starting materials in two steps[2]. Most of the *N*-protected amino acid Tmse esters can be obtained by method 1. When, owing to poor solubility, DMF has to be used, large amounts of acylurea derivatives may be formed; in such cases method 2 is preferable. Method 3 is used to prepare Trt-Cys(Trt)-OTmse, which cannot be obtained by method 1. Most of the *N*-protected amino acid Tmse esters synthesized are oily substances. They can readily be purified by chromatography on silica gel.

Cleavage. We assume that the cleavage of the Tmse group proceeds in the following way (in accordance with a suggestion made by Carpino[1] in respect of the 2-trimethylsilylethyloxycarbonyl group):

$$\text{-AA-O-}CH_2\text{-}CH_2\text{-SiMe}_3 \xrightarrow{F^{\ominus}} \text{-AA-O}^{\ominus} + CH_2{=}CH_2 + \text{F-SiMe}_3$$

The reaction is carried out under exclusion of moisture with a solution of $[R_4N]^{\oplus} F^{\ominus}$ in DMF or DMSO, R = Et or Bu (2–10 equiv. 5-60 min., 20-30°C). The cleavage of -Gly-OTmse takes longer than that of -Thr(Bu^t)-OTmse. In the presence of basic groups like Arg, a larger excess of fluoride is necessary. Special attention has been paid to the question of racemization during treatment with fluoride ion, but no evidence of racemization was found in any of the cases examined. As is well known, this side reaction does occur during alkaline hydrolysis of methylesters.

Scope and limitations. Under normal peptide-coupling and workup conditions and during hydrogenolysis of Z groups the Tmse esters are stable. When the hydrogenation is performed in MeOH, transesterification may occur under the influence of the basic NH_2 group formed. Therefore, Z should be hydrogenated either by the pH-Stat method or in isopropanol as solvent. With NaOH in organic-aqueous solution the Tmse esters are hydrolyzed at rates comparable to those of ethylesters. HCl in organic solvents cleaves the Tmse group, but the reaction is slow enough to allow selective removal of a Boc group. Example: Formation of HCl•H-Ala-Cys(Acm)-OTmse starting from the Boc derivative; yield 87% (10 equiv. HCl in ether, 45 min., room temperature). Anhydrous trifluoroacetic acid cleaves the Tmse group rapidly. Under the conditions of Tmse group cleavage by fluoride ions Boc, Bpoc, *N*- and *S*-Trt, *S*-Acm and *t*-butylether are stable. The behavior of other functional groups is indicated below.

Group	**Remarks**
Z	Formation of hydantoins[4]. Under proper conditions (minimum excess of $F^{\ominus}$, shortest time possible) the Tmse group is split with sufficient selectivity.
OMe	Very slow cleavage.
OBu^t	Very slow cleavage. Exception: -Asp(OBu^t)-AA- is cleaved to a mixture of α- and β-peptide. Under the conditions used we found 20-30% of the by-product.
OBzl	$t_{1/2}$: 30-60 min. with Boc-Gly-OBzl (4 equiv. $F^{\ominus}$, 30°C).
Asn	With C-terminal Asn formation of succinimide derivatives, cf.[5]. Negligible if Asn is not C-terminal.
R-S-S-R′	Rapid disproportionation. A selective cleavage of Tmse esters is not possible.

Applications. The usefulness of Tmse esters as a carboxyl protecting group is demonstrated in the following examples.

```
Trt Boc                           Boc Bu^t       Bu^t
 |   |                             |   |          |
Bmp-Lys-Asn-Phe-Phe-Trp-Lys-Thr-Phe-Thr-OTmse
```

Bmp = β-mercaptopropionic acid

Cleavage: 5 equiv. 0.15 *M* $[Et_4N]^{\oplus} F^{\ominus}$ in DMF, 20 min, 30°C. Precipitated with water. Yield quantitative. No racemization of Thr. The alkaline hydrolysis of the corresponding methylester takes place under drastic conditions with considerable racemization of the C-terminal Thr.

```
                                     Acm        Bu^t                   OBu^t
                                      |          |                      |
Boc-Phe-Val-Asn-Gln-His-Leu-Cys-Gly-Ser-His-Leu-Val-Glu-Ala-

Leu-OTmse = insulin B1-15.
```

Cleavage: 10 equiv. 0.15 *M* $[Et_4N]^{\oplus} F^{\ominus}$ in DMF, 10 min. 25°C. Precipitated with water at the isoelectric point. Yield quantitative. Traces of by-products. No racemization. The alkaline hydrolysis of insulin B1-16-methylester (in trifluoroethanol/water 3:2, large excess of NaOH, 50°C) showed about 7% D-Asn B3, 9% D-Cys(Acm) B7, 30% Asp instead of Asn B3.

Conclusions. 2-Trimethylsilylethyl esters are easy to prepare. They are stable under normal peptide coupling and workup conditions and during hydrogenation. The trimethylsilylethyl group is readily cleaved with a quaternary ammonium fluoride in DMF. No racemization was found in the cases examined. Some side reactions were, however, observed during the removal of the 2-trimethylsilylethyl group: peptides protected with Z are slowly transformed to hydantoins; asymmetric disulfides are rapidly disproportionated; Asp(OBu^t) placed N-terminally or in the interior of the peptide chain is subject to partial cleavage. The usefulness of the new carboxyl protecting group is demonstrated. In both examples shown, alkaline hydrolysis of the methylester was impracticable, whereas treatment of the 2-trimethylsilylethyl ester with fluoride generated the free carboxyl group in quantitative yield. Although the work reported is restricted to the chemistry of peptides, this new protecting group can also be applied in other fields of organic synthesis involving carboxylic acids.

References

1. Carpino, L. A. (1975) Lecture at CIBA-GEIGY, Basle.
2. Speier, J. L., Webster, J. A. & Barnes, G. H. (1957) *J. Amer. Chem. Soc.* **79**, 974–979.
3. According to Vowinkel, E. (1966) Chem. Ber. **99**, 1479–1484. Yield 70%. An analytical sample could be distilled, b.p. 0.09:~120°C. For preparative purposes the product is only filtered through alox in petrolether.
4. Pless, J. (1974) *J. Org. Chem.* **39**, 2644–2646.
5. König, K. & Volk, A. (1977) *Chem. Ber.* **110**, 1–11.

PHOSPHINOTHIOYL GROUPS: NEW SERIES OF ACID LABILE AMINO PROTECTING GROUPS

MASAAKI UEKI, SHIGERU IKEDA and FUMIO TONEGAWA,
Department of Chemistry, Science University of Tokyo
1–3 Kagurazaka, Shinjuku-ku, Tokyo 162, Japan

A new series of amino protecting groups which utilize the acid lability of the P-N bond was studied in order to minimize the side reactions and improve the selectivity in deprotection. Among various kinds of organo-phosphorus groups phosphinothioyl, $R_2P(S)$-, groups were selected because of easy preparation and stability to water of phosphinothioyl chlorides.[1]

Diphenylphosphinothioyl (Ppt) chloride was prepared by the Friedel-Crafts reaction using benzene and thiophosphoryl chloride.[2] Ppt-amino acids were obtained by the Schotten-Baumann type reactions as shown below.

$$C_6H_6 + P(S)Cl_3 \xrightarrow{AlCl_3} (C_6H_5)_2P(S)\text{-}Cl \quad (\text{Ppt-Cl})$$

$$\text{Ppt-Cl} + H_2NCHRCO_2^- \xrightarrow[2)\ H^+]{1)\ OH^-} \text{Ppt-NHCHRCO}_2\text{H}$$

Dimethylphosphinothioyl (Mpt) chloride was obtained by treating tetramethyldiphosphine disulfide, prepared by the reaction of the Grignard reagent with thiophosphoryl chloride, with sulfuryl chloride.[3,4] Mpt-amino acids were synthesized tentatively by alkaline hydrolysis of their esters.

$$\text{RMgX} + P(S)Cl_3 \longrightarrow R_2P(S)\text{-}P(S)R_2 \xrightarrow{SO_2Cl_2} R_2P(S)\text{-}Cl$$

$$R_2P(S)\text{-}Cl + H_2NCHRCO_2R' \longrightarrow R_2P(S)\text{-}NHCHRCO_2R' \longrightarrow R_2P(S)\text{-}NHCHRCO_2H$$

Deprotection of Phosphinothioyl Groups by Triphenylphosphine Dihydrochloride

Preliminary test using Ppt-Gly-OEt showed that the Ppt group was cleaved as fast as the *t*-butyloxycarbonyl (Boc) group by 1*N* HCl in acetic acid or 4*N* HCl in dioxane. Toward trifluoroacetic acid Ppt showed unexpected resistance, which implied the necessity of nucleophilic attack following protonation in P-N bond cleavage. Toward the final goal of selective removal between phosphinothioyl and urethane groups, we searched for a new mild reagent suitable for P-N bond cleavage. Triphenylphosphine (TPP) used as a reductant in peptide synthesis by oxidation-reduction condensation[5] is known to give dihydrochloride salt.[6] A solution of this salt could be prepared only by saturating HCl in a solution of TPP in methylene chloride. This new reagent was found to be especially useful for deprotecting phosphinothioylamino acids and peptides on a polymer support. When the extent of deprotection of Ppt-Gly-resin was measured by the Dorman titration,[7] a 1N HCl (0.5M TPP) solution in CH_2Cl_2 showed a cleaving power almost identical with that of 4N HCl in dioxane. When a 2N solution (1M TPP) was used, deprotection was complete within 5 min; but Ppt derivatives of several amino acids (Asp, Ile, Leu, Phe, Val and Tyr) showed retarded cleavage. Ppt-Leu-resin, for example, was deprotected only to the extent of 56% by 2N HCl solution (2 × 30 min). Such difficulties were not encountered with Mpt derivatives. Mpt, the most labile group in this series, could be removed within 10 min by 0.5N solution (0.25M TPP) and within 30 min by 0.25N solution (0.12M TPP). The Boc group was cleaved very slowly by these reagents. The stability of Boc was further enhanced by the addition of excess TPP.

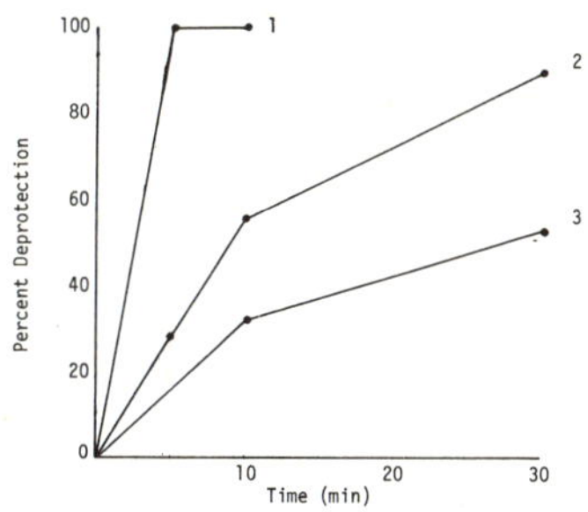

Fig. 1. The extent of deprotection of Ppt-Gly-resin at 20°C. 1) 2N HCl/1M TPP/CH_2Cl_2, 2) 4N HCl/dioxane and 3) 50% F_3CCO_2H/CH_2Cl_2.

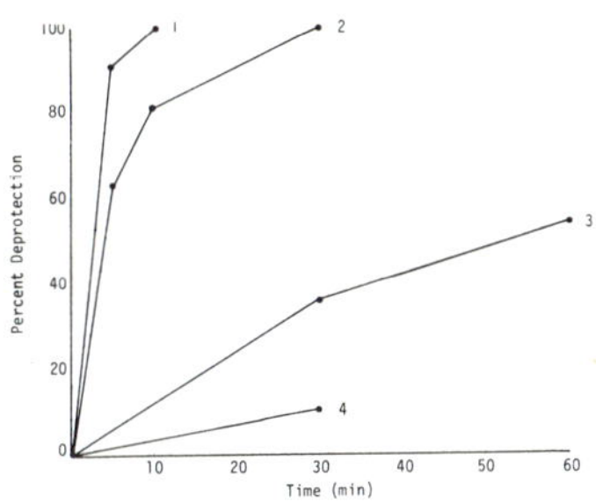

Fig. 2. Deprotection by HCl/TPP/CH_2Cl_2 of Mpt-Phe-resin, 1) 0.5N/0.25M and 2) 0.25N/0.12M, and Boc-Gly-resin, 3) 1N/0.5M and 4) 1N/1M at 20°C.

Another merit of the triphenylphosphine dihydrochloride reagent is the stability of the anchoring benzyl ester bond on the polymer support. Weak acidity of the conjugate acid of TPP (pK_a 2.3)[8] could be responsible for this stability.

Solid Phase Synthesis of Tryptophan Containing Peptides

The usefulness of the Mpt group for solid phase peptide synthesis[9] was shown in the synthesis of a tryptophan-containing peptide. Mpt-L-tryptophan, mp 130-131°C, $[\alpha]_D^{25}$ −1.2°(c 2, EtOH), was prepared by alkaline hydrolysis of its

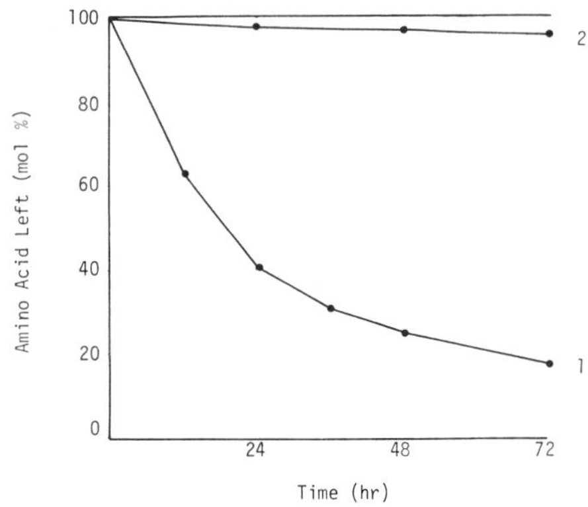

Fig. 3. Loss of amino acid from HCl•H-Gly-resin: 1) 50% F_3CCO_2H/CH_2Cl_2 and 2) 2*N* HCl/1*M* TPP/CH_2Cl_2.

methyl ester and esterified to the resin support (polystyrene-1% divinylbenzene) through its cesium salt[10] (Trp content: 0.43 meq/g). The Mpt group was removed by a once-repeated treatment for 30 min each with 0.2*N* HCl/ 0.2*M* TPP/CH_2Cl_2 at 25°C. No additional scavenger was used because triphenylphosphine was believed to have enough anti-oxidative effect. Coupling was mediated by the oxidation-reduction condensation with the use of tri(*p*-anisyl)phosphine and 2,2'-dipyridyl disulfide. After two cycles, the protected tripeptide was removed from the resin by transesterification with the use of 1*N* triethylamine in methanol[11] to afford Mpt-L-Trp-L-Trp-L-Trp-OMe in 86% yield as white amorphous powder, $[\alpha]_D^{28}$ -65.0° (c 1, EtOH).

By using 2*N* HCl/1*M* TPP/CH_2Cl_2 for deprotection Ppt-L-tryptophan could also be used for the synthesis of tryptophan containing peptides. Since the deprotection product of the Ppt group does not attack the indole nucleus, reactions proceeded smoothly without accompanying coloration. This new method could become a convenient way for the synthesis of tryptophan-containing peptides.

References

1. Neimysheva, A. A., Saushuk, V. I., Ermolaeva, M. V. & Knunyants, I. L. (1968) *Izv. Akad. Nauk SSSR, Ser. Khim.* 2222–2229.
2. Maier, L. (1964) *Helv. Chim. Acta* **47**, 120–125.
3. Parshall, G. W. (1973) in *Organic Synthesis,* Coll. Vol. 5, Baumgarten, H. E., Ed., John Wiley and Sons, New York, pp. 1016–1018.
4. Maier, L. (1961) *Chem. Ber.* **94**, 3051–3055.
5. Mukaiyama, T., Matsueda, R. & Suzuki, M. (1970) *Tetrahedron Lett.* 1901–1904.
6. van den Akker, M. & Jellinek, F. (1967) *Rec. Trav. Chim. Pays-Bas,* **86**, 897–906.
7. Dorman, L. C. (1969) *Tetrahedron Lett.* 2319–2321.
8. Goetz, H. & Sidhu, A. (1965) Justus Liebig's *Ann. Chem.* **682**, 71–74.
9. Merrifield, R. B. (1963) *J. Amer. Chem. Soc.* **85**, 2149–2154.
10. Gisin, B. F. (1973) *Helv. Chim. Acta* **56**, 1476–1482.
11. Beyerman, H. C., Hindriks, H. & de Leer, E. W. B. (1968) *Chem. Commun.* 1668.

BASE LABILITY OF REACTANTS AND PRODUCTS OF THIOLYSIS REACTION OF N^{im}-2,4-DINITROPHENYL-HISTIDINE PEPTIDES. AN EXPLANATION FOR THE pH OPTIMUM

H. JOSEPH GOREN, *Division of Medical Biochemistry, University of Calgary, Calgary, Alberta, Canada,* and MATI FRIDKIN, *Department of Organic Chemistry, Weizmann Institute of Science, Rehovot, Israel*

Since the discovery of Shaltiel[1] that *N*-2,4-dinitrophenylimidazoles (Dnp-imidazole) are readily cleaved under mild conditions, the Dnp moiety has been considered a useful blocking group for histidine in peptide synthesis.[2] Along with its ease of removal (thiolysis under mildly basic conditions) the Dnp blocking group contains the following attributes: (a) it is strongly electron withdrawing which lowers the basicity of the imidazole ring and therefore reduces side reactions, (b) it possesses acid stability and partial base stability, thus compatibility with many amino and carboxyl protecting groups, and (c) removal of Dnp moiety may be followed spectrophotometrically.[3]

When Shaltiel and Fridkin[3] followed the reaction of N^{im}-Dnp-histidine with 2-mercaptoethanol (HS-Et-OH) at 340 nm (λ_{max} for Dnp-S-Et-OH, product of thiolysis), they observed that the rate of cleavage was optimal between pH 8.5 and 9.0. Since the thiol anion of HS-Et-OH is most likely the active species in the thiolysis reaction, other reactions must be proceeding concomitantly above pH 9.0 as Dnp-S-Et-OH is being synthesized. The present report describes some base-susceptible reactions for *N*-Dnp-imidazole, HS-Et-OH, and Dnp-S-Et-OH, their mechanisms and rate constants.

Materials

2,4-Dinitrophenylmercaptan (Dnp-SH) and Dnp-S-Et-OH were prepared as previously described.[1,4] All other reagents not synthesized were of the highest quality commercially available. Syntheses (to be described elsewhere) were performed with the exclusion of light.

Methods

Spectra were run from 480 nm to 240 nm at room temperature in a Beckman Acta V Spectrophotometer in 1 cm path-length quartz cuvettes. The reference cell contained the solution solvent.

Kinetics of base hydrolysis of Dnp-S-Et-OH were under pseudo-first-order reaction conditions. The reaction was followed spectrophotometrically at 400 nm

Table I. Characterization of Synthetic Dnp-Compounds

	Yield	M p [a]	Neutral equivalent [b]	R_f [c]	Microanalyses [d] C	H	N	S
Dnp-imidazole, H_2O	50%	144-145	239(252)	0.67	42.9	3.2	22.2	
					42.9	2.4	22.4	
Dnp-S-Ac-OH [e]	80%	162-163.5	262(258)	0.45	37.2	2.3	10.8	
					37.3	2.6	10.5	
Dnp-S-S-Et-OH [e]	15%	106.4			34.8	2.9	10.1	23.2
					34.8	3.0	10.1	22.8

(a Mettler FP51-FP5, uncorrected (°C).
(b Non-aqueous titration: $HClO_4$/Ac-OH,[5] Dnp-imidazole; NaOMe,[6] Dnp-S-Ac-OH. Value in parenthesis is theoretical value.
(c Thin-layer chromatography (silica gel) in acetonitrile/water (9/1).
(d Chemalytics Inc. (Tempe, Arizona). Upper set of figures are theoretical values.
(e Dnp-S-Ac-OH, Dnp-S-thioglycollic acid; Dnp-S-S-Et-OH, Dnp-2-mercaptoethanol disulfide.

(λ_{max} of difference spectra between Dnp-S-Et-OH and Dnp-O⁻). The pseudo-first-order rate constant was determined by the initial slope method.[7]

The base hydrolysis of 2,4-dinitrothiophenol (Dnp-SH) and the oxidation of HS-Et-OH were also followed spectrophotometrically at 400 nm and 235 nm, respectively. The first-order rate constants for these reactions were determined in the usual manner.[8]

Results and Discussion

Base lability of Dnp-S-Et-OH. Scheme 1 illustrates the possible base-labile reactions.

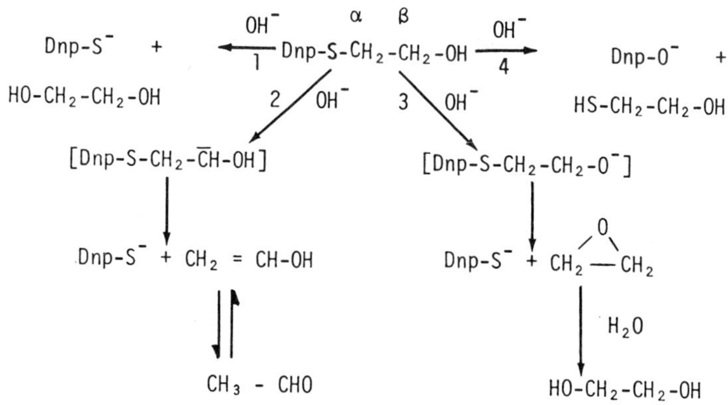

Scheme 1

Dnp-S-Ac-OH, a thioether whose susceptibility to hydroxide ion attack cannot proceed through either routes 2 or 3, was subjected to basic conditions. The following reaction scheme is derived from solution absorption spectra (Fig. 2a) when compared with the spectra of the pure compounds in solution (Fig. 1):

$$\text{Dnp-S-Ac-OH} \xrightarrow{k_1\ (OH^-)} \text{Dnp-S}^- \xrightarrow{k_2} (\text{Dnp-S})_2$$

In 0.5 M NaOH, $k_1(OH^-) > k_2$, while in 77 mM NaOH, $k_2 > k_1(OH^-)$.

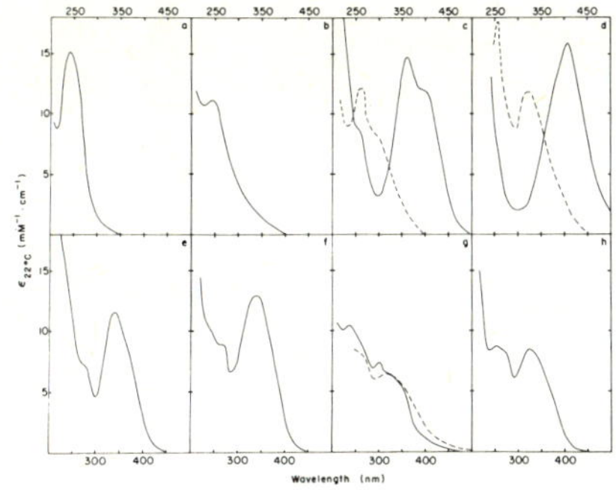

Fig. 1. Absorption spectrum of (a) Dnp-F; (b) *N*-Dnp-imidazole; (c) Dnp-OH (- -), Dnp-O$^-$ (——); (d) Dnp-SH (- -), Dnp-S$^-$ (——); (e) Dnp-S-Et-OH; (f) Dnp-S-Ac-OH; (g) $(Dnp\text{-}S)_2$ in 95% EtOH (——), and in 45% (v/v) dioxane/water freshly prepared (- -); (h) Dnp-S-S-Et-OH in 20% (v/v) dioxane/water.

In Dnp-S-EtOH the α-carbon is not as electrophilic as the α-carbon in Dnp-S-Ac-OH. Accordingly, the SN_2 reaction of route 4 proceeds rather than the SN_2 reaction of route 1 (compare spectra of Fig. 2b with 1c). Alternatively, base lability of Dnp-S-Et-OH may proceed through routes 1, 2, or 3 but the Dnp-S$^-$ that is formed is rapidly hydrolyzed to Dnp-O$^-$.

When $(Dnp\text{-}S)_2$ is dissolved in basic aqueous/organic mixtures the following reactions proceed:

$$(\text{Dnp-S})_2 \underset{k_3}{\overset{k_1}{\rightleftharpoons}} \text{Dnp-S}^- \xrightarrow{k_2(\text{OH}^-)} \text{Dnp-O}^-$$

In aqueous dioxane (Fig. 2c), hydrolysis is preferred while in aqueous ethanol re-oxidation occurs (Fig. 2d). Lowering the dioxane concentration decreases the rate of hydrolysis. The upper limit to k_2 is estimated to be 4.5 $M^{-1} \cdot \text{min}^{-1}$ (94%

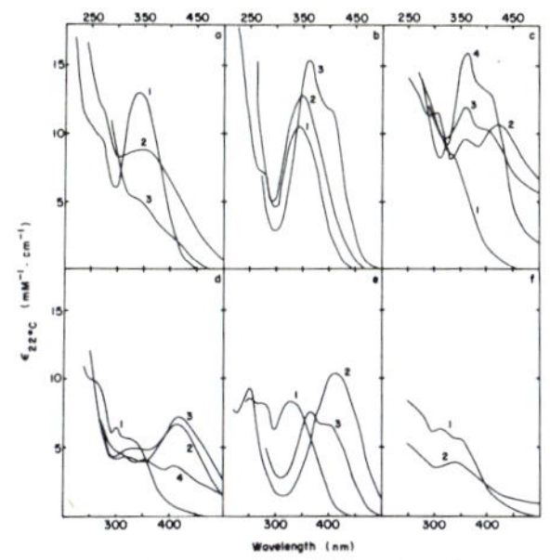

Fig. 2. Absorption spectra of: (a) Dnp-S-Ac-OH in 77 *mM* NaOH, 2% (v/v) dioxane/water, initially (1), after 2 h (2), and 20 h (3); (b) Dnp-S-Et-OH in 50 *mM* NaOH, 0.5% (v/v) EtOH/water, initially (1), after 9 min (2) and 5 h (3); (c) $(Dnp\text{-}S)_2$ in 33 *mM* NaOH, 67% (v/v) dioxane/water, initially (1), after 1 min (2), 10 min (3) and 1 h (4); (d) $(Dnp\text{-}S)_2$ in 40 *mM* NaOH, 60% (v/v) EtOH/water, initially (1), after 1 min (2), 1 h (3) and 24 h (4); (e) Dnp-S-S-Et-OH in 0.18 *M* NaOH, 20% (v/v) dioxane/water, initially (1), after 1 min (2), 1 h (3) and 24 h (4); (e) Dnp-S-S-Et-OH in 0.18 *M* NaOH, 20% (v/v) dioxane/water, initially (1), after 3 min (2), and 72 h (3); (f) $(Dnp\text{-}S)_2$ in 4% (v/v) dioxane/water, initially (1) and after 30 min (2).

(v/v) dioxane/water). The second-order rate constant for the base hydrolysis of Dnp-S-EtOH is 0.8 $M^{-1} \cdot \text{min}^{-1}$. Thus, studies with $(\text{Dnp-S})_2$ and Dnp-S-S-Et-OH (Fig. 2e) indicate that hydrolysis of Dnp-S⁻ does not occur as quickly as hydrolysis of Dnp-S-Et-OH. The hydrolysis of Dnp-S-Et-OH proceeds therefore by the S_N2 mechanism of route 4.

Base Lability of N-Dnp-imidazole and HO-Et-SH. *N*-Dnp-Imidazole forms Dnp-O⁻ slowly under basic conditions.[3] 2-Mercaptoethanol, on the other hand, is readily oxidized to the disulfide form. Oxidation was followed spectrophotometrically[9] at 4 different pH's yielding the following equation:

$$k_{ox} = \alpha \times 0.02 \ (\text{min}^{-1}) \qquad (1)$$

k_{ox} is the first order rate constant for oxidation and α is the degree of ionization of HS-Et-OH ($pK_a = 9.32$).

Conclusions

Above pH 8.5, N^{im}-Dnp-histidine peptides slowly form Dnp-O⁻; HS-Et-OH is rapidly oxidized; and Dnp-S-Et-OH hydrolyzes to form Dnp-O⁻. At pH 9.0 the first order rate constant for the latter reaction is 8×10^{-6} min^{-1}. Oxidation of HS-Et-OH is 6.3×10^{-3} min^{-1}. Thus the decrease in rate of thiolysis of *N*-Dnp-imidazoles (measured spectrophotometrically) above pH 9.0 is not due to a degradation of the product but rather due to the disappearance of one of the reactants, mercaptoethanol.

Supported by the Medical Research Council of Canada (MA 5521).

References

1. Shaltiel, S. (1967) *Biochem. Biophys. Res. Commun.* **29**, 178–183.
2. Chillemi, F. & Merrifield, R. B. (1969) *Biochemistry* **8**, 4345–4346. Nakajima, F. & Okawa, K. (1973) *Bull. Chem. Soc. Japan* **46**, 1811–1816. Goren, H. J., Katchalski-Katzir, E. & Fridkin, M. (1977) *Biopolymers* **16**, in press.
3. Shaltiel, S. & Fridkin, M. (1970) *Biochemistry* **9**, 5122–5127.
4. Sokolovsky, M. & Patchornik, A. (1964) *J. Amer. Chem. Soc.* **86**, 1860–1861.
5. Shaltiel, S. & Patchornik, A. (1963) *J. Amer. Chem. Soc.* **85**, 2799–2806.
6. Fritz, J. S. & Lisicki, N. M. (1951) *Anal. Chem.* **23**, 589–591.
7. Goren, H. J. & Fridkin, M. (1974) Europ. *J. Biochem.* **41**, 263–272.
8. Jencks, W. P. & Carriuolo, J. (1960) *J. Amer. Chem. Soc.* **82**, 1778–1785.
9. Benesch, R. E. & Benesch, R. (1955) *J. Amer. Chem. Soc.* **77**, 5877–5881.
10. Whitaker, J. R. (1972) *J. Amer. Chem. Soc.* **84**, 1900–1904.

SYNTHESIS AND USE OF PHOTOREACTIVE SULFENYL CHLORIDES

J. RAMACHANDRAN and ELEANOR CANOVA-DAVIS,
Hormone Research Laboratory, University of California, San Francisco, California 94143

The first photoaffinity labels utilized carbenes photolytically generated from diazo compounds to study the active sites of various enzymes.[1] The use of aryl azides yielding nitrenes upon photolysis was introduced recently to study a number of systems of varying complexity, such as antibody combining sites,[2] acetylcholine binding sites in the erythrocyte membrane,[3] and the opiate receptor in mouse brain.[4]

Sulfenyl chlorides have been found to be specific mild reagents for modification of tryptophan and cysteine residues of polypeptides and proteins in acidic media.[5,6] Tryptophan is converted into a derivative with a thioether function in the 2-position of the indole nucleus, and cysteine to an unsymmetrical disulfide. In aqueous alkaline solution, sulfenyl chlorides react with amino groups of amino acids and lysine residues of peptides to yield *N*-substituted derivatives. Hence, various photoaffinity labels may be obtained by the selective modification of polypeptides with phenylsulfenyl chlorides incorporating an azido group. We have prepared two new aryl sulfenyl chlorides containing photoreactive azido groups and studied their utility in the photoaffinity labeling of polypeptides.

For the synthesis of 2-nitro-4-azidophenylsulfenyl chloride (NAPS-Cl) (Figure 1) 3-nitro-4-chloroaniline was reacted with sodium sulfide and sulfur in ethanol. The resulting disulfide[7] was diazotized and allowed to react with sodium azide.[8] Chlorinolysis of the product in CCl_4 at 45°C in the presence of a catalyst[7] yielded NAPS-Cl (Infrared: CCl_4, 2140 cm^{-1}, N_3, sym. str.). A similar series of reactions with 2,4-dinitro-5-fluoroaniline yielded 2,4-dinitro-5-azidophenylsulfenyl chloride (DNAPS-Cl) (Infrared: CCl_4, 2110 cm^{-1}).

The pituitary hormone adrenocorticotropin (ACTH) was selectively modified at the 2-position of the indole side chain of the single tryptophan residue by both reagents in 90% acetic acid. The reaction with NAPS-Cl was sluggish. Figure 2 shows the carboxymethyl-cellulose chromatography of NAPS-ACTH. Prolonged reaction resulted in further modification, but the methionine residue was almost completely oxidized. On the other hand, DNAPS-Cl was highly reactive and produced quantitative modification of the tryptophan residue as indicated by amino-acid analysis, peptide mapping, and extinction coefficient at 360 nm.

Irradiation of the NAPS derivative of ACTH at 0°C with wavelengths greater than 300 nm caused spectral changes centered around 285 nm (Figure 3) indicating the photoreactivity of the modified peptide. Similar results were obtained

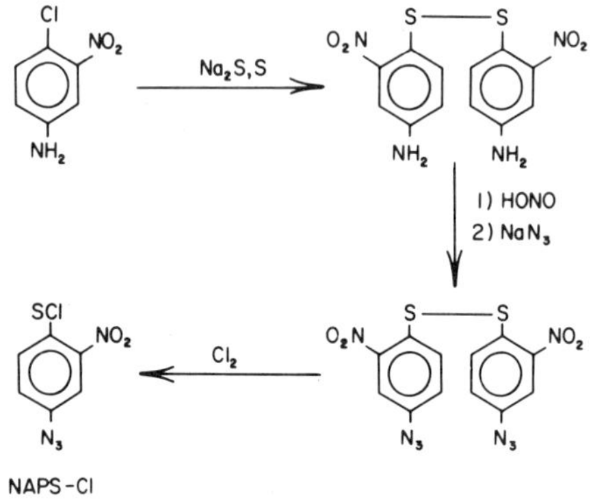

Fig. 1. Reaction scheme for the preparation of NAPS-Cl.

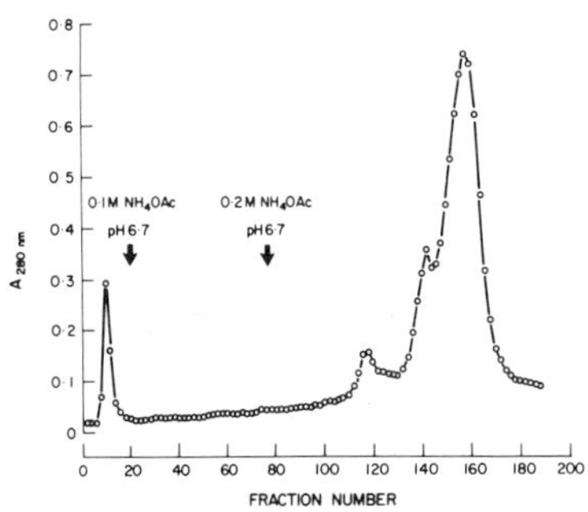

Fig. 2. CM-cellulose chromatography of NAPS-ACTH. Flow rate 60ml/h; 2ml/tube.

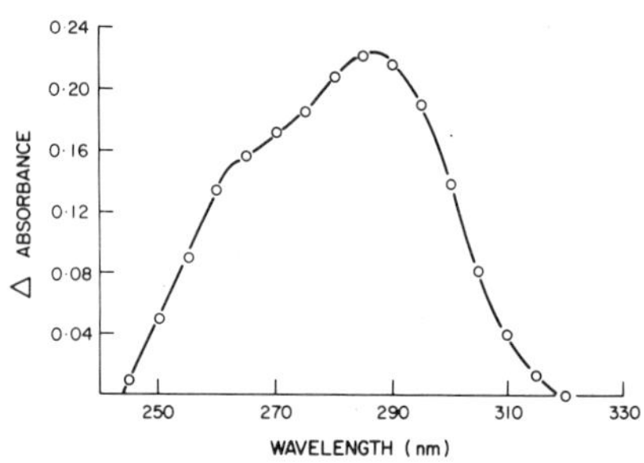

Fig. 3. Difference spectrum of photolyzed NAPS-ACTH.

with the model compound acetyl-(DNAPS)-Trp-NH_2. When a 2,4-dinitrophenylsulfenyl-ACTH derivative, which did not contain a photoreactive azido group, was irradiated, no change was evident in the difference spectrum.

Preliminary studies with a crude ACTH binding protein preparation (FI) isolated from acetone/acid extracts of whole pituitary glands[6] indicated that photolysis resulted in a covalent attachment of the NAPS-ACTH peptide. Similarly, the irradiation of the ACTH binding protein: DNAPS-ACTH complex also resulted in a covalent bond. It is known that the dinitrophenyl group can be cleaved from sulfenyl-chloride-modified-tryptophan residues by β-mercaptoethanol at pH 8.0.[9] This method was applied to attack the binding protein: NAPS-ACTH linkage. Spectral (Table I) and electrophoretic (Figure 4) analysis indicated that the ACTH moiety was indeed separated from the binding protein by this procedure.

There is the potential for reversibly cross-linking proteins by the use of these photoreactive sulfenyl chlorides. Under mildly acidic conditions, tryptophan residues, which are usually in the interior of proteins, can be selectively modified and irradiated to produce intramolecular cross-linking, providing a probe for the

Table I. Spectral Analysis of Binding Protein: (D)NAPS-ACTH Complexes

	Ratio of Absorbance 360/280	
	FI + NAPS-ACTH	FI + DNAPS-ACTH
Unphotolyzed	0.066	0.064
Photolyzed	0.104	0.168
Reduced	0.129	N.D.

The binding protein:(D)NAPS-ACTH complexes were separated on a Sephadex G-50 column (1.5 x 22.4 cm) equilibrated with 1 M formic acid to facilitate dissociation of noncovalently bound peptides.

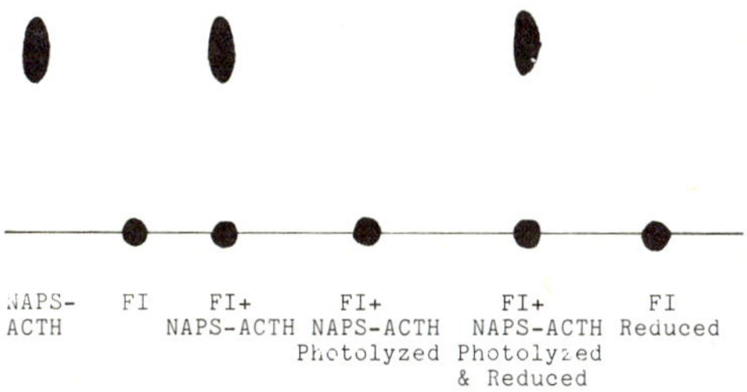

Fig. 4. Electrophoretic analysis of the binding protein: NAPS-ACTH complex. Electrophoresis was conducted at pH 3.7 (pyridine/acetic acid/water, 1:10:300; v/v) for 4 h on Whatman No. 1 paper (12 V/cm).

secondary or tertiary structure of the protein. Subsequent reduction would result in the formation of 2-thioltryptophan. In aqueous alkaline solution, the ϵ-amino groups of lysine, which are generally found on the exterior of globular proteins, could be modified and photoreacted to yield possible intermolecular cross-linkages. Since the ϵ-amino groups can be regenerated with anhydrous HCl,[10] the cross-linking is again reversible. The identification of specific receptors for ACTH on adrenocortical cells and adipocytes by photoaffinity labeling with DNAPS-ACTH is currently under investigation.

The authors thank Prof. C. H. Li for his interest. This work was supported in part by USPHS grants GM-2907 and CA-16417.

References

1. Singh, A., Thornton, E. R. & Westheimer, F. H. (1962) *J. Biol. Chem.* **237**, PC3006–3008.
2. Fleet, G. W. J., Knowles, J. R. & Porter, R. R. (1972) *Biochem. J.* **128**, 499–508.
3. Ruoho, A. E., Kiefer, H., Roeder, P. E. & Singer, S. J. (1973) *Proc. Nat. Acad. Sci. USA* **70**, 2567–2571.
4. Winter, B. & Goldstein, A. (1972) *Mol. Pharmacol.* **8**, 601–611.
5. Scoffone, E., Fontana, A. & Rocchi, R. (1968) *Biochemistry* **7**, 971–979.
6. Canova-Davis, E. & Ramachandran, J. (1976) *Biochemistry* **15**, 921–927.
7. Hubacher, M. H. (1935) in *Organic Syntheses*, Vol. XV, Noller, C. R., Ed., John Wiley and Sons, Inc., New York, pp. 45–47.
8. Wilson, D. F., Miyata, Y., Erecinska, M. & Vanderkooi, J. M. (1975) *Arch. Biochem. Biophys.* **171**, 104–107.
9. Wilchek, M. & Miron, T. (1972) *Biochem. Biophys. Res. Commun.* **47**, 1015–1020.
10. Zervas, L., Borovas, D. & Gazis, E. (1963) *J. Amer. Chem. Soc.* **85**, 3660–3666.

PEPSIN INDUCED SYNTHESIS OF PEPTIDE BONDS

A. PELLEGRINI and P. L. LUISI, *Technisch-Chemisches Laboratorium, ETH-Zentrum, Universitätstr. 6, 8092 Zürich, Switzerland*

In the past,[1-4] some investigations have been concerned with the enzymatic synthesis of peptide bonds in aqueous solution, where the chemical equilibrium is shifted towards the coupling side by choosing reagents which yield water insoluble products. Examples were described by Bergman and Fruton.[3,4] An enzymatic synthesis has some potential advantages with respect to the conventional chemical coupling procedures. Reaction conditions are very convenient (room temperature, water solution, a simple mixing of the reactants with the enzyme); the enzyme's specificity guarantees products which are optically pure and free of by-products. Also chemical protection of reactive functional side chains is unnecessary.

In view of these potentialities, we began a systematic investigation aimed at establishing whether the enzymatic approach can be useful for the synthesis of co-oligopeptides of glycine and aromatic amino acids, which until now have been prepared by us by conventional methods.[5,6] Recently, we used α-chymotrypsin to synthesize peptide bonds between an aromatic (Ar) and an aliphatic (Al) residue.[7,8] As an extension of this work, we have enzymatically coupled two aromatic amino-acid residues using pepsin as a catalyst.

Experimental

Reactions were carried out at 40°C in 0.1 *M* citrate buffer pH 4.0, containing 1 *M* NaCl, and 10% MeOH for peptides 1, 2, 3, 5 and 17% for peptide 4. For peptides 1-3 (Table I), a saturated solution of the Z-Phe-OH was mixed with 4 equivalents of solid H-Ar-Y (Y=OMe or NH_2), followed by addition of solid pepsin (up to a final concentration of 2.5 mg/ml). Peptides 4 and 5 were prepared similarly, but in the presence of undissolved Z-Ar-OH, and an 11-fold excess of the second component with respect to the total Z-Ar-OH concentration for peptide 4, and a two-fold excess for peptide 5. Reactions were protracted for several days; every other day the formed precipitate was collected until no new precipitate was formed after 3–4 days standing. The combined precipitates were washed with 3-4 ml of 1 *N* HCl, 1 *N* NaOH and H_2O, dried *in vacuo* for 2 days and crystallized from EtOAc (peptide 1), EtOH(2), and MeOH(3-5). The purity was controlled by thin layer chromatography in the systems benzene/methanol (6:1) for peptides 1 and 2; (2:1) for 3 and 4; (4:1) for 5. The products were characterized by elemental analysis, melting point, optical activity (see Table I) and mass spectroscopy. In all cases, the fragment of highest *m/e* value corresponded to the molecular ion (460, 499, 445, 484, 536, for peptides 1-5, respectively.

Table I. Characterization of Protected Dipeptides Obtained Via Enzymatic Coupling (Pepsin)[(a)]

No.	Peptides	C, H, N, in % Microanalyses	theor.	found	Melting Point °C		$\|\alpha\|^{25}$		Solvent	Chemical Yield
1	Z-Phe-Phe-OMe	C: H: N:	70.42 6.13 6.08	70.23 6.20 6.03	Lit.: Ours:	148-150[9] 132-133	Lit.: Ours:	-17.0[4] -16.6	EtOH, c 0.4	74%
2	Z-Phe-Trp-OMe	C: H: N:	69.72 5.85 8.41	69.74 5.82 8.45	Lit.: Ours:	132-133[10] 130-131	Lit.: Ours:	+39.3[5] +38.9	AcOH, c 1.5	94%
3	Z-Phe-Phe-NH_2	C: H: N:	70.09 6.11 9.43	70.04 6.28 9.32	Ours:	220-221		-11.1	AcOH, c 1.0	97%
4	Z-Trp-Phe-NH_2	C: H: N:	69.40 5.85 11.56	68.92 5.83 11.35	Ours:	212-213		-7.9	AcOH, c 1.0	47%
5	Z-Phe-Phe-Bzl	C: H: N:	73.86 6.01 5.22	73.69 6.00 5.25	Ours:	150-151		-19.1	DMF, c 1.0	90%

(a) Abbreviations: EtOH, ethanol; AcOH, acetic acid; DMF, N,N-dimethylformamide; Z, carbobenzoxy; Bzl, benzyl

Results and Discussion

The few examples reported in Table I show that the enzymatic coupling of two aromatic amino acids is possible, and with satisfactory yields. Reactions proceed with greater facility when the Z-Ar-OH partner contains a phenylalanine residue, while the H-Ar-Y component can be any aromatic residue. These observations reflect the substrate specificity of pepsin, as known from hydrolysis studies.[10] No success was obtained starting with Boc-protected aromatic amino acids. α-Chymotrypsin can effectively synthesize[7] the peptide bond using, for example, Boc-Phe-OH as a first partner (the second partner is in this case an aliphatic one). This peculiarity of pepsin may be attributed to the enzyme's specificity; however, the solubility of the Boc-components may also be important in preventing the synthesis. The synthetic approach with pepsin can in principle be extended to the preparation of compounds of the type P-$(Al)_n$-Ar-$Ar'(Al)_n'$-Y and preliminary results in our laboratory with n=1, and Al=Gly, are encouraging.

The present investigation confirms that the enzymatic synthetic approach is viable in a number of cases. The advantages mentioned in the introduction are accompanied by certain drawbacks: reactions require a long time and large reactors; the outcome of the reaction is *a priori* unpredictable, and the method limited to the preparation of insoluble products. Some of these difficulties can perhaps be overcome, and we are at the present working to generalize the method.

We are also attempting to utilize more particularly the stereospecificity of the enzyme reactions, namely the capability of pepsin to select the L-enantiomer from a racemic mixture. This characteristic, together with the almost quantitative chemical yield obtained in some cases, makes it possible to obtain an optically pure product starting from optically impure starting materials, at the same time allowing for the recovery of the unreacted D-enantiomer. For example, starting from Z-D,L-Phe-OH and H-L-Phe-NH_2, we have obtained the optically pure L-L-diastereoisomer Z-Phe-Phe-NH_2, recovering Z-D-Phe-OH with good optical purity.

References

1. Balls, A. K. & Köhler, F. (1931) *Chem. Ber.* **64**, 34–39.
2. Cohn, E. J. & Edsall, J. T. (1943) *Proteins, Amino Acids and Peptides*, Reinhold Publ. Corp., N. Y.
3. Bergman, M. & Fruton, J. S. (1941) *Adv. Enzymol.* **1**, 63–96.
4. Bergman, M. & Fruton, J. S. (1944) *Ann. N. Y. Acad. Sci.* **45**, 409–415.
5. Guarnaccia, R., Lorenzi, G. P., Rizzo, V. & Luisi, P. L. (1975) *Biopolymers* **14**, 2329–2346.
6. Guarnaccia, R., Rizzo, V., Gianola, P. & Luisi, P. L. (1976) *Biopolymers* **15**, 1103–1117.
7. Saltman, R., Vlach, D. & Luisi, P. L. (1977) *Biopolymers* **16**, 631–638.
8. Luisi, P. L., Saltman, R., Vlach, D. & Guarnaccia, R. (1977) *J. Mol. Cat.* **2**, 133–138.
9. Blaha, K. (1969) *Coll. Czech. Chem. Commun.* **34**, 4000–4005.
10. Fruton, J. (1976) *Adv. Enzymol.* **44**, 1–35.

SYNTHESIS OF PROTECTED PEPTIDE ACIDS AND ESTERS BY PHOTOSOLVOLYSIS OF 1-PEPTIDYL-5-BROMO-7-NITROINDOLINES

GILBERTO GOISSIS, BRUCE W. ERICKSON, and R. B. MERRIFIELD, *The Rockefeller University, New York, New York 10021*

Acid-labile protecting groups are commonly used to mask the α-amino and side-chain functional groups during solid-phase peptide synthesis of protected peptides.[1] Since the α-carboxyl protecting group also serves as the covalent link between the peptide and the solid support, this group should be stable to acid but be selectively cleavable by an independent reaction, such as photolysis. The 2-nitrobenzyl ester support[2] and the α-methylphenacyl ester support[3] both liberate the peptide acid on irradiation independent of solvent. More useful would be a solid support cleavable by photosolvolysis because the same peptide-resin could serve as the precursor of the peptide acid, peptide esters, and peptide amides.

Amit et al[4] found that *N*-acyl derivatives of 5-bromo-7-nitroindoline (HBni) and aromatic or simple aliphatic acids undergo efficient photosolvolysis to yield acids, esters, or amides depending on the nucleophile present during irradiation. Solvolysis of the 5-bromo-7-nitro-1-indolino (Bni) group was even observed using light above 400 nm. We have examined the usefulness of this reaction for synthesis of protected peptide acids and esters in solution.

Although direct *N*-acylation of HBni was successful using trifluoroacetic anhydride, it failed for acetic anhydride and activated Boc-amino acids due to the electron-withdrawing substituents. Thus Bni derivatives of amino acids and peptides were synthesized by the indirect scheme shown in Fig. 1. After *N*-acylation of indoline with Boc-Ala-OH or Boc-Val-OH and exchange of Boc for the acid-stable trifluoroacetyl group, the aromatic ring was brominated and nitrated. This rapid five-step scheme provided pure Tfa-aminoacyl-Bni derivatives in 50–55% overall yield. Removal of the Tfa group followed by coupling with Boc-Leu-OH or Boc-Gly-OH, respectively, furnished Boc-dipeptidyl-Bni

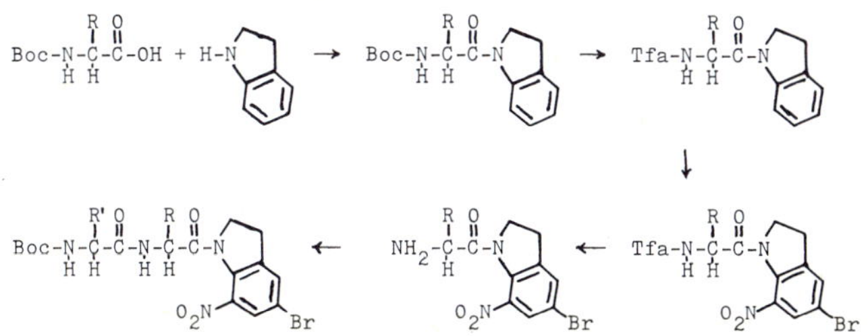

Fig. 1. Synthesis of 1-(Boc-dipeptidyl)-5-bromo-7-nitroindolines.

derivatives in 45-50% yield after preparative thin-layer chromatography (TLC). The acyl-Bni derivatives were photolabile under fluorescent light but were stable in the dark for at least several weeks.

Photolysis was conducted in Pyrex glassware through an uranium-glass filter with a medium-pressure mercury lamp. Removal of the Bni group was generally complete in 90 min as measured by release of HBni, which was monitored by TLC and ultraviolet (UV) spectroscopy at 430 nm. 1-Acetyl-5-bromo-7-nitroindoline quantitatively photosolvolyzed in 10:20:1 (v/v/v) CH_2Cl_2/dioxane/water and in 6:1 (v/v) CH_2Cl_2/methanol. Liberation of HBni during photohydrolysis was fastest during irradiation at 350–360 nm, which corresponds to the UV absorption maximum for Ac-Bni. Photolysis of Tfa-Bni, Tfa-Gly-Bni, and Ac-Gly-Bni went in 70–90% yield in these solvent mixtures. The yield of HBni for photohydrolysis of Tfa-Gly-Bni in wet 1:2 (v/v) CH_2Cl_2/dioxane increased from 45% to 62% as the water concentration was decreased from 1.7 *M* to 0.6 *M*. In the presence of 0.03 *M* trifluoroacetic acid, the yields increased from 61% to 78% as the water was decreased from 1.7 *M* to 0.06 *M*.

Photohydrolysis of Boc-Leu-Ala-Bni or Boc-Gly-Val-Bni in 10:20:1 (v/v/v) CH_2Cl_2/dioxane/water containing 0.1 *M* acetic acid went in 98–100% yield based on HBni release. Removal of Boc and ion-exchange analysis of the free dipeptide showed that the Boc-dipeptide was formed in only 80–81% yield. The 20% discrepancy is evidently due to cyclization to the diketopiperazine.

When these protected dipeptides were photolyzed in benzyl alcohol (Table I), most of the HBni was liberated by the time the starting material was consumed. But the expected benzyl ester and the free acid were both formed in significant yields. The combined yields of acid and ester from Boc-Gly-Val-Bni were about equal to the yield of liberated HBni, but 1/3 of the product obtained from Boc-Leu-Ala-Bni was ninhydrin-negative material after removal of Boc and was again probably the diketopiperazine.

The Bni protecting group was used for synthesis of Boc-Leu-Ala-Gly-Val-OH as illustrated in Fig. 2. The partially protected peptides obtained by photohydrolysis of Boc-Leu-Ala-Bni and acidolysis of Boc-Gly-Val-Bni were coupled

Table I. Photolysis of Bni Derivatives in 6:1 (v/v) CH_2Cl_2/Benzyl Alcohol

Protected dipeptide	$[CH_3CO_2H]$, M	Final yield (%)[a] HBni	Acid	Ester
Boc-Gly-Val-Bni	0	85	70	20
Boc-Gly-Val-Bni	0.1	85	56	20
Boc-Leu-Ala-Bni	0	96	29	37
Boc-Leu-Ala-Bni	0.1	96	31	25

[a] After all of the starting peptide was consumed as measured by TLC, HBni was measured by uv spectroscopy and the acid and ester were deprotected and measured with ninhydrin after ion-exchange separation.

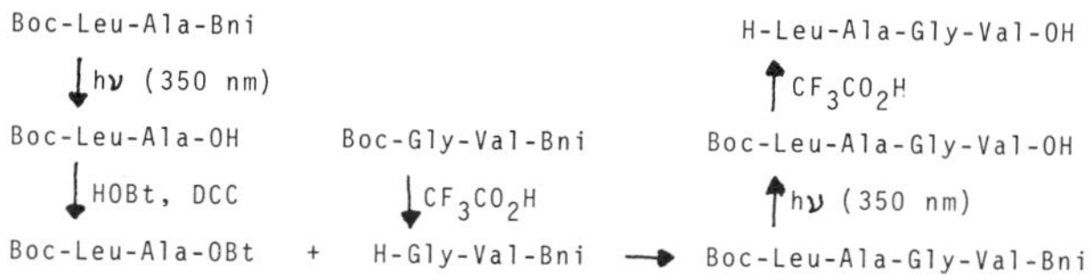

Fig. 2. Carboxyl protection by the Bni group during fragment condensation.

through the preformed benzotriazolo ester to afford the fully protected tetrapeptide in 60% overall yield after silica column chromatography. Amino-acid analysis gave 0.99 Leu, 1.00 Ala, 0.94 Gly, and 1.03 Val. Photohydrolysis of Boc-Leu-Ala-Gly-Val-Bni liberated HBni in 93% yield and afforded the Boc-tetrapeptide acid in 97% yield based on amino-acid analysis. Removal of the Boc group with trifluoroacetic acid provided free tetrapeptide in 99% yield based on ion-exchange analysis.

Amit et al.[4] observed that photosolvolysis of two derivatives of 1-benzyl-7-nitro-indoline proceeded smoothly, but that the 5-nitro and 6-nitro compounds were consumed without formation of solvolysis products. The crucial location of the nitro group *ortho* rather than *meta* or *para* to the indoline nitrogen suggests that it is not only part of the photoexcited chromophore but also participates directly in the formation of a reactive intermediate. A plausible but presently unsubstantiated mechanism involving neighboring group participation by the 7-nitro group is shown in Fig. 3. Active intermediate A would arise from an excited state by 1 → 5 transfer of the acyl group from the indoline nitrogen to one of the nitro oxygens and would subsequently undergo solvolysis.

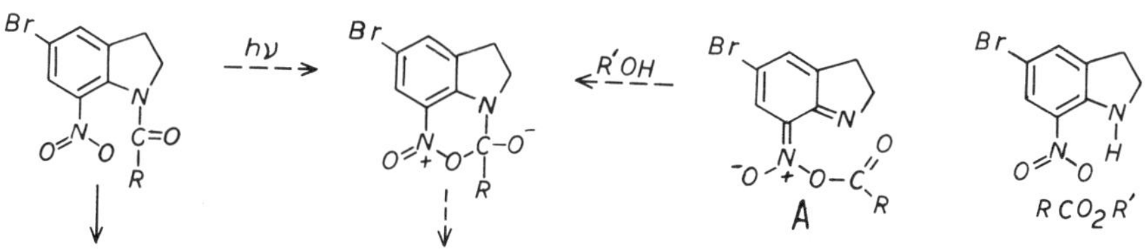

Fig. 3. A plausible mechanism for photosolvolysis of 1-peptidyl-5-bromo-7-nitroindoline derivatives.

G. G. is a postdoctoral fellow of the Research Assistance Foundation of the State of Sao Paulo (FAPESP) on leave from Escola Paulista de Medicina, Sao Paulo, Brazil. This work was supported in part by U. S. Public Health Service grants AM 01260 & HL 19795, by funds from the Hoffmann-La Roche Foundation, by American Heart Association Grant-in-Aid 74-844, and by funds contributed by the New York Heart Association.

References

1. Erickson, B. W. & Merrifield, R. B. (1976) in *The Proteins*, Neurath, H., and Hill, R. L., Eds., Academic Press, New York, 3rd Edn., Vol. II, pp. 255–527.

2. Rich, D. H. & Gurwara, S. K. (1975) *J. Amer. Chem. Soc.* 97, 1575–1579.

3. Wang, S. S. (1976) *J. Org. Chem.* **41**, 3258–3261.

4. Amit, B., Ben-Efraim, D. A. & Patchornik, A. (1976) *J. Amer. Chem. Soc.* **98**, 843–844.

IN VITRO BIOSYNTHESIS OF β-ENDORPHIN, γ-LIPOTROPIN AND β-LIPOTROPIN BY THE PARS INTERMEDIA OF BEEF PITUITARY GLANDS.

P. CRINE, S. BENJANNET, N. G. SEIDAH, M. LIS, and M. CHRÉTIEN.
Protein and Pituitary Hormone Laboratory. Clinical Research Institute of Montreal, 110 Pine Avenue West, Montreal H2W 1R7, Canada.

β-lipotropin, a 91-amino-acid pituitary peptide, comprises within its sequence the sequence of two biologically distinct peptides, namely β-melanotropin (MSH) (segment 41–58) and β-endorphin (segment 61–91). β-endorphin, β-lipotropin and γ-lipotropin (segment 1–58) have been purified from whole pituitary glands of different species including human. Immunocytofluorescence data show the presence of β-endorphin in the anterior and intermediary lobes of rat[1]. Similarly, it was shown that the major portion of opiate-like activity was found in the cells of the pars intermedia[2,3]. None of these techniques unequivocally localize the biosynthetic origin of these peptides. Recently, β-lipotropin, γ-lipotropin[4] and β-endorphin[5] were shown to be biosynthesized by the whole pituitary of beef. It was thus necessary to show that the intermediary lobe, although rich in opiate-like peptides, was also one of their biosynthetic sites. This report shows that isolated cells of the bovine pars intermedia can indeed synthesize these three peptides together with a methionyl N-terminal peptide.

The pars nervosa and pars intermedia were carefully dissected from the anterior lobe of 25 fresh beef pituitary glands. By gentle mechanical treatment the cells of the pars intermedia were then obtained, and were suspended in 10 ml of KRBG containing 0.2% bovine serum albumin. After 1 h preincubation, the cells were incubated for 3 h in a KRBG solution containing 1 *mCi* of ^{35}S-Met and 4.8 *mCi* of ^{3}H-Lys. The cells were then homogenized and extracted in 5 ml of 1 *mM* EDTA (pH 10.35) solution containing 5 mg/ml sheep pituitary fraction D[6]. After desalting, the extract was chromatographed together with 200 mg of fraction D on a CM-cellulose (CMC) column (1 × 40 cm).

Automatic degradation of the purified labeled peptides was performed on a Beckman 890 B sequencer, using 150 mmol of sperm whale apomyoglobin as carrier and 0.1 *M* Quadrol buffer[7]. The thiazolinones collected in butyl chloride were counted directly.

The CMC chromatogram of the desalted supernatant of cell homogenate is shown in Fig. 1. Four main fractions were studied: (i) fraction 59–77 corresponding to the position of carrier γ-lipotropin (ii) fraction 110–140 corresponding to the major peak of radioactivity (iii) fraction 195–235 where carrier β-lipotropin is found and (iv) fraction 236–270 from which carrier β-endorphin is recovered. Upon electrophoresis of fraction 59–77 a pure radioactive band was observed comigrating with standard γ-lipotropin at both pH 4.5 and 8.3. For a chemical proof of identity, mild (10 min) and exhaustive (24 hr) trypsin treatment was followed by direct sequencing of the digest, Fig. 2. As expected from the sequence of γ-lipotropin[6], methionine was found at the N-terminus and at cycles 7, 8 after 10 min digestion and exclusively at the N-terminus upon exhaustive tryptic digestion. This confirms the nature of this biosynthetic peptide as γ-lipotropin (β-LPH 1–58).

The electrophoresis of fraction 236–270, showed that it contained β-endorphin which was further purified on CMC as described previously[5]. A final proof of identity was provided by microsequencing which showed Met and Lys to occupy positions 5 to 9 of the sequence, respectively.

The fraction 195–235, was further purified on Sephadex G-75 followed by CM-Sephadex C-25. Electrophoresis showed a band co-migrating with β-lipotropin. However, the minute amount of material available prevented further characterization.

Finally, turning our attention to the major radioactive peak, fraction (110–140) in Fig. 1, we obtained upon electrophoresis at both pH 4.5 and 8.3 a pure material with R_f of 0.59

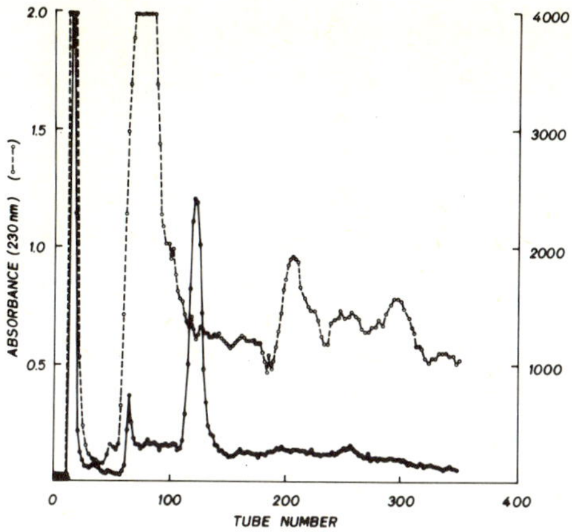

Fig. 1. CMC chromatography of labelled peptides isolated from pars intermedia of beef pituitary[5,6].

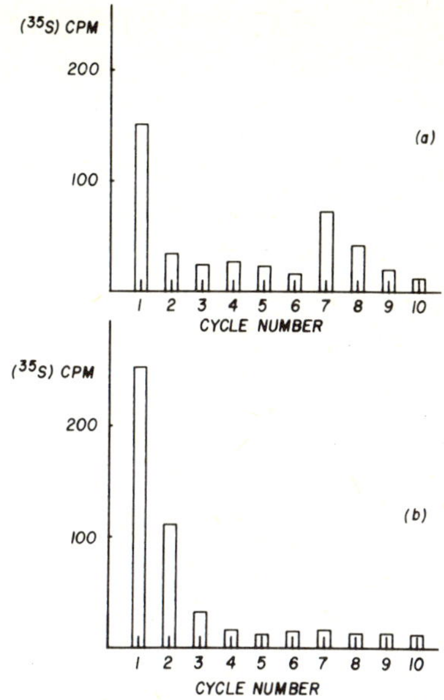

Fig. 2. Microsequence of tryptic digest of material from fraction 59.77 or 10 min (a) or 24 hours (b).

and 0.22 respectively. None of the known β-LPH fragments possessed these physical characteristics. The molecular weight was determined to be around 4000 by Sephadex G-75 chromatography. Microsequencing, Fig. 3, showed that Met occupies the N-terminal position and Lys occupies positions 6, 11, 27, 29 and 33. No known pituitary hormone of fragment known possesses this sequence.

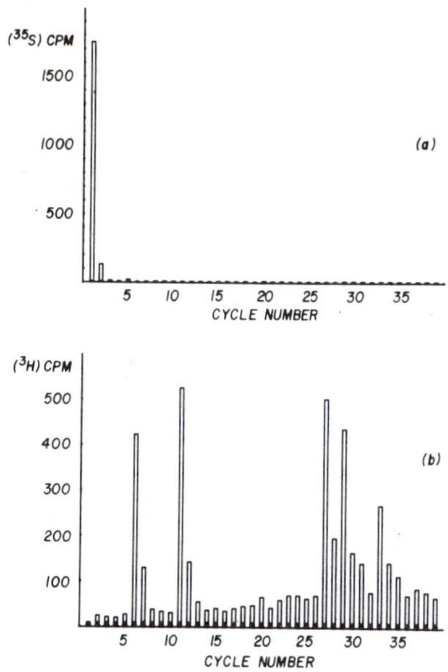

Fig. 3. Microsequencing of material in fraction 110–140 of CMC of Fig. 1.

References

1. Bloom, F., Battenberg, E., Rossier, J., Ling, N., Leppaluoto, J., Vargo, T. M. & Guillemin, R. (1977) *Life Sci* **20** 43–48.

2. Pelletier, G., Leclerc, R., Labrie, F., Côté, J., Chrétien, M. & Lis, M. (1977) *Endocrinology* **100**, 770–776.

3. Queen, G., Pinsky, C. & LaBella, F. (1976) *Biochem. Biophys. Res. Commun.* **72** 1021–1727.

4. Chrétien, M., Lis, M., Gilardeau, C. & Benjannet, S. (1976) *Can. J. Biochem.* 54 566–570.

5. Crine, P., Benjannet, S., Seidah, N. G., Lis, M. & Chrétien, M. (1977) *Proc. Nat. Acad. Sci USA* 74 1403–1406.

6. Li, C. H., Barnafi, L., Chrétien, M. & Chung, D. (1965) *Nature* **208**, 1093–1094.

7. Brauer, A. W., Margolies, M. N. & Haber, E. (1975) *Biochemistry* **14** 3029–3031.

THE SERUM THYMIC FACTOR AND ITS SYNTHESIS

E. BRICAS, J. MARTINEZ, D. BLANOT, G. AUGER, *Laboratoire des Peptides, Institut de Biochimie, Université de Paris-Sud, 91405 Orsay, France.*
M. DARDENNE, J. M. PLEAU and J. F. BACH, *INSERM U.25, Hôpital Necker 75015 Paris, France.*

The role of the thymus as an endocrine gland influencing several parameters of host immunological competence, is generally accepted[1]. During the last three years progress has been made in the isolation of highly purified thymic hormones and amino-acid sequences have been reported for three of these: thymosin α_1[2], thymopoietin II[3] and serum thymic factor (STF)[4] (Table I).

In contrast to the larger hormones obtained from thymus extracts, STF, isolated from pig blood by Bach et al.[4–6], is a nonapeptide. No homology was observed between the sequences of these peptides and STF is not a metabolite of thymopoietin II or thymosin α_1.

The thymic origin of STF was demonstrated by the rapid disappearance of this factor after thymectomy and its rapid reappearance after thymus grafting[4] (Figure 1). The purity

Table I. Amino-Acid Sequences of Three Thymic Hormones.

SERUM THYMIC FACTOR (STF) (from pig blood)

1976 - J.F. Bach et al.[5]

<Glu or Gln -Ala-Lys-Ser-Gln-Gly-Gly-Ser-Asn-OH
1 2 3 4 5 6 7 8 9

THYMOSIN α_1 (from calf thymus)

1977 - A.L. Goldstein et al.[2]

1 5 10 15 20
Ac-Ser-Asp-(Ala)$_2$-Val-Asp-Thr-(Ser)$_2$-Glu-Ile-(Thr)$_2$-Lys-Asp-Leu-Lys-Glu-(Lys)$_2$-

25 28
-Glu-(Val)$_2$-(Glu)$_2$-Ala-Glu-Asn-OH

THYMOPOIETIN II (from calf thymus)

1975 - D.H. Schlesinger and G. Goldstein[3]

1 5 10 15 20
H-Ser-Gln-Phe-Leu-Glu-Asp-Pro-Ser-Val-Leu-Thr-Lys-Glu-Lys-Leu-Lys-Ser-Glu-Leu-Val-

25 29 35 41
Ala-(Asn)$_2$-Val-Thr-Leu-Pro-Ala-Gly-Glu-Gln-Arg-Lys-Asp-Val-Tyr-Val-Gln-Leu-Tyr-Leu-

49
Glu-Thr-Leu-Thr-Ala-Val-Lys-Arg-OH

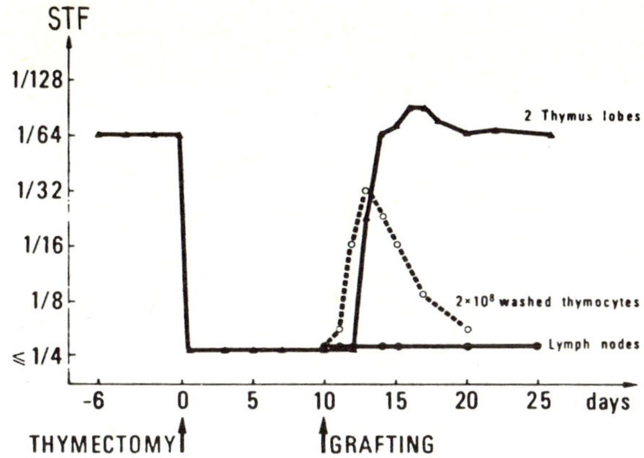

Fig. 1. Effect of thymectomy and thymus grafting on the serum thymic factor level in adult mice.

of isolated STF was assessed by TLC in different solvents and by high voltage electrophoresis[4] (Figure 2).

The reproducible stoichiometry of the amino-acid composition and the results of the sequence determination (after digestion by trypsin) by Edman's technique confirmed the purity of the isolated STF[5,6]. However two N-terminal residues are compatible with the analytical data: a pyroglutamyl or a glutaminyl residue. It is uncertain whether the pGlu residue is native or results from *in vitro* cyclization of a N-terminal Gln residue.

To confirm the proposed structure, we have carried out the synthesis of these two nonapeptides by solution methods in two different pathways of fragment condensation, see Scheme 1 as an example. Recovery was satisfactory in all steps. The majority of couplings was carried out by the mixed anhydride method. The protecting groups were cleaved from the completed nonapeptide by CF_3COOH/anisole. The synthetic <Glu'-nonapeptide behaved like the isolated natural STF on TLC in several solvent systems and in high voltage electrophoresis. The Gln'-nonapeptide was synthesized in a similar fashion.

The biological activity of the synthetic pyroglutamyl and glutaminyl nonapeptides was of the same order of magnitude as natural STF in the sheep red cell rosette assay,[7] both *in vitro* and *in vivo* (Table II). The synthetic C-terminal hexapeptide was totally inactive.

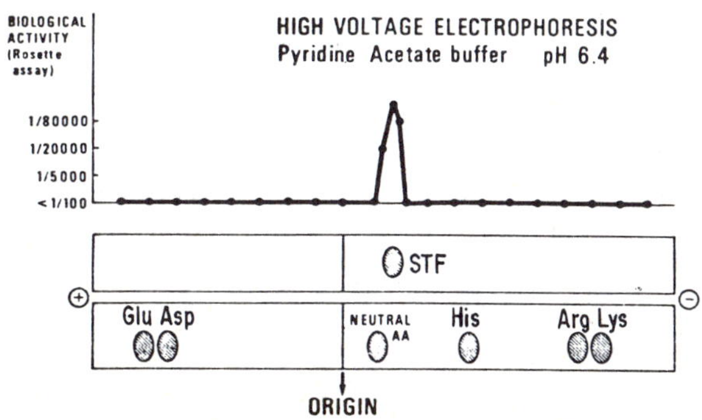

Fig. 2. Paper electrophoresis of the purified STF and biological activity of the eluted spot area.

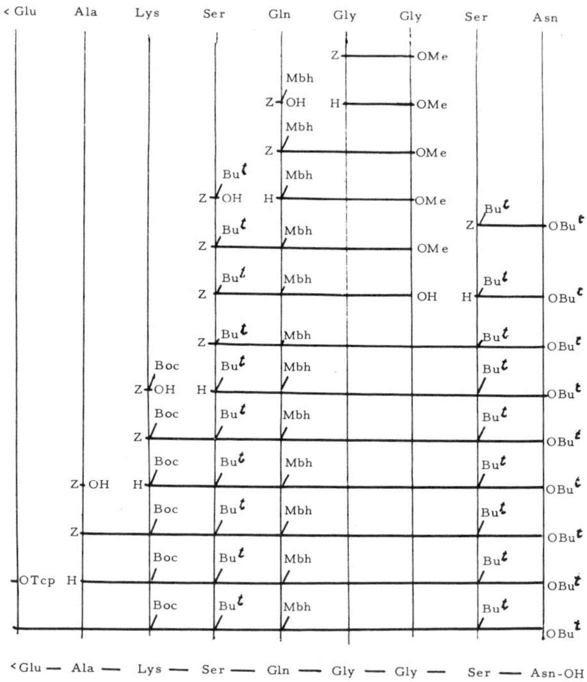

Scheme 1. Synthesis of serum thymic factor.

Table II. Biological Activities of Natural and Synthetic Serum Thymic Factor. Theta-conversion assay[7]: spleen rosette forming cells (RFC) of thymectomized adult mice in presence of azathioprine.

a. In vitro testing : minimal active concentration of thymic factor (STF).

Natural STF	<Glu-nonapeptide	Gln-nonapeptide
1-5 x 10^{-2} pg/ml	0.5 x 10^{-2} pg/ml	0.5 x 10^{-2} pg/ml
(1-5 x 10^{-14} M)	(0.5 x 10^{-14} M)	(0.5 x 10^{-14} M)

b. In vivo testing :	STF level in serum 2 hrs after injection	Spleen RFC sensitivity to azathioprine 24 hrs after injection
Normal mice (without injection)	1/128	0.5 µg
Thymectomized mice (without injection)	1/2 - 1/4	50 - 100 µg
Thymectomized mice after injection of :		
1 ng < Glupeptide	1/50 000 - 1/25 000 1/64 000 - 1/25 000	3 - 0.7 µg 0.7 - 0.3 µg
0.1 ng " "	1/50 000 - 1/12 800	0.3 - 1.5 µg
0.01 ng " "	1/12 800 - 1/12 800	3 - 3 µg
1 ng Gln-peptide	1/64 000 - 1/64 000	0.3 - 0.7 µg
0.1 ng " "	1/64 000 - 1/64 000	1.5 - 0.3 µg
10 ng C-terminal hexapeptide	inactive	50 - 100 µg

References

1. Bach, J. F. & Carnaud, C. (1976) *Prog. Allergy* **21**, 342–408.

2. Goldstein, A. L., Low, T. L. K., McAdoo, M., McClure, J., Thurman, G. B., Rossio, J., Lai, C. Y., Chaing, D., Wang, S. S., Harvey, C., Ramel, A. H. & Meienhofer, J. (1977) *Proc. Nat. Acad. Sci. USA,* **74**, 725-729.

3. Schlesinger, D. H. & Goldstein, G. (1975) *Cell,* **5**, 361–365.

4. Bach, J. F., Dardenne, M., Pleau, J. M. & Bach, M. A. (1975) *Ann. N. Y. Acad. Sci.* **249**, 186–210.

5. Bach, J. F., Dardenne, M., Pleau, J. M. & Rosa, J. (1976) *C. R. Acad. Sci. Paris,* **283D**, 1605-1607.

6. Bach, J. F., Dardenne, M., Pleau, J. M. & Rosa, J. (1977) *Nature* **266**, 55–57.

7. Dardenne, M. & Bach, J. F. (1975) in Biological Activity of Thymic Hormones, Van Bekkum, D. W., Ed., Kooyker Scientific Publication, Rotterdam, pp. 235–243.

THE ROLE OF CARBOHYDRATE IN THE STRUCTURE AND FUNCTION OF PLASMA GLYCOPROTEINS – STUDIES ON TWO INHERITED VARIANTS OF ALPHA-1-ANTITRYPSIN

CHARLES B. GLASER, ROBERT FALLAT, LUCIJA KARIC, *The Institutes of Medical Sciences, 2200 Webster Street, San Francisco, California 94115,* RICHARD STOCKERT and ANATOL G. MORELL, *Division of Genetic Medicine, Department of Medicine, Albert Einstein College of Medicine, Bronx, New York 10461*

Alpha-1-antitrypsin (AAT), the major inhibitor of proteolytic enzymes in human serum, is a glycoprotein of molecular weight ≃50,000, containing ≃12–15% carbohydrate and having an isoelectric point of 4.5-5.0. Its synthesis occurs in the liver[1] and is controlled by a pair of co-dominant allelic genes[2]. Although more than 90% of the population is found to contain proteinase inhibitor Pi type MM, a number of other variants have been found. Homozygotes for the deficiency gene, Pi type ZZ, have serum levels of AAT which are only 10–20% of the normal and these individuals are strongly disposed to lung disease[3,4].

Our goals were to establish 1) whether the low levels of circulating AAT and the deposition of AAT in the livers of ZZ individuals[5] could be accounted for by an increased catabolic rate of the Z protein; and 2) the mechanism of catabolism of AAT with particular reference to the role of the carbohydrate function. Ashwell and Morell[6] have shown that treatment of most native plasma glycoproteins with the enzyme neuraminidase will expose galactose as the terminal nonreducing sugar of the protein-linked carbohydrate chains and that this galactose will then serve as a specific determinant for the hepatic recognition, uptake and subsequent catabolism of the sialic acid deficient molecules.

Both M and Z forms of purified AAT were iodinated with $Na^{125}I$ and portions of the iodinated products were desialylated with neuraminidase. The survival time of asialo and native AAT (Fig. 1) in rats and the recovery of liver radioactivity reflects no difference between the M and Z forms. For the native proteins, ≃8% of the radioactivity was recovered in the liver after 70 minutes. The asialo forms of both M and Z protein were rapidly cleared from the circulation and within 10 minutes ≃60% of injected AAT could be recovered in the liver.

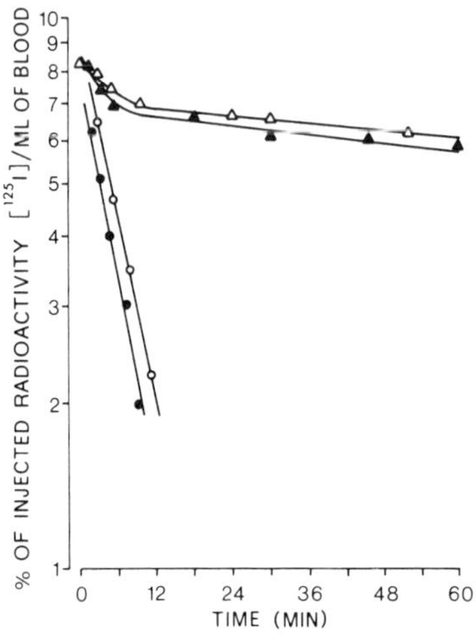

Figure 1. Short-term clearance rates of AAT (Pi types MM and ZZ) and their asialo derivatives. Male rats (200–250 g) were injected with 4.7–5.3 μg of ^{125}I-protein, diluted 1:10 with unlabelled protein, into the lateral tail vein. Blood samples were collected into weighed tubes by nicking the tail opposite the injection site and then counted for radioactivity. ▲–▲ AAT (Pi MM); △–△ AAT (Pi ZZ); ●–● asialo AAT (Pi MM); ○–○ asialo AAT (Pi ZZ).

Hepatic binding protein (HBP), specific for asialoglycoproteins was prepared from rabbit liver, bound to Sepharose, and standardized for the binding assay by established procedures[7,8]. Neither M nor Z proteins in their native forms are substrates for HBP, but the M and Z asialoproteins have strong and equal affinity.

Tritium was incorporated into terminal galactosyl residues using galactose oxidase followed by reduction with tritiated potassium borohydride[9]. None of the native forms of AAT had significant incorporation of radioactivity in the isolated neutral sugar fraction. In another experiment with the asialo derivatives, more tritium was incorporated into the M allele as compared to the Z variant. This is consistent with the presence of more total sugar chains in the M form, a finding which is also supported by our glycopeptide isolation studies with immobilized Concanavalin A. These results show that both forms of AAT have an equal rate of clearance which is dependent upon the loss of terminal sialic acid residues.

In a second study, the observed biochemical or physiological properties of the Z variant of AAT relative to the M-form (i.e., low carbohydrate content, lability, inability to efficiently be transported from the liver) were probed by near and far ultraviolet circular dichroic spectra (Fig. 2 A, B) and tryptophan fluorescence spectra. In the near ultraviolet spectra of the M-protein, the positive peaks at 259 and 265 nm and probably the negative shoulder at 269 nm are assigned to Phe (28 residues/molecule). The large negative band from 275–290 nm contains unresolved contributions from Trp and Tyr (7 residues/molecule), while the positive band at 295 nm consists of a composite of three Trp residues. In the Z protein one or more Trp chromophores are in a modified micro-environment. It is not yet possible to ascribe changes between the M and Z protein in the region 250–290 nm to specific residues. The general shift to a more negative amplitude in Z protein could reflect a Trp band, but perturbations in Tyr and Phe may also contribute. Differences in tertiary structure are also reflected in the Trp fluorescence spectra. Although the wavelength of maximum emission remains unchanged (335 nm) and indicates a moderately non-polar region for the tryptophans in both proteins, the amplitude of the emission spectra is 50% higher in the Z protein. The far ultraviolet circular dichroic spectra is almost identical for both forms of AAT, and shows an α-helix content of 45–50%.[10]

This work was supported by the National Institute of Health Grants HL-17194, 14692, AM-0159 and the American Lung Association.

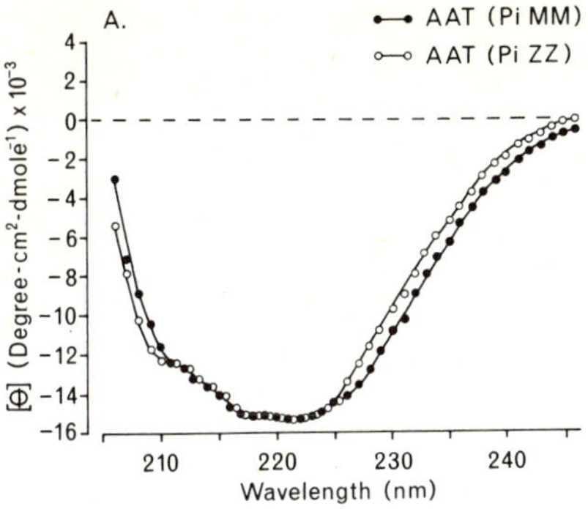

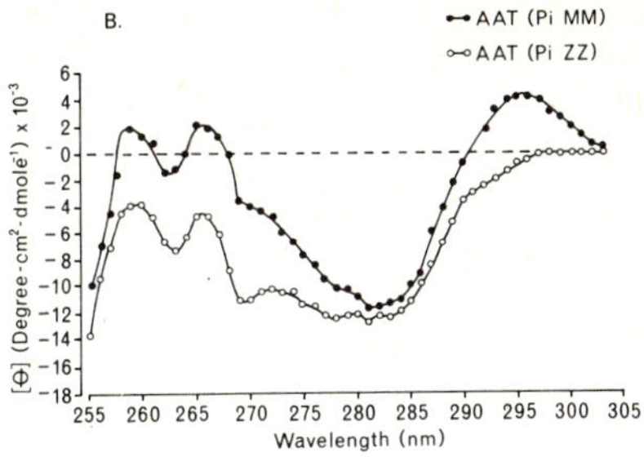

Fig. 2. (A): Circular dichroic spectra of AAT in far UV in 0.1 *M* Tris/HCl, pH 8.0. Data are expressed as mean residue ellipticity, calculated by using mean residue weight of 111 from its amino-acid composition. Protein concentrations were determined spectrophotometrically using $E^{0.1\%}_{1\ cm} = 0.50$. Each concentration value was reduced by subtracting the carbohydrate content. (B): Circular dichroic spectra of AAT in near UV has been presented in terms of molecular ellipticity. The molecular weight of 45,000 used in these calculations represents only the polypeptide portions of the molecules.

References

1. Rowley, P. T. & Miller, L. L. (1975) *Proc. Soc. Exp. Biol. Med.* **148**, 145–150.
2. Fagerhol, M. K. & Laurell, C. B. (1967) *Clin. Chim. Acta* **16**, 199–203.
3. Erikkson, S. (1965) *Acta Med. Scand. (Suppl.)* **432**, 1–85.
4. Talamo, R. C., Allen, J. D., Kahan, M. G. & Austen, K. F. (1968) *N. Engl. J. Med.* **278**, 345–351.
5. Sharp, H. L. (1976) *Gastroenterology* **70**, 611–621.
6. Ashwell, G. & Morell, A. G. (1974) *Advan. Enzymol.* **41**, 99–128.
7. Hudgin, R. L., Princer, Jr., W. E., Ashwell, G., Stockert, R. J. & Morell, A. G. (1974) *J. Biol. Chem.* **249**, 5536–5543.
8. Lunney, Y. & Ashwell, G., (1976) *Proc. Nat. Sci. USA* **73**, 341–343.
9. Morell, A. G., VanDenHamer, C. J. A., Scheinberg, I. H. & Ashwell, G. (1974) *J. Biol. Chem.* **241**, 3745–3749.
10. Chen, Y-H., Yang, J. T. & Martinez, H. M. (1972) *Biochemistry* **11**, 4120–4131.

MONITORING WITH FEEDBACK IN AUTOMATED SOLID PHASE PEPTIDE SYNTHESIS

T. CHRISTENSEN, P. VILLEMOES, and K. BRUNFELDT, *The Danish Institute of Protein Chemistry affiliated to the Danish Academy of Technical Sciences, 4, Venlighedsvej, DK-2970 Hørsholm, Denmark*

Monitoring of automated solid phase peptide synthesis can be performed by perchloric acid titration[1]. To obtain feedback control a minicomputer is now used as control unit[2,3]. A print-out of the synthetic procedure including instrumental failures and evaluation of titration data is obtained. The computer is programmed (Scheme 1) for the following decisions: 1. Whether titration data are acceptable through an evaluation of the deviation between the two last obtained values. A repetition is allowed a preset number of times until acceptable data are obtained. 2. If acceptance, the mean of the two last titrations is calculated. The value is then compared with the value after the previous coupling or deblocking cycle. If within preset limits, the synthetic procedure will continue. 3. If not, repetition of coupling or deblocking cycle will take place a preset number of times. 4. In case neither acceptable titration results nor acceptable coupling or deblocking yields are obtained, the synthesizer will be paralyzed.

The programming secures proper electrode function and correct titration procedure. Titration is stopped if a signal is not received before the delayed shut-off time has run out. By very slow addition, for example less than 0.1 ml in 5 min, the titration is brought to an end, allowing titration on compounds protected with slightly labile α-amino protection groups. $1 starts the deblocking cycle and means that one cycle will be carried out. #2 starts the titration cycle and means that two titration cycles unconditionally are carried out. /2 at the end of the titration cycle means that a maximum of two further titration cycles may be performed to try to obtain acceptable values. /1 at the end of the deblocking cycle means that if the value obtained from the succeeding titration cycles indicates an incomplete deblocking, the cycle can be repeated only once. Similarly for a coupling cycle.

Resin-bound Leu^5-enkephalin was prepared at 20°C. Coupling time 2 h. Tyr was coupled as Boc-Tyr (2-BrBzl) (Ref 4). For the time of cleavage, see Figure 1 (Fig. 1D, 1 h for the C-terminal amino acid). Cleavage from resin was performed with HBr in HOAc and 10% *m*-dimetoxybenzene in 5 h. In Fig. 1B and C with Boc-amino acids and DCC in suboptimal amounts, a pronounced decrease in the number of titratable groups after

```
$1 B1 X4 Y1 W1 X4 Y1 W1 X6 Y1 B1 X4 Y1 B1 X4 Y1 B1 X4 Y1 A1 X4 Y1
A1 X4 Y1 /1
#2 A1 X4 Y1 A1 WØ A1 X4 X2 Y1 A1 X4 Y1 A1 X4 Y1 B1 X4 Y1 B1 X4 Y1
B1 X4 Y1 B1 A1 V2 Y1 A1 X4 Y1 A1 X4 Y1 /2.
```

Scheme 1. Program for a deblocking cycle, Boc cleavage, followed by titration.

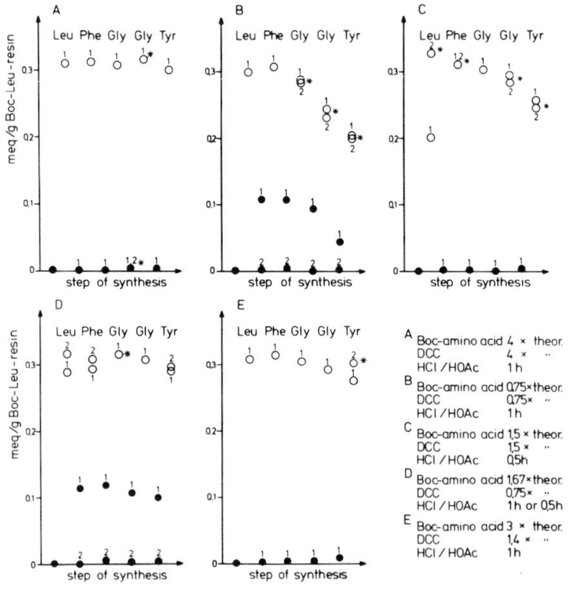

Fig. 1. Titration values during syntheses of the sequence shown. The values are means of two last values when accepted data have been obtained. o: Value after deblocking. •: Value after coupling. The indices denote the number of deblockings or couplings. * means, that the synthesizer was paralyzed because an acceptable coupling or deblocking was not achieved.

deblocking is seen. In Fig. 1B and D repetition of the coupling was necessary. It is evident that in Fig. 1B and C irreversible blocking has occurred, most likely due to the effect of free DCC in combination with incomplete and/or slow acylation. Loss of peptide from the resin can be excluded due to the presence of the expected amount of the C-terminal leucine, at the end of the synthesis (see Table I).

Table I. Amino-acid Content of the Synthesized Products Determined by Titration and Amino-acid Analysis. The values are presented as mmole per g Boc-Leu-resin, the numbers in brackets are the relative values. Ø – as some resin was lost during the cleavage procedure, it is not possible to give absolute values in this case. Calculation of the amino-acid content from the titration values is carried out by subtracting the preceding value for the coupling from the value obtained after deblocking.

Synthesis	Amino-acid	Titration	Amino acid analysis	
			Resin bound product	Cleaved crude product
A	Leu	0.306 (0.99)	0.344 (1.12)	0.323 (1.02)
	Phe	0.312 (1.01)	0.319 (1.04)	0.307 (0.97)
	Gly	0.619 (2.00)	0.614 (2.00)	0.636 (2.00)
	Tyr	0.300 (0.97)	0.274 (0.89)	0.303 (0.95)
B	Leu	0.297 (1.16)	0.342 (1.36)	0.260 (1.13)
	Phe	0.303 (1.18)	0.282 (1.12)	0.228 (0.99)
	Gly	0.513 (2.00)	0.504 (2.00)	0.461 (2.00)
	Tyr	0.196 (0.76)	0.182 (0.72)	0.182 (0.79)
C	Leu	0.326 (1.12)	0.347 (1.16)	o.266 (1.01)
	Phe	0.307 (1.05)	0.319 (1.07)	0.258 (0.98)
	Gly	0.584 (2.00)	0.596 (2.00)	0.528 (2.00)
	Tyr	0.243 (0.83)	0.233 (0.78)	o.204 (0.77)
D	Leu	0.316 (1.02)	0.340 (1.12)	0.292 (1.02)
	Phe	0.304 (0.98)	0.329 (1.o9)	0.282 (0.98)
	Gly	0.618 (2.00)	0.605 (2.00)	0.574 (2.00)
	Tyr	0.292 (0.94)	0.269 (2.00)	0.263 (0.92)
E	Leu	0.307 (1.04)	0.340 (1.11)	Ø (1.03)
	Phe	0.311 (1.06)	0.307 (1.00)	(0.94)
	Gly	0.589 (2.00)	0.615 (2.00)	(2.00)
	Tyr	0.293 (0.99)	0.263 (0.86)	(0.90)

References

1. Brunfeldt, K., Christensen, T. & Villemoes, P. (1972) *FEBS Lett* 22, 238–244.
2. Villemoes, P., Christensen, T. & Brunfeldt, K. (1976) *Hoppe-Seyler's Z. Physiol. Chem.* 357, 713–719.
3. Villemoes, P., Christensen, T. & Brunfeldt, K. To be published.
4. Salem, E., Larsen, P., Lund, E. & Schou, O. To be published.
5. Hancock, W. S., Prescott, D. J., Vagelos, P. R. & Marshall, G. R. (1973) *J. Org. Chem.* 38, 774–781.

A THREE-POINT MODEL FOR THE DIPEPTIDE SWEETENER-RECEPTOR INTERACTION

M. CHOREV, C. G. WILLSON and M. GOODMAN, *Department of Chemistry, University of California, San Diego, La Jolla, California 92093*

Since the discovery by Mazur et al.[1] that α-L-aspartyl-L-phenylalanine methyl ester is about 150 times as sweet as sucrose, much synthetic work has been undertaken to understand the molecular basis for the sweet taste.[1-6] Several models have been published to explain and predict the sweet taste of the related dipeptide derivatives. Ariyoshi's model[3] is based on Fischer projections of the structure of dipeptide derivatives. Temussi and coworkers[7] attempted to establish a receptor site based upon the most probable conformation of the dipeptide sweeteners by using pmr studies and potential energy calculations. Our group[8] viewed the dipeptide sweeteners from a topochemical standpoint. We pointed out the necessity of a planar zwitterion and a specific hydrophobic shape and size moiety.

In this paper, we extend our model to include the peptide bond in a specific location as an essential feature to elicit the sweet taste in addition to two previous requirements. Our three-point model is outlined in Figure 1.

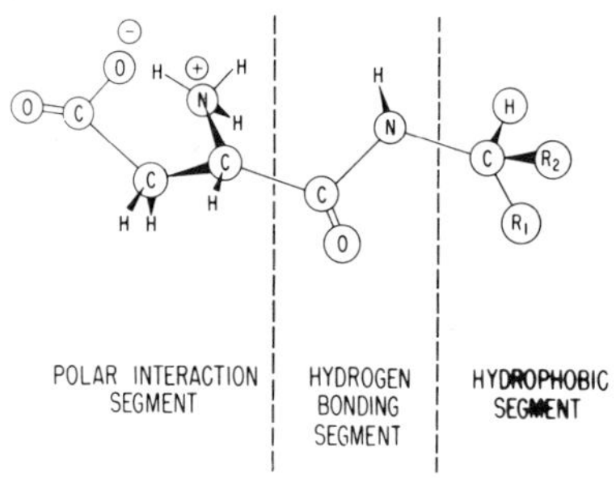

Fig. 1. Schematic representation of the three-point model.

We established the necessity of the peptide bond by synthesizing L-aspartyl-*N*-methyl-L-phenylalanine methyl ester (*1*) and L-aspartyl-L-phenyllactic acid methyl ester (*2*). Both compounds are not sweet.

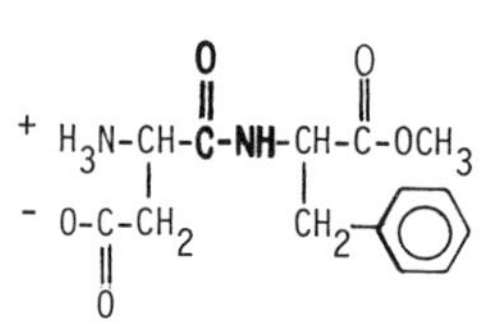

L-Asp-L-PheOMe

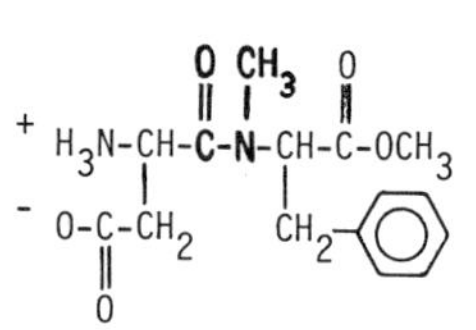

(1)

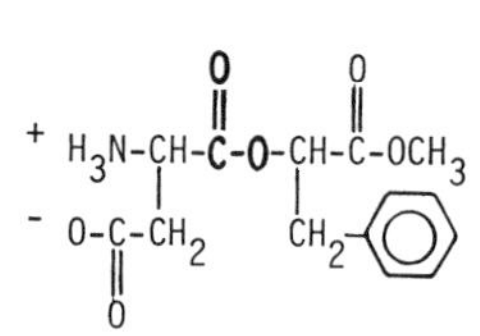

(2)

These findings have led us to consider the peptide bond as a new basic requirement, formerly ignored, to have a proper tastant-receptor interaction. The key test of our hypothesis involved the synthesis of a topochemical analog[9] of a dipeptide sweetener. This analog is essentially identical to the parent compound with respect to polarity and shape at its periphery; the only difference resides in the reversal of the peptide bond which leads to an absolute linear retro-isomer of the dipeptide sweetener.

We have synthesized the retro-isomer (*3*) of intensely sweet α-aminomalonyl-D,L-phenylalanine methyl ester (*4*).

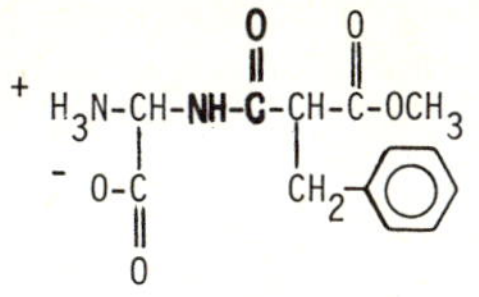

(3)

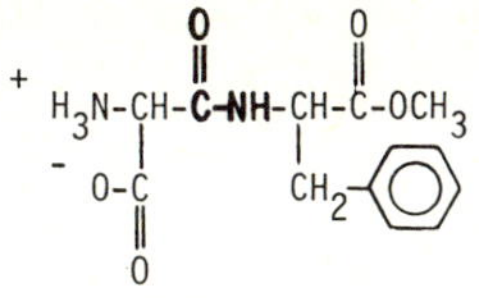

(4)

The precursor for the synthesis was *N*-benzyloxycarbonyl-α-aminomalonic acid benzyl ester hydrazide (*5*) which on treatment with NOCl/THF yielded the azide[10] (*6*) which underwent Curtius rearrangement to the corresponding isocyanate (*7*). Addition of *t*-butanol to the isocyanate (*7*) yielded *N*-benzyloxycarbonyl-*N'*-*t*-butyloxycarbonyl-α-aminoglycine benzyl ester (*8*) (Scheme 1):

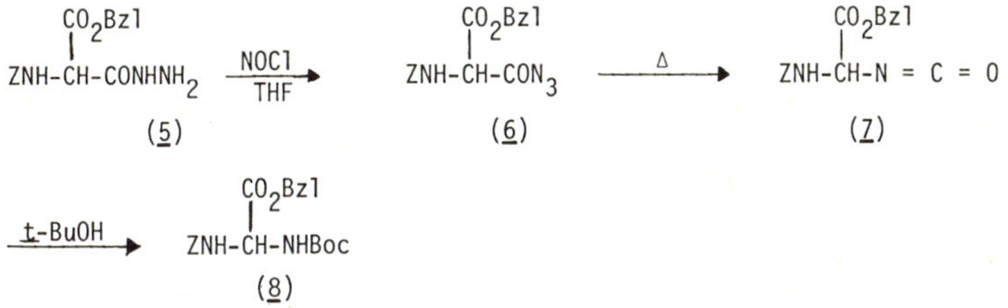

Scheme 1

The *t*-butyloxycarbonyl group was removed selectively from compound *8* by treatment with HCl/dioxane. The resulting salt was coupled immediately with 2-benzylmalonic acid mono-methyl ester using dicyclohexylcarbodiimide. The reaction yielded compound *9*, the fully protected retro-isomer of compound *4*. Hydrogenation of compound *9* at atmospheric pressure in methanol in the presence of triethylamine resulted in compound *3*, the absolute retro-isomer of the dipeptide sweetener (*4*) (Scheme 2).

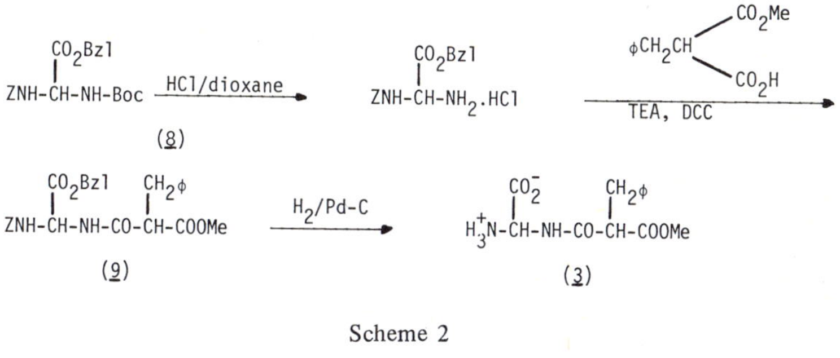

Scheme 2

Compound *3* was found *not* to be sweet; it actually possessed a slightly bitter after-taste. This provides further evidence that in addition to the other molecular features, the presence of an appropriate peptide bond is essential. The peptide bond is required not only for its polarity and hydrogen bond donor-acceptor characteristics, but also for its direction and relative position within the molecule which must be specific in order to obtain the correct interaction between the tastant and taste receptor.

Our three-point model (Fig. 1) is an extension of our two-point picture to which we have added a second polar site – the peptide bond. This is the first observation that the peptide backbone, as well as the nature and spatial arrangement of side chains and end groups, is required for biological activity.

The approaches presented in Schemes I and II for the synthesis of *gem*-diamino derivatives are part of a general methodology being developed by us.[11,12]

The authors wish to thank Mr. Fred Vernacchia for the synthesis of compound (*1*) (L-aspartyl-N-methyl-L-phenylalanine methyl ester) and Dr. Chaim Gilon for the synthesis of compound (*2*) (L-aspartyl-L-phenyllactic acid methyl ester). They also gratefully acknowledge the support of the National Institutes of Health (FD 00590).

References

1. Mazur, R. H., Schlatter, J. M. & Goldkamp, A. H. (1969) *J. Amer. Chem. Soc.* **91**, 2684-2691.

2. Mazur, R. H., Reuter, J. A., Swiatek, K. H. & Schlatter, J. M. (1973) *J. Med. Chem.* **16**, 1284–1287.

3. Ariyoshi, Y. (1976) *Agr. Biol. Chem.* **40**, 983-992.

4. Mazur, R. H., Goldkamp, A. H., James, P. A. & Schlatter, J. M. (1970) *J. Med. Chem.* **13**, 1217-1221.

5. Fujino, M., Wakimasu, M. & Tanaka, K., Aoki, H. & Nakajima, N. (1973) *Naturwissen.* **60**, 351.

6. Goodman, M., et al., unpublished data.

7. Lelj, F., Tancredi, T., Temussi, P. A., and Toniolo, C. (1976), *J. Am. Chem. Soc.*, **98**, 6669–6675.

8. Goodman, M., and Gilon, C. (1975) in Peptides 1974, Proceedings of the 13th European Peptide Symposium, Israel, Wolman, Y., Ed., John Wiley & Sons, New York; Israel Universities Press, Jerusalem, pp. 271–278.

9. Shemyakin, N. M., Ovchinnikov, Yu. A., and Ivanov, V. T. (1969), *Angew. Chem. Internat. Edit.*, 8, 492–499.

10. Honzel, J., and Rudinger, J. (1961), *Coll. Czech.*, **26**, 2333.

11. Chorev, M., Willson, C. G., and Goodman, M. (1977), submitted for publication.

12. Willson, C. G., Goodman, M., Rivier, J., and Vale, W. (1977), in *Peptides: Proceedings of the 5th American Peptide Symposium*, Goodman, M. & Meienhofer, J., Eds., John Wiley and Sons, Inc., New York, pp. 579–581.

HOW DO ENZYMES REALLY WORK?

ARIEH WARSHEL, *Department of Chemistry, University of Southern California, Los Angeles, California 90007*

Understanding of the important factors in enzymic reactions is a major problem in molecular biology. Despite extensive effort there is no quantitative explanation for the catalytic power of enzymes. Here we show quantitatively that the most important catalytic factor in enzymic reactions is electrostatic stabilization of ionic transition states. The crucial role of charge stabilization has never been demonstrated before because the available dielectric theories are inadequate for studies of electrostatic interactions at short distances.

Applying a recently developed microscopic dielectric model we show how a very small charge stabilization effect in model compounds in solution corresponds to a remarkable effect in the enzyme-active site.

In recent work[1] we developed a method for the theoretical study of enzymic reactions. The method includes the complete enzyme-substrate system and allows the evaluation of the microscopic dielectric effect inside the enzyme active site. This method was applied to the carbonium ion intermediate in the reaction of lysozyme. It was found that the charge stabilization of the carbonium transition state is a major catalytic factor. It was also concluded that the strain effect, thought to be a major catalytic factor, does not contribute significantly. However, the conclusion about the importance of charge stabilization seems to be in clear contradiction to many experiments on model compounds in solution. Such experiments have not shown any significant catalytic effect by neighboring ionized groups. The use of current theories in physical organic chemistry to extrapolate these results to low dielectric media has not indicated a large catalytic effect.

To resolve this contradiction we developed a new microscopic dielectric approach[2] which allows the accurate evaluation of the microscopic energy balance of charge separation in polar liquids. Using this approach we have shown that in polar solutions the electrostatic interaction between opposite charges is always balanced by the solvation energy (see Fig. 1). This is true even when two charges are only 3 Å apart. An experimental verification for this conclusion is found in studying the ΔpK_a of dicarboxylic acids of different lengths. This microscopic dielectric effect is due primarily to the fact that the solvent system has enough degrees of freedom, so that any electrostatic force is compensated for by reorientation and compression of the solvent dipoles.

Enzyme active sites are quite different systems. Such active sites can provide a low dielectric hydrophobic environment in which charged groups and permanent dipoles are designed to stabilize the substrate transition state. Since these groups cannot rearrange their orientation, they are able to provide a large electrostatic stabilization and to make the reaction faster than in polar solutions.

To prove our point we performed two comparative calculations[2] of the general acid catalysis of the cleavage of a glycosidic bond by lysozyme. In the first calculation we studied the reaction in aqueous solution and examined the catalytic effect of a negative

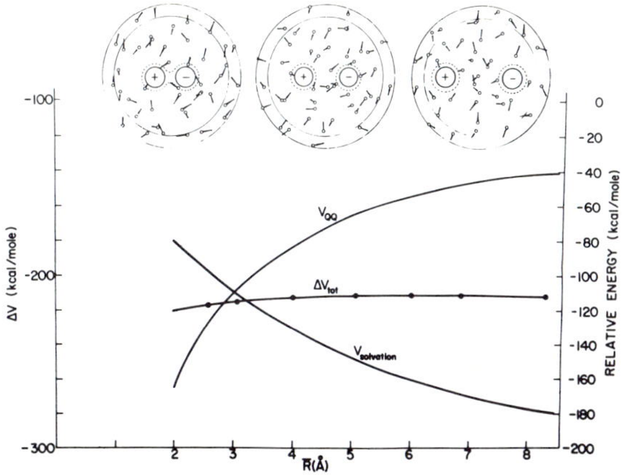

Fig. 1. Calculation of the energy balance of charge separation in polar liquids. V_{QQ}, $V_{solvation}$ and ΔV_{tot} are respectively the charge-charge interaction, the solvation energy and the total energy. More details are given in Ref. 2.

charge at 3 Å from the C_1 of the carbonium intermediate. It was found that the presence of the external charge does not lead to any significant rate enhancement. In the second calculation we took exactly the same substrate system (without any adjustable parameter) and repeated the calculation in the active site of lysozyme in the presence of the ionized Asp 52. The activation barrier in the enzyme active site was reduced by ~10 kcal/mole relative to the corresponding barrier in solution.

Considering the fact that the overall reduction in the activation barrier between reaction in aqueous solution and at the enzyme active site is 12–14 kcal/mole, we feel that charge stabilization must be the major factor in enzymic reactions

References

1. Warshel, A. & Levitt, M. (1976) *J. Mol. Biol.* **103**, 227–249.
2. Warshel, A. (1978) *J. Am. Chem. Soc.* (submitted).

THE USE OF AN ACTIVE SITE MODEL IN THE DESIGN OF SPECIFIC INHIBITORS OF ANGIOTENSIN-CONVERTING ENZYME

MIGUEL A. ONDETTI, EMILY F. SABO, KATHRYN A. LOSEE,
HONG SON CHEUNG, DAVID W. CUSHMAN, BERNARD RUBIN,
The Squibb Institute for Medical Research, Princeton, New Jersey 08540

In 1970, at the Second American Peptide Symposium[1], we reported the structure elucidation and synthesis of powerful inhibitors of angiotensin-converting enzyme (ACE) that were originally isolated from the venom of *B. jararaca*. The most potent of these inhibitors was the nonapeptide SQ 20,881, <Glu-Trp-Pro-Arg-Pro-Gln-Ile-Pro-Pro. Since then it has been shown that SQ 20,881 was a remarkably useful tool for the inhibition of the renin-angiotensin system in a large variety of animal species and in man. However, this usefulness was flawed by the lack of oral activity. Our studies with analogs and derivatives of SQ 20,881[2] did not solve this problem, but led to a better understanding of the enzyme and to the proposal that its mechanism of action was essentially similar to that of carboxypeptidase A.

Based on these similarities and the concept of byproduct inhibition introduced by Byers and Wolfenden,[3] we postulated that succinyl amino acids might function as byproduct inhibitors of ACE, since they combined in one single molecular species characteristics of the two products of peptide hydrolysis (Fig. 1). Succinyl-*L*-proline, the first derivative synthesized to prove this hypothesis, was indeed an inhibitor of ACE. Although it had a rather low potency (IC_{50} = 135 μg/ml), it was clearly a specific inhibitor of ACE, since in the smooth muscle test it was shown to inhibit the angiotensin I response and potentiate the bradykinin response without affecting those of angiotensin II or acetylcholine, a typical pattern for a specific ACE inhibitor.[4]

To increase the inhibitory activity of this prototype compound while still pursuing the concept of byproduct inhibition, varied functionality was introduced in the amino-acid and succinyl moieties. The most important conclusion derived from these studies was that substituents on the succinyl moiety increase activity only if present in the α-position with respect to the amide bond and with an absolute stereochemistry that would make them isosteric with an L-L-dipeptide. In this manner we obtained a substantial increase in inhibitory activity with the synthesis of D-2-methylsuccinyl-L-proline (IC_{50} = 12 μg/ml), a compound which was also shown to be capable of efficiently inhibiting ACE *in vivo* parenterally and, to some extent, orally.

With the introduction of these substitutions it appeared that we had driven the byproduct analog concept to its ultimate level of usefulness. We had also found that glutaryl-*L*-proline was more active than succinyl-*L*-proline (Table I, *3, 1*), an observation that was not in complete agreement with the idea of byproduct inhibition. However, a somewhat different explanation for the strong inhibitory activity of these compounds seemed possible if one hypothesizes that the active center of ACE has at least three main binding points for a substrate: namely, a cationic carboxyl-binding center; an amide recognition center, probably through a hydrogen bonding functionality; and a complexed zinc ion that polarizes the carbonyl of the scissile amide bond (Fig. 2). If a molecule could combine functionalities that can interact specifically with these binding sites, it would become a potent, but still competitive inhibitor of ACE. Succinyl and glutaryl amino acids fulfilled these requirements, if we postulated that a folded conformation of the glutaryl side chain binds more effectively to the enzyme than an extended succinyl side chain. This alternate interpretation suggests that one could increase the potency of these inhibitors by searching for moieties that could bind to the zinc atom with stronger affinity than the carboxyl residue.

This assumption led to the synthesis of acyl amino acids carrying a large variety of oxygen, nitrogen or sulfur containing functionalities on the acyl moiety (Table I, *6-16*). We found that a mercaptoalkanoyl acyl moiety could yield inhibitors that were several orders of magnitude more potent than the carboxylalkanoyl amino acids (Table I, *16, 17*). With the aid of suitably substituted mercaptans, we showed that the other two primary binding sites postulated in the hypothetical model also contribute significantly to the binding of enzyme and inhibitor (Table I, *18, 19*). In addition, other substantial interactions with the enzyme became apparent during these studies; namely, those involving the side chain of the amino-acid residue and the substitution on the acyl residue. Among all the amino-acid derivatives tested, proline or other cyclic imino acids gave the most potent inhibitors, and alkyl substitution on the α-position of the acyl moiety with the appropriate stereochemistry lowers considerably the dissociation constant of the inhibitors.

We can therefore conclude that the explanation that we had postulated for the action of the carboxyalkanoyl amino acid inhibitors of ACE – namely, a strong regio- and stereospecific interaction between several functionalities on the inhibitor molecule and binding sites on the enzyme surface – has been of great value in guiding the synthesis of novel inhibitors, and that the high potency and specificity of the inhibitors thus obtained gave indirect but strong evidence that the interactions postulated in the model are correct. The most potent of the ACE inhibitors developed so far is D-3-mercapto-2-methylpropanoyl-L-proline (SQ 14,225) which has been shown to inhibit the vasopressor response of angiotensin I in animals[4] and in man[5] when administered orally in doses ranging from 0.1 to 1 mg/kg. SQ

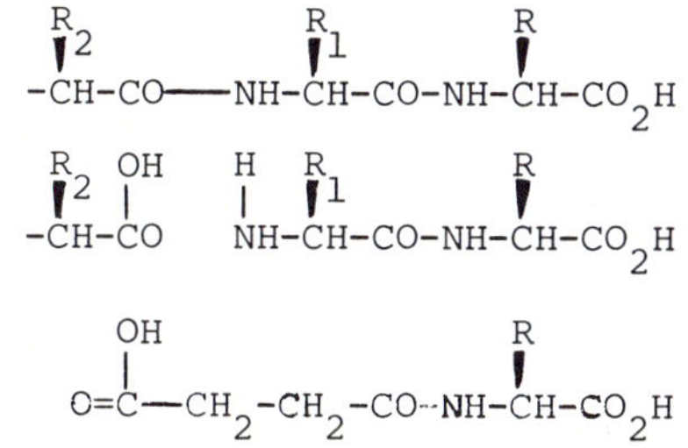

Fig. 1. Substrate, products, and inhibitors of ACE.

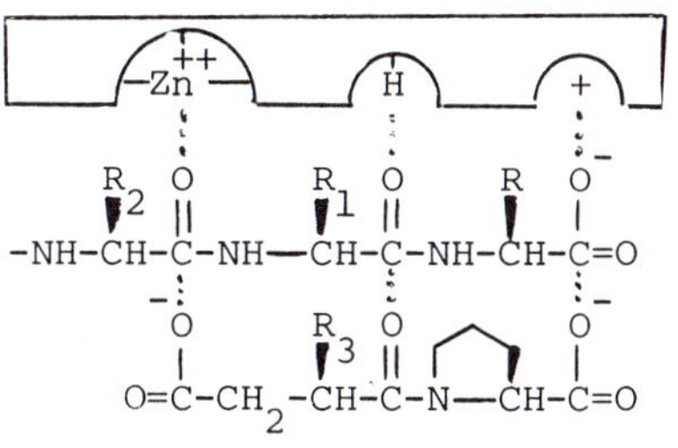

Fig. 2. Binding of substrate and inhibitors at the active site of ACE.

Table I. Activities *In Vitro* of Angiotensin-Converting Enzyme Inhibitors. IC_{50}, Concentration of Compound Producing 50 Percent Inhibition of Enzyme Activity or Agonist Effect.

No.	Structure	Angiotensin-Converting Enzyme Rabbit Lung IC_{50} (μg/ml)	Excised Guinea Pig Ileum Angiotensin I IC_{50} (μg/ml)
1	$HO_2C-CH_2-CH_2-CO$-L-Pro	135.	94.0
2	$HO_2C-CH_2-CH_2-CO$-L-Phe	145.0	>100.0
3	$HO_2C-CH_2-CH_2-CH_2-CO$-L-Pro	16.	6.6
4	$HO_2C-CH_2-CH(CH_3)-CO$-L-Pro (SQ 13,297)	12.	13.0
5	$HO_2C-CH_2-CH(CH_3)-CO$-L-Pro	340.0	>100.0
6	$CH_3-CO-CH_2-CH_2-CO$-L-Phe	600.0	>100.0
7	$CH_3-CH(OH)-CH_2-CH_2-CO$-L-Phe	>1000.	>100.0
8	$HONH-CO-CH_2-CH_2-CO$-L-Phe	19.	56.0
9	$HO_2C-CH(NH_2)-CH_2-CO$-L-Pro	950.	310.0
10	$HO_2C-CH(NH_2)-CH_2-CH_2-CO$-L-Pro	760.	>100.0
11	$NH_2-CH_2-CH_2-CO$-L-Pro	170.	>100.0
12	$NH_2-CH_2-CH_2-CH_2-CO$-L-Pro	>1000.	>100.0
13	$NH_2-C(=NH)-NH-CH_2-CH_2-CO$-L-Pro	>1000.	>100.0
14	 $CH_2-NH-CH_2CH_2-CO$-L-Pro	800.	>100.0
15	HS 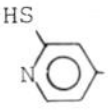 CO-L-Pro	210.	47.0
16	$HS-CH_2-CH_2-CO$-L-Pro (SQ 13,863)	0.050	0.04
17	$HS-CH_2-CH(CH_3)-CO$-L-Pro (SQ 14,225)	0.005	0.005
18	$HS-CH_2-CH_2-CH_2-CH_2-CH_2-CO_2H$	260.0	80.0
19	$HS-CH_2-CH_2-CO$-L-Pro-OEt	4.0	5.4

14,225 was also shown to have a marked oral antihypertensive activity in animal models of renovascular hypertension,[6] and it is expected to constitute a significant addition to the therapy of human hypertension.

References

1. Ondetti, M. A., Williams, N. J., Sabo, E. F., Pluscec, J., Weaver, E. R. & Kocy, O. (1972) in *Progress in Peptide Research*, Lande, S., Ed., Gordon and Breach, New York, pp. 251-259.

2. Cushman, D. W., Pluscec, J., Williams, N. J., Weaver, E. R., Sabo, E. F., Kocy, O., Cheung, H. S. & Ondetti, M. A. (1973), *Experientia*, 1032–1035.

3. Byers, L. D. & Wolfenden, R. (1973) *Biochemistry* **12**, 2070–2078.

4. Ondetti, M. A., Rubin, B. & Cushman, D. W. (1977) *Science* **196**, 441–444.

5. Furguson, R. K., Brunner, H. R., Turini, G. A., Gavras, H. & McKinstry, D. N. (1977) *Lancet*, 775-778.

6. Laffan, R. J., Goldberg, M. E., High, J. P., Schaeffer, T., Waugh, N. H. & Rubin, B. (1977), *Fed. Proc.*, **36**, 1049.

TOPOCHEMICALLY RELATED HORMONE STRUCTURES. THE SYNTHESIS OF RETRO-ANALOGS OF LRF

C. G. WILLSON, M. GOODMAN, *Department of Chemistry, University of California, La Jolla, CA 92093,* J. RIVIER and W. VALE, *The Salk Institute for Biological Studies, San Diego, CA 92112*

A retro-isomer of a peptide is defined by a reversal of the direction of the peptide bonds, while holding the rest of the structure constant. Only retro-isomers of cyclic peptides are true enantiomers of the parent compounds; they can be synthesized by inverting the absolute configuration at each chiral carbon. Retro-isomers of linear peptides are not enantiomers.[1]

The retro-isomers of several biologically active, linear peptides have been synthesized. All proved inactive[2,3] and none were reported to exhibit antagonist activity. The inactivity of these compounds may be a consequence of the fact that although retro-isomers of linear peptides have essentially identical side-chain topochemistry, their carboxyl and N-termini are necessarily reversed. Goissis et al.[4] conceived of a partial solution to the end-group-reversal problem. They synthesized a series of angiotensin analogs which were retro-isomers of angiotensin but also included a malonic acid derivative at the N-terminus, rather than a D-α-amino acid (i.e. the compounds had two carboxylic acid termini). This approach to a complete retro-isomer of a linear peptide was first described by Hayward and Morley,[2] who further proposed that the N-terminus be replaced by an α,α-diamine rather than a D-α-amino acid. We have developed a high-yield synthesis of optically active α,α-diamino compounds[5] which allowed us to apply the concept of Hayward and Morley[2] to the preparation of analogs of the luteinizing hormone-releasing factor (LRF).

The availability of α,α-diamino and α,α-dicarboxy analogs of the α-amino acids allows initiation and termination, respectively, at any residue of a peptide sequence, of a totally or partially retro-analog.

Synthesis of L-5-amino-2-pyrolidine (Apy)

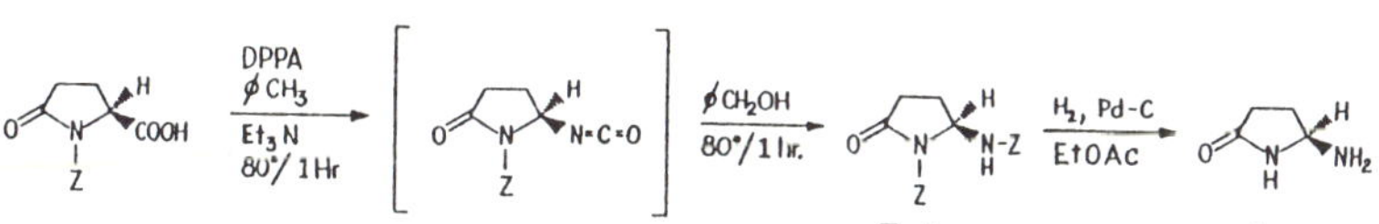

Z_2Apy Apy

The conversion of Z-<Glu to Z_2Apy) is a one-pot reaction and the product crystallizes from the cooled reaction mixture in excellent yield. Conversion to the isocyanate was accomplished by heating equimolar amounts of L-Z-<Glu, diphenylphosphoryl azide[6] (DPPA) and triethylamine in toluene at 80°C. The course of the reaction was monitored by watching the appearance of the intense isocyanate stretching absorption in the infrared. When the Curtius reaction was complete (approximately 1 hr.), a 10% molar excess of benzyl alcohol was added to the reaction mixture and heating was continued. The disappearance of the isocyanate band was monitored; when the addition was complete (approximately 1 hr.), the solution was allowed to cool, the product was isolated by filtration and recrystallized from toluene or from hexane/ethylacetate. Hydrogenation of Z_2-Apy in ethylacetate using 10% Pd/C at atmospheric pressure provided Apy as a white crystalline solid which was slightly hygroscopic (m.p. 78°C). Both Z_2-Apy and Apy were characterized by infrared and high resolution nmr spectroscopy and by mass spectrometry.

Synthesis of Boc-L-Phe-Apy. Boc-L-Phe and Apy were coupled in ethyl acetate using dicyclohexylcarbodiimide (DCC) and purified by crystallization. The crystalline product was characterized by infrared, high resolution nmr and mass spectroscopy, m.p. 105–108°C.

Synthesis of Propionyl-D-Pro-D-Arg-D-Leu-Gly-D-Tyr-D-Ser-D-Trp-D-His-OH (I) and of mono-malonamide-D-Pro-D-Arg-D-Leu-Gly-D-Tyr-D-Ser-D-Trp-OH (II). Both peptides were synthesized by the solid phase technique, purified and characterized according to published procedures.[7]

Retro-des-Gly^{10}-[Pro^{9}-NEt]-LRF (III). Apy was coupled to I using DCC/*N*-hydroxysuccinimide and the product was purified by semi-preparative high pressure liquid chromatography (HPLC).[8] Analog III showed a correct amino-acid analysis (Apy yields 2 NH_3 upon HCl hydrolysis) and was cleanly resolved from I by HPLC.

Retro-[L-Phe^{2}]-LRF (IV). Treatment of Boc-L-Phe-Apy with 4 *N* HCl/dioxane gave L-Phe-Apy which was coupled to II using DCC/*N*-hydroxysuccinimide and the crude product was purified by semi-preparative HPLC.[8] The purified analog was cleanly resolved from II on HPLC and gave a correct amino-acid analysis including 1 Phe and 3 NH_3.

Both III and IV were found to be inactive LRF agonists *in vitro* (0.01% of LRF). Compound IV was also tested for antagonism[9] activity *in vitro* and was found to be ineffective at the doses tested (IDR_{50} 3333/1). Since des-Gly^{10}-[Pro^{9}-NEt]-LRF is 3 times more potent than LRF[10] and since [D-Phe^{2}]-LRF is a relatively potent competitive inhibitor of LRF,[11] the lack of biological activity in the retro-isomers suggests that the peptide backbone[5] may be participating in the hormone-receptor interaction. One stereochemical problem exists for all retro-analogs of peptides containing proline.[3] They lack topological equivalence at the prolyl residue. Further testing of LRF analogs, containing various retro sequences not including the prolyl residue, should clarify the role of the peptide backbone in binding.

This research was supported by grants from NIH AM 15410, AM 18811, HD 09690 and National Foundation 1–411.

References

1. Shemyakin, M. M., Ovchinnikov, Yu. A. & Ivanov, V. T. (1969) *Angew. Chem. Internat. Edit.* 8, 492–499.

2. Hayward, C. F. & Morley, J. S. (1974) in *Peptides,* Y. Wolman Ed., Israel University Press, Jerusalem, pp. 287–297.

3. Rudinger, J. (1971) in *Drug Design*, Ariens, E. J., Ed., Academic Press, New York, vol. II, pp. 319–419.

4. Goissis, G., Nouailhetas, V. L. A. & Paiva, A. C. M. (1976) *J. Med. Chem.* **19**, 1287–1290.

5. Chorev, M., Willson, C. G. & Goodman, M. (1977) *Peptides: Proceedings of the 5th American Peptide Symposium,* Goodman, M. & Meienhofer, J., Eds., John Wiley and Sons, Inc., New York, pp. 572–574.

6. Shiori, T., Ninomiya, K. & Yamada, S. (1972) *J. Amer. Chem. Soc.* **94**, 6203–6205.

7. Rivier, J. E. (1974) *J. Amer. Chem. Soc.* **96**, 2986–2992.

8. Rivier, J. E., Wolbers, R. & Burgus, R. (1977) *Peptides: Proceedings of the 5th American Peptide Symposium,* Goodman, M. & Meienhofer, J., Eds., John Wiley and Sons, Inc., New York, pp. 52–55.

9. Vale, W., Grant, G., Rivier, J., Monahan, M., Amoss, M., Blackwell, R., Burgus, R. & Guillemin, R. (1972) *Science* **176**, 933-934.

10. Fujino, M., Kobayashi, S., Obayashi, M., Shinagawa, S., Fukuda, T., Kitada, C., Nakayama, R., Yanazaki, I., White, W. F. & Rippel, R. H. (1972) *Biochem. Biophys. Res. Commun.* **49**, 863-869.

11. Rees, R., Foell, T., Chai, S-Y. & Grant, N. (1974) *J. Med. Chem.* **17**, 1016–1019.

PARATHYROID HORMONE INHIBITORS: DESIGN, SYNTHESIS, AND BIOLOGICAL EVALUATION

MICHAEL ROSENBLATT, JANE E. MAHAFFEY, and JOHN T. POTTS, JR.
Endocrine Unit, Department of Medicine, Massachusetts General Hospital and Harvard Medical School, Boston, Massachusetts 02114

Two analogs of bovine parathyroid hormone (bPTH), [Nle^{8},Nle^{18},Tyr^{34}]-bPTH-(3–34)amide and [Nle^{8},Nle^{18},Nps-Trp^{23},Tyr^{34}]-bPTH-(3–34)amide[1], were synthesized by the solid phase method[2]. Although bPTH is a single-chain peptide of 84 amino acids, the structural requirements for full biological activity[3] reside within the N-terminal 34-amino-acid segment bPTH-(1-34). Investigations to determine the minimum sequence for hormonal activity revealed that deletions of amino acids from either the N- or C-terminus of the active segment result in a progressive decline in activity[4]. However, there is a striking difference in the role of the C- versus the N-terminal region of bPTH-(1–34) on receptor binding and cyclase activation. bPTH-(1–26) neither stimulates nor inhibits adenylyl-cyclase activity, thus the region 27–34 appears essential for receptor binding. N-terminal deletions also cause decline in potency but the fragment bPTH-(3–34), although not an agonist, is an antagonist and hence must occupy receptor sites without stimulating adenylyl-cyclase[5].

If positions 1 and 2 are essential for hormonal activity, then an approach to the design of antagonists is to incorporate modifications known to enhance activity in bPTH-(1–34) into analogs lacking the two N-terminal residues. Thus, receptor binding and inhibition might increase without restoration of hormonal activity. Structure-activity studies reveal two C-terminal modifications that enhance activity. Substitution[6] of Tyr for Phe at position 34 results in an analog 140% as potent as bPTH-(1–34). Conversion of the C-terminus to a carboxamide ($CONH_2$) yields nearly a 3-fold increase in activity[4]. Both modifications were incorporated into the analogs presented here (Fig. 1). Another modification was selected as a stabilizing feature. Substitution of norleucine for methionine at positions 8 and 18 was demonstrated to be well tolerated in terms of biopotency, while providing analogs that withstand oxidizing conditions with retention of activity[6]. An additional modification was undertaken based on studies of adrenocorticotropic hormone (ACTH)[7]. For ACTH, treatment with *o*-nitrophenylsulfenyl chloride to yield the Nps-tryptophan derivative yielded an inhibitor of ACTH.

The bPTH analogs lacked agonist activity but were effective inhibitors, causing 44 and 57% inhibition of bPTH activity, respectively, at concentrations 1:1 to native bPTH in the *in vitro* renal adenylyl-cyclase assay. Inhibition is proportional to the dose of inhibitor and inversely related to bPTH concentration. The analogs also are complete antagonists; that is, they can totally suppress bPTH-stimulated adenylyl-cyclase activity (Fig. 2).

Detailed study of [Nle^{8},Nle^{18},Tyr^{34}]-bPTH-(3–34)amide revealed kinetic parameters characteristic of competitive inhibition. K_i derived by Eadie-Hofstee plots was $(1.24 \pm 0.13) \times 10^{-8}$ *M*. The affinity of this analog for the *in vitro* receptor is the same order of magnitude as the native hormone, $K_m = 0.45 \times 10^{-8}$ *M* (Fig. 3).

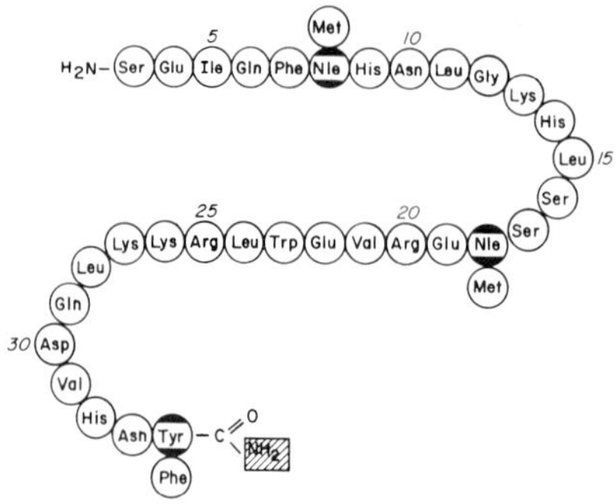

Fig. 1. The synthetic hormone antagonist [Nle^8,Nle^{18},Tyr^{34}]-bPTH-(3-34)amide. Shaded residues depict positions of substitution for the native sequence. The C-terminal COOH function has been modified to a carboxamide. An *o*-nitrophenylsulfenyl group (Nps) was linked to position 23 to form another inhibitory analog.

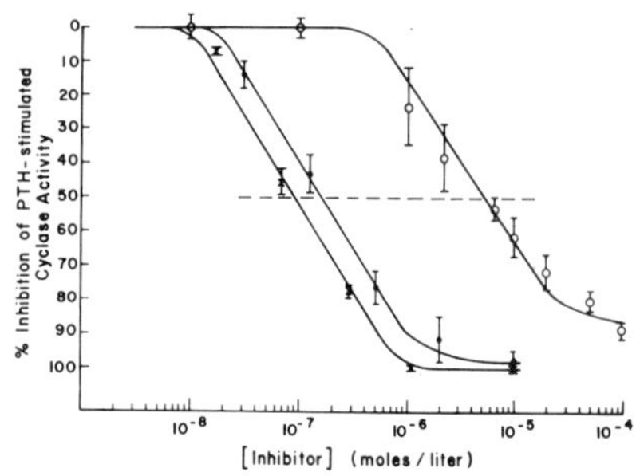

Fig. 2. Inhibition of bPTH-(1–84) stimulated generation of *c*-AMP by synthetic analogs. A near-maximal dose of bPTH-(1–84) (1.2×10^{-7} *M*) was used in the presence of variable doses of inhibitors: (o) bPTH-(3–34); (o) [Nle^8,Nle^{18},Tyr^{34}]-bPTH-(3–34)amide; (x) [Nle^8,Nle^{18},Nps-Trp^{23},Tyr^{34}]-bPTH-(3–34)amide. 50% inhibition is marked by (– – – – –).

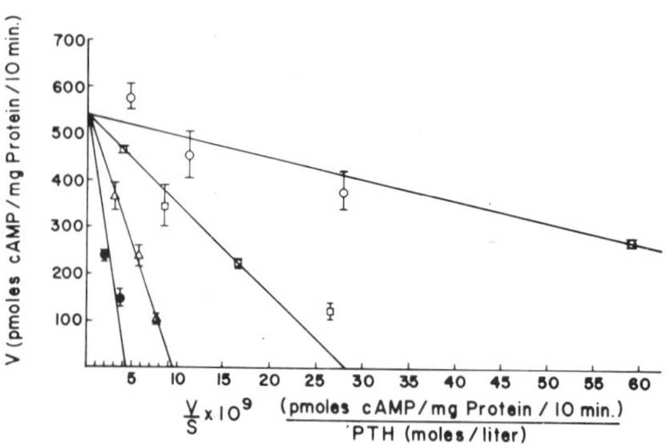

Fig. 3. Michaelis-Menten analysis of adenylyl-cyclase inhibition by Eadie-Hofstee plots of V versus V/S. Native bPTH concentration is varied in the presence of [Nle^8,Nle^{18},Tyr^{34}]-bPTH-(3–34)amide. The concentrations of inhibitors are: (○) no inhibitor; (□) 4.18×10^{-8} *M*; (Δ) 1.24×10^{-7} *M*; (○) 3.73×10^{-7} *M*. The *y*-intercept yields V_{max}. K_m = (–) slope of line (○). In the presence of inhibitor, the K_i can be derived from slope $= -q \cdot K_m$ where $q = (1 + I/K_i)$.

The biological properties of these analogs reflect current ability to design successfully and synthesize bPTH inhibitors which are, for the first time, of sufficient potency to be tested *in vivo*. These analogs represent a 60–300-fold improvement in inhibitory potency over unsubstituted bPTH-(3–34). The ability to impose activity-enhancing modifications upon an inhibitory core without conversion to partial agonism further emphasizes the dichotomy between binding and activation for this hormone. Finally, these analogs serve as models for the synthesis of PTH inhibitors of possible therapeutic significance.

This work was supported in part by grants AM04501 and AM11794 from the NIAMDD. M. R. is a recipient of a National Research Service Award.

References

1. Rosenblatt, M., Callahan, E. N., Mahaffey, J. E., Pont, A. & Potts, J. T., Jr. (1977) *J. Biol. Chem.* (in press).
2. Merrifield, R. B. (1969) *Adv. Enzymol.* **32**, 221–296.
3. Tregear, G. W., van Rietschoten, J., Greene, E., Keutmann, H. T., Niall, H. D., Reit, B., Parsons, J. A. & Potts, J. T., Jr. (1973) *Endocrinology* **93**, 1349–1353.
4. Tregear, G. W., van Rietschoten, J., Greene, E., Keutmann, H. T., Niall, H. D., Parsons, J. A. & Potts, J. T., Jr. (1974) in *Endocrinology 1973: Proceedings of the Fourth International Symposium*, S. Taylor, Ed., Heinemann, London, pp. 1–15.
5. Goltzman, D., Peytremann, A., Callahan, E., Tregear, G. W. & Potts, J. T., Jr. (1976) *J. Biol. Chem.* **250**, 3199-3203.
6. Rosenblatt, M., Goltzman, D., Keutmann, H. T., Tregear, G. W. & Potts, J. T., Jr. (1976) *J. Biol. Chem.* **251**, 159-164.
7. Ramachandran, J. & Lee, V. (1970) *Biochem. Biophys. Res. Commun.* **41**, 358–366.

SOLVENT CHANNELS IN THE CRYSTALLINE LATTICES OF ANTAMANIDE GROWN FROM POLAR AND NONPOLAR SOLVENTS

ISABELLA L. KARLE, *Laboratory for the Structure of Matter, Naval Research Laboratory, Washington, D. C. 20375*

Antamanide, a cyclic decapeptide with the sequence $Val^1Pro^2Pro^3Ala^4Phe^5Phe^6Pro^7$-$Pro^8Phe^9Phe^{10}$, binds Ca^{++}, Na^+ and Li^+ ions[1]. The complex formation was measured by the extraction of the metal picrates from aqueous solutions of antamanide with the following percentages: Na^+, 14.0; Ca^{++}, 9.7; Li^+, 0.8. [Phe^4Val^6]-antamanide, a synthetic symmetric analog, has a similar complexing behavior[1]. The schematic diagram in Fig. 1 compares the folding of the antamanide backbone in the crystals of the Li^+ and Na^+ complexes[2–4] to the elongated ring found for uncomplexed antamanide in crystals prepared from nonpolar solvents[5,6] (*n*-hexane plus methyl acetate) and from polar solvents[7] (CH_3CN, acetone and $Ca(NO_3)_2$). There is a major conformational change upon complexation with Li^+ and Na^+; however, the polarity of the solvent does not affect the conformation of the uncomplexed molecule. It is interesting to note that the uncomplexed form contains three intrinsic H_2O molecules imbedded in the interior of the peptide which form hydrogen bonds to four of the six NH moieties that are directed toward the interior of the molecule. Figure 2 shows a view into the interior of [Phe^4Val^6]-antamanide$\cdot 3H_2O$ drawn with van der Waals radii for the C, N and O atoms. The three water molecules fill the space in the interior of the ring; moreover it is evident that adjacent pyrrolidine and phenyl rings at either end of the molecule stack snugly against each other. These lipophilic groups on the exterior of the molecule play a major role in the packing in the crystal. The schematic diagram in Fig. 3(a) shows how the Pro and Phe side groups of one molecule interleave with a similar pair of an adjacent molecule to form continuous parallel stacks of lipophilic rings throughout the crystal. Between these lipophilic bands are continuous channels in which the solvent is contained.

For the crystal grown from *n*-hexane/methyl acetate, no individual atomic sites can be identified in the channel[5,6]. Rather, there is a continuous low level of electron density which indicates disordered solvent molecules, or even perhaps, molecules of solvent flowing through the crystal lattice. The diameter of the channel is large enough to accommodate

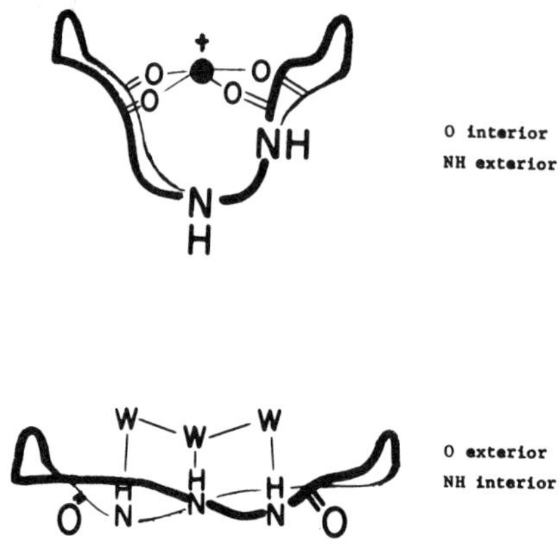

Fig. 1. (Upper) Schematic diagram of Na^+ or Li^+ complex. A fifth ligand is formed to the alkali metal ion from C_2H_5OH, CH_3CN, or $(CH_3)_2CO$. (Lower) Uncomplexed [Phe^{4}Val^6]-antamanide crystallized from polar or nonpolar solvents.

Fig. 2. [Phe^{4}Val^6]-antamanide·$3H_2O$ drawn with van der Waals radii for the C, N, and O atoms[6,8] using experimentally determined coordinates.

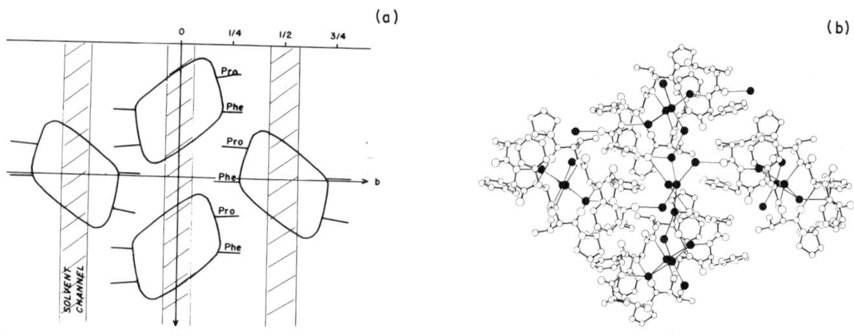

Fig. 3. (a) Schematic diagram of packing in crystal lattice[7]. (b) Actual packing[7] of [Phe^{4}Val^6]-antamanide crystallized from acetone, CH_3CN, and $Ca(NO_3)_2$. The dark circles represent the O atoms of H_2O molecules.

either *n*-hexane or methyl acetate molecules. Furthermore, the crystal is stable only in contact with the mother liquor and the lattice collapses as soon as the crystal is allowed to dry.

In the case of the crystal grown from a polar solvent composed of acetone, acetonitrile and $Ca(NO_3)_2$, the atomic sites for solvent atoms are quite distinct[7]. The actual packing diagram in Fig. 3(b), analogous to the schematic diagram in Fig. 3(a), shows the crystal from the polar solvent in which the equivalent of 12 water molecules, associated with each peptide molecule, are distributed over 16 atomic sites for the O atoms. A view of a cross-section of the channel in Fig. 4, shows that two peptide molecules form a sandwich around the channel. In the channel itself, there is an outer layer of H_2O molecules bound to the C=0 and NH groups of the peptides. These H_2O sites, of which there are 8 independent ones along the channel, are fully occupied. The core of the channel has eight independent sites

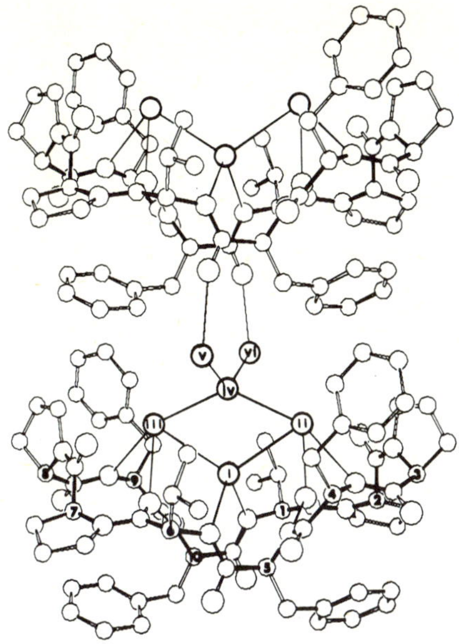

Fig. 4. A cross-section of the channel showing molecules of symmetric antamanide sandwiching the solvent channel.

that are only partially occupied, i.e. different sites are occupied in different cells of the crystal. All the partially occupied sites are at good distances for making hydrogen bonds to the H_2O in the bound layer.

The question arises as to the whereabouts of the Ca^{++} ions present in the crystallizing solution. Are they still in the mother liquor surrounding the crystal? Or is it possible that some of the sites in the inner core of the channel are partially occupied by Ca^{++} rather than by H_2O? The X-ray diffraction data cannot distinguish between the type of ion or atom occupying a site when dealing with partial occupancies, since the occupancy and the weight of an atom are not independent variables. The distances between the sites in the inner core and the H_2O in the bound layer are compatible with $Ca^{++} \cdots 0$ ligands. The speculative idea is proposed that Ca^{++} ions may be transported through such channels when rows of antamanide molecules with their lipophilic exteriors are inserted into membranes.

References

1. Wieland, Th. (1972) in *Chemistry and Biology of Peptides*, Meienhofer, J., Ed., Ann Arbor Science Publishers, Ann Arbor, MI, pp. 377–396.

2. Karle, I. L., Karle, J., Wieland, Th., Burgermeister, W., Faulstich, H. & Witkop, B. (1973) *Proc. Nat. Acad. Sci. USA* **70**, 1836–1840.

3. Karle, I. L. (1974) *J. Amer. Chem. Soc.* **96**, 4000–4006.

4. Karle, I. L. (1974) *Biochemistry* **13**, 2155–2162.

5. Karle, I. L., Karle, J., Wieland, Th., Burgermeister, W. & Witkop, B. (1976) *Proc. Nat. Acad. Sci. USA* **73**, 1782–1785.

6. Karle, I. L. (1977) *J. Amer. Chem. Soc.* 99, 5152–5157.

7. Karle, I. L. & Duesler, E. (1977) *Proc. Nat. Acad. Sci. USA* 74, 2602–2606.

8. Hanson, J. C., Ringle, W. M. & Love, W. E. (1977). *Van der Waals' Surface Program,* Department of Biophysics, Johns Hopkins University, Baltimore, MD 21218.

MONTE CARLO SIMULATION OF WATER STRUCTURE AROUND PROTEINS AND PEPTIDES

A. T. HAGLER and J. MOULT, *Chemical Physics Department Weizmann Institute of Science, Rehovot, Israel*

The solvent system which makes up the environment of biological molecules plays a central role in their structure and function[1]. In spite of this crucial importance relatively little is known about the structure and properties of such systems.[2,3] Recently it has become clear that Monte Carlo simulation of the structure of bulk water and the water around an ion can reproduce well the properties of those systems[4,5]. We have extended these methods to the calculation of the water structure around proteins, peptides and amino acids in solution. A description of the method will be given elsewhere (manuscript in preparation).

Here we describe the simulation of the solvent structure in triclinic crystals of hen egg white lysozyme[6]. Because of the large volume of solvent they contain[7], protein crystals provide good models of macromolecules in solution, and have the advantage that some checking against the experimental results is possible (Moult and Yonat, manuscript in preparation). This unit cell contains 1001 non-hydrogen atoms in the protein, 303 water molecules and eight bound nitrate ions. Figure 1 shows the average energy of the system as a function of configuration number during the simulation. The first 600,000 configurations of a total of 1,023,000 were used to randomize the system.

The nature of the water surrounding a protein and especially the existence of bound water, has been elusive[8]. To gain insight, we have compared the energy distributions of water molecules in different environments in the unit cell of the crystal and that of water in bulk water[9]. A breakdown of the energy distribution in the crystal into water-water and water-protein components (Fig. 2) shows a fairly even division between these two. In the water-water curve there is a sharp peak at zero energy produced by 9 water molecules each almost entirely isolated from the rest of the solvent. The water-protein curve has a long tail at low energies representing waters tightly bound to the protein. The interaction energies of these molecules are easily understood in terms of their environment. Three of them, for example, are situated between pairs of charges. Those most in contact with the protein move

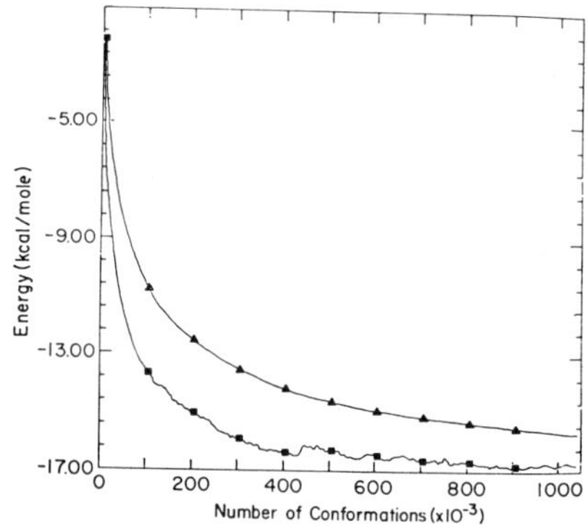

Fig. 1. Energy of the lysozyme crystal as a function of configuration number. ▲ average for the whole simulation to current point. ■ average around current point.

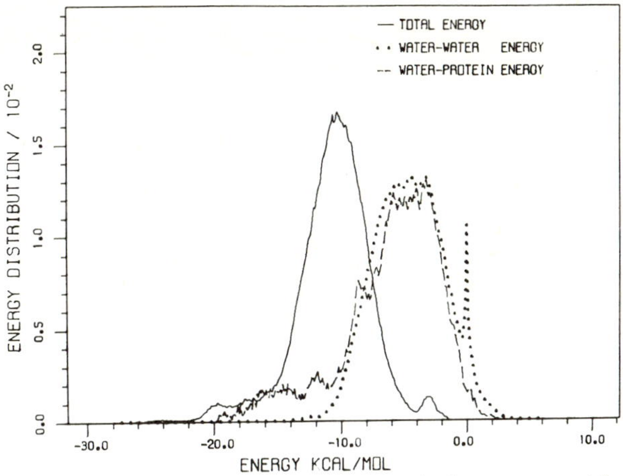

Fig. 2. Energy distribution of water in the lysozyme crystal.

least, while those far from the protein tend to move freely within their volume and are essentially disordered. In a comparison of the energy distributions for 78 of the most ordered waters, which are in contact with the protein and bulk water, the low energy tail is much more populated for these 'near' waters. Comparison of the mean positions of such ordered molecules with those observed in the X-ray work shows around half of the X-ray waters to occupy the same environmental niches as the simulated ones. A comparison of the energy distribution of the 51 waters furthest from the protein with that of bulk water shows these distributions to be similar to each other, demonstrating that these 'second shell' waters are quite like bulk water but slightly constrained by their interactions with their neighbors in contact with the protein molecules (manuscript in preparation). A breakdown of the average energies of the two systems into Coulombic and van der Waals components, (Table I) shows some interesting differences. To within the accuracy of the potentials used[10,11] the average energy in the two systems is the same, but with a slightly higher Coulombic component in the lysozyme case. This effect is much more pronounced in the water-protein component. Thus it seems that the ordering of water on the surface of the protein is greatly enhanced by the presence of charges on the protein (there are 28 distributed

Table I. Comparison of Average Energies of Water in Bulk Water and the Lysozyme Crystal (kcal/mole).

	BULK WATER	LYSOZYME CRYSTAL		
	TOTAL WATER	TOTAL WATER	WATER-WATER	WATER-PROTEIN
VAN-DER-WAALS COMPONENT	-2.28	-1.74	-1.14	-0.60
COULOMB COMPONENT	-8.06	-8.85	-3.43	-5.42
TOTAL ENERGY	-10.34	-10.59	-4.57	-6.02

over a surface of the molecule), and that this ordering may be associated with a loss of order in molecules less strongly interacting with the protein. A more detailed description of the water structure in the crystal, including an analysis of the structural features of the bound water will be given elsewhere.

References

1. Tanford, C. (1968) *Adv. Prot. Chem.* **23**, 122–275.
2. Hopfinger, A. J. (1977) *Intermolecular Interactions and Biological Organization*, John Wiley & Sons, New York.
3. Fisher, I. Z. (1969) *Statistical Theory of Liquids*, The University of Chicago Press, Chicago.
4. Watts, R. O. (1974) *Mol. Phys.* **28**, 1069–1083.
5. Fromm, J., Clementi, E. & Watts, R. O. (1975) *J. Chem. Phys.* **62**, 1388–1398.
6. Moult, J., Yonat, A., Traub, W., Smilansky, A., Podjainy, A., Rabinovich, D. & Saya, A. (1976) *J. Mol. Biol.* **100**, 179–196.
7. Matthews, B. (1968) *J. Mol. Biol.* **33**, 491–497.
8. Kuntz, I. D. & Kauzmann, W. (1974) *Adv. Prot. Chem.* **28**, 239–338.
9. Barker, J. A. & Watts, R. O. (1969) *Chem. Phys. Letts.* **3**, 144–145.
10. Hagler, A. T., Huler, E. & Lifson, S. (1974) *J. Amer. Chem. Soc.* **96**, 5319–5327.
11. Rawlinson, J. S. (1951) *Trans. Far. Soc.* **47**, 120–129.

STRUCTURE AND CONFORMATION OF NEUROPHYSINS

R. ACHER, J. CHAUVET, M. T. CHAUVET, and P. CODOGNO,
Laboratory of Biological Chemistry, University of Paris VI
96, Bd Raspail – 75006 Paris (France)

Neurophysins are small proteins which are found in the posterior pituitary gland associated with neurohypophysial hormones and which can be extracted and purified as hormone-protein complexes (review[1]). We have previously published the amino-acid sequences of ovine, bovine[2,3], and porcine[4] MSEL-neurophysins and we report now the complete amino-acid sequence of equine MSEL-neurophysin. We also describe experiments suggesting that the conformation of neurophysins is extended or extendable.

Sequence. The purification has been carried out as previously described for ox or sheep neurophysins[5]. Horse neurophysin is oxidized by performic acid in order to split the disulfide bridges. One part of the product is subjected to trypsin hydrolysis, another to hydrolysis by staphylococcal protease[6]. Resulting peptides are isolated by peptide mapping, analyzed and sequenced either directly by Edman degradation or after enzymic subcleavage under previously described conditions[2,3]. Tryptic peptides are ordered through overlapping peptides given by staphylococcal protease to determine the complete sequence (Fig. 1).

When compared with sheep, ox or pig MSEL-neurophysins, horse MSEL-neurophysin displays a 3-residue deletion at the N-terminus. Since truncated forms have also been observed in the case of ovine, bovine, and porcine proteins[3,4], this deletion is likely due to a proteolytic degradation which seems to occur in the cell. So a typical MSEL-neurophysin is a 95-residue polypeptide with 7 disulfide bridges. Fig. 1 shows that in the MSEL-neurophysin family, the substitutions from species to species virtually occur only in the

	1–25
Ovine	Ala-Met-Ser-Asp-Leu-Glu-Leu-Arg-Gln-Cys-Leu-Pro-Cys-Gly-Pro-Gly-Gly-Lys-Gly-Arg-Cys-Phe-Gly-Pro-Ser
Bovine	———————————————
Porcine	———————————————
Equine	———————————

	26–50
Ovine	Ile-Cys-Cys-Gly-Asp-Glu-Leu-Gly-Cys-Phe-Val-Gly-Thr-Ala-Glu-Ala-Leu-Arg-Cys-Gln-Glu-Glu-Ile-Tyr-Leu
Bovine	——————————————-Asn-——
Porcine	——————————————-Asn-——
Equine	——————————————-Asn-——

	51–75
Ovine	Pro-Ser-Pro-Cys-Gln-Ser-Gly-Gln-Lys-Pro-Cys-Gly-Ser-Gly-Gly-Arg-Cys-Ala-Ala-Ala-Gly-Ile-Cys-Cys-Asn
Bovine	———————————————
Porcine	———————————————
Equine	———————————————

	76–95
Ovine	Asp-Glu-Ser-Cys-Val-Thr-Glu-Pro-Glu-Cys-Arg-Glu-Gly-Ile-Gly-Phe-Pro-Arg-Arg-Val
Bovine	———————————-Ile-/-Val-———————
Porcine	———————————-Ala-Ser-——-Leu-———-Ala
Equine	———————————-Ala-——-Leu-————-Ala

Fig. 1. Comparison of amino-acid sequences of ovine, bovine, porcine, and equine MSEL-neurophysins. Sequences identical with that of ovine neurophysin are shown as solid lines. A microheterogeneity in position 89 is found for adult and fetal bovine MSEL-neurophysins.

C-terminal part of the molecule, namely in positions 89 to 95. In particular, there are four substitutions in horse MSEL-neurophysin compared to the ovine protein.

The **Conformation** of neurophysins has been studied by two approaches: (1) the accessibility to proteolytic enzymes, and (2) the accessibility to reducing agents.

(1) The native ovine MSEL-neurophysin has been hydrolyzed by trypsin at pH 7.8, by staphylococcal protease at pH 4.0, and by pepsin at pH 2. When the tryptic digest is examined by peptide mapping, the cystine-free peptides T_1, T_3, T_7, and T_8, previously characterized in the tryptic digest of the oxidized neurophysin[2,3] can be identified by their positions on the map and by their amino-acid compositions[3]. A "core" remaining near the starting point is eluted, oxidized by performic acid and the resulting peptides are isolated by fingerprinting and analyzed. The four cysteic acid-containing peptides T_2,T_4,T_5, and T_6, previously characterized in the case of trypsin hydrolysis of the oxidized molecule, are then identified. It can be concluded that trypsin cleaves "native" ovine MSEL-neurophysin in the same way as the fully unfolded oxidized polypeptide chain and that the conformation is either very flexible or non-compact.

(2) The peculiar sensitivity of some disulfide bridges of neurophysins to reduction was previously noted by others[7]. We have however observed that in the absence of urea, all the 7 disulfide bridges are reduced when a molar ratio of 2:1 of dithiothreitol to disulfide bridge is used at pH 8.1 for 2 hr. After carboxamidomethylation with iodoacetamide, the derivative is subjected to trypsin hydrolysis and resulting peptides are isolated and analyzed; the peptide map is identical with that obtained when the reduction-alkylation is carried out in 8 *M* urea. This complete reduction in the absence of urea is unusual, especially for a protein very rich in disulfide bridges. The ready reduction can be explained if the conformation is extended or rapidly extendable so that a cascade of fast reductions reaches the 7 disulfide bridges.[8]

References

1. Acher, R. (1976) *Biochemie* 58, 895–911.
2. Chauvet, M. T., Chauvet, J. & Acher, R. (1975) *FEBS Lett.* 58, 234–237.

3. Chauvet, M. T., Chauvet, J. & Acher, R. (1976) *Eur. J. Biochem.* **69**, 475–485.
4. Chauvet, M. T., Codogno, P., Chauvet, J. & Acher, R. (1976) *FEBS Lett.* **71**, 291–293.
5. Chauvet, M. T., Coffe, G., Chauvet, J. & Acher, R. (1976) *Compt. Rend. Soc. Biol.* **170**, 257-268.
6. Drapeau, G. R. & Houmard, J. (1972) *Proc. Nat. Acad. Sci. USA* **69**, 3506–3509.
7. Menendez-Botet, C. J. & Breslow, E. (1975) *Biochemistry* **14**, 3825-3835.
8. Chauvet, M. T., Chauvet, J. & Acher, R. (1971) *FEBS Lett.* **71**, 96–98.

THEORETICAL STUDIES ON PEPTIDE-SOLVENT INTERACTIONS

V. RENUGOPALAKRISHNAN* and F. JORDAN, *Dept. of Chemistry, Rutgers, The State University of New Jersey, Newark, New Jersey 07102.*
(*Present Address: Laboratory of Molecular Biophysics, Univ. of Alabama Medical Center, Box 311, University Station, Birmingham, Alabama 35294)

In an attempt to determine the peptide conformations in solution[1] we have used an approach suggested by Sinanoglu[2] in which the solvation energy is assumed to consist of two parts. In the first step a cavity is created in the liquid of appropriate dimensions to accommodate the solute. In the second step a solute molecule is placed in the cavity and interacts with its environment. The concept of continuum reaction field formulated by Onsager[3] was used in calculating the interaction energy of the solute molecule with its environment. The solvent is assumed to be a continuum possessing the macroscopic dielectric properties of the solution, which coincide with those of a pure liquid in the case of a dilute solution. The total free energy for a given conformation is assumed to be the sum of the intrinsic free energy of the solute molecule and the solvation energy. Although the above model for solvent effect seems to account satisfactorily for the effect of the bulk solvent on peptide conformations in solvents of low dielectric constant like CCl_4, we felt it necessary to include the effect of directly bound solvents like H_2O to the peptide NH and C=0 groups in the calculations[4]. Therefore we initiated a study, using quantum chemical methods, for the exploration of hydration sites in peptides in order to determine the architecture of peptide-H_2O super-molecular complex. In the present study, *N*-acetyl-L-alanine-(*N*)-methylamide was chosen as a model peptide system[1].

Atomic coordinates for the dipeptide were constructed using the bond lengths and bond angles suggested in the literature[5]. Semi-empirical, CNDO/2, and *ab initio* minimal basis set (STO-3G) wave functions were used for mapping the electrostatic potential generated by the charge distribution in the dipeptide[6]. The interaction of a unit positive charge with the electrostatic potential was obtained and the interaction energy surface was obtained as a series of isopotential curves drawn in selected planes of interest. For geometries corresponding to points of greatest interaction energy in the interaction energy surface, *ab initio* minimal basis set (STO-3G) calculations for the dipeptide-H_2O complex were performed. The interaction energy surfaces will be reported in detail elsewhere[7].

Ab initio minimal basis set calculations for the free dipeptide predict C_5 conformation to be the minimum and more stable than $C_7^{eq.}$ by 1.53 kcal/mol. $C_7^{ax.}$ conformation is predicted to be 6.43 kcal/mol. and less stable than $C_7^{eq.}$ conformation. *Ab initio* results are at variance with classical partitioned potential energy calculations reported by us earlier[1]. In the *ab initio* calculations of the dipeptide-H_2O complex, only a single molecule of H_2O was considered to interact at an instant, due to computational limitations. The intermolecular hydrogen bonds, N-H....O=C and C=O....H-O, were assumed to be linear. For C_5 and $C_7^{eq.}$

conformations, hydrogen bond energies of 2.91 and 5.97 kcal/mol., respectively, were obtained. For the $C_7^{ax.}$ conformation, *ab initio* calculations do not predict a minimum. Interaction of a single H_2O molecule was also considered with the C=O group of the dipeptide for $C_7^{eq.}$ conformer. *Ab initio* calculations predict a hydrogen bond energy of 5.41 kcal/mol. A decomposition of the hydrogen bond energies into electrostatic and delocalization energies was made using a procedure suggested by Kollman and Allen[8]. Among the two components considered, the electrostatic term is the dominant term. Details of the decomposition will be discussed elsewhere[7]. The intent of the present study is to build systematically the peptide-H_2O supermolecular complex and then apply the solvent effect model described in the introduction[1].

We thank the Center of Computer and Information Services at Rutgers University for generous allocation of computer time.

References

1. Renugopalakrishnan, V., Nir, S. & Rein, R. (1976) in *Environmental Effects on Molecular Structure and Properties,* Pullman, B., Ed., D. Reidel Publishing Co., The Netherlands, pp. 109–133.
2. Sinanoglu, O. (1968) *Molecular Associations in Biology,* Academic Press, New York, p. 427.
3. Onsager, L. (1936) *J. Amer. Chem. Soc.* **58**, 1486–1493.
4. Kuntz, I. D. & Kauzmann, W. (1974) *Adv. Protein Chem.* **28**, 239–345.
5. Ramachandran, G. N. & Sasisekharan, V. (1968), *Adv. Protein Chem.* **23**, 283–438.
6. Alagona, G., Pullman, A., Scrocco, E. & Tomasi, J. (1973) *Int. J. Pept Protein Res.* **5**, 251–259.
7. Renugopalakrishnan, V. & Jordan, F. in preparation.
8. Kollman, P. A. & Allen, L. C. (1970) *Theoret. Chim. Acta* **18**, 399–403.

SPECTROSCOPIC AND STRUCTURAL PROPERTIES OF α-CHYMOTRYPSIN IN CYCLOHEXANE

P. L. LUISI, FRANCIS J. BONNER, and CH. WALSOE, *Technisch-Chemisches Laboratorium, ETH-Zentrum, Universitätstr. 6, 8092 Zürich, Switzerland*

The interaction of amino acids, peptides and proteins with organic solvents[1–3] presents two main points of interest, (i) the influence of solvation upon structure, and (ii) the behavior of biomolecules in certain non-aqueous environments such as proteins in membranes.

Our interest was prompted by the observation that methyl-trioctyl-ammonium chloride ($^{+}NR_4Cl^{-}$) enables the transfer of amino acids, peptides and proteins from an aqueous solution to a supernatant organic solution[4,5]. Our study presents two significant new features: that the organic milieu can be a pure aprotic solvent such as cyclohexane or chloroform; and that the protein, as judged from its UV-absorption spectrum, does not appear to be denatured. The present paper deals with the further characterization of α-chymotrypsin in the cyclohexane · $^{+}NR_4Cl^{-}$ system. In particular, we will show on the basis of circular dichroic data that the main chain conformation in the cyclohexane phase is with all probability very similar to the native conformation in water. Experimental details have been described previously.[4,5]

The present technique, starting typically with a 80 μM aqueous solution of α-chymotrypsin pH 9.5 ca., (the pH being adjusted with NaOH), can transfer ca. 20 μM protein to the supernatant cyclohexane solution containing 5 mg/ml of $^{+}NR_4Cl^{-}$. We have already reported[5] that the UV spectrum of α-chymotrypsin in the water and in the cyclohexane phases is very similar, with the cyclohexane spectrum being red-shifted only by ca. 2 nm. Fig. 1 shows the CD spectrum of α-chymotrypsin in water and in the cyclohexane phase. In the near UV region (290–320 nm), where the spectral properties are largely dominated by the contribution of the 8 tryptophan residues, the form of the spectra is very similar in the two cases, the ellipticity being somewhat smaller in the cyclohexane phase. It therefore seems reasonable to assume that the conformation around tryptophan residues (and by inference the whole protein conformation) is not significantly changed upon transfer into the aprotic phase. This conclusion is reinforced by the analysis of the 200 nm region, which is dominated by the contribution of the peptide chromophore and is therefore sensitive to the overall conformation. The main peak shows a ~ 5 nm red shift with respect to the water solution, accompanied by a ~ 30% decrease in the ellipticity. This decrease may be partly interpreted on the basis of a hypochromic effect of the UV-absorption band in this region, or from microscopic refractive index or polarizability differences in the two phases. The changes taking place in the 230–280 nm region are significant, and difficult to interpret at this stage. X-Ray data indicate that α-chymotrypsin contains only a very small amount of helical structure,[6] and it is likely that the band present in water solution at 230 nm is due more to aromatic side chains (suggested also by Jirgenson[7]) than to α-helices. If this is so, then the decrease of this band in the cyclohexane phase relative to water could be ascribed to a slight local perturbation of the aromatic residues. The change in ellipticity by 260–280 nm could be due to local perturbations of one or more tyrosin residues (and/or S-S chromophores).

In general, we believe that the CD data suggest that the protein maintains in the cyclohexane phase the native tertiary structure with possible local conformation and chromophoric perturbations. Finding an explanation for this rather unexpected observation is an interesting problem.

Ultracentrifugation studies suggest that $NR_4{}^{+}Cl^{-}$ in cyclohexane builds aggregates composed of Ca. 10 molecules of ammonium salt. The system in the cyclohexane phase can be viewed as an "inverted micelle"[8], which explains the relatively small spectral and structural

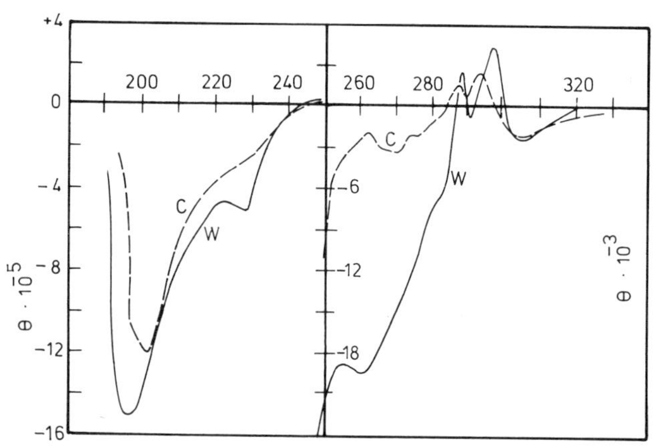

Fig. 1. Circular dichroic spectra of α-chymotrypsin in water (W) and in the cyclohexane phase (C). θ is the molar ellipticity referred to the whole macromolecule.

changes relative to water. With some water also transported into the cyclohexane, the protein could be in a rather aqueous core of the micelle and so be protected from direct interaction with the aprotic solvent. On the other hand, the fluorescence intensity of the protein in the cyclohexane phase is larger than in water (by a factor of 3 ca.), which suggests that the environment of α-chymotrypsin is not exactly the same as in bulk water. More detailed spectroscopic investigations are needed in order to assess more clearly the state of the protein in the cyclohexane phase.

References

1. Inoue, H. (1972) *Biopolymers* **11**, 737–747.
2. Kaminsky, L. S. & Davison, A. J. (1969) *FEBS Lett.* **3**, 338–340.
3. Zahler, P. & Niggli, V. (1977) *Methods in Membranol.*, in press.
4. Dossena, A., Rizzo, V., Marchelli, R., Casnati, G. & Luisi, P. L. (1976) *Biochem. Biophys. Acta* **446**, 493–505.
5. Luisi, P. L., Henninger, F., Joppich, M., Dossena, A. & Casnati, G. (1977) *Biochem. Biophys. Res. Commun.* **74**, 1384–1389.
6. Siegler, P. B., Blow, D. M., Matthews, B. W. & Henderson, R. (1968) *J. Mol. Biol.* **35**, 143–148.
7. Jirgenson, B. (1973) *Optical activity of proteins and other macromolecules.* 2nd Ed., Springer Verlag, Berlin.
8. Fendler, J. H. (1976) *Acc. Chem. Res.* 9, 153–161.

THE PREPARATION OF SEMI-SYNTHETIC HORSE HEART CYTOCHROME c

LEON E. BARSTOW, ROBERT S. YOUNG, EMEL YAKALI, *Department of Chemistry, University of Arizona, Tucson, Arizona 85721*, JOHN J. SHARP, JEAN C. O'BRIEN, PHIL BERMAN, and HENRY A. HARBURY, *Department of Biochemistry, Dartmouth Medical School, Hanover, New Hampshire 03755*

Corradin and Harbury[1] have shown that cytochrome *c* can be cleaved into heme 1–65 and non-heme 66–104 fragments by cyanogen bromide. These fragments can then be recombined to yield a fully active protein (Figure 1) that differs from native cytochrome c only at position 65[2,3]. As part of a joint effort to investigate structure-function relationships of cytochrome c, we have successfully synthesized the 66–104 non-heme fragment of horse heart cytochrome c by the solid phase method and recombined it with native heme 1–65 peptide to yield a protein indistinguishable from reconstituted native cytochrome c.

The 66–104 non-heme peptide fragment was synthesized by the Merrifield solid-phase method starting with t-Boc-L-Glutamic Acid-γ-OBzl ester resin (0.2 mmol/gram substitution) prepared according to the method of Gisin[4]. *t*-Boc-amino acids were utilized throughout the synthesis including the following with additional side chain protection: β-OBzl-Tyr, β-OBzl-Thr, β-OBzl-Asp, Asn-NP, γ-OBzl-Glu, ω-Tosyl-Arg, and ε-Z-Lys. Single couplings were used for the first nineteen amino acids, and double couplings for the remaining twenty.

Cleavage of the peptide from the resin was done according to the procedure outlined by Lenard and Robinson[5] and modified by Feinberg and Merrifield[6]. The primary difficulties with the synthesis arose from the use of HF, which destroys the peptide in cleavages lasting

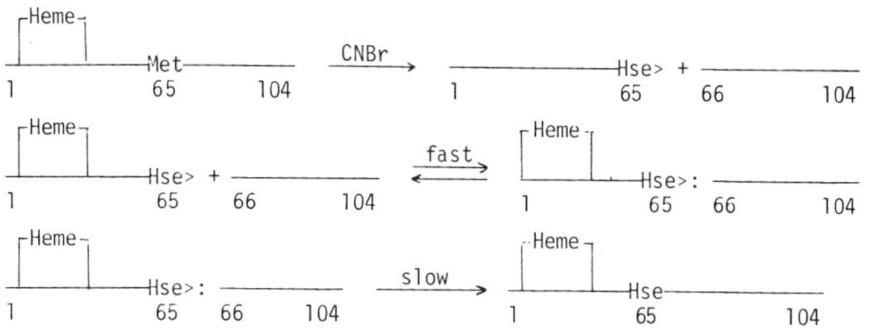

Fig. 1. Cyanogen bromide (CNBr) cleavage and reconstitution of natural cytochrome c. Hse = homoserine.

for an extended period of time and/or run at room temperature[6], and apparently fails to deblock the protecting groups in cleavages of shorter duration. Although the amino-acid analysis of peptide cleaved at 0°C for 0.5h was good, the product contained a significant quantity of insoluble peptide, indicating incomplete deblocking of protected side chains. Results of a tryptic digest[7] did not reveal the presence of the protecting groups of any one particular acid residue, and yield of recombinant with this material, while promising, was still relatively low. A second HF treatment of the cleaved peptide at 0°C for 0.5 h increased the yield of the active peptide and gave a clear separation of two peaks after chromatography on Sephadex G-50. Amino-acid analysis of this material compared favorably with isolated native 66–104.

Two other cleavage methods were attempted. Cleavage of the resin ester with HBr in TFA for an hour at 24°C followed by an HF cleavage at 0°C for 0.5 h to remove the tosyl protecting group from arginine gave peptide similar in yield and activity to that obtained after two cleavages with HF. A cleavage of the resin ester with boron Tris(trifluoroacetate) gave a very poor yield of peptide.[8]

The partially purified synthetic peptide was reconstituted with the native heme 1–65 peptide. Since the binding of the synthetic peptide to the native heme peptide is apparently selective for those molecules that have the proper sequence, the recombination step acts as an affinity technique for the peptide. The homoserine lactone at position 65 is opened by the α-amino function on the N-terminal glutamic acid of the synthetic peptide, and a covalent bond is formed between the two peptide fragments to yield an intact protein (Figure 1). Absorption spectra, CD spectra, amino-acid analysis, antibody binding and the biological activity measured by oxygen uptake (Figure 2) are identical for native reconstituted and semisynthetic cytochrome c.

The success of the reconstitution of a synthetic fragment and a native fragment offers a new technique for investigating

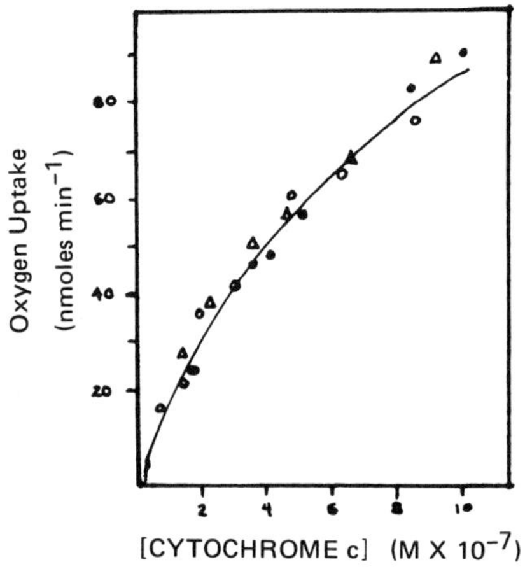

Fig. 2. Oxygen uptake (nmoles min^{-1}): • cytochrome c, o semisynthetic cytochrome c, Δ nonsynthetic, reconstituted horse cytochrome c.

structure-function relationships through the preparation of protein analogs that have a single amino acid replacement. We are currently preparing a number of analogs of horse heart cytochrome c to probe the details of the biological function of the protein.

References

1. Corradin, G. & Harbury, H. A. (1970) *Biochim. Biophys. Acta* **221**, 489–496.

2. Corradin, G. & Harbury, H. A. (1971) *Proc. Nat. Acad. Sci. USA* **68**, 3036–3039.

3. Corradin, G. & Harbury, H. A. (1974) *Biochem. Biophys. Res. Commun.* **61**, 1400–1406.

4. Gisin, B. F. (1973) *Helv. Chim. Acta* **56**, 1476–1482.

5. Lenard, J. & Robinson, A. B. (1967) *J. Amer. Chem. Soc.* **97**, 181–182.

6. Feinberg, R. S. & Merrifield, R. B. (1975) *J. Am. Chem. Soc.* **97**, 3485–3496.

7. Bodanszky, M., Ondetti, M. A., Levine, S. D. & Williams, N. J. (1967) *J. Amer. Chem. Soc.* 98, 6753–6757.

8. Bauer, W. & Pless, J. (1973) *Angew. Chem.* 85, 142–143.

FIFTH AMERICAN PEPTIDE SYMPOSIUM

La Jolla, California
June 20-24, 1977

PLANNING COMMITTEE

E. R. Blout
M. Goodman
E. Gross
E. T. Kaiser
I. Karle
D. S. Kemp
K. D. Kopple
J. Meienhofer
J. Ramachandran
R. Walter

PROGRAM COMMITTEE

Murray Goodman, Chairman

G. Fasman
E. Gross
R. Guillemin
R. Marsh
J. Meienhofer
J. T. Potts, Jr.
R. Walter

LOCAL COMMITTEE

J. Griffin
J. E. Rivier
C. J. Stewart
M. S. Verlander

SPONSORS

Abbott Laboratories
Armour Pharmaceutical Company
Ayerst Research Laboratories
Bachem Fine Chemicals, Inc.
Beckman Instruments, Inc.
Burroughs Wellcome Company
Calbiochem
Chemical Dynamics Corporation
Ciba-Geigy
Digilab, Inc.
Hoffmann-La Roche, Inc.
Hyland
Lab Systems, Inc.
Lilly Research Laboratories
McGaw Laboratories
Mead Johnson Laboratories
Merck Sharp & Dohme Research Laboratories
Pfizer Central Research
Pharmacia Fine Chemicals
Pierce Chemical Company
Sandoz, Inc.
Schering Corporation
Searle Laboratories
Shankman Laboratories
Smith, Kline and French
E. R. Squibb & Sons, Inc.
Syntex Research
UC San Diego Foundation
The Upjohn Company
Vega-Fox Biochemicals
Waters Associates
Wyeth Laboratories

This symposium was also supported by a grant from the National Institutes of Health (GM 23983).

AUTHOR INDEX

Acher, R. 588
Adawadkar, P. D. 357
Agosta, W. C. 322
Agresti, D. G. 325
Alberts, A. W. 59
Alper, J. B. 329
Andreatta, R. H. 543
Aswanikumar, S. 141
Atherton, E. 503
Auer, H. E. 353
Auger, G. 564
Austen, B. M. 103
Bach, J. F. 564
Baici, A. 395
Bara, Y. A. 343
Barabe, J. 157
Baram, T. 193
Barany, G. 488
Baron, M. H. 303
Barstow, L. E. 593
Bayer, E. 403, 514
Benedetti, E. 257
Bennett, C. D. 59
Bennett, G. W. 183
Benjannet, S. 562
Berman, P. 593
Besch, H. R., Jr. 423
Bhatnagar, P. 340
Birr, C. 510
Blanot, D. 564
Bloom, F. 63
Blumenstein, M. 179
Blundell, T. L. 136
Boggs, N. T., III 465
Böhn, L. 427
Bonner, F. J. 591
Bonora, G. M. 399, 474
Borin, G. 474
Botos, C. R. 161
Bower, J. D. 111
Bowers, C. Y. 292
Bradbury, A. F. 103
Bradbury, E. M. 427
Brendenburg, D. 136
Briand, G. 427
Bricas, E. 564
Britton, D. 107
Brunfeldt, K. 569
Buku, A. 510
Burgus, R. 52, 63
Burton, J. 525
Canova-Davis, E. 553
Caporale, L. H. 225
Cary, P. D. 427
Cellis, M. E. 197
Chang, J. K. 292
Chauvet, J. 588
Chauvet, M. T. 588
Cheung, H. S. 576
Chiao, C. Y.-C. 353
Chorev, M. 572
Chou, H. J. 213
Chou, P. Y. 284
Chretien, M. 562
Christensen, T. 569
Codogno, P. 588
Cohen, P. 176
Convert, O. 176
Corcoran, B. A. 141
Cosand, W. L. 488
Cosani, A. 368
Cowburn, D. 322
Coy, D. H. 107
Crane-Robinson, C. 427
Crine, P. 562
Currie, B. 350
Cusan, L. 107
Cushman, D. W. 576
Czerwinski, E. W. 277
D'Amour, P. 232
Danho, W. 540
Dardenne, M. 564
Davis, D. G. 215
Day, A. R. 114, 146
Deber, C. M. 357
De Coen, J. L. 88
De Loze, C. 303
Desiderio, D. M. 514
Deslauriers, R. 168
Diaconescu, C. 136
Di Bello, C. 176
Dimarchi, R. D. 480
Dockerill, S. 136
Donohue, A. M. 59
Donzel, B. 171
Drey, C. N. C. 468
Duax, W. L. 277

Dupont, A. 107
Dutta, A. S. 189
Eberle, G. 484
Eckert, H. 484
Edwardson, J. A. 183
Eisler, K. 543
Engelhard, M. 488
Englert, A. 364
Erickson, B. W. 225, 559
Fallat, R. 567
Fasman, G. D. 284
Faulstich, H. 243
Felix, A. M. 532
Feng, M.-C. 333
Fernandez, H. N. 228
Fertel, R. 107
Feurer, M. 448
Feyereisen, C. 387
Filippi, B. 474
Fillaux, F. 303
Fischer, W. 506
Fischman, A. J. 322
Flory, P. J. 288
Fontana, A. 432
Ford, J. J. 165
Frank, H. 514
Freeman, A. R. 423
Freer, R. J. 114, 146
Fridkin, M. 193, 549
Friedrich, A. 343
Friesen, H.-J. 136
Fuller, W. 471
Furr, B. J. A. 189
Garg, H. G. 477
Garner, W. H. 480
Gattner, H.-G. 136
Geisow, M. J. 103
Gibbons, W. A. 165, 329
Gilden, R. V. 239
Giles, M. B. 189
Gill, P. 107
Gisin, B. F. 215
Glaser, C. B. 567
Glickson, J. D. 325
Go, A. 333
Goissis, G. 559
Goodman, M. 171, 572, 579
Goren, H. J. 549
Gorin, F. A. 84
Gotto, A. M., Jr. 149
Grainer, C. 522
Gregerman, R. I. 213
Griffin, J. H. 176
Gross, E. 141, 205
Grzonka, Z. 153
Guest, K. P. 111
Guillemin, R. 63, 96
Gurd, F. R. N. 480
Haas, E. 361
Haber, E. 525
Hagenmaier, H. 514
Hagler, A. T. 280, 586
Haldar, J. 201
Hall, J. E. 215
Hammonds, R. G., Jr. 100
Handa, B. K. 111
Hao Li, C. 458
Harbury, H. A. 593
Hatano, M. 390
Hay, D. I. 56
Hayashi, H. 427
Hazum, E. 193
Hehlein, W. 343
Helbecque, N. 415
Henriksen, S. 63
Hermann, P. 153
Hiskey, R. G. 465
Hodges, R. S. 411
Honig, B. 280
Horváth, C. 48
Houghten, R. A. 458
Hruby, V. J. 179
Hsieh, Y. 461
Hubbach, A. 540
Hugli, T. E. 225, 228
Hui, K. Y. 461
Humblet, C. 88
Ikeda, S. 546
Irving, C. S. 419
Ivanov, V. T. 307
Jacobs, R. 56
Jeanloz, R. W. 477
Jimenez, M. H. 532
Jones, C. R. 165, 329
Jordan, F. 590
Juliano, L. 337
Kamber, B. 543
Karic, L. 567
Karle, I. L. 274, 583
Kastin, A. J. 107
Katchalski-Katzir, E. 361
Kendrick, N. E. 277

Kennedy, S. J. 423
Kessler, H. 343
Kim, S. E. 461
Klausner, Y. S. 536
Kobayashi, S. 215
Koch, Y. 193
Kojro, E. 153
Kondor, P. 343
Kopple, K. D. 333
Koshland, D. E., Jr. 117
Kovacs, J. 461
Krieger, D. 427
Krishna, R. N. 325
Krugh, T. R. 353
Kuo, M.-C. 329
Kupryszewski, G. 218
Labrie, F. 107
Lagerlund, I. 484
Lammek, B. 218
Lapidot, A. 419
Lasky, J. 350
Lazarus, H. 221
Lazarus, L. 96
Leclerc, M. 364
Lenkinski, R. E. 325
Lewis, P. N. 427
Levy, G. C. 168
Lindner, H. R. 193
Ling, N. 63, 96, 100
Lis, M. 562
Live, D. H. 44, 322
Long, M. M. 407
Lorenzi, G. P. 395
Losee, K. A. 576
Loucheux, C. 387
Loucheux-Lefebvre, M.-H. 415
Luisi, P. L. 556, 591
Madison, V. 350
Mahaffey, J. E. 581
Manavalan, P. 300
Manjula, G. 296
Manning, M. 165, 201
Marchiori, F. 474
Margolies, M. N. 525
Marlborough, D. I. 114
Marquarding, D. 484
Marshall, G. R. 84, 277
Martinez, J. 564
Mästle, W. 506
Mathews, F. S. 277
McMaster, D. 518
Meienhofer, J. 532
Meiri, T. H. 536
Merrifield, R. B. 488, 559
Metcalf, G. 111
Minick, S. 63, 96
Mojsov, S. 488
Molnár, I. 48
Molter, M. 343
Momany, F. A. 292
Moore, G. 518
Morcellet, M. 387
Morell, A. G. 567
Morell, J. L. 141
Morgan, B. A. 111
Moss, T. 427
Moult, J. 586
Muller, E. P. 522
Munekata, E. 243
Mutter, H. 403
Mutter, M. 288, 403
Naithani, V. K. 136
Nelson, B. 506
Neuenschwander, P. 395
Neugebauer, W. 218
Niall, H. D. 127
Nicholls, L. J. F. 165
Nicolas, P. 176
Nishino, N. 236
Nomori, H. 390
Nowak, J. 136
O'Brien, J. C. 593
Okamoto, K. 407
Ondetti, M. A. 576
Orlowski, R. 168
Oroszlan, S. 239
Paganetti, T. 395
Paiva, A. C. M. 337
Palacz, Z. 153
Pallai, P. 205
Palumbo, M. 368, 399
Pawlowski-Konopnicki, B. E. 353
Pease, L. G. 346
Peggion, E. 368, 399
Pellegrini, A. 556
Pert, C. B. 141
Pleau, J. M. 564
Ponnuswamy, P. K. 300
Potts, J. T., Jr. 232, 581
Powers, J. C. 236
Pownall, H. J. 149
Premilat, S. 364
Priestley, G. P. 468
Puett, D. 100
Ramachandran, J. 553

Ramakrishnan, C. 296
Rapaka, R. S. 407
Regoli, D. 157
Renugopalakrishnan, V. 590
Rich, D. H. 209, 340
Riniker, B. 543
Rink, H. 543
Rivier, J. 52, 96, 579
Rizzo, V. 395
Roderich, W. 161, 168, 325
Rodkey, J. A. 59
Roeske, R. W. 423
Rosenblatt, M. 232, 581
Rossier, J. 63
Rothe, M. 506
Rubin, B. 576
Rubinstein, M. 41
Sabo, E. F. 576
Sakakibara, S. 436
Sakarellos, C. 171
Sallay, S. I. 239
Samy, T. S. A. 221
Sander, A. 506
Sarantakis, D. 186
Sautiere, P. 427
Sawyer, W. H. 201
Scheraga, H. A. 246
Schiffmann, E. 141
Schiller, P. W. 92
Schlesinger, D. H. 56
Schneider, E. 536
Segre, G. V. 232
Seidah, N. G. 562
Sharon, R. 280
Sharp, J. J. 593
Sheppard, R. C. 503
Sieber, P. 543
Sigler, G. F. 149, 471
Skorna, G. 484
Smith, C. F. C. 111
Smith, C. W. 161
Smith, G. D. 277
Smith, I. C. P. 168
Smith, L. C. 149
Smyth, D. G. 103
Snell, C. R. 103
Snyder, S. H. 77
Soutar, A. K. 149
Sparrow, J. T. 149
Srivastava, K. S. L. 239
Stein, S. 14, 41
Steinberg, I. Z. 361
St-Pierre, S. A. 411
St-Louis, J. 157
Stevens, E. 399
Stockert, R. 567
Strauss, A. W. 59
Sun, E. 209
Teichman, J. 186
Terbojevich, M. 368
Tjoeng, F.-S. 411
Todd, C. W. 27
Tom-Kun, J. 357, 427
Toniolo, C. 399, 474
Tonegawa, F. 546
Turan, A. 201
Udenfriend, S. 14
Ueki, M. 546
Ugi, I. 484
Urban, R. 484
Urry, D. W. 407
Vale, W. 579
van Rietschoten, J. 522
Vavrek, R. 471
Vidali, G. 427
Villemoes, P. 569
Vita, C. 432
Vivas, A. 197
v. Zychlinski, H. 484
Wackerle, L. 484
Wakamiya, T. 205
Walsoe, C. 591
Wang, C.-C. 480
Wang, K.-T. 528
Warshel, A. 574
Watanabe, A. M. 423
Watson, C. 346
Weinstein, B. 539
Wengert-Müller, M. 510
Wideman, J. 41
Wieland, T. 243
Willhardt, I. 153
Willson, C. G. 572, 579
Wolbers, R. 52
Wong, C.-H. 528
Wood, S. P. 136
Wyssbrod, H. R. 322
Yakali, E. 593
Yam, C. F. 92
Yamamoto, D. M. 179
Yang, Y. C. S. 179
Yoshikawa, M. 390
Young, R. S. 593
Zahn, H. 136, 540

SUBJECT INDEX

Ac-Ala-$NHCH_3$, CD studies, 307
Accessibility of solvent to peptides, calculation of, 300
Acetimidation, use in modified myoglobin synthesis, 480
Acetyl-deoxystatin, inhibition of pepsin by, 209
Acetyl-statin, inhibition of pepsin by, 209
Acid proteases, inhibition by dextran-pepstatin conjugate, 213
ACTH, *see* adrenocorticotropin
Actinomycin D-nucleotide complexes, spectroscopic studies, 353
Active ester, water soluble, 536
Active site model, for angiotensin converting enzyme, 576
Adrenocorticotropin
  conformational analysis of, 364
  modification by sulfenyl chlorides, 553
  precursor, 63
  synthesis in aqueous solution, 536
Affinity chromatography
  of alpha-1-antitrypsin variants, 567
  of proteases A and B, 236
  of Taiwan cobra venom cardiotoxin, 528
  of thermolysin, 236
Agonist-antagonist relationships in small peptide hormones, 292
Alamethicin, synthesis of, 215
D-alloisoleucine, conformational studies of alternating co-oligomers, 395
Alpha-1-antitrypsin and variants, 567
Alternating co-oligomers of L-Ile and D-aIle, 395
Alternating liquid-solid phase synthesis of LRF, 514
Amidation of ornithine, 474
Amino acid analysis
  of endorphin analogs, 63
  of opioid peptides using fluorometric detection system, 41
Amino acid sequence of relaxin, 127
Amino acids, tetrazole analogs of, 153
α-Amino protecting groups
  biphenylisopropyloxycarbonyl, 1
  cysteic acid, 540
  dimethylphosphinothioyl, 546
  diphenylphosphinamides, 1
  isonicontinylcarbonyl, 1
  β-methylsulfonylethoxycarbonyl, 1
  of polymeric nature, 514
Aminomethyl resin, solid phase peptide synthesis on, 488
L-5-Amino-2-pyrolidine, in synthesis of retro-LRF, 579
Anaphylatoxin C3a and C5a
  bioassay for, 225
  genetic relationships, 228
  structure and functionality, 228
  synthesis of oligopeptides from (C3a), 225
Angiotensin II
  analysis of fluorescent labelled analogs, 44
  analogs of, 157
  large-scale synthesis of [Asn^1,Val^5] analog, 448
  receptor-binding studies *in vitro*, 157
Angiotensin, in brain, 77
Angiotensin-converting enzyme, inhibitors of, 576
Anionic proline-rich proteins, sequencing of, 56
Antamanide, and analogs, solvent channels in crystalline lattice of, 583
Antibiotic activity of alamethicin-γ, 215
Antibody 3368 light-chain variable region peptides, synthesis of, 525
Antigenic peptides related to RNA tumor virus group specific protein, 239
Antisera, to endorphins, 63
Antitumor protein, modification of neocarzinostatin, 221
Apamin, solid phase peptide synthesis of, 488, 522
Apolipoprotein C-1, fragments of, synthesis and activity, 149
Arginine-vasotocin, cleavage from resin, 518
Autoracemization of histidine active-ester derivatives, 461
Bacitracin, chemotactic activity of, 141
Bates' reagent, for peptide coupling, 1
Bee venom peptides, synthesis of, 488, 510, 522
O-Benzyl group, removal of, 1

N^{α}-Benzyloxycarbonyl group, cleavage in liquid ammonia, 532
Binding assay, use in opioid peptide detection, 41
Bio-assay, of enkephalin analogs, 114
Biocytin, incorporation into peptides, 1
Biological activity
  of anionic proline-rich proteins, 56
  of bPTH (28-48), 232
  of catecholamines, 197
  of dextran-pepstatin conjugate, 213
  of [$Asu^{1,7}$]-human calcitonin, 436
  of lecithin:cholesterol acyltransferase, 149
  of LRF, 292
  of modified neocarzinostatin, 221
  of neutrophils, 141
  of oligopeptides from C3a anaphylatoxin, 225
  of opioid peptides, 41, 114
  of pepsin inhibitors, 209
  of relaxin, 127
Biosynthesis of opioid peptides, 562
Biotin *p*-nitrophenyl ester, synthesis of, 1
Boc-biocytin pentachlorophenyl ester, synthesis of, 1
Bovine parathyroid hormone, synthesis of fragment (28-48), 232
Bradykinin
  low-temperature CD studies, 307
  receptor-binding studies *in vitro*, 157
  solution synthesis of, 436
Bradykinin analogs, receptor-binding studies, 157
5-Bromo-7-nitroindoline derivatives of peptides in synthesis, 559
Brownian motion, of oligopeptide chain ends, 361
S-t-Butyl protecting group, removal of, 532
Calcification, initiation of, 407
Calcitonin, [$Asu^{1,7}$]-eel, stability of, 436
Calcitonin, [$Asu^{1,7}$]-human, isolation and synthesis of, 436
Calcium ions and conformation of thermolysin fragment (121-205), 432
Calculations
  of global minimum energy conformation, 246
  of solvent accessibility to peptides, 300
Carbodiimide procedure, use in fragment condensation, 436
γ-Carboxyglutamic acid, use in peptide synthesis, 1, 465
N-Carboxyanhydrides in enkephalin synthesis, 471
Catalytic hydrogenolysis, in liquid ammonia, 532
Catecholamines, inhibition of MSH, 197
CCK, *see* cholecystokinin
Cell-cell communications system, 127
Cell wall, bacterial, study of peptidoglycan subunits of, by ^{15}N nmr, 419
^{13}C enrichment
  of human encephalitogenic (EAE) nonapeptide, 357
  of myoglobin, 480
Chain-folding, importance of β-turn in, 284
Charge-stabilization, in enzymatic reactions, 574
Charge transfer, in conformational studies of LRF analogs, 171
Chemoeffectors, interaction with receptors, 117
Chemotactically active peptides, synthesis of, 141, 146
Chemotaxis system, receptor-chemoeffector interactions in, 117
Chiral template in 4-CC peptide synthesis, 484
Cholecystokinin, synthesis of, 1
α-Chymotrypsin, conformational studies, 591
Circular dichroism, induced, of acridine orange dye in a film of poly(γ-[2-(9-carbazolyl)ethyl]-L-glutamate), 390
Circular dichroism studies
  of Ac-Ala-$NHCH_3$, 307
  of actinomycin D, interaction with nucleotides, 353
  of alpha-1-antitrypsin and variants, 567
  of bradykinin and related peptides, 307
  of α-chymotrypsin in cyclohexane, 591
  of clupeines, 474
  of co-oligopeptides of L-Ile and D-alle, 395
  of cyclic dipeptide-aromatic compound complexes in solution, 350
  of enkephalins and endorphins, 100
  of H-Gly-Pro_4-OH, 415
  of histone peptides, 427

Circular dichroism studies (Continued)
of homo-oligopeptides of L-cyclohexylalanine, L-Leu, L-Phe, 399
of oligo(Met-Val-Gly-Pro-Asn-Gly), 333
of peptides bound to polyoxyethylene, 403
of phalloidin analogs, 243
of poly(γ-[2-(9-carbazolyl)ethyl]-L-glutamate) in liquid crystalline and solid state, 390
of poly(Gly-Pro_4), 415
of poly(Met-Val-Gly-Pro-Asn-Gly), 333
of polypeptide-metal complexes, 368
of tentoxin and related cyclic tetrapeptides, 340
of tetrazole analogs of amino acids and their derivatives, 153
of tropomyosin model peptides, 417
Cis amide bond formation
in cyclic peptides, 257, 307
in linear peptides, 246, 257
Clupeines, synthesis and conformational analysis of, 474
Coil-to-helix transition, of poly[N^5-(3-hydroxypropyl)-L-glutamine], 387
Coiled-coil triple-stranded structures of collagen and collagen model peptides, 246
Complement system, activation of, 225, 228
Conformational angles
of H-bonds in peptides, 274
of protected tetrapeptides, 277
Conformational energy calculations
approximations in, 246
librational entropy in, 246
multiple-minimum problem in, 246
of cyclic peptides, 246, 296
of enkephalin analogs, 88, 246, 292
of FSH, 292
of globular proteins, 246
of gramicidin S, 246
of LH and LRF, 292
of oligopeptides, 246
of pancreatic trypsin inhibitor, 280
of peptide-solvent interactions, 590
of poly(Gly-X-Y), 246
of sequential polypeptides, 246
simulation of water structure around hen egg white lysozyme, 586
Conformational parameters
for amino acid residues in proteins, 284
for cyclic dipeptides, 257
for prediction of conformational transformations, 284
Conformational studies
Brownian motion of oligopeptide chain ends, 361
of Ac-Ala-$NHCH_3$, 307
of adrenocorticotropin, 364
of D-aIle peptides, 395
of alpha-1-antitrypsin and variants, 567
of bradykinin and related peptides, 307, 337
of α-chymotrypsin in cyclohexane, 591
of clupeines, 474
of complexes of actinomycin D with nucleotides, 353
of co-oligopeptides of L-Ile, 395
of cyclic peptides, 257, 274, 343, 346
of *cyclo*(L-Pro-D-Val) and *cyclo* (L-Pro-L-Val), 350
of elastin model polypeptides, 407
of encephalitogenic (EAE) nonapeptide, 357
of enkephalins, 329
of gallines, 474
of H-Gly-Pro_4-OH, 415
of gramicidin S in solution, 307
of β-helices in peptides and proteins, 423
of histone proteins, 427
of homo-oligopeptides of L-cyclohexylalanine, L-Leu and L-Phe, 399
of LRF analogs, 171
of neurophysins, 588
of oligo(Met-Val-Gly-Pro-Asn-Gly), 333
of oligopeptides, 361
of oxytocin and analogs, 165, 168, 322
of peptides, 307, 325
of peptides bound to polyoxyethylene, 403
of peptidoglycan subunits in bacterial cell walls, 419
of poly(γ-[2-(9-carbazolyl)ethyl]-L-glutamate), 390
of poly(Gly-Pro_4), 415
of poly[N^5-(3-hydroxypropyl)-L-glutamine], 387
of poly(Met-Val-Gly-Pro-Asn-Gly), 333

Conformational studies (Continued)
of polypeptide-ion complexes, 368
of tentoxin and related cyclic tetrapeptides, 340
of thermolysin CNBr fragments, 432
of tropomyosin model polypeptides, 411
of valinomycin in solution, 307
of vasopressin and analogs, 165
Conformational transformations, prediction of, 284
Co-oligopeptides, of L-Ile and D-alle, 395
Co(I)-phthalocyanine anion, use in cleavage of resin bound peptides, 484
Cortical adenylyl-cyclase assay for parathyroid hormone activity, 232
Corticotropin-(1-24), large scale synthesis of, 448
Counter-current distribution, in purification of peptides, 540
Coupling agents, 1,4-dinitro-1,3-butadiene-2,3-dicarboxylic anhydride, 536
Cross-linking
of proteins by photoreactive sulfenylchlorides, 553
of synthetic polypeptides of elastin, 407
Crown-ether, use in peptide cleavage from resin, 518
Crystal structure analysis of peptides, 257, 274, 277
Crystallization of insulin derivatives, 136
Cu(II) ions in conformational studies of ion-polypeptide interactions, 368
Cyclic dipeptides, conformational analysis in solid state, 257
Cyclic peptides
cis-amide, bond formation in, 307
conformational energy calculations of, 246, 296
conformational studies, 257, 307, 333
gel permeation chromatography of, 506
intramolecular H-bond studies, 274
recyclization of phalloidin analogs, 243
symmetry in, 296
X-ray studies, 257, 274
Cyclization
of resin-bound gramicidin S pentapeptide, 506
of somatostatin by $K_3Fe(CN)_6$, 532
prediction of, in polypeptides, 288
O-Cyclohexyltyrosine, use in solid phase peptide synthesis, 488
Cyclol type transformation in solid phase peptide synthesis, 510
Cyclo(Gly-Pro-Gly-D-Ala-Pro), conformational studies, 350
Cyclo[(Met-Val-Gly-Pro-Asn-Gly)$_2$] structural studies, 333
Cyclo-oligomerizations, *see* interchain reactions
N^{α}-Cyclopropylmethyl-*O*-*t*-butyltyrosine, synthesis of, 111
Cyclo(D-Val-L-Hyiv-L-Val-D-Hyiv)$_3$, conformational studies of, 307
Cyclo(Val-MeIle-D-Hyiv$_2$), conformational studies of, 307
Cyclo(Val-Sar)$_3$, preferential solution structure of, 307
Cysteic acid as amino protecting group, 540
Cysteine-containing peptides, deprotection by catalytic hydrogenolysis, 532
Cysteine protection, in solid phase peptide synthesis, 522
Cytochrome c, horse-heart, preparation of semi-synthetic protein, 593
Cytotoxicity of agarose immobilized neocarzinostatin, 221
Ddz – *see* α,α-dimethyl-3,5-dimethoxybenzyloxycarbonyl
Dehydroalanine, formation of in nisin:ring A synthesis, 205
Deprotection by HF in solution synthesis, 436
Dextran-pepstatin conjugate, synthesis and biological activity of, 213
gem-Diamino compounds, use in synthesis of retro-isomers of peptides, 572, 579
Dideoxypepstatin, inhibition of pepsin by, 209
Diketopiperazines, *see* cyclic dipeptides
1,8-Dimethylaminonaphthalene, in peptide synthesis, 1
4-Dimethylaminopyridine, active ester coupling catalyst, 1
α-α-Dimethyl-3,5-dimethoxybenzyloxycarbonyl-amino acids in peptide synthesis, 510
Dimethylphosphinothioyl chloride, for amino protection, 546
2,4-Dinitro-5-azidophenylsulfenyl chloride, for reversible cross-linking of peptides, 553

1,4 Dinitro-1,3-butadiene-2,3-dicarboxylic anhydride, synthesis and use as a coupling agent, 539
N-Dinitrophenyl-imidazole, base-susceptible reactions of, 549
Dipeptide sweeteners and analogs, synthesis of, 572
Dipeptides, resolution of unprotected stereoisomers by HPLC, 52
Dipeptidyl aminopeptidase, in peptide sequencing, 14, 27
Diphenylphosphinothioyl chloride, use for amino protection, 546
Diphenyl phosphoryl azide, use in esperic acid synthesis, 218
Dopamine, effect on MSH release, 197
Dissociation constants of, Glu derivatives, 215
Disulfide bridging in synthesis of apamin, 522
Dithiasuccinoyl amino protecting group, 488
Edman and Begg method, in sequence determination of anionic proline-rich proteins, 56
Edman degradation, automated, of preproalbumin, 59
Elastin model polypeptides, synthesis and properties, 407
Electrometric titration studies of bradykinin and related peptides, 337
Electrophoresis, use of fluorometric procedures in, 14
Encephalitogenic (EAE) nonapeptide, synthesis and conformational analysis of ^{13}C enriched, 357
β-Endorphin
  biological studies of, 63, 77
  biosynthesis of *in vitro*, 562
  blood levels in stress, 63
  precursor of, 63
  solid phase synthesis of, 503
  susceptibility to endopeptidases, 103
Endorphins
  CD studies of, 100
  isolation and purification of, 41, 63
  synthesis of, 107
Energy transfer, nonradiative, conformational analysis of human ACTH, 364
Enkephalin analogs, 77
  biological studies of, 63, 77, 111, 114
  CD studies, 100
  conformational energy calculations, 88, 246, 292
  conformational studies of, 84, 92, 329
  fluorescence studies of, 92
  hypophysiotropic activity of, 292
  inverse-γ-turn in, 111
  isolation of, 41
  localization of, 77
  nmr studies of, 88, 329
  radioimmunoassay of, 77
  relative potencies of, 96
  structure-activity relationships, 84, 111
  synthesis of, 114, 471
  topology of, 84, 92
  β-turn of type II in, 246
Enzymatic coupling of amino acids by pepsin, 556
Enzyme inhibition, by peptide hydroxamates, 236
Enzyme kinetics, inhibition of pepsin, 209
Enzyme mechanisms, model for, 574
Esperic acid analogs, synthesis and biological activity of, 218
N-Ethoxycarbonyl-L-ethoxy-1,2-dihydroquinoline, use in esperic acid synthesis, 218
Fibrous protein analogs, conformational energy calculations, 246
Fluorometric analysis
  in amino acid analyzer, 14
  in HPLC, 14
  of opioid peptides, 41
  of peptide hormones and analogs, 44
  of peptide synthesis and purification, 14
  reagents for, 14, 41
Fluorescence studies
  of lipid binding properties, 149
  of opioid peptides, 92
Folate conjugates, synthesis of, 468
Follicle-stimulating hormone, conformational energy calculations, 292
N^{α}-Formyl-Met-Leu-Phe-OH, analogs, as chemo-attractants, 146
Four-component condensation method
  in peptide synthesis, 484
  stereoselectivity in, 484
Fragment condensation
  carbodiimide procedure in, 436
  glucagon synthesis, 488
  in solid phase peptide synthesis, 488, 510, 528

Fragment condensation (Continued)
synthesis of myoglobin by, 480
Taiwan cobra venom cardiotoxin by, 528
FSH, *see* follicle-stimulating hormone
Gallins, synthesis and conformation of, 474
Gel permeation chromatography, of cyclic peptides, 506
Gene duplication, mechanism for sequence homology in hormones, 127
Genetic relationships, of anaphylatoxins C3a and C5a, 228
Global minimum energy conformation, calculation of, 246
Glucagon, synthesis of, 488
Glutaryl-L-proline, as inhibitor of angiotensin converting enzyme, 576
Glycine, function in turn formation, 280
Glycophorin, a β^{13}-helical model for, 423
Glycoproteins, role of carbohydrate in, 567
Glycotripeptides, synthesis of, 477
Gramicidin S
chemotactic activity of, 141
complex with Cu(II) ions, 368
conformational energy calculations of, 246
conformational studies of, 307
safety-catch method in synthesis of ring oligomers, 506
studies with lipid mono- and bilayers, 307
Gramicidin S analogs, structure-activity relationships, 307
Hardsphere-like calculation of peptides, 300
α-Helical region, frequency of residues in, 284
β-Helix
in co-oligomers of L-Ile and D-aIle, 395
in peptides and proteins, 423
HF, use of scavengers in deprotection by, 436
High performance liquid chromatography
coupled to mass spectrometer, 27
in fluorometric analysis, 14, 44
in monitoring of peptide synthesis, 514
of angiotensin II and analogs, 44
of fluorescent labelled oxytocin, 44
of opioid peptides, 41, 107
of oxytocin and analogs, 44
of somatostatin, 52
of somatostatin fragments, 532
of TRF, 52
of unprotected peptide stereoisomers, 52
of vasopressin and analogs, 44
racemization test by, 52
reverse phase method in, 48
Histones, complexation and conformation, 427
^{1}H nmr, *see* nmr
Homo-oligopeptides; L-cyclohexylalanine, L-Leu, and L-Phe conformation of, 399
Hormone cleavage inhibition, by bPTH (28-48), 232
Hormone evolution, mechanism of, 127
Hormone-protein interactions, 176, 179
Hormones, homology in "prepro" sequence, 127
HPLC, *see* high-performance liquid chromatography
Hybrid molecules, of relaxin and insulin, 127
Hydrophobic interactions, in reverse phase chromatography, 48
1-Hydroxybenzotriazole, as an additive in peptide synthesis, 436
2-Hydroxyethylsulfonylmethyl resin in solid phase peptide synthesis, 488
Hydroxyfatty acids in esperic acid synthesis, 218
ε-Hydroxynorleucine, incorporation into peptides, 1
N-Hydroxysuccinimide esters, use in esperic acid synthesis, 218
Hypothalamic hormones, interaction with brain-stem synaptosomes, 183
Immuno-cross-reactivity, of [Asu^{1,7}]-human-calcitonin, 436
Immunological studies, of synthetic antigenic peptides, 239
Inflammatory activity, of oligopeptides from C3a anaphylatoxin, 225
Infrared dichroism, of (Met-Val-Gly-Pro-Asn-Gly) oligomers, 333
Infrared spectroscopy
of histone peptides, 427
of homo-oligomers, 399
of *N*-methylacetamide, 303
Inhibition
of angiotensin converting enzyme, 576

Inhibition (Continued)
of acid proteases by dextran-pepstatin conjugate, 213
of thermolysin by peptide hydroxamates, 236
Insulin
structure homology with relaxin, 127
structure-activity relationship of, 136
synthesis and X-ray studies of, 136
Interchain reactions, in solid phase peptide synthesis, 506
Internal structure control in synthesis of apamin, 488
Intramolecular hydrogen bonds
in cyclic peptides, 274
in peptides, 257
Ion-binding studies of cyclic pentapeptides, 436
Ion channel, in β-helical peptides and proteins, 423
Ion-polypeptide interactions, CD studies of, 368
Ion-pair chromatography in separation of amino acids, 48
Ionic strength, effect on polypeptide conformation, 246
Ionophores, crystalline antamanide and analogs as, 583
IR Dichroism, *see* infrared dichroism
IR spectroscopy, *see* infrared spectroscopy
α-Isocyano acid derivative in 4-CC method, 484
Isocyanomethyl polystyrene resin, in solid phase peptide synthesis, 484
Isotope effect, use in racemization test, 461
Isotopic shifts in spectroscopic studies of *N*-methylacetamide, 303
Kinetic studies of base-susceptible reactions, 549
α-^{2}H-Labelled amino acids, use in racemization test, 461
meso-Lanthionine, incorporation into nisin:ring A, 205
Leu^{5}-enkephalin, synthesis by automated solid-phase technique, 569
LH, *see* luteinizing hormone
Librational entropy, in conformational energy calculations, 246
Lipid bilayer membrane, voltage-dependent conductance of, 215
Lipid binding peptides, synthesis and biological activity of, 149
β-Lipotropin
biosynthesis *in vitro*, 562
blood levels in stress, 63
human, 107
isolation of, 41
porcine, 63
precursor of, 63
subunits of, 77, 96, 100
γ-Lipotropin, biosynthesis of *in vitro*, 562
Lipotropin C fragment, *see* β-endorphin
β-LPH, *see* β-lipotropin
LRF, *see* luteinizing hormone-releasing factor
Luteinizing hormone, conformational energy of, 292
Luteinizing hormone-releasing factor
conformational energy calculations of, 292
mechanism of biological activity, 292
synthesis of, 514
Luteinizing hormone-releasing factor analogs
biological studies of, 189, 193
conformational studies, 171
retro-isomer, 579
synthesis of, 171, 189
Luteinizing hormone-releasing factor degrading enzymes, purification of, 193
Macrocyclization equilibrium constant for polypeptides, 288
Martinez and Winternitz coupling reagent, 1
Mass spectroscopy
coupling of, with HPLC, 27
of endorphin analogs, 63
use in peptide sequencing, 27
Mast-cell-degranulating peptide, solid phase synthesis of, 510
Maximum protection procedure in peptide synthesis, 436
Mean residue hydrophobicities of lipid binding peptides, 149
Melanocyte-stimulating hormone, effect of catecholamines on release of, 197
α-Melanatropin, porcine, chromatography, 48
Membrane structure, β-helical structure of proteins in, 423
2-Mercaptoethanol, kinetics of oxidation, 549

D-3-Mercapto-2-methylpropanoyl-L-proline, biological activity of, 576
Metalloprotease, inhibition by peptide hydroxamates, 236
Methionine sulfoxide, reduction of, 458
N-Methylacetamide, conformational studies of, 303
N-Methylmercaptoacetamide, reducing agent for methionine sulfoxide, 458
Microscopic dielectric effect, in enzyme reaction mechanisms, 574
Mixed anhydride fragment coupling in enkephalin synthesis, 471
Monte-Carlo simulation of water structure around hen egg white lysozyme, 586
Morphinomimetic compounds
  see also enkephalins, endorphins, opioid peptides
  biological activity of, 63, 107
MSH *see* melanocyte-stimulating hormone
Myoglobin, semisynthesis of ^{13}C enriched, 480
Neocarzinostatin, modification of, 221
Neurophysins
  interaction with oxytocin and analogs, 176, 179
  structural and conformational studies, 588
Neurotensin, biological studies of, 77
Neutron inelastic scattering spectra of *N*-methylacetamide, 303
Nisin:ring A, synthesis and biological studies of, 205
2-Nitro-4-azidophenylsulfenyl chloride, synthesis and reversible cross-linking with, 553
o-Nitro-*p*-sulfophenyl esters in peptide synthesis, 536
^{13}C Nmr studies
  of H-Gly-Pro_4-OH, 415
  of human encephalitogenic (EAE) nonapeptide, 357
  of oxytocin analogs, 168
  of peptide hormone binding to neurophysins, 176, 179
^{15}N Nmr studies
  of oxytocin analogs, 322
  of peptidoglycan subunits in bacterial cell walls, 419
Nmr studies
  of co-oligopeptides of L-Ile and D-aIle, 395
  of cyclic pentapeptides, 343
  of enkephalin analogs, 88
  of fragment complexation of histones H_3 and H_4, 427
  of H-Gly-Pro_4-OH, 415
  of peptides bound to polyoxyethylene, 403
  of protected linear somatostatin, 532
  of tentoxin and related cyclic tetrapeptides, 340
  of vasopressin analogs, 165
  relaxation time studies, 480
Nmr heteronuclear coupling studies
  of free valinomycin, 307
  of gramicidin S, 307
  of oxytocin analogs, 322
Norepinephrin
  effect on MSH release, 197
  receptor binding studies, 157
Nuclear magnetic resonance, *see* nmr
Oligopeptides
  Brownian motion of chain ends of, 361
  conformational energy calculations of, 246
  conformational studies of, 333, 395, 399
Opiate pharmacophore, topology of, 84
Opiate receptor
  fluorescence studies of, 92
  influence of ions on binding to, 77
  topography of, 92
Opioid peptides, *see also*, enkephalins, endorphins
  biosynthesis of, 562
  isolation, purification and characterization of, 41
  purification followed by fluorometric procedures, 14
  solid phase synthesis of, 503
  topology of, 84
Ornithine, transformation into arginine, 474
Orthogonal protection, in solid phase peptide synthesis, 488
Oxytocin
  conformational studies, 165
  interaction with neurophysin I, 176, 179
  quantitative fluorometric analysis, 14
Oxytocin analogs
  analysis by fluorescence, 44

Oxytocin analogs (Continued)
conformational studies of, 168, 322
synthesis and biological activity of 4-substituted, 201
Pam, *see* phenylacetamidomethyl
Pancreatic trypsin inhibitor, computed conformation for, 280
Parallel β-sheet structure, in homo-oligopeptides of, 399
Parathyroid hormone analogs, as inhibitors, 581
Partial specific volume of poly[N^5-(3-hydroxypropyl)-L-glutamine], 387
Passerini reaction, 484
Pepsin
coupling of amino acid by, 556
inhibition of, 209
Pepstatin, inhibition of pepsin by, 209
Peptide antibiotics
alamethicin, 215
esperic acid analogs, 218
Peptide conformation, *see* conformational studies and conformational energy calculations
Peptide fragment combination by 4-CC method, 484
Peptide hormones
agonist-antagonist relationships in, 292
receptor binding studies *in vitro*, 157
Peptide hydroxamates, synthesis of thermolysin inhibitors, 236
Peptide-metal complexes, structural studies in solution, 325
Peptide-solvent interactions, theoretical studies, 590
Peptidoglycan subunits in bacterial cell walls, 419
Peptide sweeteners, 3-point model for, 572
Peptide synthesis
base-susceptible reactions of *N*-dinitrophenylimidazole in, 549
coupling with 1,4-dinitro-1,3-butadiene-2,3-dicarboxylic anhydride, 539
coupling with pepsin, 556
cysteic acid as protecting group in, 540
dimethylphosphinothioyl as protecting group in, 546
4-CC method in, 484
large scale, 448
monitoring by HPLC, 514
o-nitro-*p*-sulfophenyl esters in, 536
2-trimethylsilylethyl protecting group in, 543
use of photoreactive sulfenyl chlorides, 553
Peptide synthesis, solid phase
cleavage from resin in, 518
cyclol type transformation in, 510
cyclo-oligomerization in, 506
cysteine protection in, 522
deprotection methods in, 525
fragment condensation in, 488, 510
interchain reactions in, 506
internal structure control in, 488
loss of peptide during, 525
monitoring of automated, 569
resins and supports for, 484, 488, 503, 510
side-chain protection in, 488
use of crown-ether in, 518
use of *O*-cyclohexyltyrosine in, 488
use of 1-peptidyl-5-bromo-7-nitroindolines in, 559
Petroic acid, use in synthesis of folate conjugates, 468
Phalloidin, new analogs by modifications of, 243
Phenylacetamidomethyl resin in solid phase peptide synthesis, 488
Phenacetyl support, cyclol type transformation on, 510
L-Phenylalaninol, 215
Phospholipid binding by fragments of apolipoprotein C-1 and synthetic peptides, 149
Photoaffinity labeling with photoreactive sulfenyl chlorides, 553
Photosolvolysis of 5-bromo-7-nitroindoline peptide derivatives, 559
Pituitary hormones, secretion of, 63
Pmr, *see* nmr
Poly(γ-[2-(9-carbazolyl)ethyl]-L-glutamate), CD studies of, 390
Poly(diaminobutyric acid) and derivatives, interaction with Cu(II) ions, 368
Polydimethylacrylamide support, in solid phase peptide synthesis, 503
Poly(Gly-Pro_4), conformational studies, 415
Poly(Gly-X-Y), conformational energy calculation, 246
Poly(His) interaction with Cu(II) ions, 368

Poly[N^5-(3-hydroxypropyl)-L-glutamine], coil-helix transition in, 387
Polymeric amino protection group in peptide syntheses, 514
Poly(Met-Val-Gly-Pro-Asn-Gly), conformational studies of, 333
Poly(Orn) derivatives, interaction with Cu(II) ions, 368
Polyoxyethylene, conformational studies of peptides bound to, 403
Polypeptides
  elastin-like, conformational studies, 407
  macrocyclization of, 288
  sequential, conformational studies, 246, 333, 411, 415
Preferential adsorption, to poly[N^5-(3-hydroxypropyl)-L-glutamine], 387
Preparative-HPLC, use in purification of enkephalin, 471
Prepro sequence in hormones, nature and importance of, 127
Primary structure of anaphylatoxins C3a and C5a, 228
Proinsulin, predicted conformation for, 284
Protecting groups, 488, 540, 543, 546, 559
Protein analogs, by reconstitution of synthetic and natural peptide fragments, 591
Protein folding, theoretical studies of, 280
Proteolytic enzymes
  inhibitors, predicted secondary structure, 284
  proteases A and B, purification, 236
  tetrazole derivatives as substrates, 153
Protamines, synthesis of related peptides, 474
Prothrombin, studies on, 465
Pteroylglutamates, synthesis of, 468
Pyridoxal enzyme, use of Cu(II)-peptide complexes as models for, 368
Racemization
  in peptide synthesis, 436
  kinetic studies of, 461
Racemization test by HPLC, 52
Radioimmunoassay
  of angiotensin peptides, 77
  of [$Asu^{1,7}$]-human calcitonin, 436
  of endorphins and analogs, 63, 107
  of enkephalins, 77
Raman spectra of N-methylacetamide, 303
Receptor-chemoeffector interactions of the chemotoxis system, 117
Receptor molecules, characteristics of, 127
Recyclization, method for modifying phalloidin, 243
Reduction of methionine sulfoxide, 458
Relaxation time studies by nmr, for modified myoglobin, 480
Relaxin, structure and biological activity, 127
Repetitive excess mixed anhydride-coupling method, 1
Resolution of, γ-carboxyglutamic acid derivatives, 465
Response regulators in bacteria, 117
Retro-isomers
  of LRF, 579
  of peptide sweetener, 572
Reverse phase chromatography
  hydrophobic interactions in, 48
  use in HPLC, 48
  use in resolving opioid peptides, 41
Ribonuclease S peptide, separation of tryptic digest of, 48
Ring closure probabilities for polypeptides, 288
RNA tumor virus group specific protein, synthesis of sequences related to, 239
Safety-catch method in gramicidin S synthesis, 506
Salt coupling in peptide synthesis, 468, 471
Scavengers, use in deprotection procedures, 436
Secondary structure of proteins, semiempirical predictive method, 284
Semisynthesis of myoglobin, 480
Sequence determination
  of anionic proline-rich proteins, 56
  of preproalbumin, 59
Sequence homology in hormones, 127
Serine, side reactions of, 436
Serotonin, effect on MSH release, 197
Serum thymic factor, synthesis of, 564
β-Sheet region, frequency of residues in, 284
Side-chain conformations of amino acids, studies by X-ray, 257

Side reactions
- *O*-acylation of serine, 436
- in deprotection of tyrosine, 488
- on phenacetyl support, 510

Signalling systems, in bacteria, 117
Solvent-shell model, for hydration energy calculation, 246
Somatostatin
- purification by HPLC, 52
- stepwise synthesis in solution, 532

Somatostatin carbacyclic analogs, synthesis of, 186
SRIF, *see* somatostatin
Statherin, 56
Stereoselectivity, in 4-CC peptide synthesis, 484
Stereospecificity in polypeptide-ion interactions, 368
Structure-activity relationship
- of chemo-attractant molecules, 146
- of derivatized insulin, *in vivo*, 136
- of gramicidin S, 307
- of oligopeptides from C3a anaphylatoxin, 225
- of phalloidin analogs, 243

Substance P, biological studies of, 77
Succinyl-L-proline, and analogs as inhibitors of angiotensin coverting enzyme, 576
Surface-activity of neocarzinostatin, 221
Sweetener-receptor interaction, model for, 572
Symmetry in cyclic tetrapeptides, 296
Synthesis
- of alamethicin-γ, 215
- of anaphylatoxin C3a, oligopeptides from, 225
- of 3368-antibody light-chain variable region peptides, 525
- of antigenic peptide sequences related to RNA tumor virus group specific protein, 239
- of apamin, 488, 522
- of apolipoprotein C-1 fragments, 149
- of bee venom peptides, 488
- of bPTH (28-48), 232
- of [$Asu^{1,7}$]-calcitonin, 436
- of chemotactically active peptides, 141
- of clupeines, 474
- of cytochrome c, 593
- of elastin model polypeptides, 407
- of encephalitogenic (EAE) nonapeptide, 357
- of endorphin and analogs, 107, 503
- of enkephalin analogs, 114, 471
- of esperic acid analogs, 218
- of folate conjugates, 468
- of gallines, 474
- of glucagon, 488
- of glycotripeptides, 477
- of inhibitors of bPTH, 581
- of lipid-binding peptides, 149
- of mast-call-degranulating peptide, 510
- of nisin:ring A, 205
- of [Hse^4]-oxytocin, 201
- of peptides containing γ-carboxyglutamic acid, 465
- of peptide hydroxamates, 236
- of peptide sweeteners and analogs, 572
- of pteroylglutamates, 468
- of retro-isomers of LRF analogs, 579
- of serum thymic factor, 564
- of somatostatin, 532
- of somatostatin carbacyclic analogs, 186
- of Taiwan cobra venom cardiotoxin, 528
- of vasoactive intestinal peptide, 1
- of vasopressin analogs, 161

Taiwan cobra venom cardiotoxin, synthesis of, 528
Temperature coefficients of NH chemical shifts in cyclic pentapeptides, 343
Tentoxin, conformational analysis of, 340
γ-(5-Tetrazolyl)-α-amino-L-butyric acid, synthesis of, 153
Tetrazole analogs of amino acid derivatives as proteolytic enzyme substrates, 153
Thermolysin
- conformational studies of CNBr fragments, 432
- inhibition of by peptide hydroxamates, 236
- purification by affinity chromatography, 236

Thioredoxin, predicted conformational profile for, 284
Thyrotropin releasing factor, purification by HPLC, 52
Topochemical analogs
- of dipeptide sweeteners, 572

Topochemical analogs (Continued)
of LRF, 579
Torsion modes in *N*-methylacetamide, 303
Transesterification in solid phase peptide synthesis, 518
Transition-state-analog inhibitor, biological activity of pepstatin as, 209
TRF, *see* thyrotropin releasing factor
2-Trimethylsilylethyl protecting group in peptide synthesis, 543
Triphenylphosphine dihydrochloride for specific cleavage of P-N bonds, 546
Tropomyosin, synthetic polyheptapeptides as models for, 411
Tryptophan, modification during deprotection, 436
Tuberculostatic activity of esperic acid analogs, 218
Turn formation, glycine function in, 280
β-Turns
frequency of residues in, 284
importance in chain-folding, 284
nmr studies of cyclic pentapeptides, 343
of *cyclo*(Gly-Pro-Gly-D-Ala-Pro) in solution, 346
type II′ in enkephalin, 246
γ-Turns, in cyclic pentapeptides, 343, 346
Tyrocidin, chemotactic activity of, 141
Tyrosine side-chain protecting groups, 488
Ultraviolet absorption studies
of actinomycin D-nucleotide complexes, 353
of α-chymotrypsin in cyclohexane, 591
of co-oligopeptides of L-Ile and D-aIle, 395
Valinomycin (complexed), conformational studies of, 257, 307, 325
Valinomycin (free) conformational studies of, 257, 307
Vasoactive intestinal peptide, synthesis of, 1
Vasopressin, fluorometric analysis of, 14, 44
Vasopressin analogs, conformational studies of, 161, 165
Voltage-dependent conductance in lipid bilayer membranes, 215
Woodward's reagent K, use in synthesis of glycotripeptides, 477
X-ray studies
dihedral angles from, 257
intramolecularly H-bonded conformations of peptides, 257, 274
of cyclic peptides, 257, 274
of insulin derivatives, 136
of protected tetrapeptides, 277
of side-chain conformations of amino acids, 257
of solvent channels in antamanide crystalline lattice, 583
peptide backbone conformations, 257